Bending

Normal Stress

$$\sigma = \frac{My}{I}$$

Unsymmetric Bending

$$\sigma = -\frac{M_z y}{I_z} + \frac{M_y z}{I_y}, \qquad \tan \alpha = \frac{I_z}{I_y} \tan \theta$$

Shear

Average Direct Shear Stress

$$\tau_{avg} = \frac{V}{A}$$

Transverse Shear Stress

$$\tau = \frac{VQ}{It}$$

Shear Flow

$$q = \frac{VQ}{I}$$

Stress in Thin-Walled Pressure Vessel

Cylinder

$$\sigma_1 = \frac{pr}{t}, \qquad \sigma_2 = \frac{pr}{2t}$$

Sphere

$$\sigma_1 = \sigma_2 = \frac{pr}{2t}$$

Stress Transformation Equations

$$\sigma_{x'} = \frac{\sigma_x + \sigma_y}{2} + \frac{\sigma_x - \sigma_y}{2} \cos 2\theta + \tau_{xy} \sin 2\theta$$

$$\tau_{x'y'} = -\frac{\sigma_x - \sigma_y}{2} \sin 2\theta + \tau_{xy} \cos 2\theta$$

Principal Stress

$$\tan 2\theta_p = \frac{\tau_{xy}}{(\sigma_x - \sigma_y)/2}$$

$$\sigma_{1,2} = \frac{\sigma_x + \sigma_y}{2} \pm \sqrt{\left(\frac{\sigma_x - \sigma_y}{2}\right)^2 + \tau_{xy}^2}$$

Maximum In-Plane Shear Stress

$$\tan 2\theta_s = -\frac{(\sigma_x - \sigma_y)/2}{\tau_{xy}}$$

$$\tau_{max} = \sqrt{\left(\frac{\sigma_x - \sigma_y}{2}\right)^2 + \tau_{xy}^2}$$

$$\sigma_{avg} = \frac{\sigma_x + \sigma_y}{2}$$

Absolute Maximum Shear Stress

$$\tau_{\substack{abs \\ max}} = \frac{\sigma_{max} - \sigma_{min}}{2}$$

$$\sigma_{avg} = \frac{\sigma_{max} + \sigma_{min}}{2}$$

Material Property Re...

Poisson's Ratio

$$\nu = -\frac{\epsilon_{lat}}{\epsilon_{long}}$$

Generalized Hooke's Law

$$\epsilon_x = \frac{1}{E}[\sigma_x - \nu(\sigma_y + \sigma_z)]$$

$$\epsilon_y = \frac{1}{E}[\sigma_y - \nu(\sigma_x + \sigma_z)]$$

$$\epsilon_z = \frac{1}{E}[\sigma_z - \nu(\sigma_x + \sigma_y)]$$

$$\gamma_{xy} = \frac{1}{G}\tau_{xy}, \quad \gamma_{yz} = \frac{1}{G}\tau_{yz}, \quad \gamma_{zx} = \frac{1}{G}\tau_{zx}$$

where

$$G = \frac{E}{2(1 + \nu)}$$

Relations Between w, V, M

$$\frac{dV}{dx} = w(x), \qquad \frac{dM}{dx} = V$$

Elastic Curve

$$\frac{1}{\rho} = \frac{M}{EI}$$

$$EI\frac{d^4 v}{dx^4} = w(x)$$

$$EI\frac{d^3 v}{dx^3} = V(x)$$

$$EI\frac{d^2 v}{dx^2} = M(x)$$

Buckling

Critical Axial Load

$$P_{cr} = \frac{\pi^2 EI}{(KL)^2}$$

Critical Stress

$$\sigma_{cr} = \frac{\pi^2 E}{(KL/r)^2}, \quad r = \sqrt{I/A}$$

Secant Formula

$$\sigma_{max} = \frac{P}{A}\left[1 + \frac{ec}{r^2}\sec\left(\frac{L}{2r}\sqrt{\frac{P}{EA}}\right)\right]$$

STATICS AND MECHANICS OF MATERIALS

STATICS AND MECHANICS OF MATERIALS

FIFTH EDITION

R. C. HIBBELER

PEARSON

Hoboken Boston Columbus San Francisco New York
Indianapolis London Toronto Sydney Singapore Tokyo Montréal Dubai
Madrid Hong Kong Mexico City Munich Paris Amsterdam Cape Town

Vice President and Editorial Director, ECS: *Marcia J. Horton*
Editor in Chief: *Julian Partridge*
Senior Editor: *Norrin Dias*
Editorial Assistant: *Michelle Bayman*
Program/Project Management Team Lead: *Scott Disanno*
Program Manager: *Sandra L. Rodriguez*
Project Manager: *Rose Kernan*
Director of Operations: *Nick Sklitsis*
Operations Specialist: *Maura Zaldivar-Garcia*
Field Marketing Manager: *Demetrius Hall*
Marketing Assistant: *Jon Bryant*
Cover Designer: *Black Horse Designs*
Cover Image: *Adam Lister/Getty Images*

Pearson Education Ltd., *London*
Pearson Education Singapore, Pte. Ltd
Pearson Education Canada, Inc.
Pearson Education—Japan
Pearson Education Australia PTY, Limited
Pearson Education North Asia, Ltd., *Hong Kong*
Pearson Educación de Mexico, S.A. de C.V.
Pearson Education Malaysia, Pte. Ltd.
Pearson Education, Inc., Hoboken, *New Jersey*

Library of Congress Cataloging-in-Publication Data
Hibbeler, R. C.
Statics and mechanics of materials / R.C. Hibbeler.
Fifth edition. | Hoboken : Pearson, 2016. | Includes index.
LCCN 2016013512 | ISBN 9780134382593
LCSH: Strength of materials. | Statics. | Structural analysis (Engineering)
LCC TA405 .H48 2016 | DDC 620.1/12—dc23
LC record available at http://lccn.loc.gov/2016013512

1 16

ISBN 10: 0-13-438259-5
ISBN 13: 978-0-13-438259-3

To the Student

With the hope that this work will stimulate
an interest in Engineering Mechanics and
Mechanics of Materials and provide
an acceptable guide to its understanding.

PREFACE

This book represents a combined abridged version of two of the author's books, namely *Engineering Mechanics: Statics, Fourteenth Edition* and *Mechanics of Materials, Tenth Edition.* It provides a clear and thorough presentation of both the theory and application of the important fundamental topics of these subjects, that are often used in many engineering disciplines. The development emphasizes the importance of satisfying equilibrium, compatibility of deformation, and material behavior requirements. The hallmark of the book, however, remains the same as the author's unabridged versions, and that is, strong emphasis is placed on drawing a free-body diagram, and the importance of selecting an appropriate coordinate system and an associated sign convention whenever the equations of mechanics are applied. Throughout the book, many analysis and design applications are presented, which involve mechanical elements and structural members often encountered in engineering practice.

NEW TO THIS EDITION

- **Preliminary Problems.** This new feature can be found throughout the text, and is given just before the Fundamental Problems. The intent here is to test the student's conceptual understanding of the theory. Normally the solutions require little or no calculation, and as such, these problems provide a basic understanding of the concepts before they are applied numerically. All the solutions are given in the back of the text.

- **Improved Fundamental Problems.** These problem sets are located just after the Preliminary Problems. They offer students basic applications of the concepts covered in each section, and they help provide the chance to develop their problem-solving skills before attempting to solve any of the standard problems that follow.

- **New Problems.** There are approximately 80% new problems that have been added to this edition, which involve applications to many different fields of engineering.

- **Updated Material.** Many topics in the book have been re-written in order to further enhance clarity and to be more succinct. Also, some of the artwork has been enlarged and improved throughout the book to support these changes.

• **New Layout Design.** Additional design features have been added to this edition to provide a better display of the material. Almost all the topics are presented on a one or two page spread so that page turning is minimized.

• **New Photos.** The relevance of knowing the subject matter is reflected by the real-world application of new or updated photos placed throughout the book. These photos generally are used to explain how the principles apply to real-world situations and how materials behave under load.

HALLMARK FEATURES

Besides the new features just mentioned, other outstanding features that define the contents of the text include the following.

Organization and Approach. Each chapter is organized into well-defined sections that contain an explanation of specific topics, illustrative example problems, and a set of homework problems. The topics within each section are placed into subgroups defined by boldface titles. The purpose of this is to present a structured method for introducing each new definition or concept and to make the book convenient for later reference and review.

Chapter Contents. Each chapter begins with a photo demonstrating a broad-range application of the material within the chapter. A bulleted list of the chapter contents is provided to give a general overview of the material that will be covered.

Emphasis on Free-Body Diagrams. Drawing a free-body diagram is particularly important when solving problems, and for this reason this step is strongly emphasized throughout the book. In particular, within the statics coverage some sections are devoted to show how to draw free-body diagrams. Specific homework problems have also been added to develop this practice.

Procedures for Analysis. A general procedure for analyzing any mechanics problem is presented at the end of the first chapter. Then this procedure is customized to relate to specific types of problems that are covered throughout the book. This unique feature provides the student with a logical and orderly method to follow when applying the theory. The example problems are solved using this outlined method in order to clarify its numerical application. Realize, however, that once the relevant principles have been mastered and enough confidence and judgment have been obtained, the student can then develop his or her own procedures for solving problems.

Important Points. This feature provides a review or summary of the most important concepts in a section and highlights the most significant points that should be realized when applying the theory to solve problems.

Conceptual Understanding. Through the use of photographs placed throughout the book, the theory is applied in a simplified way in order to illustrate some of its more important conceptual features and instill the physical meaning of many of the terms used in the equations. These simplified applications increase interest in the subject matter and better prepare the student to understand the examples and solve problems.

Preliminary and Fundamental Problems. These problems may be considered as extended examples, since the key equations and answers are all listed in the back of the book. Additionally, when assigned, these problems offer students an excellent means of preparing for exams, and they can be used at a later time as a review when studying for the Fundamentals of Engineering Exam.

Conceptual Problems. Throughout the text, usually at the end of each chapter, there is a set of problems that involve conceptual situations related to the application of the principles contained in the chapter. These analysis and design problems are intended to engage students in thinking through a real-life situation as depicted in a photo. They can be assigned after the students have developed some expertise in the subject matter and they work well either for individual or team projects.

Homework Problems. Apart from the Preliminary, Fundamental, and Conceptual type problems mentioned previously, other types of problems contained in the book include the following:

- **General Analysis and Design Problems.** The majority of problems in the book depict realistic situations encountered in engineering practice. Some of these problems come from actual products used in industry. It is hoped that this realism will both stimulate the student's interest in engineering mechanics and provide a means for developing the skill to reduce any such problem from its physical description to a model or symbolic representation to which the principles of mechanics may be applied.

Throughout the book, there is an approximate balance of problems using either SI of FPS units. Furthermore, in any set, an attempt has been made to arrange the problems in order of increasing difficulty, except for the end of chapter review problems, which are presented in random order. Problems that are simply indicated by a problem number have an answer given in the back of the book. However, an asterisk (*) before every fourth problem number indicates a problem without an answer.

Accuracy. In addition to the author, the text and problem solutions have been thoroughly checked for accuracy by four other parties: Scott Hendricks, Virginia Polytechnic Institute and State University; Karim Nohra, University of South Florida; Kurt Norlin, Bittner Development Group; and finally Kai Beng Yap, a practicing engineer.

CONTENTS

The book is divided into two parts, and the material is covered in the traditional manner.

Statics. The subject of statics is presented in 6 chapters. The text begins in Chapter 1 with an introduction to mechanics and a discussion of units. The notion of a vector and the properties of a concurrent force system are introduced in Chapter 2. Chapter 3 contains a general discussion of concentrated force systems and the methods used to simplify them. The principles of rigid-body equilibrium are developed in Chapter 4 and then applied to specific problems involving the equilibrium of trusses, frames, and machines in Chapter 5. Finally, topics related to the center of gravity, centroid, and moment of inertia are treated in Chapter 6.

Mechanics of Materials. This portion of the text is covered in 10 chapters. Chapter 7 begins with a formal definition of both normal and shear stress, and a discussion of normal stress in axially loaded members and average shear stress caused by direct shear; finally, normal and shear strain are defined. In Chapter 8 a discussion of some of the important mechanical properties of materials is given. Separate treatments of axial load, torsion, bending, and transverse shear are presented in Chapters 9, 10, 11, and 12, respectively. Chapter 13 provides a partial review of the material covered in the previous chapters, in which the state of stress resulting from combined loadings is discussed. In Chapter 14 the concepts for transforming stress and strain are presented. Chapter 15 provides a means for a further summary and review of previous material by covering design of beams based on allowable stress. In Chapter 16 various methods for computing deflections of beams are presented, including the method for finding the reactions on these members if they are statically indeterminate. Lastly, Chapter 17 provides a discussion of column buckling.

Sections of the book that contain more advanced material are indicated by a star (*). Time permitting, some of these topics may be included in the course. Furthermore, this material provides a suitable reference for basic principles when it is covered in other courses, and it can be used as a basis for assigning special projects.

Alternative Method for Coverage of Mechanics of Materials. It is possible to cover many of the topics in the text in several different sequences. For example, some instructors prefer to cover stress and strain transformations *first*, before discussing specific applications of axial load, torsion, bending, and shear. One possible method for doing this would be to first cover stress and strain and its transformations, Chapter 7 and Chapter 14, then Chapters 8 through 13 can be covered with no loss in continuity.

ACKNOWLEDGMENTS

Over the years, this text has been shaped by the suggestions and comments of many of my colleagues in the teaching profession. Their encouragement and willingness to provide constructive criticism are very much appreciated and it is hoped that they will accept this anonymous recognition. A note of thanks is also given to the reviewers of both my *Engineering Mechanics: Statics* and *Mechanics of Materials* texts. Their comments have guided the improvement of this book as well.

In particular, I would like to thank:

- D. Kingsbury—Arizona State University
- S. Redkar—Arizona State University
- P. Mokashi—Ohio State University
- S. Seetharaman—Ohio State University
- H. Salim—University of Missouri—Columbia

During the production process I am thankful for the assistance of Rose Kernan, my production editor for many years, and to my wife, Conny, for her help in proofreading and typing, that was needed to prepare the manuscript for publication.

I would also like to thank all my students who have used the previous edition and have made comments to improve its contents; including those in the teaching profession who have taken the time to e-mail me their comments.

I would greatly appreciate hearing from you if at any time you have any comments or suggestions regarding the contents of this edition.

Russell Charles Hibbeler
hibbeler@bellsouth.net

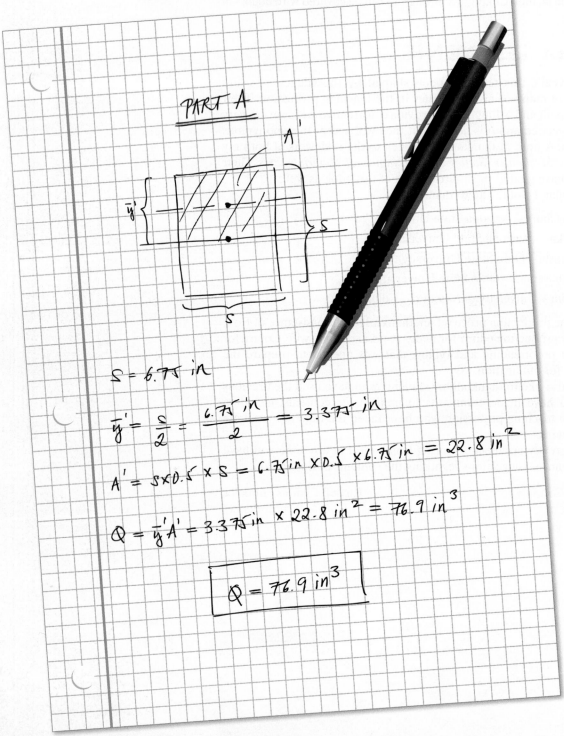

PART A

$S = 6.75$ in

$\bar{y}' = \dfrac{S}{2} = \dfrac{6.75 \text{ in}}{2} = 3.375$ in

$A' = S \times 0.5 \times S = 6.75 \text{ in} \times 0.5 \times 6.75 \text{ in} = 22.8 \text{ in}^2$

$Q = \bar{y}' A' = 3.375 \text{ in} \times 22.8 \text{ in}^2 = 76.9 \text{ in}^3$

$$\boxed{Q = 76.9 \text{ in}^3}$$

your answer specific feedback

Express your answer to three significant figures and include appropriate units.

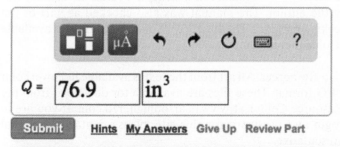

$Q =$ 76.9   in^3

Submit   **Hints** **My Answers** Give Up Review Part

Incorrect; Try Again; 5 attempts remaining

The distance between the horizontal centroidal axis of area A' and the neutral axis of the beam's cross section is half the distance between the top of the shaft and the neutral axis.

www.MasteringEngineering®.com

RESOURCES FOR INSTRUCTORS

- **MasteringEngineering.** This online Tutorial Homework program allows you to integrate dynamic homework with automatic grading and adaptive tutoring. MasteringEngineering allows you to easily track the performance of your entire class on an assignment-by-assignment basis, or the detailed work of an individual student.

- **Instructor's Solutions Manual.** An instructor's solutions manual was prepared by the author. The manual was also checked as part of the accuracy checking program. The Instructor Solutions Manual is available at www.pearsonhighered.com.

- **Presentation Resources.** All art from the text is available in PowerPoint slide and JPEG format. These files are available for download from the Instructor Resource Center at www.pearsonhighered.com. If you are in need of a login and password for this site, please contact your local Pearson representative.

- **Video Solutions.** Developed by Professor Edward Berger, Purdue University, video solutions located on the Pearson Engineering Portal offer step-by-step solution walkthroughs of representative homework problems from each section of the text. Make efficient use of class time and office hours by showing students the complete and concise problem solving approaches that they can access anytime and view at their own pace. The videos are designed to be a flexible resource to be used however each instructor and student prefers. A valuable tutorial resource, the videos are also helpful for student self-evaluation as students can pause the videos to check their understanding and work alongside the video. Access the videos at pearsonhighered.com/engineering-resources/ and follow the links for the *Statics and Mechanics of Materials* text.

RESOURCES FOR STUDENTS

- **Mastering Engineering.** Tutorial homework problems emulate the instrutor's office-hour environment.

- **Engineering Portal**—The Pearson Engineering Portal, located at pearsonhighered.com/engineering-resources/ includes opportunities for practice and review including:

- **Video Solutions**—Complete, step-by-step solution walkthroughs of representative homework problems from each section of the text. Videos offer fully worked solutions that show every step of the representative homework problems—this helps students make vital connections between concepts.

CONTENTS

mechanics of materials

STATICS AND MECHANICS OF MATERIALS

CHAPTER

CHAPTER 1

(© Andrew Peacock/Lonely Planet Images/Getty Images)

Large cranes such as this one are required to lift extremely large loads. Their design is based on the basic principles of statics and dynamics, which form the subject matter of engineering mechanics.

GENERAL PRINCIPLES

CHAPTER OBJECTIVES

- To provide an introduction to the basic quantities and idealizations of mechanics.
- To state Newton's Laws of Motion.
- To review the principles for applying the SI system of units.
- To examine the standard procedures for performing numerical calculations.
- To present a general guide for solving problems.

1.1 MECHANICS

Mechanics can be defined as that branch of the physical sciences concerned with the state of rest or motion of bodies that are subjected to the action of forces. In this book we will study two important branches of mechanics, namely, statics and mechanics of materials. These subjects form a suitable basis for the design and analysis of many types of structural, mechanical, or electrical devices encountered in engineering.

Statics deals with the equilibrium of bodies, that is, it is used to determine the forces acting either external to the body or within it that are necessary to keep the body in equilibrium. *Mechanics of materials* studies the relationships between the external loads and the distribution of internal forces acting within the body. This subject is also concerned with finding the deformations of the body, and it provides a study of the body's stability.

In this book we will first study the principles of statics, since for the design and analysis of any structural or mechanical element it is *first* necessary to determine the forces acting both on and within its various members. Once these internal forces are determined, the size of the members, their deflection, and their stability can then be determined using the fundamentals of mechanics of materials, which will be covered later.

Historical Development. The subject of statics developed very early in history because its principles can be formulated simply from measurements of geometry and force. For example, the writings of Archimedes (287–212 B.C.) deal with the principle of the lever. Studies of the pulley and inclined plane are also recorded in ancient writings—at times when the requirements for engineering were limited primarily to building construction.

The origin of mechanics of materials dates back to the beginning of the seventeenth century, when Galileo performed experiments to study the effects of loads on rods and beams made of various materials. However, at the beginning of the eighteenth century, experimental methods for testing materials were vastly improved, and at that time many experimental and theoretical studies in this subject were undertaken primarily in France, by such notables as Saint-Venant, Poisson, Lamé, and Navier.

Over the years, after many of the fundamental problems of mechanics of materials had been solved, it became necessary to use advanced mathematical and computer techniques to solve more complex problems. As a result, this subject has expanded into other areas of mechanics, such as the *theory of elasticity* and the *theory of plasticity*. Research in these fields is ongoing, in order to meet the demands for solving more advanced problems in engineering.

1.2   FUNDAMENTAL CONCEPTS

Before we begin our study, it is important to understand the definitions of certain fundamental concepts and principles.

Mass. *Mass* is a measure of a *quantity of matter* that is used to compare the action of one body with that of another. This property provides a measure of the resistance of matter to a change in velocity.

Force. In general, *force* is considered as a "push" or "pull" exerted by one body on another. This interaction can occur when there is direct contact between the bodies, such as a person pushing on a wall, or it can occur through a distance when the bodies are physically separated. Examples of the latter type include gravitational, electrical, and magnetic forces. In any case, a force is completely characterized by its magnitude, direction, and point of application.

Particle. A *particle* has a mass, but a size that can be neglected. For example, the size of the earth is insignificant compared to the size of its orbit, and therefore the earth can be modeled as a particle when studying its orbital motion. When a body is idealized as a particle, the principles of mechanics reduce to a rather simplified form since the geometry of the body *will not be involved* in the analysis of the problem.

Rigid Body. A *rigid body* can be considered as a combination of a large number of particles in which all the particles remain at a fixed distance from one another, both before and after applying a load. This model is important because the material properties of any body that is assumed to be rigid will not have to be considered when studying the effects of forces acting on the body. In most cases the actual deformations occurring in structures, machines, mechanisms, and the like are relatively small, and the rigid-body assumption is suitable for analysis.

Concentrated Force. A *concentrated force* represents the effect of a loading which is assumed to act at a point on a body. We can represent a load by a concentrated force, provided the area over which the load is applied is very small compared to the overall size of the body. An example would be the contact force between a wheel and the ground.

Three forces act on the ring. Since these forces all meet at a point, then for any force analysis, we can assume the ring to be represented as a particle.

Steel is a common engineering material that does not deform very much under load. Therefore, we can consider this railroad wheel to be a rigid body acted upon by the concentrated force of the rail.

Newton's Three Laws of Motion. Engineering mechanics is formulated on the basis of Newton's three laws of motion, the validity of which is based on experimental observation. These laws apply to the motion of a particle as measured from a *nonaccelerating* reference frame. They may be briefly stated as follows.

First Law. A particle originally at rest, or moving in a straight line with constant velocity, tends to remain in this equilibrium state provided the particle is *not* subjected to an unbalanced force, Fig. 1–1*a*.

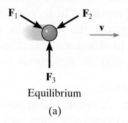

Equilibrium

(a)

Second Law. A particle acted upon by an *unbalanced force* **F** experiences an acceleration **a** that has the same direction as the force and a magnitude that is directly proportional to the force, Fig. 1–1*b*. If the particle has a mass m, this law may be expressed mathematically as

$$\mathbf{F} = m\mathbf{a} \tag{1–1}$$

F ⟶ ⬤   a ⟶

Accelerated motion

(b)

Third Law. The mutual forces of action and reaction between two particles are equal, opposite, and collinear, Fig. 1–1*c*.

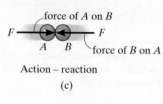

force of *A* on *B*

F — A B — F

force of *B* on *A*

Action – reaction

(c)

Fig. 1–1

Newton's Law of Gravitational Attraction. Shortly after formulating his three laws of motion, Newton postulated a law governing the gravitational attraction between any two particles. Stated mathematically,

$$F = G \frac{m_1 m_2}{r^2}$$ (1–2)

where

F = force of gravitation between the two particles

G = universal constant of gravitation; according to experimental evidence, $G = 66.73(10^{-12})$ m^3/(kg·s^2)

m_1, m_2 = mass of each of the two particles

r = distance between the two particles

Weight. According to Eq. 1–2, any two particles or bodies have a mutual attractive (gravitational) force acting between them. In the case of a particle located at or near the surface of the earth, however, the only gravitational force having any sizable magnitude is that between the earth and the particle. Consequently, this force, called the **weight**, will be the only gravitational force considered in our study of mechanics.

From Eq. 1–2, we can develop an approximate expression for finding the weight W of a particle having a mass $m_1 = m$. If we assume the earth to be a nonrotating sphere of constant density and having a mass $m_2 = M_e$, then if r is the distance between the earth's center and the particle, we have

$$W = G \frac{mM_e}{r^2}$$

Letting $g = GM_e/r^2$ yields

This astronaut's weight is diminished since she is far removed from the gravitational field of the earth. (© NikoNomad/Shutterstock)

$$\boxed{W = mg}$$ (1–3)

By comparison with $\mathbf{F} = m\mathbf{a}$, we can see that g is the acceleration due to gravity. Since it depends on r, the weight of a particle or body is *not* an absolute quantity. Instead, its magnitude is determined from where the measurement was made. For most engineering calculations, however, g is determined at sea level and at a latitude of 45°, which is considered the "standard location."

TABLE 1–1 SI System of Units				
Name	Length	Time	Mass	Force
International System of Units	meter	second	kilogram	newton*
SI	m	s	kg	$\dfrac{N}{\left(\dfrac{kg \cdot m}{s^2}\right)}$

*Derived unit.

1.3   THE INTERNATIONAL SYSTEM OF UNITS

The four basic quantities—length, time, mass, and force—are not all independent from one another; in fact, they are *related* by Newton's second law of motion, $\mathbf{F} = m\mathbf{a}$. Because of this, the *units* used to measure these quantities cannot *all* be selected arbitrarily. The equality $\mathbf{F} = m\mathbf{a}$ is maintained only if three of the four units, called **base units**, are *defined* and the fourth unit is then *derived* from the equation.

For the International System of Units, abbreviated SI after the French "Système International d'Unités," length is in meters (m), time is in seconds (s), and mass is in kilograms (kg), Table 1–1. The unit of force, called a **newton** (N), is *derived* from $\mathbf{F} = m\mathbf{a}$. Thus, 1 newton is equal to a force required to give 1 kilogram of mass an acceleration of 1 m/s^2 ($N = kg \cdot m/s^2$).

If the weight of a body located at the "standard location" is to be determined in newtons, then Eq. 1–3 must be applied. Here measurements give $g = 9.806\ 65 \text{ m/s}^2$; however, for calculations the value $g = 9.81 \text{ m/s}^2$ will be used. Thus,

$$W = mg \quad (g = 9.81 \text{ m/s}^2) \tag{1–4}$$

Therefore, a body of mass 1 kg has a weight of 9.81 N, a 2-kg body weighs 19.62 N, and so on, Fig. 1–2. Perhaps it is easier to remember that a small apple weighs one newton. Also, by comparison with the U.S. Customary system of units (FPS),

$$1 \text{ pound (lb)} = 4.448 \text{ N}$$
$$1 \text{ foot (ft)} = 0.3048 \text{ m}$$

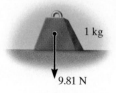

1 kg

9.81 N

Fig. 1–2

Prefixes. When a numerical quantity is either very large or very small, the units used to define its size may be modified by using a prefix. Some of the prefixes used in the SI system are shown in Table 1–2. Each represents a multiple or submultiple of a unit which, if applied successively, moves the decimal point of a numerical quantity to every third place.* For example, 4 000 000 N = 4 000 kN (kilo-newton) = 4 MN (mega-newton), or 0.005 m = 5 mm (milli-meter). Notice that the SI system does not include the multiple deca (10) or the submultiple centi (0.01), which form part of the metric system. Except for some volume and area measurements, the use of these prefixes is generally avoided in science and engineering.

TABLE 1–2   Prefixes

	Exponential Form	Prefix	SI Symbol
Multiple			
1 000 000 000	10^9	giga	G
1 000 000	10^6	mega	M
1 000	10^3	kilo	k
Submultiple			
0.001	10^{-3}	milli	m
0.000 001	10^{-6}	micro	μ
0.000 000 001	10^{-9}	nano	n

Rules for Use. Here are a few of the important rules that describe the proper use of the various SI symbols:

- Quantities defined by several units which are multiples of one another are separated by a *dot* to avoid confusion with prefix notation, as indicated by $N = kg \cdot m/s^2 = kg \cdot m \cdot s^{-2}$. Also, $m \cdot s$ (meter-second), whereas ms (milli-second).

- The exponential power on a unit having a prefix refers to both the unit *and* its prefix. For example, $\mu N^2 = (\mu N)^2 = \mu N \cdot \mu N$. Likewise, mm^2 represents $(mm)^2 = mm \cdot mm$.

- With the exception of the base unit the kilogram, in general avoid the use of a prefix in the denominator of composite units. For example, do not write N/mm, but rather kN/m; also, m/mg should be written as Mm/kg.

- When performing calculations, represent the numbers in terms of their *base or derived units* by converting all prefixes to powers of 10. The final result should then be expressed using a *single prefix*. Also, after calculation, it is best to keep numerical values between 0.1 and 1000; otherwise, a suitable prefix should be chosen. For example,

$$(50 \text{ kN})(60 \text{ nm}) = [50(10^3) \text{ N}][60(10^{-9}) \text{ m}]$$
$$= 3000(10^{-6}) \text{ N} \cdot \text{m} = 3(10^{-3}) \text{ N} \cdot \text{m} = 3 \text{ mN} \cdot \text{m}$$

*The kilogram is the only base unit that is defined with a prefix.

1.4 NUMERICAL CALCULATIONS

Numerical work in engineering practice is most often performed by using handheld calculators and computers. It is important, however, that the answers to any problem be reported with justifiable accuracy using appropriate significant figures. In this section we will discuss these topics together with some other important aspects involved in all engineering calculations.

Computers are often used in engineering for advanced design and analysis. (© Blaize Pascall/Alamy)

Dimensional Homogeneity. The terms of any equation used to describe a physical process must be ***dimensionally homogeneous***; that is, each term must be expressed in the same units. Provided this is the case, all the terms of an equation can then be combined if numerical values are substituted for the variables. Consider, for example, the equation $s = vt + \frac{1}{2} at^2$, where, in SI units, s is the position in meters, m, t is time in seconds, s, v is velocity in m/s, and a is acceleration in m/s^2. Regardless of how this equation is evaluated, it maintains its dimensional homogeneity. In the form stated, each of the three terms is expressed in meters $\left[\text{m}, (\text{m}/\cancel{s})\cancel{s}, (\text{m}/\cancel{s^2})\cancel{s^2} \right]$ or solving for a, $a = 2s/t^2 - 2v/t$, the terms are each expressed in units of m/s^2 [m/s^2, m/s^2, (m/s)/s].

Keep in mind that problems in mechanics always involve the solution of dimensionally homogeneous equations, and so this fact can then be used as a partial check for algebraic manipulations of an equation.

Significant Figures. The number of significant figures contained in any number determines the accuracy of the number. For instance, the number 4981 contains four significant figures. However, if zeros occur at the end of a whole number, it may be unclear as to how many significant figures the number represents. For example, 23 400 might have three (234), four (2340), or five (23 400) significant figures. To avoid these ambiguities, we will use ***engineering notation*** to report a result. This requires that numbers be rounded off to the appropriate number of significant digits and then expressed in multiples of (10^3), such as (10^3), (10^6), or (10^{-9}). For instance, if 23 400 has five significant figures, it is written as $23.400(10^3)$, but if it has only three significant figures, it is written as $23.4(10^3)$.

If zeros occur at the beginning of a number that is less than one, then the zeros are not significant. For example, 0.008 21 has three significant figures. Using engineering notation, this number is expressed as $8.21(10^{-3})$. Likewise, 0.000 582 can be expressed as $0.582(10^{-3})$ or $582(10^{-6})$.

Rounding Off Numbers. Rounding off a number is necessary so that the accuracy of the result will be the same as that of the problem data. As a general rule, any numerical figure ending in a number greater than five is rounded up and a number less than five is not rounded up. The rules for rounding off numbers are best illustrated by example. Suppose the number 3.5587 is to be rounded off to *three* significant figures. Because the fourth digit (8) is *greater than* 5, the third number is rounded up to 3.56. Likewise 0.5896 becomes 0.590 and 9.3866 becomes 9.39. If we round off 1.341 to three significant figures, because the fourth digit (1) is *less than* 5, then we get 1.34. Likewise 0.3762 becomes 0.376 and 9.871 becomes 9.87. There is a special case for any number that ends in a 5. As a general rule, if the digit preceding the 5 is an *even number*, then this digit is *not* rounded up. If the digit preceding the 5 is an *odd number*, then it is rounded up. For example, 75.25 rounded off to three significant digits becomes 75.2, 0.1275 becomes 0.128, and 0.2555 becomes 0.256.

Calculations. When a sequence of calculations is performed, it is best to store the intermediate results in the calculator. In other words, do not round off calculations until expressing the final result. This procedure maintains precision throughout the series of steps to the final solution. In this book we will generally round off the answers to *three significant figures* since most of the data in engineering mechanics, such as geometry and loads, may be reliably measured to this accuracy.

1.5 GENERAL PROCEDURE FOR ANALYSIS

Attending a lecture, reading this book, and studying the example problems helps, but **the most effective way of learning the principles of engineering mechanics is to *solve problems*.** To be successful at this, it is important to always present the work in a *logical* and *orderly manner*, as suggested by the following sequence of steps:

- Read the problem carefully and try to correlate the actual physical situation with the theory studied.

- Tabulate the problem data and *draw to a large scale* any necessary diagrams.

- Apply the relevant principles, generally in mathematical form. When writing any equations, be sure they are dimensionally homogeneous.

- Solve the necessary equations, and report the answer with no more than three significant figures.

- Study the answer with technical judgment and common sense to determine whether or not it seems reasonable.

When solving problems, do the work as neatly as possible. Being neat will stimulate clear and orderly thinking, and vice versa.

1

IMPORTANT POINTS

- A particle has a mass but a size that can be neglected, and a rigid body does not deform under load.

- A force is considered as a "push" or "pull" of one body on another.

- Concentrated forces are assumed to act at a point on a body.

- Newton's three laws of motion should be memorized.

- Mass is measure of a quantity of matter that does not change from one location to another. Weight refers to the gravitational attraction of the earth on a body or quantity of mass. Its magnitude depends upon the elevation at which the mass is located.

- In the SI system the unit of force, the newton, is a derived unit. The meter, second, and kilogram are base units.

- Prefixes G, M, k, m, μ, and n are used to represent large and small numerical quantities. Their exponential size should be known, along with the rules for using the SI units.

- Perform numerical calculations with several significant figures, and then report the final answer to three significant figures.

- Algebraic manipulations of an equation can be checked in part by verifying that the equation remains dimensionally homogeneous.

- Know the rules for rounding off numbers.

EXAMPLE   1.1

Convert 2 km/h to m/s. How many ft/s is this?

SOLUTION

Since 1 km = 1000 m and 1 h = 3600 s, the factors of conversion are arranged in the following order, so that a cancellation of the units can be applied:

$$2 \text{ km/h} = \frac{2 \cancel{\text{km}}}{\cancel{\text{h}}} \left(\frac{1000 \text{ m}}{\cancel{\text{km}}} \right) \left(\frac{1 \cancel{\text{h}}}{3600 \text{ s}} \right)$$

$$= \frac{2000 \text{ m}}{3600 \text{ s}} = 0.556 \text{ m/s} \qquad\qquad Ans.$$

Since 1 ft = 0.3048 m, then

$$0.556 \text{ m/s} = \left(\frac{0.556 \cancel{\text{m}}}{\text{s}} \right) \left(\frac{1 \text{ ft}}{0.3048 \cancel{\text{m}}} \right)$$

$$= 1.82 \text{ ft/s} \qquad\qquad Ans.$$

NOTE: Remember to round off the final answer to three significant figures.

EXAMPLE   1.2

Convert 300 lb · ft to appropriate SI units.

SOLUTION

Since 1 lb = 4.448 N and 1 ft = 0.3048 m, then we have

$$300 \cancel{\text{lb}} \cdot \cancel{\text{ft}} = \left(\frac{4.448 \text{ N}}{1 \cancel{\text{lb}}} \right) \left(\frac{0.3048 \text{ m}}{1 \cancel{\text{ft}}} \right)$$

$$= 407 \text{ N} \cdot \text{m} \qquad\qquad Ans.$$

1

EXAMPLE 1.3

Evaluate each of the following and express with SI units having an appropriate prefix: (a) (50 mN)(6 GN), (b) (400 mm)(0.6 MN)^2, (c) 45 MN^3/900 Gg.

SOLUTION

First convert each number to base units, perform the indicated operations, then choose an appropriate prefix.

Part (a)

$$(50 \text{ mN})(6 \text{ GN}) = \left[50(10^{-3}) \text{ N}\right]\left[6(10^{9}) \text{ N}\right]$$

$$= 300(10^{6}) \text{ N}^2$$

$$= 300(10^{6}) \text{ N}^2\left(\frac{1 \text{ kN}}{10^3 \text{ N}}\right)\left(\frac{1 \text{ kN}}{10^3 \text{ N}}\right)$$

$$= 300 \text{ kN}^2 \qquad\qquad Ans.$$

NOTE: Keep in mind the convention kN^2 = (kN)^2 = 10^6 N^2.

Part (b)

$$(400 \text{ mm})(0.6 \text{ MN})^2 = \left[400(10^{-3}) \text{ m}\right]\left[0.6(10^{6}) \text{ N}\right]^2$$

$$= \left[400(10^{-3}) \text{ m}\right]\left[0.36(10^{12}) \text{ N}^2\right]$$

$$= 144(10^{9}) \text{ m} \cdot \text{N}^2$$

$$= 144 \text{ Gm} \cdot \text{N}^2 \qquad\qquad Ans.$$

We can also write

$$144(10^{9}) \text{ m} \cdot \text{N}^2 = 144(10^{9}) \text{ m} \cdot \text{N}^2\left(\frac{1 \text{ MN}}{10^6 \text{ N}}\right)\left(\frac{1 \text{ MN}}{10^6 \text{ N}}\right)$$

$$= 0.144 \text{ m} \cdot \text{MN}^2 \qquad\qquad Ans.$$

Part (c)

$$\frac{45 \text{ MN}^3}{900 \text{ Gg}} = \frac{45(10^6 \text{ N})^3}{900(10^6) \text{ kg}}$$

$$= 50(10^{9}) \text{ N}^3/\text{kg}$$

$$= 50(10^{9}) \text{ N}^3\left(\frac{1 \text{ kN}}{10^3 \text{ N}}\right)^3 \frac{1}{\text{kg}}$$

$$= 50 \text{ kN}^3/\text{kg} \qquad\qquad Ans.$$

PROBLEMS

The answers to all but every fourth problem (asterisk) are given in the back of the book.

1–1. What is the weight in newtons of an object that has a mass of (a) 8 kg, (b) 0.04 kg, (c) 760 Mg?

1–2. Represent each of the following combinations of units in the correct SI form using an appropriate prefix: (a) kN/μs, (b) Mg/mN, (c) MN/(kg $\cdot$ ms).

1–3. Represent each of the following combinations of units in the correct SI form: (a) Mg/ms, (b) N/mm, (c) mN/(kg $\cdot$ μs).

*1–4. Convert: (a) 200 lb $\cdot$ ft to N $\cdot$ m, (b) 350 lb/ft^3 to kN/m^3, (c) 8 ft/h to mm/s. Express the result to three significant figures. Use an appropriate prefix.

1–5. Represent each of the following as a number between 0.1 and 1000 using an appropriate prefix: (a) 45 320 kN, (b) 568(10^5) mm, (c) 0.00563 mg.

1–6. Round off the following numbers to three significant figures: (a) 58 342 m, (b) 68.534 s, (c) 2553 N, (d) 7555 kg.

1–7. Represent each of the following quantities in the correct SI form using an appropriate prefix: (a) 0.000 431 kg, (b) 35.3(10^3) N, (c) 0.005 32 km.

*1–8. Represent each of the following combinations of units in the correct SI form using an appropriate prefix: (a) Mg/mm, (b) mN/μs, (c) μm $\cdot$ Mg.

1–9. Represent each of the following combinations of units in the correct SI form using an appropriate prefix: (a) m/ms, (b) μkm, (c) ks/mg, (d) km $\cdot$ μN.

1–10. Represent each of the following combinations of units in the correct SI form using an appropriate prefix: (a) GN $\cdot$ μm, (b) kg/μm, (c) N/ks^2, and (d) kN/μs.

1–11. Represent each of the following with SI units having an appropriate prefix: (a) 8653 ms, (b) 8368 N, (c) 0.893 kg.

*1–12. Evaluate each of the following to three significant figures and express each answer in SI units using an appropriate prefix: (a) (684 μm)/(43 ms), (b) (28 ms)(0.0458 Mm)/(348 mg), (c) (2.68 mm)(426 Mg).

*1–13. Convert each of the following to three significant figures. (a) 20 lb $\cdot$ ft to N $\cdot$ m, (b) 450 lb/ft^3 to kN/m^3, (c) 15 ft/h to mm/s.

1–14. Evaluate each of the following to three significant figures and express each answer in SI units using an appropriate prefix: (a) (212 mN)^2, (b) (52 800 ms)^2, (c) [548(10^6)]$^{1/2}$ ms.

1–15. Using the SI system of units, show that Eq. 1–2 is a dimensionally homogeneous equation which gives F in newtons. Determine to three significant figures the gravitational force acting between two spheres that are touching each other. The mass of each sphere is 200 kg and the radius is 300 mm.

*1–16. The *pascal* (Pa) is actually a very small unit of pressure. To show this, convert 1 Pa = 1 N/m^2 to lb/ft^2. Atmosphere pressure at sea level is 14.7 lb/in^2. How many pascals is this?

1–17. What is the weight in newtons of an object that has a mass of: (a) 10 kg, (b) 0.5 g, (c) 4.50 Mg? Express the result to three significant figures. Use an appropriate prefix.

1–18. Evaluate each of the following to three significant figures and express each answer in SI units using an appropriate prefix: (a) 354 mg(45 km)/(0.0356 kN), (b) (0.004 53 Mg)(201 ms), (c) 435 MN/23.2 mm.

1–19. A concrete column has a diameter of 350 mm and a length of 2 m. If the density (mass/volume) of concrete is 2.45 Mg/m^3, determine the weight of the column in pounds.

*1–20. Two particles have a mass of 8 kg and 12 kg, respectively. If they are 800 mm apart, determine the force of gravity acting between them. Compare this result with the weight of each particle.

1–21. If a man weighs 155 lb on earth, specify (a) his mass in kilograms, and (b) his weight in newtons. If the man is on the moon, where the acceleration due to gravity is $g_m = 5.30$ ft/s^2, determine (c) his weight in pounds, and (d) his mass in kilograms.

CHAPTER 2

(© Vasiliy Koval/Fotolia)

This electric transmission tower is stabilized by cables that exert forces on the tower at their points of connection. In this chapter we will show how to express these forces as Cartesian vectors, and then determine their resultant.

FORCE VECTORS

- To show how to add forces and resolve them into components using the Parallelogram Law.
- To express force and position as Cartesian vectors.
- To introduce the dot product in order to use it to find the angle between two vectors or the projection of one vector onto another.

2.1 SCALARS AND VECTORS

Many physical quantities in engineering mechanics are measured using either scalars or vectors.

Scalar. A *scalar* is any positive or negative physical quantity that can be completely specified by its *magnitude*. Examples of scalar quantities include length, mass, and time.

Vector. A *vector* is any physical quantity that requires both a *magnitude* and a *direction* for its complete description. Examples of vectors encountered in statics are force, position, and moment. A vector **A** is shown graphically by an arrow, Fig. 2–1. The length of the arrow represents the *magnitude* of the vector, and the angle θ between the vector and a fixed axis defines the *direction of its line of action*. The head or tip of the arrow indicates the *sense of direction* of the vector.

In print, vector quantities are represented by boldface letters such as **A**, and the magnitude of a vector is italicized, A. For handwritten work, it is often convenient to denote a vector quantity by simply drawing an arrow above it, $\vec{A}$.

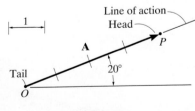

Fig. 2–1

17

2

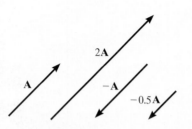

Scalar multiplication and division

Fig. 2–2

2.2 VECTOR OPERATIONS

Multiplication and Division of a Vector by a Scalar. If a vector is multiplied or divided by a positive scalar, its magnitude is increased by that amount. Multiplying or dividing by a negative scalar will also change the directional sense of the vector. Graphic examples of these operations are shown in Fig. 2–2.

Vector Addition. When adding two vectors together it is important to account for both their magnitudes and their directions. To do this we must use the ***parallelogram law***. To illustrate, the two ***component vectors*** **A** and **B** in Fig. 2–3a are added to form a ***resultant vector*** $\mathbf{R} = \mathbf{A} + \mathbf{B}$ using the following procedure:

- First join the tails of the components at a point to make them concurrent, Fig. 2–3b.

- From the head of **B**, draw a line parallel to **A**. Draw another line from the head of **A** that is parallel to **B**. These two lines intersect at point *P* to form the adjacent sides of a parallelogram.

- The diagonal of this parallelogram that extends to *P* forms **R**, which then represents the resultant vector $\mathbf{R} = \mathbf{A} + \mathbf{B}$, Fig. 2–3c.

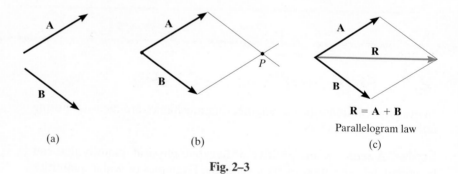

(a)   (b)   (c)

$\mathbf{R} = \mathbf{A} + \mathbf{B}$

Parallelogram law

Fig. 2–3

We can also add **B** to **A**, Fig. 2–4a, using the ***triangle rule***, which is a special case of the parallelogram law, whereby vector **B** is added to vector **A** in a "head-to-tail" fashion, i.e., by connecting the tail of **B** to the head of **A**, Fig. 2–4b. The resultant **R** extends from the tail of **A** to the head of **B**. In a similar manner, **R** can also be obtained by adding **A** to **B**, Fig. 2–4c. By comparison, it is seen that vector addition is commutative; in other words, the vectors can be added in either order, i.e., $\mathbf{R} = \mathbf{A} + \mathbf{B} = \mathbf{B} + \mathbf{A}$.

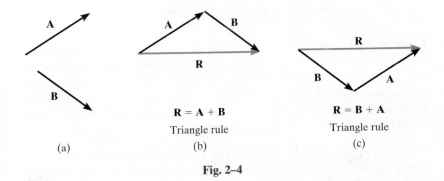

A
B
(a)
A
B
R
R = A + B
Triangle rule
(b)
R
B
A
R = B + A
Triangle rule
(c)

Fig. 2–4

As a special case, if the two vectors **A** and **B** are ***collinear***, i.e., both have the same line of action, the parallelogram law reduces to an *algebraic* or *scalar addition* $R = A + B$, as shown in Fig. 2–5.

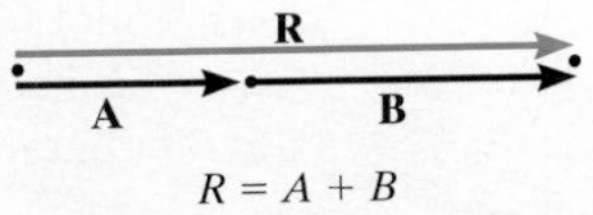

R
A
B
R = A + B

Addition of collinear vectors

Fig. 2–5

Vector Subtraction. The resultant of the *difference* between two vectors **A** and **B** of the same type may be expressed as

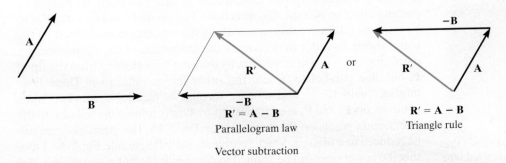

$$\mathbf{R}' = \mathbf{A} - \mathbf{B} = \mathbf{A} + (-\mathbf{B})$$

This vector sum is shown graphically in Fig. 2–6. Subtraction is therefore defined as a special case of addition, so the rules of vector addition also apply to vector subtraction.

A
B
R′
A
−B
R′ = A − B
Parallelogram law
or
−B
R′
A
R′ = A − B
Triangle rule

Vector subtraction

Fig. 2–6

2.3 VECTOR ADDITION OF FORCES

Experimental evidence has shown that a force is a vector quantity since it has a specified magnitude, direction, and sense and it adds according to the parallelogram law. Two common problems in statics involve either finding the resultant force, knowing its components, or resolving a known force into two components. We will now describe how each of these problems is solved using the parallelogram law.

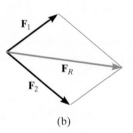

The parallelogram law must be used to determine the resultant of the two forces acting on the hook.

Finding a Resultant Force. The two component forces F_1 and F_2 acting on the pin in Fig. 2–7a are added together to form the resultant force $F_R = F_1 + F_2$, using the parallelogram law as shown in Fig. 2–7b. From this construction, or using the triangle rule, Fig. 2–7c, we can then apply the law of cosines or the law of sines to the triangle in order to obtain the magnitude of the resultant force and its direction.

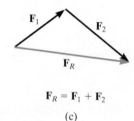

F_1

F_2

(a)

F_1

F_R

F_2

(b)

F_1

F_2

F_R

$F_R = F_1 + F_2$

(c)

Fig. 2–7

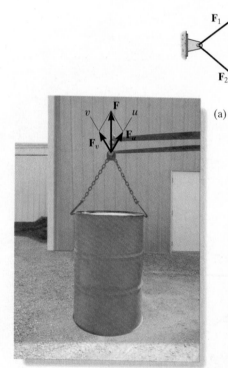

Using the parallelogram law the supporting force F can be resolved into components acting along the u and v axes.

Finding the Components of a Force. Sometimes it is necessary to resolve a force into two *components* in order to study its pulling or pushing effect in two specific directions. For example, in Fig. 2–8a, F is to be resolved into two components along the two members, defined by the u and v axes. In order to determine the magnitude of each component, a parallelogram is constructed first, by drawing lines starting from the tip of F, one line parallel to u, and the other line parallel to v. These lines intersect with the v and u axes, forming a parallelogram. The force components F_u and F_v are established by simply joining the tail of F to the intersection points on the u and v axes, Fig. 2–8b. This parallelogram can be reduced to a triangle, which represents the triangle rule, Fig. 2–8c. From this, the law of sines can be applied to determine the unknown magnitudes of the components.

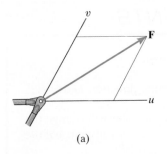
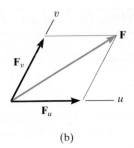
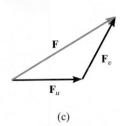

(a)   (b)   (c)

Fig. 2–8

Addition of Several Forces. If more than two forces are to be added, successive applications of the parallelogram law can be carried out in order to obtain the resultant force. For example, if three forces $\mathbf{F}_1$, $\mathbf{F}_2$, $\mathbf{F}_3$ act at a point O, Fig. 2–9, the resultant of any two of the forces is found, say, $\mathbf{F}_1 + \mathbf{F}_2$, and then this resultant is added to the third force, yielding the resultant of all three forces; i.e., $\mathbf{F}_R = (\mathbf{F}_1 + \mathbf{F}_2) + \mathbf{F}_3$. Using the parallelogram law to add more than two forces, as shown here, generally requires extensive geometric and trigonometric calculation to determine the magnitude and direction of the resultant. Instead, problems of this type are easily solved by using the "rectangular-component method," which is explained in the next section.

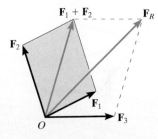

Fig. 2–9

The resultant force $\mathbf{F}_R$ on the hook requires the addition of $\mathbf{F}_1 + \mathbf{F}_2$, then this resultant is added to $\mathbf{F}_3$.

IMPORTANT POINTS

- A scalar is a positive or negative number.

- A vector is a quantity that has a magnitude, direction, and sense.

- Multiplication or division of a vector by a scalar will change the magnitude of the vector. The sense of the vector will change if the scalar is negative.

- Vectors are added or subtracted using the parallelogram law or the triangle rule.

- As a special case, if the vectors are collinear, the resultant is formed by an algebraic or scalar addition.

PROCEDURE FOR ANALYSIS

Problems that involve the addition of two forces can be solved as follows:

Parallelogram Law.

- Sketch the addition of the two "component" forces $\mathbf{F}_1$ and $\mathbf{F}_2$ according to the parallelogram law, yielding the *resultant* force $\mathbf{F}_R$ that forms the diagonal of the parallelogram, Fig. 2–10a.

- If a force $\mathbf{F}$ is to be resolved into *components* along two axes u and v, then start at the head of force $\mathbf{F}$ and construct lines parallel to the axes, thereby forming the parallelogram, Fig. 2–10b. The sides of the parallelogram represent the components, $\mathbf{F}_u$ and $\mathbf{F}_v$.

- Label all the known and unknown force magnitudes and the angles on the sketch and identify the two unknowns as the magnitude and direction of $\mathbf{F}_R$, or the magnitudes of its components.

Trigonometry.

- Redraw a half portion of the parallelogram to illustrate the triangular head-to-tail addition of the components.

- From this triangle, the magnitude of the resultant force can be determined using the law of cosines, and its direction is determined from the law of sines. The magnitudes of two force components are determined from the law of sines. The formulas are given in Fig. 2–10c.

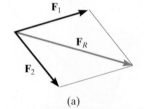

(a)

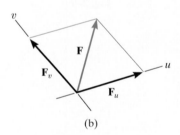

(b)

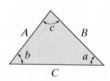

Cosine law:
$$C = \sqrt{A^2 + B^2 - 2AB \cos c}$$
Sine law:
$$\frac{A}{\sin a} = \frac{B}{\sin b} = \frac{C}{\sin c}$$

(c)

Fig. 2–10

EXAMPLE 2.1

The screw eye in Fig. 2–11a is subjected to two forces, $\mathbf{F}_1$ and $\mathbf{F}_2$. Determine the magnitude and direction of the resultant force.

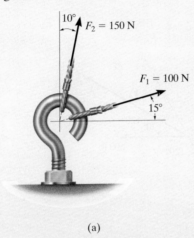

(a)

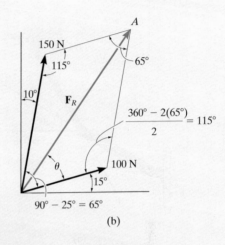

(b)

SOLUTION

Parallelogram Law. The parallelogram is formed by drawing a line from the head of $\mathbf{F}_1$ that is parallel to $\mathbf{F}_2$, and another line from the head of $\mathbf{F}_2$ that is parallel to $\mathbf{F}_1$. The resultant force $\mathbf{F}_R$ extends to where these lines intersect at point A, Fig. 2–11b. The two unknowns are the magnitude of $\mathbf{F}_R$ and the angle θ (theta).

Trigonometry. From the parallelogram, the vector triangle is shown in Fig. 2–11c. Using the law of cosines

$$F_R = \sqrt{(100 \text{ N})^2 + (150 \text{ N})^2 - 2(100 \text{ N})(150 \text{ N}) \cos 115°}$$

$$= \sqrt{10\ 000 + 22\ 500 - 30\ 000(-0.4226)} = 212.6 \text{ N}$$

$$= 213 \text{ N} \qquad\qquad Ans.$$

Applying the law of sines to determine θ,

$$\frac{150 \text{ N}}{\sin \theta} = \frac{212.6 \text{ N}}{\sin 115°} \qquad \sin \theta = \frac{150 \text{ N}}{212.6 \text{ N}} (\sin 115°)$$

$$\theta = 39.8°$$

Thus, the direction ϕ (phi) of $\mathbf{F}_R$, measured from the horizontal, is

$$\phi = 39.8° + 15.0° = 54.8° \qquad\qquad Ans.$$

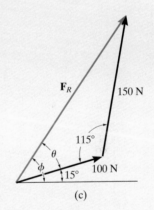

(c)

Fig. 2–11

NOTE: The results seem reasonable, since Fig. 2–11b shows $\mathbf{F}_R$ to have a magnitude larger than its components and a direction that is between them.

EXAMPLE   2.2

Resolve the horizontal 600-lb force in Fig. 2–12*a* into components acting along the *u* and *v* axes and determine the magnitudes of these components.

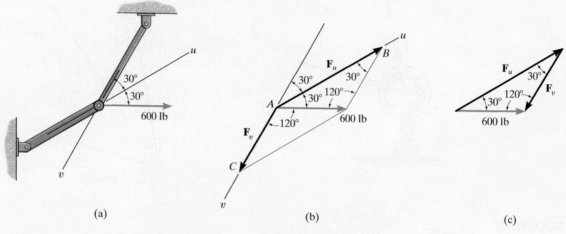

Fig. 2–12

SOLUTION

Parallelogram Law.   The parallelogram is constructed by extending a line from the *head* of the 600-lb force parallel to the *v* axis until it intersects the *u* axis at point *B*, Fig. 2–12*b*. The arrow from *A* to *B* represents $\mathbf{F}_u$. Similarly, the line extended from the head of the 600-lb force drawn parallel to the *u* axis intersects the *v* axis at point *C*, which gives $\mathbf{F}_v$.

Trigonometry.   The vector addition using the triangle rule is shown in Fig. 2–12*c*. The two unknowns are the magnitudes of $\mathbf{F}_u$ and $\mathbf{F}_v$. Applying the law of sines,

$$\frac{F_u}{\sin 120°} = \frac{600 \text{ lb}}{\sin 30°}$$

$$F_u = 1039 \text{ lb} \qquad\qquad Ans.$$

$$\frac{F_v}{\sin 30°} = \frac{600 \text{ lb}}{\sin 30°}$$

$$F_v = 600 \text{ lb} \qquad\qquad Ans.$$

NOTE: The result for F_u shows that sometimes a component can have a greater magnitude than the resultant.

EXAMPLE   2.3

Determine the magnitude of the component force **F** in Fig. 2–13a and the magnitude of the resultant force $\mathbf{F}_R$ if $\mathbf{F}_R$ is directed along the positive y axis.

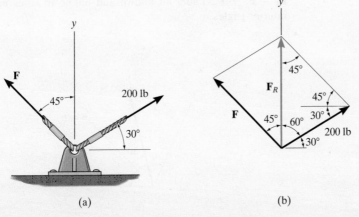

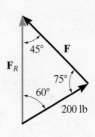

(a)   (b)   (c)

Fig. 2–13

SOLUTION

The parallelogram law of addition is shown in Fig. 2–13b, and the triangle rule is shown in Fig. 2–13c. The magnitudes of $\mathbf{F}_R$ and **F** are the two unknowns. They can be determined by applying the law of sines.

$$\frac{F}{\sin 60°} = \frac{200\ \text{lb}}{\sin 45°}$$

$$F = 245\ \text{lb} \qquad\qquad Ans.$$

$$\frac{F_R}{\sin 75°} = \frac{200\ \text{lb}}{\sin 45°}$$

$$F_R = 273\ \text{lb} \qquad\qquad Ans.$$

It is strongly suggested that you test yourself on the solutions to these examples by covering them over and then trying to draw the parallelogram law, and thinking about how the sine and cosine laws are used to determine the unknowns. Then before solving any of the problems, try to solve the Preliminary Problems and some of the Fundamental Problems given on the next pages. The solutions and answers to these are given in the back of the book. Doing this throughout the book will help immensely in developing your problem-solving skills.

PRELIMINARY PROBLEMS

Partial solutions and answers to all Preliminary Problems are given in the back of the book.

P2–1. In each case, construct the parallelogram law to show $\mathbf{F}_R = \mathbf{F}_1 + \mathbf{F}_2$. Then establish the triangle rule, where $\mathbf{F}_R = \mathbf{F}_1 + \mathbf{F}_2$. Label all known and unknown sides and internal angles.

P2–2. In each case, show how to resolve the force $\mathbf{F}$ into components acting along the u and v axes using the parallelogram law. Then establish the triangle rule to show $\mathbf{F}_R = \mathbf{F}_u + \mathbf{F}_v$. Label all known and unknown sides and interior angles.

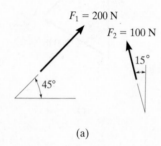

(a)

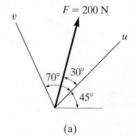

(a)

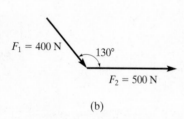

(b)

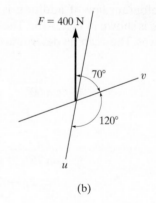

(b)

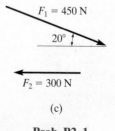

(c)

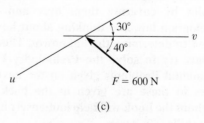

(c)

Prob. P2–1

Prob. P2–2

FUNDAMENTAL PROBLEMS

Partial solutions and answers to all Fundamental Problems are given in the back of the book.

F2–1. Determine the magnitude of the resultant force acting on the screw eye and its direction measured clockwise from the *x* axis.

Prob. F2–1

F2–2. Determine the magnitude of the resultant force.

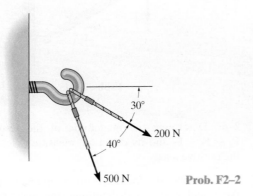

Prob. F2–2

F2–3. Determine the magnitude of the resultant force and its direction measured counterclockwise from the positive *x* axis.

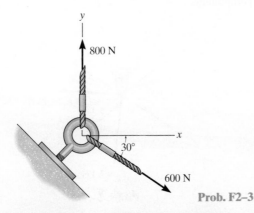

Prob. F2–3

F2–4. Resolve the 30-lb force into components along the *u* and *v* axes, and determine the magnitude of each of these components.

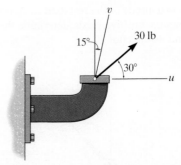

Prob. F2–4

F2–5. Resolve the force into components acting along members *AB* and *AC*, and determine the magnitude of each component.

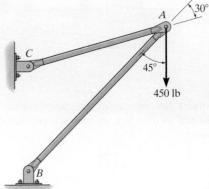

Prob. F2–5

F2–6. If force **F** is to have a component along the *u* axis of $F_u = 6$ kN, determine the magnitude of **F** and the magnitude of its component **F**$_v$ along the *v* axis.

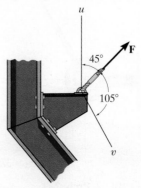

Prob. F2–6

PROBLEMS

2–1. If $\theta = 60°$ and $F = 450$ N, determine the magnitude of the resultant force and its direction, measured counterclockwise from the positive x axis.

2–2. If the magnitude of the resultant force is to be 500 N, directed along the positive y axis, determine the magnitude of force **F** and its direction θ.

*2–4. Determine the magnitudes of the two components of **F** directed along members AB and AC. Set $F = 500$ N.

2–5. Solve Prob. 2–4 with $F = 350$ lb.

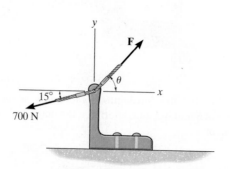

Probs. 2–1/2

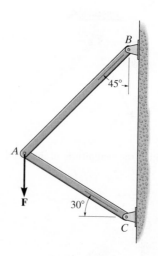

Probs. 2–4/5

2–3. Determine the magnitude of the resultant force $F_R = F_1 + F_2$ and its direction, measured counterclockwise from the positive x axis.

2–6. Determine the magnitude of the resultant force $F_R = F_1 + F_2$ and its direction, measured clockwise from the positive u axis.

2–7. Resolve the force F_1 into components acting along the u and v axes and determine the magnitudes of the components.

*2–8. Resolve the force F_2 into components acting along the u and v axes and determine the magnitudes of the components.

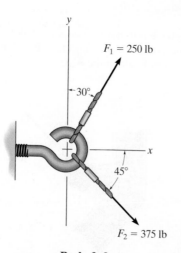

Prob. 2–3

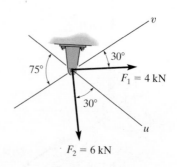

Probs. 2–6/7/8

2–9. If the resultant force acting on the support is to be 1200 lb, directed horizontally to the right, determine the force **F** in rope A and the corresponding angle θ.

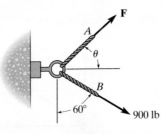

Prob. 2–9

2–10. Determine the magnitude of the resultant force and its direction, measured counterclockwise from the positive x axis.

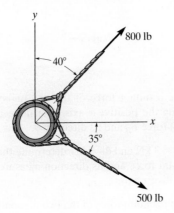

Prob. 2–10

2–11. If $\theta = 60°$, determine the magnitude of the resultant and its direction measured clockwise from the horizontal.

*2–12.** Determine the angle θ for connecting member A to the plate so that the resultant force of $\mathbf{F}_A$ and $\mathbf{F}_B$ is directed horizontally to the right. Also, what is the magnitude of the resultant force?

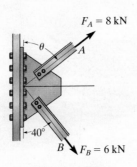

Probs. 2–11/12

2–13. The force acting on the gear tooth is $F =$ Resolve this force into two components acting along lines aa and bb.

2–14. The component of force **F** acting along line aa is required to be 30 lb. Determine the magnitude of **F** and its component along line bb.

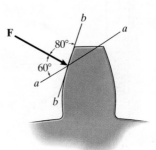

Probs. 2–13/14

2–15. Force **F** acts on the frame such that its component acting along member AB is 650 lb, directed from B towards A, and the component acting along member BC is 500 lb, directed from B towards C. Determine the magnitude of **F** and its direction θ. Set $\phi = 60°$.

*2–16.** Force **F** acts on the frame such that its component acting along member AB is 650 lb, directed from B towards A. Determine the required angle ϕ ($0° \leq \phi \leq 45°$) and the component acting along member BC. Set $F = 850$ lb and $\theta = 30°$.

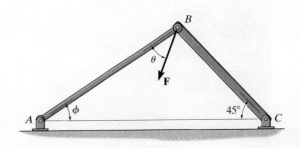

Probs. 2–15/16

2–17. If $F_1 = 30$ lb and $F_2 = 40$ lb, determine the angles θ and ϕ so that the resultant force is directed along the positive x axis and has a magnitude of $F_R = 60$ lb.

*2–20. Determine the magnitude of force **F** so that the resultant $\mathbf{F}_R$ of the three forces is as small as possible. What is the minimum magnitude of $\mathbf{F}_R$?

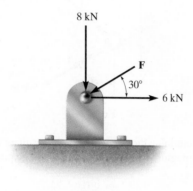

Prob. 2–20

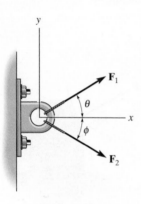

Prob. 2–17

2–21. If the resultant force of the two tugboats is 3 kN, directed along the positive x axis, determine the required magnitude of force $\mathbf{F}_B$ and its direction θ.

2–18. Determine the magnitude and direction θ of $\mathbf{F}_A$ so that the resultant force is directed along the positive x axis and has a magnitude of 1250 N.

2–22. If $F_B = 3$ kN and $\theta = 45°$, determine the magnitude of the resultant force and its direction measured clockwise from the positive x axis.

2–19. Determine the magnitude of the resultant force acting on the ring at O if $F_A = 750$ N and $\theta = 45°$. What is its direction, measured counterclockwise from the positive x axis?

2–23. If the resultant force of the two tugboats is required to be directed towards the positive x axis, and $\mathbf{F}_B$ is to be a minimum, determine the magnitude of $\mathbf{F}_R$ and $\mathbf{F}_B$ and the angle θ.

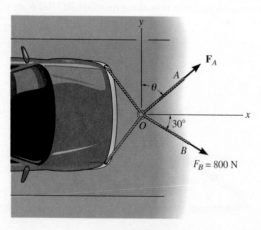

Probs. 2–18/19

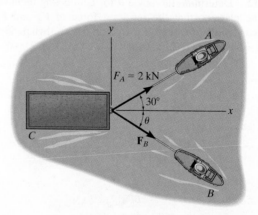

Probs. 2–21/22/23

2.4   ADDITION OF A SYSTEM OF COPLANAR FORCES

When a force is resolved into two components along the x and y axes, the components are then called **rectangular components**. For analytical work we can represent these components in one of two ways, using either scalar notation or Cartesian vector notation.

Scalar Notation. The rectangular components of force **F** shown in Fig. 2–14a are found using the parallelogram law, so that $\mathbf{F} = \mathbf{F}_x + \mathbf{F}_y$. Because these components form a right triangle, they can be determined from

$$F_x = F \cos \theta \quad \text{and} \quad F_y = F \sin \theta$$

Instead of using the angle θ, however, the direction of **F** can also be defined using a small "slope" triangle, as in the example shown in Fig. 2–15b. Since this triangle and the larger shaded triangle are similar, the proportional length of the sides gives

$$\frac{F_x}{F} = \frac{a}{c}$$

or

$$F_x = F\left(\frac{a}{c}\right)$$

and

$$\frac{F_y}{F} = \frac{b}{c}$$

or

$$F_y = -F\left(\frac{b}{c}\right)$$

Here the y component is a *negative scalar* since $\mathbf{F}_y$ is directed along the negative y axis.

It is important to keep in mind that this positive and negative scalar notation is to be used only for calculations, not for graphical representations in figures. Throughout the book, the *head of a vector arrow* in *any figure* indicates the sense of the vector *graphically*; algebraic signs are not used for this purpose. Thus, the vectors in Figs. 2–14a and 2–14b are designated by using boldface (vector) notation.* Whenever italic symbols are written near vector arrows in figures, they indicate the *magnitude* of the vector, which is *always* a *positive* quantity.

---

* Negative signs are used only in figures with boldface notation when showing equal but opposite pairs of vectors, as in Fig. 2–2.

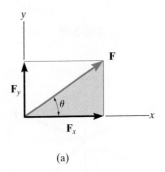

(a)

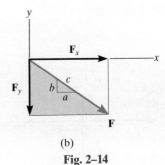

(b)

Fig. 2–14

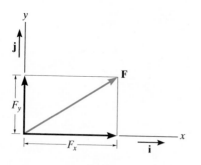

Fig. 2–15

Cartesian Vector Notation. Rather than representing the magnitude and direction of the components $\mathbf{F}_x$ and $\mathbf{F}_y$ as positive or negative scalars, we can instead consider them to be only positive scalars and thereby only report the magnitudes of the components. Their directions are then represented by the Cartesian unit vectors $\mathbf{i}$ and $\mathbf{j}$, Fig. 2–15. These are called unit vectors because they have a dimensionless magnitude of 1. By separating the magnitude and direction of each component, we can express $\mathbf{F}$ as a *Cartesian vector*.

$$\mathbf{F} = F_x \mathbf{i} + F_y \mathbf{j}$$

Coplanar Force Resultants. We can use either of the two methods just described to determine the resultant of several *coplanar forces*, i.e., forces that all lie in the same plane. To do this, each force is first resolved into its x and y components, and then the respective components are added using *scalar algebra* since they are collinear. The resultant force is then formed by adding the *resultant components* using the parallelogram law. For example, consider the three concurrent forces in Fig. 2–16a, which have x and y components shown in Fig. 2–16b. Using Cartesian vector notation, each force is first represented as a Cartesian vector, i.e.,

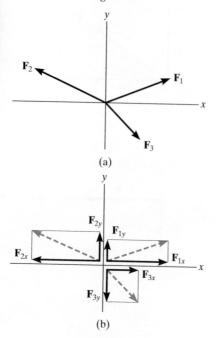

(a)

(b)

Fig. 2–16

$$\mathbf{F}_1 = F_{1x} \mathbf{i} + F_{1y} \mathbf{j}$$
$$\mathbf{F}_2 = -F_{2x} \mathbf{i} + F_{2y} \mathbf{j}$$
$$\mathbf{F}_3 = F_{3x} \mathbf{i} - F_{3y} \mathbf{j}$$

The vector resultant, Fig. 2–17c, is therefore

$$
\begin{aligned}
\mathbf{F}_R &= \mathbf{F}_1 + \mathbf{F}_2 + \mathbf{F}_3 \\
&= F_{1x}\mathbf{i} + F_{1y}\mathbf{j} - F_{2x}\mathbf{i} + F_{2y}\mathbf{j} + F_{3x}\mathbf{i} - F_{3y}\mathbf{j} \\
&= (F_{1x} - F_{2x} + F_{3x})\mathbf{i} + (F_{1y} + F_{2y} - F_{3y})\mathbf{j} \\
&= (F_R)_x \mathbf{i} + (F_R)_y \mathbf{j}
\end{aligned}
$$

If *scalar notation* is used, then indicating the positive directions of components along the x and y axes with symbolic arrows, we have

$$\xrightarrow{+} \quad (F_R)_x = F_{1x} - F_{2x} + F_{3x}$$
$$+\uparrow \quad (F_R)_y = F_{1y} + F_{2y} - F_{3y}$$

Notice that these are the *same* results as the $\mathbf{i}$ and $\mathbf{j}$ components of $\mathbf{F}_R$ determined above.

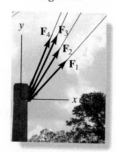

The resultant force of the four cable forces acting on the post can be determined by adding algebraically the separate x and y components of each cable force. This resultant $\mathbf{F}_R$ produces the *same pulling effect* on the post as all four cables.

In general, then, the components of the resultant force of any number of coplanar forces can be represented by the algebraic sum of the x and y components of all the forces, i.e.,

$$\begin{aligned}(F_R)_x &= \Sigma F_x \\ (F_R)_y &= \Sigma F_y\end{aligned} \qquad (2\text{–}1)$$

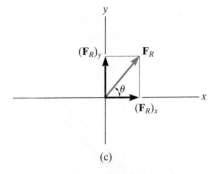

(c)

Fig. 2–16 (cont.)

Once these components are determined, they may be sketched along the x and y axes with their proper sense of direction, and the resultant force can be determined from vector addition, Fig. 2–16c. From this sketch, the magnitude of $\mathbf{F}_R$ is then found from the Pythagorean theorem; that is,

$$F_R = \sqrt{(F_R)_x^2 + (F_R)_y^2}$$

Also, the angle θ, which specifies the direction of the resultant force, is determined from trigonometry.

$$\theta = \tan^{-1}\left|\frac{(F_R)_y}{(F_R)_x}\right|$$

The above concepts are illustrated numerically in the examples which follow.

IMPORTANT POINTS

- The resultant of several coplanar forces can easily be determined if an x, y coordinate system is established and the forces are resolved into components along the axes.

- The direction of each force is specified by the angle its line of action makes with one of the axes, or by a slope triangle.

- The orientation of the x and y axes is arbitrary, and their positive direction can be specified by the Cartesian unit vectors $\mathbf{i}$ and $\mathbf{j}$.

- The x and y components of the *resultant force* are simply the algebraic addition of the components of all the coplanar forces.

- The magnitude of the resultant force is determined from the Pythagorean theorem, and when the resultant components are sketched on the x and y axes, Fig. 2–16c, the direction θ of the resultant can be determined from trigonometry.

EXAMPLE 2.4

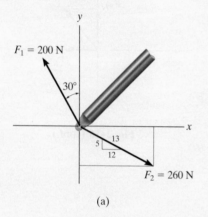

(a)

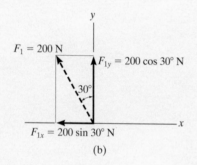

(b)

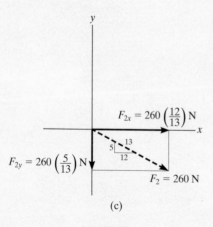

(c)

Fig. 2–17

Determine the x and y components of $\mathbf{F}_1$ and $\mathbf{F}_2$ acting on the boom shown in Fig. 2–17a. Express each force as a Cartesian vector.

SOLUTION

Scalar Notation. By the parallelogram law, $\mathbf{F}_1$ is resolved into x and y components, Fig. 2–17b. Since $\mathbf{F}_{1x}$ acts in the $-x$ direction, and $\mathbf{F}_{1y}$ acts in the $+y$ direction, we have

$$F_{1x} = -200 \sin 30° \text{ N} = -100 \text{ N} = 100 \text{ N} \leftarrow \qquad Ans.$$

$$F_{1y} = 200 \cos 30° \text{ N} = 173 \text{ N} = 173 \text{ N} \uparrow \qquad Ans.$$

The force $\mathbf{F}_2$ is resolved into its x and y components, as shown in Fig. 2-17c. From the "slope triangle" we could obtain the angle θ, e.g., $\theta = \tan^{-1}(\frac{5}{12})$, and then proceed to determine the magnitudes of the components in the same manner as for $\mathbf{F}_1$. The easier method, however, consists of using proportional parts of similar triangles, i.e.,

$$\frac{F_{2x}}{260 \text{ N}} = \frac{12}{13} \qquad F_{2x} = 260 \text{ N}\left(\frac{12}{13}\right) = 240 \text{ N}$$

Similarly,

$$F_{2y} = 260 \text{ N}\left(\frac{5}{13}\right) = 100 \text{ N}$$

Notice how the magnitude of the *horizontal component*, F_{2x}, was obtained by multiplying the force magnitude by the ratio of the *horizontal leg* of the slope triangle divided by the hypotenuse; whereas the magnitude of the *vertical component*, F_{2y}, was obtained by multiplying the force magnitude by the ratio of the *vertical leg* divided by the hypotenuse. Using scalar notation to represent the components, we have

$$F_{2x} = 240 \text{ N} = 240 \text{ N} \rightarrow \qquad Ans.$$

$$F_{2y} = -100 \text{ N} = 100 \text{ N} \downarrow \qquad Ans.$$

Cartesian Vector Notation. Having determined the magnitudes and directions of the components of each force, we can express each force as a Cartesian vector.

$$\mathbf{F}_1 = \{-100\mathbf{i} + 173\mathbf{j}\}\,\text{N} \qquad Ans.$$

$$\mathbf{F}_2 = \{240\mathbf{i} - 100\mathbf{j}\}\,\text{N} \qquad Ans.$$

EXAMPLE 2.5

The link in Fig. 2–18a is subjected to two forces $\mathbf{F}_1$ and $\mathbf{F}_2$. Determine the magnitude and direction of the resultant force.

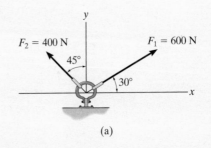

(a)

SOLUTION I

Scalar Notation. First we resolve each force into its x and y components, Fig. 2–18b, then we sum these components algebraically.

$$\xrightarrow{\,+\,} (F_R)_x = \Sigma F_x; \qquad (F_R)_x = 600 \cos 30° \text{ N} - 400 \sin 45° \text{ N}$$

$$= 236.8 \text{ N} \rightarrow$$

$$+\uparrow (F_R)_y = \Sigma F_y; \qquad (F_R)_y = 600 \sin 30° \text{ N} + 400 \cos 45° \text{ N}$$

$$= 582.8 \text{ N} \uparrow$$

The resultant force, shown in Fig. 2–18c, has a *magnitude* of

$$F_R = \sqrt{(236.8 \text{ N})^2 + (582.8 \text{ N})^2}$$

$$= 629 \text{ N} \qquad\qquad\qquad Ans.$$

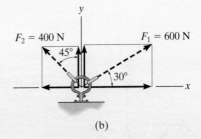

(b)

From the vector addition,

$$\theta = \tan^{-1}\left(\frac{582.8 \text{ N}}{236.8 \text{ N}}\right) = 67.9° \qquad\qquad Ans.$$

SOLUTION II

Cartesian Vector Notation. From Fig. 2–18b, each force is first expressed as a Cartesian vector.

$$\mathbf{F}_1 = \{\, 600 \cos 30°\mathbf{i} + 600 \sin 30°\mathbf{j} \,\} \text{ N}$$
$$\mathbf{F}_2 = \{\, -400 \sin 45°\mathbf{i} + 400 \cos 45°\mathbf{j} \,\} \text{ N}$$

Then,

$$\mathbf{F}_R = \mathbf{F}_1 + \mathbf{F}_2 = (600 \cos 30° \text{ N} - 400 \sin 45° \text{ N})\mathbf{i}$$
$$+ (600 \sin 30° \text{ N} + 400 \cos 45° \text{ N})\mathbf{j}$$
$$= \{\, 236.8\mathbf{i} + 582.8\mathbf{j} \,\} \text{ N}$$

The magnitude and direction of $\mathbf{F}_R$ are determined in the same manner as before.

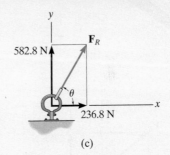

(c)

Fig. 2–18

NOTE: Comparing the two methods of solution, notice that the use of scalar notation is more efficient since the components can be found *directly*, without first having to express each force as a Cartesian vector before adding the components. Later, however, we will show that Cartesian vector analysis is very beneficial for solving three-dimensional problems.

EXAMPLE 2.6

The end of the boom O in Fig. 2–19a is subjected to the three concurrent and coplanar forces. Determine the magnitude and direction of the resultant force.

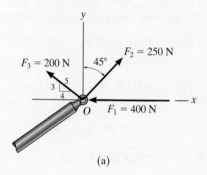

(a)

SOLUTION

Each force is resolved into its x and y components, Fig. 2–19b. Summing the x components, we have

$$\xrightarrow{+} (F_R)_x = \Sigma F_x; \qquad (F_R)_x = -400 \text{ N} + 250 \sin 45° \text{ N} - 200\left(\tfrac{4}{5}\right) \text{ N}$$

$$= -383.2 \text{ N} = 383.2 \text{ N} \leftarrow$$

Summing the y components yields

$$+\uparrow(F_R)_y = \Sigma F_y; \qquad (F_R)_y = 250 \cos 45° \text{ N} + 200\left(\tfrac{3}{5}\right) \text{ N}$$

$$= 296.8 \text{ N}\uparrow$$

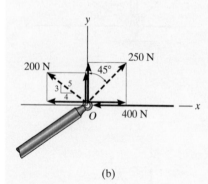

(b)

The resultant force, shown in Fig. 2–19c, has a *magnitude* of

$$F_R = \sqrt{(-383.2 \text{ N})^2 + (296.8 \text{ N})^2}$$

$$= 485 \text{ N} \qquad\qquad Ans.$$

From the vector addition in Fig. 2–19c, the direction angle θ is

$$\theta = \tan^{-1}\left(\frac{296.8}{383.2}\right) = 37.8° \qquad Ans.$$

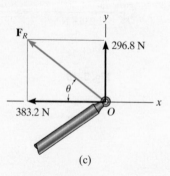

(c)

Fig. 2–19

NOTE: Application of this method is more convenient, compared to using two applications of the parallelogram law, first to add $\mathbf{F}_1$ and $\mathbf{F}_2$, then adding $\mathbf{F}_3$ to this resultant.

FUNDAMENTAL PROBLEMS

F2–7. Resolve each force into its x and y components.

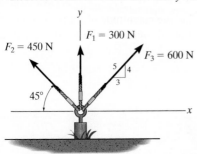

Prob. F2–7

F2–8. Determine the magnitude and direction of the resultant force.

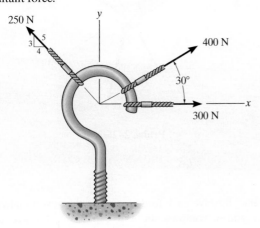

Prob. F2–8

F2–9. Determine the magnitude of the resultant force acting on the corbel and its direction θ, measured counterclockwise from the x axis.

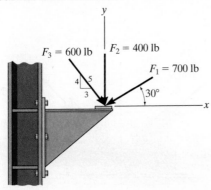

Prob. F2–9

F2–10. If the resultant force acting on the bracket is to be 750 N directed along the positive x axis, determine the magnitude of **F** and its direction θ.

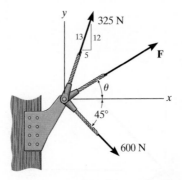

Prob. F2–10

F2–11. If the magnitude of the resultant force acting on the bracket is to be 80 lb directed along the u axis, determine the magnitude of **F** and its direction θ.

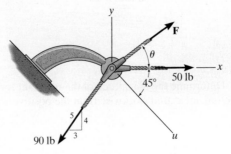

Prob. F2–11

F2–12. Determine the magnitude of the resultant force and its direction θ, measured counterclockwise from the positive x axis.

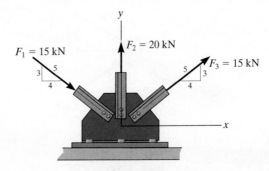

Prob. F2–12

PROBLEMS

*2–24. Determine the magnitude of the resultant force and its direction, measured counterclockwise from the positive x axis.

2–26. Resolve $\mathbf{F}_1$ and $\mathbf{F}_2$ into their x and y components.

2–27. Determine the magnitude of the resultant force and its direction measured counterclockwise from the positive x axis.

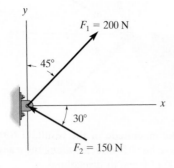

Prob. 2–24

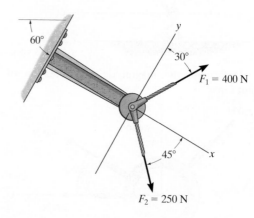

Probs. 2–26/27

2–25. Determine the magnitude of the resultant force and its direction, measured clockwise from the positive x axis.

*2–28. Resolve each force acting on the gusset plate into its x and y components, and express each force as a Cartesian vector.

2–29. Determine the magnitude of the resultant force acting on the gusset plate and its direction, measured counterclockwise from the positive x axis.

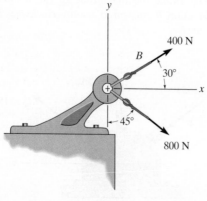

Prob. 2–25

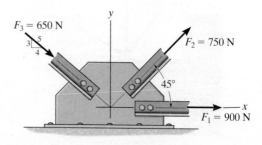

Probs. 2–28/29

2–30. Express each of the three forces acting on the support in Cartesian vector form and determine the magnitude of the resultant force and its direction, measured clockwise from positive x axis.

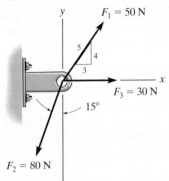

Prob. 2–30

2–31. Determine the x and y components of F_1 and F_2.

*2–32.** Determine the magnitude of the resultant force and its direction, measured counterclockwise from the positive x axis.

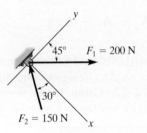

Probs. 2–31/32

2–33. Determine the magnitude of the resultant force and its direction, measured counterclockwise from the positive x axis.

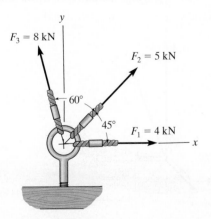

Prob. 2–33

2–34. Express F_1, F_2, and F_3 as Cartesian vectors.

2–35. Determine the magnitude of the resultant force and its direction, measured counterclockwise from the positive x axis.

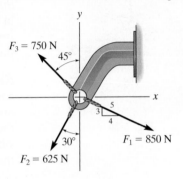

Probs. 2–34/35

*2–36.** Determine the magnitude of the resultant force and its direction, measured clockwise from the positive x axis.

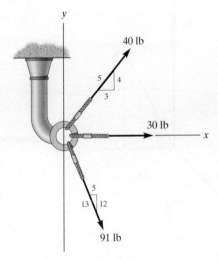

Prob. 2–36

2–37. Determine the magnitude and direction θ of the resultant force F_R. Express the result in terms of the magnitudes of the components F_1 and F_2 and the angle ϕ.

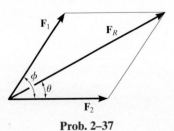

Prob. 2–37

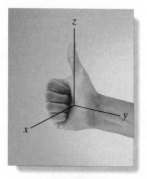

Fig. 2–20

2.5 CARTESIAN VECTORS

The operations of vector algebra, when applied to solving problems in *three dimensions*, are greatly simplified if the vectors are first represented in Cartesian vector form. In this section we will present a general method for doing this; then in the next section we will use this method for finding the resultant force of a system of concurrent forces.

Right-Handed Coordinate System. We will use a right-handed coordinate system to describe the vector algebra that follows. Specifically, a rectangular coordinate system is said to be *right handed* if the thumb of the right hand points in the direction of the positive z axis when the right-hand fingers are curled about this axis and directed from the positive x towards the positive y axis, Fig. 2–20.

Rectangular Components of a Vector. In general a vector $\mathbf{A}$ will have three rectangular components along the x, y, z coordinate axes, Fig. 2–21. These components are determined using two successive applications of the parallelogram law, that is, $\mathbf{A} = \mathbf{A}' + \mathbf{A}_z$ and then $\mathbf{A}' = \mathbf{A}_x + \mathbf{A}_y$. Combining these equations to eliminate $\mathbf{A}'$, $\mathbf{A}$ is represented by the vector sum of its *three* rectangular components,

$$\mathbf{A} = \mathbf{A}_x + \mathbf{A}_y + \mathbf{A}_z \qquad (2\text{–}2)$$

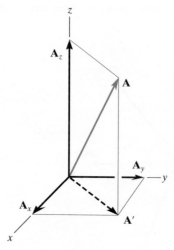

Fig. 2–21

Cartesian Vector Representation. In three dimensions, the set of Cartesian unit vectors, $\mathbf{i}, \mathbf{j}, \mathbf{k}$, is used to designate the directions of the x, y, z axes, respectively, Fig. 2–22. Using these vectors, the three components of $\mathbf{A}$ in Fig. 2–23 can be written in Cartesian vector form as

$$\boxed{\mathbf{A} = A_x\mathbf{i} + A_y\mathbf{j} + A_z\mathbf{k}} \qquad (2\text{–}3)$$

There is a distinct advantage to writing vectors in this manner. Separating the *magnitude* and *direction* of each *component vector* will simplify the operations of vector algebra, particularly in three dimensions.

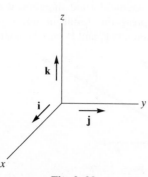

Fig. 2–22

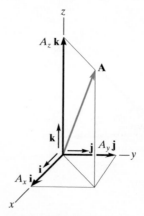

Fig. 2–23

Magnitude of a Cartesian Vector. If **A** is expressed as a Cartesian vector, then its magnitude can be determined. As shown in Fig. 2–24, from the blue right triangle, $A = \sqrt{A'^2 + A_z^2}$, and from the gray right triangle, $A' = \sqrt{A_x^2 + A_y^2}$. Combining these equations to eliminate A' yields

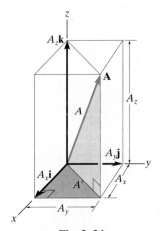

$$A = \sqrt{A_x^2 + A_y^2 + A_z^2} \qquad (2\text{–}4)$$

Fig. 2–24

Hence, the magnitude of **A** *is equal to the positive square root of the sum of the squares of the magnitudes of its components.*

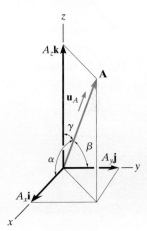

Coordinate Direction Angles. We will define the *direction* of **A** by the ***coordinate direction angles*** α (alpha), β (beta), and γ (gamma), measured between the *tail* of **A** and the *positive x, y, z* axes, Fig. 2–25. Note that regardless of where **A** is directed, each of these angles will be between 0° and 180°.

To determine α, β, and γ, consider the projection of **A** onto the *x, y, z* axes, Fig. 2–26. Referring to the three shaded right triangles shown in the figure, we have

Fig. 2–25

$$\cos\alpha = \frac{A_x}{A} \quad \cos\beta = \frac{A_y}{A} \quad \cos\gamma = \frac{A_z}{A} \qquad (2\text{–}5)$$

These numbers are known as the ***direction cosines*** of **A**. Once they have been obtained, the coordinate direction angles α, β, γ can then be determined from the inverse cosines.

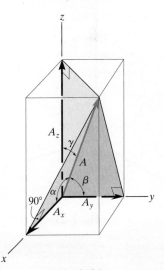

Fig. 2–26

An easy way of obtaining these direction cosines is to form a unit vector $\mathbf{u}_A$ in the direction of A, Fig. 2–25. To do this, divide $\mathbf{A}$ by its magnitude A, so that

$$\mathbf{u}_A = \frac{\mathbf{A}}{A} = \frac{A_x}{A}\mathbf{i} + \frac{A_y}{A}\mathbf{j} + \frac{A_z}{A}\mathbf{k} \qquad (2\text{–}6)$$

By comparison with Eqs. 2–5, it is seen that *the* $\mathbf{i}, \mathbf{j}, \mathbf{k}$ *components of* $\mathbf{u}_A$ *represent the direction cosines of* $\mathbf{A}$, i.e.,

$$\mathbf{u}_A = \cos\alpha\,\mathbf{i} + \cos\beta\,\mathbf{j} + \cos\gamma\,\mathbf{k} \qquad (2\text{–}7)$$

Since the magnitude of $\mathbf{u}_A$ is one, then from this equation an important relation among the direction cosines can be formulated, namely,

$$\cos^2\alpha + \cos^2\beta + \cos^2\gamma = 1 \qquad (2\text{–}8)$$

Therefore, if only *two* of the coordinate angles are known, the third angle can be found using this equation.

Finally, if the magnitude and coordinate direction angles of $\mathbf{A}$ are known, then $\mathbf{A}$ may be expressed in Cartesian vector form as

$$\begin{aligned}
\mathbf{A} &= A\mathbf{u}_A \\
&= A\cos\alpha\,\mathbf{i} + A\cos\beta\,\mathbf{j} + A\cos\gamma\,\mathbf{k} \qquad (2\text{–}9) \\
&= A_x\mathbf{i} + A_y\mathbf{j} + A_z\mathbf{k}
\end{aligned}$$

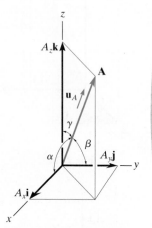

Fig. 2–25 (Repeated)

Horizontal and Vertical Angles. Sometimes the direction of $\mathbf{A}$ can be specified using a *horizontal angle* θ and a *vertical angle* ϕ (phi), such as shown in Fig. 2–27. The components of $\mathbf{A}$ can then be determined by applying trigonometry first to the light blue right triangle, which yields

$$A_z = A\cos\phi$$

and

$$A' = A\sin\phi$$

Now applying trigonometry to the dark blue right triangle,

$$A_x = A'\cos\theta = A\sin\phi\cos\theta$$

$$A_y = A'\sin\theta = A\sin\phi\sin\theta$$

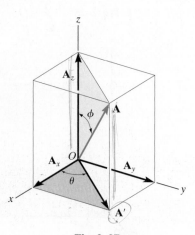

Fig. 2–27

Therefore **A** written in Cartesian vector form becomes

$$\mathbf{A} = A \sin \phi \cos \theta \, \mathbf{i} + A \sin \phi \sin \theta \, \mathbf{j} + A \cos \phi \, \mathbf{k}$$

This equation should not be memorized; rather, it is important to understand how the components were determined using trigonometry.

2.6   ADDITION OF CARTESIAN VECTORS

The addition (or subtraction) of two or more vectors is greatly simplified if the vectors are expressed in terms of their Cartesian components. For example, if $\mathbf{A} = A_x\mathbf{i} + A_y\mathbf{j} + A_z\mathbf{k}$ and $\mathbf{B} = B_x\mathbf{i} + B_y\mathbf{j} + B_z\mathbf{k}$, Fig. 2–28, then the resultant vector, **R**, has components which are the scalar sums of the **i**, **j**, **k** components of **A** and **B**, i.e.,

$$\mathbf{R} = \mathbf{A} + \mathbf{B} = (A_x + B_x)\mathbf{i} + (A_y + B_y)\mathbf{j} + (A_z + B_z)\mathbf{k}$$

If this is generalized and applied to a system of several concurrent forces, then the force resultant is the vector sum of all the forces in the system and can be written as

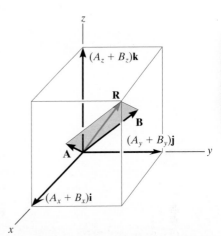

Fig. 2–28

$$\boxed{\mathbf{F}_R = \Sigma\mathbf{F} = \Sigma F_x\mathbf{i} + \Sigma F_y\mathbf{j} + \Sigma F_z\mathbf{k}} \qquad (2\text{–}10)$$

Here ΣF_x, ΣF_y, and ΣF_z represent the algebraic sums of the respective x, y, z or **i**, **j**, **k** components of each force in the system.

▶ IMPORTANT POINTS

- A Cartesian vector **A** has **i**, **j**, **k** components along the x, y, z axes. If **A** is known, its magnitude is $A = \sqrt{A_x^2 + A_y^2 + A_z^2}$.

- The direction of a Cartesian vector can be defined by the three coordinate direction angles α, β, γ, measured from the *positive* x, y, z axes to the *tail* of the vector. To find these angles, formulate a unit vector in the direction of **A**, i.e., $\mathbf{u}_A = \mathbf{A}/A$, and determine the inverse cosines of its components. Only two of these angles are independent of one another; the third angle is found from $\cos^2\alpha + \cos^2\beta + \cos^2\gamma = 1$.

- The direction of a Cartesian vector can also be specified using a horizontal angle θ and vertical angle ϕ.

Cartesian vector analysis provides a convenient method for finding both the resultant force and its components in three dimensions.

EXAMPLE   2.7

Determine the magnitude and the coordinate direction angles of the resultant force acting on the ring in Fig. 2–29a.

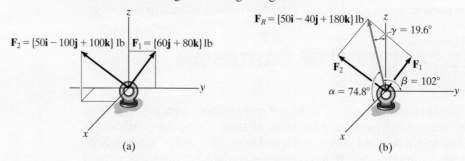

Fig. 2–29

SOLUTION

Since each force is represented in Cartesian vector form, the resultant force, shown in Fig. 2–29b, is

$$\mathbf{F}_R = \Sigma\mathbf{F} = \mathbf{F}_1 + \mathbf{F}_2 = \{60\mathbf{j} + 80\mathbf{k}\}\,\text{lb} + \{50\mathbf{i} - 100\mathbf{j} + 100\mathbf{k}\}\,\text{lb}$$
$$= \{50\mathbf{i} - 40\mathbf{j} + 180\mathbf{k}\}\,\text{lb}$$

The magnitude of $\mathbf{F}_R$ is

$$F_R = \sqrt{(50\,\text{lb})^2 + (-40\,\text{lb})^2 + (180\,\text{lb})^2} = 191.0\,\text{lb}$$
$$= 191\,\text{lb} \qquad\qquad Ans.$$

The coordinate direction angles α, β, γ are determined from the components of the unit vector acting in the direction of $\mathbf{F}_R$.

$$\mathbf{u}_{F_R} = \frac{\mathbf{F}_R}{F_R} = \frac{50}{191.0}\mathbf{i} - \frac{40}{191.0}\mathbf{j} + \frac{180}{191.0}\mathbf{k}$$

$$= 0.2617\mathbf{i} - 0.2094\mathbf{j} + 0.9422\mathbf{k}$$

so that

$$\cos\alpha = 0.2617 \qquad \alpha = 74.8° \qquad\qquad Ans.$$

$$\cos\beta = -0.2094 \qquad \beta = 102° \qquad\qquad Ans.$$

$$\cos\gamma = 0.9422 \qquad \gamma = 19.6° \qquad\qquad Ans.$$

These angles are shown in Fig. 2–29b.

NOTE: Here $\beta > 90°$ since the $\mathbf{j}$ component of $\mathbf{u}_{F_R}$ is negative. This seems reasonable considering how $\mathbf{F}_1$ and $\mathbf{F}_2$ add according to the parallelogram law.

EXAMPLE 2.8

Express the force **F** shown in Fig. 2–30a as a Cartesian vector.

SOLUTION

The angles of 60° and 45° defining the direction of **F** are *not* coordinate direction angles. Two successive applications of the parallelogram law are needed to resolve **F** into its x, y, z components. First $\mathbf{F} = \mathbf{F}' + \mathbf{F}_z$, then $\mathbf{F}' = \mathbf{F}_x + \mathbf{F}_y$, Fig. 2–30b. By trigonometry, the magnitudes of the components are

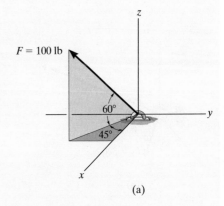

(a)

$$F_z = 100 \sin 60° \text{ lb} = 86.6 \text{ lb}$$

$$F' = 100 \cos 60° \text{ lb} = 50 \text{ lb}$$

$$F_x = F' \cos 45° = 50 \cos 45° \text{ lb} = 35.4 \text{ lb}$$

$$F_y = F' \sin 45° = 50 \sin 45° \text{ lb} = 35.4 \text{ lb}$$

Realizing that $\mathbf{F}_y$ is in the $-\mathbf{j}$ direction, we have

$$\mathbf{F} = \{35.4\mathbf{i} - 35.4\mathbf{j} + 86.6\mathbf{k}\} \text{ lb} \qquad \qquad Ans.$$

To show that the magnitude of this vector is indeed 100 lb, apply Eq. 2–4,

$$F = \sqrt{F_x^2 + F_y^2 + F_z^2}$$

$$= \sqrt{(35.4)^2 + (35.4)^2 + (86.6)^2} = 100 \text{ lb}$$

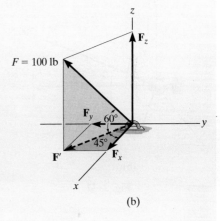

(b)

If needed, the coordinate direction angles of **F** can be determined from the components of the unit vector acting in the direction of **F**.

$$\mathbf{u} = \frac{\mathbf{F}}{F} = \frac{F_x}{F}\mathbf{i} + \frac{F_y}{F}\mathbf{j} + \frac{F_z}{F}\mathbf{k}$$

$$= \frac{35.4}{100}\mathbf{i} - \frac{35.4}{100}\mathbf{j} + \frac{86.6}{100}\mathbf{k}$$

$$= 0.354\mathbf{i} - 0.354\mathbf{j} + 0.866\mathbf{k}$$

so that

$$\alpha = \cos^{-1}(0.354) = 69.3°$$

$$\beta = \cos^{-1}(-0.354) = 111°$$

$$\gamma = \cos^{-1}(0.866) = 30.0°$$

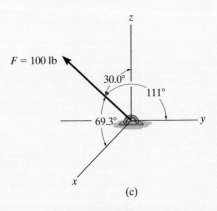

(c)

Fig. 2–30

These results are shown in Fig. 2–30c.

EXAMPLE 2.9

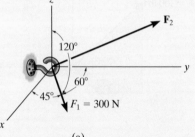

(a)

Fig. 2–31

(b)

Two forces act on the hook shown in Fig. 2–31a. Specify the magnitude of $\mathbf{F}_2$ and its coordinate direction angles so that the resultant force $\mathbf{F}_R$ acts along the positive y axis and has a magnitude of 800 N.

SOLUTION

To solve this problem, the resultant force $\mathbf{F}_R$ and its two components, $\mathbf{F}_1$ and $\mathbf{F}_2$, will each be expressed in Cartesian vector form. Then, as shown in Fig. 2–31b, it is necessary that $\mathbf{F}_R = \mathbf{F}_1 + \mathbf{F}_2$.

Applying Eq. 2–9,

$$\mathbf{F}_1 = F_1 \cos \alpha_1 \mathbf{i} + F_1 \cos \beta_1 \mathbf{j} + F_1 \cos \gamma_1 \mathbf{k}$$

$$= 300 \cos 45° \, \mathbf{i} + 300 \cos 60° \, \mathbf{j} + 300 \cos 120° \, \mathbf{k}$$

$$= \{212.1\mathbf{i} + 150\mathbf{j} - 150\mathbf{k}\} \, \text{N}$$

$$\mathbf{F}_2 = F_{2x}\mathbf{i} + F_{2y}\mathbf{j} + F_{2z}\mathbf{k}$$

Since $\mathbf{F}_R$ has a magnitude of 800 N and acts in the $+\mathbf{j}$ direction,

$$\mathbf{F}_R = (800 \text{ N})(+\mathbf{j}) = \{800\mathbf{j}\} \text{ N}$$

We require

$$\mathbf{F}_R = \mathbf{F}_1 + \mathbf{F}_2$$

$$800\mathbf{j} = 212.1\mathbf{i} + 150\mathbf{j} - 150\mathbf{k} + F_{2x}\mathbf{i} + F_{2y}\mathbf{j} + F_{2z}\mathbf{k}$$

$$800\mathbf{j} = (212.1 + F_{2x})\mathbf{i} + (150 + F_{2y})\mathbf{j} + (-150 + F_{2z})\mathbf{k}$$

To satisfy this equation the $\mathbf{i}, \mathbf{j}, \mathbf{k}$ components of $\mathbf{F}_R$ must be equal to the corresponding $\mathbf{i}, \mathbf{j}, \mathbf{k}$ components of $(\mathbf{F}_1 + \mathbf{F}_2)$. Hence,

$$0 = 212.1 + F_{2x} \qquad F_{2x} = -212.1 \text{ N}$$

$$800 = 150 + F_{2y} \qquad F_{2y} = 650 \text{ N}$$

$$0 = -150 + F_{2z} \qquad F_{2z} = 150 \text{ N}$$

The magnitude of $\mathbf{F}_2$ is thus

$$F_2 = \sqrt{(-212.1 \text{ N})^2 + (650 \text{ N})^2 + (150 \text{ N})^2}$$

$$= 700 \text{ N} \qquad\qquad\qquad Ans.$$

We can use Eq. 2–9 to determine $\alpha_2, \beta_2, \gamma_2$.

$$\cos \alpha_2 = \frac{-212.1}{700}; \qquad \alpha_2 = 108° \qquad Ans.$$

$$\cos \beta_2 = \frac{650}{700}; \qquad \beta_2 = 21.8° \qquad Ans.$$

$$\cos \gamma_2 = \frac{150}{700}; \qquad \gamma_2 = 77.6° \qquad Ans.$$

These results are shown in Fig. 2–31b.

PRELIMINARY PROBLEMS

P2–3. Sketch the following forces on the x, y, z coordinate axes. Show α, β, γ.

a) $\mathbf{F} = \{50\mathbf{i} + 60\mathbf{j} - 10\mathbf{k}\}$ kN

b) $\mathbf{F} = \{-40\mathbf{i} - 80\mathbf{j} + 60\mathbf{k}\}$ kN

P2–4. In each case, establish $\mathbf{F}$ as a Cartesian vector, and find the magnitude of $\mathbf{F}$ and the direction cosine of β.

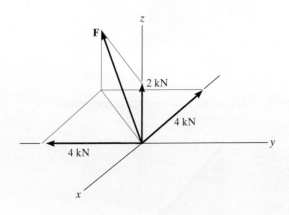

(a)

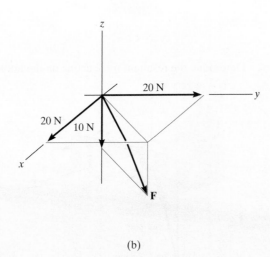

(b)

Prob. P2–4

P2–5. Show how to resolve each force into its x, y, z components. Set up the calculation used to find the magnitude of each component.

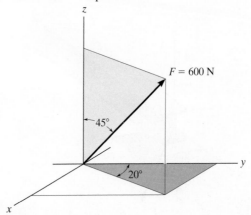

(a)

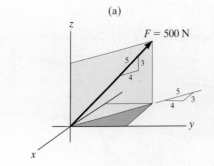

(b)

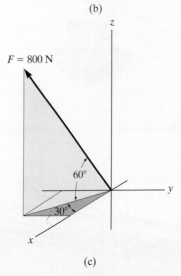

(c)

Prob. P2–5

FUNDAMENTAL PROBLEMS

F2–13. Determine the coordinate direction angles of the force.

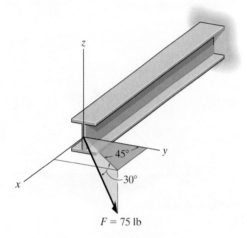

Prob. F2–13

F2–14. Express the force as a Cartesian vector.

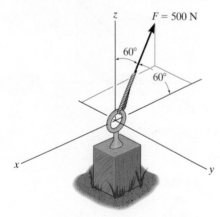

Prob. F2–14

F2–15. Express the force as a Cartesian vector.

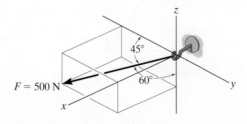

Prob. F2–15

F2–16. Express the force as a Cartesian vector.

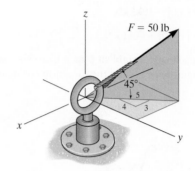

Prob. F2–16

F2–17. Express the force as a Cartesian vector.

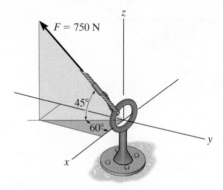

Prob. F2–17

F2–18. Determine the resultant force acting on the hook.

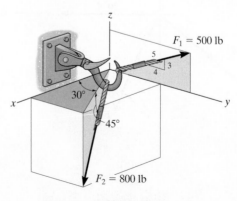

Prob. F2–18

PROBLEMS

2–38.   The force **F** has a magnitude of 80 lb. Determine the magnitudes of the x, y, z components of **F**.

*2–40.   Determine the magnitude and coordinate direction angles of the force **F** acting on the support. The component of **F** in the x–y plane is 7 kN.

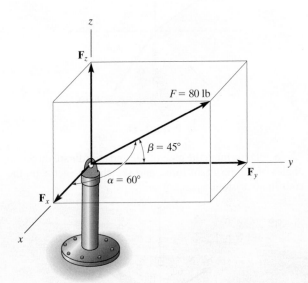

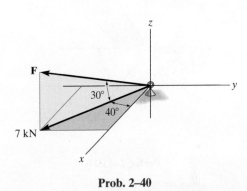

Prob. 2–40

Prob. 2–38

2–39.   The bolt is subjected to the force **F**, which has components acting along the x, y, z axes as shown. If the magnitude of **F** is 80 N, and $\alpha = 60°$ and $\gamma = 45°$, determine the magnitudes of its components.

2–41.   Determine the magnitude and coordinate direction angles of the resultant force and sketch this vector on the coordinate system.

2–42.   Specify the coordinate direction angles of $\mathbf{F}_1$ and $\mathbf{F}_2$ and express each force as a Cartesian vector.

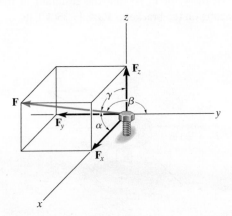

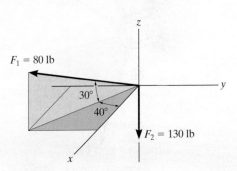

Prob. 2–39

Probs. 2–41/42

2–43. Express each force in Cartesian vector form and then determine the resultant force. Find the magnitude and coordinate direction angles of the resultant force.

*2–44. Determine the coordinate direction angles of $\mathbf{F}_1$.

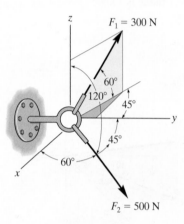

$F_1 = 300$ N

$60°$

$120°$

$45°$

$45°$

$60°$

$F_2 = 500$ N

Probs. 2–43/44

2–45. Determine the magnitude and coordinate direction angles of $\mathbf{F}_3$ so that the resultant of the three forces acts along the positive y axis and has a magnitude of 600 lb.

2–46. Determine the magnitude and coordinate direction angles of $\mathbf{F}_3$ so that the resultant of the three forces is zero.

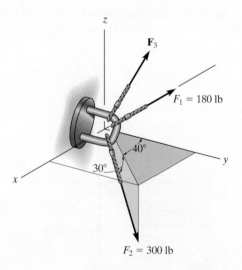

$\mathbf{F}_3$

$F_1 = 180$ lb

$40°$

$30°$

$F_2 = 300$ lb

Probs. 2–45/46

2–47. Determine the magnitude and coordinate direction angles of the resultant force, and sketch this vector on the coordinate system.

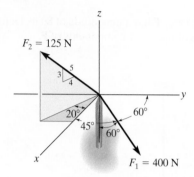

$F_2 = 125$ N

3 5

4

$20°$ $60°$

$45°$ $60°$

$F_1 = 400$ N

Prob. 2–47

*2–48. Determine the magnitude and coordinate direction angles of the resultant force, and sketch this vector on the coordinate system.

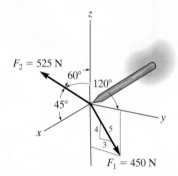

$F_2 = 525$ N

$60°$ $120°$

$45°$

4 5

3

$F_1 = 450$ N

Prob. 2–48

2–49. Determine the magnitude and coordinate direction angles $\alpha_1, \beta_1, \gamma_1$ of $\mathbf{F}_1$ so that the resultant of the three forces acting on the bracket is $\mathbf{F}_R = \{-350\mathbf{k}\}$ lb.

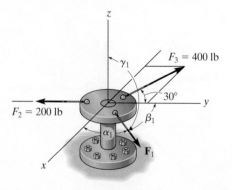

γ_1 $F_3 = 400$ lb

$F_2 = 200$ lb $30°$

β_1

α_1

$\mathbf{F}_1$

Prob. 2–49

2–50.   If the resultant force $\mathbf{F}_R$ has a magnitude of 150 lb and the coordinate direction angles shown, determine the magnitude of $\mathbf{F}_2$ and its coordinate direction angles.

2–53.   The spur gear is subjected to the two forces. Express each force as a Cartesian vector.

2–54.   The spur gear is subjected to the two forces. Determine the resultant of the two forces and express the result as a Cartesian vector.

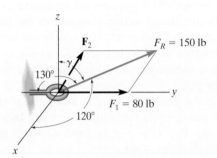

Prob. 2–50

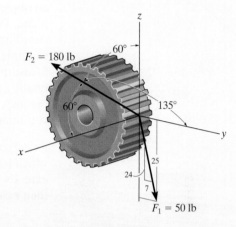

Probs. 2–53/54

2–51.   Express each force as a Cartesian vector.

*2–52.   Determine the magnitude and coordinate direction angles of the resultant force, and sketch this vector on the coordinate system.

2–55.   Determine the magnitude and coordinate direction angles of the resultant force, and sketch this vector on the coordinate system.

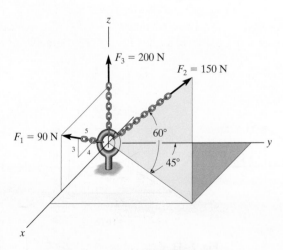

Probs. 2–51/52

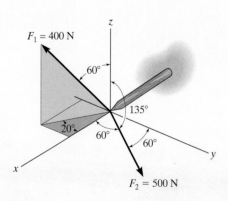

Prob. 2–55

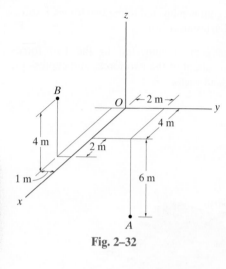

Fig. 2–32

2.7 POSITION VECTORS

In this section we will introduce the concept of a position vector. Later it will be shown that this vector is of importance in formulating a Cartesian force vector directed between two points in space.

x, y, z Coordinates. Throughout the book we will use the convention followed in many technical books, which requires the positive z axis to be directed *upward* (the zenith direction) so that it measures the height of an object or the altitude of a point. The x, y axes then lie in the horizontal plane, Fig. 2–32. Points in space are located relative to the origin of coordinates, O, by successive measurements along the x, y, z axes. For example, the coordinates of point A are obtained by starting at O and measuring $x_A = +4$ m along the x axis, $y_A = +2$ m along the y axis, and finally $z_A = -6$ m along the z axis, so that $A(4$ m, 2 m, -6 m$)$. In a similar manner, measurements along the x, y, z axes from O to B give the coordinates of B, that is, $B(6$ m, -1 m, 4 m$)$.

Position Vector. A *position vector* **r** is defined as a fixed vector which locates a point in space relative to another point. For example, if **r** extends from the origin of coordinates, O, to point $P(x, y, z)$, Fig. 2–33a, then **r** can be expressed in Cartesian vector form as

$$\mathbf{r} = x\mathbf{i} + y\mathbf{j} + z\mathbf{k}$$

Note how the head-to-tail vector addition of the three components yields vector **r**, Fig. 2–33b. Starting at the origin O, one "travels" x in the $+\mathbf{i}$ direction, then y in the $+\mathbf{j}$ direction, and finally z in the $+\mathbf{k}$ direction to arrive at point $P(x, y, z)$.

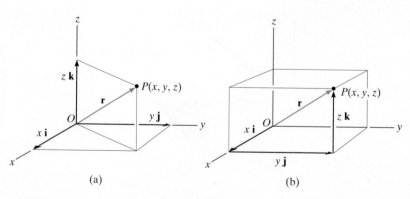

(a)   (b)

Fig. 2–33

In the more general case, the position vector may be directed from point A to point B in space. From Fig. 2–34a, by the head-to-tail vector addition, using the triangle rule, we require

$$\mathbf{r}_A + \mathbf{r} = \mathbf{r}_B$$

Solving for $\mathbf{r}$ and expressing $\mathbf{r}_A$ and $\mathbf{r}_B$ in Cartesian vector form yields

$$\mathbf{r} = \mathbf{r}_B - \mathbf{r}_A = (x_B\mathbf{i} + y_B\mathbf{j} + z_B\mathbf{k}) - (x_A\mathbf{i} + y_A\mathbf{j} + z_A\mathbf{k})$$

or

$$\mathbf{r} = (x_B - x_A)\mathbf{i} + (y_B - y_A)\mathbf{j} + (z_B - z_A)\mathbf{k} \qquad (2\text{–}11)$$

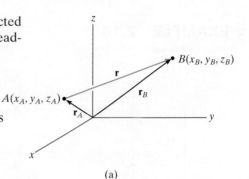

(a)

Thus, the $\mathbf{i}, \mathbf{j}, \mathbf{k}$ *components of* $\mathbf{r}$ *may be formed by taking the coordinates of the tail of the vector* $A(x_A, y_A, z_A)$ *and subtracting them from the corresponding coordinates of the head* $B(x_B, y_B, z_B)$. We can also form these components *directly*, Fig. 2–34b, by starting at A and moving through a distance of $(x_B - x_A)$ along the positive x axis $(+\mathbf{i})$, then $(y_B - y_A)$ along the positive y axis $(+\mathbf{j})$, and finally $(z_B - z_A)$ along the positive z axis $(+\mathbf{k})$ to get to B.

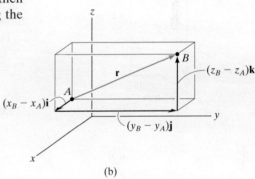

(b)

Fig. 2–34

If an x, y, z coordinate system is established, then the coordinates of two points A and B on the cable can be determined. From this the position vector $\mathbf{r}$ acting along the cable can be formulated. Its magnitude represents the distance from A to B, and its unit vector, $\mathbf{u} = \mathbf{r}/r$, gives the direction defined by α, β, γ.

EXAMPLE   2.10

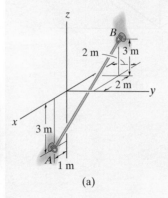

(a)

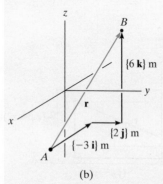

(b)

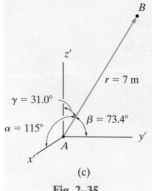

(c)

Fig. 2–35

An elastic rubber band is attached to points *A* and *B* as shown in Fig. 2–35*a*. Determine its length and its direction measured from *A* towards *B*.

SOLUTION

We first establish a position vector from *A* to *B*, Fig. 2–35*b*. In accordance with Eq. 2–11, the coordinates of the tail *A*(1 m, 0, −3 m) are subtracted from the coordinates of the head *B*(−2 m, 2 m, 3 m), which yields

$$\mathbf{r} = [-2 \text{ m} - 1 \text{ m}]\mathbf{i} + [2 \text{ m} - 0]\mathbf{j} + [3 \text{ m} - (-3 \text{ m})]\mathbf{k}$$
$$= \{-3\mathbf{i} + 2\mathbf{j} + 6\mathbf{k}\} \text{ m}$$

These components of **r** can also be determined *directly* by realizing that they represent the direction and distance one must travel along each axis in order to move from *A* to *B*, i.e., along the *x* axis {−3**i**} m, along the *y* axis {2**j**} m, and finally along the *z* axis {6**k**} m.

The length of the rubber band is therefore

$$r = \sqrt{(-3 \text{ m})^2 + (2 \text{ m})^2 + (6 \text{ m})^2} = 7 \text{ m} \qquad \text{Ans.}$$

Formulating a unit vector in the direction of **r**, we have

$$\mathbf{u} = \frac{\mathbf{r}}{r} = -\frac{3}{7}\mathbf{i} + \frac{2}{7}\mathbf{j} + \frac{6}{7}\mathbf{k}$$

The components of this unit vector give the coordinate direction angles

$$\alpha = \cos^{-1}\left(-\frac{3}{7}\right) = 115° \qquad \text{Ans.}$$

$$\beta = \cos^{-1}\left(\frac{2}{7}\right) = 73.4° \qquad \text{Ans.}$$

$$\gamma = \cos^{-1}\left(\frac{6}{7}\right) = 31.0° \qquad \text{Ans.}$$

NOTE: These angles are measured from the *positive axes* of a localized coordinate system placed at the tail of **r**, as shown in Fig. 2–35*c*.

2.8 FORCE VECTOR DIRECTED ALONG A LINE

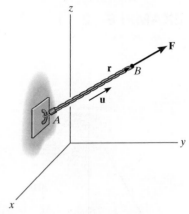

Quite often in three-dimensional statics problems, the direction of a force is specified by two points through which its line of action passes. Such a situation is shown in Fig. 2–36, where the force **F** is directed along the cord *AB*. We can formulate **F** as a Cartesian vector by realizing that it has the *same direction* and *sense* as the position vector **r** directed from point *A* to point *B* on the cord. This common direction is specified by the *unit vector* **u** = **r**/*r*, and so once **u** is determined, then

$$\mathbf{F} = F\mathbf{u} = F\left(\frac{\mathbf{r}}{r}\right) = F\left(\frac{(x_B - x_A)\mathbf{i} + (y_B - y_A)\mathbf{j} + (z_B - z_A)\mathbf{k}}{\sqrt{(x_B - x_A)^2 + (y_B - y_A)^2 + (z_B - z_A)^2}}\right)$$

Fig. 2–36

Although we have represented **F** symbolically in Fig. 2–36, note that it has *units of force*, unlike **r**, which has units of length.

The force **F** acting along the rope can be represented as a Cartesian vector by establishing *x*, *y*, *z* axes and first forming a position vector **r** along the length of the rope. Then the corresponding unit vector **u** = **r**/*r* that defines the direction of both the rope and the force can be determined. Finally, the magnitude of the force is combined with its direction, so that **F** = *F***u**.

> ## IMPORTANT POINTS

- A position vector locates one point in space relative to another point.

- The easiest way to formulate the components of a position vector is to determine the distance and direction that one must travel in the *x*, *y*, *z* directions—going from the tail to the head of the vector.

- A force **F** acting in the direction of a position vector **r** can be represented in Cartesian form if the unit vector **u** of the position vector is determined and it is multiplied by the magnitude of the force, i.e., **F** = *F***u** = *F*(**r**/*r*).

EXAMPLE 2.11

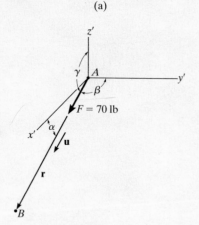

(a)

(b)

Fig. 2–37

The man shown in Fig. 2–37a pulls on the cord with a force of 70 lb. Represent this force acting on the support A as a Cartesian vector and determine its direction.

SOLUTION

Force $\mathbf{F}$ is shown in Fig. 2–37b. The *direction* of this vector, $\mathbf{u}$, is determined from the position vector $\mathbf{r}$, which extends from A to B. Rather than using the coordinates of the end points of the cord, $\mathbf{r}$ can be determined *directly* by noting in Fig. 2–37a that one must travel from A $\{-24\mathbf{k}\}$ ft, then $\{-8\mathbf{j}\}$ ft, and finally $\{12\mathbf{i}\}$ ft to get to B. Thus,

$$\mathbf{r} = \{12\mathbf{i} - 8\mathbf{j} - 24\mathbf{k}\} \text{ ft}$$

The magnitude of $\mathbf{r}$, which represents the *length* of cord AB, is

$$r = \sqrt{(12 \text{ ft})^2 + (-8 \text{ ft})^2 + (-24 \text{ ft})^2} = 28 \text{ ft}$$

Forming the unit vector that defines the direction and sense of both $\mathbf{r}$ and $\mathbf{F}$, we have

$$\mathbf{u} = \frac{\mathbf{r}}{r} = \frac{12}{28}\mathbf{i} - \frac{8}{28}\mathbf{j} - \frac{24}{28}\mathbf{k}$$

Since $\mathbf{F}$ has a *magnitude* of 70 lb and a *direction* specified by $\mathbf{u}$, then

$$\mathbf{F} = F\mathbf{u} = 70 \text{ lb}\left(\frac{12}{28}\mathbf{i} - \frac{8}{28}\mathbf{j} - \frac{24}{28}\mathbf{k}\right)$$

$$= \{30\mathbf{i} - 20\mathbf{j} - 60\mathbf{k}\} \text{ lb} \qquad \textit{Ans.}$$

The coordinate direction angles are measured between the tail of $\mathbf{r}$ (or $\mathbf{F}$) and the *positive axes* of a localized coordinate system with origin placed at A, Fig. 2–37b. From the components of the unit vector:

$$\alpha = \cos^{-1}\left(\frac{12}{28}\right) = 64.6° \qquad \textit{Ans.}$$

$$\beta = \cos^{-1}\left(\frac{-8}{28}\right) = 107° \qquad \textit{Ans.}$$

$$\gamma = \cos^{-1}\left(\frac{-24}{28}\right) = 149° \qquad \textit{Ans.}$$

NOTE: These results make sense when compared with the angles identified in Fig. 2–37b.

EXAMPLE 2.12

The roof is supported by two cables as shown in the photo. If the cables exert forces $F_{AB} = 100$ N and $F_{AC} = 120$ N on the wall hook at A as shown in Fig. 2–38a, determine the resultant force acting at A. Express the result as a Cartesian vector.

SOLUTION

The resultant force $\mathbf{F}_R$ is shown graphically in Fig. 2–38b. We can express this force as a Cartesian vector by first formulating $\mathbf{F}_{AB}$ and $\mathbf{F}_{AC}$ as Cartesian vectors and then adding their components. The directions of $\mathbf{F}_{AB}$ and $\mathbf{F}_{AC}$ are specified by forming unit vectors $\mathbf{u}_{AB}$ and $\mathbf{u}_{AC}$ along the cables. These unit vectors are obtained from the associated position vectors $\mathbf{r}_{AB}$ and $\mathbf{r}_{AC}$. With reference to Fig. 2–38a, to go from A to B, we must travel $\{-4\mathbf{k}\}$ m, and then $\{4\mathbf{i}\}$ m. Thus,

$$\mathbf{r}_{AB} = \{4\mathbf{i} - 4\mathbf{k}\} \text{ m}$$

$$r_{AB} = \sqrt{(4 \text{ m})^2 + (-4 \text{ m})^2} = 5.66 \text{ m}$$

$$\mathbf{F}_{AB} = F_{AB}\left(\frac{\mathbf{r}_{AB}}{r_{AB}}\right) = (100 \text{ N})\left(\frac{4}{5.66}\mathbf{i} - \frac{4}{5.66}\mathbf{k}\right)$$

$$\mathbf{F}_{AB} = \{70.7\mathbf{i} - 70.7\mathbf{k}\} \text{ N}$$

To go from A to C, we must travel $\{-4\mathbf{k}\}$ m, then $\{2\mathbf{j}\}$ m, and finally $\{4\mathbf{i}\}$ m. Thus,

$$\mathbf{r}_{AC} = \{4\mathbf{i} + 2\mathbf{j} - 4\mathbf{k}\} \text{ m}$$

$$r_{AC} = \sqrt{(4 \text{ m})^2 + (2 \text{ m})^2 + (-4 \text{ m})^2} = 6 \text{ m}$$

$$\mathbf{F}_{AC} = F_{AC}\left(\frac{\mathbf{r}_{AC}}{r_{AC}}\right) = (120 \text{ N})\left(\frac{4}{6}\mathbf{i} + \frac{2}{6}\mathbf{j} - \frac{4}{6}\mathbf{k}\right)$$

$$= \{80\mathbf{i} + 40\mathbf{j} - 80\mathbf{k}\} \text{ N}$$

The resultant force is therefore

$$\mathbf{F}_R = \mathbf{F}_{AB} + \mathbf{F}_{AC} = \{70.7\mathbf{i} - 70.7\mathbf{k}\} \text{ N} + \{80\mathbf{i} + 40\mathbf{j} - 80\mathbf{k}\} \text{ N}$$

$$= \{151\mathbf{i} + 40\mathbf{j} - 151\mathbf{k}\} \text{ N} \qquad \textit{Ans.}$$

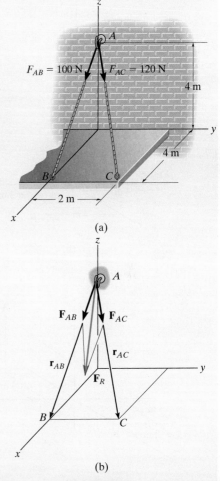

Fig. 2–38

PRELIMINARY PROBLEMS

P2–6. In each case, establish a position vector from point *A* to point *B*.

P2–7. In each case, express **F** as a Cartesian vector.

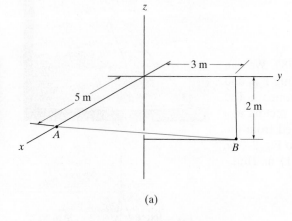

(a)

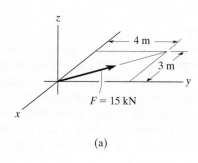

(a)

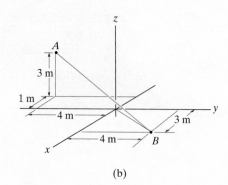

(b)

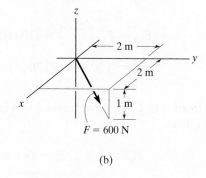

(b)

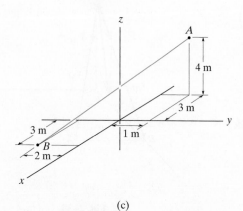

(c)

Prob. P2–6

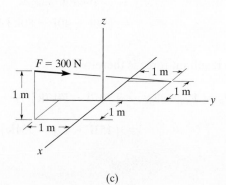

(c)

Prob. P2–7

FUNDAMENTAL PROBLEMS

F2–19. Express $\mathbf{r}_{AB}$ as a Cartesian vector, then determine its magnitude and coordinate direction angles.

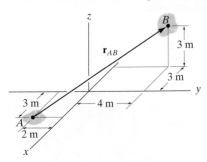

Prob. F2–19

F2–20. Determine the length of the rod and the position vector directed from A to B. What is the angle θ?

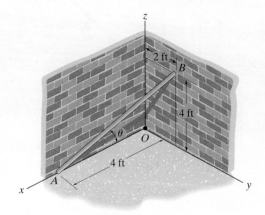

Prob. F2–20

F2–21. Express the force as a Cartesian vector.

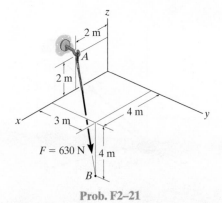

Prob. F2–21

F2–22. Express the force as a Cartesian vector.

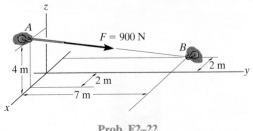

Prob. F2–22

F2–23. Determine the magnitude of the resultant force at A.

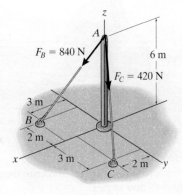

Prob. F2–23

F2–24. Determine the resultant force at A, expressed as a Cartesian vector.

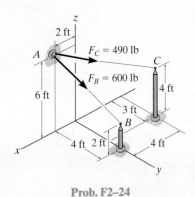

Prob. F2–24

PROBLEMS

*2–56.   Determine the length of the connecting rod AB by first formulating a position vector from A to B and then determining its magnitude.

2–58.   Express each force as a Cartesian vector, and then determine the magnitude and coordinate direction angles of the resultant force.

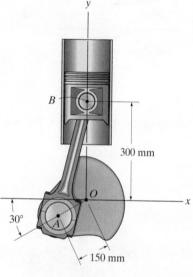

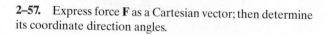

Prob. 2–56

2–57.   Express force **F** as a Cartesian vector; then determine its coordinate direction angles.

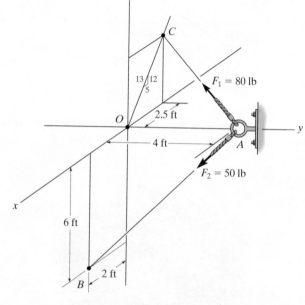

Prob. 2–58

2–59.   If $\mathbf{F} = \{350\mathbf{i} - 250\mathbf{j} - 450\mathbf{k}\}$ N and cable AB is 9 m long, determine the x, y, z coordinates of point A.

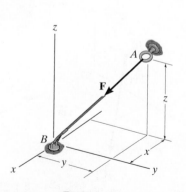

Prob. 2–57

Prob. 2–59

*2–60. The 8-m-long cable is anchored to the ground at A. If $x = 4$ m and $y = 2$ m, determine the coordinate z to the highest point of attachment along the column.

2–61. The 8-m-long cable is anchored to the ground at A. If $z = 5$ m, determine the location $+x, +y$ of the support at A. Choose a value such that $x = y$.

2–63. If $F_B = 560$ N and $F_C = 700$ N, determine the magnitude and coordinate direction angles of the resultant force acting on the flag pole.

*2–64. If $F_B = 700$ N, and $F_C = 560$ N, determine the magnitude and coordinate direction angles of the resultant force acting on the flag pole.

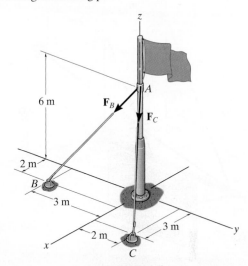

Probs. 2–63/64

2–65. The plate is suspended using the three cables which exert the forces shown. Express each force as a Cartesian vector.

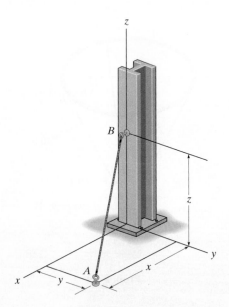

Probs. 2–60/61

2–62. Express each of the forces in Cartesian vector form and then determine the magnitude and coordinate direction angles of the resultant force.

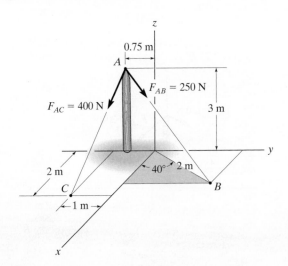

Prob. 2–62

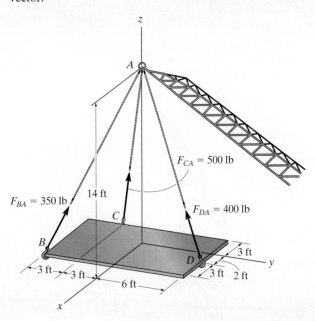

Prob. 2–65

2–66. Represent each cable force as a Cartesian vector.

2–67. Determine the magnitude and coordinate direction angles of the resultant force of the two forces acting at point A.

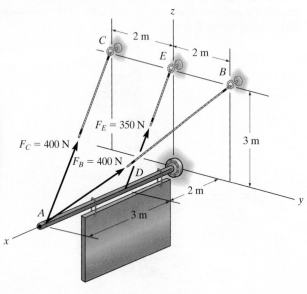

Probs. 2–66/67

*2–68. The force **F** has a magnitude of 80 lb and acts at the midpoint C of the rod. Express this force as a Cartesian vector.

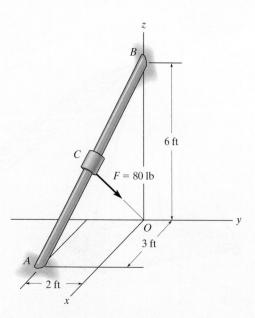

Prob. 2–68

2–69. The load at A creates a force of 60 lb in wire AB. Express this force as a Cartesian vector.

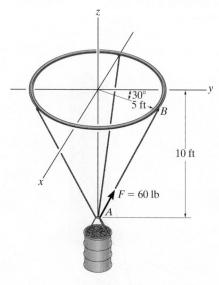

Prob. 2–69

2–70. Determine the magnitude and coordinate direction angles of the resultant force acting at point A on the post.

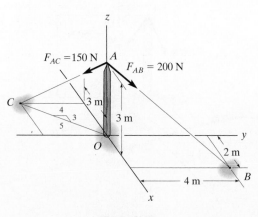

Prob. 2–70

2.9   DOT PRODUCT

Occasionally in statics one has to find the angle between two lines or the components of a force parallel and perpendicular to a line. In two dimensions, these problems can readily be solved by trigonometry since the geometry is easy to visualize. In three dimensions, however, this is often difficult, and consequently vector methods should be employed for the solution. The dot product, which defines a particular method for "multiplying" two vectors, will be used to solve the above-mentioned problems.

The **dot product** of vectors **A** and **B**, written **A** · **B**, and read "A dot B," is defined as the product of the magnitudes of **A** and **B** and the cosine of the angle θ between their tails, Fig. 2–39. Expressed in equation form,

$$\mathbf{A} \cdot \mathbf{B} = AB \cos \theta \qquad (2\text{--}12)$$

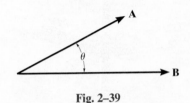

Fig. 2–39

where $0° \leq \theta \leq 180°$. The dot product is often referred to as the *scalar product* of vectors since the result is a *scalar* and not a vector.

The following three laws of operation apply.

1. Commutative law: $\mathbf{A} \cdot \mathbf{B} = \mathbf{B} \cdot \mathbf{A}$
2. Multiplication by a scalar: $a(\mathbf{A} \cdot \mathbf{B}) = (a\mathbf{A}) \cdot \mathbf{B} = \mathbf{A} \cdot (a\mathbf{B})$
3. Distributive law: $\mathbf{A} \cdot (\mathbf{B} + \mathbf{D}) = (\mathbf{A} \cdot \mathbf{B}) + (\mathbf{A} \cdot \mathbf{D})$

Cartesian Vector Formulation. If we apply Eq. 2–12, we can find the dot product for any two Cartesian unit vectors. For example, $\mathbf{i} \cdot \mathbf{i} = (1)(1) \cos 0° = 1$ and $\mathbf{i} \cdot \mathbf{j} = (1)(1) \cos 90° = 0$. If we want to find the dot product of two general vectors **A** and **B** that are expressed in Cartesian vector form, then we have

$$
\begin{aligned}
\mathbf{A} \cdot \mathbf{B} = {} & (A_x\mathbf{i} + A_y\mathbf{j} + A_z\mathbf{k}) \cdot (B_x\mathbf{i} + B_y\mathbf{j} + B_z\mathbf{k}) \\
= {} & A_xB_x\,(\mathbf{i} \cdot \mathbf{i}) + A_xB_y\,(\mathbf{i} \cdot \mathbf{j}) + A_xB_z\,(\mathbf{i} \cdot \mathbf{k}) \\
& + A_yB_x\,(\mathbf{j} \cdot \mathbf{i}) + (A_yB_y\,(\mathbf{j} \cdot \mathbf{j}) + A_yB_z\,(\mathbf{j} \cdot \mathbf{k}) \\
& + A_zB_x\,(\mathbf{k} \cdot \mathbf{i}) + A_zB_y\,(\mathbf{k} \cdot \mathbf{j}) + A_zB_z\,(\mathbf{k} \cdot \mathbf{k})
\end{aligned}
$$

Carrying out the dot-product operations, the final result becomes

$$\mathbf{A} \cdot \mathbf{B} = A_xB_x + A_yB_y + A_zB_z \qquad (2\text{--}13)$$

Thus, to determine the dot product of two Cartesian vectors, multiply their corresponding x, y, z components and sum these products algebraically. The result will be either a positive or negative *scalar*, or it could be zero.

Fig. 2–39 (Repeated)

Applications. The dot product has two important applications.

- **The angle formed between two vectors or intersecting lines.** The angle θ between the tails of vectors **A** and **B** in Fig. 2–39 can be determined from Eq. 2–12 and written as

$$\theta = \cos^{-1}\left(\frac{\mathbf{A} \cdot \mathbf{B}}{AB}\right) \quad 0° \le \theta \le 180°$$

Here $\mathbf{A} \cdot \mathbf{B}$ is found from Eq. 2–13. As a special case, if $\mathbf{A} \cdot \mathbf{B} = 0$, then $\theta = \cos^{-1} 0 = 90°$ so that **A** will be *perpendicular* to **B**.

- **The components of a vector parallel and perpendicular to a line.** The component of vector **A** parallel to or collinear with the line aa in Fig. 2–40 is defined by $A_a = A \cos \theta$. This component is sometimes referred to as the *projection* of **A** onto the line, since a *right angle* is formed in the construction. If the *direction* of the line is specified by the unit vector $\mathbf{u}_a$, and since $u_a = 1$, we can determine the magnitude of $\mathbf{A}_a$ directly from the dot product (Eq. 2–12); i.e.,

$$A_a = \mathbf{A} \cdot \mathbf{u}_a = A \cos \theta$$

Hence, the scalar projection of **A** *along a line is determined from the dot product of* **A** *and the unit vector* $\mathbf{u}_a$ *which defines the direction of the line*. Notice that if this result is positive, then $\mathbf{A}_a$ has a directional sense which is the same as $\mathbf{u}_a$; whereas if A_a is a negative scalar, then $\mathbf{A}_a$ has the opposite sense of direction to $\mathbf{u}_a$.

The component $\mathbf{A}_a$ represented as a *vector* is therefore

$$\mathbf{A}_a = A_a \mathbf{u}_a$$

The perpendicular component of **A** can also be obtained, Fig. 2–40. Since $\mathbf{A} = \mathbf{A}_a + \mathbf{A}_\perp$, then $\mathbf{A}_\perp = \mathbf{A} - \mathbf{A}_a$. There are two possible ways of obtaining $A_\perp$. One way would be to determine θ from the dot product, $\theta = \cos^{-1}(\mathbf{A} \cdot \mathbf{u}_A/A)$; then $A_\perp = A \sin \theta$. Alternatively, if A_a is known, then by the Pythagorean theorem we can also write $A_\perp = \sqrt{A^2 - A_a^2}$.

The angle θ between the rope and the beam can be determined by formulating unit vectors along the beam and rope and then using the dot product, $\mathbf{u}_b \cdot \mathbf{u}_r = (1)(1) \cos \theta$.

The projection of the cable force **F** along the beam can be determined by first finding the unit vector $\mathbf{u}_b$ that defines this direction. Then apply the dot product, $F_b = \mathbf{F} \cdot \mathbf{u}_b$.

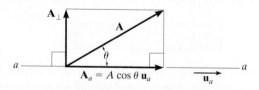

Fig. 2–40

IMPORTANT POINTS

- The dot product is used to determine the angle between two vectors or the projection of a vector in a specified direction.

- If vectors **A** and **B** are expressed in Cartesian vector form, the dot product is determined by multiplying the respective x, y, z scalar components and algebraically adding the results, i.e., $\mathbf{A} \cdot \mathbf{B} = A_x B_x + A_y B_y + A_z B_z$.

- From the definition of the dot product, the angle formed between the tails of vectors **A** and **B** is $\theta = \cos^{-1}(\mathbf{A} \cdot \mathbf{B}/AB)$.

- The magnitude of the projection of vector **A** along a line aa whose direction is specified by $\mathbf{u}_a$ is determined from the dot product, $A_a = \mathbf{A} \cdot \mathbf{u}_a$.

EXAMPLE 2.13

Determine the magnitudes of the projections of the force **F** in Fig. 2–41 onto the u and v axes.

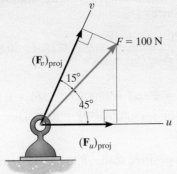

Projections of **F**
(a)

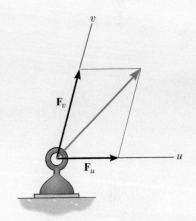

Components of **F**
(b)

Fig. 2–41

SOLUTION

Projections of Force.   The graphical representation of the *projections* is shown in Fig. 2–41a. From this figure, the magnitudes of the projections of **F** onto the u and v axes can be obtained by trigonometry:

$$(F_u)_{\text{proj}} = (100 \text{ N})\cos 45° = 70.7 \text{ N} \qquad \text{Ans.}$$
$$(F_v)_{\text{proj}} = (100 \text{ N})\cos 15° = 96.6 \text{ N} \qquad \text{Ans.}$$

NOTE: These projections *are not* equal to the magnitudes of the components of force **F** along the u and v axes found from the parallelogram law, Fig. 2–41b. They would only be equal if the u and v axes were *perpendicular* to one another.

EXAMPLE   2.14

The frame shown in Fig. 2–42a is subjected to a horizontal force $\mathbf{F} = \{300\mathbf{j}\}$ N. Determine the magnitudes of the components of this force parallel and perpendicular to member AB.

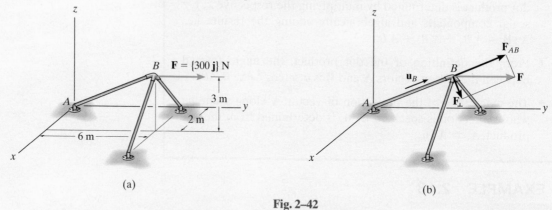

(a)

(b)

Fig. 2–42

SOLUTION

The magnitude of the projected component of $\mathbf{F}$ along AB is equal to the dot product of $\mathbf{F}$ and the unit vector $\mathbf{u}_B$, which defines the direction of AB, Fig. 2–42b. Since

$$\mathbf{u}_B = \frac{\mathbf{r}_B}{r_B} = \frac{2\mathbf{i} + 6\mathbf{j} + 3\mathbf{k}}{\sqrt{(2)^2 + (6)^2 + (3)^2}} = 0.286\mathbf{i} + 0.857\mathbf{j} + 0.429\mathbf{k}$$

then

$$\begin{aligned} F_{AB} = F\cos\theta = \mathbf{F}\cdot\mathbf{u}_B &= (300\mathbf{j})\cdot(0.286\mathbf{i} + 0.857\mathbf{j} + 0.429\mathbf{k}) \\ &= (0)(0.286) + (300)(0.857) + (0)(0.429) \\ &= 257.1 \text{ N} \end{aligned}$$

Ans.

Since the result is a positive scalar, $\mathbf{F}_{AB}$ has the same sense of direction as $\mathbf{u}_B$, Fig. 2–42b.

Expressing $\mathbf{F}_{AB}$ in Cartesian vector form, we have

$$\begin{aligned} \mathbf{F}_{AB} = F_{AB}\mathbf{u}_B &= (257.1 \text{ N})(0.286\mathbf{i} + 0.857\mathbf{j} + 0.429\mathbf{k}) \\ &= \{73.5\mathbf{i} + 220\mathbf{j} + 110\mathbf{k}\} \text{ N} \end{aligned}$$

Ans.

The perpendicular component, Fig. 2–43b, is therefore

$$\begin{aligned} \mathbf{F}_{\perp} = \mathbf{F} - \mathbf{F}_{AB} &= 300\mathbf{j} - (73.5\mathbf{i} + 220\mathbf{j} + 110\mathbf{k}) \\ &= \{-73.5\mathbf{i} + 79.6\mathbf{j} - 110\mathbf{k}\} \text{ N} \end{aligned}$$

Its magnitude can be determined either from this vector or by using the Pythagorean theorem, Fig. 2–42b:

$$\begin{aligned} F_{\perp} = \sqrt{F^2 - F_{AB}^2} &= \sqrt{(300 \text{ N})^2 - (257.1 \text{ N})^2} \\ &= 155 \text{ N} \end{aligned}$$

Ans.

EXAMPLE 2.15

The pipe in Fig. 2–43a is subjected to the force of $F = 80$ lb. Determine the angle θ between **F** and the pipe segment BA, and the projection of **F** along this segment.

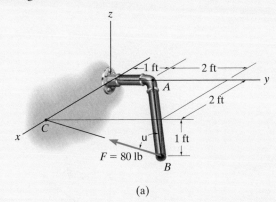

(a)

SOLUTION

Angle θ. First we will establish position vectors from B to A and B to C; Fig. 2–43b. Then we will determine the angle θ between the tails of these two vectors.

$$\mathbf{r}_{BA} = \{ -2\mathbf{i} - 2\mathbf{j} + 1\mathbf{k} \} \text{ ft}, \quad r_{BA} = 3 \text{ ft}$$
$$\mathbf{r}_{BC} = \{ -3\mathbf{j} + 1\mathbf{k} \} \text{ ft}, \quad r_{BC} = \sqrt{10} \text{ ft}$$

Thus,

$$\cos \theta = \frac{\mathbf{r}_{BA} \cdot \mathbf{r}_{BC}}{r_{BA} r_{BC}} = \frac{(-2)(0) + (-2)(-3) + (1)(1)}{3\sqrt{10}} = 0.7379$$

$$\theta = 42.5° \qquad \qquad \textit{Ans.}$$

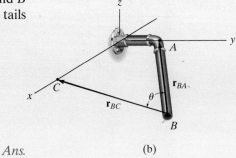

(b)

Projection of F. The projection of **F** along BA is shown in Fig. 2–43c. We must first formulate the unit vector along BA and force **F** as Cartesian vectors.

$$\mathbf{u}_{BA} = \frac{\mathbf{r}_{BA}}{r_{BA}} = \frac{(-2\mathbf{i} - 2\mathbf{j} + 1\mathbf{k})}{3} = -\frac{2}{3}\mathbf{i} - \frac{2}{3}\mathbf{j} + \frac{1}{3}\mathbf{k}$$

$$\mathbf{F} = 80 \text{ lb}\left(\frac{\mathbf{r}_{BC}}{r_{BC}}\right) = 80\left(\frac{-3\mathbf{j} + 1\mathbf{k}}{\sqrt{10}}\right) = -75.89\mathbf{j} + 25.30\mathbf{k}$$

Thus,

$$F_{BA} = \mathbf{F} \cdot \mathbf{u}_{BA} = (-75.89\mathbf{j} + 25.30\mathbf{k}) \cdot \left(-\frac{2}{3}\mathbf{i} - \frac{2}{3}\mathbf{j} + \frac{1}{3}\mathbf{k}\right)$$

$$= 0\left(-\frac{2}{3}\right) + (-75.89)\left(-\frac{2}{3}\right) + (25.30)\left(\frac{1}{3}\right)$$

$$= 59.0 \text{ lb} \qquad \qquad \textit{Ans.}$$

NOTE: Since θ has been calculated, then also, $F_{BA} = F \cos \theta = 80 \text{ lb} \cos 42.5° = 59.0 \text{ lb}.$

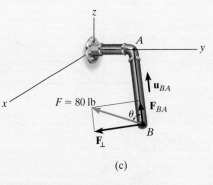

(c)

Fig. 2–43

PRELIMINARY PROBLEMS

P2–8. In each case, set up the dot product to find the angle θ. Do not calculate the result.

P2–9. In each case, set up the dot product to find the magnitude of the projection of the force **F** along a–a axes. Do not calculate the result.

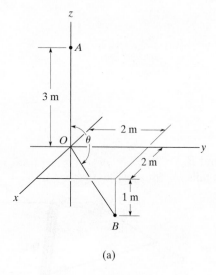

(a)

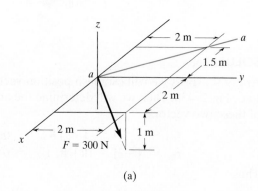

(a)

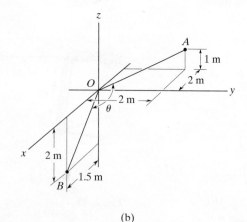

(b)

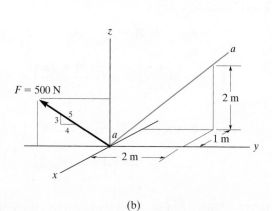

(b)

Prob. P2–8

Prob. P2–9

FUNDAMENTAL PROBLEMS

F2–25. Determine the angle θ between the force and the line AO.

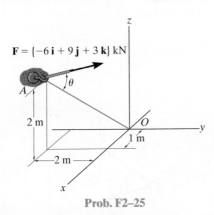

Prob. F2–25

F2–26. Determine the angle θ between the force and the line AB.

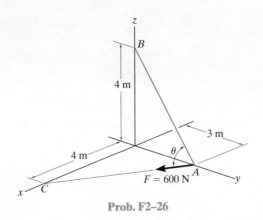

Prob. F2–26

F2–27. Determine the angle θ between the force and the line OA.

F2–28. Determine the projected component of the force along the line OA.

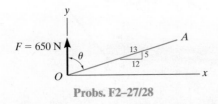

Probs. F2–27/28

F2–29. Find the magnitude of the projected component of the force along the pipe AO.

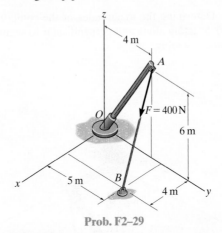

Prob. F2–29

F2–30. Determine the components of the force acting parallel and perpendicular to the axis of the pole.

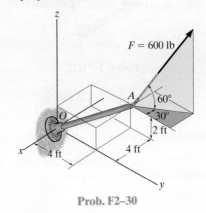

Prob. F2–30

F2–31. Determine the magnitudes of the components of the force $F = 56$ N acting along and perpendicular to line AO.

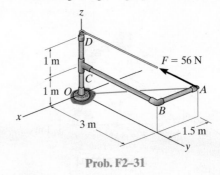

Prob. F2–31

PROBLEMS

2–71. Given the three vectors **A**, **B**, and **D**, show that
$\mathbf{A} \cdot (\mathbf{B} + \mathbf{D}) = (\mathbf{A} \cdot \mathbf{B}) + (\mathbf{A} \cdot \mathbf{D})$.

*2–72. Determine the magnitudes of the components of $F = 600$ N acting along and perpendicular to segment DE of the pipe assembly.

2–75. Determine the angle θ between the two cables.

*2–76. Determine the magnitude of the projection of the force $\mathbf{F}_1$ along cable AC.

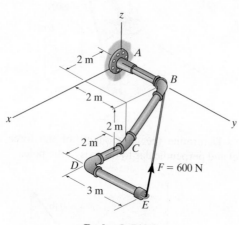

Probs. 2–71/72

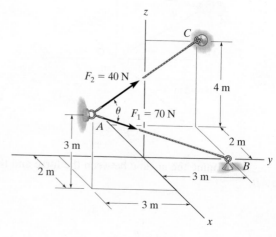

Probs. 2–75/76

2–73. Determine the angle θ between BA and BC.

2–74. Determine the magnitude of the projected component of the 3 kN force acting along axis BC of the pipe.

2–77. Determine the angle θ between the pole and the wire AB.

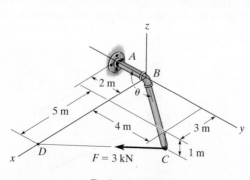

Probs. 2–73/74

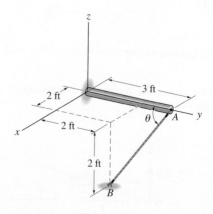

Prob. 2–77

2–78. Determine the magnitude of the projection of the force along the u axis.

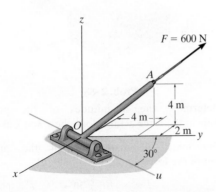

Prob. 2–78

2–79. Determine the magnitude of the projected component of the 100-lb force acting along the axis BC of the pipe.

*2–80. Determine the angle θ between pipe segments BA and BC.

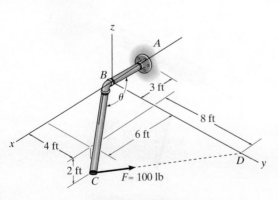

Probs. 2–79/80

2–81. Determine the angle θ between the two cables.

2–82. Determine the projected component of the force acting in the direction of cable AC. Express the result as a Cartesian vector.

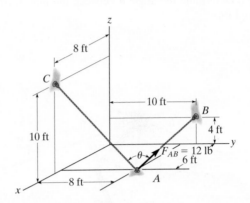

Probs. 2–81/82

2–83. Determine the angles θ and ϕ between the flag pole and the cables AB and AC.

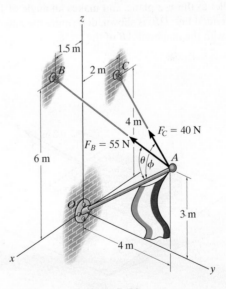

Prob. 2–83

*2–84. Determine the magnitudes of the components of **F** acting along and perpendicular to segment BC of the pipe assembly.

2–85. Determine the magnitude of the projected component of **F** along line AC. Express this component as a Cartesian vector.

2–86. Determine the angle θ between the pipe segments BA and BC.

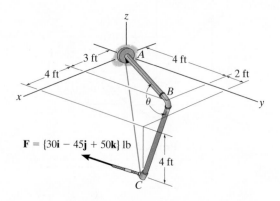

Probs. 2–84/85/86

2–87. If the force $F = 100$ N lies in the plane $DBEC$, which is parallel to the x–z plane, and makes an angle of $10°$ with the extended line DB as shown, determine the angle that **F** makes with the diagonal AB of the crate.

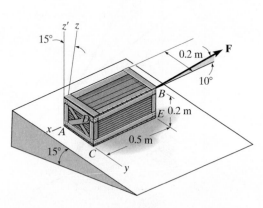

Prob. 2–87

*2–88. Determine the magnitudes of the components of the force acting parallel and perpendicular to diagonal AB of the crate.

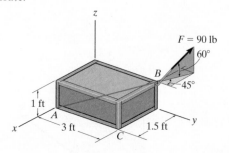

Prob. 2–88

2–89. Determine the magnitudes of the projected components of the force acting along the x and y axes.

2–90. Determine the magnitude of the projected component of the force acting along line OA.

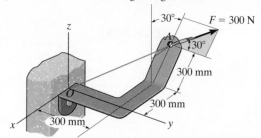

Probs. 2–89/90

2–91. Two cables exert forces on the pipe. Determine the magnitude of the projected component of $\mathbf{F}_1$ along the line of action of $\mathbf{F}_2$.

*2–92. Determine the angle θ between the two forces.

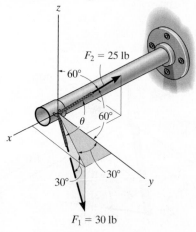

Probs. 2–91/92

CHAPTER REVIEW

A scalar is a positive or negative number; e.g., mass and temperature. A vector has a magnitude and direction, where the arrowhead represents the sense of the vector.		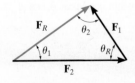
Multiplication or division of a vector by a scalar will change only the magnitude of the vector. If the scalar is negative, the sense of the vector will change so that it acts in the opposite direction.		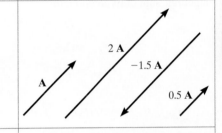
If vectors are collinear, the resultant is simply the algebraic or scalar addition.	$$R = A + B$$	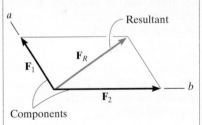
Parallelogram Law Two forces add according to the parallelogram law. The *components* form the sides of the parallelogram and the *resultant* is the diagonal. To find the components of a force along any two axes, extend lines from the head of the force, parallel to the axes, to form the components.		
Two force components can be added tip-to-tail using the triangle rule, and then the law of cosines and the law of sines can be used to calculate unknown values.	$$F_R = \sqrt{F_1^2 + F_2^2 - 2\,F_1 F_2 \cos \theta_R}$$ $$\frac{F_1}{\sin \theta_1} = \frac{F_2}{\sin \theta_2} = \frac{F_R}{\sin \theta_R}$$	

Rectangular Components: Two Dimensions

Vectors $\mathbf{F}_x$ and $\mathbf{F}_y$ are rectangular components of $\mathbf{F}$.

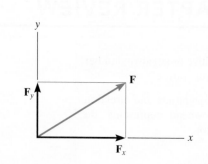

The resultant force is determined from the algebraic sum of its components.

$$(F_R)_x = \Sigma F_x$$

$$(F_R)_y = \Sigma F_y$$

$$F_R = \sqrt{(F_R)_x^2 + (F_R)_y^2}$$

$$\theta = \tan^{-1}\left|\frac{(F_R)_y}{(F_R)_x}\right|$$

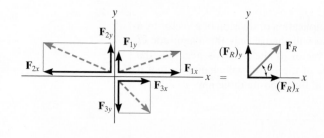

Cartesian Vectors

The unit vector $\mathbf{u}$ has a length of 1, no units, and it points in the direction of the vector $\mathbf{F}$.

$$\mathbf{u} = \frac{\mathbf{F}}{F}$$

A force can be resolved into its Cartesian components along the x, y, z axes so that $\mathbf{F} = F_x\mathbf{i} + F_y\mathbf{j} + F_z\mathbf{k}$.

The magnitude of $\mathbf{F}$ is determined from the positive square root of the sum of the squares of its components.

$$F = \sqrt{F_x^2 + F_y^2 + F_z^2}$$

The coordinate direction angles α, β, γ are determined by formulating a unit vector in the direction of $\mathbf{F}$. The x, y, z components of $\mathbf{u}$ represent $\cos\alpha, \cos\beta, \cos\gamma$.

$$\mathbf{u} = \frac{\mathbf{F}}{F} = \frac{F_x}{F}\mathbf{i} + \frac{F_y}{F}\mathbf{j} + \frac{F_z}{F}\mathbf{k}$$

$$\mathbf{u} = \cos\alpha\,\mathbf{i} + \cos\beta\,\mathbf{j} + \cos\gamma\,\mathbf{k}$$

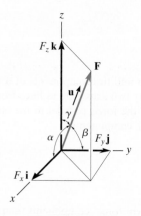

The coordinate direction angles are related, so that only two of the three angles are independent of one another.

$$\cos^2 \alpha + \cos^2 \beta + \cos^2 \gamma = 1$$

To find the resultant of a concurrent force system, express each force as a Cartesian vector and add the $\mathbf{i}, \mathbf{j}, \mathbf{k}$ components of all the forces in the system.

$$\mathbf{F}_R = \Sigma \mathbf{F} = \Sigma F_x \mathbf{i} + \Sigma F_y \mathbf{j} + \Sigma F_z \mathbf{k}$$

Position and Force Vectors

A position vector locates one point in space relative to another. The easiest way to formulate the components of a position vector is to determine the distance and direction that one must travel along the $x, y,$ and z directions—going from the tail to the head of the vector.

$$\mathbf{r} = (x_B - x_A)\mathbf{i}$$
$$+ (y_B - y_A)\mathbf{j}$$
$$+ (z_B - z_A)\mathbf{k}$$

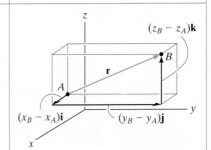

If the line of action of a force passes through points A and B, then the force acts in the same direction $\mathbf{u}$ as the position vector $\mathbf{r}$ extending from A to B. Knowing F and $\mathbf{u}$, the force can then be expressed as a Cartesian vector.

$$\mathbf{F} = F\mathbf{u} = F\left(\frac{\mathbf{r}}{r}\right)$$

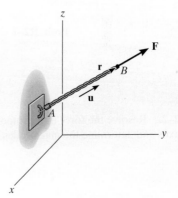

Dot Product

The dot product between two vectors $\mathbf{A}$ and $\mathbf{B}$ yields a scalar. If $\mathbf{A}$ and $\mathbf{B}$ are expressed in Cartesian vector form, then the dot product is the sum of the products of their $x, y,$ and z components.

$$\mathbf{A} \cdot \mathbf{B} = AB \cos \theta$$
$$= A_x B_x + A_y B_y + A_z B_z$$

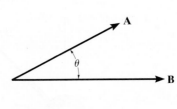

The dot product can be used to determine the angle between $\mathbf{A}$ and $\mathbf{B}$.

$$\theta = \cos^{-1}\left(\frac{\mathbf{A} \cdot \mathbf{B}}{AB}\right)$$

The dot product is also used to determine the projected component of a vector $\mathbf{A}$ onto an axis aa defined by its unit vector $\mathbf{u}_a$.

$$\mathbf{A}_a = A \cos \theta \, \mathbf{u}_a = (\mathbf{A} \cdot \mathbf{u}_a)\mathbf{u}_a$$

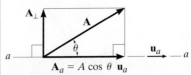

REVIEW PROBLEMS

R2–1. Determine the magnitude of the resultant force $\mathbf{F}_R$ and its direction, measured clockwise from the positive u axis.

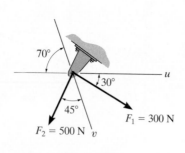

Prob. R2–1

R2–3. Determine the magnitude of the resultant force acting on the *gusset plate*.

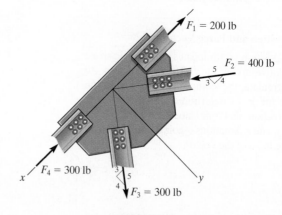

Prob. R2–3

R2–2. Resolve the force into components along the u and v axes and determine the magnitudes of these components.

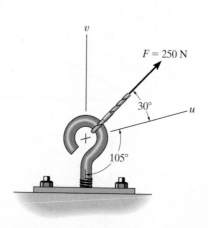

Prob. R2–2

*R2–4. The cable exerts a force of 250 lb on the crane boom as shown. Express this force as a Cartesian vector.

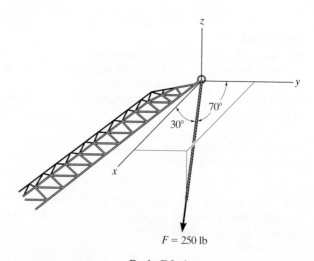

Prob. R2–4

R2–5. The cable attached to the tractor at B exerts a force of 350 lb on the framework. Express this force as a Cartesian vector.

R2–7. Determine the angle θ between the edges of the sheet-metal bracket.

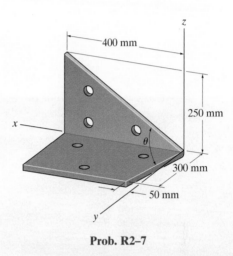

Prob. R2–7

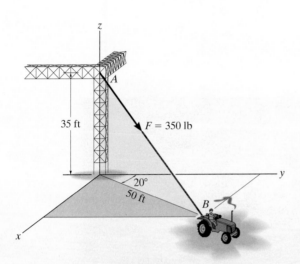

Prob. R2–5

***R2–8.** Determine the projection of the force **F** along the pole.

R2–6. Express $\mathbf{F}_1$ and $\mathbf{F}_2$ as Cartesian vectors.

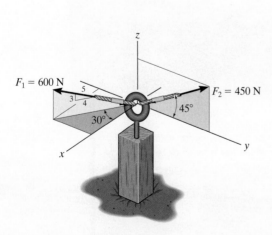

Prob. R2–6

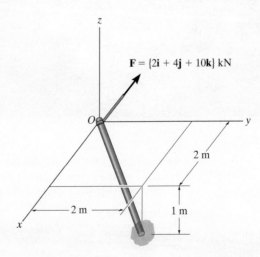

Prob. R2–8

CHAPTER 3

(© Rolf Adlercreutz/Alamy)

The force applied to this wrench will produce rotation or a tendency for rotation. This effect is called a moment, and in this chapter we will study how to determine the moment of a system of forces and calculate their resultants.

FORCE SYSTEM RESULTANTS

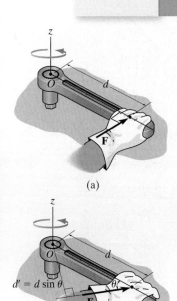

(a)

(b)

3.1 MOMENT OF A FORCE—SCALAR FORMULATION

When a force is applied to a body it will produce a tendency for the body to rotate about a point that is not on the line of action of the force. This tendency to rotate is sometimes called a *torque*, but most often it is called the moment of a force or simply the *moment*. For example, consider applying a force to the handle of the wrench used to unscrew the bolt in Fig. 3–1a. The magnitude of the moment is directly proportional to the magnitude of $\mathbf{F}$ and the perpendicular distance or moment arm d. The larger the force or the longer the moment arm, the greater the moment or turning effect. If the force $\mathbf{F}$ is applied at an angle $\theta \neq 90°$, Fig. 3–1b, then it will be more difficult to turn the bolt since the moment arm $d' = d \sin\theta$ will be smaller than d. If $\mathbf{F}$ is applied along the handle, Fig. 3–1c, its moment arm will be zero since the line of action of $\mathbf{F}$ will intersect point O (the z axis). As a result, the moment of $\mathbf{F}$ about O is also zero and no turning can occur.

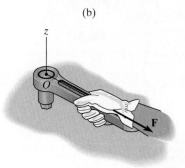

(c)

Fig. 3–1

79

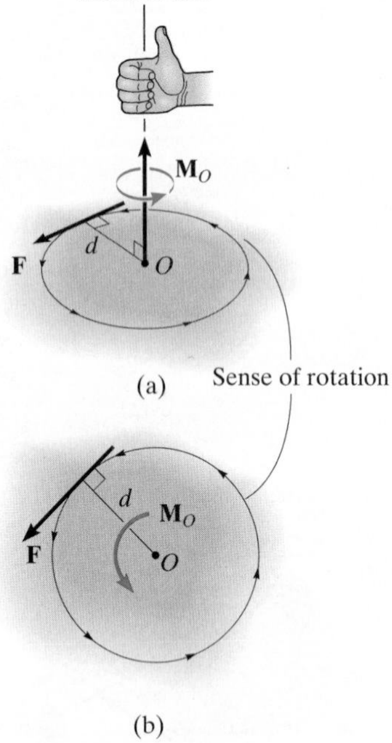

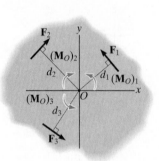

Moment axis

M_O

F

d

O

(a)   Sense of rotation

d M_O

F

O

(b)

Fig. 3–2

In general, if we consider the force **F** and point O to lie in the shaded plane shown in Fig. 3–2a, the moment $\mathbf{M}_O$ about point O, or about an axis passing through O and perpendicular to the plane, is a *vector quantity* since it has a specified magnitude and direction.

Magnitude. The magnitude of $\mathbf{M}_O$ is

$$M_O = Fd \tag{3–1}$$

where d is the **moment arm** or *perpendicular distance* from the axis at point O to the line of action of the force. Units of moment magnitude consist of force times distance, e.g., N · m or lb · ft.

Direction. The direction of $\mathbf{M}_O$ is defined by its **moment axis**, which is perpendicular to the plane that contains the force **F** and its moment arm d. The right-hand rule is used to establish the sense of direction of $\mathbf{M}_O$, where the natural curl of the fingers of the right hand, as they are drawn towards the palm, represents the rotation, or the tendency for rotation caused by the moment. Doing this, the thumb of the right hand will give the directional sense of $\mathbf{M}_O$, Fig. 3–2a. Here the moment vector is represented three-dimensionally by a curl around an arrow. In two dimensions this vector is represented only by the curl, as in Fig. 3–2b. Since this produces counterclockwise rotation, the moment vector is actually directed out of the page.

Resultant Moment. For two-dimensional problems, where all the forces lie within the x–y plane, Fig. 3–3, the resultant moment $(\mathbf{M}_R)_O$ about point O (the z axis) can be determined by *finding the algebraic sum* of the moments caused by all the forces in the system. As a convention, we will generally consider *positive moments* as *counterclockwise* since they are directed along the positive z axis (out of the page). *Clockwise moments* will be *negative*. The directional sense of each moment can be represented by a *plus* or *minus* sign. Using this sign convention, with a symbolic curl to define the positive direction, the resultant moment in Fig. 3–3 is therefore

$$\zeta + (M_R)_O = \Sigma Fd; \qquad (M_R)_O = F_1 d_1 - F_2 d_2 + F_3 d_3$$

If the numerical result of this sum is a positive scalar, $(\mathbf{M}_R)_O$ will be a counterclockwise moment (out of the page); and if the result is negative, $(\mathbf{M}_R)_O$ will be a clockwise moment (into the page).

F_2 y

$(M_O)_2$ F_1

d_2 d_1 $(M_O)_1$

$(M_O)_3$ O x

d_3

F_3

Fig. 3–3

EXAMPLE 3.1

For each case illustrated in Fig. 3–4, determine the moment of the force about point O.

SOLUTION (SCALAR ANALYSIS)
The line of action of each force is extended as a dashed line in order to establish the moment arm d. Also illustrated is the tendency of rotation of the member as caused by the force, and the orbit of the force about O is shown as a colored curl. Thus,

Fig. 3–4a $M_O = (100 \text{ N})(2 \text{ m}) = 200 \text{ N} \cdot \text{m} \,\circlearrowright$ *Ans.*

Fig. 3–4b $M_O = (50 \text{ N})(0.75 \text{ m}) = 37.5 \text{ N} \cdot \text{m} \,\circlearrowright$ *Ans.*

Fig. 3–4c $M_O = (40 \text{ lb})(4 \text{ ft} + 2 \cos 30° \text{ ft}) = 229 \text{ lb} \cdot \text{ft} \,\circlearrowright$ *Ans.*

Fig. 3–4d $M_O = (60 \text{ lb})(1 \sin 45° \text{ ft}) = 42.4 \text{ lb} \cdot \text{ft} \,\circlearrowleft$ *Ans.*

Fig. 3–4e $M_O = (7 \text{ kN})(4 \text{ m} - 1 \text{ m}) = 21.0 \text{ kN} \cdot \text{m} \,\circlearrowleft$ *Ans.*

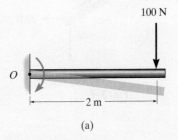

(a)

3

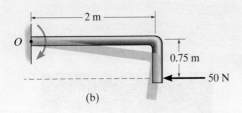

(b)

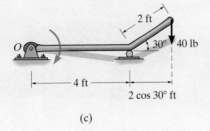

(c)

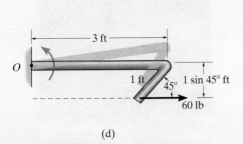

(d)

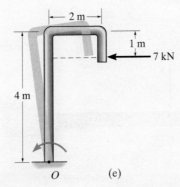

(e)

Fig. 3–4

EXAMPLE   3.2

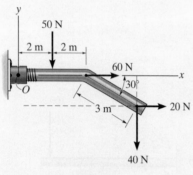

Fig. 3–5

Determine the resultant moment of the four forces acting on the rod shown in Fig. 3–5 about point O.

SOLUTION
Assuming that positive moments act in the $+\mathbf{k}$ direction, i.e., counterclockwise, we have

$$\zeta + (M_R)_o = \Sigma Fd;$$

$$(M_R)_o = -50 \text{ N}(2 \text{ m}) + 60 \text{ N}(0) + 20 \text{ N}(3 \sin 30° \text{ m})$$

$$-40 \text{ N}(4 \text{ m} + 3 \cos 30° \text{ m})$$

$$(M_R)_o = -334 \text{ N} \cdot \text{m} = 334 \text{ N} \cdot \text{m} \; \mathrel{)} \qquad\qquad\qquad Ans.$$

For this calculation, note how the moment-arm distances for the 20-N and 40-N forces are established from the extended (dashed) lines of action of each of the forces.

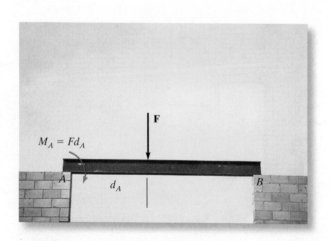

The force $\mathbf{F}$ tends to rotate the beam clockwise about its support at A with a moment $M_A = Fd_A$. The actual rotation would occur only if the support at B were removed.

The ability to remove the nail will require the moment of $\mathbf{F}_h$ about point O to be larger than the moment of the force $\mathbf{F}_n$ about O that is needed to pull the nail out.

3.2   CROSS PRODUCT

The moment of a force will be formulated using Cartesian vectors in the next section. Before doing this, however, it is first necessary to expand our knowledge of vector algebra and introduce the cross-product method of vector multiplication.

The ***cross product*** of two vectors **A** and **B** yields the vector **C**, which is written as

$$C = A \times B \qquad\qquad (3\text{–}2)$$

and is read "**C** equals **A** cross **B**."

Magnitude. The *magnitude* of **C** is defined as the product of the magnitudes of **A** and **B** and the sine of the angle θ between their tails, where $0° \le \theta \le 180°$. Thus,

$$C = AB \sin \theta$$

Direction. Vector **C** has a *direction* that is perpendicular to the plane containing **A** and **B** such that the directional sense of **C** is specified by the right-hand rule; i.e., curling the fingers of the right hand from vector **A** (cross) to vector **B**, the thumb points in the direction of **C**, as shown in Fig. 3–6.

Knowing both the magnitude and direction of **C**, we can therefore write

$$C = A \times B = (AB \sin \theta)u_C \qquad\qquad (3\text{–}3)$$

The terms of Eq. 3–3 are illustrated graphically in Fig. 3–6.

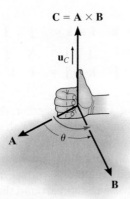

Fig. 3–6

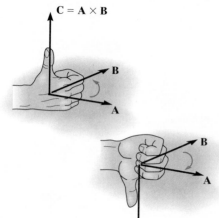

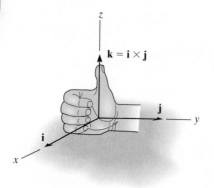

Fig. 3–7

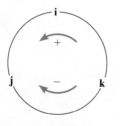

Fig. 3–8

Fig. 3–9

The following three laws of operation apply.

- The commutative law is *not* valid; i.e., $\mathbf{A} \times \mathbf{B} \neq \mathbf{B} \times \mathbf{A}$. Rather,

$$\mathbf{A} \times \mathbf{B} = -\mathbf{B} \times \mathbf{A}$$

This is shown in Fig. 3–7 by using the right-hand rule. The cross product $\mathbf{B} \times \mathbf{A}$ yields a vector that has the same magnitude but acts in the opposite sense of direction to $\mathbf{C}$; i.e., $\mathbf{B} \times \mathbf{A} = -\mathbf{C}$.

- If the cross product is multiplied by a scalar a, it obeys the associative law;

$$a(\mathbf{A} \times \mathbf{B}) = (a\mathbf{A}) \times \mathbf{B} = \mathbf{A} \times (a\mathbf{B}) = (\mathbf{A} \times \mathbf{B})a$$

This property is easily shown since the magnitude of the resultant vector ($|a| AB \sin \theta$) and its sense of direction are the same in each case.

- The vector cross product also obeys the distributive law of addition,

$$\mathbf{A} \times (\mathbf{B} + \mathbf{D}) = (\mathbf{A} \times \mathbf{B}) + (\mathbf{A} \times \mathbf{D})$$

It is important to note that *proper order* of these cross products must be maintained since they are not commutative.

Cartesian Vector Formulation. Equation 3–3 may be used to find the cross product of any pair of Cartesian unit vectors. For example, to find $\mathbf{i} \times \mathbf{j}$, the magnitude of the resultant vector is $(i)(j)(\sin 90°) = (1)(1)(1) = 1$, and its direction is determined using the right-hand rule, Fig. 3–8. Here the resultant vector points in the $+\mathbf{k}$ direction so that $\mathbf{i} \times \mathbf{j} = (1)\mathbf{k}$. In a similar manner,

$$\mathbf{i} \times \mathbf{j} = \mathbf{k} \quad \mathbf{i} \times \mathbf{k} = -\mathbf{j} \quad \mathbf{i} \times \mathbf{i} = 0$$
$$\mathbf{j} \times \mathbf{k} = \mathbf{i} \quad \mathbf{j} \times \mathbf{i} = -\mathbf{k} \quad \mathbf{j} \times \mathbf{j} = 0$$
$$\mathbf{k} \times \mathbf{i} = \mathbf{j} \quad \mathbf{k} \times \mathbf{j} = -\mathbf{i} \quad \mathbf{k} \times \mathbf{k} = 0$$

These results should *not* be memorized; rather, it should be clearly understood how each is obtained by using the right-hand rule and the definition of the cross product. A simple scheme shown in Fig. 3–9 can sometimes be helpful for obtaining the same results when the need arises. If the circle is constructed as shown, then "crossing" two unit vectors in a *counterclockwise* fashion around the circle yields the *positive* third unit vector; e.g., $\mathbf{k} \times \mathbf{i} = \mathbf{j}$. "Crossing" *clockwise*, a *negative* unit vector is obtained; e.g., $\mathbf{i} \times \mathbf{k} = -\mathbf{j}$.

Let us now consider the cross product of vectors **A** and **B** which are expressed in Cartesian vector form. We have

$$\mathbf{A} \times \mathbf{B} = (A_x\mathbf{i} + A_y\mathbf{j} + A_z\mathbf{k}) \times (B_x\mathbf{i} + B_y\mathbf{j} + B_z\mathbf{k})$$
$$= A_xB_x(\mathbf{i} \times \mathbf{i}) + A_xB_y(\mathbf{i} \times \mathbf{j}) + A_xB_z(\mathbf{i} \times \mathbf{k})$$
$$+ A_y B_x(\mathbf{j} \times \mathbf{i}) + A_yB_y(\mathbf{j} \times \mathbf{j}) + A_yB_z(\mathbf{j} \times \mathbf{k})$$
$$+ A_zB_x(\mathbf{k} \times \mathbf{i}) + A_zB_y(\mathbf{k} \times \mathbf{j}) + A_zB_z(\mathbf{k} \times \mathbf{k})$$

Carrying out the cross-product operations and combining terms yields

$$\mathbf{A} \times \mathbf{B} = (A_yB_z - A_zB_y)\mathbf{i} - (A_xB_z - A_zB_x)\mathbf{j} + (A_xB_y - A_yB_x)\mathbf{k} \quad (3\text{--}4)$$

This equation may also be written in a more compact determinant form as

$$\mathbf{A} \times \mathbf{B} = \begin{vmatrix} \mathbf{i} & \mathbf{j} & \mathbf{k} \\ A_x & A_y & A_z \\ B_x & B_y & B_z \end{vmatrix} \quad (3\text{--}5)$$

Thus, to find the cross product of **A** and **B**, it is necessary to expand a determinant whose first row of elements consists of the unit vectors **i**, **j**, and **k** and whose second and third rows represent the x, y, z components of the two vectors **A** and **B**, respectively.*

*A determinant having three rows and three columns can be expanded using three minors, each of which is multiplied by one of the three terms in the first row. There are four elements in each minor, for example,

$$\begin{vmatrix} A_{11} & A_{12} \\ A_{21} & A_{22} \end{vmatrix}$$

By *definition*, this determinant notation represents the terms $(A_{11}A_{22} - A_{12}A_{21})$, which is simply the product of the two elements intersected by the arrow slanting downward to the right $(A_{11}A_{22})$ *minus* the product of the two elements intersected by the arrow slanting downward to the left $(A_{12}A_{21})$. For a 3×3 determinant, such as Eq. 3–5, the three minors can be generated in accordance with the following scheme:

For element **i**: $\begin{vmatrix} \oplus & \mathbf{j} & \mathbf{k} \\ A_x & A_y & A_z \\ B_x & B_y & B_z \end{vmatrix} = \mathbf{i}(A_yB_z - A_zB_y)$

Remember the negative sign

For element **j**: $\begin{vmatrix} \mathbf{i} & \oplus & \mathbf{k} \\ A_x & A_y & A_z \\ B_x & B_y & B_z \end{vmatrix} = -\mathbf{j}(A_xB_z - A_zB_x)$

For element **k**: $\begin{vmatrix} \mathbf{i} & \mathbf{j} & \oplus \\ A_x & A_y & A_z \\ B_x & B_y & B_z \end{vmatrix} = \mathbf{k}(A_xB_y - A_yB_x)$

Adding these results and noting that the **j** element *must include the minus sign* yields the expanded form of **A** × **B** given by Eq. 3–4.

3.3 MOMENT OF A FORCE—VECTOR FORMULATION

The moment of a force $\mathbf{F}$ about point O, Fig. 3–10a, can be expressed using the vector cross product,

$$\mathbf{M}_O = \mathbf{r} \times \mathbf{F} \qquad (3\text{–}6)$$

Here $\mathbf{r}$ is a position vector directed *from* O to *any point* on the line of action of $\mathbf{F}$. We will now show that indeed the moment $\mathbf{M}_O$, when determined by this cross product, has the proper magnitude and direction.

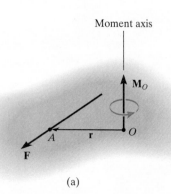

Moment axis

(a)

Magnitude. The magnitude of the cross product is defined from Eq. 3–3 as $M_O = rF \sin \theta$, where the angle θ is measured between the *tails* of $\mathbf{r}$ and $\mathbf{F}$. To establish this angle, $\mathbf{r}$ must be treated as a ***sliding vector*** so that θ can be constructed properly, Fig. 3–10b. Since the moment arm $d = r \sin \theta$, then

$$M_O = rF \sin \theta = F(r \sin \theta) = Fd$$

which agrees with Eq. 3–1.

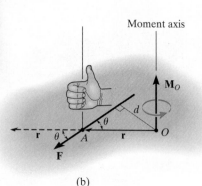

Moment axis

(b)

Fig. 3–10

Direction. The direction and sense of $\mathbf{M}_O$ in Eq. 3–6 are determined by the right-hand rule as it applies to the cross product. Thus, sliding $\mathbf{r}$ to the dashed position and curling the right-hand fingers from $\mathbf{r}$ towards $\mathbf{F}$, "$\mathbf{r}$ cross $\mathbf{F}$," the thumb is directed upward or perpendicular to the plane containing $\mathbf{r}$ and $\mathbf{F}$ and this is in the *same direction* as $\mathbf{M}_O$, the moment of the force about point O, Fig. 3–10b. Remember that the cross product does not obey the commutative law, and so the order of $\mathbf{r} \times \mathbf{F}$ must be maintained to produce the correct sense of direction for $\mathbf{M}_O$.

Principle of Transmissibility. The cross product operation is often used in three dimensions since the perpendicular distance or moment arm from point O to the line of action of the force is not needed. In other words, we can use any position vector $\mathbf{r}$ measured from point O to any point on the line of action of the force $\mathbf{F}$, Fig. 3–11. Thus,

$$\mathbf{M}_O = \mathbf{r}_1 \times \mathbf{F} = \mathbf{r}_2 \times \mathbf{F} = \mathbf{r}_3 \times \mathbf{F}$$

Since $\mathbf{F}$ can be applied at any point along its line of action and still create this *same moment* about point O, then $\mathbf{F}$ can be considered a *sliding vector*. This property is called the ***principle of transmissibility*** of a force.

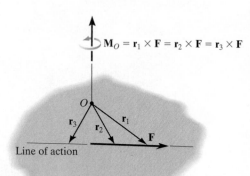

$\mathbf{M}_O = \mathbf{r}_1 \times \mathbf{F} = \mathbf{r}_2 \times \mathbf{F} = \mathbf{r}_3 \times \mathbf{F}$

Line of action

Fig. 3–11

Cartesian Vector Formulation. If we establish x, y, z coordinate axes, then the position vector $\mathbf{r}$ and force $\mathbf{F}$ can be expressed as Cartesian vectors, Fig. 3–12a. Then applying Eq. 3–5 we have

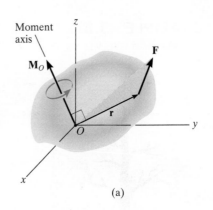

$$\mathbf{M}_O = \mathbf{r} \times \mathbf{F} = \begin{vmatrix} \mathbf{i} & \mathbf{j} & \mathbf{k} \\ r_x & r_y & r_z \\ F_x & F_y & F_z \end{vmatrix} \qquad (3\text{–}7)$$

where

r_x, r_y, r_z represent the x, y, z components of the position vector drawn from point O to *any point* on the line of action of the force

F_x, F_y, F_z represent the x, y, z components of the force vector

If the determinant is expanded, then like Eq. 3–4 we have

$$\mathbf{M}_O = (r_y F_z - r_z F_y)\mathbf{i} - (r_x F_z - r_z F_x)\mathbf{j} + (r_x F_y - r_y F_x)\mathbf{k} \qquad (3\text{–}8)$$

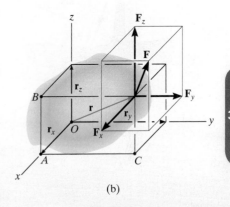

(b)

Fig. 3–12

The physical meaning of these three moment components becomes evident by studying Fig. 3–12b. For example, the $\mathbf{i}$ component of $\mathbf{M}_O$ can be determined from the moments of $\mathbf{F}_x$, $\mathbf{F}_y$, and $\mathbf{F}_z$ about the x axis. The component $\mathbf{F}_x$ does *not* create a moment or tendency to cause turning about the x axis since this force is *parallel* to the x axis. The line of action of $\mathbf{F}_y$ passes through point B, and so the magnitude of the moment of $\mathbf{F}_y$ about point A on the x axis is $r_z F_y$. By the right-hand rule this component acts in the *negative* $\mathbf{i}$ direction. Likewise, $\mathbf{F}_z$ passes through point C and so it contributes a moment component of $r_y F_z \mathbf{i}$ about the x axis. Thus, $(M_O)_x = (r_y F_z - r_z F_y)$ as shown in Eq. 3–8. As an exercise, try to establish the $\mathbf{j}$ and $\mathbf{k}$ components of $\mathbf{M}_O$ in this manner and show that indeed the expanded form of the determinant, Eq. 3–8, represents the moment of $\mathbf{F}$ about point O. Once $\mathbf{M}_O$ is determined, realize that it will always be *perpendicular* to the shaded plane containing vectors $\mathbf{r}$ and $\mathbf{F}$, Fig. 3–12a.

Resultant Moment of a System of Forces. If a body is acted upon by a system of forces, Fig. 3–13, the resultant moment of the forces about point O can be determined by vector addition of the moment of each force. This resultant can be written symbolically as

$$(\mathbf{M}_R)_o = \Sigma(\mathbf{r} \times \mathbf{F}) \qquad (3\text{–}9)$$

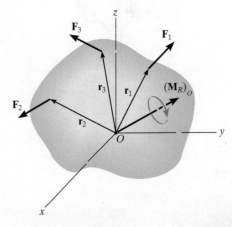

Fig. 3–13

EXAMPLE 3.3

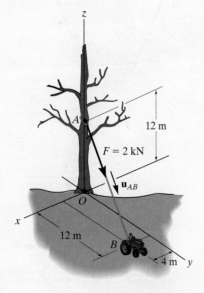

(a)

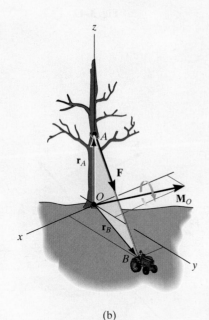

(b)

Fig. 3–14

Determine the moment produced by the force **F** in Fig. 3–14a about point O. Express the result as a Cartesian vector.

SOLUTION
As shown in Fig. 3–14b, either $\mathbf{r}_A$ or $\mathbf{r}_B$ can be used to determine the moment about point O. These position vectors are

$$\mathbf{r}_A = \{12\mathbf{k}\}\ \text{m} \qquad\qquad \mathbf{r}_B = \{4\mathbf{i} + 12\mathbf{j}\}\ \text{m}$$

Force **F** expressed as a Cartesian vector is

$$\mathbf{F} = F\mathbf{u}_{AB} = 2\ \text{kN}\left[\frac{\{4\mathbf{i} + 12\mathbf{j} - 12\mathbf{k}\}\ \text{m}}{\sqrt{(4\ \text{m})^2 + (12\ \text{m})^2 + (-12\ \text{m})^2}}\right]$$

$$= \{0.4588\mathbf{i} + 1.376\mathbf{j} - 1.376\mathbf{k}\}\ \text{kN}$$

Thus

$$\mathbf{M}_O = \mathbf{r}_A \times \mathbf{F} = \begin{vmatrix} \mathbf{i} & \mathbf{j} & \mathbf{k} \\ 0 & 0 & 12 \\ 0.4588 & 1.376 & -1.376 \end{vmatrix}$$

$$= [0(-1.376) - 12(1.376)]\mathbf{i} - [0(-1.376) - 12(0.4588)]\mathbf{j}$$
$$+ [0(1.376) - 0(0.4588)]\mathbf{k}$$

$$= \{-16.5\mathbf{i} + 5.51\mathbf{j}\}\ \text{kN}\cdot\text{m} \qquad\qquad Ans.$$

or

$$\mathbf{M}_O = \mathbf{r}_B \times \mathbf{F} = \begin{vmatrix} \mathbf{i} & \mathbf{j} & \mathbf{k} \\ 4 & 12 & 0 \\ 0.4588 & 1.376 & -1.376 \end{vmatrix}$$

$$= [12(-1.376) - 0(1.376)]\mathbf{i} - [4(-1.376) - 0(0.4588)]\mathbf{j}$$
$$+ [4(1.376) - 12(0.4588)]\mathbf{k}$$

$$= \{-16.5\mathbf{i} + 5.51\mathbf{j}\}\ \text{kN}\cdot\text{m} \qquad\qquad Ans.$$

NOTE: As shown in Fig. 3–14b, $\mathbf{M}_O$ acts perpendicular to the plane that contains **F**, $\mathbf{r}_A$, and $\mathbf{r}_B$. Had this problem been worked using $M_O = Fd$, notice the difficulty that would arise in obtaining the moment arm d.

EXAMPLE 3.4

Two forces act on the rod shown in Fig. 3–15a. Determine the resultant moment they create about the flange at O. Express the result as a Cartesian vector.

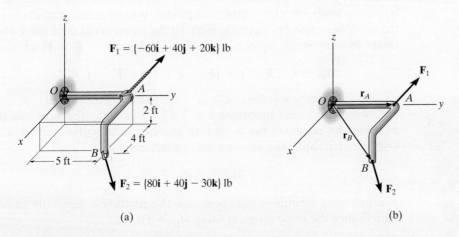

(a)

(b)

SOLUTION

Position vectors are directed from O to each force as shown in Fig. 3–15b. These vectors are

$$\mathbf{r}_A = \{5\mathbf{j}\} \text{ ft}$$

$$\mathbf{r}_B = \{4\mathbf{i} + 5\mathbf{j} - 2\mathbf{k}\} \text{ ft}$$

The resultant moment about O is therefore

$$(\mathbf{M}_R)_O = \Sigma(\mathbf{r} \times \mathbf{F})$$

$$= \mathbf{r}_A \times \mathbf{F}_1 + \mathbf{r}_B \times \mathbf{F}_2$$

$$= \begin{vmatrix} \mathbf{i} & \mathbf{j} & \mathbf{k} \\ 0 & 5 & 0 \\ -60 & 40 & 20 \end{vmatrix} + \begin{vmatrix} \mathbf{i} & \mathbf{j} & \mathbf{k} \\ 4 & 5 & -2 \\ 80 & 40 & -30 \end{vmatrix}$$

$$= [5(20) - 0(40)]\mathbf{i} - [0]\mathbf{j} + [0(40) - (5)(-60)]\mathbf{k}$$

$$\quad + [5(-30) - (-2)(40)]\mathbf{i} - [4(-30) - (-2)(80)]\mathbf{j} + [4(40) - 5(80)]\mathbf{k}$$

$$= \{30\mathbf{i} - 40\mathbf{j} + 60\mathbf{k}\} \text{ lb} \cdot \text{ft} \qquad \textit{Ans.}$$

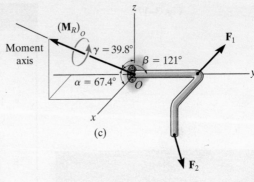

(c)

Fig. 3–15

NOTE: This result is shown in Fig. 3–15c. The coordinate direction angles were determined from the unit vector for $(\mathbf{M}_R)_O$. Realize that the two forces tend to cause the rod to rotate about the moment axis in the manner shown by the curl indicated on the moment vector.

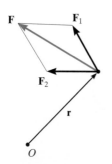

Fig. 3–16

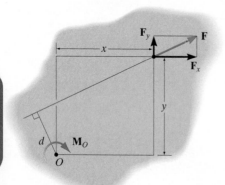

Fig. 3–17

3.4 PRINCIPLE OF MOMENTS

A concept often used in mechanics is the ***principle of moments***, which is sometimes referred to as **Varignon's theorem** since it was originally developed by the French mathematician Pierre Varignon (1654–1722). It states that *the moment of a force about a point is equal to the sum of the moments of the components of the force about the point.* This theorem can be proven easily using the vector cross product since the cross product obeys the *distributive law.* For example, consider the moments of the force **F** and two of its components about point O, Fig. 3–16. Since $\mathbf{F} = \mathbf{F}_1 + \mathbf{F}_2$ we have

$$\mathbf{M}_O = \underbrace{\mathbf{r} \times \mathbf{F}}_{\text{moment of force}} = \mathbf{r} \times (\mathbf{F}_1 + \mathbf{F}_2) = \underbrace{\mathbf{r} \times \mathbf{F}_1 + \mathbf{r} \times \mathbf{F}_2}_{\text{moment of components}}$$

For two-dimensional problems, Fig. 3–17, we can use the principle of moments by resolving the force into any two rectangular components and then determine the moment using a scalar analysis. Thus,

$$M_O = F_x y - F_y x$$

The following examples will show that this method is generally easier than finding the same moment using $M_O = Fd$.

The moment of the force about point O is $M_O = Fd$. But it is easier to find this moment using $M_O = F_x(0) + F_y r = F_y r$.

> ### ▶ IMPORTANT POINTS
>
> - The moment of a force creates the tendency of a body to turn about an axis passing through a specific point O.
>
> - Using the right-hand rule, the sense of rotation is indicated by the curl of the fingers, and the thumb produces the sense of direction of the moment.
>
> - The magnitude of the moment is determined from $M_O = Fd$, where d is called the moment arm, which represents the *perpendicular* or shortest distance from point O to the line of action of the force.
>
> - In three dimensions the vector cross product is used to determine the moment, i.e., $\mathbf{M}_O = \mathbf{r} \times \mathbf{F}$. Here $\mathbf{r}$ is directed *from* point O to *any point* on the line of action of **F**.
>
> - In two dimensions it is often easier to use the principle of moments and find the moment of the force's components about point O, rather than using $M_O = Fd$.

EXAMPLE   3.5

Determine the moment of the force in Fig. 3–18a about point O.

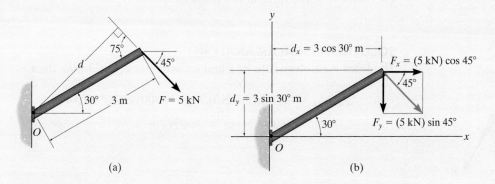

(a)   (b)

SOLUTION I

The moment arm d in Fig. 3–18a can be found from trigonometry.

$$d = (3 \text{ m}) \sin 75° = 2.898 \text{ m}$$

Thus,

$$M_O = Fd = (5 \text{ kN})(2.898 \text{ m}) = 14.5 \text{ kN} \cdot \text{m} \circlearrowright \qquad Ans.$$

Since the force tends to rotate or orbit clockwise about point O, the moment is directed into the page.

SOLUTION II

The x and y components of the force are indicated in Fig. 3–18b. Considering counterclockwise moments as positive, and applying the principle of moments, we have

$$\circlearrowleft + M_O = -F_x d_y - F_y d_x$$
$$= -(5 \cos 45° \text{ kN})(3 \sin 30° \text{ m}) - (5 \sin 45° \text{ kN})(3 \cos 30° \text{ m})$$
$$= -14.5 \text{ kN} \cdot \text{m} = 14.5 \text{ kN} \cdot \text{m} \circlearrowright \qquad Ans.$$

SOLUTION III

The x and y axes can be oriented parallel and perpendicular to the rod's axis as shown in Fig. 3–18c. Here $\mathbf{F}_x$ produces no moment about point O since its line of action passes through this point. Therefore,

$$\circlearrowleft + M_O = -F_y d_x$$
$$= -(5 \sin 75° \text{ kN})(3 \text{ m})$$
$$= -14.5 \text{ kN} \cdot \text{m} = 14.5 \text{ kN} \cdot \text{m} \circlearrowright \qquad Ans.$$

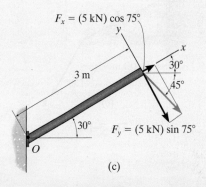

(c)

Fig. 3–18

EXAMPLE 3.6

3

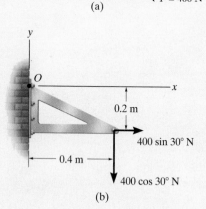

(a)

(b)

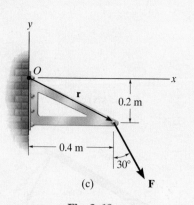

(c)

Fig. 3–19

Force **F** acts at the end of the angle bracket in Fig. 3–19a. Determine the moment of the force about point O.

SOLUTION I (SCALAR ANALYSIS)

The force is resolved into its x and y components, Fig. 3–19b, then

$$\zeta + M_O = 400 \sin 30° \text{ N}(0.2 \text{ m}) - 400 \cos 30° \text{ N}(0.4 \text{ m})$$

$$= -98.6 \text{ N} \cdot \text{m} = 98.6 \text{ N} \cdot \text{m} \, \zeta$$

or

$$\mathbf{M}_O = \{-98.6\mathbf{k}\} \text{ N} \cdot \text{m} \qquad\qquad Ans.$$

SOLUTION II (VECTOR ANALYSIS)

Using a Cartesian vector approach, the force and position vectors, Fig. 3–19c, are

$$\mathbf{r} = \{0.4\mathbf{i} - 0.2\mathbf{j}\} \text{ m}$$

$$\mathbf{F} = \{400 \sin 30°\mathbf{i} - 400 \cos 30°\mathbf{j}\} \text{ N}$$

$$= \{200.0\mathbf{i} - 346.4\mathbf{j}\} \text{ N}$$

The moment is therefore

$$\mathbf{M}_O = \mathbf{r} \times \mathbf{F} = \begin{vmatrix} \mathbf{i} & \mathbf{j} & \mathbf{k} \\ 0.4 & -0.2 & 0 \\ 200.0 & -346.4 & 0 \end{vmatrix}$$

$$= 0\mathbf{i} - 0\mathbf{j} + [0.4(-346.4) - (-0.2)(200.0)]\mathbf{k}$$

$$= \{-98.6\mathbf{k}\} \text{ N} \cdot \text{m} \qquad\qquad Ans.$$

NOTE: The scalar analysis (Solution I) provides a more *convenient method* for analysis than Solution II since the direction of the moment and the moment arm for each component force are easy to establish. For this reason, this method is generally recommended for solving problems in two dimensions, whereas a Cartesian vector analysis is generally recommended only for solving three-dimensional problems.

PRELIMINARY PROBLEMS

P3–1.   In each case, determine the moment of the force about point O.

P3–2.   In each case, set up the determinant to find the moment of the force about point P.

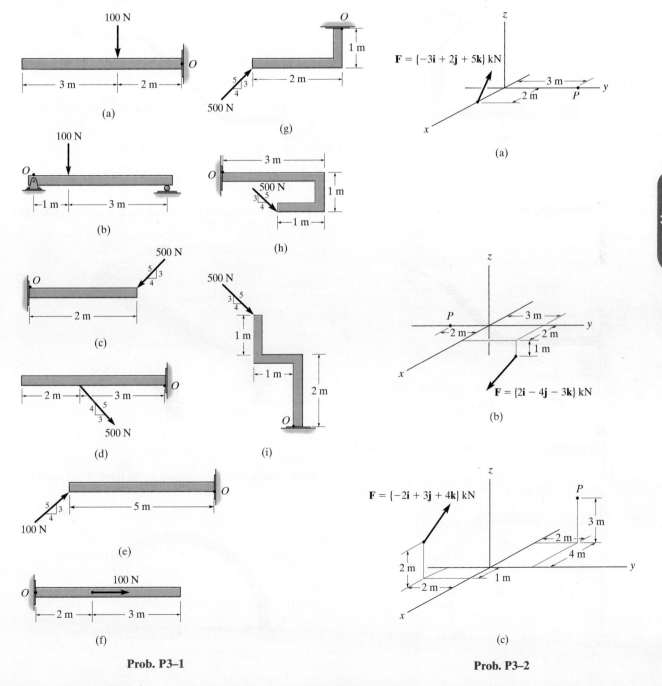

(a)

(b)

(c)

(d)

(e)

(f)

(g)

(h)

(i)

Prob. P3–1

(a)

(b)

(c)

Prob. P3–2

FUNDAMENTAL PROBLEMS

F3–1. Determine the moment of the force about point O.

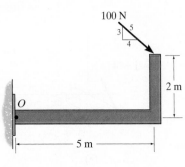

Prob. F3–1

F3–2. Determine the moment of the force about point O.

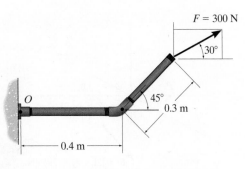

Prob. F3–2

F3–3. Determine the moment of the force about point O.

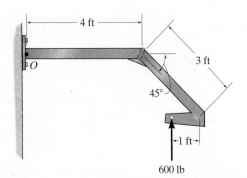

Prob. F3–3

F3–4. Determine the moment of the force about point O.

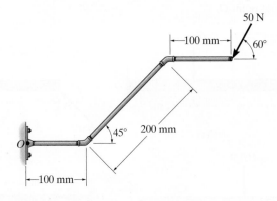

Prob. F3–4

F3–5. Determine the moment of the force about point O.

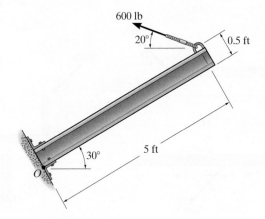

Prob. F3–5

F3–6. Determine the moment of the force about point O.

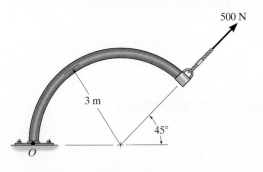

Prob. F3–6

F3–7. Determine the resultant moment produced by the forces about point O.

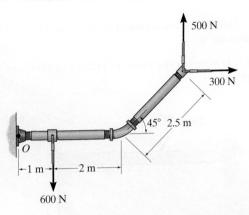

Prob. F3–7

F3–8. Determine the resultant moment produced by the forces about point O.

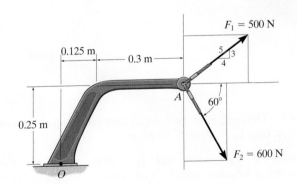

Prob. F3–8

F3–9. Determine the resultant moment produced by the forces about point O.

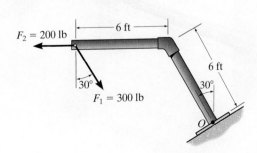

Prob. F3–9

F3–10. Determine the moment of force **F** about point O. Express the result as a Cartesian vector.

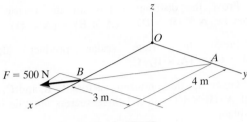

Prob. F3–10

F3–11. Determine the moment of force **F** about point O. Express the result as a Cartesian vector.

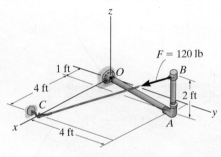

Prob. F3–11

F3–12. If $\mathbf{F}_1 = \{100\mathbf{i} - 120\mathbf{j} + 75\mathbf{k}\}$ lb and $\mathbf{F}_2 = \{-200\mathbf{i} + 250\mathbf{j} + 100\mathbf{k}\}$ lb, determine the resultant moment produced by these forces about point O. Express the result as a Cartesian vector.

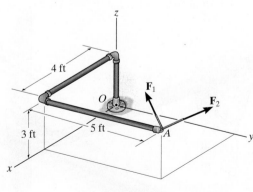

Prob. F3–12

PROBLEMS

3–1. Prove the distributive law for the vector cross product, i.e., $\mathbf{A} \times (\mathbf{B} + \mathbf{D}) = (\mathbf{A} \times \mathbf{B}) + (\mathbf{A} \times \mathbf{D})$.

3–2. Prove the triple scalar product identity $\mathbf{A} \cdot (\mathbf{B} \times \mathbf{C}) = (\mathbf{A} \times \mathbf{B}) \cdot \mathbf{C}$.

3–3. Given the three nonzero vectors $\mathbf{A}$, $\mathbf{B}$, and $\mathbf{C}$, show that if $\mathbf{A} \cdot (\mathbf{B} \times \mathbf{C}) = 0$, the three vectors *must* lie in the same plane.

*3–4. Determine the moment about point A of each of the three forces acting on the beam.

3–5. Determine the moment about point B of each of the three forces acting on the beam.

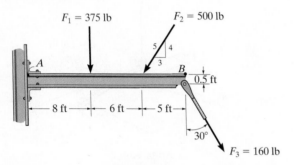

Probs. 3–4/5

3–6. The crowbar is subjected to a vertical force of $P = 25$ lb at the grip, whereas it takes a force of $F = 155$ lb at the claw to pull the nail out. Find the moment of each force about point A and determine if $\mathbf{P}$ is sufficient to pull out the nail.

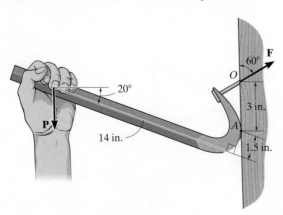

Prob. 3–6

3–7. Determine the moment of each of the three forces about point A.

*3–8. Determine the moment of each of the three forces about point B.

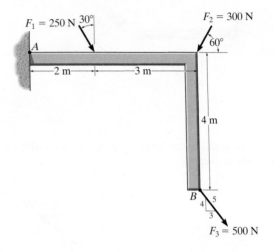

Probs. 3–7/8

3–9. Determine the moment of each force about the bolt at A. Take $F_B = 40$ lb, $F_C = 50$ lb.

3–10. If $F_B = 30$ lb and $F_C = 45$ lb, determine the resultant moment about the bolt at A.

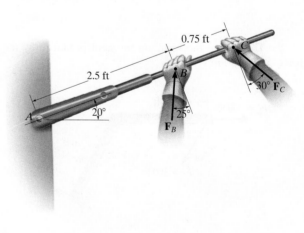

Probs. 3–9/10

3–11. The cable exerts a force of $P = 6$ kN at the end of the 8-m-long crane boom. If $\theta = 30°$, determine the placement x of the hook at B so that this force creates a maximum moment about point O. What is this moment?

*3–12. The cable exerts a force of $P = 6$ kN at the end of the 8-m-long crane boom. If $x = 10$ m, determine the angle θ of the boom so that this force creates a maximum moment about point O. What is this moment?

3–15. Two men exert forces of $F = 80$ lb and $P = 50$ lb on the ropes. Determine the moment of each force about A. Which way will the pole rotate, clockwise or counterclockwise?

*3–16. If the man at B exerts a force of $P = 30$ lb on the rope, determine the magnitude of the force **F** the man at C must exert to prevent the pole from rotating, i.e., so the resultant moment about A of both forces is zero.

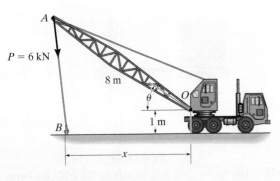

Probs. 3–11/12

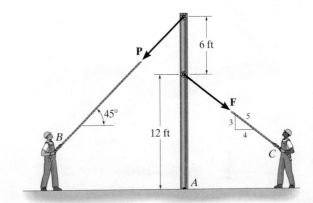

Probs. 3–15/16

3–13. The 20-N horizontal force acts on the handle of the socket wrench. What is the moment of this force about point B. Specify the coordinate direction angles α, β, γ of the moment axis.

3–14. The 20-N horizontal force acts on the handle of the socket wrench. Determine the moment of this force about point O. Specify the coordinate direction angles α, β, γ of the moment axis.

3–17. The torque wrench ABC is used to measure the moment or torque applied to a bolt when the bolt is located at A and a force is applied to the handle at C. The mechanic reads the torque on the scale at B. If an extension AO of length d is used on the wrench, determine the required scale reading if the desired torque on the bolt at O is to be M.

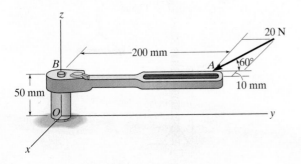

Probs. 3–13/14

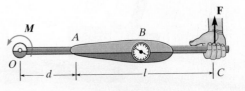

Prob. 3–17

3–18. The tongs are used to grip the ends of the drilling pipe. Determine the torque (moment) M_P that the applied force $F = 150$ lb exerts on the pipe about point P as a function of θ. Plot this moment M_P versus θ for $0° \leq \theta \leq 90°$.

3–19. The tongs are used to grip the ends of the drilling pipe. If a torque (moment) of $M_P = 800$ lb·ft is needed at P to turn the pipe, determine the cable force F that must be applied to the tongs. Set $\theta = 30°$.

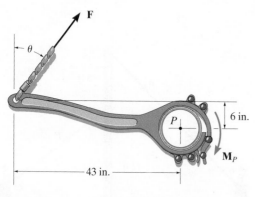

Probs. 3–18/19

*__*3–20.__ The handle of the hammer is subjected to the force of $F = 20$ lb. Determine the moment of this force about the point A.

3–21. In order to pull out the nail at B, the force $\mathbf{F}$ exerted on the handle of the hammer must produce a clockwise moment of 500 lb·in. about point A. Determine the required magnitude of force $\mathbf{F}$.

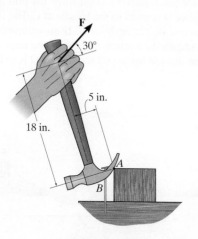

Probs. 3–20/21

3–22. Old clocks were constructed using a *fusee B* to drive the gears and watch hands. The purpose of the fusee is to increase the leverage developed by the mainspring A as it uncoils and thereby loses some of its tension. The mainspring can develop a torque (moment) $T_s = k\theta$, where $k = 0.015$ N·m/rad is the torsional stiffness and θ is the angle of twist of the spring in radians. If the torque T_f developed by the fusee is to remain constant as the mainspring winds down, and $x = 10$ mm when $\theta = 4$ rad, determine the required radius of the fusee when $\theta = 3$ rad.

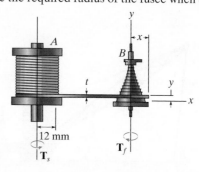

Prob. 3–22

3–23. The tower crane is used to hoist the 2-Mg load upward at constant velocity. The 1.5-Mg jib BD, 0.5-Mg jib BC, and 6-Mg counterweight C have centers of mass at G_1, G_2, and G_3, respectively. Determine the resultant moment produced by the load and the weights of the tower crane jibs about point A and about point B.

*__*3–24.__ The tower crane is used to hoist a 2-Mg load upward at constant velocity. The 1.5-Mg jib BD and 0.5-Mg jib BC have centers of mass at G_1 and G_2, respectively. Determine the required mass of the counterweight C so that the resultant moment produced by the load and the weight of the tower crane jibs about point A is zero. The center of mass for the counterweight is located at G_3.

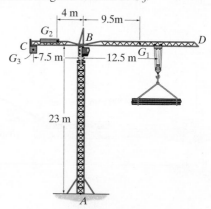

Probs. 3–23/24

3–25. If the 1500-lb boom AB, the 200-lb cage BCD, and the 175-lb man have centers of gravity located at points G_1, G_2, and G_3, respectively, determine the resultant moment produced by each weight about point A.

3–26. If the 1500-lb boom AB, the 200-lb cage BCD, and the 175-lb man have centers of gravity located at points G_1, G_2, and G_3, respectively, determine the resultant moment produced by all the weights about point A.

3–29. The force $\mathbf{F} = \{400\mathbf{i} - 100\mathbf{j} - 700\mathbf{k}\}$ lb acts at the end of the beam. Determine the moment of this force about point O.

3–30. The force $\mathbf{F} = \{400\mathbf{i} - 100\mathbf{j} - 700\mathbf{k}\}$ lb acts at the end of the beam. Determine the moment of this force about point A.

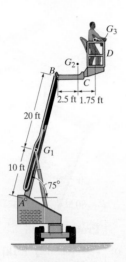

Probs. 3–25/26

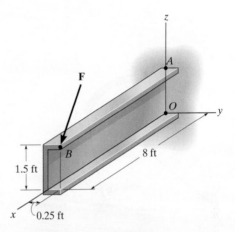

Probs. 3–29/30

3–27. Determine the moment of the force $\mathbf{F}$ about point O. Express the result as a Cartesian vector.

*3–28. Determine the moment of the force $\mathbf{F}$ about point P. Express the result as a Cartesian vector.

3–31. Determine the moment of the force $\mathbf{F}$ about point P. Express the result as a Cartesian vector.

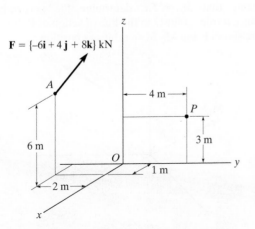

Probs. 3–27/28

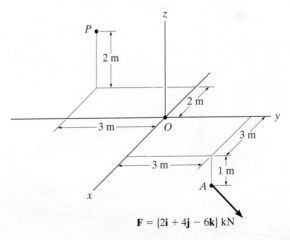

Prob. 3–31

*3–32. The curved rod has a radius of 5 ft. If a force of 60 lb acts at its end as shown, determine the moment of this force about point C.

3–33. Determine the smallest force F that must be applied along the rope in order to cause the curved rod to fail at the support C. This requires a moment of $M = 80$ lb $\cdot$ ft to be developed at C.

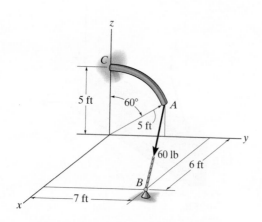

Probs. 3–32/33

3–34. A 20-N horizontal force is applied perpendicular to the handle of the socket wrench. Determine the magnitude and the coordinate direction angles of the moment created by this force about point O.

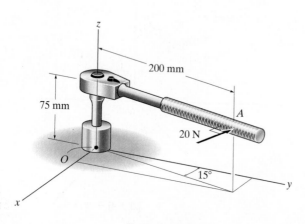

Prob. 3–34

3–35. The pipe assembly is subjected to the 80-N force. Determine the moment of this force about point A.

*3–36. The pipe assembly is subjected to the 80-N force. Determine the moment of this force about point B.

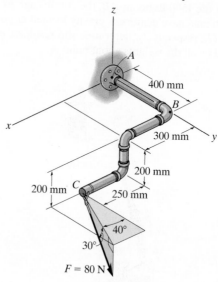

Probs. 3–35/36

3–37. A force of $\mathbf{F} = \{6\mathbf{i} - 2\mathbf{j} + 1\mathbf{k}\}$ kN produces a moment of $\mathbf{M}_O = \{4\mathbf{i} + 5\mathbf{j} - 14\mathbf{k}\}$ kN $\cdot$ m about the origin, point O. If the force acts at a point having an x coordinate of $x = 1$ m, determine the y and z coordinates. *Note:* The figure shows $\mathbf{F}$ and $\mathbf{M}_O$ in an arbitrary position.

3–38. The force $\mathbf{F} = \{6\mathbf{i} + 8\mathbf{j} + 10\mathbf{k}\}$ N creates a moment about point O of $\mathbf{M}_O = \{-14\mathbf{i} + 8\mathbf{j} + 2\mathbf{k}\}$ N $\cdot$ m. If the force passes through a point having an x coordinate of 1 m, determine the y and z coordinates of the point. Also, realizing that $M_O = Fd$, determine the perpendicular distance d from point O to the line of action of $\mathbf{F}$. *Note:* The figure shows $\mathbf{F}$ and $\mathbf{M}_O$ in an arbitrary position.

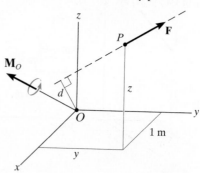

Probs. 3–37/38

3.5   MOMENT OF A FORCE ABOUT A SPECIFIED AXIS

Sometimes the moment produced by a force about a *specified axis* must be determined. For example, suppose the lug nut at O on the car tire in Fig. 3–20a needs to be loosened. The force applied to the wrench will create a tendency for the wrench and the nut to rotate about the *moment axis* passing through O; however, the nut can only rotate about the y axis. Therefore, to determine the turning effect, only the y component of the moment is needed, and the total moment produced is not important. To determine this component, we can use either a scalar or vector analysis.

Scalar Analysis. To use a scalar analysis, the moment arm, or perpendicular distance d_a from the axis to the line of action of the force, must be determined. The moment is then

$$M_a = Fd_a \qquad\qquad (3\text{--}10)$$

For example, for the lug nut in Fig. 3–20a, $d_y = d \cos \theta$, and so the moment of **F** about the y axis is $M_y = F d_y = F(d \cos \theta)$. According to the right-hand rule, $\mathbf{M}_y$ is directed along the positive y axis as shown in the figure.

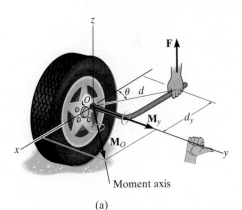

Moment axis

(a)

Fig. 3–20

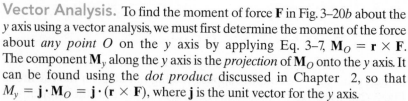

(b)

Fig. 3–20 (cont.)

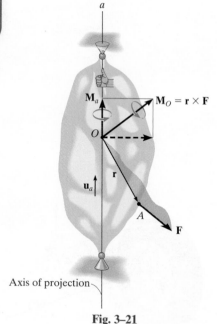

Axis of projection

Fig. 3–21

Vector Analysis. To find the moment of force **F** in Fig. 3–20b about the *y* axis using a vector analysis, we must first determine the moment of the force about *any point O* on the *y* axis by applying Eq. 3–7, $\mathbf{M}_O = \mathbf{r} \times \mathbf{F}$. The component $\mathbf{M}_y$ along the *y* axis is the *projection* of $\mathbf{M}_O$ onto the *y* axis. It can be found using the *dot product* discussed in Chapter 2, so that $M_y = \mathbf{j} \cdot \mathbf{M}_O = \mathbf{j} \cdot (\mathbf{r} \times \mathbf{F})$, where **j** is the unit vector for the *y* axis.

We can generalize this approach by letting $\mathbf{u}_a$ be the unit vector that specifies the direction of the *a* axis, shown in Fig. 3–21. Then the moment of **F** about a point *O* on the axis is $\mathbf{M}_O = \mathbf{r} \times \mathbf{F}$, and the projection of this moment onto the *a* axis is $M_a = \mathbf{u}_a \cdot (\mathbf{r} \times \mathbf{F})$. This combination is referred to as the **scalar triple product**. If the vectors are written in Cartesian form, we have

$$M_a = [u_{a_x}\mathbf{i} + u_{a_y}\mathbf{j} + u_{a_z}\mathbf{k}] \cdot \begin{vmatrix} \mathbf{i} & \mathbf{j} & \mathbf{k} \\ r_x & r_y & r_z \\ F_x & F_y & F_z \end{vmatrix}$$

$$= u_{a_x}(r_y F_z - r_z F_y) - u_{a_y}(r_x F_z - r_z F_x) + u_{a_z}(r_x F_y - r_y F_x)$$

This result can also be written in the form of a determinant, making it easier to memorize.*

$$M_a = \mathbf{u}_a \cdot (\mathbf{r} \times \mathbf{F}) = \begin{vmatrix} u_{a_x} & u_{a_y} & u_{a_z} \\ r_x & r_y & r_z \\ F_x & F_y & F_z \end{vmatrix} \qquad (3\text{–}11)$$

where

$u_{a_x}, u_{a_y}, u_{a_z}$ represent the *x, y, z* components of the unit vector defining the direction of the *a* axis

r_x, r_y, r_z represent the *x, y, z* components of the position vector extended from *any point O* on the *a* axis to *any point A* on the line of action of the force

F_x, F_y, F_z represent the *x, y, z* components of the force vector.

When M_a is evaluated from Eq. 3–11, it will yield a positive or negative scalar. The sign of this scalar indicates the sense of direction of $\mathbf{M}_a$ along the *a* axis. If it is positive, then $\mathbf{M}_a$ will have the same sense as $\mathbf{u}_a$, whereas if it is negative, then $\mathbf{M}_a$ will act opposite to $\mathbf{u}_a$. Once the *a* axis is established, point your right-hand thumb in the direction of $\mathbf{M}_a$, and the curl of your fingers will indicate the sense of twist about the axis, Fig. 3–21.

---

*Take a moment to expand this determinant, to show that it will yield the above result.

Once M_a is determined, we can then express $\mathbf{M}_a$ as a Cartesian vector, namely,

$$\mathbf{M}_a = M_a\mathbf{u}_a \qquad (3\text{--}12)$$

The examples which follow illustrate numerical applications of these concepts.

IMPORTANT POINTS

- The moment of a force about a specified axis can be determined provided the perpendicular distance d_a from the force line of action to the axis can be determined. Then $M_a = Fd_a$.

- If vector analysis is used, then $M_a = \mathbf{u}_a \cdot (\mathbf{r} \times \mathbf{F})$, where $\mathbf{u}_a$ defines the direction of the axis and $\mathbf{r}$ is extended from *any point* on the axis to *any point* on the line of action of the force.

- If M_a is calculated as a negative scalar, then the sense of direction of $\mathbf{M}_a$ is opposite to $\mathbf{u}_a$.

- The moment $\mathbf{M}_a$ expressed as a Cartesian vector is determined from $\mathbf{M}_a = M_a\mathbf{u}_a$.

EXAMPLE 3.7

Determine the resultant moment of the three forces in Fig. 3–22 about the x axis, the y axis, and the z axis.

SOLUTION
A force that is *parallel* to a coordinate axis or has a line of action that passes through the axis does *not* produce any moment or tendency for turning about the axis. Defining the positive direction of the moment of a force according to the right-hand rule, as shown in the figure, we have

$$M_x = (60 \text{ lb})(2 \text{ ft}) + (50 \text{ lb})(2 \text{ ft}) + 0 = 220 \text{ lb} \cdot \text{ft} \qquad \textit{Ans.}$$

$$M_y = 0 - (50 \text{ lb})(3 \text{ ft}) - (40 \text{ lb})(2 \text{ ft}) = -230 \text{ lb} \cdot \text{ft} \qquad \textit{Ans.}$$

$$M_z = 0 + 0 - (40 \text{ lb})(2 \text{ ft}) = -80 \text{ lb} \cdot \text{ft} \qquad \textit{Ans.}$$

The negative signs indicate that $\mathbf{M}_y$ and $\mathbf{M}_z$ act in the $-y$ and $-z$ directions, respectively.

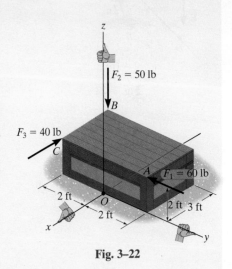

Fig. 3–22

EXAMPLE 3.8

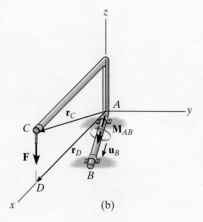

Fig. 3–23

Determine the moment $\mathbf{M}_{AB}$ produced by the force $\mathbf{F}$ in Fig. 3–23a, which tends to rotate the rod about the AB axis.

SOLUTION

A vector analysis using $M_{AB} = \mathbf{u}_B \cdot (\mathbf{r} \times \mathbf{F})$ will be considered for the solution rather than trying to find the moment arm or perpendicular distance from the line of action of $\mathbf{F}$ to the AB axis.

Unit vector $\mathbf{u}_B$ defines the direction of the AB axis of the rod, Fig. 3–23b, where

$$\mathbf{u}_B = \frac{\mathbf{r}_B}{r_B} = \frac{\{0.4\mathbf{i} + 0.2\mathbf{j}\}\ \text{m}}{\sqrt{(0.4\ \text{m})^2 + (0.2\ \text{m})^2}} = 0.8944\mathbf{i} + 0.4472\mathbf{j}$$

Vector $\mathbf{r}$ is directed from *any point* on the AB axis to *any point* on the line of action of the force. For example, position vectors $\mathbf{r}_C$ and $\mathbf{r}_D$ are suitable, Fig. 3–23b. (Although not shown, $\mathbf{r}_{BC}$ or $\mathbf{r}_{BD}$ can also be used.) For simplicity, we choose $\mathbf{r}_D$, where

$$\mathbf{r}_D = \{0.6\mathbf{i}\}\ \text{m}$$

The force is

$$\mathbf{F} = \{-300\mathbf{k}\}\ \text{N}$$

Substituting these vectors into the determinant form of the triple product and expanding, we have

$$M_{AB} = \mathbf{u}_B \cdot (\mathbf{r}_D \times \mathbf{F}) = \begin{vmatrix} 0.8944 & 0.4472 & 0 \\ 0.6 & 0 & 0 \\ 0 & 0 & -300 \end{vmatrix}$$

$$= 0.8944[0(-300) - 0(0)] - 0.4472[0.6(-300) - 0(0)] + 0[0.6(0) - 0(0)]$$

$$= 80.50\ \text{N} \cdot \text{m}$$

This positive result indicates that the sense of $\mathbf{M}_{AB}$ is in the same direction as $\mathbf{u}_B$, Fig. 3–23b.

Expressing $\mathbf{M}_{AB}$ as a Cartesian vector yields

$$\mathbf{M}_{AB} = M_{AB}\mathbf{u}_B = (80.50\ \text{N} \cdot \text{m})(0.8944\mathbf{i} + 0.4472\mathbf{j})$$

$$= \{72.0\mathbf{i} + 36.0\mathbf{j}\}\ \text{N} \cdot \text{m} \qquad\qquad Ans.$$

NOTE: If the axis AB is defined using a unit vector directed from B toward A, then in the above determinant $-\mathbf{u}_B$ would have to be used. This would lead to $M_{AB} = -80.50\ \text{N} \cdot \text{m}$. Consequently, $\mathbf{M}_{AB} = M_{AB}(-\mathbf{u}_B)$, and the same vector result would be obtained.

EXAMPLE 3.9

Determine the magnitude of the moment of force **F** about segment *OA* of the pipe assembly in Fig. 3–24*a*.

SOLUTION

The moment of **F** about the *OA* axis is determined from $M_{OA} = \mathbf{u}_{OA} \cdot (\mathbf{r} \times \mathbf{F})$, where **r** is a position vector extending from any point on the *OA* axis to any point on the line of action of **F**. As indicated in Fig. 3–24*b*, either $\mathbf{r}_{OD}$, $\mathbf{r}_{OC}$, $\mathbf{r}_{AD}$, or $\mathbf{r}_{AC}$ can be used. Here $\mathbf{r}_{OD}$ will be considered since it will simplify the calculation.

The unit vector $\mathbf{u}_{OA}$, which specifies the direction of the *OA* axis, is

$$\mathbf{u}_{OA} = \frac{\mathbf{r}_{OA}}{r_{OA}} = \frac{\{0.3\mathbf{i} + 0.4\mathbf{j}\}\ \text{m}}{\sqrt{(0.3\ \text{m})^2 + (0.4\ \text{m})^2}} = 0.6\mathbf{i} + 0.8\mathbf{j}$$

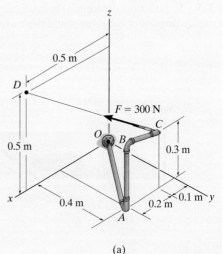

(a)

and the position vector $\mathbf{r}_{OD}$ is

$$\mathbf{r}_{OD} = \{0.5\mathbf{i} + 0.5\mathbf{k}\}\ \text{m}$$

The force **F** expressed as a Cartesian vector is

$$\mathbf{F} = F\left(\frac{\mathbf{r}_{CD}}{r_{CD}}\right)$$

$$= (300\ \text{N})\left[\frac{\{0.4\mathbf{i} - 0.4\mathbf{j} + 0.2\mathbf{k}\}\ \text{m}}{\sqrt{(0.4\ \text{m})^2 + (-0.4\ \text{m})^2 + (0.2\ \text{m})^2}}\right]$$

$$= \{200\mathbf{i} - 200\mathbf{j} + 100\mathbf{k}\}\ \text{N}$$

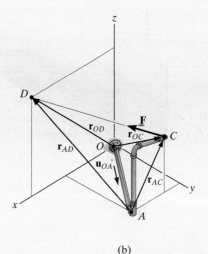

(b)

Fig. 3–24

Therefore,

$$M_{OA} = \mathbf{u}_{OA} \cdot (\mathbf{r}_{OD} \times \mathbf{F})$$

$$= \begin{vmatrix} 0.6 & 0.8 & 0 \\ 0.5 & 0 & 0.5 \\ 200 & -200 & 100 \end{vmatrix}$$

$$= 0.6[0(100) - (0.5)(-200)] - 0.8[0.5(100) - (0.5)(200)] + 0$$

$$= 100\ \text{N} \cdot \text{m} \qquad\qquad\qquad\qquad\qquad Ans.$$

PRELIMINARY PROBLEMS

P3–3. In each case, determine the resultant moment of the forces acting about the x, the y, and the z axis.

P3–4. In each case, set up the determinant needed to find the moment of the force about the a–a axes.

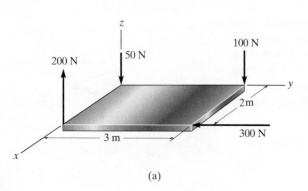

(a)

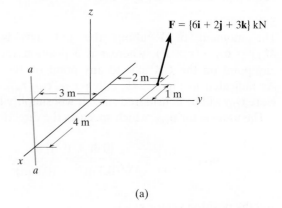

(a)

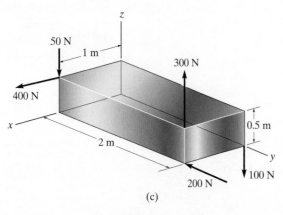

(b)

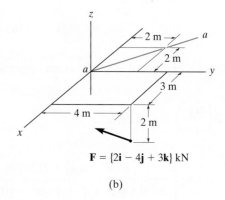

(b)

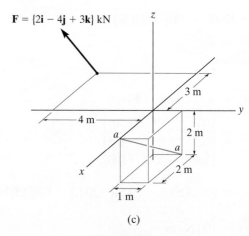

(c)

Prob. P3–3

(c)

Prob. P3–4

FUNDAMENTAL PROBLEMS

F3–13. Determine the magnitude of the moment of the force $\mathbf{F} = \{300\mathbf{i} - 200\mathbf{j} + 150\mathbf{k}\}$ N about the x axis.

F3–14. Determine the magnitude of the moment of the force $\mathbf{F} = \{300\mathbf{i} - 200\mathbf{j} + 150\mathbf{k}\}$ N about the OA axis.

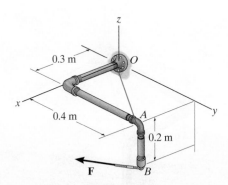

Probs. F3–13/14

F3–15. Determine the magnitude of the moment of the 200-N force about the x axis. Solve the problem using both a scalar and a vector analysis.

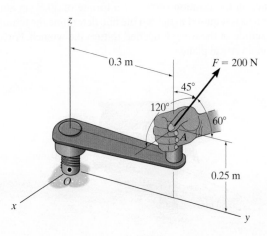

Prob. F3–15

F3–16. Determine the magnitude of the moment of the force about the y axis.
$$\mathbf{F} = \{30\mathbf{i} - 20\mathbf{j} + 50\mathbf{k}\}\ \text{N}$$

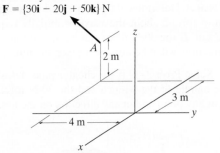

Prob. F3–16

F3–17. Determine the moment of the force $\mathbf{F} = \{50\mathbf{i} - 40\mathbf{j} + 20\mathbf{k}\}$ lb about the AB axis. Express the result as a Cartesian vector.

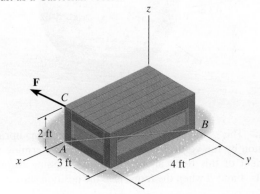

Prob. F3–17

F3–18. Determine the moment of force $\mathbf{F}$ about the x, the y, and the z axis. Solve the problem using both a scalar and a vector analysis.

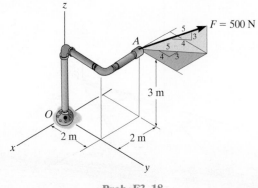

Prob. F3–18

PROBLEMS

3–39. The lug nut on the wheel of the automobile is to be removed using the wrench and applying the vertical force of $F = 30$ N at A. Determine if this force is adequate, provided 14 N · m of torque about the x axis is initially required to turn the nut. If the 30-N force can be applied at A in any other direction, will it be possible to turn the nut?

*3–40. Solve Prob. 3–39 if the cheater pipe AB is slipped over the handle of the wrench and the 30-N force can be applied at any point and in any direction on the assembly.

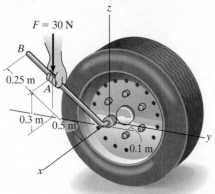

Probs. 3–39/40

3–41. The A-frame is being hoisted into an upright position by the vertical force of $F = 80$ lb. Determine the moment of this force about the y' axis passing through points A and B when the frame is in the position shown.

3–42. The A-frame is being hoisted into an upright position by the vertical force of $F = 80$ lb. Determine the moment of this force about the x axis when the frame is in the position shown.

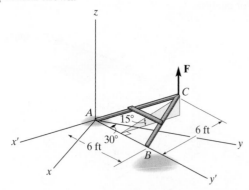

Probs. 3–41/42

3–43. Determine the magnitude of the moment of the force F about the x, the y, and the z axis. Solve the problem (a) using a Cartesian vector approach and (b) using a scalar approach.

*3–44. Determine the moment of force F about an axis extending between A and C. Express the result as a Cartesian vector.

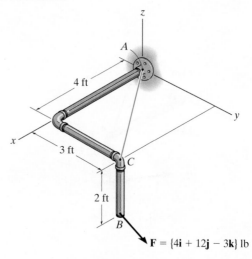

Probs. 3–43/44

3–45. The board is used to hold the end of the cross lug wrench in the position shown when the man applies a force of $F = 100$ N. Determine the magnitude of the moment produced by this force about the x axis. Force F lies in a vertical plane.

3–46. The board is used to hold the end of the cross lug wrench in the position shown. If a torque of 30 N · m about the x axis is required to tighten the nut, determine the required magnitude of the force F needed to turn the wrench. Force F lies in a vertical plane.

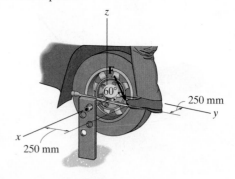

Probs. 3–45/46

3–47. The A-frame is being hoisted into an upright position by the vertical force of $F = 80$ lb. Determine the moment of this force about the y axis when the frame is in the position shown.

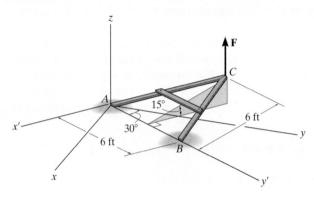

Prob. 3–47

*3–48. Determine the magnitude of the moment of the force $F = \{50\mathbf{i} - 20\mathbf{j} - 80\mathbf{k}\}$ N about member AB of the tripod.

3–49. Determine the magnitude of the moment of the force $F = \{50\mathbf{i} - 20\mathbf{j} - 80\mathbf{k}\}$ N about member BC of the tripod.

3–50. Determine the magnitude of the moment of the force $F = \{50\mathbf{i} - 20\mathbf{j} - 80\mathbf{k}\}$ N about member CA of the tripod.

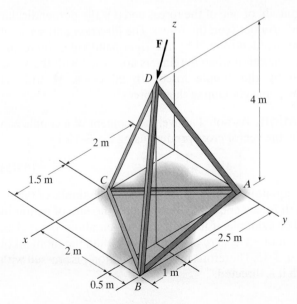

Probs. 3–48/49/50

3–51. A horizontal force of $\mathbf{F} = \{-50\mathbf{i}\}$ N is applied perpendicular to the handle of the pipe wrench. Determine the moment that this force exerts along the axis OA (z axis) of the pipe assembly. Both the wrench and pipe assembly, $OABC$, lie in the y–z plane. *Suggestion:* Use a scalar analysis.

*3–52. Determine the magnitude of the horizontal force $\mathbf{F} = -F\mathbf{i}$ acting on the handle of the pipe wrench so that this force produces a component of the moment along the OA axis (z axis) of the pipe assembly of $\mathbf{M}_z = \{4\mathbf{k}\}$ N·m. Both the wrench and the pipe assembly, $OABC$, lie in the y–z plane. *Suggestion*: Use a scalar analysis.

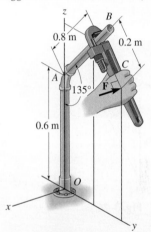

Probs. 3–51/52

3–53. Determine the moment of the force about the a–a axis of the pipe if $\alpha = 60°$, $\beta = 60°$, and $\gamma = 45°$. Also, determine the coordinate direction angles of F in order to produce the maximum moment about the a–a axis. What is this moment?

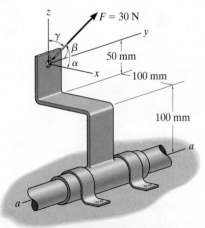

Prob. 3–53

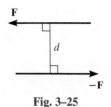

Fig. 3–25

3.6 MOMENT OF A COUPLE

A **couple** is defined as two parallel forces that have the same magnitude, but opposite directions, and are separated by a perpendicular distance d, Fig. 3–25. Since the resultant force is zero, the only effect of a couple is to produce a rotation, or if no movement is possible, there is a tendency for rotation.

The moment produced by a couple is called a *couple moment*. We can determine its value by finding the sum of the moments of both couple forces about *any* arbitrary point. For example, in Fig. 3–26, position vectors $\mathbf{r}_A$ and $\mathbf{r}_B$ are directed from point O to points A and B lying on the line of action of $-\mathbf{F}$ and $\mathbf{F}$. The couple moment determined about O is therefore

$$\mathbf{M} = \mathbf{r}_B \times \mathbf{F} + \mathbf{r}_A \times (-\mathbf{F}) = (\mathbf{r}_B - \mathbf{r}_A) \times \mathbf{F}$$

However, $\mathbf{r}_B = \mathbf{r}_A + \mathbf{r}$ or $\mathbf{r} = \mathbf{r}_B - \mathbf{r}_A$, so that

$$\mathbf{M} = \mathbf{r} \times \mathbf{F} \qquad (3\text{–}13)$$

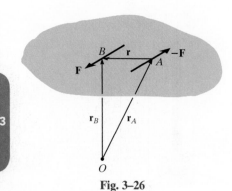

Fig. 3–26

This result indicates that a couple moment is a ***free vector***, i.e., it can act at *any point* since $\mathbf{M}$ depends *only* upon the position vector $\mathbf{r}$ directed *between* the forces and *not* the position vectors $\mathbf{r}_A$ and $\mathbf{r}_B$, directed from point O to the forces.

Scalar Formulation. The moment of a couple, Fig. 3–27, has a *magnitude* of

$$\boxed{M = Fd} \qquad (3\text{–}14)$$

where F is the magnitude of one of the forces and d is the perpendicular distance or moment arm between the forces. The *direction* and sense of the couple moment are determined by the right-hand rule, where the thumb indicates this direction when the fingers are curled with the sense of rotation caused by the couple forces. In all cases, $\mathbf{M}$ will act perpendicular to the plane containing these forces.

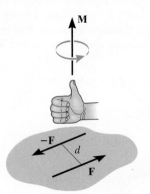

Fig. 3–27

Vector Formulation. As noted above, the moment of a couple can also be expressed by the vector cross product using Eq. 3–13, i.e.,

$$\boxed{\mathbf{M} = \mathbf{r} \times \mathbf{F}} \qquad (3\text{–}15)$$

Application of this equation is easily remembered if one thinks of taking the moments of both forces about a point lying on the line of action of one of the forces. For example, if moments are taken about point A in Fig. 3–26, the moment of $-\mathbf{F}$ is *zero* about this point, and the moment of $\mathbf{F}$ is defined from Eq. 3–15. Therefore, in the formulation $\mathbf{r}$ is crossed with the force $\mathbf{F}$ to which it is directed.

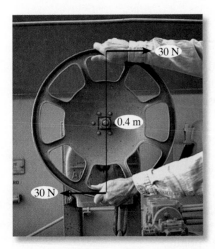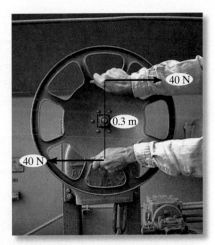

Fig. 3–28

Equivalent Couples. If two couples produce a moment with the *same magnitude and direction*, then these two couples are *equivalent*. For example, the two couples shown in Fig. 3–28 are *equivalent* because each couple moment has a magnitude of $M = 30 \text{ N}(0.4 \text{ m}) = 40 \text{ N}(0.3 \text{ m}) = 12 \text{ N} \cdot \text{m}$, and each is directed into the plane of the page. Notice that larger forces are required in the second case to create the same turning effect because the hands are placed closer together. Also, if the wheel was connected to the shaft at a point other than at its center, then the wheel would still turn when each couple is applied since this 12 N · m couple is a free vector.

Resultant Couple Moment. Since couple moments are free vectors, their resultant can be determined by moving them to a single point and using vector addition. For example, to find the resultant of couple moments $\mathbf{M}_1$ and $\mathbf{M}_2$ acting on the pipe assembly in Fig. 3–29a, we can join their tails at point O and find the resultant couple moment, $\mathbf{M}_R = \mathbf{M}_1 + \mathbf{M}_2$, as shown in Fig. 3–29b.

If more than two couple moments act on the body, we may generalize this concept and write the vector resultant as

$$\mathbf{M}_R = \Sigma(\mathbf{r} \times \mathbf{F}) \tag{3–16}$$

These concepts are illustrated numerically in the examples that follow. In general, problems projected in two dimensions should be solved using a scalar analysis since the moment arms and force components are easy to determine.

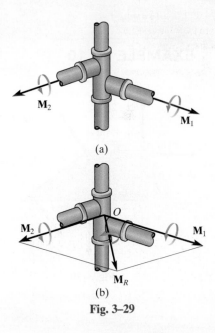

Fig. 3–29

Steering wheels on vehicles have been made smaller than on older vehicles because power steering does not require the driver to apply a large couple moment to the wheel.

IMPORTANT POINTS

- A couple moment is produced by two noncollinear forces that are equal in magnitude but opposite in direction. Its effect is to produce pure rotation, or tendency for rotation in a specified direction.

- A couple moment is a free vector, and as a result it causes the same rotational effect on a body regardless of where the couple moment is applied to the body.

- The moment of the two couple forces can be determined about *any point*. For convenience, this point is often chosen on the line of action of one of the forces in order to eliminate the moment of this force about the point.

- In three dimensions the couple moment is often determined using the vector formulation, $\mathbf{M} = \mathbf{r} \times \mathbf{F}$, where $\mathbf{r}$ is directed from *any point* on the line of action of one of the forces to *any point* on the line of action of the other force $\mathbf{F}$.

- A resultant couple moment is simply the vector sum of all the couple moments of the system.

EXAMPLE 3.10

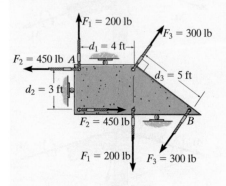

Fig. 3–30

Determine the resultant couple moment of the three couples acting on the plate in Fig. 3–30.

SOLUTION
As shown the perpendicular distances between each pair of couple forces are $d_1 = 4$ ft, $d_2 = 3$ ft, and $d_3 = 5$ ft. Considering counterclockwise couple moments as positive, we have

$$\zeta + M_R = \Sigma M; \quad M_R = -F_1 d_1 + F_2 d_2 - F_3 d_3$$

$$= -(200 \text{ lb})(4 \text{ ft}) + (450 \text{ lb})(3 \text{ ft}) - (300 \text{ lb})(5 \text{ ft})$$

$$= -950 \text{ lb} \cdot \text{ft} = 950 \text{ lb} \cdot \text{ft} \, \downarrow \qquad \textit{Ans.}$$

EXAMPLE 3.11

Determine the magnitude and direction of the couple moment acting on the gear in Fig. 3–31a.

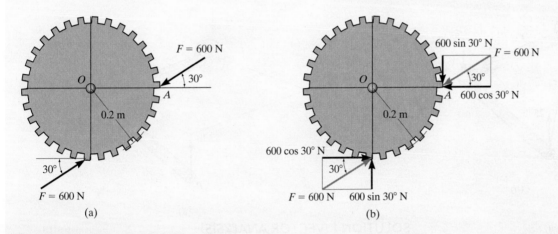

(a)   (b)

SOLUTION

The easiest solution requires resolving each force into its components as shown in Fig. 3–31b. The couple moment can be determined by summing the moments of these force components about any point, for example, the center O of the gear or point A. If we consider counterclockwise moments as positive, we have

$$\zeta + M = \Sigma M_O; \quad M = (600 \cos 30° \text{ N})(0.2 \text{ m}) - (600 \sin 30° \text{ N})(0.2 \text{ m})$$
$$= 43.9 \text{ N} \cdot \text{m} \, \zeta \qquad \qquad Ans.$$

or

$$\zeta + M = \Sigma M_A; \quad M = (600 \cos 30° \text{ N})(0.2 \text{ m}) - (600 \sin 30° \text{ N})(0.2 \text{ m})$$
$$= 43.9 \text{ N} \cdot \text{m} \, \zeta \qquad \qquad Ans.$$

This positive result indicates that **M** has a counterclockwise rotational sense, so it is directed outward, perpendicular to the page.

NOTE: The same result can also be obtained using $M = Fd$, where d is the perpendicular distance between the lines of action of the couple forces, Fig. 3–31c. However, the computation for d is more involved. Also, realize that the couple moment is a free vector and can act at any point on the gear and produce the same turning effect about point O.

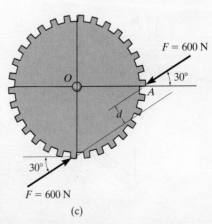

(c)

Fig. 3–31

EXAMPLE 3.12

Determine the couple moment acting on the pipe shown in Fig. 3–32a. Segment AB is directed 30° below the x–y plane.

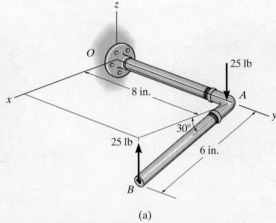

(a)

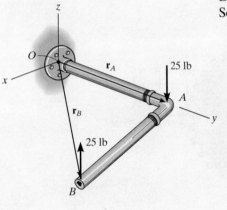

(b)

3

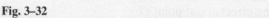

(c)

(d)

Fig. 3–32

SOLUTION I (VECTOR ANALYSIS)

The moment of the two couple forces can be found about *any point*. If point O is considered, Fig. 3–32b, we have

$$\mathbf{M} = \mathbf{r}_A \times (-25\mathbf{k}) + \mathbf{r}_B \times (25\mathbf{k})$$
$$= (8\mathbf{j}) \times (-25\mathbf{k}) + (6\cos 30°\mathbf{i} + 8\mathbf{j} - 6\sin 30°\mathbf{k}) \times (25\mathbf{k})$$
$$= -200\mathbf{i} - 129.9\mathbf{j} + 200\mathbf{i}$$
$$= \{-130\mathbf{j}\}\ \text{lb} \cdot \text{in.} \qquad\qquad Ans.$$

It is *easier* to take moments of the couple forces about a point lying on the line of action of one of the forces, e.g., point A, Fig. 3–32c. In this case the moment of the force at A is zero, so that

$$\mathbf{M} = \mathbf{r}_{AB} \times (25\mathbf{k})$$
$$= (6\cos 30°\mathbf{i} - 6\sin 30°\mathbf{k}) \times (25\mathbf{k})$$
$$= \{-130\mathbf{j}\}\ \text{lb} \cdot \text{in.} \qquad\qquad Ans.$$

SOLUTION II (SCALAR ANALYSIS)

Although this problem is shown in three dimensions, the geometry is simple enough to use the scalar equation $M = Fd$, where $d = 6\cos 30° = 5.196$ in., Fig. 3–32d. Hence, taking moments of the forces about either point A or point B yields

$$M = Fd = 25\ \text{lb}\ (5.196\ \text{in.}) = 129.9\ \text{lb} \cdot \text{in.}$$

Applying the right-hand rule, $\mathbf{M}$ acts in the $-\mathbf{j}$ direction. Thus,

$$\mathbf{M} = \{-130\mathbf{j}\}\ \text{lb} \cdot \text{in.} \qquad\qquad Ans.$$

EXAMPLE   3.13

Replace the two couples acting on the pipe assembly in Fig. 3–33a by a resultant couple moment.

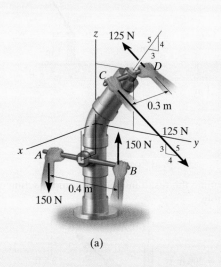

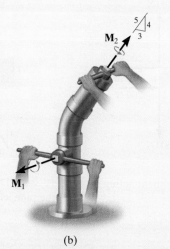

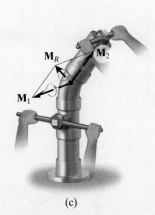

(a)   (b)   (c)

Fig. 3–33

SOLUTION *(VECTOR ANALYSIS)*
The couple moment $\mathbf{M}_1$, developed by the forces at A and B, can easily be determined from a scalar formulation.

$$M_1 = Fd = 150 \text{ N}(0.4 \text{ m}) = 60 \text{ N} \cdot \text{m}$$

By the right-hand rule, $\mathbf{M}_1$ acts in the $+\mathbf{i}$ direction, Fig. 3–33b. Hence,

$$\mathbf{M}_1 = \{60\mathbf{i}\} \text{ N} \cdot \text{m}$$

Vector analysis will be used to determine $\mathbf{M}_2$, caused by forces at C and D. If moments are calculated about point D, Fig. 3–33a, $\mathbf{M}_2 = \mathbf{r}_{DC} \times \mathbf{F}_C$, then

$$\mathbf{M}_2 = \mathbf{r}_{DC} \times \mathbf{F}_C = (0.3\mathbf{i}) \times \left[125\left(\tfrac{4}{5}\right)\mathbf{j} - 125\left(\tfrac{3}{5}\right)\mathbf{k} \right]$$
$$= (0.3\mathbf{i}) \times [100\mathbf{j} - 75\mathbf{k}] = 30(\mathbf{i} \times \mathbf{j}) - 22.5(\mathbf{i} \times \mathbf{k})$$
$$= \{22.5\mathbf{j} + 30\mathbf{k}\} \text{ N} \cdot \text{m}$$

Since $\mathbf{M}_1$ and $\mathbf{M}_2$ are free vectors, Fig. 3–33b, they may be moved to some arbitrary point and added vectorially, Fig. 3–33c. The resultant couple moment becomes

$$\mathbf{M}_R = \mathbf{M}_1 + \mathbf{M}_2 = \{60\mathbf{i} + 22.5\mathbf{j} + 30\mathbf{k}\} \text{ N} \cdot \text{m} \qquad \textit{Ans.}$$

FUNDAMENTAL PROBLEMS

F3–19. Determine the resultant couple moment acting on the beam.

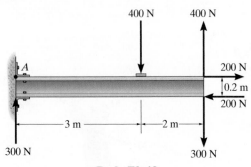

Prob. F3–19

F3–20. Determine the resultant couple moment acting on the plate.

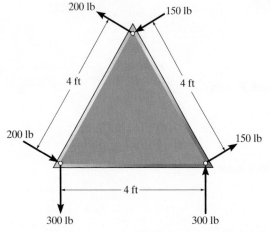

Prob. F3–20

F3–21. Determine the magnitude of **F** so that the resultant couple moment acting on the beam is 1.5 kN · m clockwise.

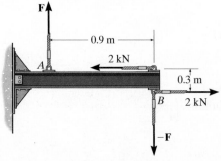

Prob. F3–21

F3–22. Determine the couple moment acting on the beam.

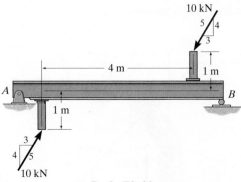

Prob. F3–22

F3–23. Determine the resultant couple moment acting on the pipe assembly.

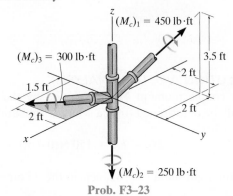

Prob. F3–23

F3–24. Determine the couple moment acting on the pipe assembly and express the result as a Cartesian vector.

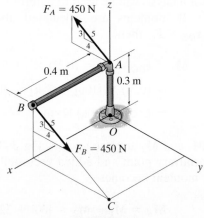

Prob. F3–24

PROBLEMS

3–54. A clockwise couple $M = 5\,\text{N}\cdot\text{m}$ is resisted by the shaft of the electric motor. Determine the magnitude of the reactive forces $-\mathbf{R}$ and $\mathbf{R}$ which act at supports A and B so that the resultant of the two couples is zero.

*3–56. If the resultant couple of the three couples acting on the triangular block is to be zero, determine the magnitude of forces $\mathbf{F}$ and $\mathbf{P}$.

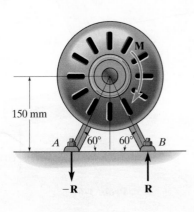

Prob. 3–54

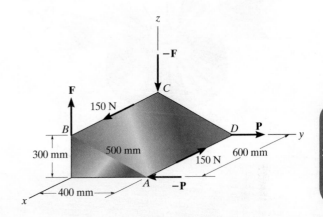

Prob. 3–56

3–55. A twist of $4\,\text{N}\cdot\text{m}$ is applied to the handle of the screwdriver. Resolve this couple moment into a pair of couple forces $\mathbf{F}$ exerted on the handle and $\mathbf{P}$ exerted on the blade.

3–57. If $F = 125\,\text{lb}$, determine the resultant couple moment.

3–58. Determine the magnitude of $\mathbf{F}$ so that the resultant couple moment is $450\,\text{lb}\cdot\text{ft}$, counterclockwise. Where on the beam does the resultant couple moment act?

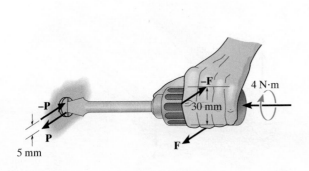

Prob. 3–55

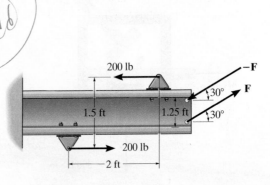

Probs. 3–57/58

3–59. Determine the magnitude and coordinate direction angles of the resultant couple moment.

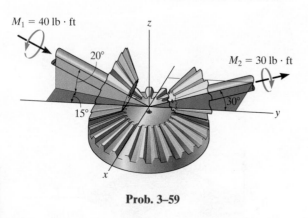

$M_1 = 40$ lb · ft

$M_2 = 30$ lb · ft

20°

15°

30°

Prob. 3–59

*3–60. Determine the required magnitude of the couple moments M_2 and M_3 so that the resultant couple moment is zero.

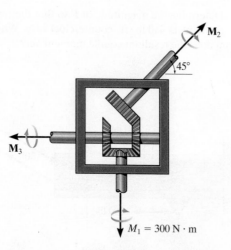

M_2

45°

M_3

$M_1 = 300$ N · m

Prob. 3–60

3–61. Determine the resultant couple moment of the two couples that act on the assembly. Specify its magnitude and coordinate direction angles.

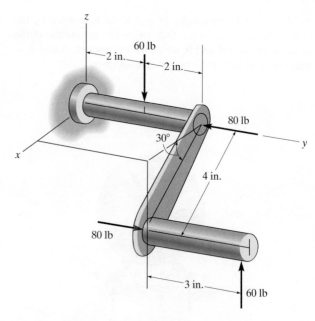

60 lb

2 in.   2 in.

80 lb

30°

4 in.

80 lb

3 in.   60 lb

Prob. 3–61

3–62. Express the moment of the couple acting on the frame in Cartesian vector form. The forces are applied perpendicular to the frame. What is the magnitude of the couple moment? Take $F = 50$ N.

3–63. In order to turn over the frame, a couple moment is applied as shown. If the component of this couple moment along the x axis is $M_x = \{-20\mathbf{i}\}$ N · m, determine the magnitude F of the couple forces.

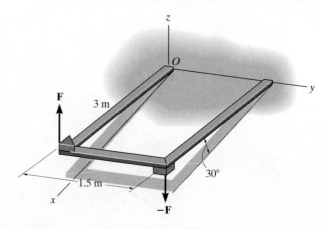

O

F

3 m

1.5 m

30°

−F

Probs. 3–62/63

*3–64. Express the moment of the couple acting on the pipe in Cartesian vector form. What is the magnitude of the couple moment? Take $F = 125$ N.

3–65. If the couple moment acting on the pipe has a magnitude of 300 N·m, determine the magnitude of the forces applied to the wrenches.

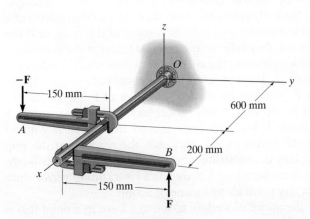

Probs. 3–64/65

3–66. If $F = 80$ N, determine the magnitude and coordinate direction angles of the couple moment. The pipe assembly lies in the x–y plane.

3–67. If the magnitude of the couple moment acting on the pipe assembly is 50 N·m, determine the magnitude of the couple forces applied to each wrench. The pipe assembly lies in the x–y plane.

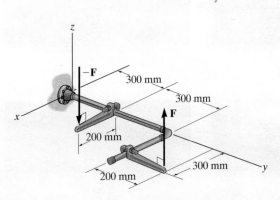

Probs. 3–66/67

*3–68. Express the moment of the couple acting on the rod in Cartesian vector form. What is the magnitude of the couple moment?

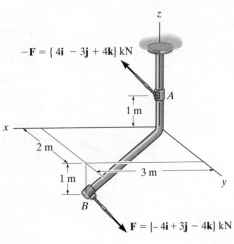

Prob. 3–68

3–69. If $F_1 = 100$ N, $F_2 = 120$ N, and $F_3 = 80$ N, determine the magnitude and coordinate direction angles of the resultant couple moment.

3–70. Determine the required magnitude of F_1, F_2, and F_3 so that the resultant couple moment is $(M_c)_R = [50\mathbf{i} - 45\mathbf{j} - 20\mathbf{k}]$ N·m.

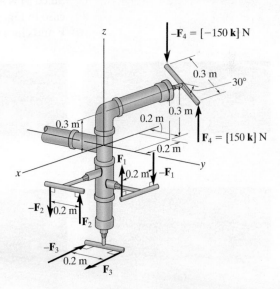

Probs. 3–69/70

3.7   SIMPLIFICATION OF A FORCE AND COUPLE SYSTEM

Sometimes it is convenient to reduce a system of forces and couple moments acting on a body to a simpler form by replacing it with an *equivalent system*, consisting of a single resultant force and a resultant couple moment. A system is equivalent if the *external effects* it produces on a body are the same as those caused by the original force and couple moment system. If the body is free to move, then the external effects of a system refer to the *translating and rotating motion* of the body, or if the body is fully supported, they refer to the *reactive forces* at the supports.

For example, consider holding the stick in Fig. 3–34a, which is subjected to the force **F** at point A. If we attach a pair of equal but opposite forces **F** and –**F** at point B, which is *on the line of action* of **F**, Fig. 3–34b, we observe that –**F** at B and **F** at A will cancel each other, leaving only **F** at B, Fig. 3–34c. Force **F** has now been moved from A to B without modifying its *external effects* on the stick; i.e., the reaction at the grip remains the same. This demonstrates the *principle of transmissibility*, which states that a force acting on a body (stick) is a *sliding vector* since it can be applied at any point along its line of action.

We can also use the above procedure to move a force to a point that is *not* on the line of action of the force. If **F** is applied perpendicular to the stick, as in Fig. 3–35a, then we can attach a pair of equal but opposite forces **F** and –**F** to B, Fig. 3–35b. Force **F** is now applied at B, and the other two forces, **F** at A and –**F** at B, form a couple that produces the couple moment $M = Fd$, Fig. 3–35c. Therefore, the force **F** can be moved from A to B provided a couple moment **M** is *added* to maintain an equivalent system. This couple moment is determined by taking the moment of **F** about B. Since **M** is actually a *free vector*, it can act at any point on the stick. In each case in Fig. 3–35 the systems are equivalent. This causes a downward force **F** and clockwise couple moment $M = Fd$ to be felt at the grip.

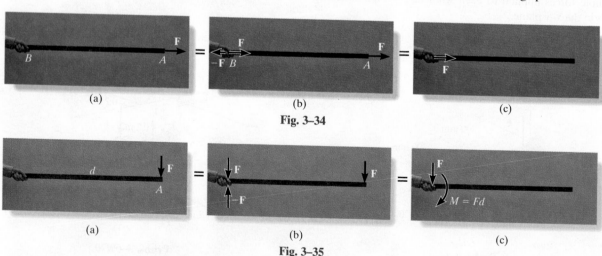

(a)   (b)   (c)

Fig. 3–34

(a)   (b)   (c)

Fig. 3–35

System of Forces and Couple Moments. Using this method, a system of several forces and couple moments acting on a body can be reduced to an equivalent single resultant force acting at a point O and a resultant couple moment. For example, in Fig. 3–36a, O is not on the line of action of $\mathbf{F}_1$, and so this force can be moved to point O provided a couple moment $(\mathbf{M}_O)_1 = \mathbf{r}_1 \times \mathbf{F}$ is added to the body. Similarly, the couple moment $(\mathbf{M}_O)_2 = \mathbf{r}_2 \times \mathbf{F}_2$ should be added to the body when we move $\mathbf{F}_2$ to point O. Finally, since the couple moment $\mathbf{M}$ is a free vector, it can just be moved to point O. Doing this, we obtain the equivalent system shown in Fig. 3–36b, which produces the same external effects on the body as that of the force and couple system shown in Fig. 3–36a. If we sum the forces and couple moments, we obtain the resultant force $\mathbf{F}_R = \mathbf{F}_1 + \mathbf{F}_2$ and the resultant couple moment $(\mathbf{M}_R)_O = \mathbf{M} + (\mathbf{M}_O)_1 + (\mathbf{M}_O)_2$, Fig. 3–36$c$.

Notice that $\mathbf{F}_R$ is independent of the location of point O since it is simply a summation of the forces. However, $(\mathbf{M}_R)_O$ depends upon this location since the moments $\mathbf{M}_1$ and $\mathbf{M}_2$ are determined using the position vectors $\mathbf{r}_1$ and $\mathbf{r}_2$, which extend from O to each force. Also note that $(\mathbf{M}_R)_O$ is a free vector and can act at *any point* on the body, although point O is generally chosen as its point of application.

We can now generalize the above method of reducing a force and couple system to an equivalent resultant force $\mathbf{F}_R$ acting at point O and a resultant couple moment $(\mathbf{M}_R)_O$ by using the following two equations.

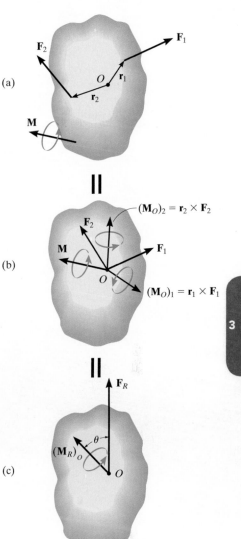

$$\begin{aligned} \mathbf{F}_R &= \Sigma \mathbf{F} \\ (\mathbf{M}_R)_O &= \Sigma \mathbf{M}_O + \Sigma \mathbf{M} \end{aligned} \qquad (3\text{–}17)$$

The first equation states that the resultant force of the system is equivalent to the sum of all the forces; and the second equation states that the resultant couple moment of the system is equivalent to the sum of all the couple moments $\Sigma \mathbf{M}$ plus the moments of all the forces about point O, $\Sigma \mathbf{M}_O$. If the force system lies in the x–y plane and any couple moments are perpendicular to this plane, then the above equations reduce to the following three scalar equations.

Fig. 3–36

$$\begin{aligned} (F_R)_x &= \Sigma F_x \\ (F_R)_y &= \Sigma F_y \\ (M_R)_O &= \Sigma M_O + \Sigma M \end{aligned} \qquad (3\text{–}18)$$

Here the resultant force is determined from the vector sum of its two components $(F_R)_x$ and $(F_R)_y$.

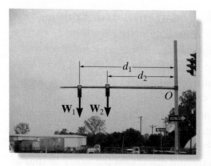

The weights of these traffic lights can be replaced by their equivalent resultant force $W_R = W_1 + W_2$ and a couple moment $(M_R)_O = W_1 d_1 + W_2 d_2$ at the support, O. In both cases the support must provide the same resistance to translation and rotation in order to keep the member in the horizontal position.

PROCEDURE FOR ANALYSIS

The following points should be kept in mind when simplifying a force and couple moment system to an equivalent resultant force and couple system.

- Establish the coordinate axes with the origin located at point O and the axes having a selected orientation.

Force Summation.

- If the force system is *coplanar*, resolve each force into its x and y components. If a component is directed along the positive x or y axis, it represents a positive scalar; whereas if it is directed along the negative x or y axis, it is a negative scalar.

- In three dimensions, represent each force as a Cartesian vector before summing the forces.

Moment Summation.

- When determining the moments of a *coplanar* force system about point O, it is generally advantageous to use the principle of moments, i.e., determine the moments of the components of each force, rather than the moment of the force itself.

- In three dimensions use the vector cross product to determine the moment of each force about point O. Here the position vectors extend from O to *any point* on the line of action of each force.

EXAMPLE 3.14

Replace the force and couple system shown in Fig. 3–37a by an equivalent resultant force and couple moment acting at point O.

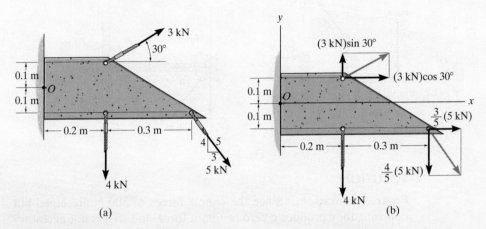

(a)   (b)

SOLUTION

Force Summation. The 3 kN and 5 kN forces are resolved into their x and y components as shown in Fig. 3–37b. We have

$$\xrightarrow{+} (F_R)_x = \Sigma F_x; \qquad (F_R)_x = (3 \text{ kN}) \cos 30° + \left(\tfrac{3}{5}\right)(5 \text{ kN}) = 5.598 \text{ kN} \rightarrow$$

$$+\uparrow (F_R)_y = \Sigma F_y; \qquad (F_R)_y = (3 \text{ kN}) \sin 30° - \left(\tfrac{4}{5}\right)(5 \text{ kN}) - 4 \text{ kN} = -6.50 \text{ kN} = 6.50 \text{ kN}\downarrow$$

Using the Pythagorean theorem, Fig. 3–37c, the magnitude of $\mathbf{F}_R$ is

$$F_R = \sqrt{(F_R)_x^2 + (F_R)_y^2} = \sqrt{(5.598 \text{ kN})^2 + (6.50 \text{ kN})^2} = 8.58 \text{ kN} \quad Ans.$$

Its direction θ is

$$\theta = \tan^{-1}\left(\frac{(F_R)_y}{(F_R)_x}\right) = \tan^{-1}\left(\frac{6.50 \text{ kN}}{5.598 \text{ kN}}\right) = 49.3° \qquad Ans.$$

Moment Summation. Referring to Fig. 3–37b, we have

$$\zeta + (M_R)_O = \Sigma M_O;$$

$$(M_R)_O = (3 \text{ kN}) \sin 30°(0.2 \text{ m}) - (3 \text{ kN}) \cos 30°(0.1 \text{ m}) + \left(\tfrac{3}{5}\right)(5 \text{ kN})(0.1 \text{ m})$$

$$- \left(\tfrac{4}{5}\right)(5 \text{ kN})(0.5 \text{ m}) - (4 \text{ kN})(0.2 \text{ m})$$

$$= -2.46 \text{ kN} \cdot \text{m} = 2.46 \text{ kN} \cdot \text{m}\, \zeta \qquad Ans.$$

This clockwise moment is shown in Fig. 3–37c.

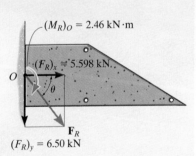

(c)

Fig. 3–37

NOTE: Realize that the resultant force and couple moment in Fig. 3–37c will produce the same external effects or reactions at the wall support as those produced by the force system, Fig. 3–37a.

EXAMPLE   3.15

Replace the force and couple system acting on the member in Fig. 3–38a by an equivalent resultant force and couple moment acting at point O.

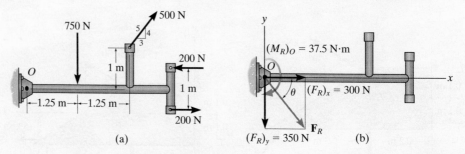

(a)

$(M_R)_O = 37.5$ N·m

$(F_R)_x = 300$ N

$(F_R)_y = 350$ N

F_R

(b)

Fig. 3–38

SOLUTION

Force Summation.   Since the couple forces of 200 N are equal but opposite, they produce a zero resultant force, and so it is not necessary to consider them in the force summation. The 500-N force is resolved into its x and y components, thus,

$$\xrightarrow{+}(F_R)_x = \Sigma F_x;\ (F_R)_x = \left(\tfrac{3}{5}\right)(500\text{ N}) = 300\text{ N} \rightarrow$$

$$+\uparrow(F_R)_y = \Sigma F_y;\ (F_R)_y = (500\text{ N})\left(\tfrac{4}{5}\right) - 750\text{ N} = -350\text{ N} = 350\text{ N}\downarrow$$

From Fig. 3–15b, the magnitude of $\mathbf{F}_R$ is

$$F_R = \sqrt{(F_R)_x^2 + (F_R)_y^2}$$

$$= \sqrt{(300\text{ N})^2 + (350\text{ N})^2} = 461\text{ N} \qquad Ans.$$

And the angle θ is

$$\theta = \tan^{-1}\left(\frac{(F_R)_y}{(F_R)_x}\right) = \tan^{-1}\left(\frac{350\text{ N}}{300\text{ N}}\right) = 49.4° \qquad Ans.$$

Moment Summation.   Since the couple moment is a free vector, it can act at any point on the member. Referring to Fig. 3–38a, we have

$$\zeta + (M_R)_O = \Sigma M_O + \Sigma M$$

$$(M_R)_O = (500\text{ N})\left(\tfrac{4}{5}\right)(2.5\text{ m}) - (500\text{ N})\left(\tfrac{3}{5}\right)(1\text{ m})$$

$$- (750\text{ N})(1.25\text{ m}) + 200\text{ N}\cdot\text{m}$$

$$= -37.5\text{ N}\cdot\text{m} = 37.5\text{ N}\cdot\text{m}\ \rangle \qquad Ans.$$

This clockwise moment is shown in Fig. 3–38b.

EXAMPLE 3.16

The member is subjected to a couple moment $\mathbf{M}$ and forces $\mathbf{F}_1$ and $\mathbf{F}_2$ in Fig. 3–39a. Replace this system by an equivalent resultant force and couple moment acting at its base, point O.

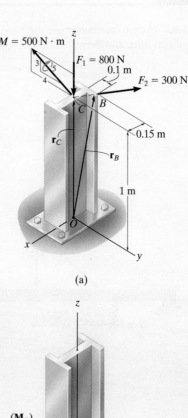

(a)

SOLUTION (VECTOR ANALYSIS)

The three-dimensional aspects of the problem can be simplified by using a Cartesian vector analysis. Expressing the forces and couple moment as Cartesian vectors, we have

$$\mathbf{F}_1 = \{-800\mathbf{k}\} \text{ N}$$

$$\mathbf{F}_2 = (300 \text{ N})\mathbf{u}_{CB}$$

$$= (300 \text{ N})\left(\frac{\mathbf{r}_{CB}}{r_{CB}}\right)$$

$$= 300 \text{ N}\left[\frac{\{-0.15\mathbf{i} + 0.1\mathbf{j}\} \text{ m}}{\sqrt{(-0.15 \text{ m})^2 + (0.1 \text{ m})^2}}\right] = \{-249.6\mathbf{i} + 166.4\mathbf{j}\} \text{ N}$$

$$\mathbf{M} = -500\left(\tfrac{4}{5}\right)\mathbf{j} + 500\left(\tfrac{3}{5}\right)\mathbf{k} = \{-400\mathbf{j} + 300\mathbf{k}\} \text{ N} \cdot \text{m}$$

Force Summation.

$$\mathbf{F}_R = \Sigma\mathbf{F}; \qquad \mathbf{F}_R = \mathbf{F}_1 + \mathbf{F}_2 = -800\mathbf{k} - 249.6\mathbf{i} + 166.4\mathbf{j}$$

$$= \{-250\mathbf{i} + 166\mathbf{j} - 800\mathbf{k}\} \text{ N} \qquad\qquad Ans.$$

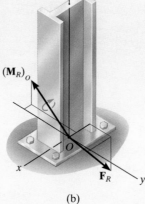

(b)

Fig. 3–39

Moment Summation.

$$(\mathbf{M}_R)_o = \Sigma\mathbf{M} + \Sigma\mathbf{M}_O$$

$$(\mathbf{M}_R)_o = \mathbf{M} + \mathbf{r}_C \times \mathbf{F}_1 + \mathbf{r}_B \times \mathbf{F}_2$$

$$(\mathbf{M}_R)_o = (-400\mathbf{j} + 300\mathbf{k}) + (1\mathbf{k}) \times (-800\mathbf{k}) + \begin{vmatrix} \mathbf{i} & \mathbf{j} & \mathbf{k} \\ -0.15 & 0.1 & 1 \\ -249.6 & 166.4 & 0 \end{vmatrix}$$

$$= (-400\mathbf{j} + 300\mathbf{k}) + (0) + (-166.4\mathbf{i} - 249.6\mathbf{j})$$

$$= \{-166\mathbf{i} - 650\mathbf{j} + 300\mathbf{k}\} \text{ N} \cdot \text{m} \qquad\qquad Ans.$$

The results are shown in Fig. 3–39b.

PRELIMINARY PROBLEM

P3–5. In each case, determine the x and y components of the resultant force and the resultant couple moment at point O.

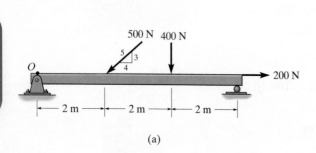

(a)

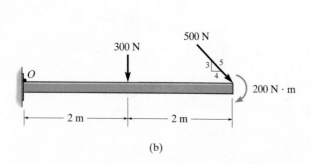

(b)

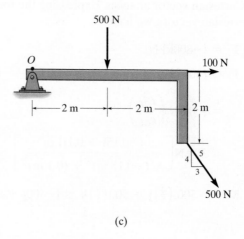

(c)

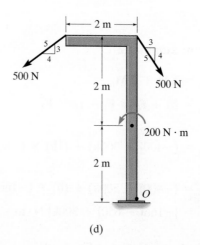

(d)

Prob. P3–5

FUNDAMENTAL PROBLEMS

F3–25. Replace the loading by an equivalent resultant force and couple moment acting at point A.

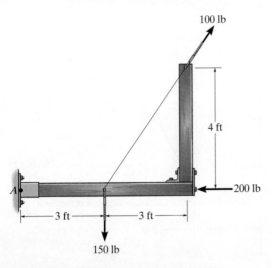

Prob. F3–25

F3–26. Replace the loading by an equivalent resultant force and couple moment acting at point A.

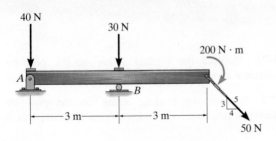

Prob. F3–26

F3–27. Replace the loading by an equivalent resultant force and couple moment acting at point A.

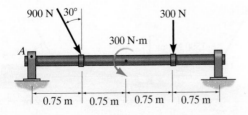

Prob. F3–27

F3–28. Replace the loading by an equivalent resultant force and couple moment acting at point A.

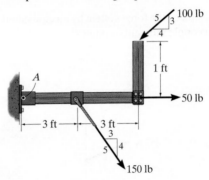

Prob. F3–28

F3–29. Replace the loading by an equivalent resultant force and couple moment acting at point O.

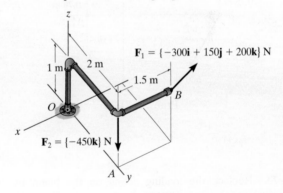

Prob. F3–29

F3–30. Replace the loading by an equivalent resultant force and couple moment acting at point O.

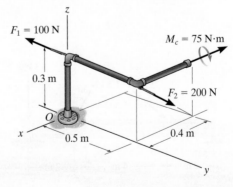

Prob. F3–30

PROBLEMS

3–71. Replace the force system by an equivalent resultant force and couple moment at point O.

***3–72.** Replace the force system by an equivalent resultant force and couple moment at point P.

3–75. Replace the loading acting on the beam by an equivalent resultant force and couple moment at point O.

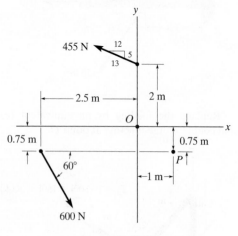

Probs. 3–71/72

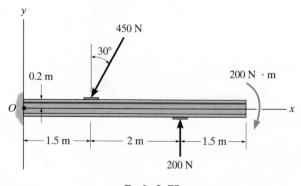

Prob. 3–75

3–73. Replace the loading acting on the beam by an equivalent force and couple moment at point A.

3–74. Replace the loading acting on the beam by an equivalent force and couple moment at point B.

***3–76.** Replace the loading acting on the post by an equivalent resultant force and couple moment at point A.

3–77. Replace the loading acting on the post by an equivalent resultant force and couple moment at point B.

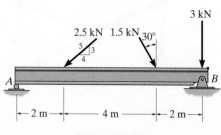

Probs. 3–73/74

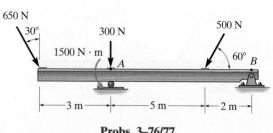

Probs. 3–76/77

3–78.   Replace the loading acting on the post by a resultant force and couple moment at point O.

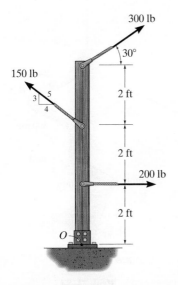

Prob. 3–78

3–79.   Replace the loading acting on the frame by an equivalent resultant force and couple moment acting at point A.

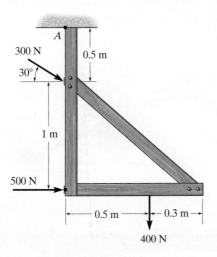

Prob. 3–79

***3–80.**   The forces $\mathbf{F}_1 = \{-4\mathbf{i} + 2\mathbf{j} - 3\mathbf{k}\}$ kN and $\mathbf{F}_2 = \{3\mathbf{i} - 4\mathbf{j} - 2\mathbf{k}\}$ kN act on the end of the beam. Replace these forces by an equivalent force and couple moment acting at point O.

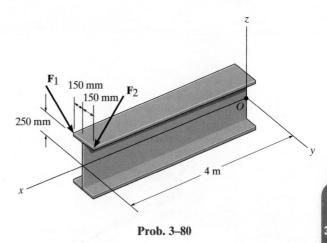

Prob. 3–80

3–81.   A biomechanical model of the lumbar region of the human trunk is shown. The forces acting in the four muscle groups consist of $F_R = 35$ N for the rectus, $F_O = 45$ N for the oblique, $F_L = 23$ N for the lumbar latissimus dorsi, and $F_E = 32$ N for the erector spinae. These loadings are symmetric with respect to the y–z plane. Replace this system of parallel forces by an equivalent force and couple moment acting at the spine, point O. Express the results in Cartesian vector form.

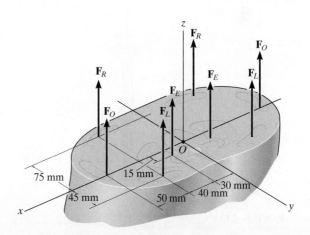

Prob. 3–81

3–82. Replace the loading by an equivalent resultant force and couple moment at point O. Take $F_3 = \{-200i + 500j - 300k\}$ N.

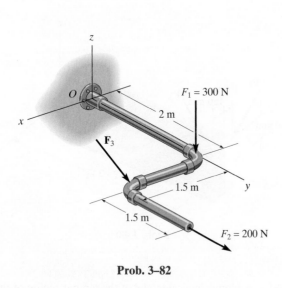

Prob. 3–82

3–83. Replace the loading by an equivalent resultant force and couple moment at point O.

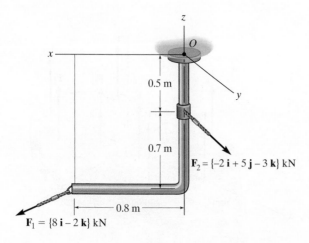

Prob. 3–83

*3–84. Replace the force of $F = 80$ N acting on the pipe assembly by an equivalent resultant force and couple moment at point A.

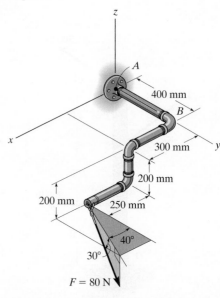

Prob. 3–84

3–85. The belt passing over the pulley is subjected to forces F_1 and F_2, each having a magnitude of 40 N. F_1 acts in the $-k$ direction. Replace these forces by an equivalent force and couple moment at point A. Express the result in Cartesian vector form. Set $\theta = 0°$ so that F_2 acts in the $-j$ direction.

3–86. The belt passing over the pulley is subjected to two forces F_1 and F_2, each having a magnitude of 40 N. F_1 acts in the $-k$ direction. Replace these forces by an equivalent force and couple moment at point A. Express the result in Cartesian vector form. Take $\theta = 45°$.

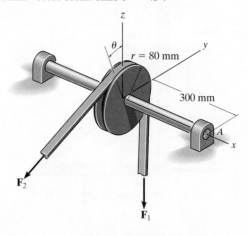

Probs. 3–85/86

3.8   FURTHER SIMPLIFICATION OF A FORCE AND COUPLE SYSTEM

In the preceding section, we developed a way to reduce a force and couple moment system acting on a rigid body into an equivalent resultant force $\mathbf{F}_R$ acting at a specific point O and a resultant couple moment $(\mathbf{M}_R)_O$. The force system can be further reduced to an equivalent single resultant force provided the lines of action of $\mathbf{F}_R$ and $(\mathbf{M}_R)_O$ are *perpendicular* to each other. This occurs when the force system is either concurrent, coplanar, or parallel.

Concurrent Force System. Since a ***concurrent force system*** is one in which the lines of action of all the forces intersect at a common point O, Fig. 3–40a, then the force system produces no moment about this point. As a result, the equivalent system can be represented by a single resultant force $\mathbf{F}_R = \Sigma\mathbf{F}$ acting at O, Fig. 3–40b.

Coplanar Force System. In the case of a ***coplanar force system***, the lines of action of all the forces lie in the same plane, Fig. 3–41a, and so the resultant force $\mathbf{F}_R = \Sigma\mathbf{F}$ of this system also lies in this plane. Furthermore, the moment of each of the forces about any point O is directed perpendicular to this plane. Thus, the resultant moment $(\mathbf{M}_R)_O$ and resultant force $\mathbf{F}_R$ will be *mutually perpendicular*, Fig. 3–41b. The resultant moment can be replaced by moving the resultant force $\mathbf{F}_R$ a perpendicular or moment arm distance d away from point O such that $\mathbf{F}_R$ produces the *same moment* $(\mathbf{M}_R)_O$ about point O, Fig. 3–41c. This distance d can be determined from the scalar equation $(M_R)_O = F_R d = \Sigma M_O$ or $d = (M_R)_O / F_R$.

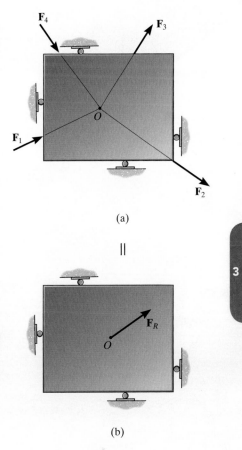

(a)

||

(b)

Fig. 3–40

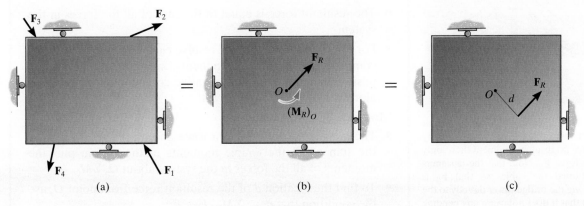

(a)   (b)   (c)

Fig. 3–41

(a)

(b)

(c)

Fig. 3–42

Parallel Force System. The ***parallel force system*** shown in Fig. 3–42a consists of forces that are all parallel to the z axis. Thus, the resultant force $\mathbf{F}_R = \Sigma \mathbf{F}$ at point O must also be parallel to this axis, Fig. 3–42b. The moment produced by each force lies in the plane of the plate, and so the resultant couple moment, $(\mathbf{M}_R)_O$, will also lie in this plane, along the moment axis a since $\mathbf{F}_R$ and $(\mathbf{M}_R)_O$ are mutually perpendicular. As a result, the force system can be further reduced to an equivalent single resultant force $\mathbf{F}_R$, acting through point P located on the perpendicular b axis, Fig. 3–42c. The distance d along this axis from point O requires $(M_R)_O = F_R d = \Sigma M_O$ or $d = \Sigma M_O / F_R$.

PROCEDURE FOR ANALYSIS

The technique used to reduce a coplanar or parallel force system to a single resultant force follows a similar procedure outlined in the previous section.

- Establish the x, y, z axes and locate the resultant force $\mathbf{F}_R$ an arbitrary distance away from the origin of the coordinates.

Force Summation.

- The resultant force is equal to the sum of all the forces in the system.

- For a coplanar force system, resolve each force into its x and y components. Positive components are directed along the positive x and y axes, and negative components are directed along the negative x and y axes.

Moment Summation.

- The moment of the resultant force about point O is equal to the sum of all the couple moments in the system plus the moments of all the forces in the system about O, ΣM_O.

- To find the location d of the resultant force from point O, use the condition that $d = \Sigma M_O / F_R$.

The four cable forces are all concurrent at point O on this bridge tower. Consequently they produce no resultant moment there, only a resultant force $\mathbf{F}_R$. Note that the designers have positioned the cables so that $\mathbf{F}_R$ is directed *along* the bridge tower directly to the support, so that it does not cause any bending of the tower.

EXAMPLE 3.17

Replace the force and couple moment system acting on the beam in Fig. 3–43a by an equivalent resultant force, and find where its line of action intersects the beam, measured from point O.

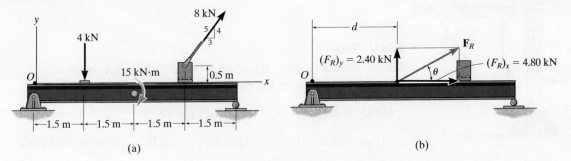

Fig. 3–43

SOLUTION

Force Summation. Summing the force components,

$$\xrightarrow{+}(F_R)_x = \Sigma F_x; \qquad (F_R)_x = 8 \text{ kN}\left(\tfrac{3}{5}\right) = 4.80 \text{ kN} \rightarrow$$

$$+\uparrow(F_R)_y = \Sigma F_y; \qquad (F_R)_y = -4 \text{ kN} + 8 \text{ kN}\left(\tfrac{4}{5}\right) = 2.40 \text{ kN}\uparrow$$

From Fig. 3–44b, the magnitude of F_R is

$$F_R = \sqrt{(4.80 \text{ kN})^2 + (2.40 \text{ kN})^2} = 5.37 \text{ kN} \qquad Ans.$$

The angle θ is

$$\theta = \tan^{-1}\left(\frac{2.40 \text{ kN}}{4.80 \text{ kN}}\right) = 26.6° \qquad Ans.$$

Moment Summation. We must equate the moment of F_R about point O in Fig. 3–43b to the sum of the moments of the force and couple moment system about point O in Fig. 3–43a. Since the line of action of $(F_R)_x$ passes through point O, *only $(F_R)_y$ produces a moment* about this point. Thus,

$$\zeta+ (M_R)_O = \Sigma M_O; \qquad 2.40 \text{ kN}(d) = -(4 \text{ kN})(1.5 \text{ m}) - 15 \text{ kN}\cdot\text{m}$$
$$-\left[8 \text{ kN}\left(\tfrac{3}{5}\right)\right](0.5 \text{ m}) + \left[8 \text{ kN}\left(\tfrac{4}{5}\right)\right](4.5 \text{ m})$$

$$d = 2.25 \text{ m} \qquad Ans.$$

EXAMPLE 3.18

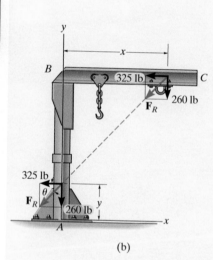

(a)

(b)

Fig. 3–44

The jib crane shown in Fig. 3–44a is subjected to three coplanar forces. Replace this loading by an equivalent resultant force and specify where the resultant's line of action intersects the column AB and boom BC.

SOLUTION

Force Summation. Resolving the 250-lb force into x and y components and summing the force components yields

$\xrightarrow{+} (F_R)_x = \Sigma F_x;$ $(F_R)_x = -250 \text{ lb}\left(\frac{3}{5}\right) - 175 \text{ lb} = -325 \text{ lb} = 325 \text{ lb} \leftarrow$

$+\uparrow (F_R)_y = \Sigma F_y;$ $(F_R)_y = -250 \text{ lb}\left(\frac{4}{5}\right) - 60 \text{ lb} = -260 \text{ lb} = 260 \text{ lb}\downarrow$

As shown by the vector addition in Fig. 3–44b,

$$F_R = \sqrt{(325 \text{ lb})^2 + (260 \text{ lb})^2} = 416 \text{ lb} \qquad Ans.$$

$$\theta = \tan^{-1}\left(\frac{260 \text{ lb}}{325 \text{ lb}}\right) = 38.7° \nearrow \qquad Ans.$$

Moment Summation. Moments will be summed about point A. Assuming the line of action of $\mathbf{F}_R$ intersects AB at a distance y from A, Fig. 3–44b, we have

$\zeta + (M_R)_A = \Sigma M_A;$ $325 \text{ lb}\,(y) + 260 \text{ lb}\,(0)$

$= 175 \text{ lb}\,(5 \text{ ft}) - 60 \text{ lb}\,(3 \text{ ft}) + 250 \text{ lb}\left(\frac{3}{5}\right)(11 \text{ ft}) - 250 \text{ lb}\left(\frac{4}{5}\right)(8 \text{ ft})$

$$y = 2.29 \text{ ft} \qquad Ans.$$

By the principle of transmissibility, $\mathbf{F}_R$ can also be placed at a distance x where it intersects BC, Fig. 3–44b. In this case we have

$\zeta + (M_R)_A = \Sigma M_A;$ $325 \text{ lb}\,(11 \text{ ft}) - 260 \text{ lb}\,(x)$

$= 175 \text{ lb}\,(5 \text{ ft}) - 60 \text{ lb}\,(3 \text{ ft}) + 250 \text{ lb}\left(\frac{3}{5}\right)(11 \text{ ft}) - 250 \text{ lb}\left(\frac{4}{5}\right)(8 \text{ ft})$

$$x = 10.9 \text{ ft} \qquad Ans.$$

EXAMPLE   3.19

The slab in Fig. 3–45a is subjected to four parallel forces. Determine the magnitude and direction of a resultant force equivalent to the given force system, and locate its point of application on the slab.

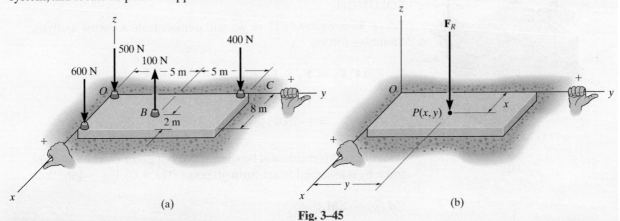

(a)   (b)

Fig. 3–45

SOLUTION *(SCALAR ANALYSIS)*

Force Summation.   From Fig. 3–45a, the resultant force is

$+\downarrow F_R = \Sigma F;$ $F_R = 600 \text{ N} - 100 \text{ N} + 400 \text{ N} + 500 \text{ N}$

$= 1400 \text{ N} \downarrow$ *Ans.*

Moment Summation.   We require the moment about the x axis of the resultant force, Fig. 3–45b, to be equal to the sum of the moments about the x axis of all the forces in the system, Fig. 3–45a. The moment arms are determined from the y coordinates, since these coordinates represent the *perpendicular distances* from the x axis to the lines of action of the forces. Using the right-hand rule, we have

$(M_R)_x = \Sigma M_x;$

$-(1400 \text{ N})y = 600 \text{ N}(0) + 100 \text{ N}(5 \text{ m}) - 400 \text{ N}(10 \text{ m}) + 500 \text{ N}(0)$

$-1400y = -3500$ $y = 2.50 \text{ m}$ *Ans.*

In a similar manner, a moment equation can be written about the y axis using moment arms defined by the x coordinates of each force.

$(M_R)_y = \Sigma M_y;$

$(1400 \text{ N})x = 600 \text{ N}(8 \text{ m}) - 100 \text{ N}(6 \text{ m}) + 400 \text{ N}(0) + 500 \text{ N}(0)$

$1400x = 4200$

$x = 3 \text{ m}$ *Ans.*

NOTE: A force of $F_R = 1400 \text{ N}$ placed at point $P(3.00 \text{ m}, 2.50 \text{ m})$ on the slab, Fig. 3–45b, is therefore equivalent to the parallel force system acting on the slab in Fig. 3–45a.

EXAMPLE 3.20

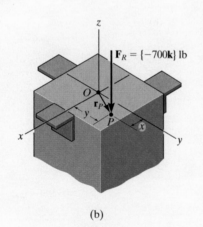

(a)

(b)

Fig. 3–46

Replace the force system in Fig. 3–46a by an equivalent resultant force and specify its point of application on the pedestal.

SOLUTION

Force Summation. Here we will demonstrate a vector analysis. Summing forces,

$$\mathbf{F}_R = \Sigma\mathbf{F}; \ \mathbf{F}_R = \mathbf{F}_A + \mathbf{F}_B + \mathbf{F}_C$$

$$= \{-300\mathbf{k}\}\,\text{lb} + \{-500\mathbf{k}\}\,\text{lb} + \{100\mathbf{k}\}\,\text{lb}$$

$$= \{-700\mathbf{k}\}\,\text{lb} \qquad\qquad Ans.$$

Location. Moments will be summed about point O. The resultant force $\mathbf{F}_R$ is assumed to act through point $P\,(x, y, 0)$, Fig. 3–46b. Thus

$$(\mathbf{M}_R)_O = \Sigma\mathbf{M}_O;$$

$$\mathbf{r}_P \times \mathbf{F}_R = (\mathbf{r}_A \times \mathbf{F}_A) + (\mathbf{r}_B \times \mathbf{F}_B) + (\mathbf{r}_C \times \mathbf{F}_C)$$

$$(x\mathbf{i} + y\mathbf{j}) \times (-700\mathbf{k}) = [(4\mathbf{i}) \times (-300\mathbf{k})]$$

$$+ [(-4\mathbf{i} + 2\mathbf{j}) \times (-500\mathbf{k})] + [(-4\mathbf{j}) \times (100\mathbf{k})]$$

$$-700x(\mathbf{i} \times \mathbf{k}) - 700y(\mathbf{j} \times \mathbf{k}) = -1200(\mathbf{i} \times \mathbf{k}) + 2000(\mathbf{i} \times \mathbf{k})$$

$$- 1000(\mathbf{j} \times \mathbf{k}) - 400(\mathbf{j} \times \mathbf{k})$$

$$700x\mathbf{j} - 700y\mathbf{i} = 1200\mathbf{j} - 2000\mathbf{j} - 1000\mathbf{i} - 400\mathbf{i}$$

Equating the $\mathbf{i}$ and $\mathbf{j}$ components,

$$-700y = -1400 \qquad\qquad (1)$$

$$y = 2\,\text{in.} \qquad\qquad Ans.$$

$$700x = -800 \qquad\qquad (2)$$

$$x = -1.14\,\text{in.} \qquad\qquad Ans.$$

The negative sign indicates that the x coordinate of point P is negative.

NOTE: As demonstrated in Example 3.19, it is also possible to establish Eq. 1 and 2 directly by summing moments about the x and y axes. Using the right-hand rule, we have

$$(M_R)_x = \Sigma M_x; \qquad -700y = -100\,\text{lb}(4\,\text{in.}) - 500\,\text{lb}(2\,\text{in.})$$

$$(M_R)_y = \Sigma M_y; \qquad 700x = 300\,\text{lb}(4\,\text{in.}) - 500\,\text{lb}(4\,\text{in.})$$

PRELIMINARY PROBLEMS

P3–6. In each case, determine the x and y components of the resultant force and specify the distance where this force acts from point O.

P3–7. In each case, determine the resultant force and specify its coordinates x and y where it acts on the x–y plane.

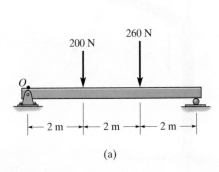

(a)

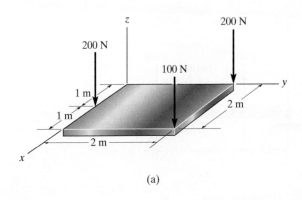

(a)

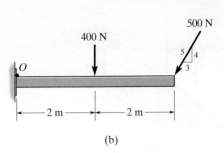

(b)

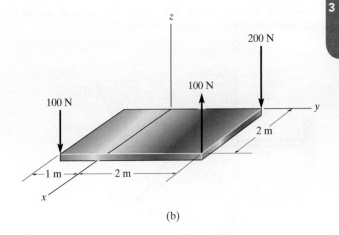

(b)

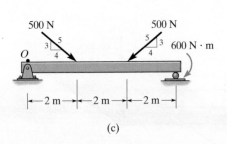

(c)

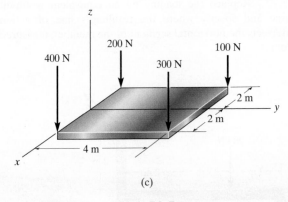

(c)

Prob. P3–6

Prob. P3–7

FUNDAMENTAL PROBLEMS

F3–31. Replace the loading by an equivalent resultant force and specify where the resultant's line of action intersects the beam, measured from O.

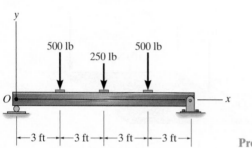

Prob. F3–31

F3–32. Replace the loading by an equivalent resultant force and specify where the resultant's line of action intersects the member, measured from A.

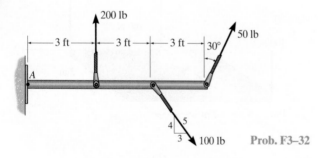

Prob. F3–32

F3–33. Replace the loading by an equivalent resultant force and specify where the resultant's line of action intersects the horizontal segment of the member, measured from A.

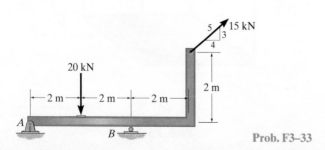

Prob. F3–33

F3–34. Replace the loading by an equivalent resultant force and specify where the resultant's line of action intersects the member AB, measured from A.

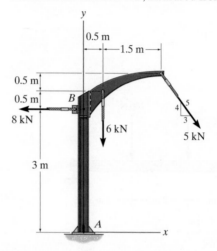

Prob. F3–34

F3–35. Replace the loading by an equivalent single resultant force and specify the x and y coordinates of its line of action.

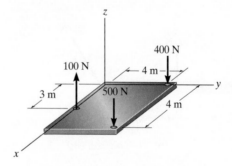

Prob. F3–35

F3–36. Replace the loading by an equivalent single resultant force and specify the x and y coordinates of its line of action.

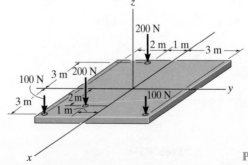

Prob. F3–36

PROBLEMS

3–87. The weights of the various components of the truck are shown. Replace this system of forces by an equivalent resultant force and specify its location, measured from *B*.

*3–88. The weights of the various components of the truck are shown. Replace this system of forces by an equivalent resultant force and specify its location, measured from point *A*.

3–91. Replace the loading by a single resultant force. Specify where the force acts, measured from end *A*.

*3–92. Replace the loading by a single resultant force. Specify where the force acts, measured from *B*.

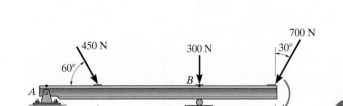

Probs. 3–91/92

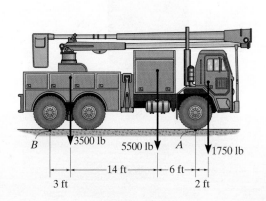

Probs. 3–87/88

3–93. Replace the loading by a single resultant force. Specify where its line of action intersects a vertical line along member *AB*, measured from *A*.

3–89. Replace the three forces acting on the shaft by a single resultant force. Specify where the force acts, measured from end *A*.

3–90. Replace the three forces acting on the shaft by a single resultant force. Specify where the force acts, measured from end *B*.

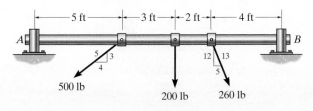

Probs. 3–89/90

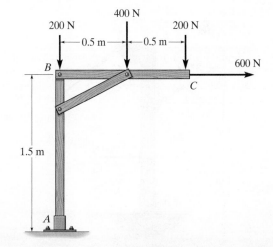

Prob. 3–93

3–94. Replace the loading on the frame by a single resultant force. Specify where its line of action intersects a vertical line along member AB, measured from A.

3–95. Replace the loading on the frame by a single resultant force. Specify where its line of action intersects a horizontal line along member CB, measured from end C.

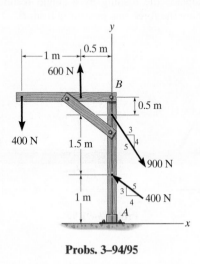

Probs. 3–94/95

***3–96.** Replace the loading acting on the post by a resultant force, and specify where its line of action intersects the post AB, measured from point A.

3–97. Replace the loading acting on the post by a resultant force, and specify where its line of action intersects the post AB, measured from point B.

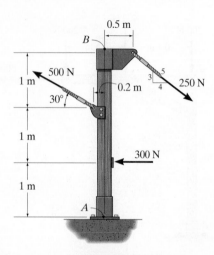

Probs. 3–96/97

3–98. Replace the parallel force system acting on the plate by a resultant force and specify its location on the x–z plane.

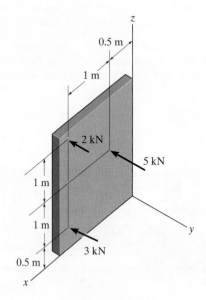

Prob. 3–98

3–99. Replace the loading acting on the frame by an equivalent resultant force and specify where the resultant's line of action intersects member AB, measured from A.

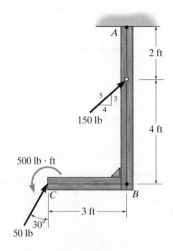

Prob. 3–99

*3–100. Replace the loading acting on the frame by an equivalent resultant force and specify where the resultant's line of action intersects member BC, measured from B.

3–102. Determine the magnitudes of $\mathbf{F}_A$ and $\mathbf{F}_B$ so that the resultant force passes through point O.

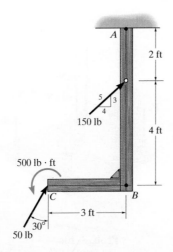

Prob. 3–100

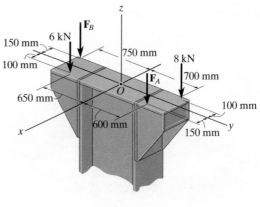

Prob. 3–102

3–101. If $F_A = 7$ kN and $F_B = 5$ kN, represent the force system by a resultant force, and specify its location on the x–y plane.

3–103. The tube supports the four parallel forces. Determine the magnitudes of forces $\mathbf{F}_C$ and $\mathbf{F}_D$ acting at C and D so that the equivalent resultant force of the force system acts through the midpoint O of the tube.

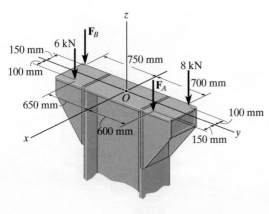

Prob. 3–101

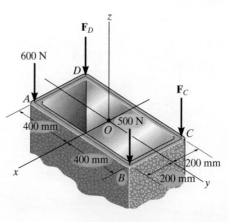

Prob. 3–103

*3–104. The building slab is subjected to four parallel column loadings. Determine the equivalent resultant force and specify its location (x, y) on the slab. Take $F_1 = 8$ kN and $F_2 = 9$ kN.

3–106. If $F_A = 40$ kN and $F_B = 35$ kN, determine the magnitude of the resultant force and specify the location of its point of application (x, y) on the slab.

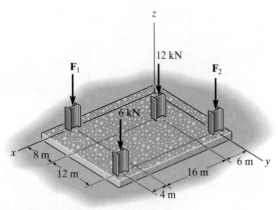

Prob. 3–104

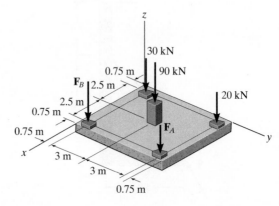

Prob. 3–106

3–105. The building slab is subjected to four parallel column loadings. Determine F_1 and F_2 if the resultant force acts through point $(12 \text{ m}, 10 \text{ m})$.

3–107. If the resultant force is required to act at the center of the slab, determine the magnitude of the column loadings F_A and F_B and the magnitude of the resultant force.

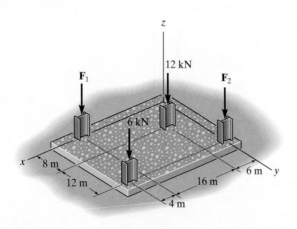

Prob. 3–105

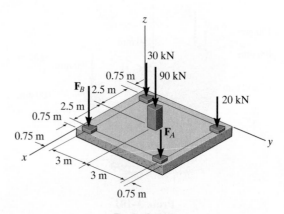

Prob. 3–107

3.9   REDUCTION OF A SIMPLE DISTRIBUTED LOADING

Sometimes a body may be subjected to a loading that is distributed over its surface. For example, wind on the face of a sign, water within a tank, or sand on the floor of a storage container all exert *distributed loadings*. The pressure caused by a loading at each point on the surface represents the intensity of the loading. It is measured using pascals, Pa (or N/m^2) in SI units, or lb/ft^2 in the U.S. Customary system.

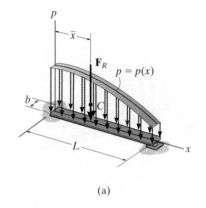

(a)

Loading Along a Single Axis.

The most common type of distributed pressure loading is represented along a single axis. For example, consider the beam (or plate) in Fig. 3–47a that has a constant width and is subjected to a pressure loading that varies only along the x axis. This loading can be described by the function $p = p(x)\ N/m^2$. Since it contains only one variable, x, we can represent it as a *coplanar distributed load*. To do so, we must multiply it by the width b m of the beam, so that $w(x) = p(x)b\ N/m$, Fig. 3–47b. Using the methods of Sec. 3–8, we can replace this coplanar parallel force system with a single equivalent resultant force $\mathbf{F}_R$, Fig. 3–47c.

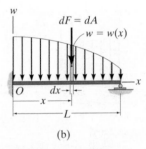

(b)

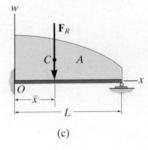

(c)

Fig. 3–47

Magnitude of Resultant Force.

The magnitude of $\mathbf{F}_R$ is equivalent to the sum of all the forces in the system, $F_R = \Sigma F$. In this case integration must be used since there is an infinite number of parallel forces $d\mathbf{F}$ acting on the beam, Fig. 3–47b. Each $d\mathbf{F}$ is acting on an element of length dx, and since $w(x)$ is a force per unit length, then $dF = w(x)\,dx = dA$ where dA is the colored *differential area* under the loading curve. For the entire length L,

$$+\downarrow F_R = \Sigma F; \qquad \boxed{F_R = \int_L w(x)\,dx = \int_A dA = A} \qquad (3\text{–}19)$$

Therefore, *the magnitude of the resultant force is equal to the area A under the loading diagram*, Fig. 3–47c.

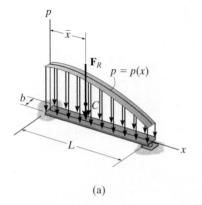

(a)

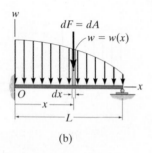

(b)

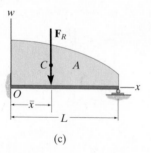

(c)

Fig. 3–47 (Repeated)

The pile of brick creates an approximate triangular distributed loading on the board.

Location of Resultant Force. The location $\bar{x}$ of $\mathbf{F}_R$ can be determined by equating the moments of the force resultant and the parallel force distribution about point O (the y axis), $(M_R)_O = \Sigma M_O$. Since dF produces a moment of $x \, dF = xw(x) \, dx$ about O, Fig. 3–47b, then for the entire length L, Fig. 3–47c, we have

$$\zeta + (M_R)_O = \Sigma M_O; \qquad\qquad -\bar{x}F_R = -\int_L xw(x) \, dx$$

Solving for $\bar{x}$, using Eq. 3–19, we have

$$\bar{x} = \frac{\displaystyle\int_L xw(x) \, dx}{\displaystyle\int_L w(x) \, dx} = \frac{\displaystyle\int_A x \, dA}{\displaystyle\int_A dA} \qquad (3\text{--}20)$$

This coordinate $\bar{x}$ locates the geometric center or **centroid** of the *area* under the distributed loading. *In other words, the line of action of the resultant force passes through the centroid C (geometric center) of the area under the loading diagram*, Fig. 3–47c. When the distributed-loading diagram is in the shape of a rectangle, triangle, or some other simple geometric form, then the centroid location for such common shapes does not have to be determined from the above equation. Rather it can be obtained directly from the tabulation given on the inside back cover.

Once $\bar{x}$ is determined, $\mathbf{F}_R$ by symmetry passes through point $(\bar{x}, 0)$ on the surface of the beam, Fig. 3–47a, and so in three dimensions *the resultant force has a magnitude equal to the volume under the loading curve $p = p(x)$ and a line of action which passes through the centroid (geometric center) of this volume.*

IMPORTANT POINTS

- Coplanar distributed loadings are defined by using a loading function $w = w(x)$ that indicates the intensity of the loading along the length of a member. This intensity is measured in N/m or lb/ft.

- The external effects caused by a coplanar distributed load acting on a member can be represented by a resultant force.

- This resultant force is equivalent to the *area* under the loading diagram, and has a line of action that passes through the *centroid* or geometric center of this area.

EXAMPLE 3.21

Determine the magnitude and location of the equivalent resultant force acting on the shaft in Fig. 3–48a.

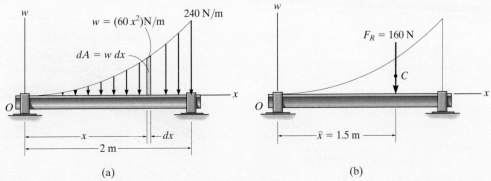

(a)

(b)

Fig. 3–48

SOLUTION

Since $w = w(x)$ is given, this problem will be solved by integration.

The differential element has an area $dA = w\,dx = 60x^2\,dx$. Applying Eq. 3–19,

$$+\downarrow F_R = \Sigma F;$$

$$F_R = \int_A dA = \int_0^{2\,m} 60x^2\,dx = 60\left(\frac{x^3}{3}\right)\Big|_0^{2\,m} = 60\left(\frac{2^3}{3} - \frac{0^3}{3}\right)$$

$$= 160\,\text{N} \hspace{4cm} Ans.$$

The location $\bar{x}$ of $\mathbf{F}_R$ measured from O, Fig. 3–48b, is determined from Eq. 3–20.

$$\bar{x} = \frac{\int_A x\,dA}{\int_A dA} = \frac{\int_0^{2\,m} x(60x^2)\,dx}{160\,\text{N}} = \frac{60\left(\frac{x^4}{4}\right)\Big|_0^{2\,m}}{160\,\text{N}} = \frac{60\left(\frac{2^4}{4} - \frac{0^4}{4}\right)}{160\,\text{N}}$$

$$= 1.5\,\text{m} \hspace{4cm} Ans.$$

NOTE: These results can be checked by using the table in Appendix B, where for the exparabolic area of length a, height b, and shape shown in Fig. 3–48a, we have

$$A = \frac{ab}{3} = \frac{2\,\text{m}(240\,\text{N/m})}{3} = 160\,\text{N and } \bar{x} = \frac{3}{4}a = \frac{3}{4}(2\,\text{m}) = 1.5\,\text{m}$$

EXAMPLE   3.22

A distributed loading of $p = (800x)$ Pa acts over the top surface of the beam shown in Fig. 3–49a. Determine the magnitude and location of the equivalent resultant force.

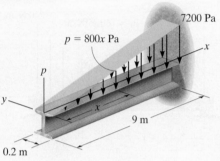

(a)

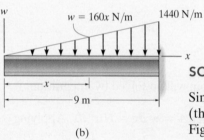

(b)

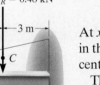

(c)

Fig. 3–49

SOLUTION

Since the loading intensity is uniform along the width of the beam (the y axis), the loading can be viewed in two dimensions as shown in Fig. 3–49b. Here

$$w = (800x \text{ N/m}^2)(0.2 \text{ m})$$
$$= (160x) \text{ N/m}$$

At $x = 9$ m, $w = 1440$ N/m. We may again apply Eqs. 3–19 and 3–20 as in the previous example; however, here it is easier to find the area and its centroid using Appendix B.

The magnitude of the resultant force is equivalent to the area of the triangle.

$$F_R = \tfrac{1}{2}(9 \text{ m})(1440 \text{ N/m}) = 6480 \text{ N} = 6.48 \text{ kN} \qquad \textit{Ans.}$$

The line of action of $\mathbf{F}_R$ passes through the *centroid C* of this triangle. Hence,

$$\bar{x} = 9 \text{ m} - \tfrac{1}{3}(9 \text{ m}) = 6 \text{ m} \qquad \textit{Ans.}$$

The results are shown in Fig. 3–49c.

NOTE: We can also view the resultant $\mathbf{F}_R$ as *acting* through the *centroid* of the *volume* of the loading diagram $p = p(x)$ in Fig. 3–49a. Then $\mathbf{F}_R$ intersects the x–y plane at the point (6 m, 0). Furthermore, the magnitude of $\mathbf{F}_R$ is equal to the volume under this loading diagram; i.e.,

$$F_R = V = \tfrac{1}{2}(7200 \text{ N/m}^2)(9 \text{ m})(0.2 \text{ m}) = 6.48 \text{ kN} \qquad \textit{Ans.}$$

EXAMPLE 3.23

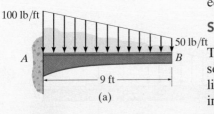

100 lb/ft

50 lb/ft

A

B

9 ft

(a)

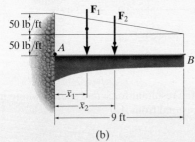

F_1 F_2

50 lb/ft

50 lb/ft A

B

$\bar{x}_1$

$\bar{x}_2$

9 ft

(b)

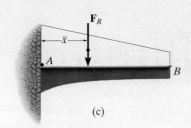

F_R

$\bar{x}$

A

B

(c)

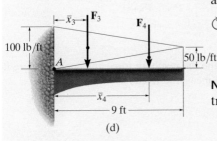

$\bar{x}_3$ F_3

F_4

100 lb/ft

A

50 lb/ft

$\bar{x}_4$

9 ft

(d)

Fig. 3–50

The granular material exerts the distributed loading on the beam as shown in Fig. 3–50a. Determine the magnitude and location of the equivalent resultant of this load.

SOLUTION

The area of the loading diagram is a *trapezoid*, and therefore the solution can be obtained directly from the formulas for a trapezoid listed in Appendix B. Since these formulas are not easily remembered, instead we will solve this problem by using "composite" areas. Here we will divide the loading into a rectangular and a triangular loading as shown in Fig. 3–50b. The magnitude of the force represented by each of these loadings is equal to its associated *area*,

$$F_1 = \tfrac{1}{2}(9 \text{ ft})(50 \text{ lb/ft}) = 225 \text{ lb}$$
$$F_2 = (9 \text{ ft})(50 \text{ lb/ft}) = 450 \text{ lb}$$

The lines of action of these parallel forces act through the respective *centroids* of their associated areas and therefore intersect the beam at

$$\bar{x}_1 = \tfrac{1}{3}(9 \text{ ft}) = 3 \text{ ft}$$
$$\bar{x}_2 = \tfrac{1}{2}(9 \text{ ft}) = 4.5 \text{ ft}$$

The two parallel forces $\mathbf{F}_1$ and $\mathbf{F}_2$ can be reduced to a single resultant $\mathbf{F}_R$. The magnitude of $\mathbf{F}_R$ is

$$+\!\downarrow F_R = \Sigma F; \qquad F_R = 225 + 450 = 675 \text{ lb} \qquad \textit{Ans.}$$

We can find the location of $\mathbf{F}_R$ with reference to point A, Fig. 3–50b and Fig. 3–50c. We require

$$\zeta + (M_R)_A = \Sigma M_A; \quad \bar{x}(675) = 3(225) + 4.5(450)$$
$$\bar{x} = 4 \text{ ft} \qquad \textit{Ans.}$$

NOTE: The trapezoidal area in Fig. 3–50a can also be divided into two triangular areas as shown in Fig. 3–50d. In this case

$$F_3 = \tfrac{1}{2}(9 \text{ ft})(100 \text{ lb/ft}) = 450 \text{ lb}$$
$$F_4 = \tfrac{1}{2}(9 \text{ ft})(50 \text{ lb/ft}) = 225 \text{ lb}$$

and

$$\bar{x}_3 = \tfrac{1}{3}(9 \text{ ft}) = 3 \text{ ft}$$
$$\bar{x}_4 = 9 \text{ ft} - \tfrac{1}{3}(9 \text{ ft}) = 6 \text{ ft}$$

Using these results, show again that $F_R = 675$ lb and $\bar{x} = 4$ ft.

FUNDAMENTAL PROBLEMS

F3–37. Determine the resultant force and specify where it acts on the beam, measured from A.

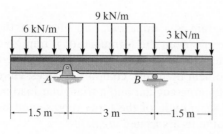

9 kN/m
6 kN/m
3 kN/m

├─1.5 m─┼────3 m────┼─1.5 m─┤

Prob. F3–37

F3–38. Determine the resultant force and specify where it acts on the beam, measured from A.

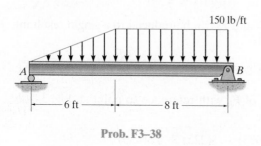

150 lb/ft

├──6 ft──┼───8 ft───┤

Prob. F3–38

F3–39. Determine the resultant force and specify where it acts on the beam, measured from A.

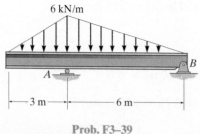

6 kN/m

├─3 m─┼──6 m──┤

Prob. F3–39

F3–40. Determine the resultant force and specify where it acts on the beam, measured from A.

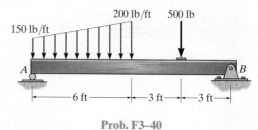

200 lb/ft   500 lb
150 lb/ft

├──6 ft──┼─3 ft─┼─3 ft─┤

Prob. F3–40

F3–41. Determine the resultant force and specify where it acts on the beam, measured from A.

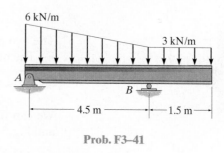

6 kN/m
3 kN/m

├───4.5 m───┼─1.5 m─┤

Prob. F3–41

F3–42. Determine the resultant force and specify where it acts on the beam, measured from A.

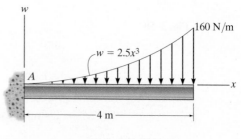

w
160 N/m
$w = 2.5x^3$
x

├────4 m────┤

Prob. F3–42

PROBLEMS

*3–108. Replace the loading by an equivalent resultant force and couple moment acting at point O.

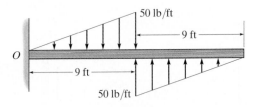

Prob. 3–108

3–109. Replace the distributed loading with an equivalent resultant force, and specify its location on the beam, measured from point O.

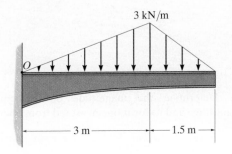

Prob. 3–109

3–110. Replace the loading by an equivalent resultant force and specify its location on the beam, measured from A.

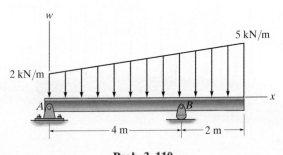

Prob. 3–110

3–111. Currently eighty-five percent of all neck injuries are caused by rear-end car collisions. To alleviate this problem, an automobile seat restraint has been developed that provides additional pressure contact with the cranium. During dynamic tests the distribution of load on the cranium has been plotted and shown to be parabolic. Determine the equivalent resultant force and its location, measured from point A.

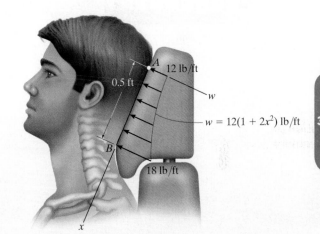

Prob. 3–111

*3–112. Replace the distributed loading by an equivalent resultant force, and specify its location on the beam, measured from the pin at A.

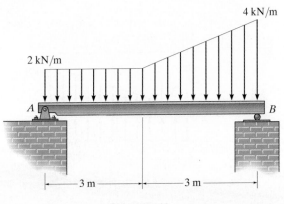

Prob. 3–112

3–113. Replace the distributed loading by an equivalent resultant force and specify where its line of action intersects a horizontal line along member AB, measured from A.

3–114. Replace the distributed loading by an equivalent resultant force and specify where its line of action intersects a vertical line along member BC, measured from C.

*3–116. Determine the equivalent resultant force and couple moment at point O.

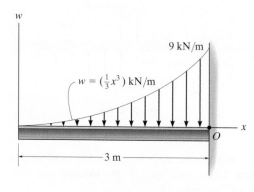

Prob. 3–116

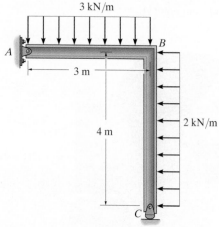

Probs. 3–113/114

3–117. Determine the magnitude of the equivalent resultant force and its location, measured from point O.

3–115. Determine the length b of the triangular load and its position a on the beam so that the equivalent resultant force is zero and the resultant couple moment is 8 kN · m clockwise.

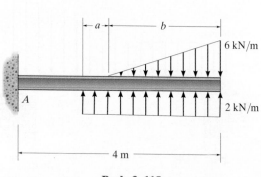

Prob. 3–115

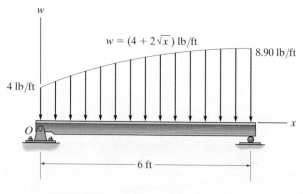

Prob. 3–117

CHAPTER REVIEW

Moment of Force — Scalar Definition

A force produces a turning effect or moment about a point O that does not lie on the force's line of action. In scalar form, the moment *magnitude* is the product of the force and the moment arm or perpendicular distance from point O to the line of action of the force.

$$M_O = Fd$$

The *direction* of the moment is defined using the right-hand rule. $\mathbf{M}_O$ always acts along an axis perpendicular to the plane containing $\mathbf{F}$ and d, and passes through the point O.

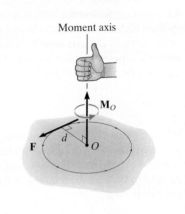

Principle of Moments

Rather than finding d, it is normally easier to resolve the force into its x and y components, determine the moment of each component about the point, and then sum the results. This is called the principle of moments.

$$M_O = Fd = F_x y - F_y x$$

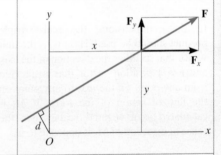

Moment of a Force — Vector Definition

Since three-dimensional geometry is generally more difficult to visualize, the vector cross product should be used to determine the moment. Here $\mathbf{M}_O = \mathbf{r} \times \mathbf{F}$, where $\mathbf{r}$ is a position vector that extends from point O to any point A, B, or C on the line of action of $\mathbf{F}$.

$$\mathbf{M}_O = \mathbf{r}_A \times \mathbf{F} = \mathbf{r}_B \times \mathbf{F} = \mathbf{r}_C \times \mathbf{F}$$

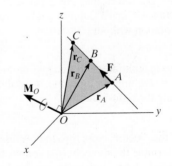

If the position vector $\mathbf{r}$ and force $\mathbf{F}$ are expressed as Cartesian vectors, then the cross product can be evaluated from the expansion of a determinant.

$$\mathbf{M}_O = \mathbf{r} \times \mathbf{F} = \begin{vmatrix} \mathbf{i} & \mathbf{j} & \mathbf{k} \\ r_x & r_y & r_z \\ F_x & F_y & F_z \end{vmatrix}$$

Moment about an Axis

If the moment of a force **F** is to be determined about a specific axis a, then for a scalar solution the moment arm, or shortest distance d_a from the line of action of the force to the axis must be used. This distance is perpendicular to both the axis and the line of action of the force.

$$M_a = F d_a$$

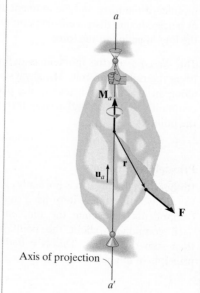

When the line of action of **F** intersects the axis then the moment of **F** about the axis is zero. Also, when the line of action of **F** is parallel to the axis, the moment of **F** about the axis is zero.

In three dimensions, the scalar triple product should be used. Here $\mathbf{u}_a$ is the unit vector that specifies the direction of the axis, and **r** is a position vector that is directed from any point on the axis to any point on the line of action of the force. If M_a is calculated as a negative scalar, then the sense of direction of $\mathbf{M}_a$ is opposite to $\mathbf{u}_a$.

$$M_a = \mathbf{u}_a \cdot (\mathbf{r} \times \mathbf{F}) = \begin{vmatrix} u_{a_x} & u_{a_y} & u_{a_z} \\ r_x & r_y & r_z \\ F_x & F_y & F_z \end{vmatrix}$$

Axis of projection

Couple Moment

A couple consists of two equal but opposite forces that act a perpendicular distance d apart. Couples tend to produce a rotation without translation.

The magnitude of the couple moment is $M = Fd$, and its direction is established using the right-hand rule.

$$M = Fd$$

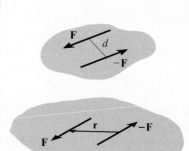

If the vector cross product is used to determine the moment of a couple, then **r** extends from any point on the line of action of one of the forces to any point on the line of action of the other force **F** that is used in the cross product.

$$\mathbf{M} = \mathbf{r} \times \mathbf{F}$$

Simplification of a Force and Couple System

Any system of forces and couples can be reduced to a single resultant force and resultant couple moment acting at a point O. The resultant force is the sum of all the forces in the system, $\mathbf{F}_R = \Sigma\mathbf{F}$, and the resultant couple moment is equal to the sum of the couple moments and the moments of all the forces about point O. $(\mathbf{M}_R)_O = \Sigma\mathbf{M} + \Sigma\mathbf{M}_O$.

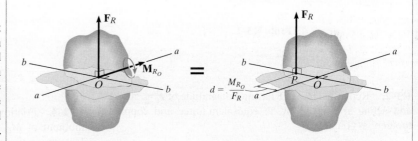

Further simplification to a single resultant force is possible, provided the force system is concurrent, coplanar, or parallel. To find the location of the resultant force from point O, it is necessary to equate the moment of the resultant force about the point to the moment of the forces and couples in the system about the same point.

Coplanar Distributed Loading

A simple distributed loading can be represented by its resultant force, which is equivalent to the *area* under the loading curve. This resultant has a line of action that passes through the *centroid* or the geometric center of the area under the loading diagram.

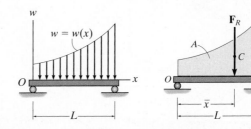

REVIEW PROBLEMS

R3–1. The boom has a length of 30 ft, a weight of 800 lb, and mass center at G. If the maximum moment that can be developed by a motor at A is $M = 20(10^3)$ lb·ft, determine the maximum load W, having a mass center at G', that can be lifted.

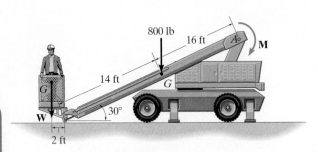

Prob. R3–1

R3–3. The hood of the automobile is supported by the strut AB, which exerts a force of $F = 24$ lb on the hood. Determine the moment of this force about the hinged axis y.

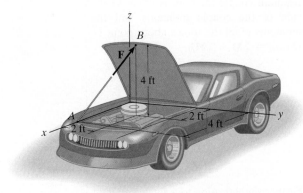

Prob. R3–3

R3–2. Replace the force $\mathbf{F}$ having a magnitude of $F = 50$ lb and acting at point A by an equivalent force and couple moment at point C.

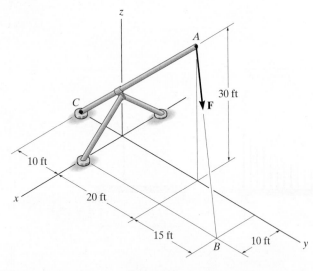

Prob. R3–2

*R3–4. Friction on the concrete surface creates a couple moment of $M_O = 100$ N·m on the blades of the trowel. Determine the magnitude of the couple forces so that the resultant couple moment on the trowel is zero. The forces lie in a horizontal plane and act perpendicular to the handle of the trowel.

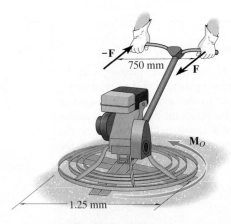

Prob. R3–4

R3–5. Replace the force and couple system by an equivalent force and couple moment at point P.

R3–7. The building slab is subjected to four parallel column loadings. Determine the equivalent resultant force and specify its location (x, y) on the slab. Take $F_1 = 30$ kN, $F_2 = 40$ kN.

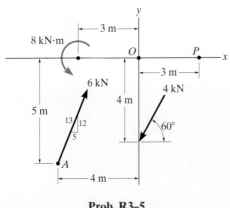

Prob. R3–5

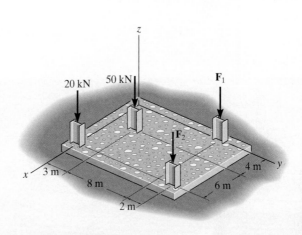

Prob. R3–7

R3–6. Replace the force system acting on the frame by a resultant force, and specify where its line of action intersects member AB, measured from point A.

*R3–8. Replace the distributed loading by an equivalent resultant force, and specify its location on the beam, measured from the pin at C.

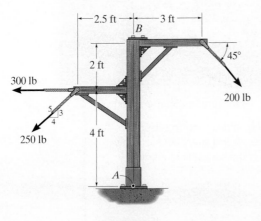

Prob. R3–6

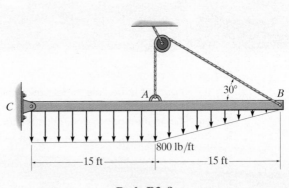

Prob. R3–8

CHAPTER 4

(© YuryZap/Shutterstock)

It is important to be able to determine the forces in the cables used to support this boom to ensure that it does not fail. In this chapter we will study how to apply equilibrium methods to determine the forces acting on the supports of a rigid body such as this.

EQUILIBRIUM OF A RIGID BODY

CHAPTER OBJECTIVES

- To develop the equations of equilibrium.
- To introduce the concept of the free-body diagram.
- To show how to solve rigid-body equilibrium problems in two and three dimensions.

4.1 CONDITIONS FOR RIGID-BODY EQUILIBRIUM

In this section, we will develop both the necessary and sufficient conditions for the equilibrium of the rigid body shown in Fig. 4–1a. This body is subjected to an external force and couple moment system that is the result of the effects of gravitational, electrical, magnetic, or contact forces caused by supports or adjacent bodies. The internal forces caused by interactions between particles within the body are not shown in this figure, because these forces occur in equal but opposite collinear pairs and hence will cancel out, a consequence of Newton's third law.

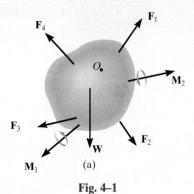

(a)

Fig. 4–1

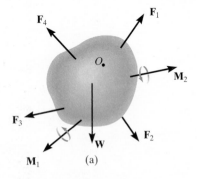

(a)

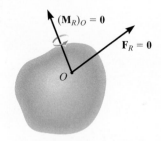

$(\mathbf{M}_R)_O = \mathbf{0}$

$\mathbf{F}_R = \mathbf{0}$

O

(b)

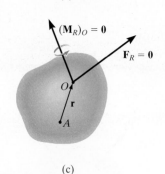

$(\mathbf{M}_R)_O = \mathbf{0}$

$\mathbf{F}_R = \mathbf{0}$

O

$\mathbf{r}$

A

(c)

Fig. 4–1 (cont.)

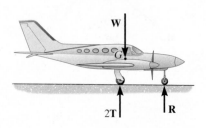

Fig. 4–2

Using the methods of the previous chapter, the force and couple moment system acting on a body can be reduced to an equivalent resultant force and resultant couple moment at any arbitrary point O on or off the body, Fig. 4–1b. If these two resultants are both equal to zero, then the body is said to be in *equilibrium*, which means it is at rest or will move with constant velocity. Mathematically, the equilibrium of a body is expressed as

$$\mathbf{F}_R = \Sigma\mathbf{F} = \mathbf{0}$$
$$(\mathbf{M}_R)_O = \Sigma\mathbf{M}_O = \mathbf{0} \qquad (4\text{–}1)$$

The first of these equations states that the sum of the forces acting on the body is equal to *zero*. The second equation states that the sum of the moments of all the forces in the system about point O, added to all the couple moments, is equal to *zero*. These two equations are not only necessary for equilibrium, they are also sufficient. To show this, consider summing moments about some other point, such as point A in Fig. 4–1c. We require

$$\Sigma\mathbf{M}_A = \mathbf{r}\times\mathbf{F}_R + (\mathbf{M}_R)_O = \mathbf{0}$$

Since $\mathbf{r}\neq\mathbf{0}$, this equation is satisfied if Eqs. 4–1 are satisfied, namely $\mathbf{F}_R = \mathbf{0}$ and $(\mathbf{M}_R)_O = \mathbf{0}$.

When applying the equations of equilibrium, we will assume that the body remains rigid. In reality, all bodies deform when subjected to loads; however, most engineering materials such as steel and concrete are very stiff and so their deformation is usually very small. Therefore, when applying the equations of equilibrium, we can generally assume that the body will remain *rigid* and *not deform* under the applied load without introducing any significant error. This way the direction of the applied forces and their moment arms with respect to a fixed reference remain the same both before and after a load is applied.

EQUILIBRIUM IN TWO DIMENSIONS

In the first part of the chapter, we will consider the case where the force system acting on a rigid body lies in or may be projected onto a *single* plane and, furthermore, any couple moments acting on the body are directed perpendicular to this plane. This type of force and couple system is often referred to as a two-dimensional or *coplanar* force system. For example, the airplane in Fig. 4–2 has a plane of symmetry through its center axis, and so the loads acting on the airplane are symmetrical with respect to this plane. Thus, each of the two wing tires will support the same load $\mathbf{T}$, which is represented on the side (two-dimensional) view of the plane as $2\mathbf{T}$.

4.2   FREE-BODY DIAGRAMS

Successful application of the equations of equilibrium, which will be discussed in Sec. 4.3, requires a complete specification of *all* the known and unknown external forces that act *on* the body. The best way to account for these forces is to draw a ***free-body diagram*** of the body. This diagram is a sketch of the outlined shape of the body, which represents it as being *isolated* or "free" from its surroundings, i.e., a "free body." On this sketch it is necessary to show *all* the forces and couple moments that the supports and the surroundings exert *on the body*, so that these effects can be accounted for when the equations of equilibrium are applied. *A thorough understanding of how to draw a free-body diagram is of primary importance for solving problems in both statics and mechanics of materials.*

Support Reactions. Before presenting a formal procedure as to how to draw a free-body diagram, we will first consider the various types of reactions that occur at supports and at points of contact between bodies subjected to coplanar force systems. As a general rule,

- A support prevents the translation of a body by exerting a force on the body.

- A support prevents the rotation of a body by exerting a couple moment on the body.

For example, let us consider three ways in which a horizontal member, such as a beam, is supported at its end. One method consists of a *roller* or cylinder, Fig. 4–3a. Since this support only prevents the beam from *translating* in the vertical direction, the roller will only exert a *force* on the beam in this direction, Fig. 4–3b.

The beam can be supported in a more restrictive manner by using a *pin*, Fig. 4–3c. The pin passes through a hole in the beam and two leaves which are fixed to the ground. Here the pin can prevent *translation* of the beam in *any direction* ϕ, Fig. 4–3d, and so the pin must exert a *force* $\mathbf{F}$ on the beam in the opposite direction. For purposes of analysis, it is generally easier to represent this resultant force $\mathbf{F}$ by its two rectangular components $\mathbf{F}_x$ and $\mathbf{F}_y$, Fig. 4–3e. Once F_x and F_y are known, then F and ϕ can be calculated.

The most restrictive way to support the beam would be to use a *fixed support* as shown in Fig. 4–3f. This support will prevent both *translation and rotation* of the beam. As a result, a *force and couple moment* must be developed on the beam at its point of connection, Fig. 4–3g. Like the case of the pin, the force is usually represented by its rectangular components $\mathbf{F}_x$ and $\mathbf{F}_y$.

Table 4–1 lists other common types of supports for bodies subjected to coplanar force systems. (In all cases the angle θ is assumed to be known.) Carefully study each of the symbols used to represent these supports and the types of reactions they exert on their contacting members.

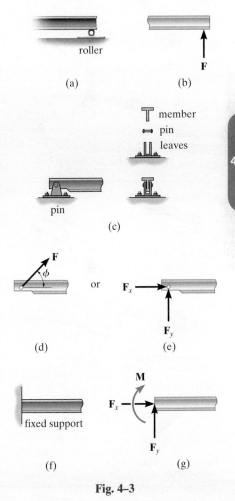

Fig. 4–3

TABLE 4–1 Supports for Rigid Bodies Subjected to Two-Dimensional Force Systems

Types of Connection	Reaction	Number of Unknowns
(1) θ cable	θ **F**	One unknown. The reaction is a tension force which acts away from the member in the direction of the cable.
(2) θ weightless link	θ **F** or θ **F**	One unknown. The reaction is a force which acts along the axis of the link.
(3) θ roller	θ **F**	One unknown. The reaction is a force which acts perpendicular to the surface at the point of contact.
(4) θ rocker	θ **F**	One unknown. The reaction is a force which acts perpendicular to the surface at the point of contact.
(5) θ smooth contacting surface	θ **F**	One unknown. The reaction is a force which acts perpendicular to the surface at the point of contact.
(6) θ roller or pin in confined smooth slot	**F** θ or **F** θ	One unknown. The reaction is a force which acts perpendicular to the slot.
(7) θ member pin connected to collar on smooth rod	θ or θ **F**	One unknown. The reaction is a force which acts perpendicular to the rod.

continued

TABLE 4–1 Continued

Types of Connection	Reaction	Number of Unknowns
(8) smooth pin or hinge	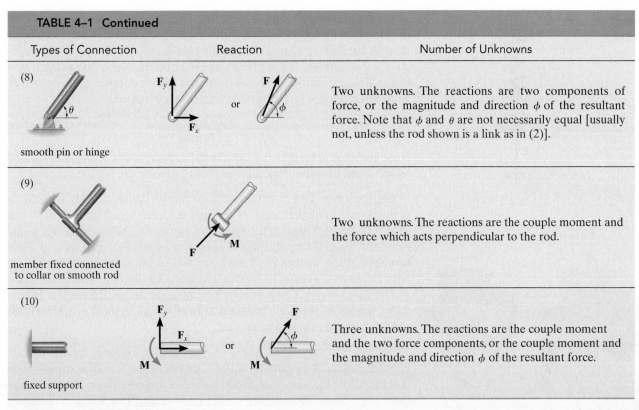	Two unknowns. The reactions are two components of force, or the magnitude and direction ϕ of the resultant force. Note that ϕ and θ are not necessarily equal [usually not, unless the rod shown is a link as in (2)].
(9) member fixed connected to collar on smooth rod		Two unknowns. The reactions are the couple moment and the force which acts perpendicular to the rod.
(10) fixed support		Three unknowns. The reactions are the couple moment and the two force components, or the couple moment and the magnitude and direction ϕ of the resultant force.

Typical examples of actual supports are shown in the following sequence of photos. The numbers refer to the connection types in Table 4–1.

The cable exerts a force on the bracket in the direction of the cable. (1)

The rocker support for this bridge girder allows horizontal movement so the bridge is free to expand and contract due to a change in temperature. (4)

This concrete girder rests on the ledge that is assumed to act as a smooth contacting surface. (5)

Typical pin support for a beam. (8)

The floor beams of this building are welded together and thus form fixed connections. (10)

Fig. 4–4

Springs.

If a *linear elastic spring* as in Fig. 4–4 is used to support a body, the length of the spring will change in direct proportion to the force acting on it. A characteristic that defines the "elasticity" of a spring is the *spring constant* or *stiffness k*. Specifically, the magnitude of force developed by a linear elastic spring which has a stiffness k, and is deformed (elongated or compressed) a distance s measured from its unloaded position, is

$$F = ks \qquad (4\text{–}2)$$

Note that s is determined from the difference in the spring's deformed length l and its undeformed length l_0, i.e., $s = l - l_0$.

Weight and the Center of Gravity.

When a body is within a gravitational field, then each of its particles has a specified weight. It was shown in Sec. 3.8 that such a system of forces can be reduced to a single resultant force acting through a specified point. We refer to this force resultant as the *weight* **W** of the body and to the location of its point of application as the *center of gravity*. The methods used for its determination will be developed in Chapter 6. In the examples and problems that follow, if the weight of the body is important for the analysis, this force will be reported in the problem statement.

Internal Forces.

As stated in Sec. 4.1, the internal forces that act between adjacent particles in a body always occur in collinear pairs such that they have the same magnitude and act in opposite directions (Newton's third law). Since these forces cancel each other, they will not create an *external effect* on the body. It is for this reason that the internal forces should *not* be included on the free-body diagram if the entire body is to be considered. For example, the engine shown in Fig. 4–5a has a free-body diagram shown in Fig. 4–5b. The internal forces between all its connected parts, such as the screws and bolts, will cancel out. Only the external forces **T₁** and **T₂** exerted by the chains and the engine weight **W** are shown on the free-body diagram.

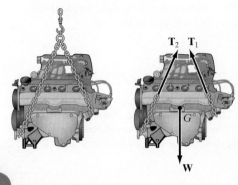

(a)   (b)

Fig. 4–5

Idealized Models.

When an engineer performs a force analysis of any object, he or she must consider a corresponding analytical or idealized *model* that gives results that approximate as closely as possible the actual situation. To do this, careful choices have to be made so that selection of the type of supports, the material behavior, and the object's dimensions can be justified. This way one can feel confident that any design or analysis will yield results which can be trusted. In complex cases this process may require developing several different models of the object that must be analyzed. However, in any case, this selection process requires both skill and experience.

The following two cases illustrate what is required to develop a proper model. In Fig. 4–6a, the steel beam is to be used to support the three roof joists of a building. For a force analysis it is reasonable to assume the material (steel) is rigid since only very small deflections will occur when the beam is loaded. A bolted connection at A will allow for any slight rotation that occurs here when the load is applied, and so a *pin* can be considered for this support. At B a *roller* can be considered since this support offers no resistance to horizontal movement. A building code is used to specify the roof loading so that the joist loads **F** can be calculated. These forces are intented to be larger than any actual loading on the beam since they account for extreme loading cases and for any dynamic or vibrational effects. Finally, the weight of the beam is generally neglected when it is small compared to the load the beam supports. The idealized model of the beam is therefore shown with average dimensions a, b, c, and d in Fig. 4–6b.

As a second case, consider the lift boom in Fig. 4–7a. By inspection, it is supported by a pin at A and by the hydraulic cylinder BC, which can be approximated as a weightless link. The material can be assumed rigid, and with its density known, the weight of the boom and the location of its center of gravity G are determined. When a design loading **P** is specified, the idealized model shown in Fig. 4–7b can be used for a force analysis. Average dimensions (not shown) are used to specify the location of the loads and the supports.

Idealized models of specific objects will be given in some of the examples throughout the text. In all cases, it should be realized that each represents the reduction of a practical situation using simplifying assumptions like the ones illustrated here.

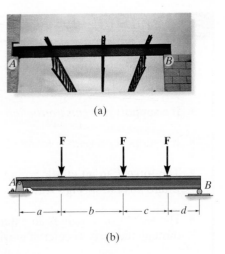

(a)

(b)

Fig. 4–6

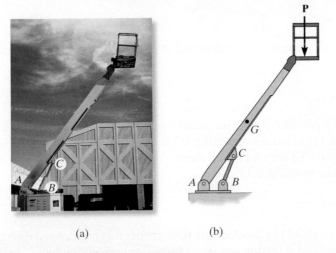

(a)   (b)

Fig. 4–7

IMPORTANT POINTS

- No equilibrium problem should be solved without *first drawing the free-body diagram*, so as to account for all the forces and couple moments that act on the body.

- If a support *prevents translation* of a body, then the support exerts a *force* on the body.

- If a support *prevents rotation* of a body, then the support exerts a *couple moment* on the body.

- The force F in an elastic spring is related to the extension or compression of the spring using $F = ks$, where k is the spring's stiffness.

- The weight of a body is an external force, and its effect is represented by a single resultant force acting through the body's center of gravity G.

- Internal forces are never shown on the free-body diagram since they occur in equal but opposite collinear pairs and therefore cancel out.

- *Couple moments* can be placed anywhere on the free-body diagram since they are *free vectors*. Forces can act at any point along their lines of action since they are *sliding vectors*.

PROCEDURE FOR ANALYSIS

To construct a free-body diagram for a rigid body or any group of bodies considered as a single system, the following steps should be performed:

Draw Outlined Shape.

Imagine the body to be *isolated* or cut "free" from its constraints and connections and draw (sketch) its outlined shape.

Show All Forces and Couple Moments.

Identify all the known and unknown *external forces* and couple moments that *act on the body*. Those generally encountered are due to (1) applied loadings, (2) reactions occurring at the supports or at points of contact with other bodies, and (3) the weight of the body. To account for all these effects, it may help to trace over the boundary, carefully noting each force or couple moment acting on it.

Identify Each Loading and Give Dimensions.

The forces and couple moments that are known should be labeled with their proper magnitudes and directions. Letters are used to represent the magnitudes and direction angles of forces and couple moments that are unknown. Finally, indicate the dimensions of the body necessary for calculating the moments of forces.

EXAMPLE 4.1

The sphere in Fig. 4–8a has a mass of 6 kg and is supported as shown. Draw a free-body diagram of the sphere, the cord CE, and the knot at C.

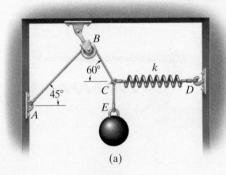

(a)

$\mathbf{F}_{CE}$ (Force of cord CE acting on sphere)

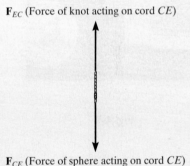

58.9 N (Weight or gravity acting on sphere)

(b)

SOLUTION

Sphere. By inspection, there are only two forces acting on the sphere, namely, its weight, 6 kg (9.81 m/s²) = 58.9 N, and the force of cord CE. The free-body diagram is shown in Fig. 4–8b.

Cord CE. When the cord CE is isolated from its surroundings, its free-body diagram shows only two forces acting on it, namely, the force of the sphere and the force of the knot, Fig. 4–8c. Notice that $\mathbf{F}_{CE}$ shown here is equal but opposite to that shown in Fig. 4–8b, a consequence of Newton's third law of action–reaction. Also, $\mathbf{F}_{CE}$ and $\mathbf{F}_{EC}$ pull on the cord and keep it in tension so that it doesn't collapse. For equilibrium, $F_{CE} = F_{EC}$.

$\mathbf{F}_{EC}$ (Force of knot acting on cord CE)

$\mathbf{F}_{CE}$ (Force of sphere acting on cord CE)

(c)

Knot. The knot at C is subjected to three forces, Fig. 4–8d. They are caused by the cords CBA and CE and the spring CD. As required, the free-body diagram shows all these forces labeled with their magnitudes and directions. It is important to recognize that the weight of the sphere does not directly act on the knot. Instead, the cord CE subjects the knot to this force.

$\mathbf{F}_{CBA}$ (Force of cord CBA acting on knot)

60°

C

$\mathbf{F}_{CD}$ (Force of spring acting on knot)

$\mathbf{F}_{CE}$ (Force of cord CE acting on knot)

(d)

Fig. 4–8

EXAMPLE 4.2

Draw the free-body diagram of the foot lever shown in Fig. 4–9a. The operator applies a vertical force to the pedal so that the spring is stretched 1.5 in. and the force in the short link at B is 20 lb.

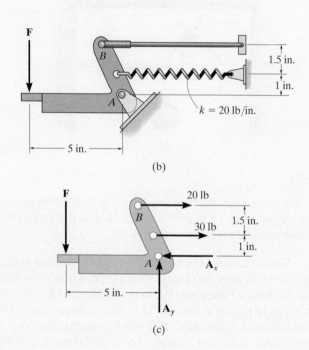

(b)

(a)

Fig. 4–9

(c)

4

SOLUTION

By inspection of the photo, the lever is loosely bolted to the frame at A. The rod at B is pinned at its ends and acts as a "short link." After making the proper measurements, the idealized model of the lever is shown in Fig. 4–9b. From this, the free-body diagram is shown in Fig. 4–9c. The pin support at A exerts force components A_x and A_y on the lever. The link at B exerts a force of 20 lb, acting in the direction of the link. In addition the spring also exerts a horizontal force on the lever. If the stiffness is measured and found to be $k = 20$ lb/in., then since the stretch $s = 1.5$ in., using Eq. 4–2, $F_s = ks = 20$ lb/in. (1.5 in.) $= 30$ lb. Finally, the operator's shoe applies a vertical force of **F** on the pedal. The dimensions of the lever are also shown on the free-body diagram, since this information will be useful when calculating the moments of the forces. As usual, the senses of the unknown forces at A have been assumed. The correct senses will become apparent after solving the equilibrium equations.

EXAMPLE 4.3

Two smooth pipes, each having a mass of 300 kg, are supported by the forked tines of the tractor in Fig. 4–10a. Draw the free-body diagrams for each pipe and both pipes together.

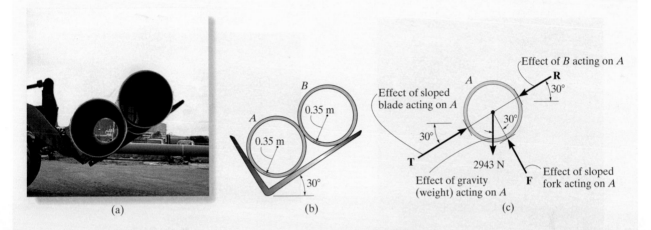

(a)

(b)

(c)

SOLUTION

The idealized model from which we must draw the free-body diagrams is shown in Fig. 4–10b. Here the pipes are identified, the dimensions have been added, and the physical situation is reduced to its simplest form.

The free-body diagram of pipe A is shown in Fig. 4–10c. Its weight is $W = 300(9.81) \text{ N} = 2943 \text{ N}$. Assuming all contacting surfaces are *smooth*, the reactive forces $\mathbf{T}, \mathbf{F}, \mathbf{R}$ act in a direction *normal* to the tangent at their surfaces of contact.

The free-body diagram of pipe B is shown in Fig. 4–10d. Can you identify each of the three forces acting *on this pipe*? Note that $\mathbf{R}$ representing the force of A on B, Fig. 4–10d, is equal and opposite to $\mathbf{R}$ representing the force of B on A, Fig. 4–10c.

The free-body diagram of both pipes combined ("system") is shown in Fig. 4–10e. Here the contact force $\mathbf{R}$, which acts between A and B, is considered an *internal* force and hence is not shown on the free-body diagram. That is, it represents a pair of equal but opposite collinear forces which cancel each other.

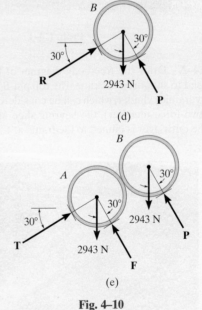

(d)

(e)

Fig. 4–10

CONCEPTUAL PROBLEMS

C4–1. Draw the free-body diagram of the uniform trash bucket which has a significant weight. It is pinned at A and rests against the smooth horizontal member at B. Show your result in side view. Label any necessary dimensions.

Prob. C4–1

C4–2. Draw the free-body diagram of the outrigger ABC used to support a backhoe. The top pin B is connected to the hydraulic cylinder, which can be considered to be a short link (two-force member), the bearing shoe at A is smooth, and the outrigger is pinned to the frame at C.

Prob. C4–2

C4–3. Draw the free-body diagram of the wing on the passenger plane. The weights of the engine and wing are significant. The tires at B are smooth.

Prob. C4–3

C4–4. Draw the free-body diagram of the wheel and member ABC used as part of the landing gear on a jet plane. The hydraulic cylinder AD acts as a two-force member, and there is a pin connection at B.

Prob. C4–4

4.3   EQUATIONS OF EQUILIBRIUM

In Sec. 4.1 we developed the two equations which are both necessary and sufficient for the equilibrium of a rigid body, namely, $\Sigma \mathbf{F} = \mathbf{0}$ and $\Sigma \mathbf{M}_O = \mathbf{0}$. When the body is subjected to a system of forces, which all lie in the x–y plane, then the forces can be resolved into their x and y components. The conditions for equilibrium in two dimensions then become

$$\Sigma F_x = 0$$
$$\Sigma F_y = 0 \tag{4–3}$$
$$\Sigma M_O = 0$$

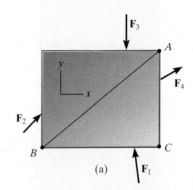

(a)

Here ΣF_x and ΣF_y represent, respectively, the algebraic sums of the x and y components of all the forces acting on the body, and ΣM_O represents the algebraic sum of the couple moments and the moments of all the force components about the z axis that pass through the arbitrary point O.

Alternative Sets of Equilibrium Equations. Although Eqs. 4–3 are *most often* used for solving coplanar equilibrium problems, two *alternative* sets of three independent equilibrium equations may also be used. One such set is

$$\Sigma F_x = 0$$
$$\Sigma M_A = 0 \tag{4–4}$$
$$\Sigma M_B = 0$$

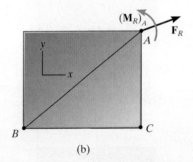

(b)

When using these equations, it is required that the line passing through points A and B not be parallel to the y axis. To show that these equations provide the *conditions* for equilibrium, consider the free-body diagram of the plate in Fig. 4–11a. Using the methods of Sec. 3.7, all the forces on the free-body diagram are first replaced by an equivalent resultant force $\mathbf{F}_R = \Sigma \mathbf{F}$, and a resultant couple moment $\left(\mathbf{M}_R\right)_A = \Sigma \mathbf{M}_A$, Fig. 4–11$b$. If $\Sigma M_A = 0$ is satisfied, then $\left(\mathbf{M}_R\right)_A = \mathbf{0}$. If $\Sigma F_x = 0$ is satisfied, then $\mathbf{F}_R$ must have *no component* along the x axis, and therefore $\mathbf{F}_R$ must be parallel to the y axis, Fig. 4–11c. Finally, if $\Sigma M_B = 0$, where B does not lie on the line of action of $\mathbf{F}_R$, then $\mathbf{F}_R = \mathbf{0}$ and therefore the body in Fig. 4–11a must be in equilibrium.

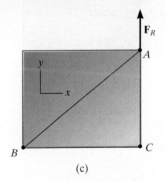

(c)

Fig. 4–11

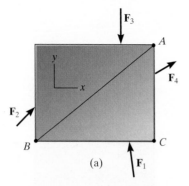

(a)

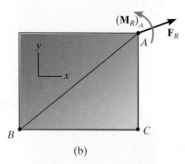

(b)

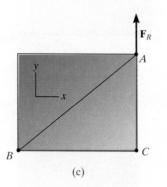

(c)

Fig. 4–11 (Repeated)

A second alternative set of equilibrium equations is

$$\Sigma M_A = 0$$
$$\Sigma M_B = 0 \qquad\qquad (4\text{–}5)$$
$$\Sigma M_C = 0$$

Here it is necessary that points A, B, and C do not lie on the same line. To show that these equations, when satisfied, ensure equilibrium, consider again the free-body diagram in Fig. 4–11b. If $\Sigma M_A = 0$ is to be satisfied, then $(\mathbf{M}_R)_A = \mathbf{0}$. If $\Sigma M_C = 0$ is satisfied, then the line of action of $\mathbf{F}_R$ passes through point C, Fig. 4–11c. Finally, if $\Sigma M_B = 0$ is satisfied, then $\mathbf{F}_R = \mathbf{0}$, and so the plate in Fig. 4–11a must be in equilibrium.

▶ PROCEDURE FOR ANALYSIS

Coplanar force equilibrium problems can be solved using the following procedure.

Free-Body Diagram.

- Establish the x, y coordinate axes in any suitable orientation.

- Draw an outlined shape of the body.

- Show all the forces and couple moments acting on the body.

- Label all the loadings and specify their directions relative to the x or y axis. The sense of a force or couple moment having an *unknown* magnitude but known line of action can be *assumed*.

- Indicate the dimensions of the body necessary for calculating the moments of forces.

Equations of Equilibrium.

- Apply the moment equation of equilibrium, $\Sigma M_O = 0$, about a point O that lies at the intersection of the lines of action of two unknown forces. In this way, the moments of these unknowns are zero about O, and a *direct solution* for the third unknown can be determined.

- When applying the force equilibrium equations, $\Sigma F_x = 0$ and $\Sigma F_y = 0$, orient the x and y axes along lines that will provide the simplest resolution of the forces into their x and y components.

- If the solution of the equilibrium equations yields a negative scalar for a force or couple moment magnitude, this indicates that the sense is opposite to that which was assumed on the free-body diagram.

EXAMPLE 4.4

Determine the tension in cables BA and BC necessary to support the 60-kg cylinder in Fig. 4–12a.

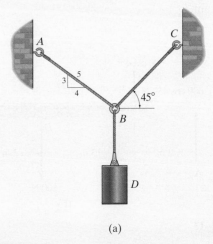

(a)

$T_{BD} = 60\ (9.81)$ N

$60\ (9.81)$ N

(b)

SOLUTION

Free-Body Diagram. Due to equilibrium, the weight of the cylinder causes the tension in cable BD to be $T_{BD} = 60(9.81)$ N, Fig. 4–12b. The forces in cables BA and BC can be determined by investigating the equilibrium of ring B. Its free-body diagram is shown in Fig. 4–12c. The magnitudes of $\mathbf{T}_A$ and $\mathbf{T}_C$ are unknown, but their directions are known.

Equations of Equilibrium. Applying the equations of equilibrium along the x and y axes, we have

$$\xrightarrow{+} \Sigma F_x = 0; \qquad T_C \cos 45^\circ - \left(\tfrac{4}{5}\right) T_A = 0 \qquad (1)$$

$$+\uparrow \Sigma F_y = 0; \qquad T_C \sin 45^\circ + \left(\tfrac{3}{5}\right) T_A - 60(9.81)\ \text{N} = 0 \qquad (2)$$

Equation (1) can be written as $T_A = 0.8839 T_C$. Substituting this into Eq. (2) yields

$$T_C \sin 45^\circ + \left(\tfrac{3}{5}\right)(0.8839 T_C) - 60(9.81)\text{N} = 0$$

So that

$$T_C = 475.66\ \text{N} = 476\ \text{N} \qquad \qquad Ans.$$

Substituting this result into either Eq. (1) or Eq. (2), we get

$$T_A = 420\ \text{N} \qquad \qquad Ans.$$

NOTE: The accuracy of these results, of course, depends on the accuracy of the data, i.e., measurements of geometry and loads. For most engineering work involving a problem such as this, the data as measured to three significant figures would be sufficient.

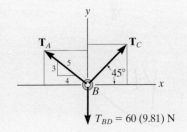

(c)

Fig. 4–12

EXAMPLE 4.5

Determine the horizontal and vertical components of reaction on the beam caused by the pin at B and the rocker at A as shown in Fig. 4–13a. Neglect the weight of the beam.

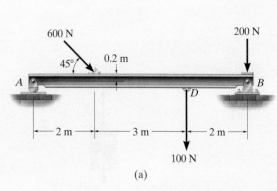

(a)

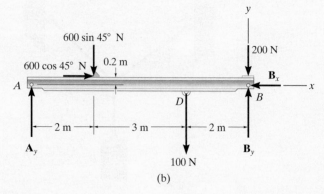

(b)

Fig. 4–13

SOLUTION

Free-Body Diagram. Identify each of the forces shown on the free-body diagram in Fig. 4–13b. Here the 600-N force is represented by its x and y components.

Equations of Equilibrium. Summing forces in the x direction yields

$$\xrightarrow{+} \Sigma F_x = 0; \qquad 600 \cos 45° \text{ N} - B_x = 0$$

$$B_x = 424 \text{ N} \qquad\qquad Ans.$$

A direct solution for $\mathbf{A}_y$ can be obtained by applying the moment equation about point B.

$$\zeta + \Sigma M_B = 0; \qquad 100 \text{ N} (2 \text{ m}) + (600 \sin 45° \text{ N})(5 \text{ m})$$

$$- (600 \cos 45° \text{ N})(0.2 \text{ m}) - A_y(7 \text{ m}) = 0$$

$$A_y = 319 \text{ N} \qquad\qquad Ans.$$

Summing forces in the y direction, using this result, gives

$$+\uparrow \Sigma F_y = 0; \qquad 319 \text{ N} - 600 \sin 45° \text{ N} - 100 \text{ N} - 200 \text{ N} + B_y = 0$$

$$B_y = 405 \text{ N} \qquad\qquad Ans$$

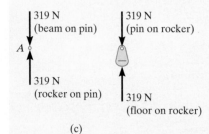

(c)

NOTE: The support forces in Fig. 4–13b are caused by the pins that *act on the beam.* The opposite forces act on the pins. For example, Fig. 4–13c shows the equilibrium of the pin at A and the rocker.

EXAMPLE   4.6

The cord shown in Fig. 4–14a supports a force of 100 lb and wraps over the frictionless pulley. Determine the tension in the cord at C and the horizontal and vertical components of reaction at pin A.

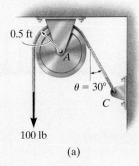

0.5 ft

A

$\theta = 30°$

C

100 lb

(a)

Fig. 4–14

SOLUTION

Free-Body Diagrams.   The free-body diagrams of the cord and pulley are shown in Fig. 4–14b. Note that the principle of action, equal but opposite reaction must be carefully observed when drawing each of these diagrams: the cord exerts an unknown load distribution p on the pulley at the contact surface, whereas the pulley exerts an equal but opposite effect on the cord. For the solution, however, it is simpler to *combine* the free-body diagrams of the pulley and this portion of the cord, so that the distributed load becomes *internal* to this "system" and is therefore eliminated from the analysis, Fig. 4–14c.

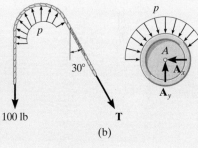

(b)

Equations of Equilibrium.   Summing moments about point A to eliminate $\mathbf{A}_x$ and $\mathbf{A}_y$, Fig. 4–14c, we have

$\zeta + \Sigma M_A = 0;$ $100 \text{ lb } (0.5 \text{ ft}) - T(0.5 \text{ ft}) = 0$

$$T = 100 \text{ lb} \qquad\qquad Ans.$$

Using this result,

$\xrightarrow{+} \Sigma F_x = 0;$ $-A_x + 100 \sin 30° \text{ lb} = 0$

$$A_x = 50.0 \text{ lb} \qquad\qquad Ans.$$

$+\uparrow \Sigma F_y = 0;$ $A_y - 100 \text{ lb} - 100 \cos 30° \text{ lb} = 0$

$$A_y = 187 \text{ lb} \qquad\qquad Ans.$$

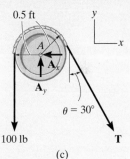

(c)

NOTE: It is seen that the tension in the cord remains *constant* as the cord passes over the pulley. (This of course is true for *any angle* θ at which the cord is directed and for *any radius r* of the pulley.)

EXAMPLE 4.7

The member shown in Fig. 4–15a is pin connected at A and rests against a smooth support at B. Determine the horizontal and vertical components of reaction at the pin A.

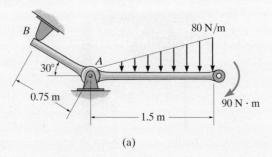

(a)

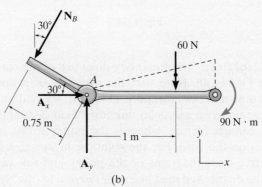

(b)

Fig. 4–15

SOLUTION

Free-Body Diagram. As shown on the free-body diagram, Fig. 4–15b, the reaction N_B must be perpendicular to the member at B. Also, horizontal and vertical components of reaction are represented at A.

Equations of Equilibrium. Summing moments about A, we obtain a direct solution for N_B,

$$\zeta + \Sigma M_A = 0; \quad -90 \text{ N} \cdot \text{m} - 60 \text{ N}(1 \text{ m}) + N_B(0.75 \text{ m}) = 0$$

$$N_B = 200 \text{ N}$$

Using this result,

$$\xrightarrow{+} \Sigma F_x = 0; \qquad A_x - 200 \sin 30° \text{ N} = 0$$

$$A_x = 100 \text{ N} \qquad\qquad Ans.$$

$$+\uparrow \Sigma F_y = 0; \qquad A_y - 200 \cos 30° \text{ N} - 60 \text{ N} = 0$$

$$A_y = 233 \text{ N} \qquad\qquad Ans.$$

4.4 TWO- AND THREE-FORCE MEMBERS

The solutions to some equilibrium problems can be simplified by recognizing members that are subjected to only two or three forces.

Two-Force Members. As the name implies, a *two-force member* has forces applied at only two points on the member. An example of a two-force member is shown in Fig. 4–16a. To satisfy force equilibrium, $\mathbf{F}_A$ and $\mathbf{F}_B$ must be equal in magnitude, $F_A = F_B = F$, but opposite in direction ($\Sigma\mathbf{F} = \mathbf{0}$), Fig. 4–16b. Furthermore, moment equilibrium requires that $\mathbf{F}_A$ and $\mathbf{F}_B$ share the same line of action, which can only happen if they are directed along the line joining points A and B ($\Sigma M_A = 0$ or $\Sigma M_B = 0$), Fig. 4–16c. Therefore, for any two-force member to be in equilibrium, the two forces acting on the member *must have the same magnitude, act in opposite directions, and have the same line of action, directed along the line joining the two points where these forces act.*

The hydraulic cylinder AB is a typical example of a two-force member since it is pin connected at its ends and, provided its weight is neglected, only the resultant pin forces act on this member.

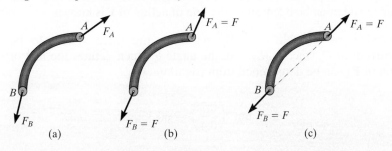

Two-force member

Fig. 4–16

The boom-and-bucket on this lift is a three-force member, provided its weight is neglected. Here the lines of action of the weight of the worker, $\mathbf{W}$, and the force of the two-force member (hydraulic cylinder) at B, $\mathbf{F}_B$, intersect at O. For moment equilibrium, the resultant force at the pin A, $\mathbf{F}_A$, must also be directed towards O.

Three-Force Members. If a member is subjected to only *three forces*, it is called a *three-force member*. Moment equilibrium can be satisfied only if the three forces form a *concurrent* or *parallel* force system. To illustrate, consider the member in Fig. 4–17a subjected to the three forces $\mathbf{F}_1$, $\mathbf{F}_2$, and $\mathbf{F}_3$. If the lines of action of $\mathbf{F}_1$ and $\mathbf{F}_2$ intersect at point O, then the line of action of $\mathbf{F}_3$ must *also* pass through point O so that the forces satisfy $\Sigma M_O = 0$. As a special case, if the three forces are all parallel, Fig. 4–17b, the location of the point of intersection, O, will approach infinity.

The link used for this railroad car brake is a three-force member. Since the force $\mathbf{F}_B$ in the tie rod at B and $\mathbf{F}_C$ from the link at C are parallel, then for equilibrium the resultant force $\mathbf{F}_A$ at the pin A must also be parallel with these two forces.

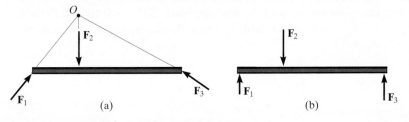

Three-force member

Fig. 4–17

EXAMPLE 4.8

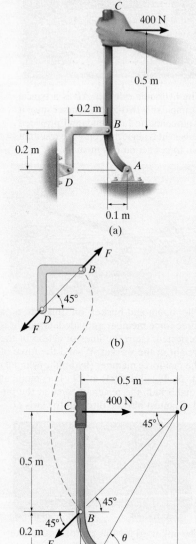

The lever *ABC* is pin supported at *A* and connected to a short link *BD* as shown in Fig. 4–18a. If the weight of the members is negligible, determine the force of the pin on the lever at *A*.

SOLUTION

Free-Body Diagrams. As shown in Fig. 4–18b, the short link *BD* is a *two-force member*, so the *resultant forces* from the pins *D* and *B* must be equal, opposite, and collinear. Although the magnitude of the force is unknown, the line of action is known since it passes through *B* and *D*.

Lever *ABC* is a *three-force member*, and therefore, in order to satisfy moment equilibrium, the three nonparallel forces acting on it must be concurrent at *O*, Fig. 4–18c. Note that the force **F** on the lever at *B* is equal but opposite to the force **F** acting at *B* on the link. Why? The distance *CO* must be 0.5 m since the line of action of **F** is known.

Equations of Equilibrium. By requiring the force system to be concurrent at *O*, since $\Sigma M_O = 0$, the angle θ which defines the line of action of $\mathbf{F}_A$ can be determined from trigonometry,

$$\theta = \tan^{-1}\left(\frac{0.7}{0.4}\right) = 60.3°$$

Using the *x*, *y* axes and applying the force equilibrium equations,

$$\xrightarrow{+} \Sigma F_x = 0; \qquad F_A \cos 60.3° - F \cos 45° + 400 \text{ N} = 0$$

$$+\uparrow \Sigma F_y = 0; \qquad F_A \sin 60.3° - F \sin 45° = 0$$

Solving, we get

$$F_A = 1.07 \text{ kN} \qquad\qquad\qquad Ans.$$

$$F = 1.32 \text{ kN}$$

NOTE: We can also solve this problem by representing the force at *A* by its two components $\mathbf{A}_x$ and $\mathbf{A}_y$ and applying $\Sigma M_A = 0$ to get *F*, then $\Sigma F_x = 0$, $\Sigma F_y = 0$ to get A_x and A_y. Once A_x and A_y are determined, we can get F_A and θ.

Fig. 4–18

PRELIMINARY PROBLEM

P4–1.   Draw the free-body diagram of each object.

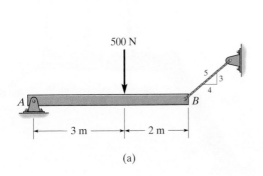

(a)

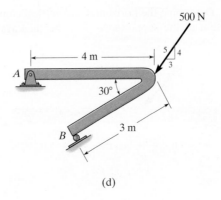

(d)

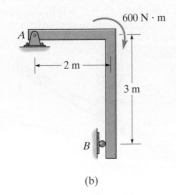

(b)

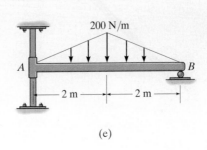

(e)

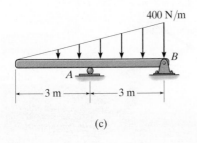

(c)

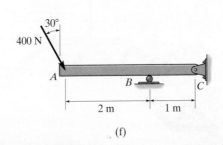

(f)

Prob. P4–1

FUNDAMENTAL PROBLEMS

All problem solutions must include an FBD.

F4–1.   Determine the horizontal and vertical components of reaction at the supports. Neglect the thickness of the beam.

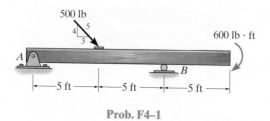

Prob. F4–1

F4–2.   Determine the horizontal and vertical components of reaction at the pin A and the reaction on the beam at C.

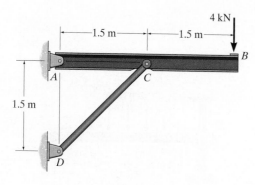

Prob. F4–2

F4–3.   The truss is supported by a pin at A and a roller at B. Determine the support reactions.

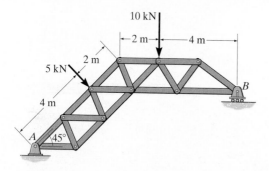

Prob. F4–3

F4–4.   Determine the components of reaction at the fixed support A. Neglect the thickness of the beam.

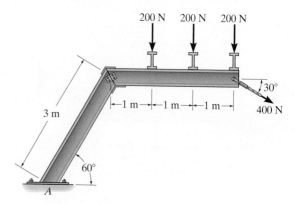

Prob. F4–4

F4–5.   The 25-kg bar has a center of mass at G. If it is supported by a smooth peg at C, a roller at A, and cord AB, determine the reactions at these supports.

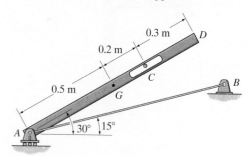

Prob. F4–5

F4–6.   Determine the reactions at the smooth contact points A, B, and C on the bar.

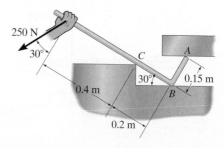

Prob. F4–6

PROBLEMS

All problem solutions must include an FBD.

4–1. Determine the components of the support reactions at the fixed support *A* on the cantilevered beam.

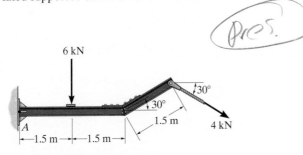

Prob. 4–1

4–2. Determine the reactions at the supports.

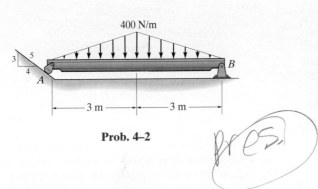

Prob. 4–2

4–3. Determine the horizontal and vertical components of reaction of the pin *A* and the reaction of the rocker *B* on the beam.

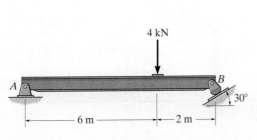

Prob. 4–3

*4–4. Determine the reactions at the supports.

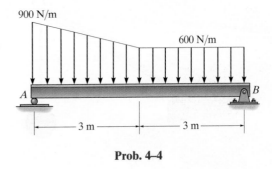

Prob. 4–4

4–5. Determine the reactions at the supports.

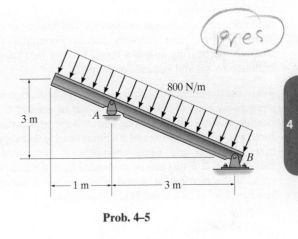

Prob. 4–5

4–6. Determine the reactions at the supports.

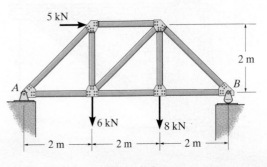

Prob. 4–6

4–7. Determine the magnitude of force at the pin A and in the cable BC needed to support the 500-lb load. Neglect the weight of the boom AB.

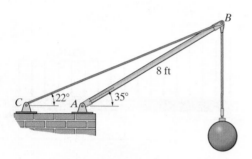

Prob. 4–7

4–8. The dimensions of a jib crane are given in the figure. If the crane has a mass of 800 kg and a center of mass at G, and the maximum rated force at its end is F = 15 kN, determine the reactions at its bearings. The bearing at A is a journal bearing and supports only a horizontal force, whereas the bearing at B is a thrust bearing that supports both horizontal and vertical components.

4–9. The dimensions of a jib crane are given in the figure. The crane has a mass of 800 kg and a center of mass at G. The bearing at A is a journal bearing and can support a horizontal force, whereas the bearing at B is a thrust bearing that supports both horizontal and vertical components. Determine the maximum load F that can be suspended from the end of the crane if the bearings at A and B can sustain a maximum resultant load of 24 kN and 34 kN, respectively.

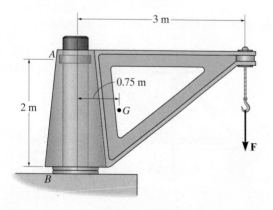

Probs. 4–8/9

4–10. The smooth pipe rests against the opening at the points of contact A, B, and C. Determine the reactions at these points needed to support the force of 300 N. Neglect the pipe's thickness.

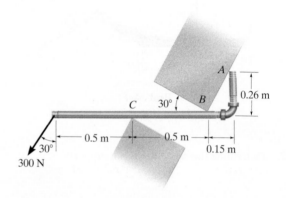

Prob. 4–10

4–11. The beam is horizontal and the springs are unstretched when there is no load on the beam. Determine the angle of tilt of the beam when the load is applied.

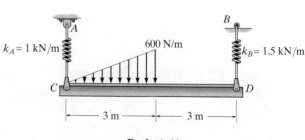

Prob. 4–11

*4–12. The 10-kg uniform rod is pinned at end A. If it is subjected to a couple moment of 50 N · m, determine the smallest angle θ for equilibrium. The spring is unstretched when $\theta = 0°$, and has a stiffness of $k = 60$ N/m.

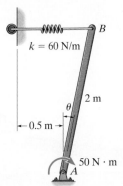

k = 60 N/m

B

2 m

θ

0.5 m

50 N · m

A

Prob. 4–12

4–13. The man uses the hand truck to move material up the step. If the truck and its contents have a mass of 50 kg with center of gravity at G, determine the normal reaction on both wheels and the magnitude and direction of the minimum force required at the grip B needed to lift the load.

0.4 m

B

0.5 m

0.2 m

G

0.4 m

60°

0.4 m

A

0.1 m

Prob. 4–13

4–14. Three uniform books, each having a weight W and length a, are stacked as shown. Determine the maximum distance d that the top book can extend out from the bottom one so the stack does not topple over.

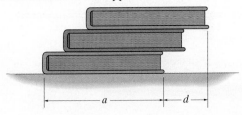

a

d

Prob. 4–14

4–15. Determine the reactions at the pin A and the tension in cord BC. Set $F = 40$ kN. Neglect the thickness of the beam.

*4–16. If rope BC will fail when the tension becomes 50 kN, determine the greatest vertical load F that can be applied to the beam at B. What is the magnitude of the reaction at A for this loading? Neglect the thickness of the beam.

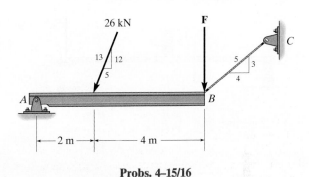

26 kN

F

13 / 12

5

A

B

5 3

4

2 m

4 m

C

Probs. 4–15/16

4–17. The rigid metal strip of negligible weight is used as part of an electromagnetic switch. If the stiffness of the springs at A and B is $k = 5$ N/m and the strip is originally horizontal when the springs are unstretched, determine the smallest force F needed to close the contact gap at C.

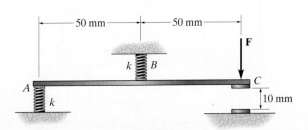

50 mm

50 mm

F

k

B

A

k

C

10 mm

Prob. 4–17

4–18. The rigid metal strip of negligible weight is used as part of an electromagnetic switch. Determine the maximum stiffness k of the springs at A and B so that the contact at C closes when the vertical force developed there is $F = 0.5$ N. Originally the strip is horizontal as shown.

*4–20. The uniform beam has a weight W and length l and is supported by a pin at A and a cable BC. Determine the horizontal and vertical components of reaction at A and the tension in the cable necessary to hold the beam in the position shown.

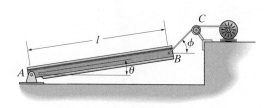

Prob. 4–20

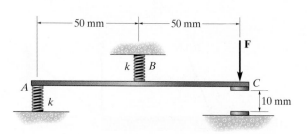

Prob. 4–18

4–21. A boy stands out at the end of the diving board, which is supported by two springs A and B, each having a stiffness of $k = 15$ kN/m. In the position shown the board is horizontal. If the boy has a mass of 40 kg, determine the angle of tilt which the board makes with the horizontal after he jumps off. Neglect the weight of the board and assume it is rigid.

4–19. The cantilever footing is used to support a wall near its edge A so that it causes a uniform soil pressure under the footing. Determine the uniform distribution loads, w_A and w_B, measured in lb/ft at pads A and B, necessary to support the wall forces of 8000 lb and 20 000 lb.

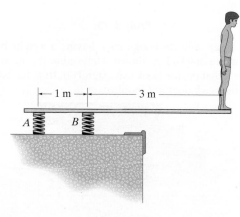

Prob. 4–21

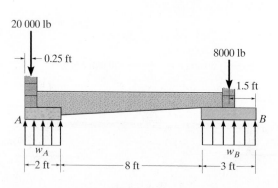

Prob. 4–19

4–22. The beam is subjected to the two concentrated loads. Assuming that the foundation exerts a linearly varying load distribution on its bottom, determine the load intensities w_1 and w_2 for equilibrium in terms of the parameters shown.

*4–24. Determine the distance d for placement of the load **P** for equilibrium of the smooth bar when it is held in the position θ as shown. Neglect the weight of the bar.

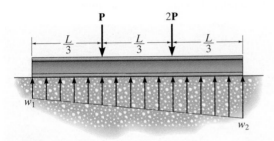

Prob. 4–22

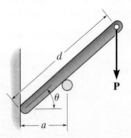

Prob. 4–24

4–23. The rod supports a weight of 200 lb and is pinned at its end A. If it is subjected to a couple moment of 100 lb·ft, determine the angle θ for equilibrium. The spring has an unstretched length of 2 ft and a stiffness of $k = 50$ lb/ft.

4–25. If $d = 1$ m, and $\theta = 30°$, determine the normal reaction at the smooth supports and the required distance a for the placement of the roller if $P = 600$ N. Neglect the weight of the bar.

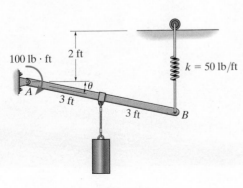

Prob. 4–23

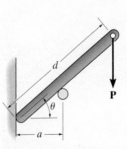

Prob. 4–25

CONCEPTUAL PROBLEMS

C4–5. The tie rod is used to support this overhang at the entrance of a building. If it is pin connected to the building wall at A and to the center of the overhang B, determine if the force in the rod will increase, decrease, or remain the same if (a) the support at A is moved to a lower position D, and (b) the support at B is moved to the outer position C. Explain your answer with an equilibrium analysis, using dimensions and loads. Assume the overhang is pin supported from the building wall.

Prob. C4–5

C4–6. The man attempts to pull the four wheeler up the incline and onto the trailer. From the position shown, is it more effective to pull on the rope at A, or would it be better to pull on the rope at B? Draw a free-body diagram for each case, and do an equilibrium analysis to explain your answer. Use appropriate numerical values to do your calculations.

Prob. C4–6

C4–7. Like all aircraft, this jet plane rests on three wheels. Why not use an additional wheel at the tail for better support? (Can you think of any other reason for not including this wheel?) If there was a fourth tail wheel, draw a free-body diagram of the plane from a side (2 D) view, and show why one would not be able to determine all the wheel reactions using the equations of equilibrium.

Prob. C4–7

C4–8. Where is the best place to arrange most of the logs in the wheelbarrow so that it minimizes the amount of force on the backbone of the person transporting the load? Do an equilibrium analysis to explain your answer.

Prob. C4–8

EQUILIBRIUM IN THREE DIMENSIONS

4.5 FREE-BODY DIAGRAMS

The first step in solving three-dimensional equilibrium problems, as in the case of two dimensions, is to draw a free-body diagram. Before we can do this, however, it is first necessary to discuss the types of reactions that can occur at the supports.

Support Reactions. The reactive forces and couple moments acting at various types of supports and connections, when the members are viewed in three dimensions, are listed in Table 4–2. It is important to recognize the symbols used to represent each of these supports and to understand clearly how the forces and couple moments are developed. As in the two-dimensional case:

- A support prevents the translation of a body by exerting a force on the body.

- A support prevents the rotation of a body by exerting a couple moment on the body.

For example, in Table 4–2, item (4), the ball-and-socket joint prevents any translation of the connecting member; therefore, a force must act on the member at the point of connection. This force has three components having unknown magnitudes, F_x, F_y, F_z. Provided these components are known, one can obtain the magnitude of force, $F = \sqrt{F_x^2 + F_y^2 + F_z^2}$, and the force's orientation defined by its coordinate direction angles α, β, γ, Eqs. 2–5.* Since the connecting member is allowed to rotate freely about *any* axis, no couple moment is resisted by a ball-and-socket joint.

Notice that the *single* bearing supports in items (5) and (7), the *single* pin (8), and the *single* hinge (9) are shown to resist both force *and* couple-moment components. If, however, these supports are used with *other* bearings, pins, or hinges to hold a rigid body in equilibrium and these supports are *properly aligned* when connected to the body, then the *force reactions* at these supports *alone* are adequate for supporting the body. In other words, the couple moments will not develop since the body is *prevented* from rotating by the other supports. The reason for this should become clear after studying the examples which follow.

---

* The three unknowns may also be represented as an unknown force magnitude F and two unknown coordinate direction angles. The third direction angle is obtained using the identity $\cos^2 \alpha + \cos^2 \beta + \cos^2 \gamma = 1$, Eq. 2–8.

TABLE 4–2 Supports for Rigid Bodies Subjected to Three-Dimensional Force Systems

Types of Connection	Reaction	Number of Unknowns
(1) cable	$\mathbf{F}$	One unknown. The reaction is a force which acts away from the member in the known direction of the cable.
(2) smooth surface support	$\mathbf{F}$	One unknown. The reaction is a force which acts perpendicular to the surface at the point of contact.
(3) roller	$\mathbf{F}$	One unknown. The reaction is a force which acts perpendicular to the surface at the point of contact.
(4) ball and socket	$\mathbf{F}_z$ $\mathbf{F}_y$ $\mathbf{F}_x$	Three unknowns. The reactions are three rectangular force components.
(5) single journal bearing	$\mathbf{M}_z$ $\mathbf{F}_z$ $\mathbf{M}_x$ $\mathbf{F}_x$	Four unknowns. The reactions are two force and two couple-moment components which act perpendicular to the shaft. Note: The couple moments are *generally not applied* if the body is supported elsewhere. See the examples.

continued

TABLE 4–2 Continued

Types of Connection	Reaction	Number of Unknowns
(6) single journal bearing with square shaft	M_z F_z M_y M_x F_x	Five unknowns. The reactions are two force and three couple-moment components. *Note*: The couple moments *are generally not applied* if the body is supported elsewhere. See the examples.
(7) single thrust bearing	M_z F_y F_z M_x F_x	Five unknowns. The reactions are three force and two couple-moment components. *Note*: The couple moments *are generally not applied* if the body is supported elsewhere. See the examples.
(8) single smooth pin	M_z F_z F_y M_y F_x	Five unknowns. The reactions are three force and two couple-moment components. *Note*: The couple moments *are generally not applied* if the body is supported elsewhere. See the examples.
(9) single hinge	M_z F_z F_y F_x M_x	Five unknowns. The reactions are three force and two couple-moment components. *Note*: The couple moments *are generally not applied* if the body is supported elsewhere. See the examples.
(10) fixed support	M_z F_z F_x F_y M_y M_x	Six unknowns. The reactions are three force and three couple-moment components.

Typical examples of actual supports that are referenced to Table 4–2 are shown in the following sequence of photos.

The journal bearings support the ends of the shaft. (5)

This ball-and-socket joint provides a connection for a member of an earth grader to its frame. (4)

This thrust bearing is used to support the drive shaft on a machine. (7)

This pin is used to support the end of the strut used on a tractor. (8)

Free-Body Diagrams. The general procedure for establishing the free-body diagram of a rigid body has been outlined in Sec. 4.2. Essentially it requires first "isolating" the body by drawing its outlined shape. This is followed by a careful *labeling* of *all* the forces and couple moments with reference to an established x, y, z coordinate system. As a general rule, show the unknown components of reaction as acting on the free-body diagram in the *positive sense*. In this way, if any negative values are obtained, they will indicate that the components act in the negative coordinate directions.

EXAMPLE 4.9

Consider the two rods and plate, along with their associated free-body diagrams, shown in Fig. 4–19. The x, y, z axes are established on each diagram and the unknown reaction components are indicated in the *positive sense*. The weight is neglected.

SOLUTION

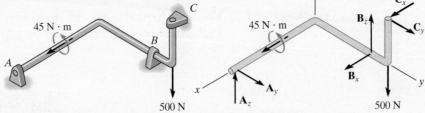

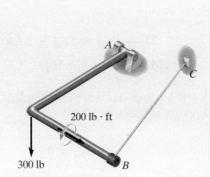

Properly aligned journal bearings at A, B, C.

The force reactions developed by the bearings are *sufficient* for equilibrium since they prevent the shaft from rotating about each of the coordinate axes. No couple moments at each bearing are developed.

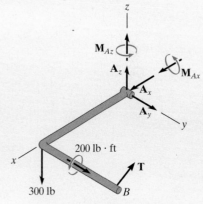

Pin at A and cable BC.

Moment components are developed by the pin on the rod to prevent rotation about the x and z axes.

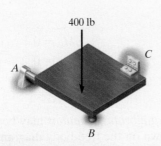

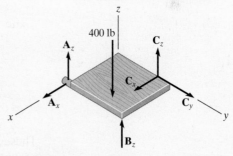

Properly aligned journal bearing at A and hinge at C. Roller at B.

Only force reactions are developed by the bearing and hinge on the plate to prevent rotation about each coordinate axis. No moments are developed at the hinge or bearing.

Fig. 4–19

4.6 EQUATIONS OF EQUILIBRIUM

As stated in Sec. 4.1, the conditions for equilibrium of a rigid body subjected to a three-dimensional force system require that both the *resultant* force and *resultant* couple moment acting on the body be equal to *zero*.

Vector Equations of Equilibrium. The two conditions for equilibrium of a rigid body may be expressed mathematically in vector form as

$$\Sigma \mathbf{F} = \mathbf{0}$$
$$\Sigma \mathbf{M}_O = \mathbf{0} \qquad\qquad (4\text{--}6)$$

where $\Sigma \mathbf{F}$ is the vector sum of all the external forces acting on the body, and $\Sigma \mathbf{M}_O$ is the sum of the couple moments and the moments of all the forces about any point O located either on or off the body.

Scalar Equations of Equilibrium. If all the external forces and couple moments are expressed in Cartesian vector form and substituted into Eqs. 4–6, we have

$$\Sigma \mathbf{F} = \Sigma F_x \mathbf{i} + \Sigma F_y \mathbf{j} + \Sigma F_z \mathbf{k} = \mathbf{0}$$
$$\Sigma \mathbf{M}_O = \Sigma M_x \mathbf{i} + \Sigma M_y \mathbf{j} + \Sigma M_z \mathbf{k} = \mathbf{0}$$

Since the $\mathbf{i}, \mathbf{j},$ and $\mathbf{k}$ components are independent from one another, then these equations are satisfied provided

$$\Sigma F_x = 0$$
$$\Sigma F_y = 0$$
$$\Sigma F_z = 0 \qquad\qquad (4\text{--}7a)$$

and

$$\Sigma M_x = 0$$
$$\Sigma M_y = 0$$
$$\Sigma M_z = 0 \qquad\qquad (4\text{--}7b)$$

These *six scalar equilibrium equations* may be used to solve for at most six unknowns shown on the free-body diagram. Equations 4–7a require the sum of the external force components acting in the x, y, z directions to be zero, and Eqs. 4–7b require the sum of the moment components about the x, y, z axes to be zero.

IMPORTANT POINTS

- Always draw the free-body diagram first when solving any equilibrium problem.

- If a support *prevents translation* of a body, then the support exerts a *force* on the body.

- If a support *prevents rotation* of a body, then the support exerts a *couple moment* on the body.

PROCEDURE FOR ANALYSIS

Three-dimensional equilibrium problems for a rigid body can be solved using the following procedure.

Free-Body Diagram.

- Draw an outlined shape of the body.
- Show all the forces and couple moments acting on the body.
- Establish the origin of the x, y, z axes at a convenient point and orient the axes so that they are parallel to as many of the external forces and moments as possible.
- Label all the loadings and specify their directions. In general, show all the *unknown* components having a *positive sense* along the x, y, z axes.
- Indicate the dimensions of the body necessary for calculating the moments of forces.

Equations of Equilibrium.

- If the x, y, z force and moment components seem easy to determine, then apply the six scalar equations of equilibrium; otherwise use the vector equations.
- It is not necessary that the set of axes chosen for force summation coincide with the set of axes chosen for moment summation.
- Choose the direction of an axis for moment summation such that it intersects the lines of action of as many unknown forces as possible. Realize that the moments of forces passing through points on this axis, and the moments of forces which are parallel to the axis, will then be zero.
- If the solution of the equilibrium equations yields a negative scalar for a force or couple moment magnitude, it indicates that the sense is opposite to that assumed on the free-body diagram.

EXAMPLE 4.10

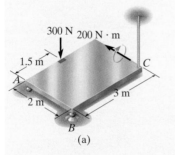

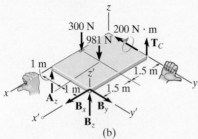

(a)

(b)

Fig. 4–20

The homogeneous plate shown in Fig. 4–20a has a mass of 100 kg and is subjected to a force and couple moment along its edges. If it is supported in the horizontal plane by a roller at A, a ball-and-socket joint at B, and a cord at C, determine the components of reaction at these supports.

SOLUTION (SCALAR ANALYSIS)

Free-Body Diagram.   There are five unknown reactions acting on the plate, as shown in Fig. 4–20b. Each of these reactions is assumed to act in a positive coordinate direction.

Equations of Equilibrium.   Since the three-dimensional geometry is rather simple, a *scalar analysis* provides a *direct solution* to this problem. A force summation along each axis yields

$$\Sigma F_x = 0; \qquad B_x = 0 \qquad\qquad\qquad\qquad\qquad\qquad\text{Ans.}$$
$$\Sigma F_y = 0; \qquad B_y = 0 \qquad\qquad\qquad\qquad\qquad\qquad\text{Ans.}$$
$$\Sigma F_z = 0; \qquad A_z + B_z + T_C - 300\text{ N} - 981\text{ N} = 0 \qquad (1)$$

Recall that the moment of a force about an axis is equal to the product of the force magnitude .and the perpendicular distance (moment arm) from the line of action of the force to the axis. Also, forces that are parallel to an axis or pass through it create no moment about the axis. Hence, summing moments about the positive x and y axes, we have

$$\Sigma M_x = 0; \qquad T_C(2\text{ m}) - 981\text{ N}(1\text{ m}) + B_z(2\text{ m}) = 0 \qquad (2)$$
$$\Sigma M_y = 0;$$

$$-300\text{ N}(1.5\text{ m}) + 981\text{ N}(1.5\text{ m}) - B_z(3\text{ m}) - A_z(3\text{ m}) - 200\text{ N}\cdot\text{m} = 0 \qquad (3)$$

The components of the force at B can be eliminated if moments are summed about the x' and y' axes. We obtain

$$\Sigma M_{x'} = 0; \qquad 981\text{ N}(1\text{ m}) + 300\text{ N}(2\text{ m}) - A_z(2\text{ m}) = 0 \qquad (4)$$
$$\Sigma M_{y'} = 0;$$

$$-300\text{ N}(1.5\text{ m}) - 981\text{ N}(1.5\text{ m}) - 200\text{ N}\cdot\text{m} + T_C(3\text{ m}) = 0 \qquad (5)$$

Solving Eqs. (1) through (3) or the more convenient Eqs. (1), (4), and (5) yields

$$A_z = 790\text{ N} \quad B_z = -217\text{ N} \quad T_C = 707\text{ N} \qquad\qquad\text{Ans.}$$

The negative sign indicates that $\mathbf{B}_z$ acts downward.

NOTE: The solution of this problem does not require a summation of moments about the z axis. The plate is partially constrained because the supports cannot prevent it from turning about the z axis if a force is applied to it in the x–y plane.

EXAMPLE   4.11

Determine the components of reaction that the ball-and-socket joint at A, the smooth journal bearing at B, and the roller support at C exert on the rod assembly in Fig. 4–21a.

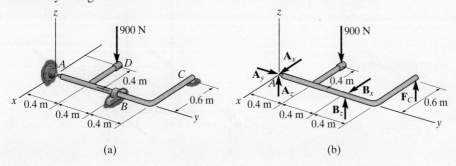

(a)   (b)

Fig. 4–21

SOLUTION

Free-Body Diagram.   As shown on the free-body diagram, Fig. 4–21b, the reactive forces of the supports will prevent the assembly from rotating about each coordinate axis, and so the journal bearing at B only exerts reactive forces on the member.

Equations of Equilibrium.   A direct solution for A_y can be obtained by summing forces along the y axis.

$$\Sigma F_y = 0; \qquad A_y = 0 \hspace{4cm} Ans.$$

The force F_C can be determined directly by summing moments about the y axis.

$$\Sigma M_y = 0; \qquad F_C(0.6 \text{ m}) - 900 \text{ N}(0.4 \text{ m}) = 0$$
$$F_C = 600 \text{ N} \hspace{4cm} Ans.$$

Using this result, B_z can be determined by summing moments about the x axis.

$$\Sigma M_x = 0; \qquad B_z(0.8 \text{ m}) + 600 \text{ N}(1.2 \text{ m}) - 900 \text{ N}(0.4 \text{ m}) = 0$$
$$B_z = -450 \text{ N} \hspace{4cm} Ans.$$

The negative sign indicates that $\mathbf{B}_z$ acts downward. The force $\mathbf{B}_x$ can be found by summing moments about the z axis.

$$\Sigma M_z = 0; \qquad -B_x(0.8 \text{ m}) = 0 \quad B_x = 0 \hspace{2.5cm} Ans.$$

Thus,

$$\Sigma F_x = 0; \qquad A_x + 0 = 0 \qquad A_x = 0 \hspace{3cm} Ans.$$

Finally, using the results of B_z and F_C,

$$\Sigma F_z = 0; \qquad A_z + (-450 \text{ N}) + 600 \text{ N} - 900 \text{ N} = 0$$
$$A_z = 750 \text{ N} \hspace{4cm} Ans.$$

PRELIMINARY PROBLEMS

P4–2.   Draw the free-body diagram of each object.

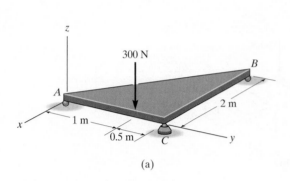

(a)

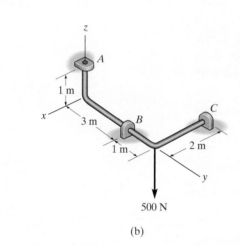

(b)

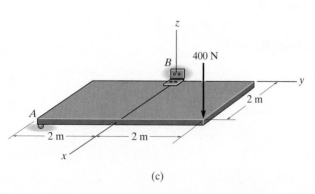

(c)

Prob. P4–2

P4–3.   In each case, write the moment equations about the x, y, and z axes.

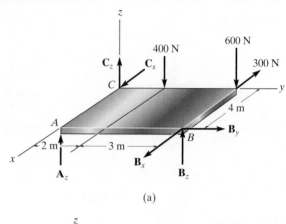

(a)

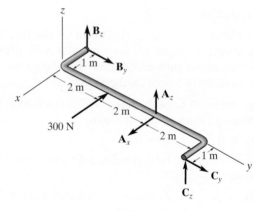

(b)

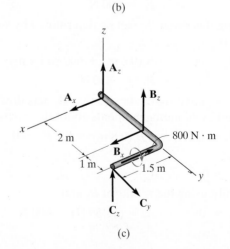

(c)

Prob. P4–3

FUNDAMENTAL PROBLEMS

All problem solutions must include an FBD.

F4–7. The uniform plate has a weight of 500 lb. Determine the tension in each of the supporting cables.

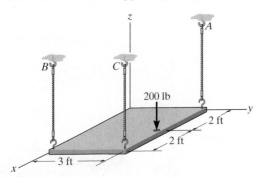

Prob. F4–7

F4–8. Determine the reactions at the roller support A, the ball-and-socket joint D, and the tension in cable BC for the plate.

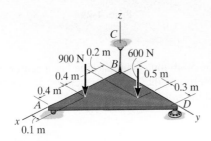

Prob. F4–8

F4–9. The rod is supported by smooth journal bearings at A, B, and C. Determine the reactions at these supports.

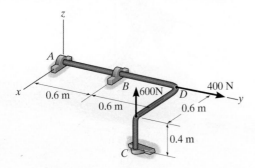

Prob. F4–9

F4–10. Determine the support reactions at the smooth journal bearings A, B, and C of the pipe assembly.

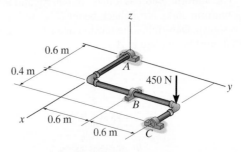

Prob. F4–10

F4–11. Determine the force developed in the short link BD, and the tension in the cords CE and CF, and the reactions of the ball-and-socket joint A on the block.

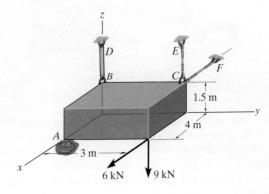

Prob. F4–11

F4–12. Determine the components of reaction that the thrust bearing A and cable BC exert on the bar.

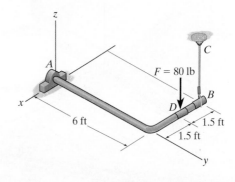

Prob. F4–12

PROBLEMS

All problem solutions must include an FBD.

4–26. The uniform load has a mass of 600 kg and is lifted using a uniform 30-kg strongback beam *BAC* and the four ropes as shown. Determine the tension in each rope and the force that must be applied at *A*.

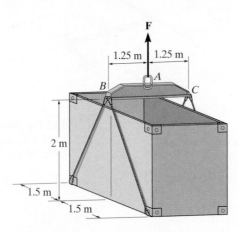

Prob. 4–26

4–27. Due to an unequal distribution of fuel in the wing tanks, the centers of gravity for the airplane fuselage *A* and wings *B* and *C* are located as shown. If these components have weights $W_A = 45\,000$ lb, $W_B = 8000$ lb, and $W_C = 6000$ lb, determine the normal reactions of the wheels *D*, *E*, and *F* on the ground.

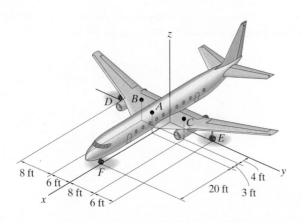

Prob. 4–27

*4–28.** Determine the components of reaction at the fixed support *A*. The 400 N, 500 N, and 600 N forces are parallel to the *x*, *y*, and *z* axes, respectively.

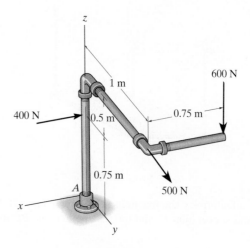

Prob. 4–28

4–29. The 50-lb mulching machine has a center of gravity at *G*. Determine the vertical reactions at the wheels *C* and *B* and the smooth contact point *A*.

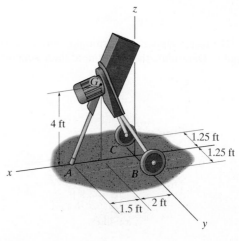

Prob. 4–29

4–30. The smooth uniform rod AB is supported by a ball-and-socket joint at A, the wall at B, and cable BC. Determine the components of reaction at A, the tension in the cable, and the normal reaction at B if the rod has a mass of 20 kg.

*4–32. The 100-lb door has its center of gravity at G. Determine the components of reaction at hinges A and B if hinge B resists only forces in the x and y directions and A resists forces in the x, y, z directions.

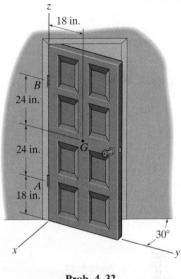

Prob. 4–32

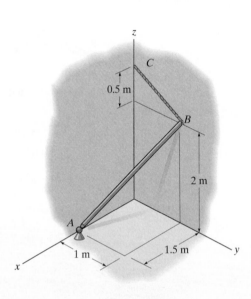

Prob. 4–30

4–33. Determine the tension in each cable and the components of reaction at D needed to support the load.

4–31. The uniform concrete slab has a mass of 2400 kg. Determine the tension in each of the three parallel supporting cables.

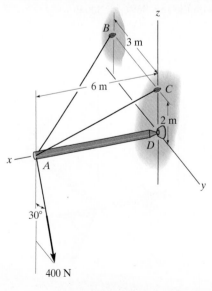

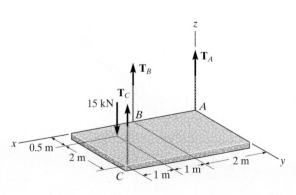

Prob. 4–31

Prob. 4–33

4–34. The bent rod is supported at A, B, and C by smooth journal bearings. Calculate the x, y, z components of reaction at the bearings if the rod is subjected to forces $F_1 = 300$ lb and $F_2 = 250$ lb. $\mathbf{F}_1$ lies in the y–z plane. The bearings are in proper alignment and exert only force reactions on the rod.

*4–36.** The bar AB is supported by two smooth collars. At A the connection is a ball-and-socket joint and at B it is a rigid attachment. If a 50-lb load is applied to the bar, determine the x, y, z components of reaction at A and B.

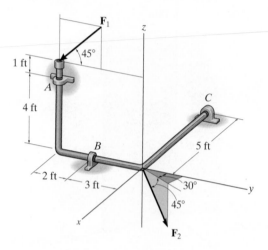

Prob. 4–34

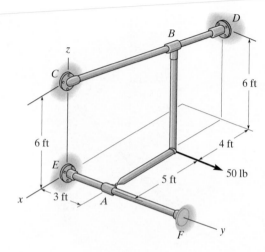

Prob. 4–36

4–35. The bent rod is supported at A, B, and C by smooth journal bearings. Determine the magnitude of $\mathbf{F}_2$ which will cause the reaction $\mathbf{C}_y$ at the bearing C to be equal to zero. The bearings are in proper alignment and exert only force reactions on the rod. Set $F_1 = 300$ lb.

4–37. The rod has a weight of 6 lb/ft. If it is supported by a ball-and-socket joint at C and a journal bearing at D, determine the x, y, z components of reaction at these supports and the moment M that must be applied along the axis of the rod to hold it in the position shown.

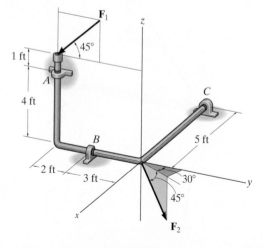

Prob. 4–35

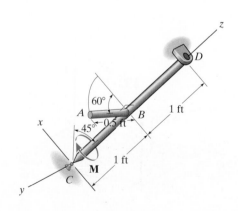

Prob. 4–37

4–38. The sign has a mass of 100 kg with center of mass at G. Determine the x, y, z components of reaction at the ball-and-socket joint A and the tension in wires BC and BD.

*4–40. Both pulleys are fixed to the shaft and as the shaft turns with constant angular velocity, the power of pulley A is transmitted to pulley B. Determine the horizontal tension $\mathbf{T}$ in the belt on pulley B and the x, y, z components of reaction at the journal bearing C and thrust bearing D if $\theta = 45°$. The bearings are in proper alignment and exert only force reactions on the shaft.

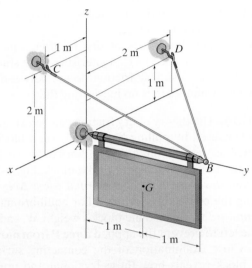

Prob. 4–38

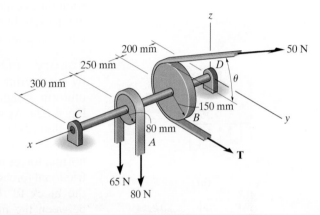

Prob. 4–40

4–39. Both pulleys are fixed to the shaft and as the shaft turns with constant angular velocity, the power of pulley A is transmitted to pulley B. Determine the horizontal tension $\mathbf{T}$ in the belt on pulley B and the x, y, z components of reaction at the journal bearing C and thrust bearing D if $\theta = 0°$. The bearings are in proper alignment and exert only force reactions on the shaft.

4–41. Member AB is supported by a cable BC and at A by a *square* rod which fits loosely through the square hole at the end collar of the member as shown. Determine the x, y, z components of reaction at A and the tension in the cable needed to hold the 800-lb cylinder in equilibrium.

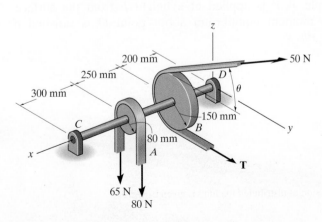

Prob. 4–39

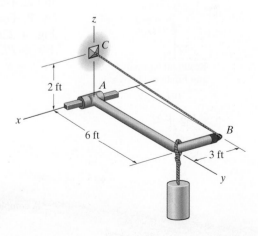

Prob. 4–41

4.7   CHARACTERISTICS OF DRY FRICTION

Friction is a force that resists the movement between two contacting surfaces that slide relative to one another. This force always acts *tangent* to the surface at the points of contact, and is directed so as to oppose the possible or existing motion between the surfaces.

In this section, we will study the effects of **dry friction**, which is sometimes called *Coulomb friction* since its characteristics were studied extensively by C.A. Coulomb in 1781. Dry friction occurs between the contacting surfaces of bodies when there is no lubricating fluid.

Theory of Dry Friction.

The theory of dry friction can be explained by considering the effects caused by pulling horizontally on a block of uniform weight **W** which is resting on a rough horizontal surface, Fig. 4–22a. As shown on the free-body diagram of the block, Fig. 4–22b, the floor exerts an uneven *distribution* of both *normal force* ΔN_n and *frictional force* ΔF_n along the contacting surface.* For equilibrium, the normal forces must act *upward* to balance the block's weight **W**, and the frictional forces act to the left to prevent the applied force **P** from moving the block to the right. Close examination of the contacting surfaces between the floor and block reveals how these frictional and normal forces develop, Fig. 4–22c. It can be seen that many microscopic irregularities exist between the two surfaces and, as a result, reactive forces ΔR_n are developed at each point of contact. As shown, each reactive force contributes both a frictional component ΔF_n and a normal component ΔN_n.

Equilibrium.

The effect of the *distributed* normal and frictional loadings is indicated by their *resultants* **N** and **F** on the free-body diagram shown in Fig. 4–22d. Notice that **N** acts a distance x to the right of the line of action of **W**, Fig. 4–22d. This location of the normal force distribution in Fig. 4–22b is necessary in order to balance the "tipping effect" caused by **P**. For example, if **P** is applied at a height h from the surface, Fig. 4–22d, then moment equilibrium about point O is satisfied if $Wx = Ph$ or $x = Ph/W$.

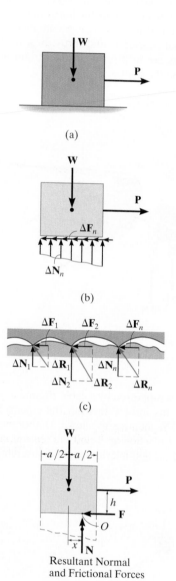

(a)

(b)

(c)

Resultant Normal
and Frictional Forces

(d)

Fig. 4–22

---

* A complete discussion of distributed loadings is given in Sec. 3.9.

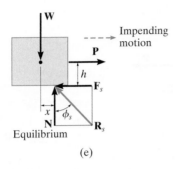

Fig. 4–22 (cont.)

Impending Motion. In cases where the surfaces of contact are rather "slippery," the frictional force **F** may *not* be great enough to balance **P**, and consequently the block will tend to slip. In other words, as P is slowly increased, F correspondingly increases until it attains a certain *maximum value* F_s, called the **limiting static frictional force**, Fig. 4–22e. When this value is reached, the block is in *unstable equilibrium* since any further increase in P will cause the block to move. Experimentally, it has been determined that F_s is *directly proportional* to the resultant normal force N. Expressed mathematically,

$$F_s = \mu_s N \qquad (4\text{–}8)$$

where the constant of proportionality, μ_s (mu "sub" s), is called the **coefficient of static friction**.

Thus, when the block is on the *verge of sliding*, the normal force **N** and frictional force $\mathbf{F}_s$ combine to create a resultant $\mathbf{R}_s$, Fig. 4–22e. The angle ϕ_s (phi) that $\mathbf{R}_s$ makes with **N** is called the **angle of static friction**. From the figure,

$$\phi_s = \tan^{-1}\left(\frac{F_s}{N}\right) = \tan^{-1}\left(\frac{\mu_s N}{N}\right) = \tan^{-1}\mu_s$$

Typical values for μ_s are given in Table 4–3. As indicated, these values will vary since experimental testing was done under variable conditions of roughness and cleanliness of the contacting surfaces. For applications, therefore, it is important that both caution and judgment be exercised when selecting a coefficient of friction for a given set of conditions. When a more accurate calculation of F_s is required, the coefficient of friction should be determined directly by an experiment that involves the two contacting materials.

Table 4–3 Typical Values for	
Contact Materials	Coefficient of Static Friction (μ_s)
Metal on ice	0.03–0.05
Wood on wood	0.30–0.70
Leather on wood	0.20–0.50
Leather on metal	0.30–0.60
Copper on copper	0.74–1.21

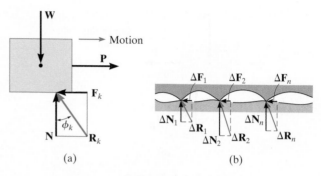

Fig. 4–23

Motion. If the magnitude of **P** acting on the block is increased so that it becomes slightly greater than F_s, the frictional force at the contacting surface will drop to a smaller value F_k, called the **kinetic frictional force**. The block will then begin to slide with increasing speed, Fig. 4–23a. As this occurs, the block will "ride" on top of the peaks at the points of contact, as shown in Fig. 4–23b. The continued breakdown of the nonrigid surfaces is the dominant mechanism creating kinetic friction.

Experiments with sliding blocks indicate that the magnitude of the kinetic friction force is directly proportional to the magnitude of the resultant normal force, expressed mathematically as

$$F_k = \mu_k N \qquad (4-9)$$

Here the constant of proportionality, μ_k, is called the **coefficient of kinetic friction**. Typical values for μ_k are approximately 25 percent *smaller* than those listed in Table 4–3 for μ_s.

As shown in Fig. 4–23a, in this case, the resultant force at the surface of contact, $\mathbf{R}_k$, has a line of action defined by ϕ_k. This angle is referred to as the **angle of kinetic friction**, where

$$\phi_k = \tan^{-1}\left(\frac{F_k}{N}\right) = \tan^{-1}\left(\frac{\mu_k N}{N}\right) = \tan^{-1} \mu_k$$

By comparison, $\phi_s \geq \phi_k$.

Characteristics of Dry Friction. As a result of *experiments* that pertain to the foregoing discussion, we can state the following rules which apply to bodies subjected to dry friction.

- The frictional force acts *tangent* to the contacting surfaces in a direction *opposed* to the *motion* or tendency for motion of one surface relative to another.

- The maximum static frictional force F_s that can be developed is independent of the area of contact between the surfaces, provided the normal pressure is not very low nor great enough to severely deform or crush the surfaces between the bodies.

- The maximum static frictional force is generally greater than the kinetic frictional force for any two surfaces of contact. However, if one of the bodies is moving with a *very low velocity* over the surface of another, F_k becomes approximately equal to F_s, i.e., $\mu_s \approx \mu_k$.

- When *slipping* at the surface of contact is *about to occur*, the maximum static frictional force is proportional to the normal force, such that $F_s = \mu_s N$.

- When *slipping* at the surface of contact is *occurring*, the kinetic frictional force is proportional to the normal force, such that $F_k = \mu_k N$.

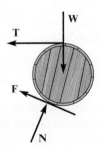

Some objects, such as this barrel, may not be on the verge of slipping, and therefore the friction force **F** must be determined strictly from the equations of equilibrium.

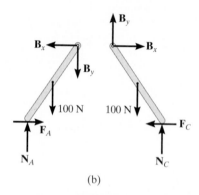

$\mu_A = 0.3$ $\mu_C = 0.5$

(a)

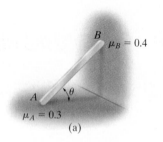

$\mu_B = 0.4$

$\mu_A = 0.3$

(a)

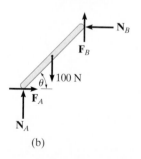

(b)

Fig. 4–25

B

$B_x \leftarrow$ $\uparrow B_y$ $\rightarrow B_x$

$\downarrow B_y$

100 N   100 N

$\leftarrow F_C$

$\uparrow F_A$ $\uparrow N_C$

N_A

(b)

Fig. 4–24

4.8   PROBLEMS INVOLVING DRY FRICTION

If a rigid body is in equilibrium when it is subjected to a system of forces that includes the effect of friction, the force system must satisfy not only the equations of equilibrium but *also* the laws that govern the frictional forces.

Types of Friction Problems. In general, there are three types of static problems involving dry friction. They can easily be classified once free-body diagrams are drawn and the total number of unknowns are identified and compared with the total number of available equilibrium equations.

No Apparent Impending Motion. Problems in this category are strictly equilibrium problems, which require the number of unknowns to be *equal* to the number of available equilibrium equations. Once the frictional forces are determined from the solution, however, their numerical values must be checked to be sure they satisfy the inequality $F \le \mu_s N$; otherwise, slipping will occur and the body will not remain in equilibrium. A problem of this type is shown in Fig. 4–24a. Here we must determine the frictional forces at A and C to check if the equilibrium position of the two-member frame can be maintained. If the members are uniform and have known weights of 100 N each, then the free-body diagrams are as shown in Fig. 4–24b. There are six unknown force components which can be determined *strictly* from the six equilibrium equations (three for each member). Once F_A, N_A, F_C, and N_C are determined, then the members will remain in equilibrium provided $F_A \le 0.3N_A$ and $F_C \le 0.5N_C$ are satisfied.

Impending Motion at All Points of Contact. In this case the total number of unknowns will *equal* the total number of available equilibrium equations *plus* the total number of available frictional equations, $F = \mu N$. When *motion is impending* at the points of contact, then $F_s = \mu_s N$; whereas if the body is *slipping*, then $F_k = \mu_k N$. For example, consider the problem of finding the smallest angle θ at which the 100-N bar in Fig. 4–25a can be placed against the wall without slipping. The free-body diagram of the bar is shown in Fig. 4–25b. Here the *five* unknowns are determined from the *three* equilibrium equations and *two* static frictional equations which apply at *both* points of contact, so that $F_A = 0.3N_A$ and $F_B = 0.4N_B$.

Impending Motion at Some Points of Contact. For these types of problems, the number of unknowns will be *less* than the number of available equilibrium equations plus the number of available frictional equations or conditional equations for tipping. As a result, several possibilities for motion or impending motion will exist and the problem will involve a determination of the kind of motion which actually occurs. For example, consider the two-member frame in Fig. 4–26a. In this problem we wish to determine the horizontal force P needed to cause movement. If each member has a weight of 100 N, then the free-body diagrams are as shown in Fig. 4–26b. There are *seven* unknowns. For a unique solution we must satisfy the *six* equilibrium equations (three for each member) and only *one* of two possible static frictional equations. This means that as P increases it will either cause slipping at A and no slipping at C, so that $F_A = 0.3N_A$ and $F_C \leq 0.5N_C$, or slipping will occur at C and no slipping at A, in which case $F_C = 0.5N_C$ and $F_A \leq 0.3N_A$. The actual situation can be determined by calculating P for each case, and then choosing the case for which P is *smaller*. If in both cases the *same value* for P is calculated, which would be highly improbable, then slipping at both points occurs simultaneously; i.e., the *seven unknowns* would have to satisfy *eight equations*.

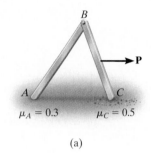

(a)

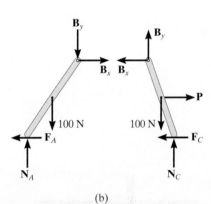

(b)

Fig. 4–26

Equilibrium Versus Frictional Equations. Whenever we solve a problem such as the one in Fig. 4-24, where the friction force F is to be an "equilibrium force" and satisfies the inequality $F < \mu_s N$, then we can assume the sense of direction of F on the free-body diagram. The correct sense is made known *after* solving the equations of equilibrium for F. If F is a negative scalar the sense of $\mathbf{F}$ is the reverse of that which was assumed. This convenience of *assuming* the sense of $\mathbf{F}$ is possible because the equilibrium equations equate to zero the *components of vectors* acting in the *same direction*. However, in cases where the frictional equation $F = \mu N$ is used in the solution of a problem, this convenience of *assuming* the sense of $\mathbf{F}$ is *lost*, since the frictional equation relates only the *magnitudes* of two *perpendicular* vectors. Consequently, $\mathbf{F}$ *must always* be shown acting with its *correct sense* on the free-body diagram, *whenever* the frictional equation is used for the solution of a problem.

IMPORTANT POINTS

- Friction is a tangential force that resists the movement of one surface relative to another.

- If no sliding occurs, the maximum value for the friction force is equal to the product of the coefficient of static friction and the normal force at the surface, $F_s = \mu_s N$.

- If sliding occurs, then the friction force is the product of the coefficient of kinetic friction and the normal force at the surface, $F_k = \mu_k N$.

- There are three types of static friction problems. Each of these problems is analyzed by first drawing the necessary free-body diagrams, and then applying the equations of equilibrium, while satisfying either the conditions of friction or the possibility of tipping.

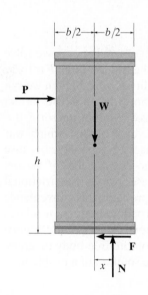

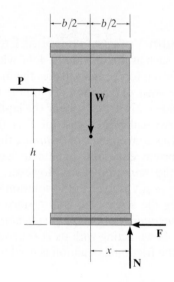

Consider pushing on the uniform crate that has a weight W and sits on the rough surface. As shown on the first free-body diagram, if the magnitude of **P** is small, the crate will remain in equilibrium. As P increases the crate will either be on the verge of slipping on the surface ($F = \mu_s N$), or if the surface is very rough (large μ_s), then the resultant normal force will shift to the corner, $x = b/2$, as shown on the second free-body diagram. At this point the crate will begin to tip over. The crate also has a greater chance of tipping if **P** is applied at a greater height h above the surface, or if its width b is smaller.

PROCEDURE FOR ANALYSIS

Equilibrium problems involving dry friction can be solved using the following procedure.

Free-Body Diagrams.

- Draw the necessary free-body diagrams, and unless it is stated in the problem that impending motion or slipping occurs, *always* show the frictional forces as unknowns (i.e., *do not assume $F = \mu N$*).

- Determine the number of unknowns and compare this with the number of available equilibrium equations.

- If there are more unknowns than equations of equilibrium, it will be necessary to apply the frictional equation at some, if not all, points of contact to obtain the extra equations needed for a complete solution.

- If the equation $F = \mu N$ is to be used, it will be necessary to show **F** acting in the correct sense of direction on the free-body diagram.

Equations of Equilibrium and Friction.

- Apply the equations of equilibrium and the necessary frictional equations (or conditional equations if tipping is possible) and solve for the unknowns.

- If the problem involves a three-dimensional force system such that it becomes difficult to obtain the force components or the necessary moment arms, apply the equations of equilibrium using Cartesian vectors.

4

EXAMPLE 4.12

The uniform crate shown in Fig. 4–27a has a mass of 20 kg. If a force $P = 80$ N is applied to the crate, determine if it remains in equilibrium. The coefficient of static friction is $\mu_s = 0.3$.

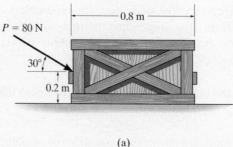

(a)

Fig. 4–27

SOLUTION

Free-Body Diagram.   As shown in Fig. 4–27b, the *resultant* normal force $\mathbf{N}_C$ must act a distance x from the crate's centerline in order to counteract the tipping effect caused by $\mathbf{P}$. There are *three unknowns*, F, N_C, and x, which can be determined strictly from the *three* equations of equilibrium.

Equations of Equilibrium.

$$\xrightarrow{+} \Sigma F_x = 0; \qquad 80 \cos 30° \text{ N} - F = 0$$

$$+\uparrow \Sigma F_y = 0; \qquad -80 \sin 30° \text{ N} + N_C - 196.2 \text{ N} = 0$$

$$\zeta + \Sigma M_O = 0; \; 80 \sin 30° \text{ N}(0.4 \text{ m}) - 80 \cos 30° \text{ N}(0.2 \text{ m}) + N_C(x) = 0$$

Solving,

$$F = 69.3 \text{ N}$$
$$N_C = 236.2 \text{ N}$$
$$x = -0.00908 \text{ m} = -9.08 \text{ mm}$$

Since x is negative it indicates the *resultant* normal force acts (slightly) to the *left* of the crate's centerline. Also, the *maximum* frictional force which can be developed at the surface of contact is $F_{\text{max}} = \mu_s N_C = 0.3(236.2 \text{ N}) = 70.9$ N. Since $F = 69.3$ N < 70.9 N, the crate will *not slip*, although it is very close to doing so.

196.2 N

$P = 80$ N |—0.4 m—|—0.4 m—|

30°

0.2 m

O

$\mathbf{F}$

|—x—|

$\mathbf{N}_C$

(b)

4

EXAMPLE   4.13

It is observed that when the bed of the dump truck is raised to an angle of $\theta = 25°$ the vending machines will begin to slide off the bed, Fig. 4–28a. Determine the coefficient of static friction between a vending machine and the surface of the truckbed.

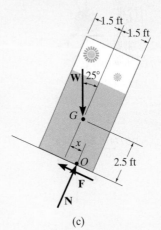

(a)

SOLUTION
An idealized model of a vending machine resting on the truckbed is shown in Fig. 4–28b. The dimensions have been measured and the center of gravity has been located. We will assume that the vending machine weighs W.

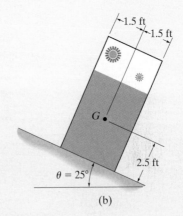

(b)

Free-Body Diagram.   As shown in Fig. 4–28c, the dimension x is used to locate the position of the resultant normal force N. There are four unknowns, $N, F, \mu_s,$ and x.

Equations of Equilibrium.

$$+\searrow\Sigma F_x = 0; \qquad\qquad W \sin 25° - F = 0 \qquad\qquad (1)$$

$$+\nearrow\Sigma F_y = 0; \qquad\qquad N - W \cos 25° = 0 \qquad\qquad (2)$$

$$\zeta+\Sigma M_O = 0; \quad -W \sin 25°(2.5 \text{ ft}) + W \cos 25°(x) = 0 \qquad (3)$$

Since slipping impends at $\theta = 25°$, using Eqs. 1 and 2, we have

$$F_s = \mu_s N; \qquad W \sin 25° = \mu_s(W \cos 25°)$$

$$\mu_s = \tan 25° = 0.466 \qquad\qquad\qquad Ans.$$

The angle of $\theta = 25°$ is referred to as the *angle of repose*, and by comparison, it is equal to the angle of static friction, $\theta = \phi_s$. Since θ is independent of the weight of the vending machine, knowing θ provides a convenient method for determining the coefficient of static friction.

NOTE: From Eq. 3, we find $x = 1.17$ ft. Since 1.17 ft $<$ 1.5 ft, indeed the vending machine will slip down the truckbed and not tip over.

(c)

Fig. 4–28

EXAMPLE 4.14

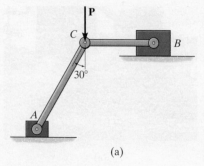

(a)

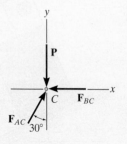

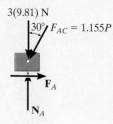

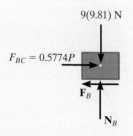

(b)

Fig. 4–29

Blocks A and B have a mass of 3 kg and 9 kg, respectively, and are connected to the weightless links shown in Fig. 4–29a. Determine the largest vertical force $\mathbf{P}$ that can be applied to the pin C without causing any movement. The coefficient of static friction between the blocks and the contacting surfaces is $\mu_s = 0.3$.

SOLUTION

Free-Body Diagram. The links are two-force members and so the free-body diagrams of pin C and blocks A and B are shown in Fig. 4–29b. Since the horizontal component of $\mathbf{F}_{AC}$ tends to move block A to the left, $\mathbf{F}_A$ must act to the right. Similarly, $\mathbf{F}_B$ must act to the left to oppose the tendency of motion of block B to the right, caused by $\mathbf{F}_{BC}$. There are seven unknowns and six available force equilibrium equations, two for the pin and two for each block, so that *only one* friction equation is needed.

Equations of Equilibrium and Friction. The force in links AC and BC can be related to P by considering the equilibrium of pin C.

$$+\uparrow \Sigma F_y = 0; \qquad F_{AC} \cos 30° - P = 0; \qquad F_{AC} = 1.155P$$

$$\xrightarrow{+} \Sigma F_x = 0; \qquad 1.155P \sin 30° - F_{BC} = 0; \qquad F_{BC} = 0.5774P$$

Using the result for F_{AC}, for block A,

$$\xrightarrow{+} \Sigma F_x = 0; \qquad F_A - 1.155P \sin 30° = 0; \quad F_A = 0.5774P \qquad (1)$$

$$+\uparrow \Sigma F_y = 0; \qquad N_A - 1.155P \cos 30° - 3(9.81 \text{ N}) = 0;$$

$$N_A = P + 29.43 \text{ N} \qquad (2)$$

Using the result for F_{BC}, for block B,

$$\xrightarrow{+} \Sigma F_x = 0; \qquad (0.5774P) - F_B = 0; \qquad F_B = 0.5774P \qquad (3)$$

$$+\uparrow \Sigma F_y = 0; \qquad N_B - 9(9.81) \text{ N} = 0; \qquad N_B = 88.29 \text{ N}$$

Movement of the system may be caused by the initial slipping of *either* block A or block B. If we *assume* that block A slips first, then

$$F_A = \mu_s N_A = 0.3 N_A \qquad (4)$$

Substituting Eqs. 1 and 2 into Eq. 4,

$$0.5774P = 0.3(P + 29.43)$$

$$P = 31.8 \text{ N} \qquad \qquad \textit{Ans.}$$

Substituting this result into Eq. 3, we obtain $F_B = 18.4$ N. Since the maximum static frictional force at B is $(F_B)_{max} = \mu_s N_B = 0.3(88.29 \text{ N}) = 26.5 \text{ N} > F_B$, block B will not slip. Thus, the above assumption is correct. Notice that if the inequality were not satisfied, we would have to assume slipping of block B ($F_B = 0.3 N_B$) and then solve for P.

PRELIMINARY PROBLEMS

P4–4. Determine the friction force at the surface of contact.

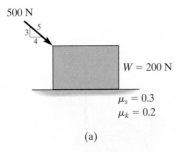

500 N

$W = 200$ N

$\mu_s = 0.3$
$\mu_k = 0.2$

(a)

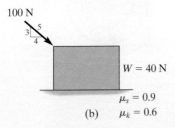

100 N

$W = 40$ N

$\mu_s = 0.9$
(b) $\mu_k = 0.6$

Prob. P4–4

P4–5. Determine the couple moment M needed to cause impending motion of the cylinder.

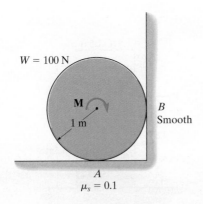

$W = 100$ N

M

1 m

B
Smooth

A
$\mu_s = 0.1$

Prob. P4–5

P4–6. Determine the force P to move block B.

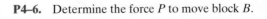
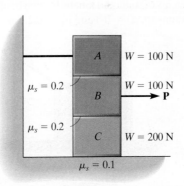

A $W = 100$ N

$\mu_s = 0.2$ $W = 100$ N
B **P**

$\mu_s = 0.2$
C $W = 200$ N

$\mu_s = 0.1$

Prob. P4–6

P4–7. Determine the force P needed to cause impending motion of the block.

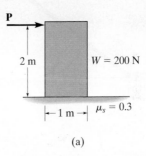

P

2 m $W = 200$ N

$\mu_s = 0.3$
⊢1 m⊣

(a)

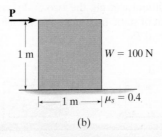

P

1 m $W = 100$ N

⊢1 m⊣ $\mu_s = 0.4$

(b)

Prob. P4–7

FUNDAMENTAL PROBLEMS

All problem solutions must include FBDs.

F4–13. Determine the friction developed between the 50-kg crate and the ground if a) $P = 200$ N, and b) $P = 400$ N. The coefficients of static and kinetic friction between the crate and the ground are $\mu_s = 0.3$ and $\mu_k = 0.2$.

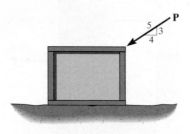

Prob. F4–13

F4–14. Determine the minimum force P to prevent the 30-kg rod AB from sliding. The contact surface at B is smooth, whereas the coefficient of static friction between the rod and the wall at A is $\mu_s = 0.2$.

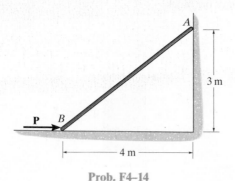

Prob. F4–14

F4–15. Determine the maximum force P that can be applied without causing the two 50-kg crates to move. The coefficient of static friction between each crate and the ground is $\mu_s = 0.25$.

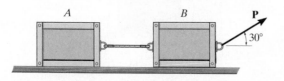

Prob. F4–15

F4–16. If the coefficient of static friction at contact points A and B is $\mu_s = 0.3$, determine the maximum force P that can be applied without causing the 100-kg spool to move.

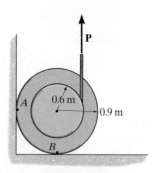

Prob. F4–16

F4–17. Determine the maximum force P that can be applied without causing movement of the 250-lb crate that has a center of gravity at G. The coefficient of static friction at the floor is $\mu_s = 0.4$.

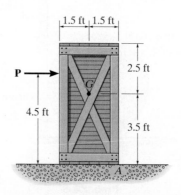

Prob. F4–17

F4–18. Determine the minimum coefficient of static friction between the uniform 50-kg spool and the wall so that the spool does not slip.

F4–20. If the coefficient of static friction at all contacting surfaces is μ_s, determine the inclination θ at which the identical blocks, each of weight W, begin to slide.

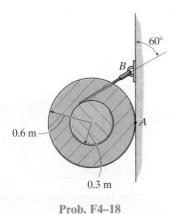

Prob. F4–18

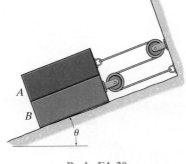

Prob. F4–20

F4–19. Blocks A, B, and C have weights of 50 N, 25 N, and 15 N, respectively. Determine the smallest horizontal force P that will cause impending motion. The coefficient of static friction between A and B is $\mu_s = 0.3$, between B and C, $\mu_s' = 0.4$, and between block C and the ground, $\mu_s'' = 0.35$.

F4–21. Blocks A and B have a mass of 7 kg and 10 kg, respectively. Using the coefficients of static friction indicated, determine the largest force P which can be applied to the cord without causing motion. There are pulleys at C and D.

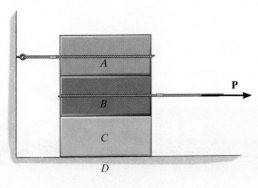

Prob. F4–19

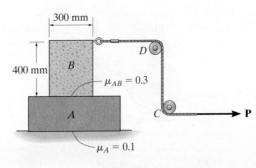

Prob. F4–21

PROBLEMS

All problem solutions must include FBDs.

4–42. Determine the maximum force P the connection can support so that no slipping occurs between the plates. There are four bolts used for the connection and each is tightened so that it is subjected to a tension of 4 kN. The coefficient of static friction between the plates is $\mu_s = 0.4$.

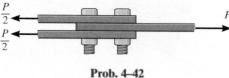

Prob. 4–42

4–43. The tractor exerts a towing force $T = 400$ lb. Determine the normal reactions at each of the two front and two rear tires and the tractive frictional force **F** on each rear tire needed to pull the load forward at constant velocity. The tractor has a weight of 7500 lb and a center of gravity located at G_T. An additional weight of 600 lb is added to its front having a center of gravity at G_A. Take $\mu_s = 0.4$. The front wheels are free to roll.

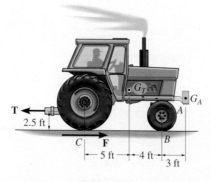

Prob. 4–43

*4–44.** The mine car and its contents have a total mass of 6 Mg and a center of gravity at G. If the coefficient of static friction between the wheels and the tracks is $\mu_s = 0.4$ when the wheels are locked, find the normal force acting on the front wheels at B and the rear wheels at A when the brakes at both A and B are locked. Does the car move?

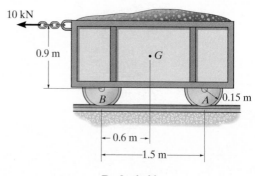

Prob. 4–44

4–45. The winch on the truck is used to hoist the garbage bin onto the bed of the truck. If the loaded bin has a weight of 8500 lb and center of gravity at G, determine the force in the cable needed to begin the lift. The coefficients of static friction at A and B are $\mu_A = 0.3$ and $\mu_B = 0.2$, respectively. Neglect the height of the support at A.

Prob. 4–45

4-46.   The automobile has a mass of 2 Mg and center of mass at G. Determine the towing force $\mathbf{F}$ required to move the car if the back brakes are locked, and the front wheels are free to roll. Take $\mu_s = 0.3$.

4-47.   The automobile has a mass of 2 Mg and center of mass at G. Determine the towing force $\mathbf{F}$ required to move the car. Both the front and rear brakes are locked. Take $\mu_s = 0.3$.

4-49.   The block brake consists of a pin-connected lever and friction block at B. The coefficient of static friction between the wheel and the lever is $\mu_s = 0.3$, and a torque of 5 N·m is applied to the wheel. Determine if the brake can hold the wheel stationary when the force applied to the lever is (a) $P = 30$ N, (b) $P = 70$ N.

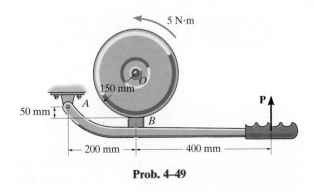

Prob. 4-49

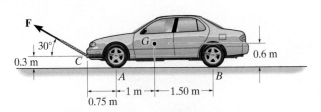

Probs. 4-46/47

***4-48.**   The block brake consists of a pin-connected lever and friction block at B. The coefficient of static friction between the wheel and the lever is $\mu_s = 0.3$, and a torque of 5 N·m is applied to the wheel. Determine if the brake can hold the wheel stationary when the force applied to the lever is (a) $P = 30$ N, (b) $P = 70$ N.

4-50.   The pipe of weight W is to be pulled up the inclined plane of slope α using a force $\mathbf{P}$. If $\mathbf{P}$ acts at an angle ϕ, show that for slipping $P = W \sin(\alpha + \theta)/\cos(\phi - \theta)$, where θ is the angle of static friction; $\theta = \tan^{-1} \mu_s$.

4-51.   Determine the angle ϕ at which the applied force $\mathbf{P}$ should act on the pipe so that the magnitude of $\mathbf{P}$ is as small as possible for pulling the pipe up the incline. What is the corresponding value of P? The pipe weighs W and the slope α is known. Express the answer in terms of the angle of kinetic friction, $\theta = \tan^{-1} \mu_k$.

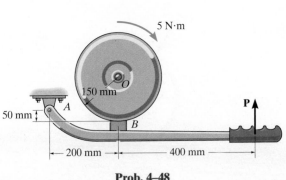

Prob. 4-48

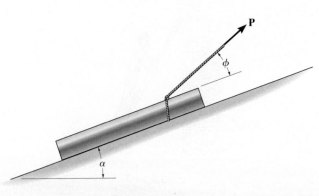

Probs. 4-50/51

*4–52. The log has a coefficient of static friction of $\mu_s = 0.3$ with the ground and a weight of 40 lb/ft. If a man can pull on the rope with a maximum force of 80 lb, determine the greatest length l of log he can drag.

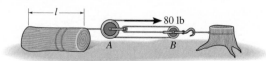

Prob. 4–52

4–53. The 180-lb man climbs up the ladder and stops at the position shown after he senses that the ladder is on the verge of slipping. Determine the inclination θ of the ladder if the coefficient of static friction between the friction pad A and the ground is $\mu_s = 0.4$. Assume the wall at B is smooth. The center of gravity for the man is at G. Neglect the weight of the ladder.

4–54. The 180-lb man climbs up the ladder and stops at the position shown after he senses that the ladder is on the verge of slipping. Determine the coefficient of static friction between the friction pad at A and ground if the inclination of the ladder is $\theta = 60°$ and the wall at B is smooth. The center of gravity for the man is at G. Neglect the weight of the ladder.

Probs. 4–53/54

4–55. The spool of wire having a weight of 300 lb rests on the ground at B and against the wall at A. Determine the force P required to begin pulling the wire horizontally off the spool. The coefficient of static friction between the spool and its points of contact is $\mu_s = 0.25$.

*4–56. The spool of wire having a weight of 300 lb rests on the ground at B and against the wall at A. Determine the normal force acting on the spool at A if $P = 300$ lb. The coefficient of static friction between the spool and the ground at B is $\mu_s = 0.35$. The wall at A is smooth.

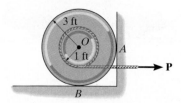

Probs. 4–55/56

4–57. The ring has a mass of 0.5 kg and is resting on the surface of the table. To move the ring a normal force **P** from the finger is exerted on it as shown. Determine its magnitude when the ring is on the verge of slipping at A. The coefficient of static friction at A is $\mu_A = 0.2$ and at B, $\mu_B = 0.3$.

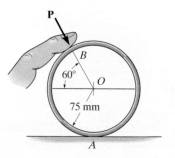

Prob. 4–57

4–58. Determine the smallest force P that must be applied in order to cause the 150-lb uniform crate to move. The coefficient of static friction between the crate and the floor is $\mu_s = 0.5$.

4–59. The man having a weight of 200 lb pushes horizontally on the crate. If the coefficient of static friction between the 450-lb crate and the floor is $\mu_s = 0.3$ and between his shoes and the floor is $\mu'_s = 0.6$, determine if he can move the crate.

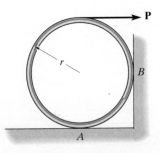

Probs. 4–58/59

*4–60.** The uniform hoop of weight W is subjected to the horizontal force P. Determine the coefficient of static friction between the hoop and the surface at A and B if the hoop is on the verge of rotating.

4–61. Determine the maximum horizontal force $\mathbf{P}$ that can be applied to the 30-lb hoop without causing it to rotate. The coefficient of static friction between the hoop and the surfaces A and B is $\mu_s = 0.2$. Take $r = 300$ mm.

Probs. 4–60/61

4–62. Determine the minimum force P needed to push the tube E up the incline. The coefficients of static friction at the contacting surfaces are $\mu_A = 0.2$, $\mu_B = 0.3$, and $\mu_C = 0.4$. The 100-kg roller and 40-kg tube each have a radius of 150 mm.

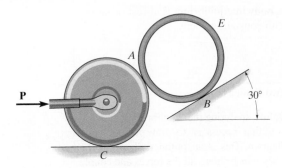

Prob. 4–62

4–63. The coefficients of static and kinetic friction between the drum and brake bar are $\mu_s = 0.4$ and $\mu_k = 0.3$, respectively. If $M = 50\ \mathrm{N \cdot m}$ and $P = 85$ N, determine the horizontal and vertical components of reaction at the pin O. Neglect the weight and thickness of the brake. The drum has a mass of 25 kg.

*4–64.** The coefficient of static friction between the drum and brake bar is $\mu_s = 0.4$. If the moment $M = 35\ \mathrm{N \cdot m}$, determine the smallest force P that needs to be applied to the brake bar in order to prevent the drum from rotating. Also determine the corresponding horizontal and vertical components of reaction at pin O. Neglect the weight and thickness of the brake bar. The drum has a mass of 25 kg.

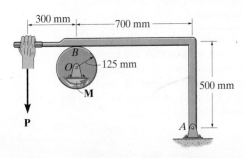

Probs. 4–63/64

CHAPTER REVIEW

Equilibrium A body in equilibrium is at rest or moves with constant velocity.	$\Sigma F = 0$ $\Sigma M = 0$	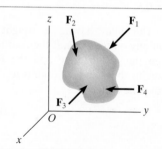
Two Dimensions Before analyzing the equilibrium of a body, it is first necessary to draw its free-body diagram. This is an outlined shape of the body, which shows all the forces and couple moments that act on it. Couple moments can be placed anywhere on the free-body diagram since they are free vectors. Forces can act at any point along their line of action since they are sliding vectors. Angles used to resolve forces, and dimensions used to take moments of the forces, should also be shown on the free-body diagram. Remember that a support will exert a force on the body in a particular direction if it prevents translation of the body in that direction, and it will exert a couple moment on the body if it prevents rotation. The three scalar equations of equilibrium can be applied when solving problems in two dimensions, since the geometry is easy to visualize.	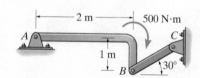 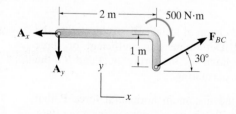 $\Sigma F_x = 0$ $\Sigma F_y = 0$ $\Sigma M_O = 0$	
For the most direct solution, try to sum forces along an axis that will eliminate as many unknown forces as possible. Sum moments about a point A that passes through the line of action of as many unknown forces as possible.		

Three Dimensions

In three dimensions, it is often advantageous to use a Cartesian vector analysis when applying the equations of equilibrium. To do this, first express each known and unknown force and couple moment shown on the free-body diagram as a Cartesian vector. Then set the force summation equal to zero. Take moments about a point O that lies on the line of action of as many unknown force components as possible. From point O direct position vectors to each force, and then use the cross product to determine the moment of each force.

The six scalar equations of equilibrium are established by setting the respective $\mathbf{i}$, $\mathbf{j}$, and $\mathbf{k}$ components of these force and moment summations equal to zero.

$$\Sigma \mathbf{F} = 0$$

$$\Sigma \mathbf{M}_O = 0$$

$$\Sigma F_x = 0$$

$$\Sigma F_y = 0$$

$$\Sigma F_z = 0$$

$$\Sigma M_x = 0$$

$$\Sigma M_y = 0$$

$$\Sigma M_z = 0$$

Dry Friction

Frictional forces exist between two rough surfaces of contact. These forces act on a body so as to oppose its motion or tendency of motion.

A static frictional force has a maximum value of $F_s = \mu_s N$, where μ_s is the *coefficient of static friction*. In this case, motion between the contacting surfaces is impending.

If slipping occurs, then the friction force remains essentially constant and equal to $F_k = \mu_k N$. Here μ_k is the *coefficient of kinetic friction*.

The solution of a problem involving friction requires first drawing the free-body diagram of the body. If the unknowns cannot be determined strictly from the equations of equilibrium, and the possibility of slipping occurs, then the friction equation should be applied at the appropriate points of contact in order to complete the solution.

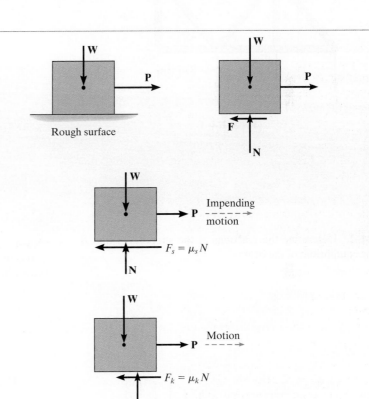

Rough surface

REVIEW PROBLEMS

All problem solutions must include an FBD.

R4–1. If the roller at B can sustain a maximum load of 3 kN, determine the largest magnitude of each of the three forces **F** that can be supported by the truss.

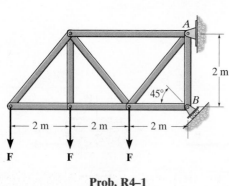

Prob. R4–1

R4–3. Determine the normal reaction at the roller A and horizontal and vertical components at pin B for equilibrium of the member.

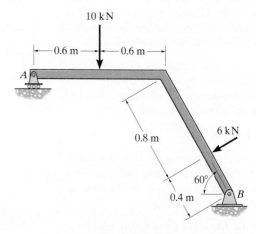

Prob. R4–3

R4–2. Determine the reactions at the supports A and B for equilibrium of the beam.

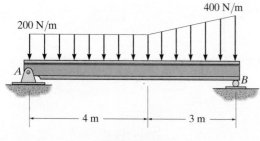

Prob. R4–2

*R4–4. Determine the horizontal and vertical components of reaction at the pin A and the reaction at the roller B on the lever.

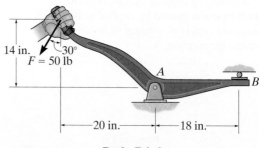

Prob. R4–4

R4–5. Determine the x, y, z components of reaction at the fixed wall A. The 150-N force is parallel to the z axis and the 200-N force is parallel to the y axis.

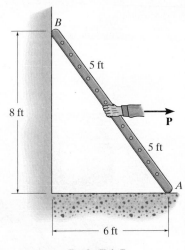

Prob. R4–5

R4–7. The uniform 20-lb ladder rests on the rough floor for which the coefficient of static friction is $\mu_s = 0.4$ and against the smooth wall at B. Determine the horizontal force P the man must exert on the ladder in order to cause it to move.

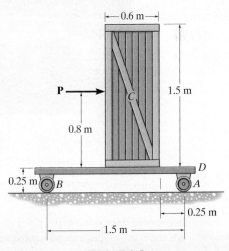

Prob. R4–7

R4–6. A vertical force of 80 lb acts on the crankshaft. Determine the horizontal equilibrium force **P** that must be applied to the handle and the x, y, z components of reaction at the journal bearing A and thrust bearing B. The bearings are properly aligned and exert only force reactions on the shaft.

*R4–8. The uniform 60-kg crate C rests uniformly on a 10-kg dolly D. If the front wheels of the dolly at A are locked to prevent rolling while the wheels at B are free to roll, determine the maximum force **P** that may be applied without causing motion of the crate. The coefficient of static friction between the wheels and the floor is $\mu_f = 0.35$ and between the dolly and the crate, $\mu_d = 0.5$.

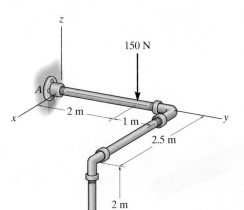

Prob. R4–6

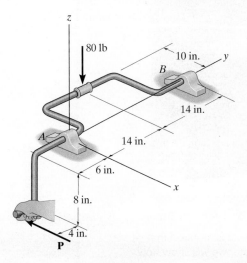

Prob. R4–8

CHAPTER 5

(© Tim Scrivener/Alamy)

In order to design the many parts of this boom assembly it is required that we know the forces that they must support. In this chapter we will show how to analyze such structures using the equations of equilibrium.

STRUCTURAL ANALYSIS

CHAPTER OBJECTIVES

■ To show how to determine the forces in the members of a truss using the method of joints and the method of sections.

■ To analyze the forces acting on the members of a frame or machine composed of pin-connected members.

5.1 SIMPLE TRUSSES

A *truss* is a structure composed of slender members joined together at their end points. The members commonly used in construction consist of wooden struts or metal bars. In particular, *planar trusses* lie in a single plane and are often used to support roofs and bridges. The truss shown in Fig. 5–1a is an example of a typical roof-supporting truss. Here, the roof load is transmitted to the truss *at the joints* by means of a series of purlins. Since this loading acts in the same plane as the truss, Fig. 5–1b, the analysis of the forces developed in the truss members will be two-dimensional.

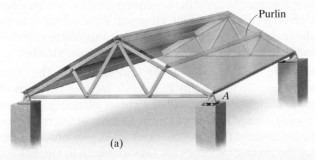

(a)

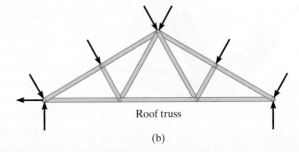

Roof truss

(b)

Fig. 5–1

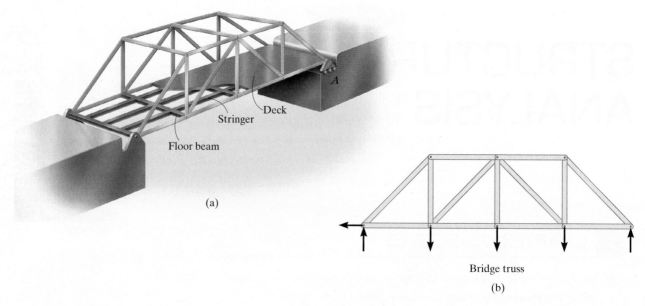

(a)

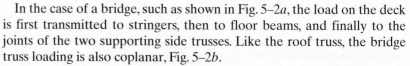

Bridge truss

(b)

Fig. 5–2

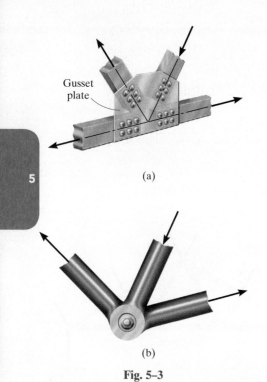

(a)

(b)

Fig. 5–3

In the case of a bridge, such as shown in Fig. 5–2a, the load on the deck is first transmitted to stringers, then to floor beams, and finally to the joints of the two supporting side trusses. Like the roof truss, the bridge truss loading is also coplanar, Fig. 5–2b.

When bridge or roof trusses extend over large distances, a rocker or roller is commonly used for supporting one end, for example, joint A in Figs. 5–1a and 5–2a. This type of support allows freedom for expansion or contraction of the members due to a change in temperature or application of loads.

Assumptions for Design. To design both the members and the connections of a truss, it is first necessary to determine the *force* developed in each member when the truss is subjected to a given loading. To do a force analysis we will make two important assumptions:

- *All loadings are applied at the joints.* In most situations, such as for bridge and roof trusses, this assumption is true. Frequently the weight of the members is neglected because the force supported by each member is usually much larger than its weight. However, if the weight is to be included in the analysis, it is generally satisfactory to apply it as a vertical force, with half of its magnitude applied at each end of the member.

- *The members are joined together by smooth pins.* The joint connections are usually formed by bolting or welding the ends of the members to a common plate, called a *gusset plate*, Fig. 5–3a, or by simply passing a large bolt or pin through each of the members, Fig. 5–3b. We can assume these connections act as pins provided the centerlines of the joining members are *concurrent*, as in Fig. 5–3.

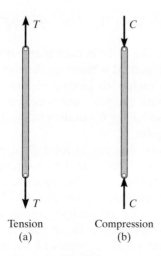

Tension
(a)

Compression
(b)

Fig. 5–4

Because of these two assumptions, *each truss member will act as a two-force member*, and therefore the force acting at each end of the member will be directed along the axis of the member. If the force tends to *elongate* the member, it is a ***tensile force*** (T), Fig. 5–4*a*; whereas if it tends to *shorten* the member, it is a ***compressive force*** (C), Fig. 5–4*b*. In the actual design of a truss it is important to state whether the force is tensile or compressive. Often, compression members must be made *thicker* than tension members because of the buckling or sudden collapse that can occur when a member is in compression.

Simple Truss. If three members are pin connected at their ends, they form a *triangular truss* that will be *rigid*, Fig. 5–5. Attaching two more members and connecting them to a new joint *D* forms a larger truss, Fig. 5–6. This procedure can be repeated as many times as desired to form an even larger truss, and by doing this one forms a ***simple truss***.

The use of metal gusset plates in the construction of these Warren trusses is clearly evident.

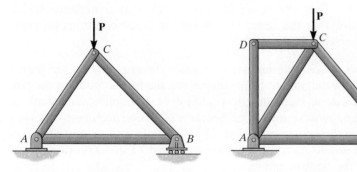

Fig. 5–5   **Fig. 5–6**

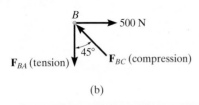

(a)

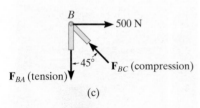

(b)

(c)

Fig. 5–7

The forces in the members of this simple roof truss can be determined using the method of joints.

5.2   THE METHOD OF JOINTS

One way to determine the force in each member of a truss is to use the *method of joints*. This method is based on the fact that if the entire truss is in equilibrium, then each of its joints is also in equilibrium. Therefore, if the free-body diagram of each joint is drawn, the force equilibrium equations, $\Sigma F_x = 0$ and $\Sigma F_y = 0$, can then be used to obtain the member forces acting on each joint.

For example, consider the pin at joint B of the truss in Fig. 5–7a. As shown on its free-body diagram, Fig. 5–7b, three forces act on the pin, namely, the 500-N force and the forces exerted by members BA and BC. Here, $\mathbf{F}_{BA}$ is "pulling" on the pin, which means that member BA is in *tension*; whereas $\mathbf{F}_{BC}$ is "pushing" on the pin, and consequently member BC is in *compression*. These effects can also be seen by isolating the joint with small segments of the member connected to the pin, Fig. 5–7c. The pushing or pulling on these small segments indicates the effect on the members being either in compression or tension.

When using the method of joints, always start at a joint having *at least one known force and at most two unknown forces*, as in Fig. 5–7b. In this way, application of $\Sigma F_x = 0$ and $\Sigma F_y = 0$ yields two algebraic equations which can be solved for the two unknowns. When applying these equations, the correct sense of an unknown member force can be determined using one of two possible methods.

- The *correct* sense of direction of an unknown member force can, in many cases, be determined "by inspection." For example, $\mathbf{F}_{BC}$ in Fig. 5–7b must push on the pin (compression) since its horizontal component, $F_{BC} \sin 45°$, must balance the 500-N force ($\Sigma F_x = 0$). Likewise, $\mathbf{F}_{BA}$ is a tensile force since it balances the vertical component, $F_{BC} \cos 45°$ ($\Sigma F_y = 0$). In more complicated cases, the sense of an unknown member force can be *assumed*; then, after applying the equilibrium equations, the assumed sense can be verified from the numerical results. A *positive* scalar indicates that the sense is *correct*, whereas a *negative* scalar indicates that the sense shown on the free-body diagram must be *reversed*.

- *Always assume* the *unknown member forces* acting on the joint's free-body diagram to be in *tension*; i.e., the forces "pull" on the pin. If this is done, then numerical solution of the equilibrium equations will yield *positive scalars for members in tension and negative scalars for members in compression*. Once an unknown member force is found, use its *correct* magnitude and sense (T or C) on subsequent joint free-body diagrams.

IMPORTANT POINTS

- Simple trusses are composed of triangular elements. The members are assumed to be pin connected at their ends and the loads applied at the joints.

- If a truss is in equilibrium, then each of its joints is in equilibrium. The internal forces in the members become external forces when the free-body diagram of each joint of the truss is drawn. A force pulling on a joint is caused by tension in a member, and a force pushing on a joint is caused by compression.

PROCEDURE FOR ANALYSIS

The following procedure provides a means for analyzing a truss using the method of joints.

- Draw the free-body diagram of a joint having at least one known force and at most two unknown forces. If this joint is at one of the supports, then it may be necessary first to calculate the external reactions at the support.

- Orient the x and y axes such that the forces on the free-body diagram can be easily resolved into their x and y components, and then apply the two force equilibrium equations $\Sigma F_x = 0$ and $\Sigma F_y = 0$. Solve for the two unknown member forces and verify their correct sense.

- Using the calculated results, continue to analyze each of the other joints. Remember that a member in *compression* "pushes" on the joint and a member in *tension* "pulls" on the joint.

EXAMPLE 5.1

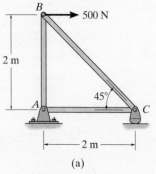

(a)

(b)

(c)

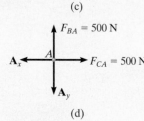

(d)

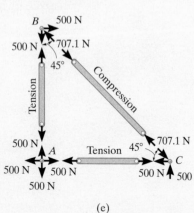

(e)

Fig. 5–8

Determine the force in each member of the truss shown in Fig. 5–8*a* and indicate whether the members are in tension or compression.

SOLUTION

We will begin our analysis at joint B since there are one known force and two unknown forces there.

Joint B. The free-body diagram is shown in Fig. 5–8*b*. Applying the equations of equilibrium, we have

$$\xrightarrow{+}\ \Sigma F_x = 0; \quad 500\ \text{N} - F_{BC} \sin 45° = 0 \qquad F_{BC} = 707.1\ \text{N (C)} \quad Ans.$$
$$+\uparrow \Sigma F_y = 0; \quad F_{BC} \cos 45° - F_{BA} = 0 \qquad F_{BA} = 500\ \text{N (T)} \quad Ans.$$

Now that the force in member BC has been calculated, we can proceed to analyze joint C to determine the force in member CA and the support reaction at the rocker.

Joint C. From the free-body diagram, Fig. 5–8*c*, we have

$$\xrightarrow{+}\ \Sigma F_x = 0; \quad -F_{CA} + 707.1 \cos 45°\ \text{N} = 0 \quad F_{CA} = 500\ \text{N (T)} \quad Ans.$$
$$+\uparrow \Sigma F_y = 0; \quad C_y - 707.1 \sin 45°\ \text{N} = 0 \qquad C_y = 500\ \text{N} \qquad Ans.$$

Joint A. Although it is not necessary, we can determine the components of the support reactions at joint A using the results of F_{CA} and F_{BA}. From the free-body diagram, Fig. 5–8*d*, we have

$$\xrightarrow{+}\ \Sigma F_x = 0; \quad 500\ \text{N} - A_x = 0 \qquad A_x = 500\ \text{N}$$
$$+\uparrow \Sigma F_y = 0; \quad 500\ \text{N} - A_y = 0 \qquad A_y = 500\ \text{N}$$

NOTE: The results of this analysis are summarized in Fig. 5–8*e*. Here the free-body diagram of each joint (or pin) shows the effects of all the connected members and external forces applied to the joint, whereas the free-body diagram of each member shows only the effects of the joints on the member.

EXAMPLE   5.2

Determine the forces acting in all the members of the truss shown in Fig. 5–9a, and indicate whether the members are in tension or compression.

SOLUTION
By inspection, there are more than two unknowns at each joint. As a result, the support reactions on the truss must first be determined. Show that they have been correctly calculated on the free-body diagram in Fig. 5–9b. We can now begin the analysis at joint C.

Joint C.   From the free-body diagram, Fig. 5–9c,

$$\xrightarrow{+} \Sigma F_x = 0; \quad -F_{CD} \cos 30° + F_{CB} \sin 45° = 0$$
$$+\uparrow \Sigma F_y = 0; \quad 1.5 \text{ kN} + F_{CD} \sin 30° - F_{CB} \cos 45° = 0$$

These two equations must be solved *simultaneously* for each of the two unknowns. A more *direct solution* can be obtained by applying a force summation along an axis that is *perpendicular* to the direction of the other unknown force. For example, summing forces along the y' axis, which is perpendicular to the direction of $\mathbf{F}_{CD}$, Fig. 5–9d, yields a *direct solution* for F_{CB}.

$$+\nearrow \Sigma F_{y'} = 0; \quad 1.5 \cos 30° \text{ kN} - F_{CB} \sin 15° = 0$$
$$F_{CB} = 5.019 \text{ kN} = 5.02 \text{ kN (C)} \qquad Ans.$$

Then,

$$+\searrow \Sigma F_{x'} = 0;$$
$$-F_{CD} + 5.019 \cos 15° - 1.5 \sin 30° = 0; \quad F_{CD} = 4.10 \text{ kN (T)} \quad Ans.$$

Joint D.   We can now proceed to analyze joint D. The free-body diagram is shown in Fig. 5–9e.

$$\xrightarrow{+} \Sigma F_x = 0; \quad -F_{DA} \cos 30° + 4.10 \cos 30° \text{ kN} = 0$$
$$F_{DA} = 4.10 \text{ kN} \quad \text{(T)} \qquad Ans.$$

$$+\uparrow \Sigma F_y = 0; \quad F_{DB} - 2(4.10 \sin 30° \text{ kN}) = 0$$
$$F_{DB} = 4.10 \text{ kN} \quad \text{(T)} \qquad Ans.$$

NOTE: The force in the last member, BA, can be obtained from joint B or joint A. As an exercise, draw the free-body diagram of joint B, sum the forces in the horizontal direction, and show that $F_{BA} = 0.776$ kN (C).

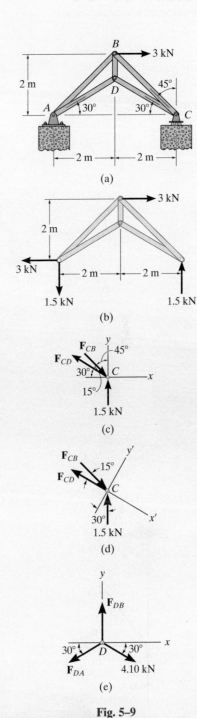

Fig. 5–9

EXAMPLE 5.3

Determine the force in each member of the truss shown in Fig. 5–10a. Indicate whether the members are in tension or compression.

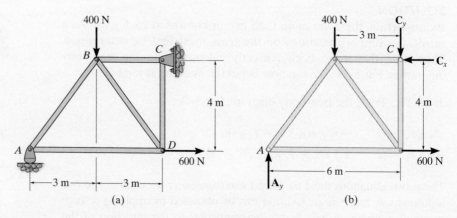

Fig. 5–10

SOLUTION

Support Reactions. No joint can be analyzed until the support reactions are determined, because each joint has at least three unknown forces acting on it. A free-body diagram of the entire truss is given in Fig. 5–10b. Applying the equations of equilibrium, we have

$$\xrightarrow{+} \Sigma F_x = 0; \qquad 600\ \text{N} - C_x = 0 \qquad\qquad\qquad C_x = 600\ \text{N}$$
$$\zeta + \Sigma M_C = 0; \qquad -A_y(6\ \text{m}) + 400\ \text{N}(3\ \text{m}) + 600\ \text{N}(4\ \text{m}) = 0$$
$$A_y = 600\ \text{N}$$
$$+\uparrow \Sigma F_y = 0; \qquad 600\ \text{N} - 400\ \text{N} - C_y = 0 \qquad\qquad C_y = 200\ \text{N}$$

The analysis can now start at either joint A or C. The choice is arbitrary since there are one known and two unknown member forces acting on the pin at each of these joints.

Joint A. (Fig. 5–10c). As shown on the free-body diagram, $\mathbf{F}_{AB}$ is assumed to be compressive and $\mathbf{F}_{AD}$ is tensile. Applying the equations of equilibrium, we have

$$+\uparrow \Sigma F_y = 0; \qquad 600\ \text{N} - \tfrac{4}{5}F_{AB} = 0 \qquad F_{AB} = 750\ \text{N} \quad (C) \qquad Ans.$$
$$\xrightarrow{+} \Sigma F_x = 0; \qquad F_{AD} - \tfrac{3}{5}(750\ \text{N}) = 0 \qquad F_{AD} = 450\ \text{N} \quad (T) \qquad Ans.$$

Joint D.   (Fig. 5–10d). Using the result for F_{AD} and summing forces in the horizontal direction, we have

$\xrightarrow{+} \Sigma F_x = 0;$ $-450 \text{ N} + \frac{3}{5}F_{DB} + 600 \text{ N} = 0$ $F_{DB} = -250 \text{ N}$

The negative sign indicates that $\mathbf{F}_{DB}$ acts in the *opposite sense* to that shown in Fig. 5–10d.* Hence,

$$F_{DB} = 250 \text{ N (T)} \qquad\qquad Ans.$$

To determine $\mathbf{F}_{DC}$, we can either correct the sense of $\mathbf{F}_{DB}$ on the free-body diagram, and then apply $\Sigma F_y = 0$, or apply this equation and retain the negative sign for F_{DB}, i.e.,

$+\uparrow \Sigma F_y = 0;$ $-F_{DC} - \frac{4}{5}(-250 \text{ N}) = 0$ $F_{DC} = 200 \text{ N}$ (C)   *Ans.*

Joint C.   (Fig. 5–10e).

$\xrightarrow{+} \Sigma F_x = 0;$ $F_{CB} - 600 \text{ N} = 0$ $F_{CB} = 600 \text{ N}$ (C)   *Ans.*
$+\uparrow \Sigma F_y = 0;$ $200 \text{ N} - 200 \text{ N} \equiv 0$ (check)

NOTE: The analysis is summarized in Fig. 5–10f, which shows the free-body diagram for each joint and member.

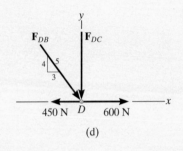

(d)

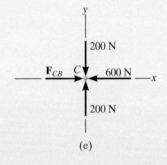

(e)

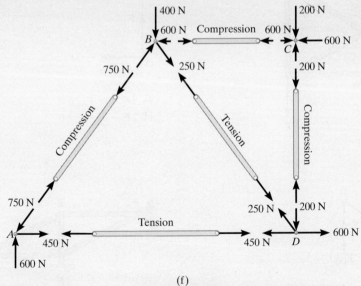

(f)

Fig. 5–10 (cont.)

*The proper sense could have been determined by inspection, prior to applying $\Sigma F_x = 0$.

5.3   ZERO-FORCE MEMBERS

Truss analysis using the method of joints is greatly simplified if we can first identify those members which support *no loading*. These ***zero-force members*** are used to increase the stability of the truss during construction and to provide added support if the loading is changed.

The zero-force members of a truss can generally be found by inspection of each of the joints. For example, consider the truss shown in Fig. 5–11*a*. If a free-body diagram of the pin at joint *A* is drawn, Fig. 5–11*b*, it is seen that members *AB* and *AF* are zero-force members. (We could not come to this conclusion if we had considered the free-body diagrams of joints *F* or *B*, simply because there are five unknowns at each of these joints.) In a similar manner, consider the free-body diagram of joint *D*, Fig. 5–11*c*. Here again it is seen that *DC* and *DE* are zero-force members. To summarize, then, *if only two non-collinear members form a truss joint and no external load or support reaction is applied to the joint, the two members must be zero-force members.* The load on the truss in Fig. 5–11*a* is therefore actually supported by only five members, as shown in Fig. 5–11*d*.

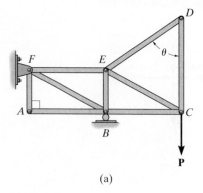

(a)

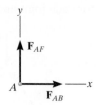

$$\overset{+}{\rightarrow} \Sigma F_x = 0; \quad F_{AB} = 0$$
$$+\uparrow \Sigma F_y = 0; \quad F_{AF} = 0$$

(b)

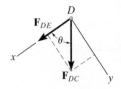

$$+\searrow \Sigma F_y = 0; F_{DC} \sin \theta = 0; \quad F_{DC} = 0 \text{ since } \sin \theta \neq 0$$
$$+\swarrow \Sigma F_x = 0; F_{DE} + 0 = 0; \quad F_{DE} = 0$$

(c)

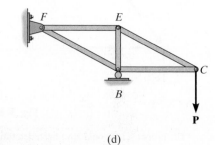

(d)

Fig. 5–11

Now consider the truss shown in Fig. 5–12a. The free-body diagram of the pin at joint D is shown in Fig. 5–12b. By orienting the y axis along members DC and DE and the x axis along member DA, it is seen that DA is a zero-force member. From the free-body diagram of joint C, Fig. 5–12c, it can be seen that this is also the case for member CA. In general then, *if three members form a truss joint for which two of the members are collinear, the third member is a zero-force member provided no external force or support reaction has a component that acts along this member*. The truss shown in Fig. 5–12d is therefore suitable for supporting the load **P**.

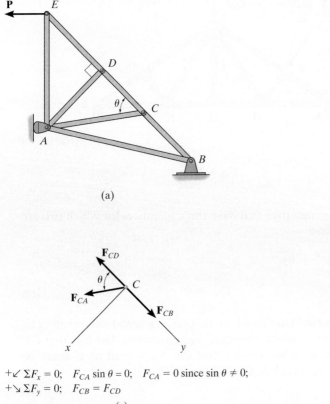

(a)

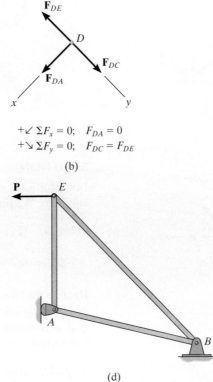

$+\swarrow \Sigma F_x = 0; \quad F_{DA} = 0$

$+\searrow \Sigma F_y = 0; \quad F_{DC} = F_{DE}$

(b)

$+\swarrow \Sigma F_x = 0; \quad F_{CA} \sin \theta = 0; \quad F_{CA} = 0 \text{ since } \sin \theta \neq 0;$

$+\searrow \Sigma F_y = 0; \quad F_{CB} = F_{CD}$

(c)

(d)

Fig. 5–12

▶ IMPORTANT POINT

- Zero-force members support no load; however, they are necessary for stability, and are available when additional loadings are applied to the joints of the truss. These members can usually be identified by inspection. They occur at joints where only *two members* are connected and no external load acts along either member. Also, at joints having two collinear members, a third member will be a zero-force member if no external force components act along this member.

EXAMPLE 5.4

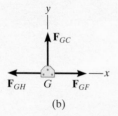

(b)

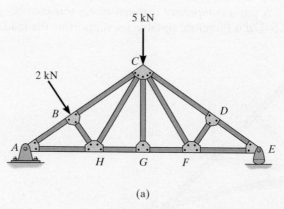

(a)

Fig. 5–13

Determine all the zero-force members of the *Fink roof truss* shown in Fig. 5–13a. Assume all joints are pin connected.

SOLUTION

Look for joint geometries that have three members for which two are collinear. We have

Joint G. (Fig. 5–13b).

$$+\uparrow \Sigma F_y = 0; \qquad\qquad F_{GC} = 0 \qquad\qquad Ans.$$

Realize that we could not conclude that GC is a zero-force member by considering joint C, where there are five unknowns. The fact that GC is a zero-force member means that the 5-kN load at C must be supported by members CB, CH, CF, and CD.

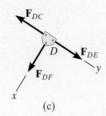

(c)

Joint D. (Fig. 5–13c).

$$+\swarrow \Sigma F_x = 0; \qquad\qquad F_{DF} = 0 \qquad\qquad Ans.$$

Joint F. (Fig. 5–13d).

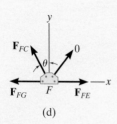

(d)

$$+\uparrow \Sigma F_y = 0; \quad F_{FC}\cos\theta = 0 \quad \text{Since } \theta \neq 90°, \quad F_{FC} = 0 \qquad Ans.$$

NOTE: If joint B is analyzed, Fig. 5–13e,

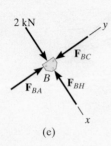

(e)

$$+\searrow \Sigma F_x = 0; \qquad 2 \text{ kN} - F_{BH} = 0 \qquad F_{BH} = 2 \text{ kN} \quad (C)$$

Also, F_{HC} must satisfy $\Sigma F_y = 0$, Fig. 5–13f, and therefore HC is *not* a zero-force member.

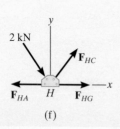

(f)

PRELIMINARY PROBLEMS

P5–1. In each case, calculate the support reactions and then draw the free-body diagrams of joints A, B, and C of the truss.

P5–2. Identify the zero-force members in each truss.

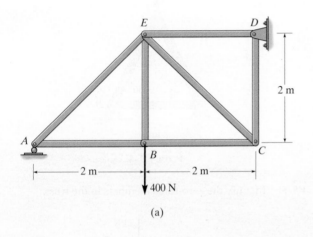

(a)

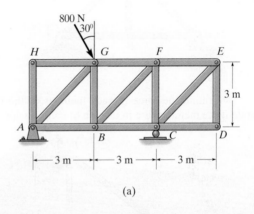

(a)

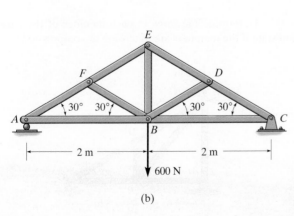

(b)

Prob. P5–1

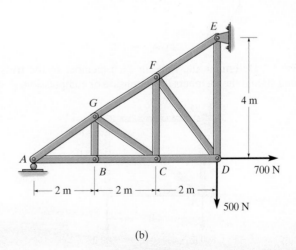

(b)

Prob. P5–2

FUNDAMENTAL PROBLEMS

All problem solutions must include FBDs.

F5–1. Determine the force in each member of the truss and state if the members are in tension or compression.

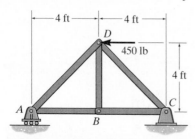

Prob. F5–1

F5–2. Determine the force in each member of the truss and state if the members are in tension or compression.

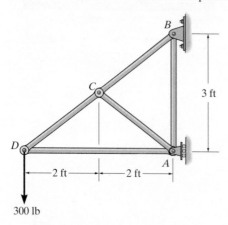

Prob. F5–2

F5–3. Determine the force in each member of the truss and state if the members are in tension or compression.

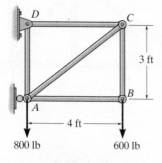

Prob. F5–3

F5–4. Determine the greatest load P that can be applied to the truss so that none of the members are subjected to a force exceeding either 2 kN in tension or 1.5 kN in compression.

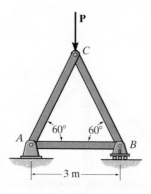

Prob. F5–4

F5–5. Identify the zero-force members in the truss.

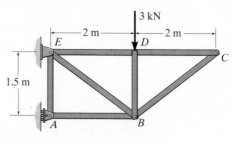

Prob. F5–5

F5–6. Determine the force in each member of the truss and state if the members are in tension or compression.

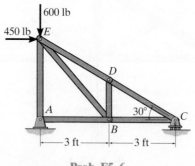

Prob. F5–6

PROBLEMS

All problem solutions must include FBDs.

5–1. Determine the force in each member of the truss and state if the members are in tension or compression. Set $P_1 = 20$ kN, $P_2 = 10$ kN.

5–2. Determine the force in each member of the truss and state if the members are in tension or compression. Set $P_1 = 45$ kN, $P_2 = 30$ kN.

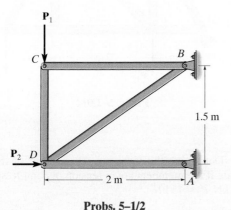

Probs. 5–1/2

5–3. Determine the force in each member of the truss and state if the members are in tension or compression.

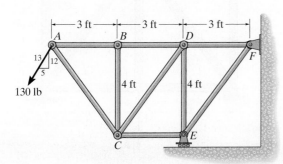

Prob. 5–3

*5–4. Determine the force in each member of the truss and state if the members are in tension or compression.

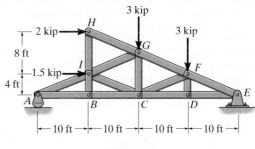

Prob. 5–4

5–5. Determine the force in each member of the truss, and state if the members are in tension or compression. Set $\theta = 0°$.

5–6. Determine the force in each member of the truss, and state if the members are in tension or compression. Set $\theta = 30°$.

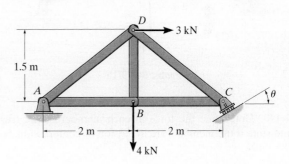

Probs. 5–5/6

5–7. Determine the force in each member of the truss and state if the members are in tension or compression.

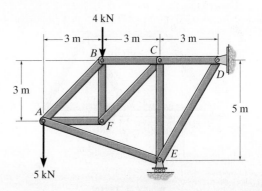

Prob. 5–7

5

*5–8. Determine the force in each member of the truss in terms of the load P and state if the members are in tension or compression.

5–9. Members AB and BC can each support a maximum compressive force of 800 lb, and members AD, DC, and BD can support a maximum tensile force of 1500 lb. If $a = 10$ ft, determine the greatest load P the truss can support.

5–10. Members AB and BC can each support a maximum compressive force of 800 lb, and members AD, DC, and BD can support a maximum tensile force of 2000 lb. If $a = 6$ ft, determine the greatest load P the truss can support.

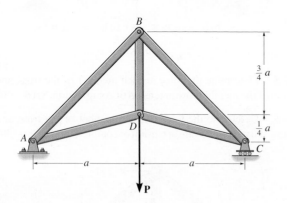

Probs. 5–8/9/10

5–11. Determine the force in each member of the truss and state if the members are in tension or compression. Set $P = 8$ kN.

*5–12. If the maximum force that any member can support is 8 kN in tension and 6 kN in compression, determine the maximum force P that can be supported at joint D.

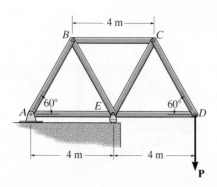

Probs. 5–11/12

5–13. Determine the force in each member of the truss and state if the members are in tension or compression. Set $P_1 = 10$ kN, $P_2 = 8$ kN.

5–14. Determine the force in each member of the truss and state if the members are in tension or compression. Set $P_1 = 8$ kN, $P_2 = 12$ kN.

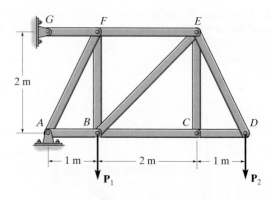

Probs. 5–13/14

5–15. Determine the force in each member of the truss and state if the members are in tension or compression. Set $P_1 = 9$ kN, $P_2 = 15$ kN.

*5–16. Determine the force in each member of the truss and state if the members are in tension or compression. Set $P_1 = 30$ kN, $P_2 = 15$ kN.

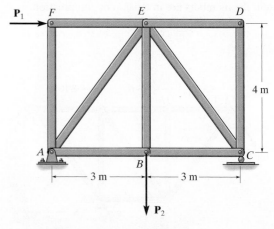

Probs. 5–15/16

5.4 THE METHOD OF SECTIONS

When we need to find the force in only a few members of a truss, we can analyze the truss using the *method of sections*. It is based on the principle that if the truss is in equilibrium then any part of the truss is also in equilibrium. For example, consider the two truss members shown in Fig. 5–14. If the forces within the members are to be determined, then an imaginary section, indicated by the blue line, can be used to cut each member into two parts and thereby "expose" each internal force as "external" to the free-body diagrams of the parts shown on the right. Clearly, it can be seen that equilibrium requires that the member in tension (T) be subjected to a "pull," whereas the member in compression (C) is subjected to a "push."

The method of sections can also be used to "cut" or section the members of an entire truss. If the section passes through the truss and the free-body diagram of either of its two parts is drawn, we can then apply the equations of equilibrium to that part to determine the member forces at the section. Since only *three* independent equilibrium equations ($\Sigma F_x = 0$, $\Sigma F_y = 0$, $\Sigma M_O = 0$) can be applied to the free-body diagram of any part, then we should try to select a section that, in general, passes through not more than *three* members in which the forces are unknown. For example, consider the truss in Fig. 5–15a. If the forces in members BC, GC, and GF are to be determined, then section aa would be appropriate. The free-body diagrams of the two parts are shown in Figs. 5–15b and 5–15c. Notice that the direction of each member force is specified from the *geometry* of the truss, since the force in a member is along its axis. Also, the member forces acting on one part of the truss are equal but opposite to those acting on the other part—Newton's third law. Members BC and GC are assumed to be in *tension* since they are subjected to a "pull," whereas GF is in *compression* since it is subjected to a "push."

The three unknown member forces $\mathbf{F}_{BC}$, $\mathbf{F}_{GC}$, and $\mathbf{F}_{GF}$ can be obtained by applying the three equilibrium equations to the free-body diagram in Fig. 5–15b. If, however, the free-body diagram in Fig. 5–15c is considered, then the three support reactions $\mathbf{D}_x$, $\mathbf{D}_y$, and $\mathbf{E}_x$ will have to be known, because only three equations of equilibrium are available. (This, of course, is done in the usual manner by considering a free-body diagram of the *entire truss*.)

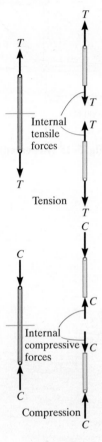

Fig. 5–14

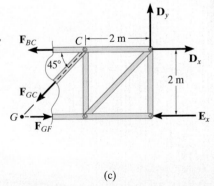

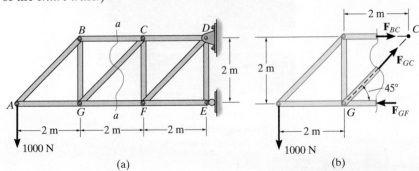

(a)   (b)   (c)

Fig. 5–15

2 m

$\mathbf{F}_{BC}$ C

2 m

$\mathbf{F}_{GC}$

45°

G $\mathbf{F}_{GF}$

2 m

1000 N

(b)

Fig. 5–15 (Repeated)

When applying the equilibrium equations, we should carefully consider ways of writing the equations so as to yield a *direct solution* for each of the unknowns, rather than having to solve simultaneous equations. For example, using the part in Fig. 5–15b and summing moments about C will yield a direct solution for $\mathbf{F}_{GF}$ since $\mathbf{F}_{BC}$ and $\mathbf{F}_{GC}$ create zero moment about C. Likewise, $\mathbf{F}_{BC}$ can be directly obtained by summing moments about G. Finally, $\mathbf{F}_{GC}$ can be found directly from a force summation in the vertical direction since $\mathbf{F}_{GF}$ and $\mathbf{F}_{BC}$ have no vertical components. This ability to *determine directly* the force in a particular truss member is one of the main advantages of using the method of sections.*

As in the method of joints, there are two ways in which we can determine the correct sense of an unknown member force:

- The correct sense can in many cases be determined "by inspection." For example, $\mathbf{F}_{BC}$ is shown as a tensile force in Fig. 5–15b since moment equilibrium about G requires that $\mathbf{F}_{BC}$ create a moment opposite to that of the 1000-N force. Also, $\mathbf{F}_{GC}$ is tensile since its vertical component must balance the 1000-N force which acts downward. In more complicated cases, the sense of an unknown force may be *assumed*. If the solution yields a *negative* scalar, it indicates that the force's sense of direction is *opposite* to that shown on the free-body diagram.

- *Always assume* that the unknown member forces at the section are *tensile* forces, i.e., "pulling" on the member. By doing this, the numerical solution of the equilibrium equations will yield *positive scalars for members in tension and negative scalars for members in compression.*

*If the method of joints were used to determine, say, the force in member GC, it would be necessary to analyze joints A, B, and G in sequence.

The forces in selected members of this Pratt truss can readily be determined using the method of sections.

IMPORTANT POINT

- If a truss is in equilibrium, then each of its parts is in equilibrium. The internal forces in the members become external forces when the free-body diagram of a part of the truss is drawn. A force pulling on a member causes tension in the member, and a force pushing on a member causes compression.

Simple trusses are often used in the construction of large cranes in order to reduce the weight of the boom and tower.

PROCEDURE FOR ANALYSIS

The forces in the members of a truss may be determined by the method of sections using the following procedure.

Free-Body Diagram.

- Make a decision as to how to "cut" or section the truss through the members where forces are to be determined.

- Before isolating any part of the truss, it may first be necessary to determine the truss's support reactions that act on the part. Once this is done then the three equilibrium equations will be available to solve for member forces at the section.

- Draw the free-body diagram of that part of the sectioned truss which has the least number of forces acting on it.

Equations of Equilibrium.

- Moments should be summed about a point that lies at the intersection of the lines of action of two unknown forces, so that the third unknown force can be determined directly from the moment equation.

- If two of the unknown forces are *parallel*, forces may be summed *perpendicular* to their direction in order to *directly* determine the third unknown force.

5

EXAMPLE 5.5

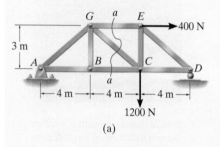

(a)

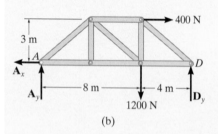

(b)

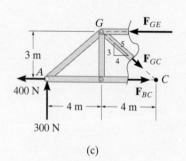

(c)

Fig. 5–16

Determine the force in members GE, GC, and BC of the truss shown in Fig. 5–16a. Indicate whether the members are in tension or compression.

SOLUTION

Section aa in Fig. 5–16a has been chosen since it cuts through the *three* members whose forces are to be determined. In order to use the method of sections, however, it is *first* necessary to determine the external reactions at A or D. Why? A free-body diagram of the entire truss is shown in Fig. 5–16b. Applying the equations of equilibrium, we have

$$\xrightarrow{+} \Sigma F_x = 0; \qquad 400\ \text{N} - A_x = 0 \qquad\qquad A_x = 400\ \text{N}$$

$$\zeta + \Sigma M_A = 0; \qquad -1200\ \text{N}(8\ \text{m}) - 400\ \text{N}(3\ \text{m}) + D_y(12\ \text{m}) = 0$$

$$D_y = 900\ \text{N}$$

$$+\uparrow \Sigma F_y = 0; \qquad A_y - 1200\ \text{N} + 900\ \text{N} = 0 \qquad A_y = 300\ \text{N}$$

Free-Body Diagram. For the analysis the free-body diagram of the left part of the sectioned truss will be used, since it involves the least number of forces, Fig. 5–16c.

Equations of Equilibrium. Summing moments about point G eliminates $\mathbf{F}_{GE}$ and $\mathbf{F}_{GC}$ and yields a direct solution for F_{BC}.

$$\zeta + \Sigma M_G = 0; \quad -300\ \text{N}(4\ \text{m}) - 400\ \text{N}(3\ \text{m}) + F_{BC}(3\ \text{m}) = 0$$

$$F_{BC} = 800\ \text{N} \quad (\text{T}) \qquad\qquad Ans.$$

In the same manner, summing moments about point C we obtain a direct solution for F_{GE}.

$$\zeta + \Sigma M_C = 0; \quad -300\ \text{N}(8\ \text{m}) + F_{GE}(3\ \text{m}) = 0$$

$$F_{GE} = 800\ \text{N} \quad (\text{C}) \qquad\qquad Ans.$$

Since $\mathbf{F}_{BC}$ and $\mathbf{F}_{GE}$ have no vertical components, summing forces in the y direction directly yields F_{GC}, i.e.,

$$+\uparrow \Sigma F_y = 0; \qquad 300\ \text{N} - \tfrac{3}{5}F_{GC} = 0$$

$$F_{GC} = 500\ \text{N} \quad (\text{T}) \qquad\qquad Ans.$$

NOTE: Here it is possible to tell, by inspection, the proper direction for each unknown member force. For example, $\Sigma M_C = 0$ requires $\mathbf{F}_{GE}$ to be *compressive* because it must balance the moment of the 300-N force about C.

EXAMPLE 5.6

Determine the force in member *CF* of the truss shown in Fig. 5–17*a*. Indicate whether the member is in tension or compression. Assume each member is pin connected.

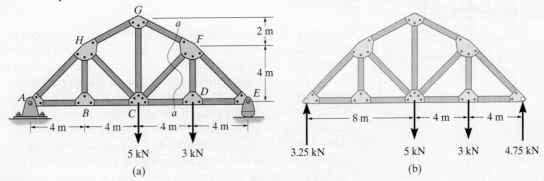

(a)   (b)

Fig. 5–17

SOLUTION

Free-Body Diagram.   Section *aa* in Fig. 5–17*a* will be used since this section will "expose" the internal force in member *CF* as "external" on the free-body diagram of either the right or left portion of the truss. It is first necessary, however, to determine the support reactions on either the left or right side. Verify the results shown on the free-body diagram in Fig. 5–17*b*.

The free-body diagram of the right part of the truss, which is the easiest to analyze, is shown in Fig. 5–17*c*. There are three unknowns, F_{FG}, F_{CF}, and F_{CD}.

Equations of Equilibrium.   We will apply the moment equation about point *O* in order to eliminate the two unknowns F_{FG} and F_{CD}. The location of point *O*, measured from *E*, can be determined from proportional triangles, i.e., $4/(4 + x) = 6/(8 + x)$, $x = 4$ m. Or, stated in another manner, since the slope of member *GF* has a drop of 2 m to a horizontal distance of 4 m, and *FD* is 4 m, Fig. 5–17*c*, then from *D* to *O* the distance must be 8 m.

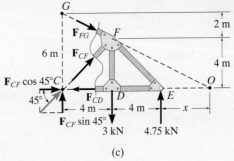

(c)

An easy way to determine the moment of $\mathbf{F}_{CF}$ about point *O* is to use the principle of transmissibility and slide $\mathbf{F}_{CF}$ to point *C*, and then resolve $\mathbf{F}_{CF}$ into its two rectangular components. We have

$\zeta + \Sigma M_O = 0;$

$$-F_{CF} \sin 45°(12 \text{ m}) + (3 \text{ kN})(8 \text{ m}) - (4.75 \text{ kN})(4 \text{ m}) = 0$$

$$F_{CF} = 0.589 \text{ kN} \text{(C)} \qquad \qquad Ans.$$

EXAMPLE 5.7

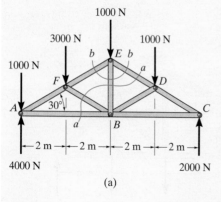

(a)

Determine the force in member *EB* of the roof truss shown in Fig. 5–18*a*. Indicate whether the member is in tension or compression.

SOLUTION

Free-Body Diagrams.   By the method of sections, any imaginary section that cuts through *EB* will also have to cut through three other members for which the forces are unknown. For example, section *aa* cuts through *ED*, *EB*, *FB*, and *AB*. If a free-body diagram of the left part of this section is considered, Fig. 5–18*b*, it is possible to obtain $\mathbf{F}_{ED}$ by summing moments about *B* to eliminate the other three unknowns; however, $\mathbf{F}_{EB}$ cannot be determined from the remaining two equilibrium equations.

One possible way of obtaining $\mathbf{F}_{EB}$ is first to determine $\mathbf{F}_{ED}$ from section *aa*, then use this result on section *bb*, Fig. 5–18*a*, which is shown in Fig. 5–18*c*. Here the force system is concurrent and our sectioned free-body diagram is the same as the free-body diagram for the joint at *E*.

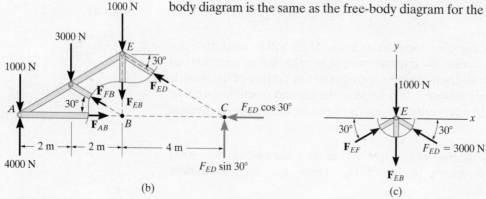

(b)

(c)

Fig. 5–18

Equations of Equilibrium.   In order to determine the moment of $\mathbf{F}_{ED}$ about point *B*, Fig. 5–18*b*, we will use the principle of transmissibility and slide this force to point *C* and then resolve it into its rectangular components as shown. Therefore,

$$\zeta + \Sigma M_B = 0; \quad 1000 \text{ N}(4 \text{ m}) + 3000 \text{ N}(2 \text{ m}) - 4000 \text{ N}(4 \text{ m})$$

$$+ F_{ED} \sin 30°(4 \text{ m}) = 0$$

$$F_{ED} = 3000 \text{ N} \quad (\text{C})$$

Considering now the free-body diagram of section *bb*, Fig. 5–18*c*, we have

$$\xrightarrow{+} \Sigma F_x = 0; \quad F_{EF} \cos 30° - 3000 \cos 30° \text{ N} = 0$$

$$F_{EF} = 3000 \text{ N} \quad (\text{C})$$

$$+\uparrow \Sigma F_y = 0; \quad 2(3000 \sin 30° \text{ N}) - 1000 \text{ N} - F_{EB} = 0$$

$$F_{EB} = 2000 \text{ N} \quad (\text{T}) \qquad \textit{Ans.}$$

FUNDAMENTAL PROBLEMS

All problem solutions must include FBDs.

F5–7. Determine the force in members BC, CF, and FE and state if the members are in tension or compression.

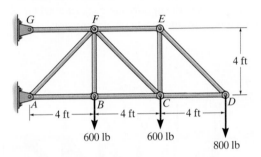

600 lb 600 lb 800 lb

Prob. F5–7

F5–8. Determine the force in members LK, KC, and CD of the Pratt truss and state if the members are in tension or compression.

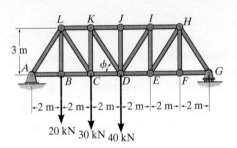

20 kN 30 kN 40 kN

Prob. F5–8

F5–9. Determine the force in members KJ, KD, and CD of the Pratt truss and state if the members are in tension or compression.

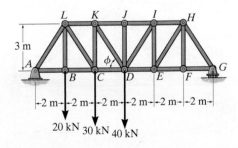

20 kN 30 kN 40 kN

Prob. F5–9

F5–10. Determine the force in members EF, CF, and BC of the truss and state if the members are in tension or compression.

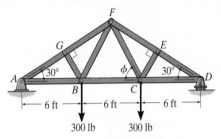

300 lb 300 lb

Prob. F5–10

F5–11. Determine the force in members GF, GD, and CD of the truss and state if the members are in tension or compression.

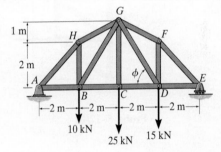

10 kN 25 kN 15 kN

Prob. F5–11

F5–12. Determine the force in members DC, HI, and JI of the truss and state if the members are in tension or compression. *Suggestion:* Use the sections shown.

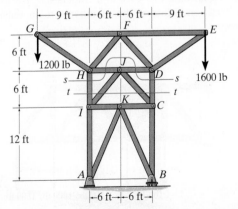

Prob. F5–12

PROBLEMS

All problem solutions must include FBDs.

5–17. Determine the force in members DC, HC, and HI of the truss and state if the members are in tension or compression.

5–18. Determine the force in members ED, EH, and GH of the truss and state if the members are in tension or compression.

5–21. Determine the force in members CD, CJ, KJ, and DJ of the truss which serves to support the deck of a bridge. State if these members are in tension or compression.

5–22. Determine the force in members EI and JI of the truss which serves to support the deck of a bridge. State if these members are in tension or compression.

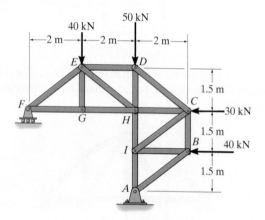

Probs. 5–17/18

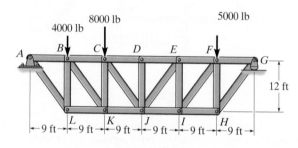

Probs. 5–21/22

5–23. The *Howe truss* is subjected to the loading shown. Determine the force in members GF, CD, and GC and state if the members are in tension or compression.

5–19. Determine the force in members HG, HE, and DE of the truss and state if the members are in tension or compression.

*5–20. Determine the force in members CD, HI, and CH of the truss and state if the members are in tension or compression.

*5–24. The *Howe truss* is subjected to the loading shown. Determine the force in members GH, BC, and BG of the truss and state if the members are in tension or compression.

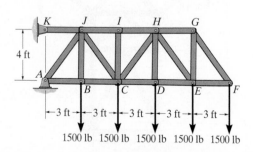

Probs. 5–19/20

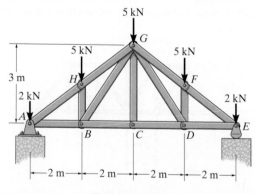

Probs. 5–23/24

5–25. Determine the force in members *EF, CF,* and *BC,* and state if the members are in tension or compression.

5–26. Determine the force in members *AF, BF,* and *BC,* and state if the members are in tension or compression.

5–29. Determine the force in members *BC, HC,* and *HG.* After the truss is sectioned use a single equation of equilibrium for the calculation of each force. State if these members are in tension or compression.

5–30. Determine the force in members *CD, CF,* and *CG* and state if these members are in tension or compression.

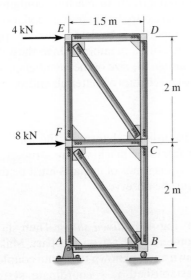

Probs. 5–25/26

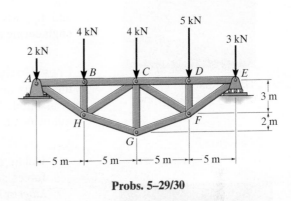

Probs. 5–29/30

5–27. Determine the force in members *EF, BE, BC,* and *BF* of the truss and state if these members are in tension or compression. Set $P_1 = 9$ kN, $P_2 = 12$ kN, and $P_3 = 6$ kN.

*5–28. Determine the force in members *BC, BE,* and *EF* of the truss and state if these members are in tension or compression. Set $P_1 = 6$ kN, $P_2 = 9$ kN, and $P_3 = 12$ kN.

5–31. Determine the force developed in members *FE, EB,* and *BC* of the truss and state if these members are in tension or compression.

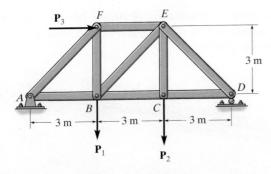

Probs. 5–27/28

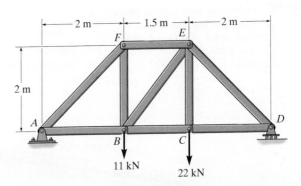

Prob. 5–31

5.5   FRAMES AND MACHINES

Frames and machines are two types of structures which are often composed of pin-connected *multiforce members*, i.e., members that are subjected to more than two forces. *Frames* are used to support loads, whereas *machines* contain moving parts and are designed to transmit and alter the effect of forces. Provided a frame or machine contains no more supports or members than are necessary to prevent its collapse, then the forces acting at the joints and supports can be determined by applying the equations of equilibrium to each of its members. Once these forces are obtained, it is then possible to *design* the size of the members, connections, and supports using the theory of mechanics of materials and an appropriate engineering design code.

This crane is a typical example of a framework.

Free-Body Diagrams. In order to determine the forces acting at the joints and supports of a frame or machine, the structure must be disassembled and the free-body diagrams of its parts must be drawn. The following important points *must* be observed:

- Isolate each part by drawing its *outlined shape*. Then show all the forces and/or couple moments that act on the part. Make sure to *label* or *identify* each known and unknown force and couple moment with reference to an established *x, y* coordinate system. Also, indicate any dimensions used for taking moments. As usual, the sense of an unknown force or couple moment can be assumed.

- Identify all the two-force members in the structure and represent their free-body diagrams as having two equal but opposite collinear forces acting at their points of application. (See Sec. 4.4.) By doing this, we can avoid solving an unnecessary number of equilibrium equations.

- Forces common to any two *contacting* members act with equal magnitudes but opposite sense on the free-body diagrams of the respective members.

Common tools such as these pliers act as simple machines. Here the applied force on the handles creates a much larger force at the jaws.

The following two examples graphically illustrate how to draw the free-body diagrams of a dismembered frame or machine. In all cases, the weight of the members is neglected.

EXAMPLE 5.8

For the frame shown in Fig. 5–19a, draw the free-body diagram of (a) each member, (b) the pins at B and A, and (c) the two members connected together.

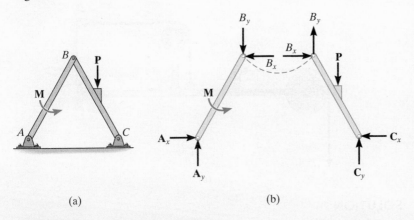

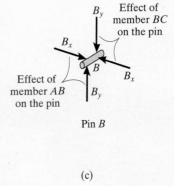

(a)   (b)   (c)

SOLUTION

Part (a). By inspection, members BA and BC are *not* two-force members. Instead, as shown on the free-body diagrams, Fig. 5–19b, BC is subjected to a force from each of the pins at B and C and the external force **P**. Likewise, AB is subjected to a force from each of the pins at A and B and the external couple moment **M**. The pin forces are represented by their x and y components.

Part (b). The pin at B is subjected to only *two forces*, i.e., the force of member BC and the force of member AB. For *equilibrium* these forces (or their respective components) must be equal but opposite, Fig. 5–19c. Notice that Newton's third law is applied between the pin and its connected members, i.e., the effect of the pin on the two members, Fig. 5–19b, and the equal but opposite effect of the two members on the pin, Fig. 5–19c. In the same manner, there are three forces on pin A, Fig. 5–19d, caused by the force components of member AB and each of the two pin leaves.

Part (c). The free-body diagram of both members connected together, yet removed from the supporting pins at A and C, is shown in Fig. 5–19e. The force components $\mathbf{B}_x$ and $\mathbf{B}_y$ are *not shown* on this diagram since they are *internal* forces (Fig. 5–19b) and therefore cancel out. Also, to be consistent when later applying the equilibrium equations, the unknown force components at A and C in Fig. 5–19e must act in the *same sense* as those shown in Fig. 5–19b.

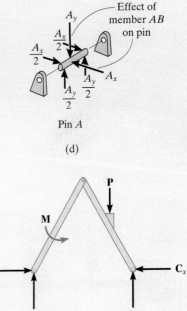

(d)

(e)

Fig. 5–19

EXAMPLE 5.9

For the frame shown in Fig. 5–20a, draw the free-body diagrams of (a) the entire frame including the pulleys and cords, (b) the frame without the pulleys and cords, and (c) each of the pulleys.

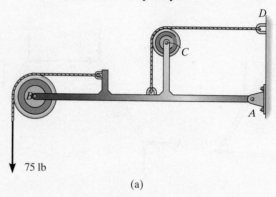

(a)

SOLUTION

Part (a). When the entire frame including the pulleys and cords is considered, the interactions at the points where the pulleys and cords are connected to the frame become pairs of *internal* forces which cancel each other and therefore are not shown on the free-body diagram, Fig. 5–20b.

Part (b). When the cords and pulleys are removed, their effect *on the frame* must be shown, Fig. 5–20c.

Part (c). The force components B_x, B_y, C_x, C_y of the pins on the pulleys are equal but opposite to the force components exerted by the pins on the frame, Fig. 5–20c.

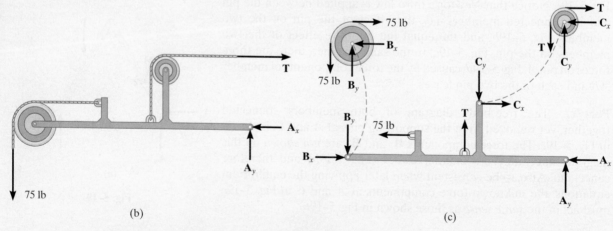

(b)

(c)

Fig. 5–20

PROCEDURE FOR ANALYSIS

The joint reactions on frames or machines (structures) composed of multiforce members can be determined using the following procedure.

Free-Body Diagram.

- Draw the free-body diagram of the entire frame or machine, a portion of it, or each of its members. The choice should be made so that it leads to the most direct solution of the problem.

- When the free-body diagram of a group of members of a frame or machine is drawn, the forces between the connected parts of this group are internal forces and are not shown on the free-body diagram of the group.

- Forces common to two members which are in contact act with equal magnitude but opposite sense on the respective free-body diagrams of the members.

- A two-force member, regardless of its shape, has equal but opposite collinear forces acting at the ends of the member.

- In many cases it is possible to tell by inspection the proper sense of the unknown forces acting on a member; however, if this seems difficult, the sense can be assumed.

- Remember that a couple moment is a free vector and can act at any point on the free-body diagram. Also, a force is a sliding vector and can act at any point along its line of action.

Equations of Equilibrium.

- Count the number of unknowns and compare it to the total number of equilibrium equations that are available. In two dimensions, there are three equilibrium equations that can be written for each member.

- Sum moments about a point that lies at the intersection of the lines of action of as many of the unknown forces as possible.

- If the solution of a force or couple moment is found to be a negative scalar, it means the sense of the force is the reverse of that shown on the free-body diagram.

5

EXAMPLE 5.10

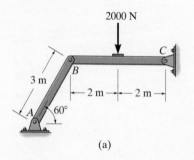

(a)

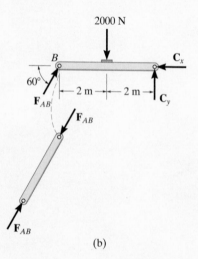

(b)

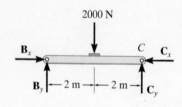

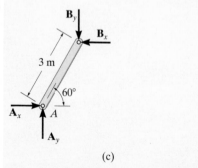

(c)

Fig. 5–21

Determine the horizontal and vertical components of force which the pin at C exerts on member BC of the frame in Fig. 5–21a.

SOLUTION I

Free-Body Diagrams. By inspection it can be seen that AB is a two-force member. The free-body diagrams are shown in Fig. 5–21b.

Equations of Equilibrium. The *three unknowns* can be determined by applying the three equations of equilibrium to member CB.

$$\zeta+\Sigma M_C = 0; \quad 2000 \text{ N}(2 \text{ m}) - (F_{AB} \sin 60°)(4 \text{ m}) = 0 \quad F_{AB} = 1154.7 \text{ N}$$

$$\overset{+}{\to}\Sigma F_x = 0; \quad 1154.7 \cos 60° \text{ N} - C_x = 0 \quad C_x = 577 \text{ N} \qquad \textit{Ans.}$$

$$+\uparrow\Sigma F_y = 0; \quad 1154.7 \sin 60° \text{ N} - 2000 \text{ N} + C_y = 0$$

$$C_y = 1000 \text{ N} \qquad \textit{Ans.}$$

SOLUTION II

Free-Body Diagrams. If one does not recognize that AB is a two-force member, then more work is involved in solving this problem. The free-body diagrams are shown in Fig. 5–21c.

Equations of Equilibrium. The *six unknowns* are determined by applying the three equations of equilibrium to each member.

Member AB

$$\zeta+\Sigma M_A = 0; \quad B_x(3 \sin 60° \text{ m}) - B_y(3 \cos 60° \text{ m}) = 0 \qquad (1)$$

$$\overset{+}{\to}\Sigma F_x = 0; \quad A_x - B_x = 0 \qquad (2)$$

$$+\uparrow\Sigma F_y = 0; \quad A_y - B_y = 0 \qquad (3)$$

Member BC

$$\zeta+\Sigma M_C = 0; \quad 2000 \text{ N}(2 \text{ m}) - B_y(4 \text{ m}) = 0 \qquad (4)$$

$$\overset{+}{\to}\Sigma F_x = 0; \quad B_x - C_x = 0 \qquad (5)$$

$$+\uparrow\Sigma F_y = 0; \quad B_y - 2000 \text{ N} + C_y = 0 \qquad (6)$$

The results for C_x and C_y can be determined by solving these equations in the following sequence: 4, 1, 5, then 6. The results are

$$B_y = 1000 \text{ N}$$

$$B_x = 577 \text{ N}$$

$$C_x = 577 \text{ N} \qquad \textit{Ans.}$$

$$C_y = 1000 \text{ N} \qquad \textit{Ans.}$$

By comparison, Solution I is simpler since the requirement that F_{AB} in Fig. 5–21b be equal, opposite, and collinear at the ends of member AB automatically satisfies Eqs. (1), (2), and (3) above, and therefore eliminates the need to write these equations. *As a result, save yourself some time and effort by always identifying the two-force members before starting the analysis!*

EXAMPLE 5.11

The compound beam shown in Fig. 5–22a is pin connected at B. Determine the components of reaction at its supports. Neglect its weight and thickness.

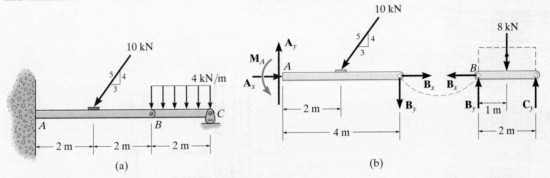

Fig. 5–22

SOLUTION

Free-Body Diagrams. By inspection, if we consider a free-body diagram of the *entire beam ABC*, there will be three unknown reactions at A and one at C. These four unknowns cannot all be obtained from the three available equations of equilibrium, and so for the solution it will become necessary to dismember the beam into its two segments, as shown in Fig. 5–22b.

Equations of Equilibrium. The six unknowns are determined as follows:

Segment BC

$$\xrightarrow{+} \Sigma F_x = 0; \quad B_x = 0$$

$$\zeta + \Sigma M_B = 0; \quad -8 \text{ kN}(1 \text{ m}) + C_y(2 \text{ m}) = 0$$

$$+\uparrow \Sigma F_y = 0; \quad B_y - 8 \text{ kN} + C_y = 0$$

Segment AB

$$\xrightarrow{+} \Sigma F_x = 0; \quad A_x - (10 \text{ kN})\left(\tfrac{3}{5}\right) + B_x = 0$$

$$\zeta + \Sigma M_A = 0; \quad M_A - (10 \text{ kN})\left(\tfrac{4}{5}\right)(2 \text{ m}) - B_y(4 \text{ m}) = 0$$

$$+\uparrow \Sigma F_y = 0; \quad A_y - (10 \text{ kN})\left(\tfrac{4}{5}\right) - B_y = 0$$

Solving each of these equations successively, using previously calculated results, we obtain

$$A_x = 6 \text{ kN} \quad A_y = 12 \text{ kN} \qquad M_A = 32 \text{ kN} \cdot \text{m} \qquad \textit{Ans.}$$

$$B_x = 0 \qquad B_y = 4 \text{ kN}$$

$$C_y = 4 \text{ kN} \qquad\qquad\qquad\qquad\qquad\qquad \textit{Ans.}$$

EXAMPLE 5.12

The 500-kg elevator car in Fig. 5–23a is being hoisted at constant speed by motor A using the pulley system shown. Determine the force developed in the two cables.

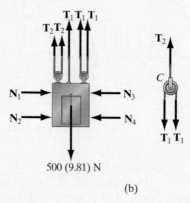

(a)

(b)

Fig. 5–23

SOLUTION

Free-Body Diagrams. We can solve this problem using the free-body diagrams of the elevator car and pulley C, Fig. 5–23b. The tensile forces developed in the two cables are denoted as T_1 and T_2.

Equations of Equilibrium. For pulley C,

$$+\uparrow\Sigma F_y = 0; \qquad T_2 - 2T_1 = 0 \qquad \text{or} \qquad T_2 = 2T_1 \tag{1}$$

For the elevator car,

$$+\uparrow\Sigma F_y = 0; \qquad 3T_1 + 2T_2 - 500(9.81)\ \text{N} = 0 \tag{2}$$

Substituting Eq. (1) into Eq. (2) yields

$$3T_1 + 2(2T_1) - 500(9.81)\ \text{N} = 0$$

$$T_1 = 700.71\ \text{N} = 701\ \text{N} \qquad\qquad Ans.$$

Substituting this result into Eq. (1),

$$T_2 = 2(700.71)\ \text{N} = 1401\ \text{N} = 1.40\ \text{kN} \qquad\qquad Ans.$$

EXAMPLE 5.13

The two planks in Fig. 5–24a are connected together by cable BC and a smooth spacer DE. Determine the reactions at the smooth supports A and F, and also find the force developed in the cable and spacer.

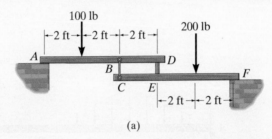

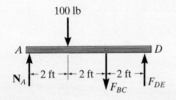

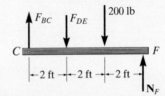

(b)

Fig. 5–24

SOLUTION

Free-Body Diagrams. The free-body diagram of each plank is shown in Fig. 5–24b. It is important to apply Newton's third law to the interaction forces as shown.

Equations of Equilibrium. For plank AD,

$$\zeta + \Sigma M_A = 0; \quad F_{DE}(6 \text{ ft}) - F_{BC}(4 \text{ ft}) - 100 \text{ lb } (2 \text{ ft}) = 0$$

For plank CF,

$$\zeta + \Sigma M_F = 0; \quad F_{DE}(4 \text{ ft}) - F_{BC}(6 \text{ ft}) + 200 \text{ lb } (2 \text{ ft}) = 0$$

Solving simultaneously,

$$F_{DE} = 140 \text{ lb} \quad F_{BC} = 160 \text{ lb} \qquad\qquad Ans.$$

Using these results, for plank AD,

$$+\uparrow \Sigma F_y = 0; \quad N_A + 140 \text{ lb} - 160 \text{ lb} - 100 \text{ lb} = 0$$

$$N_A = 120 \text{ lb} \qquad\qquad Ans.$$

And for plank CF,

$$+\uparrow \Sigma F_y = 0; \quad N_F + 160 \text{ lb} - 140 \text{ lb} - 200 \text{ lb} = 0$$

$$N_F = 180 \text{ lb} \qquad\qquad Ans.$$

5

PRELIMINARY PROBLEM

P5–3. In each case, identify any two-force members, and then draw the free-body diagrams of each member of the frame.

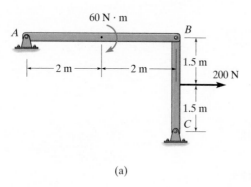

(a)

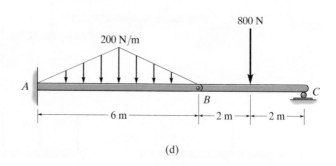

(d)

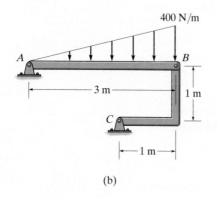

(b)

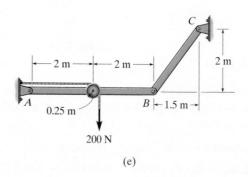

(e)

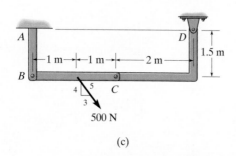

(c)

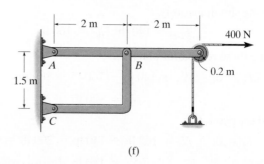

(f)

Prob. P5–3

FUNDAMENTAL PROBLEMS

All problem solutions must include FBDs.

F5–13. Determine the force P needed to hold the 60-lb weight in equilibrium.

F5–15. If a 100-N force is applied to the handles of the pliers, determine the clamping force exerted on the smooth pipe B and the magnitude of the resultant force that one of the members exerts on pin A.

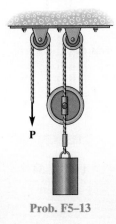

Prob. F5–13

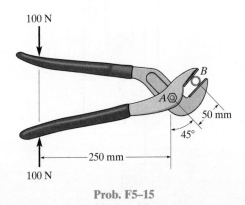

Prob. F5–15

F5–14. Determine the horizontal and vertical components of reaction at pin C.

F5–16. Determine the horizontal and vertical components of reaction at pin C.

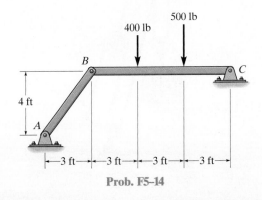

Prob. F5–14

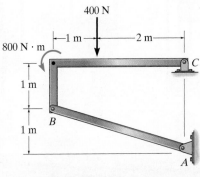

Prob. F5–16

PROBLEMS

All problem solutions must include FBDs.

*5–32. Determine the force **P** required to hold the 100-lb weight in equilibrium.

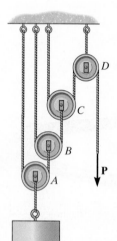

Prob. 5–32

5–33. In each case, determine the force **P** required to maintain equilibrium of the 100-lb block.

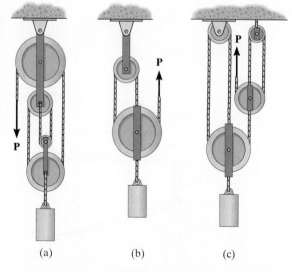

(a)   (b)   (c)

Prob. 5–33

5–34. Determine the force **P** required to hold the 50-kg block in equilibrium.

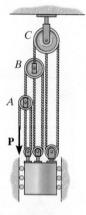

Prob. 5–34

5–35. Determine the force **P** required to hold the 150-kg crate in equilibrium.

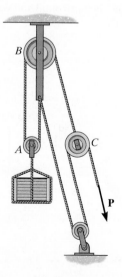

Prob. 5–35

*5–36. Determine the reactions at the supports A, C, and E of the compound beam.

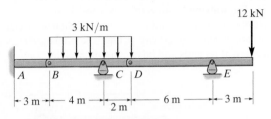

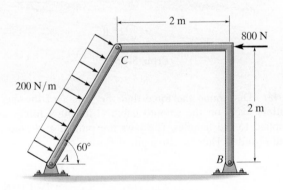

Prob. 5–36

5–37. Determine the resultant force at pins A, B, and C on the three-member frame.

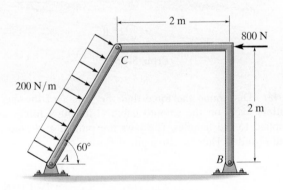

Prob. 5–37

5–38. Determine the reactions at the supports at A, E, and B of the compound beam.

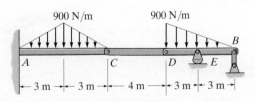

Prob. 5–38

5–39. The wall crane supports a load of 700 lb. Determine the horizontal and vertical components of reaction at the pins A and D. Also, what is the force in the cable at the winch W?

*5–40. The wall crane supports a load of 700 lb. Determine the horizontal and vertical components of reaction at the pins A and D. Also, what is the force in the cable at the winch W? The jib ABC has a weight of 100 lb and member BD has a weight of 40 lb. Each member is uniform and has a center of gravity at its center.

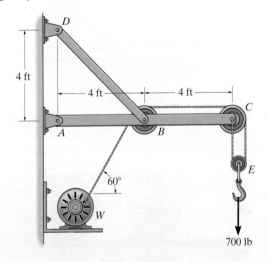

Probs. 5–39/40

5–41. Determine the horizontal and vertical components of force which the pins at A and B exert on the frame.

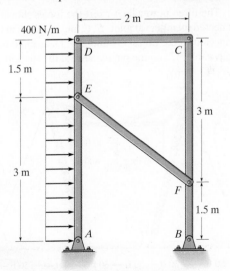

Prob. 5–41

5–42. Determine the force in members FD and DB of the frame. Also, find the horizontal and vertical components of reaction the pin at C exerts on member ABC and member EDC.

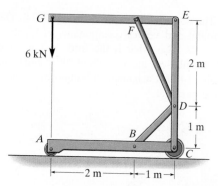

Prob. 5–42

5–43. Determine the force that the smooth 20-kg cylinder exerts on members AB and CDB. Also, what are the horizontal and vertical components of reaction at pin A?

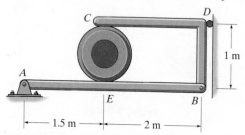

Prob. 5–43

***5–44.** The three power lines exert the forces shown on the pin-connected members at joints B, C, and D, which in turn are pin connected to the poles AH and EG. Determine the force in the guy cable AI and the pin reaction at the support H.

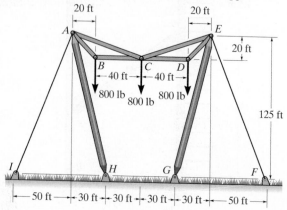

Prob. 5–44

5–45. The pumping unit is used to recover oil. When the walking beam ABC is horizontal, the force acting in the wireline at the well head is 250 lb. Determine the torque **M** which must be exerted by the motor in order to overcome this load. The horse-head C weighs 60 lb and has a center of gravity at G_C. The walking beam ABC has a weight of 130 lb and a center of gravity at G_B, and the counterweight has a weight of 200 lb and a center of gravity at G_W. The pitman, AD, is pin connected at its ends and has negligible weight.

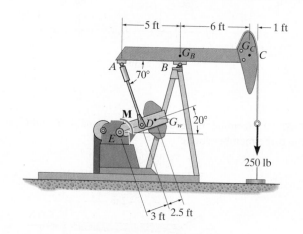

Prob. 5–45

5–46. Determine the force that the jaws J of the metal cutters exert on the smooth cable C if 100-N forces are applied to the handles. The jaws are pinned at E and A, and D and B. There is also a pin at F.

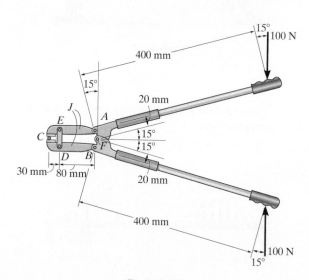

Prob. 5–46

5–47. The machine shown is used for forming metal plates. It consists of two toggles *ABC* and *DEF*, which are operated by the hydraulic cylinder *H*. The toggles push the movable bar *G* forward, pressing the plate *p* into the cavity. If the force which the plate exerts on the head is *P* = 12 kN, determine the force *F* in the hydraulic cylinder when θ = 30°.

5–49. The pipe cutter is clamped around the pipe *P*. If the wheel at *A* exerts a normal force of $F_A = 80$ N on the pipe, determine the normal forces of wheels *B* and *C* on the pipe. Also calculate the pin reaction on the wheel at *C*. The three wheels each have a radius of 7 mm and the pipe has an outer radius of 10 mm.

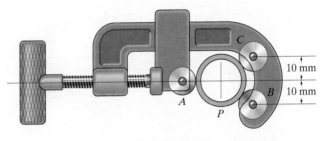

Prob. 5–49

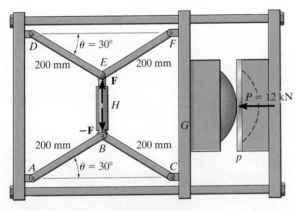

Prob. 5–47

5–50. Determine the force created in the hydraulic cylinders *EF* and *AD* in order to hold the shovel in equilibrium. The shovel load has a mass of 1.25 Mg and a center of gravity at *G*. All joints are pin connected.

*5–48. Determine the horizontal and vertical components of force which pin *C* exerts on member *ABC*. The 600-N force is applied to the pin.

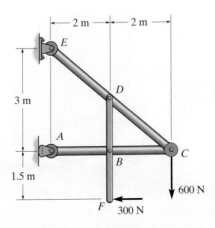

Prob. 5–48

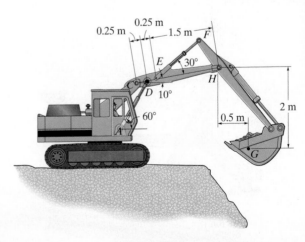

Prob. 5–50

5–51. The hydraulic crane is used to lift the 1400-lb load. Determine the force in the hydraulic cylinder AB and the force in links AC and AD when the load is held in the position shown.

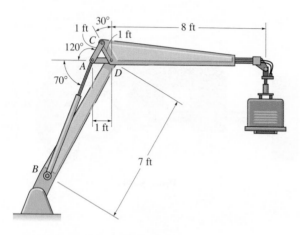

Prob. 5–51

***5–52.** Determine force **P** on the cable if the spring is compressed 25 mm when the mechanism is in the position shown. The spring has a stiffness of $k = 6$ kN/m.

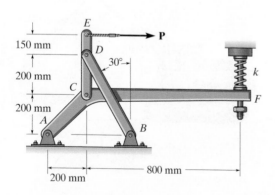

Prob. 5–52

5–53. If $d = 0.75$ ft and the spring has an unstretched length of 1 ft, determine the force F required for equilibrium.

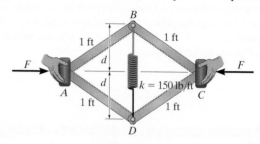

Prob. 5–53

5–54. If a force of $F = 50$ lb is applied to the pads at A and C, determine the smallest dimension d required for equilibrium if the spring has an unstretched length of 1 ft.

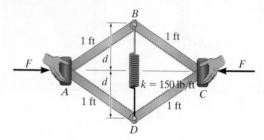

Prob. 5–54

5–55. The skid-steer loader has a mass of 1.18 Mg, and in the position shown the center of mass is at G_1. If there is a 300-kg stone in the bucket, with center of mass at G_2, determine the reactions of each pair of wheels A and B on the ground and the force in the hydraulic cylinder CD and at the pin E. There is a similar linkage on each side of the loader.

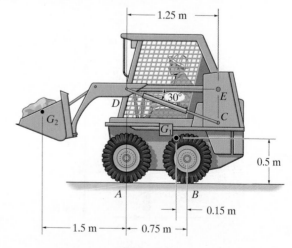

Prob. 5–55

*5–56. Determine the force **P** on the cable if the spring is compressed 0.5 in. when the mechanism is in the position shown. The spring has a stiffness of $k = 800$ lb/ft.

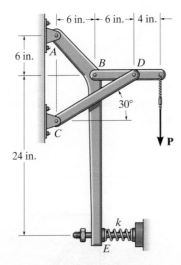

Prob. 5–56

5–57. The spring has an unstretched length of 0.3 m. Determine the angle θ for equilibrium if the uniform bars each have a mass of 20 kg.

5–58. The spring has an unstretched length of 0.3 m. Determine the mass m of each uniform bar if each angle $\theta = 30°$ for equilibrium.

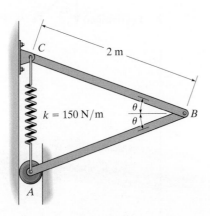

Probs. 5–57/58

5–59. The piston C moves vertically between the two smooth walls. If the spring has a stiffness of $k = 15$ lb/in., and is unstretched when $\theta = 0°$, determine the couple moment that must be applied to AB to hold the mechanism in equilibrium when $\theta = 30°$.

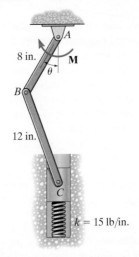

Prob. 5–59

*5–60. The platform scale consists of a combination of third and first class levers so that the load on one lever becomes the effort that moves the next lever. Through this arrangement, a small weight can balance a massive object. If $x = 450$ mm, determine the required mass of the counterweight S required to balance the load L having a mass of 90 kg.

5–61. The platform scale consists of a combination of third and first class levers so that the load on one lever becomes the effort that moves the next lever. Through this arrangement, a small weight can balance a massive object. If $x = 450$ mm, and the mass of the counterweight S is 2 kg, determine the mass of the load L required to maintain the balance.

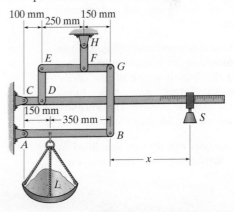

Probs. 5–60/61

5

CHAPTER REVIEW

Simple Truss

A simple truss consists of triangular elements connected together by pinned joints. The forces within its members can be determined by assuming the members are all two-force members, connected concurrently at each joint. The members are either in tension or compression, or carry no force.

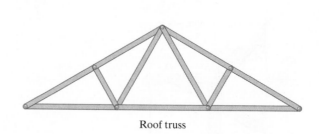

Roof truss

Method of Joints

The method of joints states that if a truss is in equilibrium, then each of its joints is also in equilibrium. For a plane truss, the concurrent force system at each joint must satisfy force equilibrium.

$$\Sigma F_x = 0$$

$$\Sigma F_y = 0$$

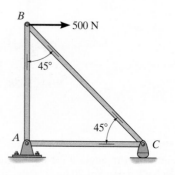

To obtain a numerical solution for the forces in the members, select a joint that has a free-body diagram with at most two unknown forces and one known force. (This may require first finding the reactions at the supports.)

Once a member force is determined, use its value and apply it to an adjacent joint.

Remember that forces that *pull* on the joint are *tensile forces*, and those that *push* on the joint are *compressive forces*.

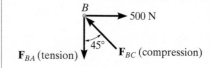

To avoid a simultaneous solution of two equations, set one of the coordinate axes along the line of action of one of the unknown forces and sum forces perpendicular to this axis. This will allow a direct solution for the other unknown.

The analysis can also be simplified by first identifying all the zero-force members.

Method of Sections

The method of sections states that if a truss is in equilibrium, then each part of the truss is also in equilibrium. Pass a section through the truss and the member whose force is to be determined. Then draw the free-body diagram of the sectioned part having the least number of forces on it.

Sectioned members subjected to *pulling* are in *tension*, and those that are subjected to *pushing* are in *compression*.

Three equations of equilibrium are available to determine the unknowns.

If possible, sum forces in a direction that is perpendicular to two of the three unknown forces. This will yield a direct solution for the third force.

Sum moments about the point where the lines of action of two of the three unknown forces intersect, so that the third unknown force can be determined directly.

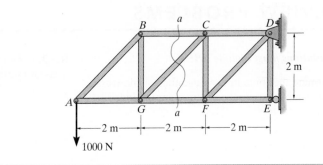

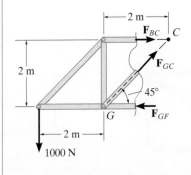

$$\Sigma F_x = 0$$
$$\Sigma F_y = 0$$
$$\Sigma M_O = 0$$
$$+\uparrow \Sigma F_y = 0$$
$$-1000 \text{ N} + F_{GC} \sin 45° = 0$$
$$F_{GC} = 1.41 \text{ kN (T)}$$
$$\zeta + \Sigma M_C = 0$$
$$1000 \text{ N}(4 \text{ m}) - F_{GF}(2 \text{ m}) = 0$$
$$F_{GF} = 2 \text{ kN (C)}$$

Frames and Machines

Frames and machines are structures that contain one or more multiforce members, that is, members with three or more forces or couples acting on them. Frames are designed to support loads, and machines transmit and alter the effect of forces.

The forces acting at the joints of a frame or machine can be determined by drawing the free-body diagrams of each of its members or parts. The principle of action–reaction should be carefully observed when indicating these forces on the free-body diagram of each adjacent member or pin. For a coplanar force system, there are three equilibrium equations available for each member.

To simplify the analysis, be sure to recognize all two-force members. They have equal but opposite collinear forces at their ends.

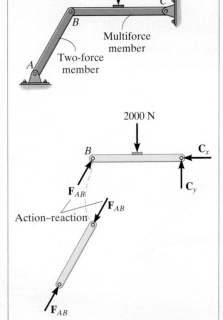

REVIEW PROBLEMS

All problem solutions must include FBDs.

R5–1.   Determine the force in each member of the truss and state if the members are in tension or compression.

R5–3.   Determine the force in member *GJ* and *GC* of the truss and state if the members are in tension or compression.

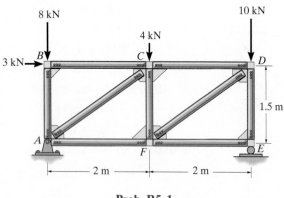

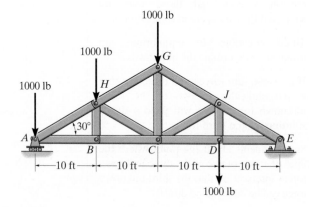

Prob. R5–1

Prob. R5–3

R5–2.   Determine the force in each member of the truss and state if the members are in tension or compression.

*R5–4.   Determine the force in members *GF*, *FB*, and *BC* of the *Fink truss* and state if the members are in tension or compression.

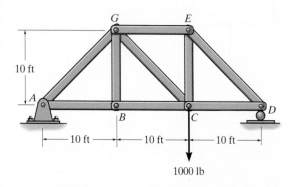

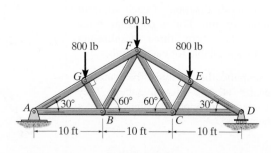

Prob. R5–2

Prob. R5–4

R5–5. Determine the horizontal and vertical components of force that the pins A and B exert on the two-member frame.

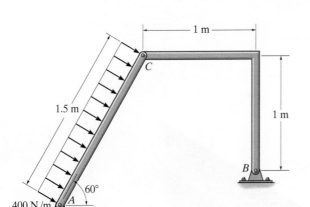

Prob. R5–5

R5–6. Determine the horizontal and vertical components of force that the pins A and C exert on the two-member frame.

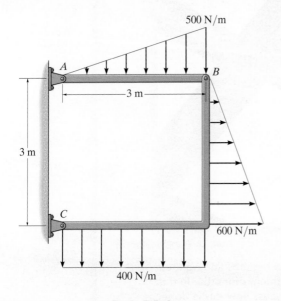

Prob. R5–6

R5–7. The three pin-connected members shown in the *top view* support a downward force of 60 lb at G. If only vertical forces are supported at the connections B, C, E and pad supports A, D, F, determine the reactions at each pad.

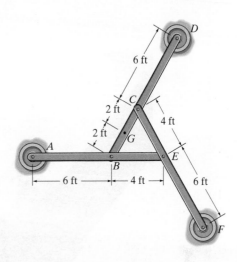

Prob. R5–7

*R5–8. Determine the resultant forces at pins B and C on member ABC of the four-member frame.

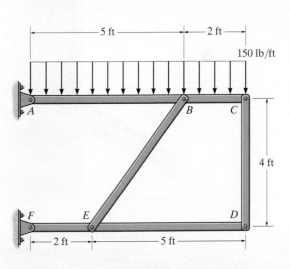

Prob. R5–8

CHAPTER 6

(© Michael N. Paras/AGE Fotostock/Alamy)

The design of these structural members requires finding their centroid and calculating their moment of inertia. In this chapter we will discuss how this is done.

CENTER OF GRAVITY, CENTROID, AND MOMENT OF INERTIA

CHAPTER OBJECTIVES

■ To show how to determine the location of the center of gravity and centroid for a body of arbitrary shape and for a composite body.

■ To present a method for finding the resultant of a general distributed loading.

■ To show how to determine the moment of inertia of an area.

6.1 CENTER OF GRAVITY AND THE CENTROID OF A BODY

In this section we will show how to locate the center of gravity for a body of arbitrary shape, and then we will show that the centroid of the body can be found using this same method.

Center of Gravity. A body is composed of an infinite number of particles of differential size, and so if the body is located within a gravitational field, then each of these particles will have a weight dW. These weights will form an approximately parallel force system, and the resultant of this system is the total weight of the body, which passes through a single point called the *center of gravity*, G.

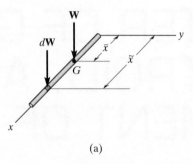

(a)

Fig. 6–1

To show how to determine the location of the center of gravity, consider the rod in Fig. 6–1a, where the segment having the weight dW is located at the arbitrary position $\tilde{x}$. Using the methods outlined in Sec. 3.8, the total weight of the rod is the sum of the weights of all of its particles, that is

$$+\downarrow F_R = \Sigma F_z; \qquad\qquad W = \int dW$$

The location of the center of gravity, measured from the y axis, is determined by equating the moment of W about the y axis, Fig. 6–1b, to the sum of the moments of the weights of all its particles about this same axis. Therefore,

$$(M_R)_y = \Sigma M_y; \qquad\qquad \bar{x}W = \int \tilde{x}dW$$

$$\bar{x} = \frac{\displaystyle\int \tilde{x}\,dW}{\displaystyle\int dW}$$

In a similar manner, if the body represents a plate, Fig. 6–1b, then a moment balance about the x and y axes would be required to determine the location $(\bar{x}, \bar{y})$ of point G. Finally we can generalize this idea to a three-dimensional body, Fig. 6–1c, and perform a moment balance about each of the three axes to locate G for any rotated position of the axes. This results in the following equations.

$$\bar{x} = \frac{\displaystyle\int \tilde{x}\,dW}{\displaystyle\int dW} \qquad \bar{y} = \frac{\displaystyle\int \tilde{y}\,dW}{\displaystyle\int dW} \qquad \bar{z} = \frac{\displaystyle\int \tilde{z}\,dW}{\displaystyle\int dW} \qquad (6\text{–}1)$$

where

$\bar{x}, \bar{y}, \bar{z}$ are the coordinates of the center of gravity G.
$\tilde{x}, \tilde{y}, \tilde{z}$ are the coordinates of an arbitrary particle in the body.

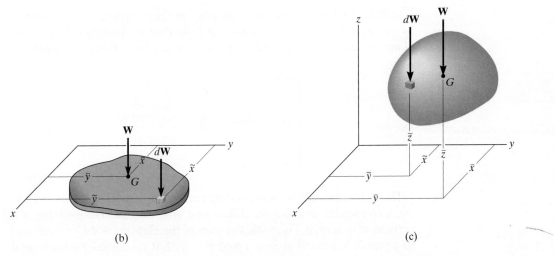

(b)

(c)

Fig. 6–1 (cont.)

Centroid of a Volume. If the body in Fig. 6–2 is made of the same material, then its specific weight γ (gamma) will be constant throughout the body. Therefore, a differential element of volume dV will have a weight $dW = \gamma \, dV$. Substituting this into Eqs. 6–1 and canceling out γ, we obtain formulas that locate the **centroid** *C* or geometric center of the body; namely

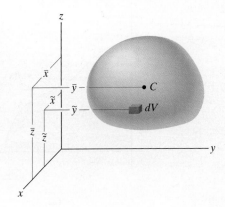

$$\bar{x} = \frac{\displaystyle\int_V \tilde{x} \, dV}{\displaystyle\int_V dV} \qquad \bar{y} = \frac{\displaystyle\int_V \tilde{y} \, dV}{\displaystyle\int_V dV} \qquad \bar{z} = \frac{\displaystyle\int_V \tilde{z} \, dV}{\displaystyle\int_V dV} \qquad (6\text{–}2)$$

Fig. 6–2

Since these equations represent a balance of the moments of the volume of the body, then if the volume possesses two planes of symmetry, its centroid will lie along the line of intersection of these two planes. For example, the cone in Fig. 6–3 has a centroid that lies on the *y* axis so that $\bar{x} = \bar{z} = 0$. To find the location $\bar{y}$ of the centroid, we can use the second of Eqs. 6–2. Here a single integration is possible if we choose a differential element represented by a *thin disk* having a thickness dy and radius $r = z$. Its volume is $dV = \pi r^2 \, dy = \pi z^2 \, dy$ and its centroid is at $\tilde{x} = 0, \tilde{y} = y, \tilde{z} = 0$.

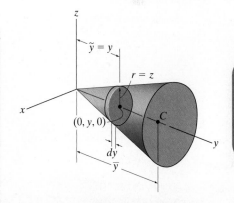

Fig. 6–3

6

Centroid of an Area. If an area lies in the x–y plane and is bounded by the curve $y = f(x)$, as shown in Fig. 6–4a, then its centroid will be in this plane and can be determined from integrals similar to Eqs. 6–2, namely,

$$\bar{x} = \frac{\int_A \tilde{x}\, dA}{\int_A dA} \qquad \bar{y} = \frac{\int_A \tilde{y}\, dA}{\int_A dA} \tag{6–3}$$

These integrals can be evaluated by performing a *single integration* if we use a *rectangular strip* for the differential area element. For example, if a vertical strip is used, Fig. 6–4b, the area of the element is $dA = y\, dx$, and its centroid is located at $\tilde{x} = x$ and $\tilde{y} = y/2$. If we consider a horizontal strip, Fig. 6–4c, then $dA = x\, dy$, and its centroid is located at $\tilde{x} = x/2$ and $\tilde{y} = y$.

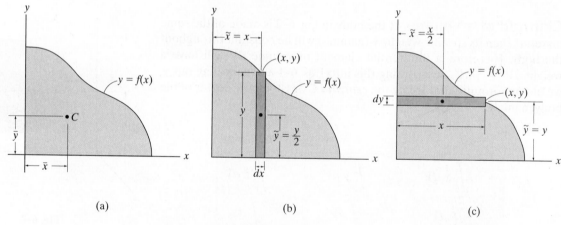

(a)   (b)   (c)

Fig. 6–4

IMPORTANT POINTS

- The centroid represents the geometric center of a body. This point coincides with the center of gravity only if the material composing the body is uniform or homogeneous.

- Formulas used to locate the center of gravity or the centroid represent a balance between the sum of moments of all the parts of the system and the moment of the "resultant" for the system.

PROCEDURE FOR ANALYSIS

The center of gravity or centroid of an object or shape can be determined by single integrations using the following procedure.

Differential Element.

- Select an appropriate coordinate system, specify the coordinate axes, and then choose a differential element for integration.

- For areas the differential element is generally a *rectangle* of area dA, having a finite length and differential width.

- For volumes the differential element can be a circular *disk* of volume dV, having a finite radius and differential thickness.

- Locate the element so that it touches the arbitrary point (x, y, z) on the curve that defines the boundary of the shape.

Size and Moment Arms.

- Express the area dA or volume dV of the element in terms of the coordinates describing the curve.

- Express the moment arms $\tilde{x}, \tilde{y}, \tilde{z}$ for the centroid or center of gravity of the element in terms of the coordinates describing the curve.

Integrations.

- Substitute the formulations for $\tilde{x}, \tilde{y}, \tilde{z}$ and dA or dV into the appropriate equations (Eqs. 6–1 through 6–3).

- Express the function in the integrand in terms of the same variable as the differential thickness of the element.

- The limits of the integral are defined from the two extreme locations of the element's differential thickness, so that when the elements are "summed" or the integration performed, the entire region is covered.[*]

[*]Some formulas for integration are given in Appendix A.

6

EXAMPLE 6.1

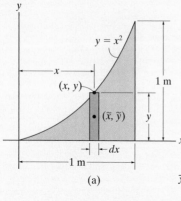

(a)

Fig. 6–5

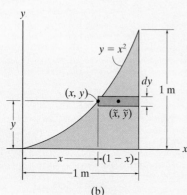

(b)

Locate the centroid of the area shown in Fig. 6–5a.

SOLUTION I

Differential Element. A differential element of thickness dx is shown in Fig. 6–5a. The element intersects the curve at the *arbitrary point* (x, y), and so it has a height y.

Area and Moment Arms. The area of the element is $dA = y\,dx$, and its centroid is located at $\tilde{x} = x$, $\tilde{y} = y/2$.

Integrations. Applying Eqs. 6–3 and integrating with respect to x yields

$$\bar{x} = \frac{\displaystyle\int_A \tilde{x}\,dA}{\displaystyle\int_A dA} = \frac{\displaystyle\int_0^{1\,m} xy\,dx}{\displaystyle\int_0^{1\,m} y\,dx} = \frac{\displaystyle\int_0^{1\,m} x^3\,dx}{\displaystyle\int_0^{1\,m} x^2\,dx} = \frac{0.250}{0.333} = 0.75\ m \qquad Ans.$$

$$\bar{y} = \frac{\displaystyle\int_A \tilde{y}\,dA}{\displaystyle\int_A dA} = \frac{\displaystyle\int_0^{1\,m} (y/2)y\,dx}{\displaystyle\int_0^{1\,m} y\,dx} = \frac{\displaystyle\int_0^{1\,m} (x^2/2)x^2\,dx}{\displaystyle\int_0^{1\,m} x^2\,dx} = \frac{0.100}{0.333} = 0.3\ m \qquad Ans.$$

SOLUTION II

Differential Element. The differential element of thickness dy is shown in Fig. 6–5b. The element intersects the curve at the *arbitrary point* (x, y), and so it has a length $(1 - x)$.

Area and Moment Arms. The area of the element is $dA = (1 - x)\,dy$, and its centroid is located at

$$\tilde{x} = x + \left(\frac{1 - x}{2}\right) = \frac{1 + x}{2}, \quad \tilde{y} = y$$

Integrations. Applying Eqs. 6–3 and integrating with respect to y, we obtain

$$\bar{x} = \frac{\displaystyle\int_A \tilde{x}\,dA}{\displaystyle\int_A dA} = \frac{\displaystyle\int_0^{1\,m} [(1 + x)/2](1 - x)\,dy}{\displaystyle\int_0^{1\,m} (1 - x)\,dy} = \frac{\dfrac{1}{2}\displaystyle\int_0^{1\,m} (1 - y)\,dy}{\displaystyle\int_0^{1\,m} (1 - \sqrt{y})\,dy} = \frac{0.250}{0.333} = 0.75\ m \qquad Ans.$$

$$\bar{y} = \frac{\displaystyle\int_A \tilde{y}\,dA}{\displaystyle\int_A dA} = \frac{\displaystyle\int_0^{1\,m} y(1 - x)\,dy}{\displaystyle\int_0^{1\,m} (1 - x)\,dy} = \frac{\displaystyle\int_0^{1\,m} (y - y^{3/2})\,dy}{\displaystyle\int_0^{1\,m} (1 - \sqrt{y})\,dy} = \frac{0.100}{0.333} = 0.3\ m \qquad Ans.$$

NOTE: Plot these results and notice that they seem reasonable. Also, by comparison, elements of thickness dx offer a simpler solution.

EXAMPLE 6.2

Locate the centroid of the semi-elliptical area shown in Fig. 6–6a.

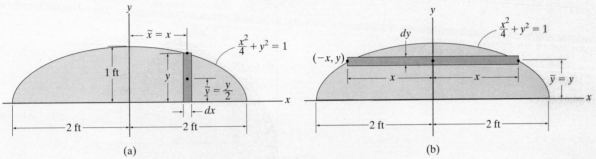

Fig. 6–6

SOLUTION I

Differential Element.   The rectangular differential element parallel to the y axis shown shaded in Fig. 6–6a will be considered. This element has a thickness of dx and a height of y.

Area and Moment Arms.   Thus, the area is $dA = y\,dx$, and its centroid is located at $\tilde{x} = x$ and $\tilde{y} = y/2$.

Integration.   Since the area is symmetrical about the y axis,

$$\bar{x} = 0 \qquad \qquad \textit{Ans.}$$

Applying the second of Eqs. 6–3 with $y = \sqrt{1 - \dfrac{x^2}{4}}$, we have

$$\bar{y} = \frac{\displaystyle\int_A \tilde{y}\,dA}{\displaystyle\int_A dA} = \frac{\displaystyle\int_{-2\,\text{ft}}^{2\,\text{ft}} \frac{y}{2}(y\,dx)}{\displaystyle\int_{-2\,\text{ft}}^{2\,\text{ft}} y\,dx} = \frac{\dfrac{1}{2}\displaystyle\int_{-2\,\text{ft}}^{2\,\text{ft}}\left(1 - \frac{x^2}{4}\right)dx}{\displaystyle\int_{-2\,\text{ft}}^{2\,\text{ft}}\sqrt{1 - \frac{x^2}{4}}\,dx} = \frac{4/3}{\pi} = 0.424\ \text{ft} \qquad \textit{Ans.}$$

SOLUTION II

Differential Element.   The shaded rectangular differential element of thickness dy and length $2x$ will be considered, Fig. 6–6b.

Area and Moment Arms.   The area is $dA = 2x\,dy$, and its centroid is at $\tilde{x} = 0$ and $\tilde{y} = y$.

Integration.   Applying the second of Eqs. 6–3, with $x = 2\sqrt{1 - y^2}$, we have

$$\bar{y} = \frac{\displaystyle\int_A \tilde{y}\,dA}{\displaystyle\int_A dA} = \frac{\displaystyle\int_0^{1\,\text{ft}} y(2x\,dy)}{\displaystyle\int_0^{1\,\text{ft}} 2x\,dy} = \frac{\displaystyle\int_0^{1\,\text{ft}} 4y\sqrt{1 - y^2}\,dy}{\displaystyle\int_0^{1\,\text{ft}} 4\sqrt{1 - y^2}\,dy} = \frac{4/3}{\pi}\ \text{ft} = 0.424\ \text{ft} \qquad \textit{Ans.}$$

6

EXAMPLE   6.3

Locate the $\bar{y}$ centroid for the paraboloid of revolution, shown in Fig. 6–7.

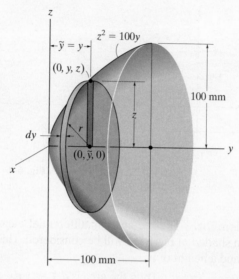

Fig. 6–7

SOLUTION

Differential Element.   An element having the shape of a *thin disk* is chosen. This element has a thickness dy, it intersects the curve at the *arbitrary point* $(0, y, z)$, and so its radius is $r = z$.

Volume and Moment Arm.   The volume of the element is $dV = (\pi z^2)\, dy$, and its centroid is located at $\tilde{y} = y$.

Integration.   Applying the second of Eqs. 6–2 and integrating with respect to y yields

$$\bar{y} = \frac{\displaystyle\int_V \tilde{y}\, dV}{\displaystyle\int_V dV} = \frac{\displaystyle\int_0^{100\text{ mm}} y(\pi z^2)\, dy}{\displaystyle\int_0^{100\text{ mm}} (\pi z^2)\, dy} = \frac{100\pi \displaystyle\int_0^{100\text{ mm}} y^2\, dy}{100\pi \displaystyle\int_0^{100\text{ mm}} y\, dy} = 66.7 \text{ mm} \qquad Ans.$$

PRELIMINARY PROBLEM

P6–1.   In each case, use the element shown and specify $\tilde{x}, \tilde{y},$ and dA.

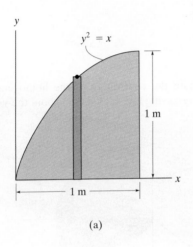

(a)

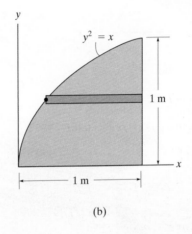

(b)

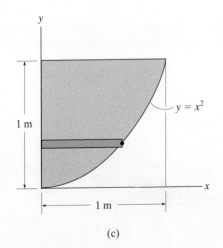

(c)

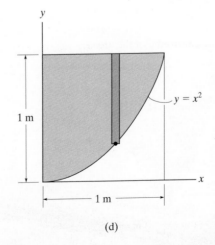

(d)

Prob. P6–1

FUNDAMENTAL PROBLEMS

F6–1. Determine the centroid $(\bar{x}, \bar{y})$ of the area.

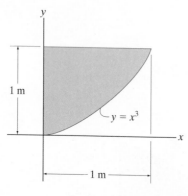

$y = x^3$

1 m

1 m

Prob. F6–1

F6–2. Determine the centroid $(\bar{x}, \bar{y})$ of the area.

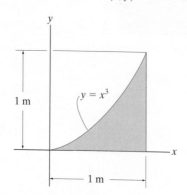

$y = x^3$

1 m

1 m

Prob. F6–2

F6–3. Determine the centroid $\bar{y}$ of the area.

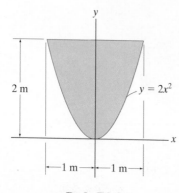

2 m

$y = 2x^2$

1 m   1 m

Prob. F6–3

F6–4. Locate the center of gravity $\bar{x}$ of the straight rod if its weight per unit length is given by $W = W_0(1 + x^2/L^2)$.

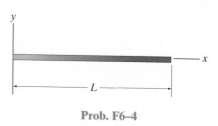

L

Prob. F6–4

F6–5. Locate the centroid $\bar{y}$ of the homogeneous solid formed by revolving the shaded area about the y axis.

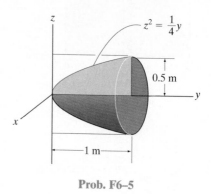

$z^2 = \frac{1}{4}y$

0.5 m

1 m

Prob. F6–5

F6–6. Locate the centroid $\bar{z}$ of the homogeneous solid formed by revolving the shaded area about the z axis.

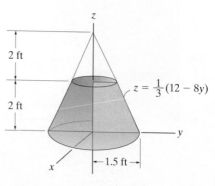

2 ft

2 ft

$z = \frac{1}{3}(12 - 8y)$

1.5 ft

Prob. F6–6

PROBLEMS

6–1. Locate the centroid $\bar{x}$ of the area.

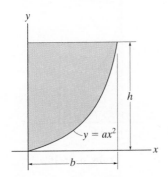

Prob. 6–1

6–2. Locate the centroid of the area.

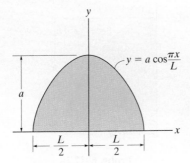

Prob. 6–2

6–3. Locate the centroid $\bar{x}$ of the area.

*6–4. Locate the centroid $\bar{y}$ of the area.

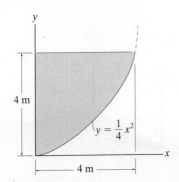

Probs. 6–3/4

6–5. Locate the centroid $\bar{x}$ of the area.

6–6. Locate the centroid $\bar{y}$ of the area.

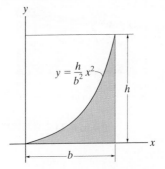

Probs. 6–5/6

6–7. Locate the centroid $\bar{x}$ of the area.

*6–8. Locate the centroid $\bar{y}$ of the area.

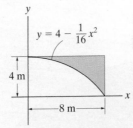

Probs. 6–7/8

6–9. Locate the centroid $\bar{x}$ of the area. Solve the problem by evaluating the integrals using Simpson's rule.

6–10. Locate the centroid $\bar{y}$ of the area. Solve the problem by evaluating the integrals using Simpson's rule.

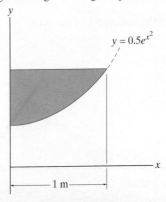

Probs. 6–9/10

6–11. Locate the centroid $\bar{y}$ of the area.

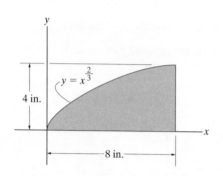

Prob. 6–11

*6–12. Locate the centroid $\bar{x}$ of the area.

6–13. Locate the centroid $\bar{y}$ of the area.

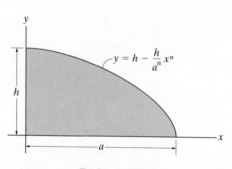

Probs. 6–12/13

6–14. Locate the centroid $\bar{y}$ of the area.

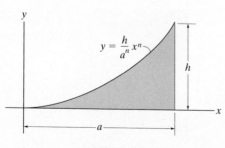

Prob. 6–14

6–15. Locate the centroid $\bar{x}$ of the area.

*6–16. Locate the centroid $\bar{y}$ of the area.

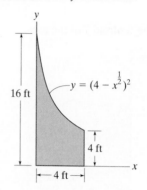

Probs. 6–15/16

6–17. Locate the centroid $\bar{x}$ of the area.

6–18. Locate the centroid $\bar{y}$ of the area.

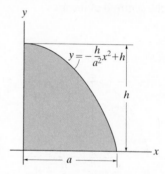

Probs. 6–17/18

6–19. The plate has a thickness of 0.25 ft and a specific weight of $\gamma = 180\,\text{lb/ft}^3$. Determine the location of its center of gravity. Also, find the tension in each of the cords used to support it.

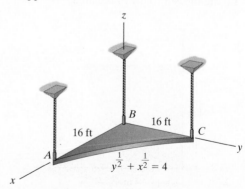

Prob. 6–19

*6–20. Locate the centroid $\bar{x}$ of the area.

6–21. Locate the centroid $\bar{y}$ of the area.

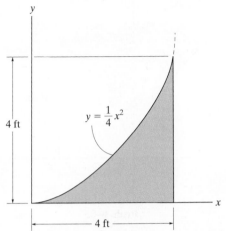

Probs. 6–20/21

6–22. Locate the centroid $\bar{x}$ of the area.

6–23. Locate the centroid $\bar{y}$ of the area.

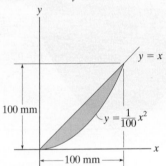

Probs. 6–22/23

*6–24. Locate the centroid $\bar{x}$ of the area.

6–25. Locate the centroid $\bar{y}$ of the area.

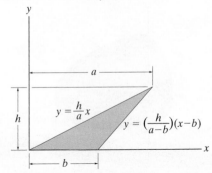

Probs. 6–24/25

6–26. Locate the centroid $\bar{x}$ of the area.

6–27. Locate the centroid $\bar{y}$ of the area.

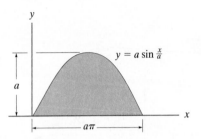

Probs. 6–26/27

*6–28. The steel plate is 0.3 m thick and has a density of 7850 kg/m^3. Determine the location of its center of gravity. Also find the reactions at the pin and roller support.

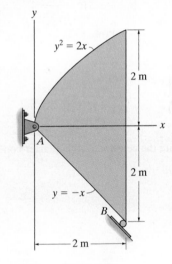

Prob. 6–28

6–29. Locate the centroid $\bar{x}$ of the area.

6–30. Locate the centroid $\bar{y}$ of the area.

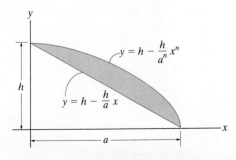

Probs. 6–29/30

6

6–31. Locate the centroid $\bar{y}$ of the solid.

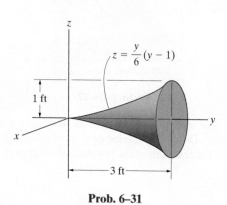

$z = \dfrac{y}{6}(y-1)$

1 ft

3 ft

Prob. 6–31

*6–32. Locate the centroid of the quarter-cone.

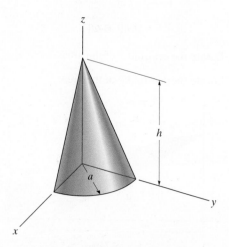

h

a

Prob. 6–32

6–33. Locate the centroid $\bar{z}$ of the solid.

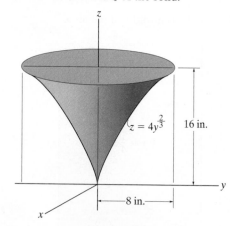

$z = 4y^{\frac{2}{3}}$ 16 in.

—8 in.—

Prob. 6–33

6–34. Locate the centroid $\bar{z}$ of the volume.

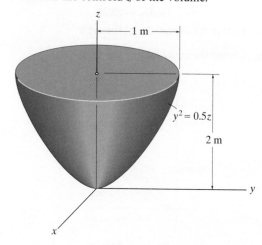

—1 m—

$y^2 = 0.5z$

2 m

Prob. 6–34

6–35. Locate the centroid of the ellipsoid of revolution.

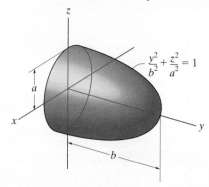

$\dfrac{y^2}{b^2} + \dfrac{z^2}{a^2} = 1$

a

b

Prob. 6–35

6.2 COMPOSITE BODIES

A *composite body* consists of a series of connected "simpler" shaped bodies, which may be rectangular, triangular, semicircular, etc. Such a body, as shown in Fig. 6–8, can often be divided up into its composite parts and, provided the *weight* and location of the center of gravity of each of these parts are known, we can then eliminate the need for integration to determine the center of gravity for the body. The method for doing this follows the same procedure outlined in Sec. 6.1, and so formulas analogous to Eqs. 6–1 result. Here, however, we have a finite number of weights, and so the equations become

$$\overline{x} = \frac{\Sigma \tilde{x} W}{\Sigma W} \qquad \overline{y} = \frac{\Sigma \tilde{y} W}{\Sigma W} \qquad \overline{z} = \frac{\Sigma \tilde{z} W}{\Sigma W} \qquad (6\text{--}4)$$

where

$\overline{x}, \overline{y}, \overline{z}$ represent the coordinates of the center of gravity G of the composite body

$\tilde{x}, \tilde{y}, \tilde{z}$ represent the coordinates of the center of gravity of each composite part of the body

ΣW is the sum of the weights of all the composite parts of the body, or simply the total weight of the body

When the body has a *constant density or specific weight*, the center of gravity *coincides* with the centroid of the body. The centroid for composite lines, areas, and volumes can then be found using relations analogous to Eqs. 6–4; however, the Ws are replaced by Ls, As, and Vs, respectively. Centroids for areas and volumes that often make up a composite body are given in Appendix B.

Fig. 6–8

In order to determine the force required to tip over this concrete barrier, it is first necessary to determine the location of its center of gravity G. This point will lie on the vertical axis of symmetry.

PROCEDURE FOR ANALYSIS

The location of the center of gravity of a body or the centroid of a composite geometrical object represented by an area or volume can be determined using the following procedure.

Composite Parts.

- Using a sketch, divide the body or object into a finite number of composite parts that have simpler shapes.
- If a composite body has a *hole*, or a geometric region having no material, then consider the composite body without the hole and consider the hole as an *additional* composite part having *negative* weight or size.

Moment Arms.

- Establish the coordinate axes on the sketch and determine the coordinates $\tilde{x}, \tilde{y}, \tilde{z}$ of the center of gravity or centroid of each part.

Summations.

- Determine $\bar{x}, \bar{y}, \bar{z}$ by applying the center of gravity equations, Eqs. 6–4, or the analogous centroid equations.
- If an object is *symmetrical* about an axis, the centroid of the object lies on this axis.

If desired, the calculations can be arranged in tabular form, as indicated in the following examples.

The center of gravity of this water tank can be determined by dividing it into composite parts and applying Eqs. 6–4.

EXAMPLE   6.4

Locate the centroid of the plate area shown in Fig. 6–9a.

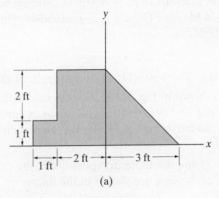

(a)

Fig. 6–9

SOLUTION

Composite Parts.   The plate is divided into three segments as shown in Fig. 6–9b. Here the area of the small rectangle ③ is considered "negative" since it must be subtracted from the larger area ②.

Moment Arms.   The location of the centroid of each segment is shown in the figure. Note that the $\tilde{x}$ coordinates of ② and ③ are *negative*.

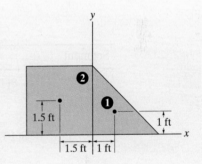

Summations.   Taking the data from Fig. 6–9b, the calculations are tabulated as follows:

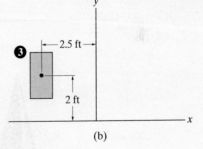

Segment	A (ft^2)	$\tilde{x}$ (ft)	$\tilde{y}$ (ft)	$\tilde{x}A$ (ft^3)	$\tilde{y}A$ (ft^3)
1	$\frac{1}{2}(3)(3) = 4.5$	1	1	4.5	4.5
2	$(3)(3) = 9$	−1.5	1.5	−13.5	13.5
3	$-(2)(1) = -2$	−2.5	2	5	−4
	$\Sigma A = 11.5$			$\Sigma\tilde{x}A = -4$	$\Sigma\tilde{y}A = 14$

Thus,

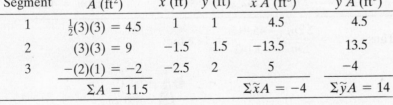

$$\bar{x} = \frac{\Sigma\tilde{x}A}{\Sigma A} = \frac{-4}{11.5} = -0.348 \text{ ft} \qquad Ans.$$

$$\bar{y} = \frac{\Sigma\tilde{y}A}{\Sigma A} = \frac{14}{11.5} = 1.22 \text{ ft} \qquad Ans.$$

(b)

NOTE: If these results are plotted in Fig. 6–9a, the location of point C seems reasonable.

6

EXAMPLE 6.5

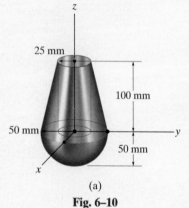

(a)

Fig. 6–10

Locate the center of gravity of the assembly shown in Fig. 6–10a. The conical frustum has a density of $\rho_c = 8 \text{ Mg/m}^3$, and the hemisphere has a density of $\rho_h = 4 \text{ Mg/m}^3$. There is a 25-mm-radius cylindrical hole in the center of the frustum.

SOLUTION

Composite Parts.   The assembly can be thought of as consisting of four segments as shown in Fig. 6–10b. For the calculations, ③ and ④ must be considered as "negative" segments in order that the four segments, when added together, yield the total composite shape shown in Fig. 6–10a.

Moment Arm.   Using the table in Appendix B, the calculations for the centroid $\tilde{z}$ of each piece are shown in the figure.

Summations.   Because of *symmetry*,

$$\bar{x} = \bar{y} = 0 \qquad \qquad Ans.$$

Since $W = mg$, and g is constant, the third of Eqs. 6–4 becomes $\bar{z} = \Sigma\tilde{z}m/\Sigma m$. The mass of each piece can be calculated from $m = \rho V$. Also, $1 \text{ Mg/m}^3 = 10^{-6} \text{ kg/mm}^3$, so that

Segment	m (kg)	$\tilde{z}$ (mm)	$\tilde{z}m$ (kg·mm)
1	$8(10^{-6})\left(\frac{1}{3}\right)\pi(50)^2(200) = 4.189$	50	209.440
2	$4(10^{-6})\left(\frac{2}{3}\right)\pi(50)^3 = 1.047$	-18.75	-19.635
3	$-8(10^{-6})\left(\frac{1}{3}\right)\pi(25)^2(100) = -0.524$	$100 + 25 = 125$	-65.450
4	$-8(10^{-6})\pi(25)^2(100) = -1.571$	50	-78.540
	$\Sigma m = 3.142$		$\Sigma\tilde{z}m = 45.815$

Thus, $\qquad \bar{z} = \dfrac{\Sigma\tilde{z}m}{\Sigma m} = \dfrac{45.815}{3.142} = 14.6 \text{ mm}$ $\qquad\qquad$ *Ans.*

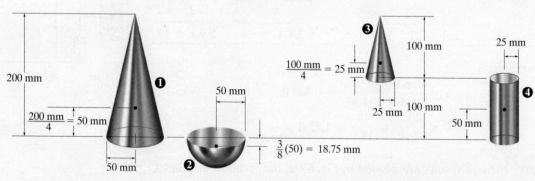

(b)

FUNDAMENTAL PROBLEMS

F6–7. Locate the centroid $(\bar{x}, \bar{y}, \bar{z})$ of the wire bent in the shape shown.

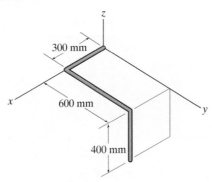

Prob. F6–7

F6–8. Locate the centroid $\bar{y}$ of the beam's cross-sectional area.

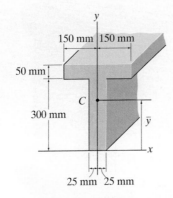

Prob. F6–8

F6–9. Locate the centroid $\bar{y}$ of the beam's cross-sectional area.

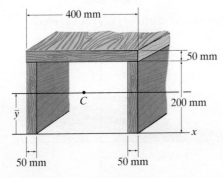

Prob. F6–9

F6–10. Locate the centroid $(\bar{x}, \bar{y})$ of the cross-sectional area.

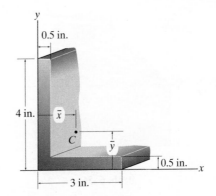

Prob. F6–10

F6–11. Locate the center of gravity $(\bar{x}, \bar{y}, \bar{z})$ of the homogeneous solid block.

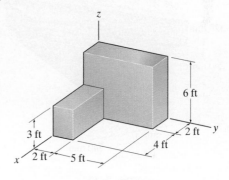

Prob. F6–11

F6–12. Locate the center of gravity $(\bar{x}, \bar{y}, \bar{z})$ of the homogeneous solid block.

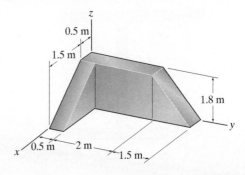

Prob. F6–12

6

PROBLEMS

*6–36. Locate the centroid $(\bar{x}, \bar{y})$ of the area.

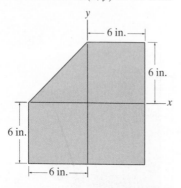

Prob. 6–36

6–37. Locate the centroid $\bar{y}$ for the beam's cross-sectional area.

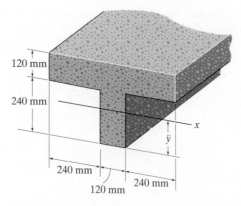

Prob. 6–37

6–38. Locate the centroid $\bar{y}$ of the beam having the cross-sectional area shown.

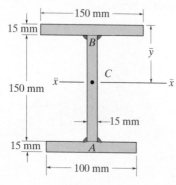

Prob. 6–38

6–39. Locate the centroid $(\bar{x}, \bar{y})$ of the area.

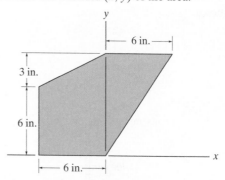

Prob. 6–39

*6–40. Locate the centroid $\bar{y}$ of the beam's cross-sectional area. Neglect the size of the corner welds at A and B for the calculation.

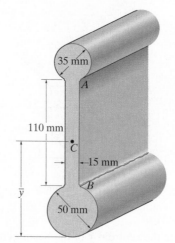

Prob. 6–40

6–41. Locate the centroid $(\bar{x}, \bar{y})$ of the area.

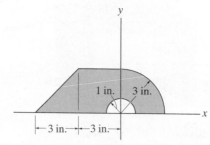

Prob. 6–41

6–42. Locate the centroid $(\bar{x}, \bar{y})$ of the area.

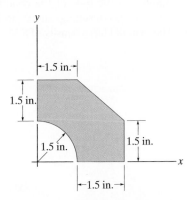

Prob. 6–42

6–43. Locate the centroid $\bar{y}$ of the cross-sectional area of the beam. The beam is symmetric with respect to the y axis.

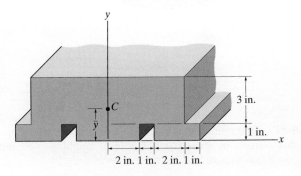

Prob. 6–43

*6–44. Locate the centroid $\bar{y}$ of the cross-sectional area of the beam constructed from a channel and a plate. Assume all corners are square and neglect the size of the weld at A.

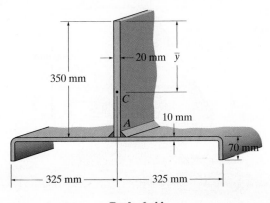

Prob. 6–44

6–45. A triangular plate made of homogeneous material has a constant thickness that is very small. If it is folded over as shown, determine the location $\bar{y}$ of the plate's center of gravity G.

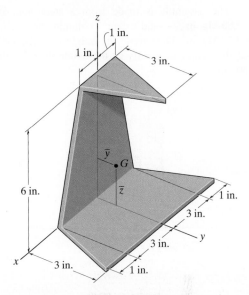

Prob. 6–45

6–46. A triangular plate made of homogeneous material has a constant thickness that is very small. If it is folded over as shown, determine the location $\bar{z}$ of the plate's center of gravity G.

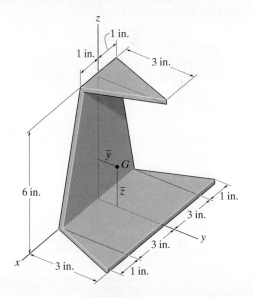

Prob. 6–46

6

6–47. The assembly is made from a steel hemisphere, $\rho_{st} = 7.80 \text{ Mg/m}^3$, and an aluminum cylinder, $\rho_{al} = 2.70 \text{ Mg/m}^3$. Determine the center of gravity of the assembly if the height of the cylinder is $h = 200$ mm.

*6–48. The assembly is made from a steel hemisphere, $\rho_{st} = 7.80 \text{ Mg/m}^3$, and an aluminum cylinder, $\rho_{al} = 2.70 \text{ Mg/m}^3$. Determine the height h of the cylinder so that the center of gravity of the assembly is located at $\bar{z} = 160$ mm.

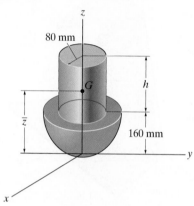

Probs. 6–47/48

6–49. The car rests on four scales and in this position the scale readings of both the front and rear tires are shown by F_A and F_B. When the rear wheels are elevated to a height of 3 ft above the front scales, the new readings of the front wheels are also recorded. Use this data to calculate the location $\bar{x}$ and $\bar{y}$ to the center of gravity G of the car. The tires each have a diameter of 1.98 ft.

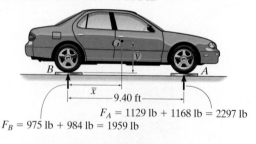

$$F_A = 1129 \text{ lb} + 1168 \text{ lb} = 2297 \text{ lb}$$
$$F_B = 975 \text{ lb} + 984 \text{ lb} = 1959 \text{ lb}$$

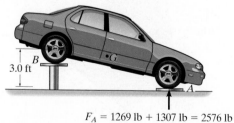

$$F_A = 1269 \text{ lb} + 1307 \text{ lb} = 2576 \text{ lb}$$

Prob. 6–49

6–50. Determine the distance h to which a 100-mm-diameter hole must be bored into the base of the cone so that the center of gravity of the resulting shape is located at $\bar{z} = 115$ mm. The material has a density of 8 Mg/m³.

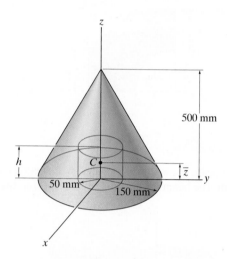

Prob. 6–50

6–51. Determine the distance $\bar{z}$ to the centroid of the shape that consists of a cone with a hole of height $h = 50$ mm bored into its base.

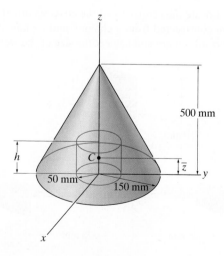

Prob. 6–51

*6–52. Locate the center of gravity $\bar{z}$ of the assembly. The cylinder and the cone are made from materials having densities of 5 Mg/m³ and 9 Mg/m³, respectively.

6–54. The assembly consists of a 20-in. wooden dowel rod and a tight-fitting steel collar. Determine the distance $\bar{x}$ to its center of gravity if the specific weights of the materials are $\gamma_w = 150$ lb/ft³ and $\gamma_{st} = 490$ lb/ft³. The radii of the dowel and collar are shown.

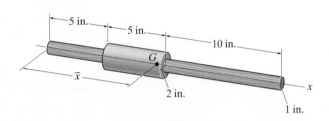

Prob. 6–54

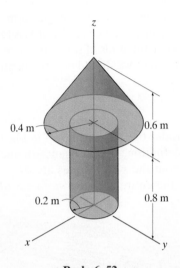

Prob. 6–52

6–53. Major floor loadings in a shop are caused by the weights of the objects shown. Each force acts through its respective center of gravity G. Locate the center of gravity $(\bar{x}, \bar{y})$ of all these components.

6–55. The composite plate is made from both steel (A) and brass (B) segments. Determine the weight and location $(\bar{x}, \bar{y}, \bar{z})$ of its center of gravity G. Take $\rho_{st} = 7.85$ Mg/m³, and $\rho_{br} = 8.74$ Mg/m³.

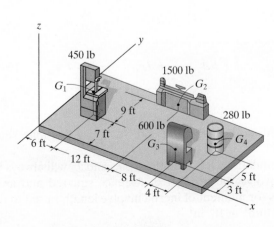

Prob. 6–53

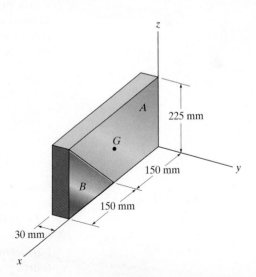

Prob. 6–55

6

6.3 MOMENTS OF INERTIA FOR AREAS

In the first few sections of this chapter, we determined the centroid for an area by considering the first moment of the area about an axis; that is, for the computation we had to evaluate an integral of the form $\int x\, dA$. An integral of the second moment of an area, such as $\int x^2\, dA$, is referred to as the *moment of inertia* for the area. Integrals of this form arise in formulas used in mechanics of materials, and so we should become familiar with the methods used for their computation.

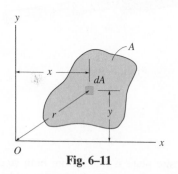

Fig. 6–11

Moment of Inertia. Consider the area A, shown in Fig. 6–11, which lies in the $x-y$ plane. By definition, the moments of inertia of the differential planar area dA about the x and y axes are $dI_x = y^2\, dA$ and $dI_y = x^2 dA$, respectively. For the entire area the *moments of inertia* are determined by integration; i.e.,

$$I_x = \int_A y^2\, dA$$
$$I_y = \int_A x^2\, dA$$

(6–5)

We can also formulate this quantity for dA about the "pole" O or z axis, Fig. 6–11. This is referred to as the *polar moment of inertia*. It is defined as $dJ_O = r^2 dA$, where r is the perpendicular distance from the pole (z axis) to the element dA. For the entire area the *polar moment of inertia* is

$$J_O = \int_A r^2 dA = I_x + I_y$$

(6–6)

Notice that this relation between J_O and I_x, I_y is possible since $r^2 = x^2 + y^2$, Fig. 6–11.

From the above formulations it is seen that I_x, I_y, and J_O will always be positive since they involve the product of distance squared and area. Furthermore, the units for moment of inertia involve length raised to the fourth power, e.g., m^4, mm^4, or ft^4, in^4.

6.4 PARALLEL-AXIS THEOREM FOR AN AREA

If the moment of inertia is known about an axis passing through the centroid of an area, then the *parallel-axis theorem* can be used to find the moment of inertia of the area about *any axis* that is parallel to the centroidal axis. To develop this theorem, consider finding the moment of inertia about the x axis of the shaded area shown in Fig. 6–12.

If we choose a differential element dA located at an arbitrary distance y' from the *centroidal x'* axis, then the distance between the parallel x and x' axes is d_y, and so the moment of inertia of dA about the x axis is $dI_x = (y' + d_y)^2\, dA$. For the entire area,

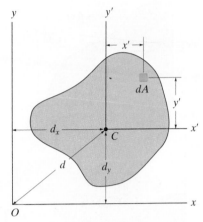

Fig. 6–12

$$I_x = \int_A (y' + d_y)^2\, dA$$

$$= \int_A y'^2\, dA + 2d_y \int_A y'\, dA + d_y^2 \int_A dA$$

The first integral represents the moment of inertia of the area about the centroidal axis, $\bar{I}_{x'}$. The second integral is zero since the x' axis passes through the area's centroid C; i.e., $\int y'\, dA = \bar{y}' \int dA = 0$ since $\bar{y}' = 0$. Since the third integral represents the area A, the final result is therefore

$$\boxed{I_x = \bar{I}_{x'} + Ad_y^2} \qquad (6\text{–}7)$$

A similar expression can be written for I_y; i.e.,

$$\boxed{I_y = \bar{I}_{y'} + Ad_x^2} \qquad (6\text{–}8)$$

And finally, for the polar moment of inertia, since $\bar{J}_C = \bar{I}_{x'} + \bar{I}_{y'}$ and $d^2 = d_x^2 + d_y^2$, we have

$$\boxed{J_O = \bar{J}_C + Ad^2} \qquad (6\text{–}9)$$

The form of each of these three equations states that *the moment of inertia for an area about an axis is equal to its moment of inertia about a parallel axis passing through the area's centroid, plus the product of the area and the square of the perpendicular distance between the axes.*

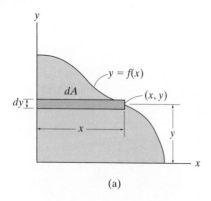

(a)

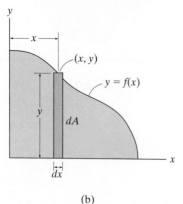

(b)

Fig. 6–13

> # PROCEDURE FOR ANALYSIS

In most cases the moment of inertia can be determined using a single integration. The following procedure shows two ways in which this can be done.

- If the curve defining the boundary of the area is expressed as $y = f(x)$, then select a rectangular differential element such that it has a finite length and differential width.

- The element should be located so that it intersects the curve at the *arbitrary point* (x, y).

Case 1:

- Orient the element so that its length is *parallel* to the axis about which the moment of inertia is calculated. This situation occurs when the rectangular element shown in Fig. 6–13a is used to determine I_x for the area. Here the entire element is at a distance y from the x axis since it has a thickness dy. Thus $I_x = \int y^2 \, dA$. To find I_y, the element is oriented as shown in Fig. 6–13b. This element lies at the *same* distance x from the y axis so that $I_y = \int x^2 \, dA$.

Case 2:

- The length of the element can be oriented *perpendicular* to the axis about which the moment of inertia is calculated; however, Eq. 6–7 *does not apply* since all points on the element will *not* lie at the same moment-arm distance from the axis. For example, if the rectangular element in Fig. 6–13a is used to determine I_y, it will first be necessary to calculate the moment of inertia of the *element* about an axis parallel to the y axis that passes through the element's centroid, and then determine the moment of inertia of the *element* about the y axis using the parallel-axis theorem. Integration of this result will yield I_y. The details are given in Example 6–7.

6

EXAMPLE 6.6

Determine the moment of inertia for the rectangular area shown in Fig. 6–14 with respect to (a) the centroidal x' axis, (b) the axis x_b passing through the base of the rectangle, and (c) the pole or z' axis perpendicular to the x'–y' plane and passing through the centroid C.

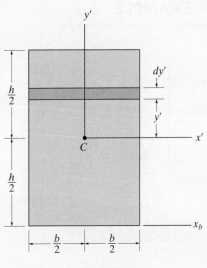

SOLUTION (CASE 1)

Part (a).   The differential element shown in Fig. 6–14 is chosen for integration. Because of its location and orientation, the *entire element* is at a distance y' from the x' axis. Here it is necessary to integrate from $y' = -h/2$ to $y' = h/2$. Since $dA = b\,dy'$, then

$$\bar{I}_{x'} = \int_A y'^2\,dA = \int_{-h/2}^{h/2} y'^2(b\,dy') = b\int_{-h/2}^{h/2} y'^2\,dy'$$

$$\bar{I}_{x'} = \frac{1}{12}bh^3 \qquad\qquad Ans.$$

Fig. 6–14

Part (b).   The moment of inertia about an axis passing through the base of the rectangle can be obtained by using the above result and applying the parallel-axis theorem, Eq. 6–7.

$$I_{x_b} = \bar{I}_{x'} + Ad_y^2$$

$$= \frac{1}{12}bh^3 + bh\left(\frac{h}{2}\right)^2 = \frac{1}{3}bh^3 \qquad\qquad Ans.$$

Part (c).   To obtain the polar moment of inertia about point C, we must first obtain $\bar{I}_{y'}$, which may be found by interchanging the dimensions b and h in the result of part (a), i.e.,

$$\bar{I}_{y'} = \frac{1}{12}hb^3$$

Using Eq. 6–9, the polar moment of inertia about C is therefore

$$\bar{J}_C = \bar{I}_{x'} + \bar{I}_{y'} = \frac{1}{12}bh(h^2 + b^2) \qquad\qquad Ans.$$

6

EXAMPLE 6.7

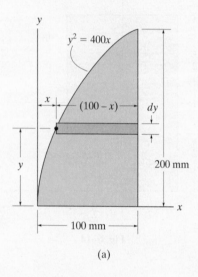

(a)

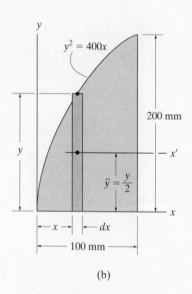

(b)

Fig. 6–15

Determine the moment of inertia of the shaded area shown in Fig. 6–15a about the x axis.

SOLUTION I (CASE 1)

A differential element that is *parallel* to the x axis, Fig. 6–15a, is chosen for integration. It intersects the curve at the *arbitrary point* (x, y). Since this element has a thickness dy and intersects the curve at the *arbitrary point* (x, y), its area is $dA = (100 - x)\, dy$. Furthermore, the element lies at the same distance y from the x axis. Hence, integrating with respect to y, from $y = 0$ to $y = 200$ mm, we have

$$I_x = \int_A y^2\, dA = \int_0^{200\,mm} y^2 (100 - x)\, dy$$

$$= \int_0^{200\,mm} y^2 \left(100 - \frac{y^2}{400}\right) dy = \int_0^{200\,mm} \left(100 y^2 - \frac{y^4}{400}\right) dy$$

$$= 107(10^6)\ \text{mm}^4 \qquad\qquad\qquad Ans.$$

SOLUTION II (CASE 2)

A differential element *parallel* to the y axis, Fig. 6–15b, is chosen for integration. It intersects the curve at the *arbitrary point* (x, y). In this case, all points of the element do *not* lie at the same distance from the x axis, and therefore the parallel-axis theorem must be used to determine the *moment of inertia of the element* with respect to this axis. For a rectangle having a base b and height h, the moment of inertia about its centroidal axis has been determined in part (a) of Example 6–6. There it was found that $\bar{I}_{x'} = \frac{1}{12} bh^3$. For the differential element shown in Fig. 6–15b, $b = dx$ and $h = y$, and so $d\bar{I}_{x'} = \frac{1}{12} dx\, y^3$. Since the centroid of the element is $\tilde{y} = y/2$ from the x axis, the moment of inertia of the element about this axis is

$$dI_x = d\bar{I}_{x'} + dA\, \tilde{y}^2 = \frac{1}{12} dx\, y^3 + y\, dx \left(\frac{y}{2}\right)^2 = \frac{1}{3} y^3\, dx$$

(This result can also be concluded from part (b) of Example 6–6.) Integrating with respect to x, from $x = 0$ to $x = 100$ mm, yields

$$I_x = \int dI_x = \int_0^{100\,mm} \frac{1}{3} y^3\, dx = \int_0^{100\,mm} \frac{1}{3} (400x)^{3/2}\, dx$$

$$= 107(10^6)\ \text{mm}^4 \qquad\qquad\qquad Ans.$$

FUNDAMENTAL PROBLEMS

F6–13. Determine the moment of inertia of the area about the x axis.

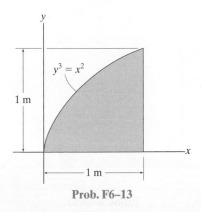

Prob. F6–13

F6–14. Determine the moment of inertia of the area about the x axis.

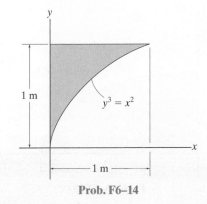

Prob. F6–14

F6–15. Determine the moment of inertia of the area about the y axis.

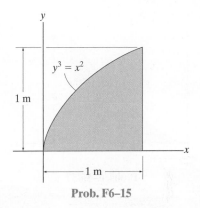

Prob. F6–15

F6–16. Determine the moment of inertia of the area about the y axis.

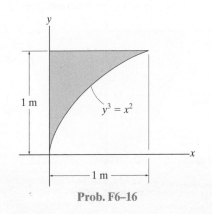

Prob. F6–16

6

PROBLEMS

*6–56. Determine the moment of inertia of the area about the x axis.

6–57. Determine the moment of inertia of the area about the y axis.

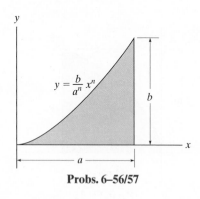

Probs. 6–56/57

6–58. Determine the moment of inertia for the area about the x axis.

6–59. Determine the moment of inertia for the area about the y axis.

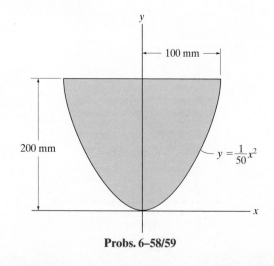

Probs. 6–58/59

*6–60. Determine the moment of inertia for the area about the x axis.

6–61. Determine the moment of inertia for the area about the y axis.

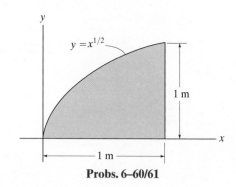

Probs. 6–60/61

6–62. Determine the moment of inertia for the area about the x axis.

6–63. Determine the moment of inertia for the area about the y axis.

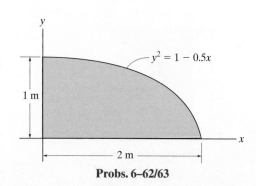

Probs. 6–62/63

*6–64. Determine the moment of inertia of the area about the x axis. Solve the problem in two ways, using rectangular differential elements: (a) having a thickness dx and (b) having a thickness of dy.

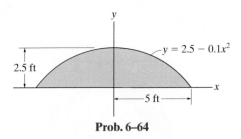

Prob. 6–64

6–65. Determine the moment of inertia of the area about the x axis.

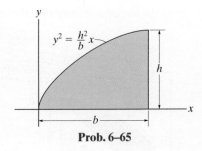

Prob. 6–65

6–66. Determine the moment of inertia for the area about the x axis.

6–67. Determine the moment of inertia for the area about the y axis.

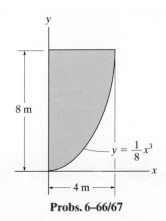

Probs. 6–66/67

*6–68. Determine the moment of inertia about the x axis.

6–69. Determine the moment of inertia about the y axis.

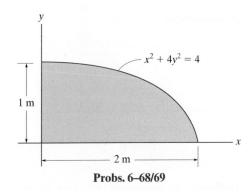

Probs. 6–68/69

6–70. Determine the moment of inertia for the area about the x axis.

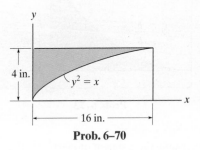

Prob. 6–70

6–71. Determine the moment of inertia for the area about the y axis.

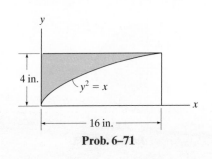

Prob. 6–71

6

*6–72. Determine the moment of inertia for the area about the x axis.

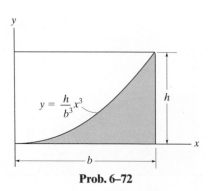

Prob. 6–72

6–73. Determine the moment of inertia for the area about the y axis.

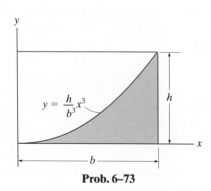

Prob. 6–73

6–74. Determine the moment of inertia for the area about the x axis.

6–75. Determine the moment of inertia for the area about the y axis.

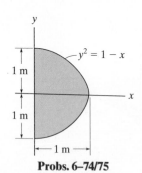

Probs. 6–74/75

*6–76. Determine the moment of inertia for the area about the x axis.

6–77. Determine the moment of inertia for the area about the y axis.

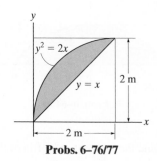

Probs. 6–76/77

6–78. Determine the moment of inertia for the area about the x axis.

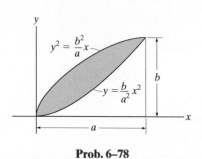

Prob. 6–78

6–79. Determine the moment of inertia for the area about the y axis.

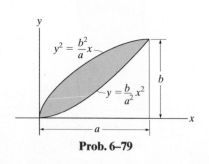

Prob. 6–79

6.5 MOMENTS OF INERTIA FOR COMPOSITE AREAS

The moment of inertia of a composite area that consists of a series of connected "simpler" parts or shapes can be determined about any axis provided the moment of inertia of each of its parts is known or can be determined about the axis. The following procedure outlines a method for doing this.

▶ PROCEDURE FOR ANALYSIS

Composite Parts.

- Using a sketch, divide the area into its composite parts and indicate the perpendicular distance from the centroid of each part to the axis.

Parallel-Axis Theorem.

- If the centroidal axis for each part does not coincide with the axis, the parallel-axis theorem, $I = \bar{I} + Ad^2$, must be used to determine the moment of inertia of the part about the axis. For the calculation of $\bar{I}$, use Appendix B.

Summation.

- The moment of inertia of the entire area about the axis is determined by summing the results of its composite parts about this axis.

- If a composite part has an empty region or hole, its moment of inertia is found by subtracting the moment of inertia of the hole from the moment of inertia of the entire part including the hole.

To design or analyze this T-beam, it is necessary to locate the centroidal axis of its cross-sectional area, and then find the moment of inertia of the area about this axis.

EXAMPLE 6.8

Determine the moment of inertia of the area shown in Fig. 6–16a about the x axis.

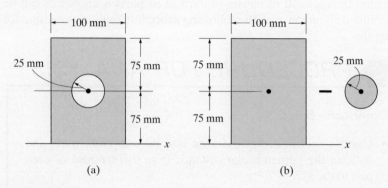

Fig. 6–16

SOLUTION

Composite Parts. The area can be obtained by *subtracting* the circle from the rectangle shown in Fig. 6–16b. The centroid of each area is shown in the figure.

Parallel-Axis Theorem. The moments of inertia about the x axis are determined using the parallel-axis theorem and the moment of inertia formulas for circular and rectangular areas, $I_x = \frac{1}{4}\pi r^4$ and $I_x = \frac{1}{12}bh^3$, found in Appendix B.

Circle

$$I_x = \bar{I}_{x'} + Ad_y^2$$

$$= \frac{1}{4}\pi(25)^4 + \pi(25)^2(75)^2 = 11.4(10^6) \text{ mm}^4$$

Rectangle

$$I_x = \bar{I}_{x'} + Ad_y^2$$

$$= \frac{1}{12}(100)(150)^3 + (100)(150)(75)^2 = 112.5(10^6) \text{ mm}^4$$

Summation. The moment of inertia for the area is therefore

$$I_x = -11.4(10^6) + 112.5(10^6)$$

$$= 101(10^6) \text{ mm}^4 \qquad\qquad Ans.$$

EXAMPLE   6.9

Determine the moments of inertia for the cross-sectional area of the member shown in Fig. 6–17a about the x and y centroidal axes.

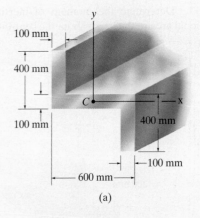

(a)

SOLUTION

Composite Parts.   The cross section can be subdivided into the three rectangular areas A, B, and D shown in Fig. 6–17b. For the calculation, the centroid of each of these rectangles is located in the figure.

Parallel-Axis Theorem.   From the table in Appendix B, or from Example 6–6, the moment of inertia of a rectangle about its centroidal axis is $\bar{I} = \frac{1}{12}bh^3$. Hence, using the parallel-axis theorem for rectangles A and D, the calculations are as follows:

Rectangles A and D

$$I_x = \bar{I}_{x'} + Ad_y^2 = \frac{1}{12}(100)(300)^3 + (100)(300)(200)^2$$

$$= 1.425(10^9) \text{ mm}^4$$

$$I_y = \bar{I}_{y'} + Ad_x^2 = \frac{1}{12}(300)(100)^3 + (100)(300)(250)^2$$

$$= 1.90(10^9) \text{ mm}^4$$

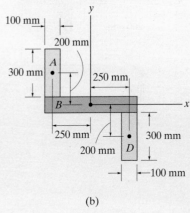

(b)

Fig. 6–17

Rectangle B

$$I_x = \frac{1}{12}(600)(100)^3 = 0.05(10^9) \text{ mm}^4$$

$$I_y = \frac{1}{12}(100)(600)^3 = 1.80(10^9) \text{ mm}^4$$

Summation.   The moments of inertia for the entire cross section are thus

$$I_x = 2[1.425(10^9)] + 0.05(10^9)$$
$$= 2.90(10^9) \text{ mm}^4 \qquad \qquad Ans.$$

$$I_y = 2[1.90(10^9)] + 1.80(10^9)$$
$$= 5.60(10^9) \text{ mm}^4 \qquad \qquad Ans.$$

6

FUNDAMENTAL PROBLEMS

F6–17. Determine the moment of inertia of the cross-sectional area of the beam about the centroidal x and y axes.

F6–19. Determine the moment of inertia of the cross-sectional area of the channel with respect to the y axis.

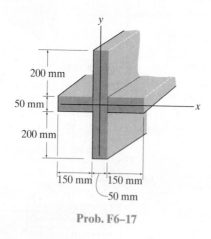

200 mm

50 mm

200 mm

150 mm   150 mm

50 mm

Prob. F6–17

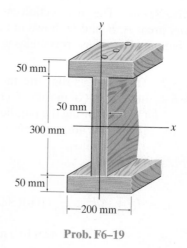

50 mm

50 mm

300 mm

50 mm

200 mm

Prob. F6–19

F6–18. Determine the moment of inertia of the cross-sectional area of the beam about the centroidal x and y axes.

F6–20. Determine the moment of inertia of the cross-sectional area of the T-beam with respect to the x' axis passing through the centroid of the cross section.

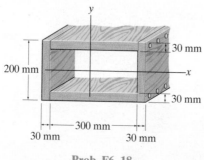

200 mm

30 mm

30 mm

300 mm

30 mm

30 mm

Prob. F6–18

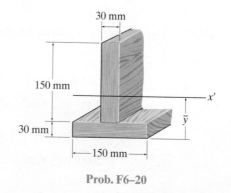

30 mm

150 mm

30 mm

150 mm

Prob. F6–20

PROBLEMS

*6–80. Determine the moment of inertia of the composite area about the x axis.

6–81. Determine the moment of inertia of the composite area about the y axis.

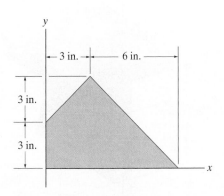

3 in.   6 in.

3 in.

3 in.

Probs. 6–80/81

6–82. The polar moment of inertia for the area is $J_C = 642\ (10^6)\ \text{mm}^4$, about the z' axis passing through the centroid C. The moment of inertia about the y' axis is $264\ (10^6)\ \text{mm}^4$, and the moment of inertia about the x axis is $938\ (10^6)\ \text{mm}^4$. Determine the area A.

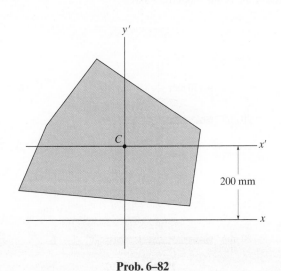

200 mm

Prob. 6–82

6–83. Determine the location $\bar{y}$ of the centroid of the cross-sectional area of the channel, and then calculate the moment of inertia of this area about this axis.

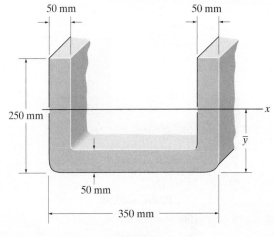

50 mm   50 mm

250 mm

50 mm

350 mm

Prob. 6–83

*6–84. Determine $\bar{y}$, which locates the centroidal axis x' for the cross-sectional area of the T-beam, and then find the moments of inertia $I_{x'}$ and $I_{y'}$.

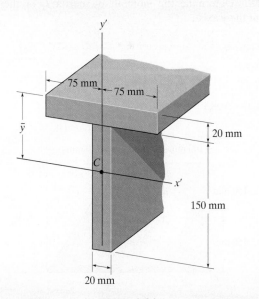

75 mm   75 mm

20 mm

150 mm

20 mm

Prob. 6–84

6–85. Determine the moment of inertia of the cross-sectional area of the beam about the x axis.

6–86. Determine the moment of inertia of the cross-sectional area of the beam about the y axis.

6–89. Determine the moment of inertia of the cross-sectional area of the beam about the y axis.

6–90. Determine $\bar{y}$, which locates the centroidal axis x' for the cross-sectional area of the T-beam, and then find the moment of inertia about the x' axis.

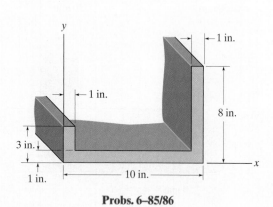

Probs. 6–85/86

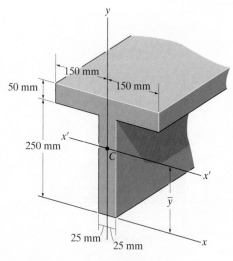

Probs. 6–89/90

6–87. Determine the moment of inertia I_x of the area about the x axis.

*6–88. Determine the moment of inertia I_x of the area about the y axis.

6–91. Determine the moment of inertia of the cross-sectional area of the beam about the x axis.

*6–92. Determine the moment of inertia of the cross-sectional area of the beam about the y axis.

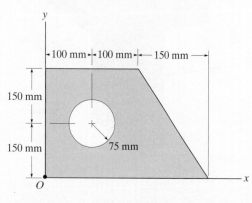

Probs. 6–87/88

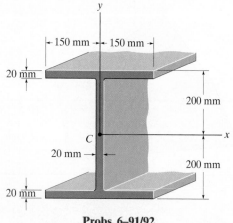

Probs. 6–91/92

CHAPTER REVIEW

Center of Gravity and Centroid

The *center of gravity* G represents a point where the weight of the body can be considered concentrated. The distance from an axis to this point can be determined from a balance of moments, which requires that the moment of the weight of all the particles of the body about this axis must equal the moment of the entire weight of the body about the axis.

$$\bar{x} = \frac{\int \tilde{x} \, dW}{\int dW}$$

$$\bar{y} = \frac{\int \tilde{y} \, dW}{\int dW}$$

$$\bar{z} = \frac{\int \tilde{z} \, dW}{\int dW}$$

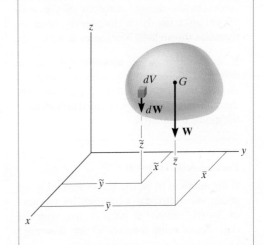

The *centroid* is the location of the geometric center for the body. It is determined in a similar manner, using a moment balance of geometric elements such as area or volume segments. For bodies having an arbitrary shape, moments are summed (integrated) using differential elements.

$$\bar{x} = \frac{\int_A \tilde{x} \, dA}{\int_A dA} \qquad \bar{y} = \frac{\int_A \tilde{y} \, dA}{\int_A dA} \qquad \bar{z} = \frac{\int_A \tilde{z} \, dA}{\int_A dA}$$

$$\bar{x} = \frac{\int_V \tilde{x} \, dV}{\int_V dV} \qquad \bar{y} = \frac{\int_V \tilde{y} \, dV}{\int_V dV} \qquad \bar{z} = \frac{\int_V \tilde{z} \, dV}{\int_V dV}$$

Composite Body

If the body is a composite of several shapes, each having a known location for its center of gravity or centroid, then the location of the center of gravity or centroid of the body can be determined from a discrete summation using its composite parts.

$$\bar{x} = \frac{\Sigma \tilde{x} W}{\Sigma W}$$

$$\bar{y} = \frac{\Sigma \tilde{y} W}{\Sigma W}$$

$$\bar{z} = \frac{\Sigma \tilde{z} W}{\Sigma W}$$

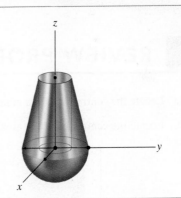

6

Area Moment of Inertia

The *area and polar moments of inertia* represent the second moment of the area about an axis. It is frequently used in formulas throughout mechanics of materials.

If the area shape is irregular but can be described mathematically, then a differential element must be selected and integration over the entire area must be performed to determine the moment of inertia.

$$I_x = \int_A y^2 \, dA$$

$$I_y = \int_A x^2 \, dA$$

$$J_O = \int_A r^2 \, dA$$

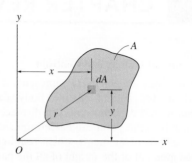

Parallel-Axis Theorem

If the moment of inertia for an area is known about its centroidal axis, then its moment of inertia about a parallel axis can be determined using the parallel-axis theorem.

$$I = \bar{I} + Ad^2$$

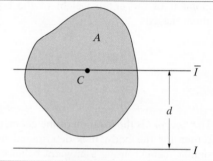

Composite Area

If an area is a composite of common shapes, then its moment of inertia is equal to the algebraic sum of the moments of inertia of each of its parts.

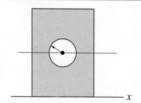

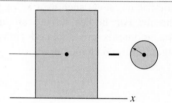

REVIEW PROBLEMS

R6–1.   Locate the centroid $\bar{x}$ of the area.

R6–2.   Locate the centroid $\bar{y}$ of the area.

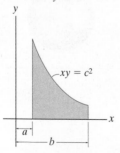

Probs. R6–1/2

R6–3.   Locate the centroid of the rod.

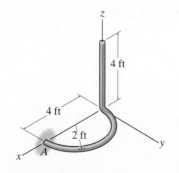

Prob. R6–3

*R6–4. Locate the centroid $\bar{y}$ of the cross-sectional area of the beam.

R6–6. Determine the area moment of inertia of the area about the y axis.

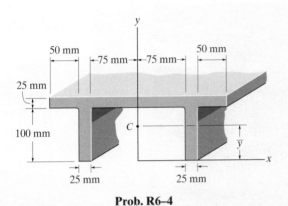

Prob. R6–4

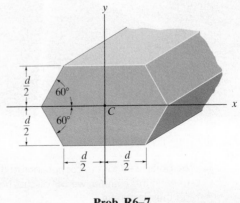

Prob. R6–6

R6–7. Determine the area moment of inertia of the cross-sectional area of the beam about the x axis which passes through the centroid C.

R6–5. Determine the moment of inertia for the area about the x axis.

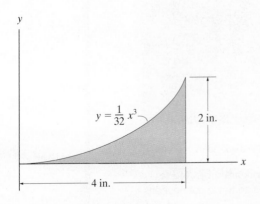

Prob. R6–5

Prob. R6–7

CHAPTER 7

(© Jack Sullivan/Alamy)

The bolts used for the connections of this steel framework are subjected to stress. In this chapter, we will discuss how engineers design these connections and their fasteners.

STRESS AND STRAIN

CHAPTER OBJECTIVES

- To show how to use the method of sections for determining the internal loadings in a member.
- To introduce the concepts of normal and shear stress, and to use them in the analysis and design of members subject to axial load and direct shear.
- To define normal and shear strain, and show how they can be determined for various types of problems.

7.1 INTRODUCTION

Mechanics of materials is a study of the internal effects of stress and strain in a solid body that is subjected to an external loading. Stress is associated with the strength of the material from which the body is made, while strain is a measure of the deformation of the body. A thorough understanding of the fundamentals of this subject is of vital importance for the design of any machine or structure, because many of the formulas and rules of design cited in engineering codes are based upon the principles of this subject.

7.2   INTERNAL RESULTANT LOADINGS

When applying the methods of mechanics of materials, statics along with the method of sections is used to determine the resultant loadings that act within a body. For example, consider the body shown in Fig. 7–1a, which is held in equilibrium by the four external forces.* In order to obtain the internal loadings acting on a specific region within the body, it is necessary to pass an imaginary section or "cut" through the region where the internal loadings are to be determined. The two parts of the body are then separated, and a free-body diagram of one of the parts is drawn, Fig. 7–1b. Here there is actually a distribution of internal force acting on the "exposed" area of the section. These forces actually represent the effects of the top section of the body acting on its bottom section.

Although the exact distribution of this internal loading may be *unknown*, we can find its resultants $\mathbf{F}_R$ and $(\mathbf{M}_R)_O$ at any specific point O on the sectioned area, Fig. 7–1c, by applying the equations of equilibrium to the bottom part of the body. It will be shown later in the book that point O is most often chosen at the *centroid* of the sectioned area, and so we will always choose this location for O, unless otherwise stated. Also, if a member is long and slender, as in the case of a rod or beam, the section to be considered is generally taken *perpendicular* to the longitudinal axis of the member. This section is referred to as the ***cross section***.

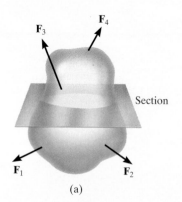

In order to design the horizontal members of this building frame, it is first necessary to find the internal loadings at various points along their length.

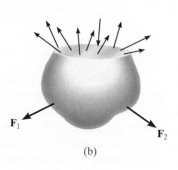

(a)

(b)

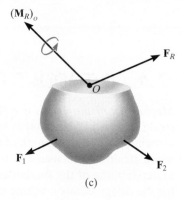

(c)

Fig. 7–1

---

*The body's weight is not shown, since it is assumed to be quite small, and therefore negligible compared with the other loads.

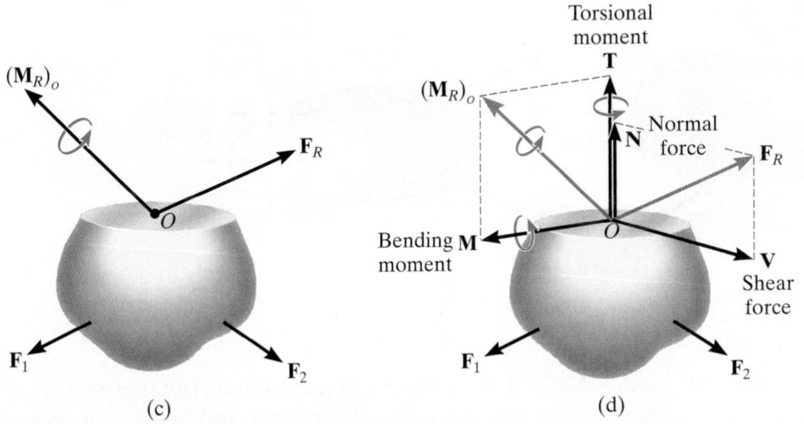

(c)

(d)

Fig. 7–1 (cont.)

Three Dimensions. For later application of the formulas of mechanics of materials, we will consider the components of $\mathbf{F}_R$ and $(\mathbf{M}_R)_o$ acting both normal and perpendicular to the sectioned area, Fig. 7–1d. Four different types of resultant loadings can then be defined as follows:

Normal force, N. This force acts perpendicular to the area. It is developed whenever the external loads tend to push or pull on the two segments of the body.

Shear force, V. The shear force lies in the plane of the area and it is developed when the external loads tend to cause the two segments of the body to slide over one another.

Torsional moment or torque, T. This effect is developed when the external loads tend to twist one segment of the body with respect to the other about an axis perpendicular to the area.

Bending moment, M. The bending moment is caused by the external loads that tend to bend the body about an axis lying within the plane of the area.

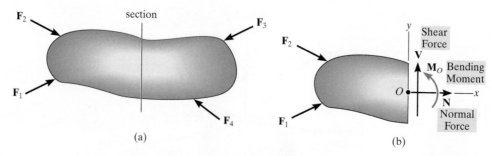

Fig. 7–2

Coplanar Loadings. If the body is subjected to a *coplanar system of forces*, Fig. 7–2a, then only normal-force, shear-force, and bending-moment components will exist at the section, Fig. 7–2b. If we use the x, y, z coordinate axes, as shown on the left segment, then **N** can be obtained by applying $\Sigma F_x = 0$, and **V** can be obtained from $\Sigma F_y = 0$. Finally, the bending moment $\mathbf{M}_O$ can be determined by summing moments about point O (the z axis), $\Sigma M_O = 0$, in order to eliminate the moments caused by the unknowns **N** and **V**.

▶ IMPORTANT POINTS

- *Mechanics of materials* is a study of the relationship between the external loads applied to a body and the stress and strain caused by the internal loads within the body.

- External forces can be applied to a body as *distributed* or *concentrated surface loadings*, or as *body forces* that act throughout the volume of the body.

- Linear distributed loadings produce a *resultant force* having a *magnitude* equal to the *area* under the load diagram, and having a *location* that passes through the *centroid* of this area.

- A support produces a *force* in a particular direction on its attached member if it *prevents translation* of the member in that direction, and it produces a *couple moment* on the member if it *prevents rotation*.

- The equations of equilibrium $\Sigma \mathbf{F} = \mathbf{0}$ and $\Sigma \mathbf{M} = \mathbf{0}$ must be satisfied in order to prevent a body from translating with accelerated motion and from rotating.

- The method of sections is used to determine the internal resultant loadings acting on the surface of a sectioned body. In general, these resultants consist of a normal force, shear force, torsional moment, and bending moment.

PROCEDURE FOR ANALYSIS

The resultant *internal* loadings at a point located on the section of a body can be obtained using the method of sections. This requires the following steps.

Support Reactions.

- When the body is sectioned, decide which segment of the body is to be considered. If the segment has a support or connection to another body, then *before* the body is sectioned, it will be necessary to determine the reactions acting on the chosen segment. To do this, draw the free-body diagram of the *entire body* and then apply the necessary equations of equilibrium to obtain these reactions.

Free-Body Diagram.

- Keep all external distributed loadings, couple moments, torques, and forces in their *exact locations*, before passing the section through the body at the point where the resultant internal loadings are to be determined.

- Draw a free-body diagram of one of the "cut" segments and indicate the unknown resultants **N, V, M**, and **T** at the section. These resultants are normally placed at the point representing the geometric center or *centroid* of the sectioned area.

- If the member is subjected to a *coplanar* system of forces, only **N, V**, and **M** act at the centroid.

- Establish the x, y, z coordinate axes with origin at the centroid and show the resultant internal loadings acting along the axes.

Equations of Equilibrium.

- Moments should be summed at the section, about each of the coordinate axes where the resultants act. Doing this eliminates the unknown forces **N** and **V** and allows a direct solution for **M** and **T**.

- If the solution of the equilibrium equations yields a negative value for a resultant, the *directional sense* of the resultant is *opposite* to that shown on the free-body diagram.

The following examples illustrate this procedure numerically and also provide a review of some of the important principles of statics.

EXAMPLE 7.1

Determine the resultant internal loadings acting on the cross section at C of the cantilevered beam shown in Fig. 7–3a.

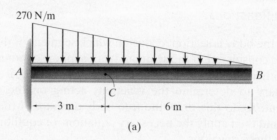

(a)

Fig. 7–3

SOLUTION

Support Reactions. The support reactions at A do not have to be determined if segment CB is considered.

Free-Body Diagram. The free-body diagram of segment CB is shown in Fig. 7–3b. It is important to keep the distributed loading on the segment until *after* the section is made. Only then should this loading be replaced by a single resultant force. Notice that the intensity of the distributed loading at C is found by proportion, i.e., from Fig. 7–3a, $w/6$ m $= (270$ N/m$)/9$ m, $w = 180$ N/m. The magnitude of the resultant of the distributed load is equal to the area under the loading curve (triangle) and acts through the centroid of this area. Thus, $F = \frac{1}{2}(180$ N/m$)(6$ m$) = 540$ N, which acts $\frac{1}{3}(6$ m$) = 2$ m from C as shown in Fig. 7–3b.

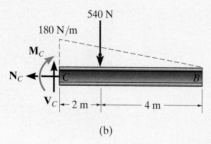

(b)

Equations of Equilibrium. Applying the equations of equilibrium we have

$$\xrightarrow{+} \Sigma F_x = 0; \qquad\qquad -N_C = 0$$
$$N_C = 0 \qquad\qquad Ans.$$
$$+\uparrow \Sigma F_y = 0; \qquad\qquad V_C - 540 \text{ N} = 0$$
$$V_C = 540 \text{ N} \qquad\qquad Ans.$$
$$\zeta + \Sigma M_C = 0; \qquad\qquad -M_C - 540 \text{ N}(2 \text{ m}) = 0$$
$$M_C = -1080 \text{ N} \cdot \text{m} \qquad\qquad Ans.$$

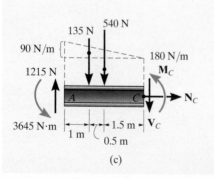

(c)

The negative sign indicates that $\mathbf{M}_C$ acts in the opposite direction to that shown on the free-body diagram. Try solving this problem using segment AC, by first checking the support reactions at A, which are given in Fig. 7–3c.

EXAMPLE 7.2

The 500-kg engine is suspended from the crane boom in Fig. 7–4a. Determine the resultant internal loadings acting on the cross section of the boom at point E.

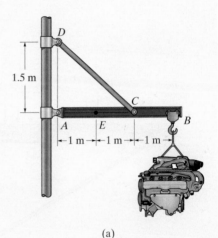

(a)

SOLUTION

Support Reactions. We will consider segment AE of the boom, so we must first determine the pin reactions at A. Since member CD is a two-force member, it acts like a cable, and therefore exerts a force F_{CD} having a known direction. The free-body diagram of the boom is shown in Fig. 7–4b. Applying the equations of equilibrium,

$$\zeta + \Sigma M_A = 0; \qquad F_{CD}\left(\tfrac{3}{5}\right)(2 \text{ m}) - [500(9.81) \text{ N}](3 \text{ m}) = 0$$

$$F_{CD} = 12\,262.5 \text{ N}$$

$$\xrightarrow{+} \Sigma F_x = 0; \qquad A_x - (12\,262.5 \text{ N})\left(\tfrac{4}{5}\right) = 0$$

$$A_x = 9810 \text{ N}$$

$$+\uparrow \Sigma F_y = 0; \qquad -A_y + (12\,262.5 \text{ N})\left(\tfrac{3}{5}\right) - 500(9.81) \text{ N} = 0$$

$$A_y = 2452.5 \text{ N}$$

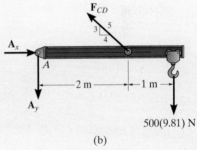

(b)

Free-Body Diagram. The free-body diagram of segment AE is shown in Fig. 7–4c.

Equations of Equilibrium.

$$\xrightarrow{+} \Sigma F_x = 0; \qquad N_E + 9810 \text{ N} = 0$$

$$N_E = -9810 \text{ N} = -9.81 \text{ kN} \qquad Ans.$$

$$+\uparrow \Sigma F_y = 0; \qquad -V_E - 2452.5 \text{ N} = 0$$

$$V_E = -2452.5 \text{ N} = -2.45 \text{ kN} \qquad Ans.$$

$$\zeta + \Sigma M_E = 0; \qquad M_E + (2452.5 \text{ N})(1 \text{ m}) = 0$$

$$M_E = -2452.5 \text{ N} \cdot \text{m} = -2.45 \text{ kN} \cdot \text{m} \qquad Ans.$$

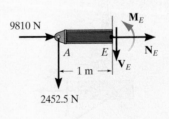

(c)

Fig. 7–4

EXAMPLE 7.3

Determine the resultant internal loadings acting on the cross section at C of the beam shown in Fig. 7–5a.

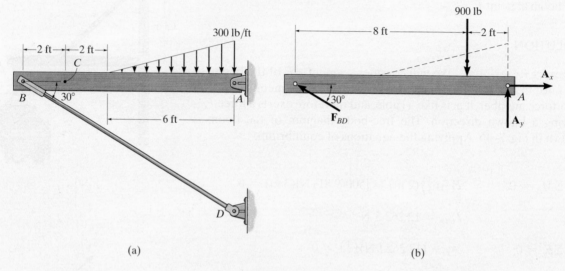

(a)

(b)

Fig. 7–5

SOLUTION

Support Reactions. Here we will consider segment BC, but first we must find the force components at pin A. The free-body diagram of the *entire* beam is shown in Fig. 7–5b. Since member BD is a two-force member, like member CD in Example 7.2, the force at B has a known direction, Fig. 7–5b. We have

$$\zeta + \Sigma M_A = 0; \quad (900 \text{ lb})(2 \text{ ft}) - (F_{BD} \sin 30°) \, 10 \text{ ft} = 0 \quad F_{BD} = 360 \text{ lb}$$

Free-Body Diagram. Using this result, the free-body diagram of segment BC is shown in Fig. 7–5c.

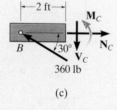

(c)

Equations of Equilibrium.

$$\xrightarrow{+} \Sigma F_x = 0; \qquad N_C - (360 \text{ lb}) \cos 30° = 0$$

$$N_C = 312 \text{ lb} \qquad \qquad Ans.$$

$$+\uparrow \Sigma F_y = 0; \qquad (360 \text{ lb}) \sin 30° - V_C = 0$$

$$V_C = 180 \text{ lb} \qquad \qquad Ans.$$

$$\zeta + \Sigma M_C = 0; \qquad M_C - (360 \text{ lb}) \sin 30°(2 \text{ ft}) = 0$$

$$M_C = 360 \text{ lb} \cdot \text{ft} \qquad \qquad Ans.$$

EXAMPLE   7.4

Determine the resultant internal loadings acting on the cross section at B of the pipe shown in Fig. 7–6a. End A is subjected to a vertical force of 50 N, a horizontal force of 30 N, and a couple moment of 70 N · m. Neglect the pipe's mass.

SOLUTION

The problem can be solved by considering segment AB, so we do not need to calculate the support reactions at C.

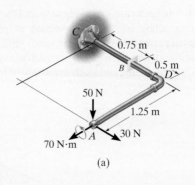

(a)

Free-Body Diagram. The free-body diagram of segment AB is shown in Fig. 7–6b, where the x, y, z axes are established at B. The resultant force and moment components at the section are assumed to act in the *positive coordinate directions* and to pass through the *centroid* of the cross-sectional area at B.

Equations of Equilibrium. Applying the six scalar equations of equilibrium, we have*

$\Sigma F_x = 0;$ $\qquad\qquad\qquad\qquad\qquad\qquad (F_B)_x = 0$ $\qquad$ *Ans.*

$\Sigma F_y = 0;$ $\qquad (F_B)_y + 30\text{ N} = 0$ $\qquad (F_B)_y = -30\text{ N}$ $\qquad$ *Ans.*

$\Sigma F_z = 0;$ $\qquad (F_B)_z - 50\text{ N} = 0$ $\qquad (F_B)_z = 50\text{ N}$ $\qquad$ *Ans.*

$\Sigma (M_B)_x = 0;$ $\qquad (M_B)_x + 70\text{ N} \cdot \text{m} - 50\text{ N}\,(0.5\text{ m}) = 0$

$\qquad\qquad\qquad\qquad (M_B)_x = -45\text{ N} \cdot \text{m}$ $\qquad$ *Ans.*

$\Sigma (M_B)_y = 0;$ $\qquad (M_B)_y + 50\text{ N}\,(1.25\text{ m}) = 0$

$\qquad\qquad\qquad\qquad (M_B)_y = -62.5\text{ N} \cdot \text{m}$ $\qquad$ *Ans.*

$\Sigma (M_B)_z = 0;$ $\qquad (M_B)_z + (30\text{ N})(1.25) = 0$

$\qquad\qquad\qquad\qquad (M_B)_z = -37.5\text{ N} \cdot \text{m}$ $\qquad$ *Ans.*

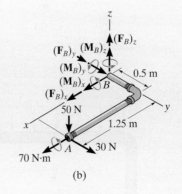

(b)

Fig. 7–6

NOTE: What do the negative signs for $(F_B)_y$, $(M_B)_x$, $(M_B)_y$, and $(M_B)_z$ indicate? The normal force $N_B = |(F_B)_y| = 30$ N, whereas the shear force is $V_B = \sqrt{(0)^2 + (50)^2} = 50$ N. Also, the torsional moment is $T_B = |(M_B)_y| = 62.5$ N · m, and the bending moment is $M_B = \sqrt{(45)^2 + (37.5)^2} = 58.6$ N · m.

*The *magnitude* of each moment about the x, y, or z axis is equal to the magnitude of each force times the perpendicular distance from the axis to the line of action of the force. The *direction* of each moment is determined using the right-hand rule, with positive moments (thumb) directed along the positive coordinate axes.

PRELIMINARY PROBLEM

P7–1. In each case, explain how to find the resultant internal loading acting on the cross section at point A. Draw all necessary free-body diagrams, and indicate the relevant equations of equilibrium. Do not calculate values. The lettered dimensions, angles, and loads are assumed to be known.

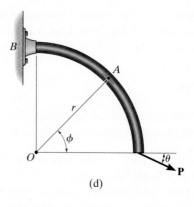

(d)

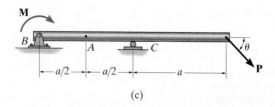

(a)

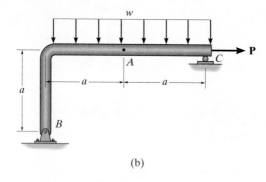

(b)

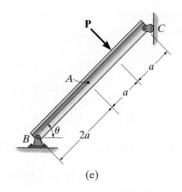

(e)

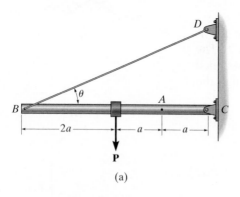

(c)

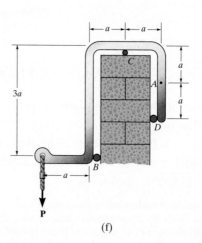

(f)

Prob. P7–1

FUNDAMENTAL PROBLEMS

F7–1. Determine the internal normal force, shear force, and bending moment at point C in the beam.

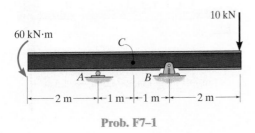

Prob. F7–1

F7–2. Determine the internal normal force, shear force, and bending moment at point C in the beam.

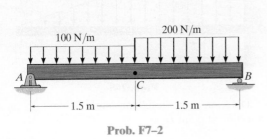

Prob. F7–2

F7–3. Determine the internal normal force, shear force, and bending moment at point C in the beam.

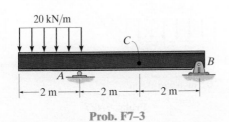

Prob. F7–3

F7–4. Determine the internal normal force, shear force, and bending moment at point C in the beam.

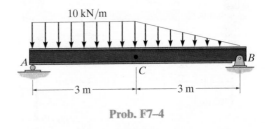

Prob. F7–4

F7–5. Determine the internal normal force, shear force, and bending moment at point C in the beam.

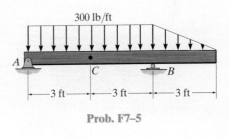

Prob. F7–5

F7–6. Determine the internal normal force, shear force, and bending moment at point C in the beam.

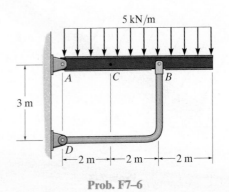

Prob. F7–6

PROBLEMS

7–1. The shaft is supported by a smooth thrust bearing at *B* and a journal bearing at *C*. Determine the resultant internal loadings acting on the cross section at *E*.

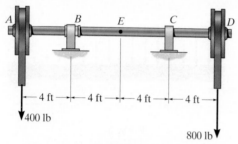

Prob. 7–1

7–2. Determine the resultant internal normal and shear force in the member at (a) section *a–a* and (b) section *b–b*, each of which passes through point *A*. The 500-lb load is applied along the centroidal axis of the member.

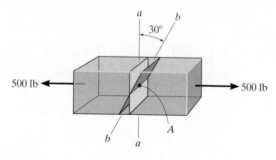

Prob. 7–2

7–3. Determine the resultant internal loadings acting on section *b–b* through the centroid, point *C* on the beam.

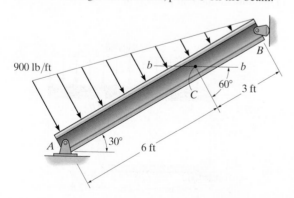

Prob. 7–3

***7–4.** The shaft is supported by a smooth thrust bearing at *A* and a smooth journal bearing at *B*. Determine the resultant internal loadings acting on the cross section at *C*.

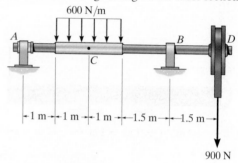

Prob. 7–4

7–5. Determine the resultant internal loadings acting on the cross section at point *B*.

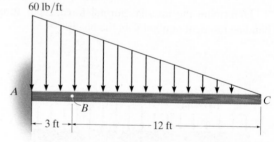

Prob. 7–5

7–6. Determine the resultant internal loadings on the cross section at point *D*.

7–7. Determine the resultant internal loadings at cross sections at points *E* and *F* on the assembly.

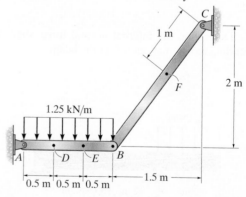

Probs. 7–6/7

*7–8. The beam supports the distributed load shown. Determine the resultant internal loadings acting on the cross section at point C. Assume the reactions at the supports A and B are vertical.

7–9. The beam supports the distributed load shown. Determine the resultant internal loadings acting on the cross section at point D. Assume the reactions at the supports A and B are vertical.

7–11. Determine the resultant internal loadings acting on the cross sections at points D and E of the frame.

*7–12. Determine the resultant internal loadings acting on the cross sections at points F and G of the frame.

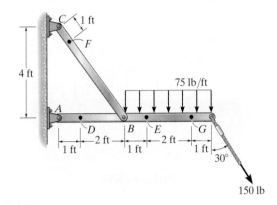

Probs. 7–11/12

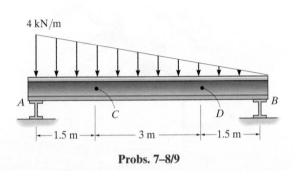

Probs. 7–8/9

7–10. The boom DF of the jib crane and the column DE have a uniform weight of 50 lb/ft. If the hoist and load weigh 300 lb, determine the resultant internal loadings in the crane on cross sections at points A, B, and C.

7–13. The blade of the hacksaw is subjected to a pretension force of $F = 100$ N. Determine the resultant internal loadings acting on section a–a at point D.

7–14. The blade of the hacksaw is subjected to a pretension force of $F = 100$ N. Determine the resultant internal loadings acting on section b–b at point D.

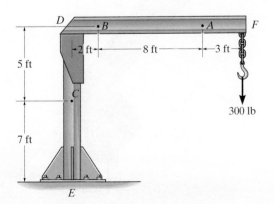

Prob. 7–10

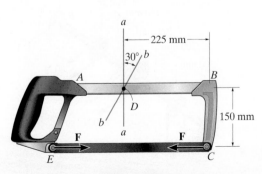

Probs. 7–13/14

7–15. The beam supports the triangular distributed load shown. Determine the resultant internal loadings on the cross section at point C. Assume the reactions at the supports A and B are vertical.

*7–16. The beam supports the distributed load shown. Determine the resultant internal loadings on the cross sections at points D and E. Assume the reactions at the supports A and B are vertical.

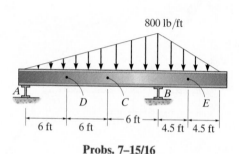

Probs. 7–15/16

7–17. The shaft is supported at its ends by two bearings A and B and is subjected to the forces applied to the pulleys fixed to the shaft. Determine the resultant internal loadings acting on the cross section at point D. The 400-N forces act in the −z direction and the 200-N and 80-N forces act in the +y direction. The journal bearings at A and B exert only y and z components of force on the shaft.

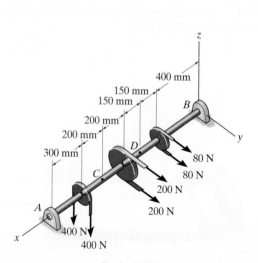

Prob. 7–17

7–18. The shaft is supported at its ends by two bearings A and B and is subjected to the forces applied to the pulleys fixed to the shaft. Determine the resultant internal loadings acting on the cross section at point C. The 400-N forces act in the −z direction and the 200-N and 80-N forces act in the +y direction. The journal bearings at A and B exert only y and z components of force on the shaft.

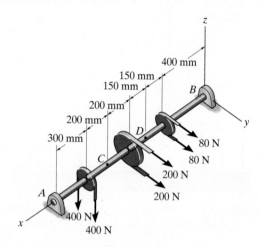

Prob. 7–18

7–19. The hand crank that is used in a press has the dimensions shown. Determine the resultant internal loadings acting on the cross section at point A if a vertical force of 50 lb is applied to the handle as shown. Assume the crank is fixed to the shaft at B.

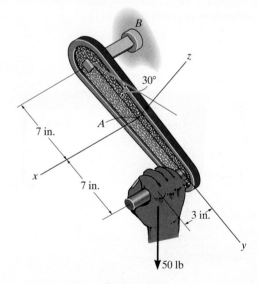

Prob. 7–19

7

*7–20.** Determine the resultant internal loadings acting on the cross section at point C in the beam. The load D has a mass of 300 kg and is being hoisted by the motor M with constant velocity.

7–21. Determine the resultant internal loadings acting on the cross section at point E. The load D has a mass of 300 kg and is being hoisted by the motor M with constant velocity.

*7–24.** Determine the resultant internal loadings acting on the cross section at point C. The cooling unit has a total weight of 52 kip and a center of gravity at G.

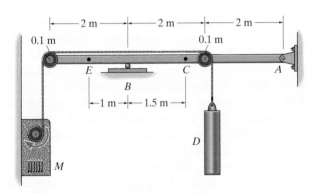

Probs. 7–20/21

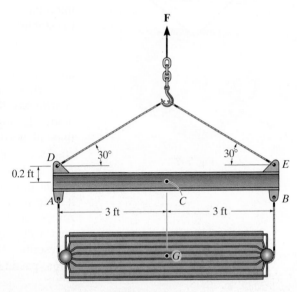

Prob. 7–24

7–22. The metal stud punch is subjected to a force of 120 N on the handle. Determine the magnitude of the reactive force at the pin A and in the short link BC. Also, determine the resultant internal loadings acting on the cross section at point D.

7–23. Determine the resultant internal loadings acting on the cross section at point E of the handle arm, and on the cross section of the short link BC.

7–25. Determine the resultant internal loadings acting on the cross section at points B and C of the curved member.

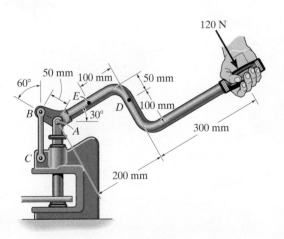

Probs. 7–22/23

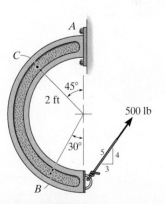

Prob. 7–25

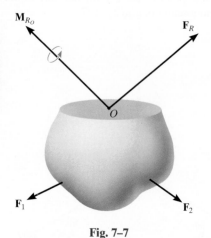

Fig. 7–7

7.3 STRESS

It was stated in Section 7.2 that the force and moment acting at a specified point O on the sectioned area of the body, Fig. 7–7, represents the resultant effects of the *distribution of loading* that acts over the sectioned area, Fig. 7–8a. Obtaining this *distribution* is of primary importance in mechanics of materials. To solve this problem it is first necessary to establish the concept of stress.

We begin by considering the sectioned area to be subdivided into small areas, such as ΔA shown in Fig. 7–8a. As we reduce ΔA to a smaller and smaller size, we will make two assumptions regarding the properties of the material. We will consider the material to be ***continuous***, that is, to consist of a *continuum* or uniform distribution of matter having no voids. Also, the material must be ***cohesive***, meaning that all portions of it are connected together, without having breaks, cracks, or separations. A typical finite yet very small force $\Delta \mathbf{F}$, acting on ΔA, is shown in Fig. 7–8a. This force, like all the others, will have a unique direction, but to compare it with all the other forces, we will replace it by its *three components*, namely, $\Delta \mathbf{F}_x$, $\Delta \mathbf{F}_y$, and $\Delta \mathbf{F}_z$. As ΔA approaches zero, so do $\Delta \mathbf{F}$ and its components; however, the quotient of the force and area will approach a finite limit. This quotient is called ***stress***, and it describes the *intensity of the internal force* acting on a *specific plane* (area) passing through a point.

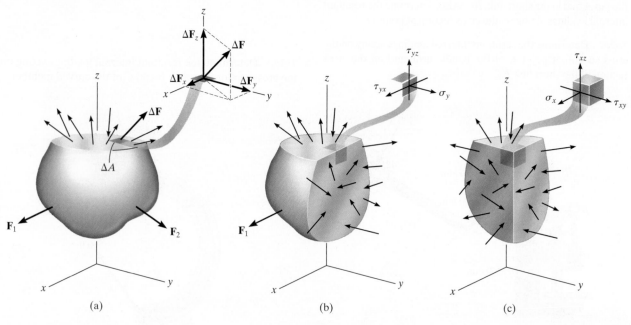

Fig. 7–8

Normal Stress. The *intensity* of the force acting normal to ΔA is referred to as the **normal stress**, σ (sigma). Since $\Delta \mathbf{F}_z$ is normal to the area then

$$\sigma_z = \lim_{\Delta A \to 0} \frac{\Delta F_z}{\Delta A} \qquad (7\text{--}1)$$

If the normal force or stress "pulls" on ΔA as shown in Fig. 7–8a, it is *tensile stress*, whereas if it "pushes" on ΔA it is *compressive stress*.

Shear Stress. The intensity of force acting tangent to ΔA is called the **shear stress**, τ (tau). Here we have two shear stress components,

$$\tau_{zx} = \lim_{\Delta A \to 0} \frac{\Delta F_x}{\Delta A}$$

$$\qquad (7\text{--}2)$$

$$\tau_{zy} = \lim_{\Delta A \to 0} \frac{\Delta F_y}{\Delta A}$$

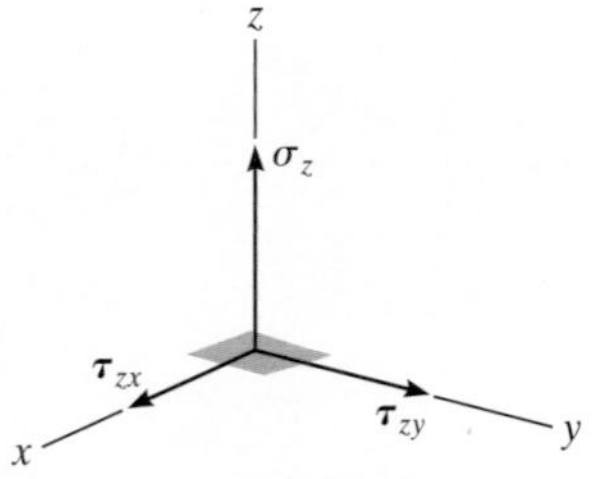

The subscript notation z specifies the orientation of the area ΔA, Fig. 7–9, and x and y indicate the axes along which each shear stress acts.

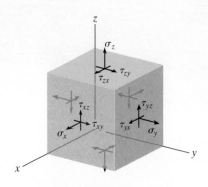

Fig. 7–9

General State of Stress. If the body is further sectioned by planes parallel to the x–z plane, Fig. 7–8b, and the y–z plane, Fig. 7–8c, we can then "cut out" a cubic volume element of material that represents the **state of stress** acting around a chosen point in the body. This state of stress is then characterized by three components acting on each face of the element, Fig. 7–10.

Units. Since stress represents a force per unit area, in the International Standard or SI system, the magnitudes of both normal and shear stress are specified in the base units of newtons per square meter (N/m^2). This combination of units is called a pascal ($1 \text{ Pa} = 1 \text{ N/m}^2$), and because it is rather small, prefixes such as kilo- (10^3), symbolized by k, mega- (10^6), symbolized by M, or giga- (10^9), symbolized by G, are used in engineering to represent larger, more realistic values of stress.* In the Foot-Pound-Second system of units, engineers usually express stress in pounds per square inch (psi) or kilopounds per square inch (ksi), where 1 kilopound (kip) = 1000 lb.

Fig. 7–10

---

*Sometimes stress is expressed in units of N/mm^2, where $1 \text{ mm} = 10^{-3} \text{ m}$. However, in the SI system, prefixes are not allowed in the denominator of a fraction, and therefore it is better to use the equivalent $1 \text{ N/mm}^2 = 1 \text{ MN/m}^2 = 1 \text{ MPa}$.

7.4   AVERAGE NORMAL STRESS IN AN AXIALLY LOADED BAR

We will now determine the average stress distribution acting over the cross-sectional area of an axially loaded bar such as the one shown in Fig. 7–11a. Specifically, the **cross section** is the section taken *perpendicular* to the longitudinal axis of the bar, and since the bar is prismatic all cross sections are the same throughout its length. Provided the material of the bar is both **homogeneous** and **isotropic**, that is, it has the same physical and mechanical properties throughout its volume, and it has the same properties in all directions, then when the load P is applied to the bar through the centroid of its cross-sectional area, the bar will deform uniformly throughout the central region of its length, Fig. 7–11b.

Realize that many engineering materials may be approximated as being both homogeneous and isotropic. Steel, for example, contains thousands of randomly oriented crystals in each cubic millimeter of its volume, and since most objects made of this material have a physical size that is very much larger than a single crystal, the above assumption regarding the material's composition is quite realistic.

Note that **anisotropic materials**, such as wood, have different properties in different directions; and although this is the case, if the grains of wood are oriented along the bar's axis (as for instance in a typical wood board), then the bar will also deform uniformly when subjected to the axial load P.

Average Normal Stress Distribution. If we pass a section through the bar, and separate it into two parts, then equilibrium requires the resultant normal force N at the section to be equal to P, Fig. 7–11c. And because the material undergoes a *uniform* deformation, it is necessary that the cross section be subjected to a *constant normal stress distribution*.

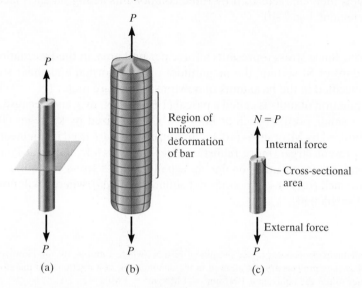

Region of uniform deformation of bar

$N = P$

Internal force

Cross-sectional area

External force

(a)   (b)   (c)

Fig. 7–11

As a result, each small area ΔA on the cross section is subjected to a force $\Delta N = \sigma \, \Delta A$, Fig. 7–11d, and the *sum* of these forces acting over the entire cross-sectional area must be equivalent to the internal resultant force **P** at the section. If we let $\Delta A \rightarrow dA$ and therefore $\Delta N \rightarrow dN$, then, recognizing σ is *constant*, we have

$$+\uparrow F_{Rz} = \Sigma F_z; \qquad \int dN = \int_A \sigma \, dA$$

$$N = \sigma A$$

$$\boxed{\sigma = \frac{N}{A}} \qquad (7\text{–}3)$$

Here

σ = average normal stress at any point on the cross-sectional area

N = *internal resultant normal force*, which acts through the *centroid* of the cross-sectional area. N is determined using the method of sections and the equations of equilibrium, where for this case $N = P$.

A = cross-sectional area of the bar where σ is determined

Equilibrium. The stress distribution in Fig. 7–11 indicates that only a normal stress exists on any small volume element of material located at each point on the cross section. Thus, if we consider vertical equilibrium of an element of material and then apply the equation of force equilibrium to its free-body diagram, Fig. 7–12,

$$\Sigma F_z = 0; \qquad \sigma(\Delta A) - \sigma'(\Delta A) = 0$$

$$\sigma = \sigma'$$

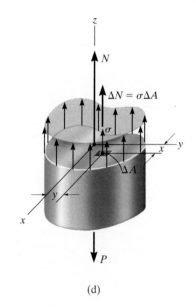

(d)

Fig. 7–11 (cont.)

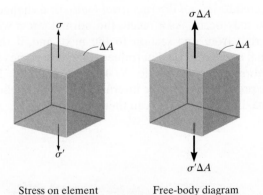

Stress on element   Free-body diagram

Fig. 7–12

7

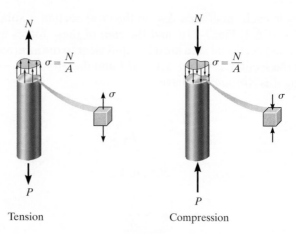

$$\sigma = \frac{N}{A}$$ $$\sigma$$ $$\sigma = \frac{N}{A}$$ $$\sigma$$

Tension   Compression

Fig. 7–13

In other words, the normal stress components on the element must be equal in magnitude but opposite in direction. Under this condition the material is subjected to **uniaxial stress**, and this analysis applies to members subjected to either tension or compression, as shown in Fig. 7–13.

   Although we have developed this analysis for *prismatic* bars, this assumption can be relaxed somewhat to include bars that have a *slight taper*. For example, it can be shown, using the more exact analysis of the theory of elasticity, that for a tapered bar of rectangular cross section, where the angle between two adjacent sides is 15°, the average normal stress, as calculated by $\sigma = N/A$, is only 2.2% *less* than its value found from the theory of elasticity.

Maximum Average Normal Stress. For our analysis, both the internal force N and the cross-sectional area A were *constant* along the longitudinal axis of the bar, and as a result the normal stress $\sigma = N/A$ is also *constant* throughout the bar's length. Occasionally, however, the bar may be subjected to *several external axial loads*, or a change in its cross-sectional area may occur. As a result, the normal stress within the bar may be different from one section to the next, and, if the *maximum average normal stress* is to be determined, then it becomes important to find the location where the ratio N/A is a *maximum*. Example 7.5 illustrates the procedure. Once the internal loading throughout the bar is known, the maximum ratio N/A can then be identified.

This steel tie rod is used as a hanger to suspend a portion of a staircase, and as a result it is subjected to tensile stress.

▶ IMPORTANT POINTS

- When a body subjected to external loads is sectioned, there is a distribution of force acting over the sectioned area which holds each segment of the body in equilibrium. The intensity of this internal force at a point in the body is referred to as *stress*.

- Stress is the limiting value of force per unit area, as the area approaches zero. For this definition, the material is considered to be continuous and cohesive.

- The magnitude of the stress components at a point depends upon the type of loading acting on the body, and the orientation of the element at the point.

- When a prismatic bar is made of homogeneous and isotropic material, and is subjected to an axial force acting through the centroid of the cross-sectional area, then the center region of the bar will deform uniformly. As a result, the material will be subjected *only to normal stress*. This stress is uniform or *averaged* over the cross-sectional area.

▶ PROCEDURE FOR ANALYSIS

The equation $\sigma = N/A$ gives the *average* normal stress on the cross-sectional area of a member when the section is subjected to an internal resultant normal force **N**. Application of this equation requires the following steps.

Internal Loading.

- Section the member *perpendicular* to its longitudinal axis at the point where the normal stress is to be determined, and draw the free-body diagram of one of the segments. Apply the force equation of equilibrium to obtain the *internal axial force* **N** at the section.

Average Normal Stress.

- Determine the member's cross-sectional area at the section and calculate the average normal stress $\sigma = N/A$.

- It is suggested that σ be shown acting on a small volume element of the material located at a point on the section where stress is calculated. To do this, first draw σ on the face of the element coincident with the sectioned area A. Here σ acts in the *same direction* as the internal force **N** since all the normal stresses on the cross section develop this resultant. The normal stress σ on the opposite face of the element acts in the opposite direction.

EXAMPLE 7.5

The bar in Fig. 7–14a has a constant width of 35 mm and a thickness of 10 mm. Determine the maximum average normal stress in the bar when it is subjected to the loading shown.

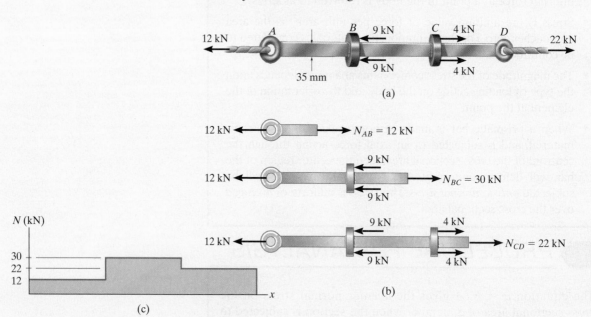

(a)

(b)

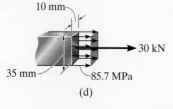

(c)

(d)

Fig. 7–14

SOLUTION

Internal Loading.   By inspection, the internal axial forces in regions *AB, BC,* and *CD* are all constant yet have different magnitudes. Using the method of sections, these loadings are shown on the free-body diagrams of the left segments shown in Fig. 7–14b.* The **normal force diagram**, which represents these results graphically, is shown in Fig. 7–14c. The largest loading is in region *BC,* where $N_{BC} = 30$ kN. Since the cross-sectional area of the bar is *constant,* the largest average normal stress also occurs within this region of the bar.

Average Normal Stress.   Applying Eq. 7–3, we have

$$\sigma_{BC} = \frac{N_{BC}}{A} = \frac{30(10^3)\ \text{N}}{(0.035\ \text{m})(0.010\ \text{m})} = 85.7\ \text{MPa} \qquad Ans.$$

The stress distribution acting on an arbitrary cross section of the bar within region *BC* is shown in Fig. 7–14d.

*Show that you get these *same results* using the right segments.

EXAMPLE   7.6

The 80-kg lamp is supported by two rods AB and BC as shown in Fig. 7–15a. If AB has a diameter of 10 mm and BC has a diameter of 8 mm, determine the average normal stress in each rod.

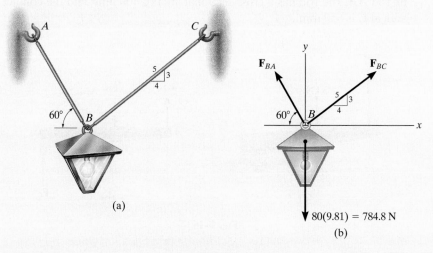

(a)

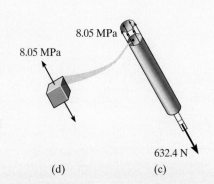

$80(9.81) = 784.8$ N

(b)

Fig. 7–15

SOLUTION

Internal Loading.   We must first determine the axial force in each rod. A free-body diagram of the lamp is shown in Fig. 7–15b. Applying the equations of force equilibrium,

$$\xrightarrow{+} \Sigma F_x = 0; \quad F_{BC}\left(\tfrac{4}{5}\right) - F_{BA}\cos 60° = 0$$

$$+\uparrow \Sigma F_y = 0; \quad F_{BC}\left(\tfrac{3}{5}\right) + F_{BA}\sin 60° - 784.8 \text{ N} = 0$$

$$F_{BC} = 395.2 \text{ N}, \quad F_{BA} = 632.4 \text{ N}$$

By Newton's third law of action, equal but opposite reaction, these forces subject the rods to tension throughout their length.

Average Normal Stress.   Applying Eq. 7–3,

$$\sigma_{BC} = \frac{F_{BC}}{A_{BC}} = \frac{395.2 \text{ N}}{\pi(0.004 \text{ m})^2} = 7.86 \text{ MPa} \qquad \textit{Ans.}$$

$$\sigma_{BA} = \frac{F_{BA}}{A_{BA}} = \frac{632.4 \text{ N}}{\pi(0.005 \text{ m})^2} = 8.05 \text{ MPa} \qquad \textit{Ans.}$$

The average normal stress distribution acting over a cross section of rod AB is shown in Fig. 7–15c, and at a point on this cross section, an element of material is stressed as shown in Fig. 7–15d.

8.05 MPa

8.05 MPa

632.4 N

(d)

(c)

EXAMPLE 7.7

Member AC shown in Fig. 7–16a is subjected to a vertical force of 3 kN. Determine the position x of this force so that the average compressive stress at the smooth support C is equal to the average tensile stress in the tie rod AB. The rod has a cross-sectional area of 400 mm² and the contact area at C is 650 mm².

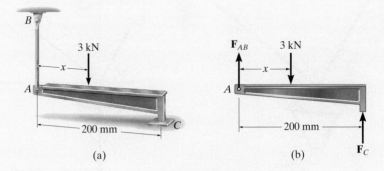

(a)   (b)

Fig. 7–16

SOLUTION

Internal Loading. The forces at A and C can be related by considering the free-body diagram of member AC, Fig. 7–16b. There are three unknowns, namely, F_{AB}, F_C, and x. To solve we will work in units of newtons and millimeters.

$$+\uparrow\Sigma F_y = 0; \qquad\qquad F_{AB} + F_C - 3000 \text{ N} = 0 \qquad (1)$$

$$\zeta+\Sigma M_A = 0; \qquad\qquad -3000 \text{ N}(x) + F_C(200 \text{ mm}) = 0 \qquad (2)$$

Average Normal Stress. A necessary third equation can be written that requires the tensile stress in the bar AB and the compressive stress at C to be equivalent, i.e.,

$$\sigma = \frac{F_{AB}}{400 \text{ mm}^2} = \frac{F_C}{650 \text{ mm}^2}$$

$$F_C = 1.625 F_{AB}$$

Substituting this into Eq. 1, solving for F_{AB}, then solving for F_C, we obtain

$$F_{AB} = 1143 \text{ N}$$

$$F_C = 1857 \text{ N}$$

The position of the applied load is determined from Eq. 2,

$$x = 124 \text{ mm} \qquad\qquad\qquad Ans.$$

As required, $0 < x < 200$ mm.

7.5   AVERAGE SHEAR STRESS

Shear stress has been defined in Section 7.3 as the stress component that acts *in the plane* of the sectioned area. To show how this stress can develop, consider the effect of applying a force **F** to the bar in Fig. 7–17a. If **F** is large enough, it can cause the material of the bar to deform and fail along the planes identified by *AB* and *CD*. A free-body diagram of the unsupported center segment of the bar, Fig. 7–17b, indicates that the shear force $V = F/2$ must be applied at each section to hold the segment in equilibrium. The ***average shear stress*** distributed over each sectioned area that develops this shear force is defined by

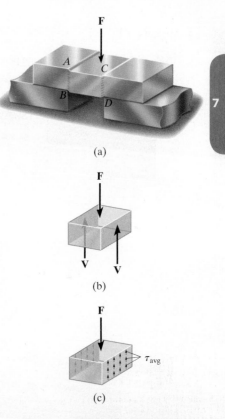

(a)

(b)

$$\tau_{avg} = \frac{V}{A} \qquad (7\text{–}4)$$

Here

τ_{avg} = average shear stress at the section, which is assumed to be the *same* at each point on the section

V = internal resultant shear force on the section determined from the equations of equilibrium

A = area of the section

(c)

Fig. 7–17

The distribution of average shear stress acting over the sections is shown in Fig. 7–17c. Notice that τ_{avg} is in the *same direction* as **V**, since the shear stress must create associated forces, all of which contribute to the internal resultant force **V**.

The loading case discussed here is an example of **simple or direct shear**, since the shear is caused by the *direct action* of the applied load **F**. This type of shear often occurs in various types of simple connections that use bolts, pins, welding material, etc. In all these cases, however, application of Eq. 7–4 is *only approximate*. A more precise investigation of the shear-stress distribution over the section often reveals that much larger shear stresses occur in the material than those predicted by this equation. Although this may be the case, application of Eq. 7–4 is generally acceptable for many problems involving the design or analysis of small elements. For example, engineering codes allow its use for determining the size or cross section of fasteners such as bolts, and for obtaining the bonding strength of glued joints subjected to shear loadings.

The pin *A* used to connect the linkage of this tractor is subjected to *double shear* because shearing stresses occur on the surface of the pin at *B* and *C*. See Fig. 7–19c.

7

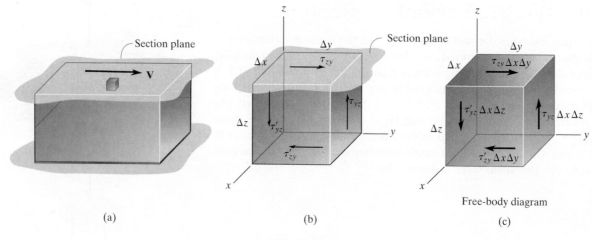

(a)

(b)

Free-body diagram

(c)

Fig. 7–18

Pure shear

(d)

Shear Stress Equilibrium. Let us consider the block in Fig. 7–18a, which has been sectioned and is subjected to the internal shear force V. A volume element taken at a point located on its surface will be subjected to a direct shear stress τ_{xy}, as shown in Fig. 7–18b. However, force and moment equilibrium of this element will also require shear stress to be developed on three other sides of the element. To show this, it is first necessary to draw the free-body diagram of the element, Fig. 7–18c. Then force equilibrium in the y direction requires

$$\Sigma F_y = 0; \qquad \tau_{zy}(\overset{\text{stress area}}{\overbrace{\Delta x\,\Delta y}}) - \tau'_{zy}\,\Delta x\,\Delta y = 0$$

$$\tau_{zy} = \tau'_{zy}$$

In a similar manner, force equilibrium in the z direction yields $\tau_{yz} = \tau'_{yz}$. Finally, taking moments about the x axis,

$$\Sigma M_x = 0; \qquad -\tau_{zy}(\Delta x\,\Delta y)\,\Delta z + \tau_{yz}(\Delta x\,\Delta z)\,\Delta y = 0$$

$$\tau_{zy} = \tau_{yz}$$

In other words,

$$\tau_{zy} = \tau'_{zy} = \tau_{yz} = \tau'_{yz} = \tau$$

and so, **all four shear stresses must have equal magnitude and be directed either toward or away from each other at opposite edges of the element**, Fig. 7–18d. This is referred to as the **complementary property of shear**, and the element in this case is subjected to *pure shear*.

IMPORTANT POINTS

- If two parts are *thin or small* when joined together, the applied loads may cause shearing of the material with negligible bending. If this is the case, it is generally assumed that an *average shear stress* acts over the cross-sectional area.

- When shear stress τ acts on a plane, then equilibrium of a volume element of material at a point on the plane requires associated shear stress of the same magnitude act on the three other sides of the element.

PROCEDURE FOR ANALYSIS

The equation $\tau_{avg} = V/A$ is used to determine the *average shear stress* in the material. Application requires the following steps.

Internal Shear.

- Section the member at the point where the average shear stress is to be determined.

- Draw the necessary free-body diagram, and calculate the internal shear force **V** acting at the section that is necessary to hold the part in equilibrium.

Average Shear Stress.

- Determine the sectioned area A, and then calculate the average shear stress $\tau_{avg} = V/A$.

- It is suggested that τ_{avg} be shown on a small volume element of material located at a point on the section where it is determined. To do this, first draw τ_{avg} on the face of the element, coincident with the sectioned area A. This stress acts in the *same direction* as **V**. The shear stresses acting on the three adjacent planes can then be drawn in their appropriate directions following the scheme shown in Fig. 7–18d.

EXAMPLE 7.8

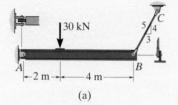

(a)

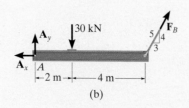

(b)

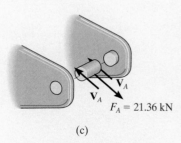

(c)

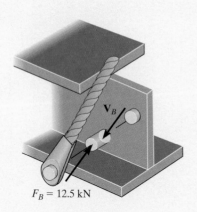

(d)

Fig. 7–19

Determine the average shear stress in the 20-mm-diameter pin at A and the 30-mm-diameter pin at B that support the beam in Fig. 7–19a.

SOLUTION

Internal Loadings. The forces on the pins can be obtained by considering the equilibrium of the beam, Fig. 7–19b.

$$\zeta + \Sigma M_A = 0;$$
$$F_B\left(\frac{4}{5}\right)(6\text{ m}) - 30\text{ kN}(2\text{ m}) = 0 \qquad F_B = 12.5\text{ kN}$$

$$\xrightarrow{+} \Sigma F_x = 0; \quad (12.5\text{ kN})\left(\frac{3}{5}\right) - A_x = 0 \qquad A_x = 7.50\text{ kN}$$

$$+\uparrow \Sigma F_y = 0; \quad A_y + (12.5\text{ kN})\left(\frac{4}{5}\right) - 30\text{ kN} = 0 \qquad A_y = 20\text{ kN}$$

Thus, the resultant force acting on pin A is

$$F_A = \sqrt{A_x^2 + A_y^2} = \sqrt{(7.50\text{ kN})^2 + (20\text{ kN})^2} = 21.36\text{ kN}$$

The pin at A is supported by two fixed "leaves" and so the free-body diagram of the center segment of the pin shown in Fig. 7–19c has *two* shearing surfaces between the beam and each leaf. Since the force of the beam (21.36 kN) acting on the pin is supported by shear force on each of two surfaces, it is called **double shear**. Thus,

$$V_A = \frac{F_A}{2} = \frac{21.36\text{ kN}}{2} = 10.68\text{ kN}$$

In Fig. 7–19a, note that pin B is subjected to **single shear**, which occurs on the section between the cable and beam, Fig. 7–19d. For this pin segment,

$$V_B = F_B = 12.5\text{ kN}$$

Average Shear Stress.

$$(\tau_A)_{avg} = \frac{V_A}{A_A} = \frac{10.68(10^3)\text{ N}}{\dfrac{\pi}{4}(0.02\text{ m})^2} = 34.0\text{ MPa} \qquad Ans.$$

$$(\tau_B)_{avg} = \frac{V_B}{A_B} = \frac{12.5(10^3)\text{ N}}{\dfrac{\pi}{4}(0.03\text{ m})^2} = 17.7\text{ MPa} \qquad Ans.$$

EXAMPLE 7.9

If the wood joint in Fig. 7–20a has a thickness of 150 mm, determine the average shear stress along shear planes a–a and b–b of the connected member. For each plane, represent the state of stress on an element of the material.

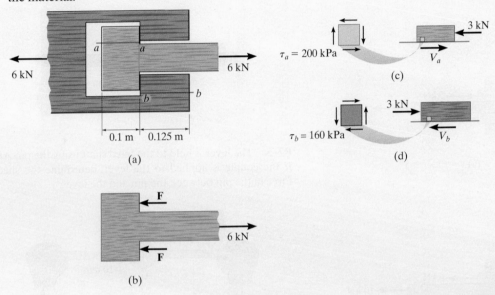

Fig. 7–20

SOLUTION

Internal Loadings. Referring to the free-body diagram of the member, Fig. 7–20b,

$$\xrightarrow{+} \Sigma F_x = 0; \quad 6 \text{ kN} - F - F = 0 \quad F = 3 \text{ kN}$$

Now consider the equilibrium of segments cut across shear planes a–a and b–b, shown in Figs. 7–20c and 7–20d.

$$\xrightarrow{+} \Sigma F_x = 0; \quad V_a - 3 \text{ kN} = 0 \quad V_a = 3 \text{ kN}$$

$$\xrightarrow{+} \Sigma F_x = 0; \quad 3 \text{ kN} - V_b = 0 \quad V_b = 3 \text{ kN}$$

Average Shear Stress.

$$(\tau_a)_{avg} = \frac{V_a}{A_a} = \frac{3(10^3) \text{ N}}{(0.1 \text{ m})(0.15 \text{ m})} = 200 \text{ kPa} \qquad Ans.$$

$$(\tau_b)_{avg} = \frac{V_b}{A_b} = \frac{3(10^3) \text{ N}}{(0.125 \text{ m})(0.15 \text{ m})} = 160 \text{ kPa} \qquad Ans.$$

The state of stress on elements located on sections a–a and b–b is shown in Figs. 7–20c and 7–20d, respectively.

PRELIMINARY PROBLEMS

P7–2. In each case, determine the largest internal shear force resisted by the bolt. Include all necessary free-body diagrams.

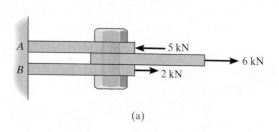

(a)

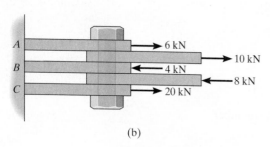

(b)

Prob. P7–2

P7–3. Determine the largest internal normal force in the bar.

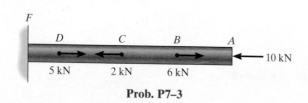

Prob. P7–3

P7–4. Determine the internal normal force at section A if the rod is subjected to the external uniformly distributed loading along its length of 8 kN/m.

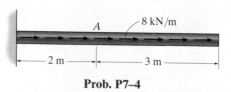

Prob. P7–4

P7–5. The lever is held to the fixed shaft using the pin AB. If the couple is applied to the lever, determine the shear force in the pin between the pin and the lever.

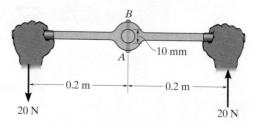

Prob. P7–5

P7–6. The single-V butt joint transmits the force of 5 kN from one bar to the other. Determine the resultant normal and shear force components on the face of the weld, section AB.

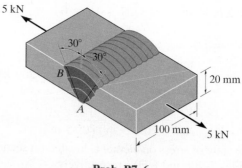

Prob. P7–6

FUNDAMENTAL PROBLEMS

F7–7. The uniform beam is supported by two rods AB and CD that have cross-sectional areas of 10 mm² and 15 mm², respectively. Determine the intensity w of the distributed load so that the average normal stress in each rod does not exceed 300 kPa.

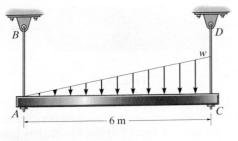

Prob. F7–7

F7–8. Determine the average normal stress on the cross section. Sketch the normal stress distribution over the cross section.

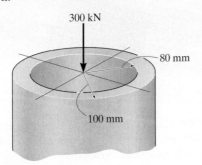

Prob. F7–8

F7–9. Determine the average normal stress on the cross section. Sketch the normal stress distribution over the cross section.

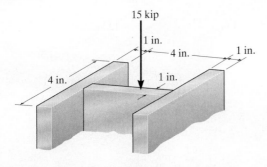

Prob. F7–9

F7–10. If the 600-kN force acts through the centroid of the cross section, determine the location $\bar{y}$ of the centroid and the average normal stress on the cross section. Also, sketch the normal stress distribution over the cross section.

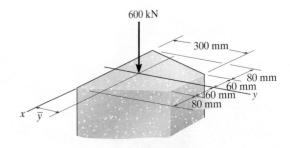

Prob. F7–10

F7–11. Determine the average normal stress at points A, B, and C. The diameter of each segment is indicated in the figure.

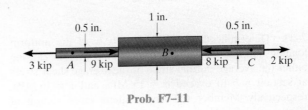

Prob. F7–11

F7–12. Determine the average normal stress in rod AB if the load has a mass of 50 kg. The diameter of rod AB is 8 mm.

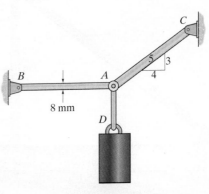

Prob. F7–12

PROBLEMS

7–26. The supporting wheel on a scaffold is held in place on the leg using a 4-mm-diameter pin. If the wheel is subjected to a normal force of 3 kN, determine the average shear stress in the pin. Assume the pin only supports the vertical 3-kN load.

3 kN

Prob. 7–26

7–27. Determine the largest intensity w of the uniform loading that can be applied to the frame without causing either the average normal stress or the average shear stress at section b–b to exceed $\sigma = 15$ MPa and $\tau = 16$ MPa, respectively. Member CB has a square cross section of 30 mm on each side.

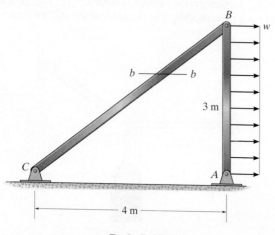

Prob. 7–27

*7–28. The bar has a cross-sectional area A and is subjected to the axial load P. Determine the average normal and average shear stresses acting over the shaded section, which is oriented at θ from the horizontal. Plot the variation of these stresses as a function of θ $(0 \leq \theta \leq 90°)$.

Prob. 7–28

7–29. The small block has a thickness of 0.5 in. If the stress distribution at the support developed by the load varies as shown, determine the force **F** applied to the block, and the distance d to where it is applied.

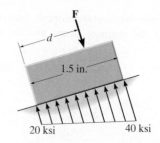

Prob. 7–29

7–30. If the material fails when the average normal stress reaches 120 psi, determine the largest centrally applied vertical load **P** the block can support.

7–31. If the block is subjected to a centrally applied force of $P = 6$ kip, determine the average normal stress in the material. Show the stress acting on a differential volume element of the material.

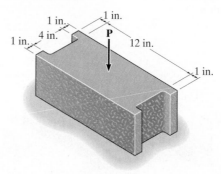

Probs. 7–30/31

*7–32. The plate has a width of 0.5 m. If the stress distribution at the support varies as shown, determine the force **P** applied to the plate and the distance d to where it is applied.

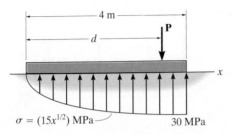

$\sigma = (15x^{1/2})$ MPa

30 MPa

Prob. 7–32

7–33. The board is subjected to a tensile force of 200 lb. Determine the average normal and average shear stress in the wood fibers, which are oriented along plane a–a at $20°$ with the axis of the board.

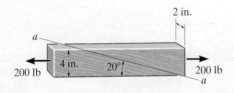

2 in.

a

4 in. 200 lb

$20°$

200 lb

a

Prob. 7–33

7–34. The boom has a uniform weight of 600 lb and is hoisted into position using the cable BC. If the cable has a diameter of 0.5 in., plot the average normal stress in the cable as a function of the boom position θ for $0° \le \theta \le 90°$.

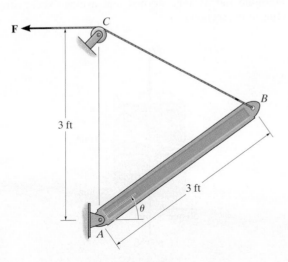

Prob. 7–34

7–35. Determine the average normal stress in each of the 20-mm-diameter bars of the truss. Set $P = 40$ kN.

*7–36. If the average normal stress in each of the 20-mm-diameter bars is not allowed to exceed 150 MPa, determine the maximum force **P** that can be applied to joint C.

7–37. Determine the maximum average shear stress in pin A of the truss. A horizontal force of $P = 40$ kN is applied to joint C. Each pin has a diameter of 25 mm and is subjected to double shear.

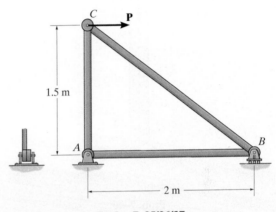

Probs. 7–35/36/37

7–38. If $P = 5$ kN, determine the average shear stress in the pins at A, B, and C. All pins are in double shear, and each has a diameter of 18 mm.

7–39. Determine the maximum magnitude P of the loads the beam can support if the average shear stress in each pin is not to exceed 80 MPa. All pins are in double shear, and each has a diameter of 18 mm.

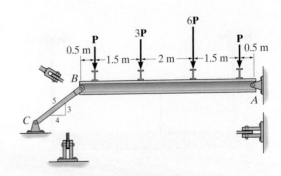

Probs. 7–38/39

*7–40. The column is made of concrete having a density of 2.30 Mg/m³. At its top *B* it is subjected to an axial compressive force of 15 kN. Determine the average normal stress in the column as a function of the distance *z* measured from its base.

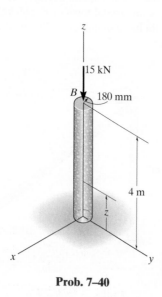

Prob. 7–40

7–41. The beam is supported by two rods *AB* and *CD* that have cross-sectional areas of 12 mm² and 8 mm², respectively. If *d* = 1 m, determine the average normal stress in each rod.

7–42. The beam is supported by two rods *AB* and *CD* that have cross-sectional areas of 12 mm² and 8 mm², respectively. Determine the position *d* of the 6-kN load so that the average normal stress in each rod is the same.

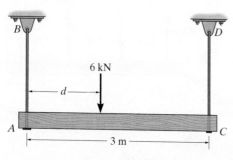

Probs. 7–41/42

7–43. If *P* = 15 kN, determine the average shear stress in the pins at *A*, *B*, and *C*. All pins are in double shear, and each has a diameter of 18 mm.

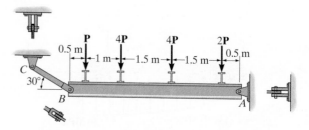

Prob. 7–43

*7–44. The railcar docklight is supported by the $\frac{1}{8}$-in.-diameter pin at *A*. If the lamp weighs 4 lb, and the extension arm *AB* has a weight of 0.5 lb/ft, determine the average shear stress in the pin needed to support the lamp. *Hint:* The shear force in the pin is caused by the couple moment required for equilibrium at *A*.

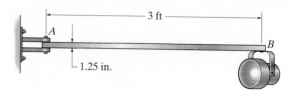

Prob. 7–44

7–45. The plastic block is subjected to an axial compressive force of 600 N. Assuming that the caps at the top and bottom distribute the load uniformly throughout the block, determine the average normal and average shear stress acting along section *a–a*.

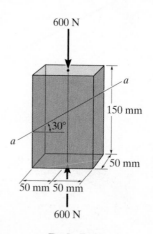

Prob. 7–45

7–46. The two steel members are joined together using a 30° scarf weld. Determine the average normal and average shear stress resisted in the plane of the weld.

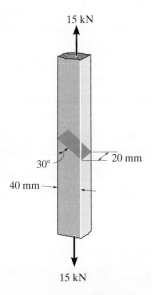

15 kN

20 mm

30°

40 mm

15 kN

Prob. 7–46

7–49. The two members used in the construction of an aircraft fuselage are joined together using a 30° fish-mouth weld. Determine the average normal and average shear stress on the plane of each weld. Assume each inclined plane supports a horizontal force of 400 lb.

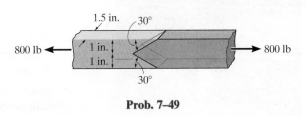

1.5 in.   30°

800 lb

1 in.
1 in.

800 lb

30°

Prob. 7–49

7–50. The 2-Mg concrete pipe has a center of mass at point G. If it is suspended from cables AB and AC, determine the average normal stress in the cables. The diameters of AB and AC are 12 mm and 10 mm, respectively.

7–51. The 2-Mg concrete pipe has a center of mass at point G. If it is suspended from cables AB and AC, determine the diameter of cable AB so that the average normal stress in this cable is the same as in the 10-mm-diameter cable AC.

7–47. The bar has a cross-sectional area of $400(10^{-6})$ m². If it is subjected to a triangular axial distributed loading along its length which is 0 at $x = 0$ and 9 kN/m at $x = 1.5$ m, and to two concentrated loads as shown, determine the average normal stress in the bar as a function of x for $0 \leq x < 0.6$ m.

*7–48. The bar has a cross-sectional area of $400(10^{-6})$ m². If it is subjected to a uniform axial distributed loading along its length of 9 kN/m, and to two concentrated loads as shown, determine the average normal stress in the bar as a function of x for 0.6 m $< x \leq 1.5$ m.

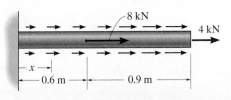

8 kN

4 kN

x

0.6 m   0.9 m

Probs. 7–47/48

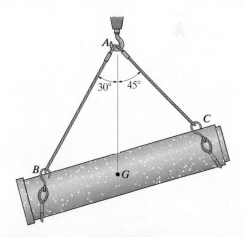

A

30°   45°

C

B   •G

Probs. 7–50/51

7

7.6   ALLOWABLE STRESS DESIGN

To ensure the safety of a structural or mechanical member, it is necessary to restrict the applied load to one that is *less than* the load the member can fully support. There are many reasons for doing this.

- The intended measurements of a structure or machine may not be exact, due to errors in fabrication or in the assembly of its component parts.

- Unknown vibrations, impact, or accidental loadings can occur that may not be accounted for in the design.

- Atmospheric corrosion, decay, or weathering tend to cause materials to deteriorate during service.

- Some materials, such as wood, concrete, or fiber-reinforced composites, can show high variability in mechanical properties.

One method of specifying the allowable load for a member is to use a number called the ***factor of safety*** (F.S.). It is a ratio of the failure load F_{fail} to the allowable load F_{allow},

Cranes are often supported using bearing pads to give them stability. Care must be taken not to crush the supporting surface, due to the large bearing stress developed between the pad and the surface.

$$\text{F.S.} = \frac{F_{\text{fail}}}{F_{\text{allow}}} \tag{7-5}$$

Here F_{fail} is found from experimental testing of the material.

If the load applied to the member is *linearly related* to the stress developed within the member, as in the case of $\sigma = N/A$ and $\tau_{\text{avg}} = V/A$, then we can also express the factor of safety as a ratio of the failure stress σ_{fail} (or τ_{fail}) to the *allowable stress* σ_{allow} (or τ_{allow}). Here the area A will cancel, and so,

$$\text{F.S.} = \frac{\sigma_{\text{fail}}}{\sigma_{\text{allow}}} \tag{7-6}$$

or

$$\text{F.S.} = \frac{\tau_{\text{fail}}}{\tau_{\text{allow}}} \tag{7-7}$$

Specific values of F.S. depend on the types of materials to be used and the intended purpose of the structure or machine, while accounting for the previously mentioned uncertainties. For example, the F.S. used in the design of aircraft-or space-vehicle components may be close to 1 in order to reduce the weight of the vehicle. Or, in the case of a nuclear power plant, the factor of safety for some of its components may be as high as 3 due to uncertainties in loading or material behavior. Whatever the case, the factor of safety or the allowable stress for a specific case can be found in design codes and engineering handbooks. Design that is based on an allowable stress limit is called ***allowable stress design*** (ASD). Using this method will ensure a balance between both public and environmental safety on the one hand and economic considerations on the other.

Simple Connections. By making simplifying assumptions regarding the behavior of the material, the equations $\sigma = N/A$ and $\tau_{avg} = V/A$ can often be used to analyze or design a simple connection or mechanical element. For example, if a member is subjected to normal force at a section, its required area at the section is determined from

$$A = \frac{N}{\sigma_{allow}} \qquad (7\text{–}8)$$

or if the section is subjected to an average shear force, then the required area at the section is

$$A = \frac{V}{\tau_{allow}} \qquad (7\text{–}9)$$

Three examples of where the above equations apply are shown in Fig. 7–21. The first figure shows the normal stress acting on the bottom of a base plate. This compressive stress caused by one surface that bears against another is often called ***bearing stress***.

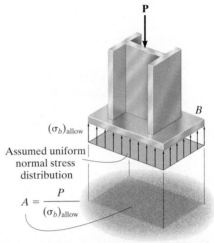

$(\sigma_b)_{allow}$

Assumed uniform normal stress distribution

$$A = \frac{P}{(\sigma_b)_{allow}}$$

The area of the column base plate B is determined from the allowable bearing stress for the concrete.

Assumed uniform shear stress τ_{allow}

$$l = \frac{P}{\tau_{allow}\pi d}$$

d

The embedded length l of this rod in concrete can be determined using the allowable shear stress of the bonding glue.

The area of the bolt for this lap joint is determined from the shear stress, which is largest between the plates.

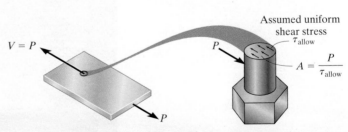

$V = P$

Assumed uniform shear stress τ_{allow}

$$A = \frac{P}{\tau_{allow}}$$

Fig. 7–21

IMPORTANT POINT

- Design of a member for strength is based on selecting an allowable stress that will enable it to safely support its intended load. Since there are many unknown factors that can influence the actual stress in a member, then depending upon the intended use of the member, a *factor of safety* is applied to obtain the allowable load the member can support.

PROCEDURE FOR ANALYSIS

When solving problems using the average normal and shear stress equations, a careful consideration should first be made as to choose the section over which the critical stress is acting. Once this section is determined, the member must then be designed to have a sufficient area at the section to resist the stress that acts on it. This area is determined using the following steps.

Internal Loading.

- Section the member through the area and draw a free-body diagram of a segment of the member. The internal resultant force at the section is then determined using the equations of equilibrium.

Required Area.

- Provided the allowable stress is known or can be determined, the required area needed to sustain the load at the section is then determined from $A = P/\sigma_{\text{allow}}$ or $A = V/\tau_{\text{allow}}$.

Appropriate factors of safety must be considered when designing cranes and cables used to transfer heavy loads.

EXAMPLE 7.10

The control arm is subjected to the loading shown in Fig. 7–22a. Determine to the nearest $\frac{1}{4}$ in. the required diameters of the steel pins at A and C if the factor of safety for shear is F.S. = 1.5 and the failure shear stress is $\tau_{fail} = 12$ ksi.

SOLUTION

Pin Forces.   A free-body diagram of the arm is shown in Fig. 7–22b. For equilibrium we have

$$\zeta + \Sigma M_C = 0; \quad F_{AB}(8 \text{ in.}) - 3 \text{ kip } (3 \text{ in.}) - 5 \text{ kip } \left(\tfrac{3}{5}\right)(5 \text{ in.}) = 0$$

$$F_{AB} = 3 \text{ kip}$$

$$\xrightarrow{+} \Sigma F_x = 0; \quad -3 \text{ kip} - C_x + 5 \text{ kip } \left(\tfrac{4}{5}\right) = 0 \quad C_x = 1 \text{ kip}$$

$$+\uparrow \Sigma F_y = 0; \quad C_y - 3 \text{ kip} - 5 \text{ kip}\left(\tfrac{3}{5}\right) = 0 \quad C_y = 6 \text{ kip}$$

The pin at C resists the resultant force at C, which is

$$F_C = \sqrt{(1 \text{ kip})^2 + (6 \text{ kip})^2} = 6.083 \text{ kip}$$

Allowable Shear Stress.   We have

$$\text{F.S.} = \frac{\tau_{fail}}{\tau_{allow}}; \quad 1.5 = \frac{12 \text{ ksi}}{\tau_{allow}}; \quad \tau_{allow} = 8 \text{ ksi}$$

Pin A.   This pin is subjected to *single shear*, Fig. 7–22c, so that

$$A = \frac{V}{\tau_{allow}}; \quad \pi\left(\frac{d_A}{2}\right)^2 = \frac{3 \text{ kip}}{8 \text{ kip/in}^2}; \quad d_A = 0.691 \text{ in.}$$

Use $d_A = \tfrac{3}{4}$ in.   *Ans.*

Pin C.   Since this pin is subjected to *double shear*, a shear force of 3.041 kip acts over its cross-sectional area *between* the arm and each supporting leaf for the pin, Fig. 7–22d. We have

$$A = \frac{V}{\tau_{allow}}; \quad \pi\left(\frac{d_C}{2}\right)^2 = \frac{3.041 \text{ kip}}{8 \text{ kip/in}^2}; \quad d_C = 0.696 \text{ in.}$$

Use $d_C = \tfrac{3}{4}$ in.   *Ans.*

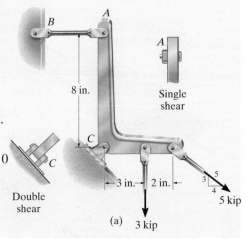

(a)

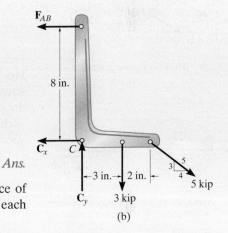

(b)

3 kip

3 kip

Pin at A

(c)

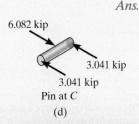

6.082 kip

3.041 kip

3.041 kip

Pin at C

(d)

Fig. 7–22

EXAMPLE 7.11

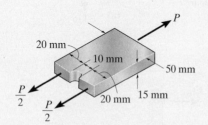

(a)

Determine the largest load P that can be applied to the bars of the lap joint shown in Fig. 7–23a. The bolt has a diameter of 10 mm and an allowable shear stress of 80 MPa. Each plate has an allowable tensile stress of 50 MPa, an allowable bearing stress of 80 MPa, and an allowable shear stress of 30 MPa.

SOLUTION

To solve the problem we will determine P for each possible failure condition; then we will choose the smallest value of P. Why?

Failure of the Plate in Tension.   If the plate fails in tension, it will do so at its smallest cross section, Fig. 7–23b.

$$(\sigma_{allow})_t = \frac{N}{A}; \qquad 50(10^6)\,\text{N/m}^2 = \frac{P}{2(0.02\,\text{m})(0.015\,\text{m})}$$

$$P = 30\,\text{kN}$$

Failure of the Plate by Bearing.   A free-body diagram of the top plate, Fig. 7–23c, shows that the bolt will exert a complicated distribution of stress on the plate along the curved central area of contact with the bolt.* To simplify the analysis for small connections having pins or bolts such as this, design codes allow the *projected area* of the bolt to be used when calculating the bearing stress. Therefore,

$$(\sigma_{allow})_b = \frac{N}{A}; \qquad 80(10^6)\text{N/m}^2 = \frac{P}{(0.01\,\text{m})(0.015\,\text{m})}$$

$$P = 12\,\text{kN}$$

Failure of plate in tension

(b)

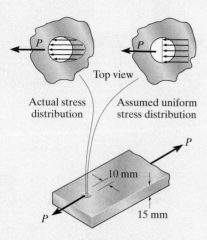

Failure of plate in bearing caused by bolt

(c)

Fig. 7–23

*The material strength of a bolt or pin is generally greater than that of the plate material, so bearing failure of the member is of greater concern.

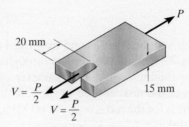

$$V = \frac{P}{2}$$

$$V = \frac{P}{2}$$

Failure of plate by shear

(d)

Failure of the Plate by Shear.   There is the possibility for the bolt to tear through the plate along the section shown on the free-body diagram in Fig. 7–23d. Here the shear is $V = P/2$, and so

$$(\tau_{\text{allow}})_p = \frac{V}{A}; \quad 30(10^6) \text{ N/m}^2 = \frac{P/2}{(0.02 \text{ m})(0.015 \text{ m})}$$

$$P = 18 \text{ kN}$$

Failure of the Bolt by Shear.   The bolt can fail in shear along the plane between the plates. The free-body diagram in Fig. 7–23e indicates that $V = P$, so that

$$(\tau_{\text{allow}})_b = \frac{V}{A}; \quad 80(10^6) \text{ N/m}^2 = \frac{P}{\pi(0.005 \text{ m})^2}$$

$$P = 6.28 \text{ kN}$$

Comparing the above results, the largest allowable load for the connections depends upon the bolt shear. Therefore,

$$P = 6.28 \text{ kN} \qquad\qquad Ans.$$

$$V = P$$

Failure of bolt by shear

(e)

Fig. 7–23 (cont.)

EXAMPLE 7.12

The suspender rod is supported at its end by a fixed-connected circular disk as shown in Fig. 7–24a. If the rod passes through a 40-mm-diameter hole, determine the minimum required diameter of the rod and the minimum thickness of the disk needed to support the 20-kN load. The allowable normal stress for the rod is $\sigma_{allow} = 60$ MPa, and the allowable shear stress for the disk is $\tau_{allow} = 35$ MPa.

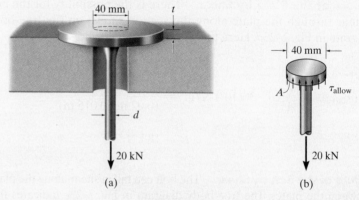

Fig. 7–24

SOLUTION

Diameter of Rod. By inspection, the axial force in the rod is 20 kN. Thus the required cross-sectional area of the rod is

$$A = \frac{N}{\sigma_{allow}}; \qquad \frac{\pi}{4}d^2 = \frac{20(10^3)\ \text{N}}{60(10^6)\ \text{N/m}^2}$$

so that

$$d = 0.0206\ \text{m} = 20.6\ \text{mm} \qquad \qquad Ans.$$

Thickness of Disk. As shown on the free-body diagram in Fig. 7–24b, the material at the sectioned area of the disk must resist *shear stress* to prevent movement of the disk through the hole. If this shear stress is *assumed* to be uniformly distributed over the sectioned area, then, since $V = 20$ kN, we have

$$A = \frac{V}{\tau_{allow}}; \qquad 2\pi\,(0.02\ \text{m})\,(t) = \frac{20(10^3)\ \text{N}}{35(10^6)\ \text{N/m}^2}$$

$$t = 4.55(10^{-3})\ \text{m} = 4.55\ \text{mm} \qquad \qquad Ans.$$

EXAMPLE 7.13

The shaft shown in Fig. 7–25a is supported by the collar at C, which is attached to the shaft and located on the right side of the bearing at B. Determine the largest value of P for the axial forces at E and F so that the bearing stress on the collar does not exceed an allowable stress of $(\sigma_b)_{\text{allow}} = 75$ MPa and the average normal stress in the shaft does not exceed an allowable stress of $(\sigma_t)_{\text{allow}} = 55$ MPa.

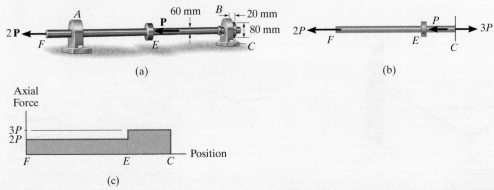

(a)

(b)

Axial Force

(c)

Fig. 7–25

SOLUTION

To solve the problem we will determine P for each possible failure condition. Then we will choose the *smallest* value. Why?

Normal Stress. Using the method of sections, the axial load within region FE of the shaft is 2P, whereas the *largest* axial force, 3P, occurs within region EC, Fig. 7–25b. The variation of the internal loading is clearly shown on the normal-force diagram, Fig. 7–25c. Since the cross-sectional area of the entire shaft is constant, region FC is subjected to the maximum averge normal stress. Applying Eq. 7–8, we have

$$A = \frac{N}{\sigma_{\text{allow}}}; \qquad \pi(0.03 \text{ m})^2 = \frac{3P}{55(10^6) \text{ N/m}^2}$$
$$P = 51.8 \text{ kN} \qquad\qquad Ans.$$

Bearing Stress. As shown on the free-body diagram in Fig. 7–25d, the collar at C must resist the load of 3P, which acts over a bearing area of $A_b = [\pi(0.04 \text{ m})^2 - \pi(0.03 \text{ m})^2] = 2.199(10^{-3}) \text{ m}^2$. Thus,

$$A = \frac{N}{\sigma_{\text{allow}}}; \qquad 2.199(10^{-3}) \text{ m}^2 = \frac{3P}{75(10^6) \text{ N/m}^2}$$
$$P = 55.0 \text{ kN}$$

(d)

By comparsion, the largest load that can be applied to the shaft is P = 51.8 kN, since any load larger than this will cause the allowable normal stress in the shaft to be exceeded.

NOTE: Here we have not considered a possible shear failure of the collar as in Example 7.12.

FUNDAMENTAL PROBLEMS

7

F7–13. Rods AC and BC are used to suspend the 200-kg mass. If each rod is made of a material for which the average normal stress can not exceed 150 MPa, determine the minimum required diameter of each rod to the nearest mm.

F7–15. Determine the maximum average shear stress developed in each 3/4-in.-diameter bolt.

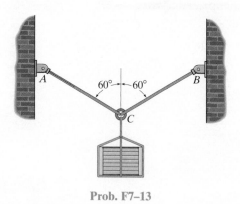

Prob. F7–13

Prob. F7–15

F7–14. The pin at A has a diameter of 0.25 in. If it is subjected to double shear, determine the average shear stress in the pin.

F7–16. If each of the three nails has a diameter of 4 mm and can withstand an average shear stress of 60 MPa, determine the maximum allowable force **P** that can be applied to the board.

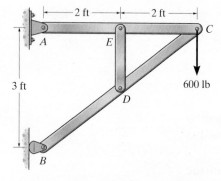

Prob. F7–14

Prob. F7–16

F7–17. The strut is glued to the horizontal member at surface AB. If the strut has a thickness of 25 mm and the glue can withstand an average shear stress of 600 kPa, determine the maximum force **P** that can be applied to the strut.

F7–19. If the eyebolt is made of a material having a yield stress of $\sigma_Y = 250$ MPa, determine the minimum required diameter d of its shank. Apply a factor of safety F.S. = 1.5 against yielding.

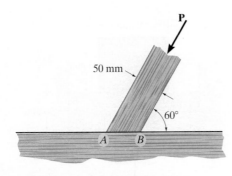

Prob. F7–17

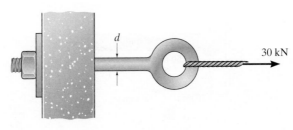

Prob. F7–19

F7–18. Determine the maximum average shear stress developed in the 30-mm-diameter pin.

F7–20. If the bar assembly is made of a material having a yield stress of $\sigma_Y = 50$ ksi, determine the minimum required dimensions h_1 and h_2 to the nearest $1/8$ in. Apply a factor of safety F.S. = 1.5 against yielding. Each bar has a thickness of 0.5 in.

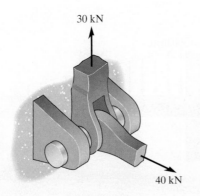

Prob. F7–18

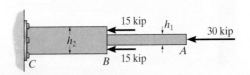

Prob. F7–20

7

F7–21. Determine the maximum force **P** that can be applied to the rod if it is made of material having a yield stress of $\sigma_Y = 250$ MPa. Consider the possibility that failure occurs in the rod and at section *a–a*. Apply a factor of safety of F.S. = 2 against yielding.

F7–23. If the bolt head and the supporting bracket are made of the same material having a failure shear stress of $\tau_{fail} = 120$ MPa, determine the maximum allowable force **P** that can be applied to the bolt so that it does not pull through the plate. Apply a factor of safety of F.S. = 2.5 against shear failure.

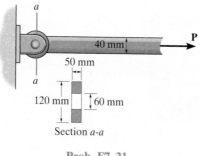

Section *a-a*

Prob. F7–21

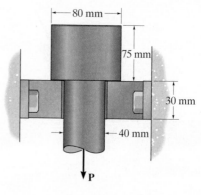

Prob. F7–23

F7–22. The pin is made of a material having a failure shear stress of $\tau_{fail} = 100$ MPa. Determine the minimum required diameter of the pin to the nearest mm. Apply a factor of safety of F.S. = 2.5 against shear failure.

F7–24. Six nails are used to hold the hanger at *A* against the column. Determine the minimum required diameter of each nail to the nearest 1/16 in. if it is made of material having $\tau_{fail} = 16$ ksi. Apply a factor of safety of F.S. = 2 against shear failure.

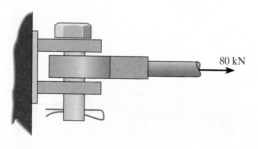

Prob. F7–22

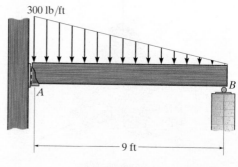

Prob. F7–24

PROBLEMS

*7–52. If A and B are both made of wood and are $\frac{3}{8}$ in. thick, determine to the nearest $\frac{1}{4}$ in. the smallest dimension h of the vertical segment so that it does not fail in shear. The allowable shear stress for the segment is $\tau_{allow} = 300$ psi.

7–54. The connection is made using a bolt and nut and two washers. If the allowable bearing stress of the washers on the boards is $(\sigma_b)_{allow} = 2$ ksi, and the allowable tensile stress within the bolt shank S is $(\sigma_t)_{allow} = 18$ ksi, determine the maximum allowable tension in the bolt shank. The bolt shank has a diameter of 0.31 in., and the washers have an outer diameter of 0.75 in. and inner diameter (hole) of 0.50 in.

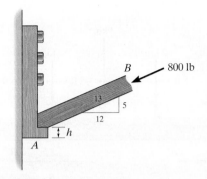

Prob. 7–52

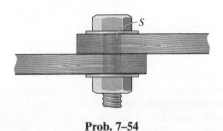

Prob. 7–54

7–53. The lever is attached to the shaft A using a key that has a width d and length of 25 mm. If the shaft is fixed and a vertical force of 200 N is applied perpendicular to the handle, determine the dimension d if the allowable shear stress for the key is $\tau_{allow} = 35$ MPa.

7–55. The tension member is fastened together using *two* bolts, one on each side of the member as shown. Each bolt has a diameter of 0.3 in. Determine the maximum load P that can be applied to the member if the allowable shear stress for the bolts is $\tau_{allow} = 12$ ksi and the allowable average normal stress is $\sigma_{allow} = 20$ ksi.

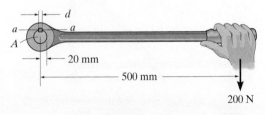

Prob. 7–53

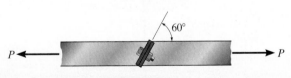

Prob. 7–55

*7–56. The steel swivel bushing in the elevator control of an airplane is held in place using a nut and washer as shown in Fig. (*a*). Failure of the washer *A* can cause the push rod to separate as shown in Fig. (*b*). If the maximum average shear stress is $\tau_{max} = 21$ ksi, determine the force **F** that must be applied to the bushing. The washer is $\frac{1}{16}$ in. thick.

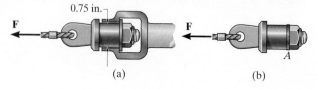

(a)   (b)

Prob. 7–56

7–57. The spring mechanism is used as a shock absorber for a load applied to the drawbar *AB*. Determine the force in each spring when the 50-kN force is applied. Each spring is originally unstretched and the drawbar slides along the smooth guide posts *CG* and *EF*. The ends of all springs are attached to their respective members. Also, what is the required diameter of the shank of bolts *CG* and *EF* if the allowable stress for the bolts is $\sigma_{allow} = 150$ MPa?

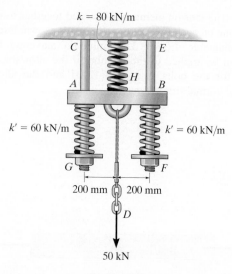

Prob. 7–57

7–58. Determine the size of *square* bearing plates *A′* and *B′* required to support the loading. Take $P = 1.5$ kip. Dimension the plates to the nearest $\frac{1}{2}$ in. The reactions at the supports are vertical and the allowable bearing stress for the plates is $(\sigma_b)_{allow} = 400$ psi.

7–59. Determine the maximum load **P** that can be applied to the beam if the bearing plates *A′* and *B′* have square cross sections of 2 in. × 2 in. and 4 in. × 4 in., respectively, and the allowable bearing stress for the material is $(\sigma_b)_{allow} = 400$ psi.

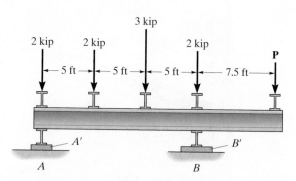

Probs. 7–58/59

*7–60. Determine the required diameter of the pins at *A* and *B* to the nearest $\frac{1}{16}$ in. if the allowable shear stress for the material is $\tau_{allow} = 6$ ksi. Pin *A* is subjected to double shear, whereas pin *B* is subjected to single shear.

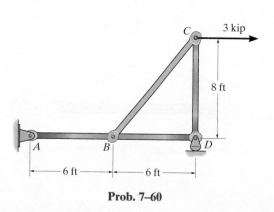

Prob. 7–60

7–61.   If the allowable tensile stress for wires AB and AC is $\sigma_{\text{allow}} = 200$ MPa, determine the required diameter of each wire if the applied load is $P = 6$ kN.

7–62.   If the allowable tensile stress for wires AB and AC is $\sigma_{\text{allow}} = 180$ MPa, and wire AB has a diameter of 5 mm and AC has a diameter of 6 mm, determine the greatest force P that can be applied to the chain.

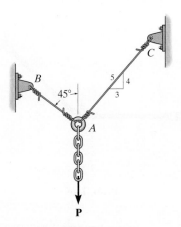

Probs. 7–61/62

7–63.   The cotter is used to hold the two rods together. Determine the smallest thickness t of the cotter and the smallest diameter d of the rods. All parts are made of steel for which the failure normal stress is $\sigma_{\text{fail}} = 500$ MPa and the failure shear stress is $\tau_{\text{fail}} = 375$ MPa. Use a factor of safety of $(\text{F.S.})_t = 2.50$ in tension and $(\text{F.S.})_s = 1.75$ in shear.

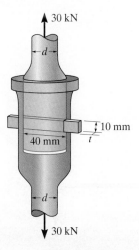

Prob. 7–63

*7–64.   Determine the required diameter of the pins at A and B if the allowable shear stress for the material is $\tau_{\text{allow}} = 100$ MPa. Both pins are subjected to double shear.

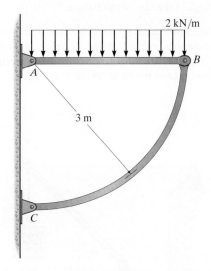

Prob. 7–64

7–65.   The steel pipe is supported on the circular base plate and concrete pedestal. If the thickness of the pipe is $t = 5$ mm and the base plate has a radius of 150 mm, determine the factors of safety against failure of the steel and concrete. The applied force is 500 kN, and the normal failure stresses for steel and concrete are $(\sigma_{\text{fail}})_{\text{st}} = 350$ MPa and $(\sigma_{\text{fail}})_{\text{con}} = 25$ MPa, respectively.

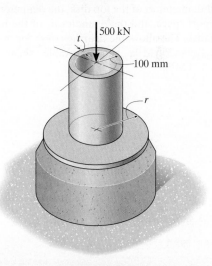

Prob. 7–65

7–66. The boom is supported by the winch cable that has a diameter of 0.25 in. and an allowable normal stress of $\sigma_{allow} = 24$ ksi. Determine the greatest weight of the crate that can be supported without causing the cable to fail if $\phi = 30°$. Neglect the size of the winch.

7–67. The boom is supported by the winch cable that has an allowable normal stress of $\sigma_{allow} = 24$ ksi. If it supports the 5000 lb crate when $\phi = 20°$, determine the smallest diameter of the cable to the nearest $\frac{1}{16}$ in.

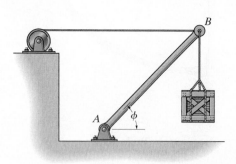

Probs. 7–66/67

7–69. The two aluminum rods support the vertical force of $P = 20$ kN. Determine their required diameters if the allowable tensile stress for the aluminum is $\sigma_{allow} = 150$ MPa.

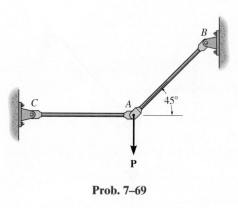

Prob. 7–69

***7–68.** The assembly consists of three disks A, B, and C that are used to support the load of 140 kN. Determine the smallest diameter d_1 of the top disk, the diameter d_2 within the support space, and the diameter d_3 of the hole in the bottom disk. The allowable bearing stress for the material is $(\sigma_b)_{allow} = 350$ MPa and allowable shear stress is $\tau_{allow} = 125$ MPa.

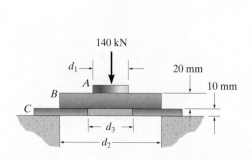

Prob. 7–68

7–70. The two aluminum rods AB and AC have diameters of 10 mm and 8 mm, respectively. Determine the largest vertical force $\mathbf{P}$ that can be supported. The allowable tensile stress for the aluminum is $\sigma_{allow} = 150$ MPa.

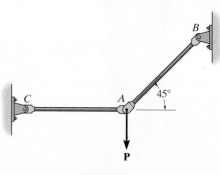

Prob. 7–70

7.7   DEFORMATION

Whenever a force is applied to a body, it will tend to change the body's shape and size. These changes are referred to as ***deformation***, and they may be either highly visible or practically unnoticeable. For example, a rubber band will undergo a very large deformation when stretched, whereas only slight deformations of structural members occur when a building is occupied by people walking about. Deformation of a body can also occur when the temperature of the body is changed. A typical example is the thermal expansion or contraction of a roof caused by the weather.

In a general sense, the deformation of a body will not be uniform throughout its volume, and so the change in geometry of any line segment within the body may vary substantially along its length. Hence, to study deformational changes in a more uniform manner, we will consider line segments that are very short and located in the neighborhood of a point. Realize, however, that these changes will also depend on the orientation of the line segment at the point. For example, a line segment may elongate if it is oriented in one direction, whereas it may contract if it is oriented in another direction.

Note the before and after positions of three diffferent line segments on this rubber membrane which is subjected to tension. The vertical line is lengthened, the horizontal line is shortened, and the inclined line changes its length and rotates.

7.8 STRAIN

In order to describe the deformation of a body by changes in lengths of line segments and changes in the angles between them, we will develop the concept of strain. Strain is actually measured by experiment, and once the strain is obtained, it will be shown in the next chapter how it can be related to the stress acting within the body.

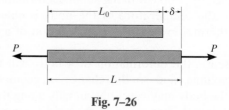

Fig. 7–26

Normal Strain. If an axial load P is applied to the bar in Fig. 7–26, it will change the bar's length L_0 to a length L. We will define the ***average normal strain*** ϵ (epsilon) of the bar as the change in its length δ (delta) $= L - L_0$ divided by its original length, that is

$$\epsilon_{\text{avg}} = \frac{L - L_0}{L_0} \tag{7–10}$$

The normal strain at a point in a body of arbitary shape is defined in a similar manner. For example, consider the very small line segment Δs located at the point, Fig. 7–27. After deformation it becomes $\Delta s'$, and the change in its length is therefore $\Delta s' - \Delta s$. As $\Delta s \to 0$, in the limit the normal strain at the point is therefore

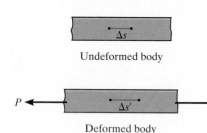

Undeformed body

Deformed body

Fig. 7–27

$$\epsilon = \lim_{\Delta s \to 0} \frac{\Delta s' - \Delta s}{\Delta s} \tag{7–11}$$

In both cases ϵ (or ϵ_{avg}) is a change in length per unit length, and it is positive when the initial line elongates, and negative when the line contracts.

Units. As shown, normal strain is a *dimensionless quantity*, since it is a ratio of two lengths. However, it is sometimes stated in terms of a ratio of length units. If the SI system is used, where the basic unit for length is the meter (m), then since ϵ is generally very small, for most engineering applications, measurements of strain will be in micrometers per meter (μm/m), where 1μm $= 10^{-6}$ m. In the Foot-Pound-Second system, strain is often stated in units of inches per inch (in./in.), and for experimental work, strain is sometimes expressed as a percent. For example, a normal strain of $480(10^{-6})$ can be reported as $480(10^{-6})$ in./in., 480μm/m, or 0.0480%. Or one can state the strain as simply 480μ (480 "micros").

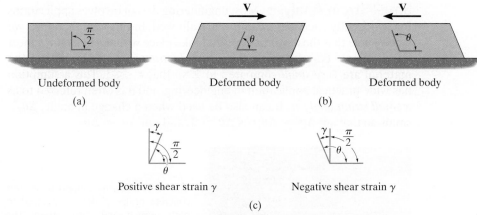

Positive shear strain γ Negative shear strain γ

(c)

Fig. 7–28

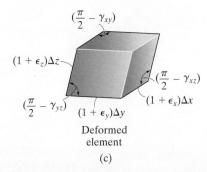

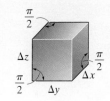

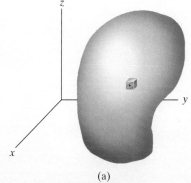

Shear Strain.
Deformations not only cause line segments to elongate or contract, but they also cause them to change direction. If we select two line segments that are originally perpendicular to one another, then the *change in angle* that occurs between them is referred to as **shear strain**. This angle is denoted by γ (gamma) and is always measured in radians (rad), which are dimensionless. For example, consider the two perpendicular line segments at a point in the block shown in Fig. 7–28a. If an applied loading causes the block to deform as shown in Fig. 7–28b, so that the angle between the line segments becomes θ, then the shear strain at the point becomes

$$\gamma = \frac{\pi}{2} - \theta \qquad (7\text{–}12)$$

Notice that if θ is smaller than $\pi/2$, Fig. 7–28c, then the shear strain is *positive*, whereas if θ is larger than $\pi/2$, then the shear strain is *negative*.

Cartesian Strain Components.
We can generalize our definitions of normal and shear strain and consider the undeformed element at a point in a body, Fig. 7–29a. Since the element's dimensions are very small, its deformed shape will become a parallelepiped, Fig. 7–29b. Here the *normal strains* change the sides of the element to

$$(1 + \epsilon_x)\Delta x \qquad (1 + \epsilon_y)\Delta y \qquad (1 + \epsilon_z)\Delta z$$

which produces a *change in the volume of the element*. And the *shear strain* changes the angles between the sides of the element to

$$\frac{\pi}{2} - \gamma_{xy} \qquad \frac{\pi}{2} - \gamma_{yz} \qquad \frac{\pi}{2} - \gamma_{xz}$$

which produces a *change in the shape of the element*.

(a)

Undeformed element

(b)

Deformed element

(c)

Fig. 7–29

Small Strain Analysis. Most engineering design involves applications for which only *small deformations* are allowed. In this text, therefore, we will assume that the deformations that take place within a body are almost infinitesimal. For example, the *normal strains* occurring within the material are *very small* compared to 1, so that $\epsilon \ll 1$. This assumption has wide practical application in engineering, and it is often referred to as a *small strain analysis*. It can also be used when a change in angle, $\Delta\theta$, is small, so that $\sin \Delta\theta \approx \Delta\theta$, $\cos \Delta\theta \approx 1$, and $\tan \Delta\theta \approx \Delta\theta$.

The rubber bearing support under this concrete bridge girder is subjected to both normal and shear strain. The normal strain is caused by the weight and bridge loads on the girder, and the shear strain is caused by the horizontal movement of the girder due to temperature changes.

IMPORTANT POINTS

- Loads will cause all material bodies to deform and, as a result, points in a body will undergo *displacements or changes in position*.

- *Normal strain* is a measure per unit length of the elongation or contraction of a small line segment in the body, whereas *shear strain* is a measure of the change in angle that occurs between two small line segments that are originally perpendicular to one another.

- The state of strain at a point is characterized by six strain components: three normal strains ϵ_x, ϵ_y, ϵ_z and three shear strains γ_{xy}, γ_{yz}, γ_{xz}. These components all depend upon the original orientation of the line segments and their location in the body.

- Strain is the geometrical quantity that is measured using experimental techniques. Once obtained, the stress in the body can then be determined from material property relations, as discussed in the next chapter.

- Most engineering materials undergo very small deformations, and so the normal strain $\epsilon \ll 1$. This assumption of "small strain analysis" allows the calculations for normal strain to be simplified, since first-order approximations can be made about its size.

EXAMPLE 7.14

Determine the average normal strains in the two wires in Fig. 7–30 if the ring at A moves to A'.

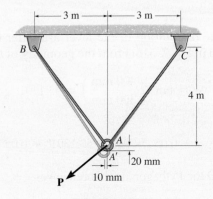

Fig. 7–30

SOLUTION

Geometry. The original length of each wire is

$$L_{AB} = L_{AC} = \sqrt{(3\text{ m})^2 + (4\text{ m})^2} = 5\text{ m}$$

The final lengths are

$$L_{A'B} = \sqrt{(3\text{ m} - 0.01\text{ m})^2 + (4\text{ m} + 0.02\text{ m})^2} = 5.01004\text{ m}$$

$$L_{A'C} = \sqrt{(3\text{ m} + 0.01\text{ m})^2 + (4\text{ m} + 0.02\text{ m})^2} = 5.02200\text{ m}$$

Average Normal Strain.

$$\epsilon_{AB} = \frac{L_{A'B} - L_{AB}}{L_{AB}} = \frac{5.01004\text{ m} - 5\text{ m}}{5\text{ m}} = 2.01(10^{-3})\text{ m/m} \quad Ans.$$

$$\epsilon_{AC} = \frac{L_{A'C} - L_{AC}}{L_{AC}} = \frac{5.02200\text{ m} - 5\text{ m}}{5\text{ m}} = 4.40(10^{-3})\text{ m/m} \quad Ans.$$

EXAMPLE 7.15

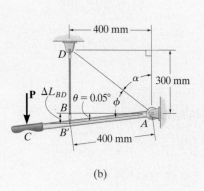

(a)

(b)

Fig. 7–31

When force **P** is applied to the rigid lever arm ABC in Fig. 7–31a, the arm rotates counterclockwise about pin A through an angle of $0.05°$. Determine the normal strain in wire BD.

SOLUTION I

Geometry. The orientation of the lever arm after it rotates about point A is shown in Fig. 7–31b. From the geometry of this figure,

$$\alpha = \tan^{-1}\left(\frac{400 \text{ mm}}{300 \text{ mm}}\right) = 53.1301°$$

Then

$$\phi = 90° - \alpha + 0.05° = 90° - 53.1301° + 0.05° = 36.92°$$

For triangle ABD the Pythagorean theorem gives

$$L_{AD} = \sqrt{(300 \text{ mm})^2 + (400 \text{ mm})^2} = 500 \text{ mm}$$

Using this result and applying the law of cosines to triangle $AB'D$,

$$\begin{aligned} L_{B'D} &= \sqrt{L_{AD}^2 + L_{AB'}^2 - 2(L_{AD})(L_{AB'})\cos\phi} \\ &= \sqrt{(500 \text{ mm})^2 + (400 \text{ mm})^2 - 2(500 \text{ mm})(400 \text{ mm})\cos 36.92°} \\ &= 300.3491 \text{ mm} \end{aligned}$$

Normal Strain.

$$\begin{aligned} \epsilon_{BD} &= \frac{L_{B'D} - L_{BD}}{L_{BD}} \\ &= \frac{300.3491 \text{ mm} - 300 \text{ mm}}{300 \text{ mm}} = 0.00116 \text{ mm/mm} \quad \textit{Ans.} \end{aligned}$$

SOLUTION II

Since the strain is small, this same result can be obtained by approximating the elongation of wire BD as ΔL_{BD}, shown in Fig. 7–31b. Here,

$$\Delta L_{BD} = \theta L_{AB} = \left[\left(\frac{0.05°}{180°}\right)(\pi \text{ rad})\right](400 \text{ mm}) = 0.3491 \text{ mm}$$

Therefore,

$$\epsilon_{BD} = \frac{\Delta L_{BD}}{L_{BD}} = \frac{0.3491 \text{ mm}}{300 \text{ mm}} = 0.00116 \text{ mm/mm} \quad \textit{Ans.}$$

EXAMPLE 7.16

The plate shown in Fig. 7–32a is fixed connected along AB and held in the horizontal guides at its top and bottom, AD and BC. If its right side CD is given a uniform horizontal displacement of 2 mm, determine (a) the average normal strain along the diagonal AC, and (b) the shear strain at E relative to the x, y axes.

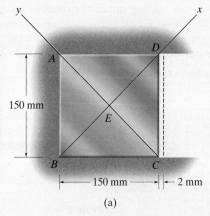

(a)

SOLUTION

Part (a). When the plate is deformed, the diagonal AC becomes AC', Fig. 7–32b. The lengths of diagonals AC and AC' can be found from the Pythagorean theorem. We have

$$AC = \sqrt{(0.150 \text{ m})^2 + (0.150 \text{ m})^2} = 0.21213 \text{ m}$$

$$AC' = \sqrt{(0.150 \text{ m})^2 + (0.152 \text{ m})^2} = 0.21355 \text{ m}$$

Therefore the average normal strain along AC is

$$(\epsilon_{AC})_{\text{avg}} = \frac{AC' - AC}{AC} = \frac{0.21355 \text{ m} - 0.21213 \text{ m}}{0.21213 \text{ m}}$$

$$= 0.00669 \text{ mm/mm} \qquad Ans.$$

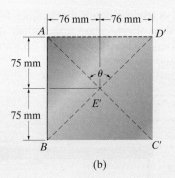

(b)

Fig. 7–32

Part (b). To find the shear strain at E relative to the x and y axes, which are 90° apart, it is necessary to find the change in the angle at E. The angle θ after deformation, Fig. 7–32b, is

$$\tan\left(\frac{\theta}{2}\right) = \frac{76 \text{ mm}}{75 \text{ mm}}$$

$$\theta = 90.759° = \left(\frac{\pi}{180°}\right)(90.759°) = 1.58404 \text{ rad}$$

Applying Eq. 7–12, the shear strain at E is therefore the change in the angle AED,

$$\gamma_{xy} = \frac{\pi}{2} - 1.58404 \text{ rad} = -0.0132 \text{ rad} \qquad Ans.$$

The *negative sign* indicates that the once 90° angle becomes larger.

NOTE: If the x and y axes were horizontal and vertical at point E, then the 90° angle between these axes would not change due to the deformation, and so $\gamma_{xy} = 0$ at point E.

PRELIMINARY PROBLEMS

P7–7. A loading causes the member to deform into the dashed shape. Explain how to determine the normal strains ϵ_{CD} and ϵ_{AB}. The displacement Δ and the lettered dimensions are known.

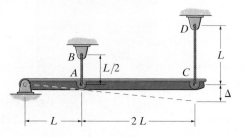

Prob. P7–7

P7–8. A loading causes the member to deform into the dashed shape. Explain how to determine the normal strains ϵ_{CD} and ϵ_{AB}. The displacement Δ and the lettered dimensions are known.

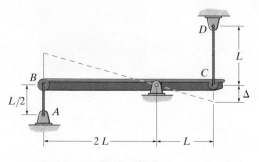

Prob. P7–8

P7–9. A loading causes the wires to elongate into the dashed shape. Explain how to determine the normal strain ϵ_{AB} in wire AB. The displacement Δ and the distances between all lettered points are known.

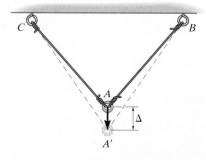

Prob. P7–9

P7–10. A loading causes the block to deform into the dashed shape. Explain how to determine the strains ϵ_{AB}, ϵ_{AC}, ϵ_{BC}, $(\gamma_A)_{xy}$. The angles and distances between all lettered points are known.

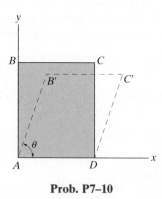

Prob. P7–10

P7–11. A loading causes the block to deform into the dashed shape. Explain how to determine the strains $(\gamma_A)_{xy}$, $(\gamma_B)_{xy}$. The angles and distances between all lettered points are known.

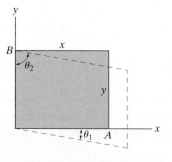

Prob. P7–11

FUNDAMENTAL PROBLEMS

F7–25. When force **P** is applied to the rigid arm ABC, point B displaces vertically downward through a distance of 0.2 mm. Determine the normal strain in wire CD.

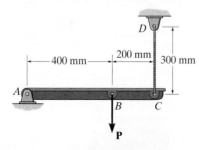

Prob. F7–25

F7–26. If the force **P** causes the rigid arm ABC to rotate clockwise about pin A through an angle of 0.02°, determine the normal strain in wires BD and CE.

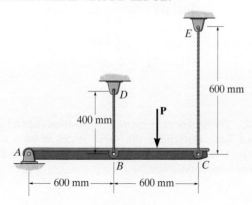

Prob. F7–26

F7–27. The rectangular plate is deformed into the shape of a parallelogram shown by the dashed line. Determine the average shear strain at corner A with respect to the x and y axes.

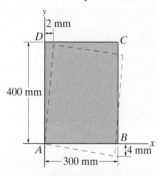

Prob. F7–27

F7–28. The triangular plate is deformed into the shape shown by the dashed line. Determine the normal strain developed along edge BC and the average shear strain at corner A with respect to the x and y axes.

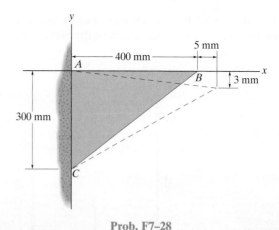

Prob. F7–28

F7–29. The square plate is deformed into the shape shown by the dashed line. Determine the average normal strain along diagonal AC and the shear strain at point E with respect to the x and y axes.

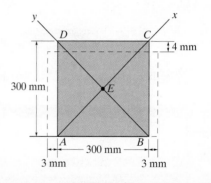

Prob. F7–29

PROBLEMS

7–71. An air-filled rubber ball has a diameter of 6 in. If the air pressure within the ball is increased until the diameter becomes 7 in., determine the average normal strain in the rubber.

*7–72. A thin strip of rubber has an unstretched length of 15 in. If it is stretched around a pipe having an outer diameter of 5 in., determine the average normal strain in the strip.

7–73. If the load **P** on the beam causes the end C to be displaced 10 mm downward, determine the normal strain in wires CE and BD.

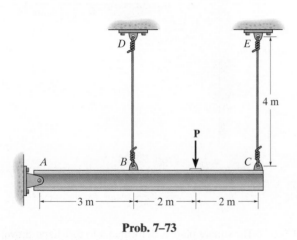

Prob. 7–73

7–74. The force applied at the handle of the rigid lever causes the lever to rotate clockwise about the pin B through an angle of 2°. Determine the average normal strain in each wire. The wires are unstretched when the lever is in the horizontal position.

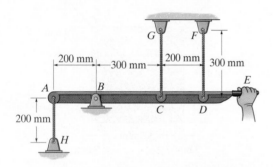

Prob. 7–74

7–75. The rectangular plate is subjected to the deformation shown by the dashed lines. Determine the average shear strain γ_{xy} in the plate.

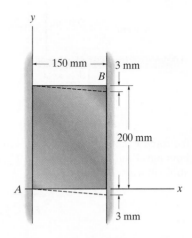

Prob. 7–75

*7–76. The square deforms into the position shown by the dashed lines. Determine the shear strain at each of its corners, A, B, C, and D, relative to the x, y axes. Side $D'B'$ remains horizontal.

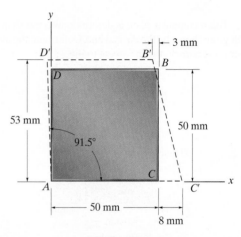

Prob. 7–76

7-77. The pin-connected rigid rods AB and BC are inclined at $\theta = 30°$ when they are unloaded. When the force **P** is applied θ becomes 30.2°. Determine the average normal strain in wire AC.

*7-80.** Determine the shear strain γ_{xy} at corners A and B if the plastic distorts as shown by the dashed lines.

7-81. Determine the shear strain γ_{xy} at corners D and C if the plastic distorts as shown by the dashed lines.

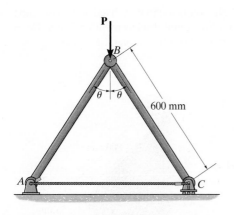

Prob. 7-77

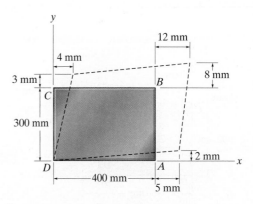

Probs. 7-80/81

7-78. The wire AB is unstretched when $\theta = 45°$. If a load is applied to the bar AC, which causes θ to become 47°, determine the normal strain in the wire.

7-79. If a load applied to the bar AC causes point A to be displaced to the right by an amount ΔL, determine the normal strain in the wire AB. Originally, $\theta = 45°$.

7-82. The material distorts into the dashed position shown. Determine the average normal strains ϵ_x, ϵ_y and the shear strain γ_{xy} at A, and the average normal strain along line BE.

7-83. The material distorts into the dashed position shown. Determine the average normal strains along the diagonals AD and CF.

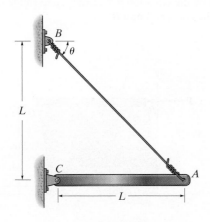

Probs. 7-78/79

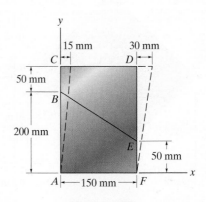

Probs. 7-82/83

*7–84.   Determine the shear strain γ_{xy} at corners A and B if the plastic distorts as shown by the dashed lines.

7–85.   Determine the shear strain γ_{xy} at corners D and C if the plastic distorts as shown by the dashed lines.

7–86.   Determine the average normal strain that occurs along the diagonals AC and DB.

*7–88.   The triangular plate is fixed at its base, and its apex A is given a horizontal displacement of 5 mm. Determine the shear strain, γ_{xy}, at A.

7–89.   The triangular plate is fixed at its base, and its apex A is given a horizontal displacement of 5 mm. Determine the average normal strain ϵ_x along the x axis.

7–90.   The triangular plate is fixed at its base, and its apex A is given a horizontal displacement of 5 mm. Determine the average normal strain $\epsilon_{x'}$ along the x' axis.

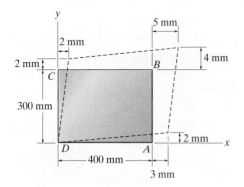

Probs. 7–84/85/86

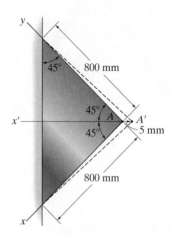

Probs. 7–88/89/90

7–87.   The corners of the square plate are given the displacements indicated. Determine the average normal strains ϵ_x and ϵ_y along the x and y axes.

7–91.   The polysulfone block is glued at its top and bottom to the rigid plates. If a tangential force, applied to the top plate, causes the material to deform so that its sides are described by the equation $y = 3.56x^{1/4}$, determine the shear strain at the corners A and B.

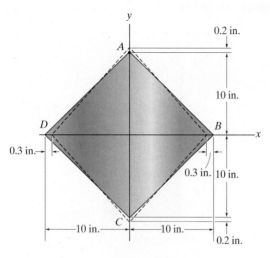

Prob. 7–87

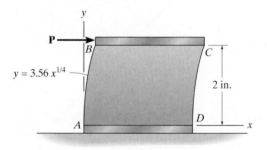

Prob. 7–91

CONCEPTUAL PROBLEMS

C7–1. Hurricane winds have caused the failure of this highway sign. Assuming the wind creates a uniform pressure on the sign of 2 kPa, use reasonable dimensions for the sign and determine the resultant shear and moment at each of the two connections where the failure occurred.

Prob. C7–1

C7–2. High-heel shoes can often do damage to soft wood or linoleum floors. Using a reasonable weight and dimensions for the heel of a regular shoe and a high-heel shoe, determine the bearing stress under each heel if the weight is transferred down only to the heel of one shoe.

Prob. C7–2

C7–3. Here is an example of the single shear failure of a bolt. Using appropriate free-body diagrams, explain why the bolt failed along the section between the plates, and not along some intermediate section such as *a–a*.

Prob. C7–3

C7–4. The vertical load on the hook is 1000 lb. Draw the appropriate free-body diagrams and determine the maximum average shear force on the pins at *A*, *B*, and *C*. Note that due to symmetry four wheels are used to support the loading on the railing.

Prob. C7–4

CHAPTER REVIEW

The internal loadings in a body consist of a normal force, shear force, bending moment, and torsional moment. They represent the resultants of both a normal and shear stress distribution that act over the cross section. To obtain these resultants, use the method of sections and the equations of equilibrium.	$\Sigma F_x = 0$ $\Sigma F_y = 0$ $\Sigma F_z = 0$ $\Sigma M_x = 0$ $\Sigma M_y = 0$ $\Sigma M_z = 0$	
If a bar is made from homogeneous isotropic material and it is subjected to a series of external axial loads that pass through the centroid of the cross section, then a uniform normal stress distribution will act over the cross section. This average normal stress can be determined from $\sigma = P/A$, where P is the internal axial load at the section.	$\sigma = \dfrac{P}{A}$	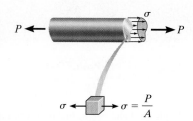
The average shear stress can be determined using $\tau_{\text{avg}} = V/A$, where V is the shear force acting on the cross section. This formula is often used to find the average shear stress in fasteners or in parts used for connections.	$\tau_{\text{avg}} = \dfrac{V}{A}$	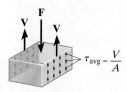

The design of any simple connection requires that the average stress along any cross section not exceed an allowable stress of σ_{allow} or τ_{allow}. These values are reported in codes and are considered safe on the basis of experiments or through experience. Sometimes a factor of safety is reported provided the failure stress is known.	$$\text{F.S.} = \frac{\sigma_{fail}}{\sigma_{allow}} = \frac{\tau_{fail}}{\tau_{allow}}$$	
Deformation is defined as the change in the shape and size of a body. It causes line segments to change length and orientation.		
Normal strain ϵ the change in length per unit length of a line segment. If ϵ is positive, the line segment elongates. If it is negative, the line segment contracts.	$$\epsilon_{avg} = \frac{\Delta s' - \Delta s}{\Delta s}$$	
Shear strain γ is a measure of the change in angle made between two line segments that are originally perpendicular to one another.	$$\gamma = \frac{\pi}{2} - \theta$$	
Strain is dimensionless; however, ϵ is sometimes reported in in./in., mm/mm, and γ is in radians.		

REVIEW PROBLEMS

R7–1. The beam AB is pin supported at A and supported by a cable BC. A *separate* cable CG is used to hold up the frame. If AB weighs 120 lb/ft and the column FC has a weight of 180 lb/ft, determine the resultant internal loadings acting on cross sections located at points D and E. Neglect the thickness of both the beam and column in the calculation.

R7–3. Determine the required thickness of member BC and the diameter of the pins at A and B if the allowable normal stress for member BC is $\sigma_{allow} = 29$ ksi, and the allowable shear stress for the pins is $\tau_{allow} = 10$ ksi.

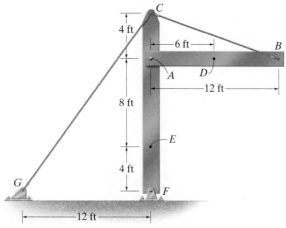

Prob. R7–1

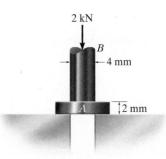

Prob. R7–3

R7–2. The long bolt passes through the 30-mm-thick plate. If the force in the bolt shank is 8 kN, determine the average normal stress in the shank, the average shear stress along the cylindrical area of the plate defined by the section lines a–a, and the average shear stress in the bolt head along the cylindrical area defined by the section lines b–b.

*****R7–4.** The circular punch B exerts a force of 2 kN on the top of the plate A. Determine the average shear stress in the plate due to this loading.

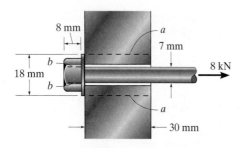

Prob. R7–2

Prob. R7–4

R7–5. Determine the average punching shear stress the circular shaft creates in the metal plate through section AC and BD. Also, what is the bearing stress developed on the surface of the plate under the shaft?

R7–7. The square plate is deformed into the shape shown by the dashed lines. If DC has a normal strain $\epsilon_x = 0.004$, DA has a normal strain $\epsilon_y = 0.005$ and at D, $\gamma_{xy} = 0.02$ rad, determine the average normal strain along diagonal CA.

*R7–8. The square plate is deformed into the shape shown by the dashed lines. If DC has a normal strain $\epsilon_x = 0.004$, DA has a normal strain $\epsilon_y = 0.005$ and at D, $\gamma_{xy} = 0.02$ rad, determine the shear strain at point E with respect to the x' and y' axes.

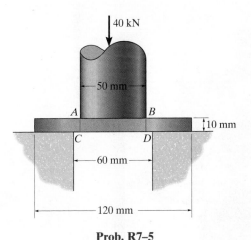

Prob. R7–5

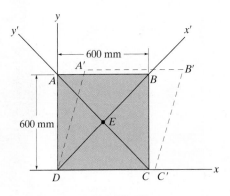

Probs. R7–7/8

R7–6. The bearing pad consists of a 150 mm by 150 mm block of aluminum that supports a compressive load of 6 kN. Determine the average normal and shear stress acting on the plane through section a–a. Show the results on a differential volume element located on the plane.

R7–9. The rubber block is fixed along edge AB, and edge CD is moved so that the vertical displacement of any point in the block is given by $v(x) = (v_0/b^3)\, x^3$. Determine the shear strain γ_{xy} at points $(b/2, a/2)$ and (b, a).

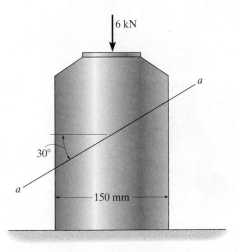

Prob. R7–6

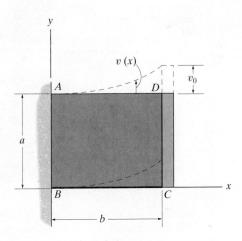

Prob. R7–9

CHAPTER 8

(© Tom Wang/Alamy)

Horizontal ground displacements caused by an earthquake produced fracture of this concrete column. The material properties of the steel and concrete must be determined so that engineers can properly design the column to resist the loadings that caused this failure.

MECHANICAL PROPERTIES OF MATERIALS

■ To show how stress can be related to strain by using experimental methods to determine the stress–strain diagram for a particular material.

■ To discuss the properties of the stress–strain diagram for materials commonly used in engineering.

8.1 THE TENSION AND COMPRESSION TEST

The strength of a material depends on its ability to sustain a load without undue deformation or failure. This strength is inherent in the material itself and must be determined by *experiment*. One of the most important tests to perform in this regard is the ***tension or compression test***. Once this test is performed, we can then determine the relationship between the average normal stress and average normal strain in many engineering materials such as metals, ceramics, polymers, and composites.

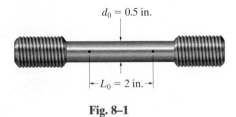

$d_0 = 0.5$ in.

$L_0 = 2$ in.

Fig. 8–1

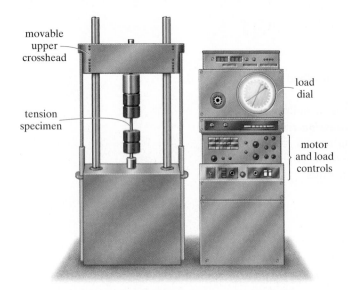

Typical steel specimen with attached strain gage

To perform a tension or compression test, a specimen of the material is made into a "standard" shape and size, Fig. 8–1. As shown it has a constant circular cross section with enlarged ends, so that when tested, failure will occur somewhere within the central region of the specimen. Before testing, two small punch marks are sometimes placed along the specimen's uniform length. Measurements are taken of both the specimen's initial cross-sectional area, A_0, and the **gage-length** distance L_0 between the punch marks. For example, when a metal specimen is used in a tension test, it generally has an initial diameter of $d_0 = 0.5$ in. (13 mm) and a gage length of $L_0 = 2$ in. (51 mm), Fig. 8–1. A testing machine like the one shown in Fig. 8–2 is then used to stretch the specimen at a very slow, constant rate until it fails. The machine is designed to read the load required to maintain this uniform stretching.

At frequent intervals, data is recorded of the applied load P. Also, the elongation $\delta = L - L_0$ between the punch marks on the specimen may be measured, using either a caliper or a mechanical or optical device called an **extensometer**. Rather than taking this measurement and then calculating the strain, it is also possible to read the normal strain *directly* on the specimen by using an **electrical-resistance strain gage**, which looks like the one shown in Fig. 8–3. As shown in the adjacent photo, the gage is cemented to the specimen along its length, so that it becomes an integral part of the specimen. When the specimen is strained in the direction of the gage, both the wire and specimen will experience the same deformation or strain. By measuring the change in the electrical resistance of the wire, the gage may then be calibrated to directly read the normal strain in the specimen.

movable upper crosshead

load dial

tension specimen

motor and load controls

Fig. 8–2

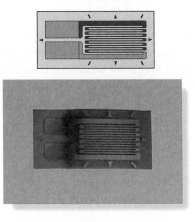

Electrical-resistance strain gage

Fig. 8–3

8.2   THE STRESS–STRAIN DIAGRAM

Once the stress and strain data from the test are known, then the results can be plotted to produce a curve called the **stress–strain diagram**. This diagram is very useful since it applies to a specimen of the material made of *any* size. There are two ways in which the stress–strain diagram is normally described.

Conventional Stress–Strain Diagram. The **nominal** or **engineering stress** is determined by dividing the applied load P by the specimen's *original* cross-sectional area A_0. This calculation assumes that the stress is *constant* over the cross section and throughout the gage length. We have

$$\sigma = \frac{P}{A_0} \tag{8–1}$$

Likewise, the **nominal** or **engineering strain** is found directly from the strain gage reading, or by dividing the change in the specimen's gage length, δ, by the specimen's *original gage length* L_0. Thus,

$$\epsilon = \frac{\delta}{L_0} \tag{8–2}$$

When these values of σ and ϵ are plotted, where the vertical axis is the stress and the horizontal axis is the strain, the resulting curve is called a **conventional stress–strain diagram**. A typical example of this curve is shown in Fig. 8–4. Realize, however, that two stress–strain diagrams for a particular material will be quite similar, but will never be exactly the same. This is because the results actually depend upon such variables as the material's composition, microscopic imperfections, the way the specimen is manufactured, the rate of loading, and the temperature during the time of the test.

From the curve in Fig. 8–4, we can identify four different regions in which the material behaves in a unique way, depending on the amount of strain induced in the material.

Elastic Behavior. The initial region of the curve, indicated in light orange, is referred to as the elastic region. Here the curve is a *straight line* up to the point where the stress reaches the **proportional limit**, σ_{pl}. When the stress slightly exceeds this value, the curve bends until the stress reaches an elastic limit. For most materials, these points are very close, and therefore it becomes rather difficult to distinguish their exact values. What makes the elastic region unique, however, is that after reaching σ_Y, if the load is removed, the specimen will recover its original shape. In other words, no damage will be done to the material.

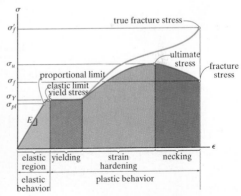

Conventional and true stress–strain diagram for ductile material (steel) (not to scale)

Fig. 8–4

Because the curve is a straight line up to σ_{pl}, any increase in stress will cause a proportional increase in strain. This fact was discovered in 1676 by Robert Hooke, using springs, and is known as **Hooke's law**. It is expressed mathematically as

$$\sigma = E\epsilon \qquad (8\text{–}3)$$

Here E represents the constant of proportionality, which is called the **modulus of elasticity** or **Young's modulus**, named after Thomas Young, who published an account of it in 1807.

As noted in Fig. 8–4, the modulus of elasticity represents the *slope* of the straight line portion of the curve. Since strain is dimensionless, from Eq. 8–3, E will have the same units as stress, such as psi, ksi, or pascals.

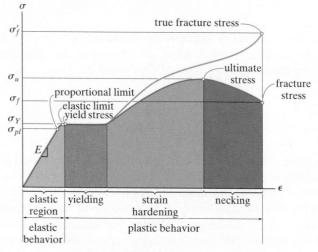

Conventional and true stress–strain diagram
for ductile material (steel) (not to scale)

Fig. 8–4 (Repeated)

Yielding. A slight increase in stress above the elastic limit will result in a breakdown of the material and cause it to *deform permanently*. This behavior is called **yielding**, and it is indicated by the rectangular dark orange region in Fig. 8–4. The stress that causes yielding is called the **yield stress** or **yield point**, σ_Y, and the deformation that occurs is called **plastic deformation**. Although not shown in Fig. 8–4, for low-carbon steels or those that are hot rolled, the yield point is often distinguished by two values. The **upper yield point** occurs first, followed by a sudden decrease in load-carrying capacity to a **lower yield point**. Once the yield point is reached, then as shown in Fig. 8–4, *the specimen will continue to elongate (strain) **without** any increase in load*. When the material behaves in this manner, it is often referred to as being **perfectly plastic**.

Strain Hardening. When yielding has ended, any load causing an increase in stress will be supported by the specimen, resulting in a curve that rises continuously but becomes flatter until it reaches a maximum stress referred to as the ***ultimate stress***, σ_u. The rise in the curve in this manner is called ***strain hardening***, and it is identified in Fig. 8–4 as the region in light green.

Necking. Up to the ultimate stress, as the specimen elongates, its cross-sectional area will decrease in a fairly *uniform* manner over the specimen's entire gage length. However, just after reaching the ultimate stress, the cross-sectional area will then begin to decrease in a *localized* region of the specimen, and so it is here where the stress begins to increase. As a result, a constriction or "neck" tends to form with further elongation, Fig. 8–5a. This region of the curve due to necking is indicated in dark green in Fig. 8–4. Here the stress–strain diagram tends to curve downward until the specimen breaks at the ***fracture stress***, σ_f, Fig. 8–5b.

Typical necking pattern which has occurred on this steel specimen just before fracture.

True Stress–Strain Diagram. Instead of always using the *original* cross-sectional area A_0 and specimen length L_0 to calculate the (engineering) stress and strain, we could have used the *actual* cross-sectional area A and specimen length L at the *instant* the load is measured. The values of stress and strain found from these measurements are called ***true stress*** and ***true strain***, and a plot of their values is called the ***true stress–strain diagram***. When this diagram is plotted, it has a form shown by the upper blue curve in Fig. 8–4. Note that the conventional and true σ–ϵ diagrams are practically coincident when the strain is small. The differences begin to appear in the strain-hardening range, where the magnitude of strain becomes more significant. From the conventional σ–ϵ diagram, the specimen appears to support a *decreasing* stress (or load), since A_0 is constant, $\sigma = N/A_0$. In fact, the true σ–ϵ diagram shows the area A within the necking region is always *decreasing* until fracture, σ_f', and so the material *actually* sustains *increasing stress*, since $\sigma = N/A$.

Although there is this divergence between these two diagrams, we can neglect this effect since most engineering design is done only within the elastic range. This will generally restrict the deformation of the material to very small values, and when the load is removed the material will restore itself to its original shape. The conventional stress–strain diagram can be used in the elastic region because the true strain up to the elastic limit is small enough, so that the error in using the engineering values of σ and ϵ is very small (about 0.1%) compared with their true values.

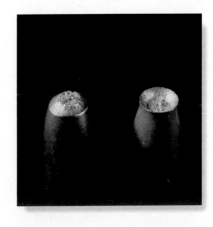

This steel specimen clearly shows the necking that occurred just before the specimen failed. This resulted in the formation of a "cup-cone" shape at the fracture location, which is characteristic of ductile materials.

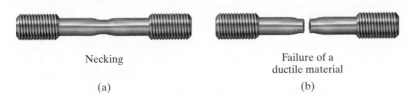

Necking	Failure of a ductile material
(a)	(b)

Fig. 8–5

Steel. A typical conventional stress–strain diagram for a mild steel specimen is shown in Fig. 8–6. In order to enhance the details, the elastic region of the curve has been shown in green using an exaggerated strain scale, also shown in green. Following this curve, as the load (stress) is increased, the proportional limit is reached at $\sigma_{pl} = 35$ ksi (241 MPa), where $\epsilon_{pl} = 0.0012$ in./in. When the load is further increased, the stress reaches an upper yield point of $(\sigma_Y)_u = 38$ ksi (262 MPa), followed by a drop in stress to a lower yield point of $(\sigma_Y)_l = 36$ ksi (248 MPa). The end of yielding occurs at a strain of $\epsilon_Y = 0.030$ in./in., which is 25 times greater than the strain at the proportional limit! Continuing, the specimen undergoes strain hardening until it reaches the ultimate stress of $\sigma_u = 63$ ksi (434 MPa); then it begins to neck down until fracture occurs, at $\sigma_f = 47$ ksi (324 MPa). By comparison, the strain at failure, $\epsilon_f = 0.380$ in./in., is 317 times greater than ϵ_{pl}!

Since $\sigma_{pl} = 35$ ksi and $\epsilon_{pl} = 0.0012$ in./in., we can determine the modulus of elasticity. From Hooke's law, it is

$$E = \frac{\sigma_{pl}}{\epsilon_{pl}} = \frac{35 \text{ ksi}}{0.0012 \text{ in./in.}} = 29(10^3) \text{ ksi}$$

Although steel alloys have different carbon contents, most grades of steel, from the softest rolled steel to the hardest tool steel, have about this same modulus of elasticity, as shown in Fig. 8–7.

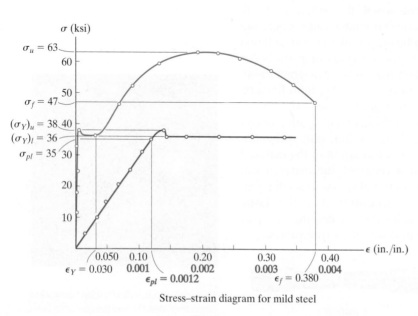

Stress–strain diagram for mild steel

Fig. 8–6

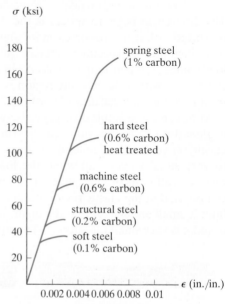

Fig. 8–7

8.3 STRESS–STRAIN BEHAVIOR OF DUCTILE AND BRITTLE MATERIALS

Materials can be classified as either being ductile or brittle, depending on their stress–strain characteristics.

Ductile Materials. Any material that can be subjected to large strains before it fractures is called a **ductile material**. Mild steel, as discussed previously, is a typical example. Engineers often choose ductile materials for design because these materials are capable of absorbing shock or energy, and if they become overloaded, they will usually exhibit large deformation before failing.

One way to specify the ductility of a material is to report its percent elongation or percent reduction in area at the time of fracture. The **percent elongation** is the specimen's fracture strain expressed as a percent. Thus, if the specimen's original gage length is L_0 and its length at fracture is L_f, then

$$\text{Percent elongation} = \frac{L_f - L_0}{L_0}(100\%) \qquad (8\text{--}4)$$

For example, as in Fig. 8–6, since $\epsilon_f = 0.380$, this value would be 38% for a mild steel specimen.

The **percent reduction in area** is another way to specify ductility. It is defined within the region of necking as follows:

$$\text{Percent reduction of area} = \frac{A_0 - A_f}{A_0}(100\%) \qquad (8\text{--}5)$$

Here A_0 is the specimen's original cross-sectional area and A_f is the area of the neck at fracture. Mild steel has a typical value of 60%.

Besides steel, other metals such as brass, molybdenum, and zinc may also exhibit ductile stress–strain characteristics similar to steel, whereby they undergo elastic stress–strain behavior, yielding at constant stress, strain hardening, and finally necking until fracture. In most metals and some plastics, however, constant yielding will *not occur* beyond the elastic range. One metal where this is the case is aluminum, Fig. 8–8. Actually, this metal often does not have a well-defined *yield point*, and consequently it is standard practice to define a **yield strength** using a graphical procedure called the **offset method**. Normally for structural design a 0.2% strain (0.002 in./in.) is chosen, and from this point on the ϵ axis a line parallel to the initial straight line portion of the stress–strain diagram is drawn. The point where this line intersects the curve defines the yield strength. From the graph, the yield strength is $\sigma_{YS} = 51$ ksi (352 MPa).

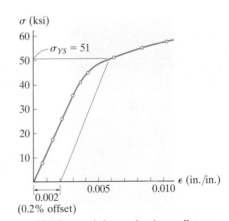

Yield strength for an aluminum alloy

Fig. 8–8

8

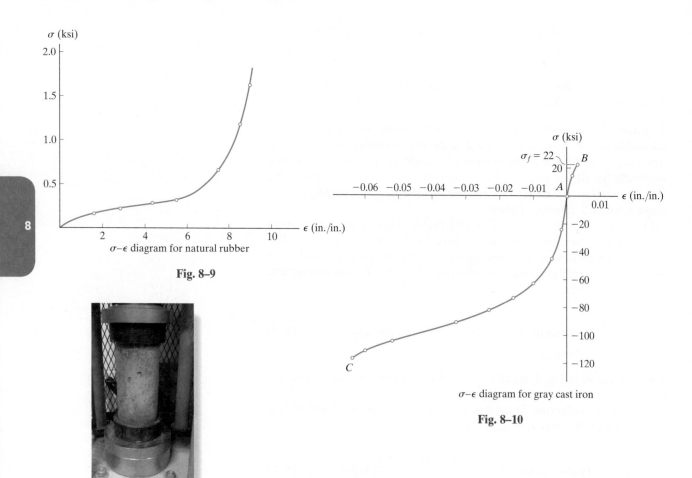

σ–ϵ diagram for natural rubber

Fig. 8–9

σ–ϵ diagram for gray cast iron

Fig. 8–10

Concrete used for structural purposes must be tested in compression to be sure it reaches its ultimate design stress after curing for 30 days.

Realize that the yield strength is not a physical property of the material, since it is a stress that causes a *specified* permanent strain in the material. In this text, however, we will assume that the yield strength, yield point, elastic limit, and proportional limit all *coincide* unless otherwise stated. An exception would be natural rubber, which in fact does not even have a proportional limit, since stress and strain are *not* linearly related. Instead, as shown in Fig. 8–9, this material, which is known as a polymer, exhibits *nonlinear elastic behavior*.

Wood is a material that is often moderately ductile, and as a result it is usually designed to respond only to elastic loadings. The strength characteristics of wood vary greatly from one species to another, and for each species they depend on the moisture content, age, and the size and arrangement of knots in the wood. Since wood is a fibrous material, its tensile or compressive characteristics parallel to its grain will differ greatly from these characteristics perpendicular to its grain. Specifically, wood splits easily when it is loaded in tension perpendicular to its grain, and consequently tensile loads are almost always intended to be applied parallel to the grain of wood members.

Tension failure of
a brittle material

(a)

Compression causes
material to bulge out

(b)

Fig. 8–11

σ (ksi)

$(\sigma_t)_{max} = 0.4$

-0.0025 -0.0015 -0.0005

ϵ (in./in.)

0   0.0005

$(\sigma_c)_{max} = 5$

$\sigma-\epsilon$ diagram for typical concrete mix

Fig. 8–12

8

Brittle Materials. Materials that exhibit little or no yielding before failure are referred to as **brittle materials**. Gray cast iron is an example, having a stress–strain diagram in tension as shown by the curve AB in Fig. 8–10. Here fracture at $\sigma_f = 22$ ksi (152 MPa) occurred due to a microscopic crack, which then spread rapidly across the specimen, causing complete fracture. Since the appearance of initial cracks in a specimen is quite random, brittle materials do not have a well-defined tensile fracture stress. Instead the *average* fracture stress from a set of observed tests is generally reported. A typical failed specimen is shown in Fig. 8–11a.

Compared with their behavior in tension, brittle materials exhibit a much higher resistance to axial compression, as evidenced by segment AC of the gray cast iron curve in Fig. 8–10. For this case any cracks or imperfections in the specimen tend to close up, and as the load increases the material will generally bulge or become barrel shaped as the strains become larger, Fig. 8–11b.

Like gray cast iron, concrete is classified as a brittle material, and it also has a low strength capacity in tension. The characteristics of its stress–strain diagram depend primarily on the mix of concrete (water, sand, gravel, and cement) and the time and temperature of curing. A typical example of a "complete" stress–strain diagram for concrete is given in Fig. 8–12. By inspection, its maximum compressive strength is about 12.5 times greater than its tensile strength, $(\sigma_c)_{max} = 5$ ksi (34.5 MPa) versus $(\sigma_t)_{max} = 0.40$ ksi (2.76 MPa). For this reason, concrete is almost always reinforced with steel bars or rods whenever it is designed to support tensile loads.

It can generally be stated that most materials exhibit both ductile and brittle behavior. For example, steel has brittle behavior when it contains a high carbon content, and it is ductile when the carbon content is reduced. Also, at low temperatures materials become harder and more brittle, whereas when the temperature rises they become softer and more ductile. This effect is shown in Fig. 8–13 for a methacrylate plastic.

Steel rapidly loses its strength when heated. For this reason engineers often require main structural members to be insulated in case of fire.

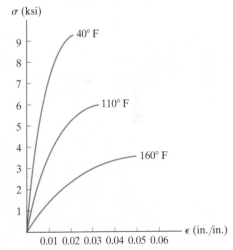

σ (ksi)

40° F

110° F

160° F

ϵ (in./in.)

0.01 0.02 0.03 0.04 0.05 0.06

$\sigma-\epsilon$ diagrams for a methacrylate plastic

Fig. 8–13

Stiffness. The modulus of elasticity is a mechanical property that indicates the **stiffness** of a material. Materials that are very stiff, such as steel, have large values of $E[E_{st} = 29(10^3)$ ksi or 200 GPa], whereas spongy materials such as vulcanized rubber have low values $[E_r = 0.10$ ksi or 0.69 MPa]. Values of E for commonly used engineering materials are often tabulated in engineering codes and reference books. Representative values are also listed on the inside back cover.

The modulus of elasticity is one of the most important mechanical properties used in the development of equations presented in this text. It must always be remembered, though, that E, through the application of Hooke's law, Eq. 8–3, can be used only if a material has *linear elastic behavior*. Also, if the stress in the material is *greater* than the proportional limit, the stress–strain diagram ceases to be a straight line, and so Hooke's law is no longer valid.

Strain Hardening. If a specimen of ductile material, such as steel, is loaded into the *plastic region* and then unloaded, *elastic strain is recovered* as the material returns to its equilibrium state. The *plastic strain remains*, however, and as a result the material will be subjected to a **permanent set**. For example, a wire when bent (plastically) will spring back a little (elastically) when the load is removed; however, it will not fully return to its original position. This behavior is illustrated on the stress–strain diagram shown in Fig. 8–14a. Here the specimen is loaded beyond its yield point A to point A'. Since interatomic forces have to be overcome to elongate the specimen *elastically*, then these same forces pull the atoms back together when the load is removed, Fig. 8–14a. Consequently, the modulus of elasticity, E, is the same, and therefore the slope of line $O'A'$ is the same as line OA. With the load removed, the permanent set is OO'.

If the load is reapplied, the atoms in the material will again be displaced until yielding occurs at or near the stress A', and the stress–strain diagram continues along the same path as before, Fig. 8–14b. Although this new stress–strain diagram, defined by $O'A'B$, now has a *higher* yield point (A'), a consequence of strain hardening, it also has *less ductility*, or a smaller plastic region, than when it was in its original state.

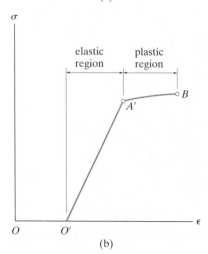

(a)

(b)

Fig. 8–14

This pin was made of a hardened steel alloy, that is, one having a high carbon content. It failed due to brittle fracture.

8.4   STRAIN ENERGY

As a material is deformed by an external load, the load will do external work, which in turn will be stored in the material as internal energy. This energy is related to the strains in the material, and so it is referred to as **strain energy**. To show how to calculate strain energy, consider a small volume element of material taken from a tension test specimen, Fig. 8–15a. It is subjected to the uniaxial stress σ. This stress develops a force $\Delta F = \sigma \, \Delta A = \sigma (\Delta x \, \Delta y)$ on the top and bottom faces of the element, which causes the element to undergo a vertical displacement $\epsilon \, \Delta z$, Fig. 8–15b. By definition, **work** is determined by the product of a force and displacement in the direction of the force. Here the force is increased uniformly from *zero* to its final magnitude ΔF when the displacement $\epsilon \, \Delta z$ occurs, and so during the displacement the work done on the element by the force is equal to the *average* force magnitude $(\Delta F/2)$ times the displacement $\epsilon \, \Delta z$. The conservation of energy requires this "external work" on the element to be equivalent to the "internal work" or strain energy stored in the element, assuming that no energy is lost in the form of heat. Consequently, the strain energy is $\Delta U = \left(\tfrac{1}{2}\Delta F\right) \epsilon \, \Delta z = \left(\tfrac{1}{2} \sigma \, \Delta x \, \Delta y\right) \epsilon \, \Delta z$. Since the volume of the element is $\Delta V = \Delta x \, \Delta y \, \Delta z$, then $\Delta U = \tfrac{1}{2}\sigma\epsilon \, \Delta V$.

For engineering applications, it is often convenient to specify the strain energy per unit volume of material. This is called the **strain energy density**, and it can be expressed as

$$u = \frac{\Delta U}{\Delta V} = \frac{1}{2}\sigma\epsilon \qquad (8\text{–}6)$$

Finally, if the material behavior is *linear elastic*, then Hooke's law applies, $\sigma = E\epsilon$, and therefore we can express the **elastic strain energy density** in terms of the uniaxial stress σ as

$$u = \frac{1}{2}\frac{\sigma^2}{E} \qquad (8\text{–}7)$$

Modulus of Resilience. When the stress in a material reaches the proportional limit, the strain energy density, as calculated by Eq. 8–6 or 8–7, is referred to as the **modulus of resilience**. It is

$$u_r = \frac{1}{2}\sigma_{pl}\,\epsilon_{pl} = \frac{1}{2}\frac{\sigma_{pl}^2}{E} \qquad (8\text{–}8)$$

Here u_r is equivalent to the shaded *triangular area* under the elastic region of the stress–strain diagram, Fig. 8–16a. Physically the modulus of resilience represents the largest amount of strain energy per unit volume the material can absorb without causing any permanent damage to the material. Certainly this property becomes important when designing bumpers or shock absorbers.

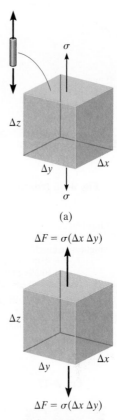

(a)

Free-body diagram

(b)

Fig. 8–15

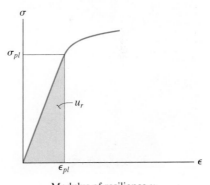

Modulus of resilience u_r

(a)

Fig. 8–16

σ

Modulus of toughness u_t

(b)

Fig. 8–16 (cont.)

σ

hard steel
(0.6% carbon)
highest strength

structural steel
(0.2% carbon)
toughest

soft steel
(0.1% carbon)
most ductile

ϵ

Fig. 8–17

This nylon specimen exhibits a high degree of toughness as noted by the large amount of necking that has occurred just before fracture.

Modulus of Toughness. Another important property of a material is its **modulus of toughness**, u_t. This quantity represents the *entire area* under the stress–strain diagram, Fig. 8–16b, and therefore it indicates the maximum amount of strain energy per unit volume the material can absorb just before it fractures. Certainly this becomes important when designing members that may be accidentally overloaded. By alloying metals, engineers can change their resilience and toughness. For example, by changing the percentage of carbon in steel, the resulting stress–strain diagrams in Fig. 8–17 show how its resilience and toughness can be changed.

IMPORTANT POINTS

- A *conventional stress–strain diagram* is important in engineering since it provides a means for obtaining data about a material's tensile or compressive strength without regard for the material's physical size or shape.

- *Engineering stress and strain* are calculated using the *original* cross-sectional area and gage length of the specimen.

- A *ductile material*, such as mild steel, has four distinct behaviors as it is loaded. They are *elastic behavior, yielding, strain hardening,* and *necking.*

- A material is *linear elastic* if the stress is proportional to the strain within the elastic region. This behavior is described by *Hooke's law*, $\sigma = E\epsilon$, where the *modulus of elasticity E* is the slope of the line.

- Important points on the stress–strain diagram are the *proportional limit, elastic limit, yield stress, ultimate stress,* and *fracture stress.*

- The *ductility* of a material can be specified by the specimen's *percent elongation* or the *percent reduction in area.*

- If a material does not have a distinct yield point, a *yield strength* can be specified using a graphical procedure such as the *offset method.*

- *Brittle materials,* such as gray cast iron, have very little or no yielding and so they can fracture suddenly.

- *Strain hardening* is used to establish a higher yield point for a material. This is done by straining the material beyond the elastic limit, then releasing the load. The modulus of elasticity remains the same; however, the material's ductility *decreases.*

- *Strain energy* is energy stored in a material due to its deformation. This energy per unit volume is called *strain energy density.* If it is measured up to the proportional limit, it is referred to as the *modulus of resilience,* and if it is measured up to the point of fracture, it is called the *modulus of toughness.* It can be determined from the area under the $\sigma-\epsilon$ diagram.

EXAMPLE 8.1

A tension test for a steel alloy results in the stress–strain diagram shown in Fig. 8–18. Calculate the modulus of elasticity and the yield strength based on a 0.2% offset. Identify on the graph the ultimate stress and the fracture stress.

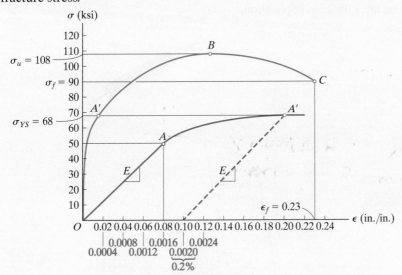

Fig. 8–18

SOLUTION

Modulus of Elasticity. We must calculate the *slope* of the initial straight-line portion of the graph. Using the magnified curve and scale shown in green, this line extends from point O to point A, which has coordinates of approximately (0.0016 in./in., 50 ksi). Therefore,

$$E = \frac{50 \text{ ksi}}{0.0016 \text{ in./in.}} = 31.2 \, (10^3) \text{ ksi} \qquad \text{Ans.}$$

Note that the equation of line OA is thus $\sigma = 31.2\,(10^3)\epsilon$.

Yield Strength. For a 0.2% offset, we begin at a strain of 0.2% or 0.0020 in./in. and graphically extend a (dashed) line parallel to OA until it intersects the σ–ϵ curve at A'. The yield strength is approximately

$$\sigma_{YS} = 68 \text{ ksi} \qquad \text{Ans.}$$

Ultimate Stress. This is defined by the peak of the σ–ϵ graph, point B in Fig. 8–18.

$$\sigma_u = 108 \text{ ksi} \qquad \text{Ans.}$$

Fracture Stress. When the specimen is strained to its maximum of $\epsilon_f = 0.23$ in./in., it fractures at point C. Thus,

$$\sigma_f = 90 \text{ ksi} \qquad \text{Ans.}$$

EXAMPLE 8.2

The stress–strain diagram for an aluminum alloy that is used for making aircraft parts is shown in Fig. 8–19. If a specimen of this material is stressed to $\sigma = 600$ MPa, determine the permanent set that remains in the specimen when the load is released. Also, find the modulus of resilience both before and after the load application.

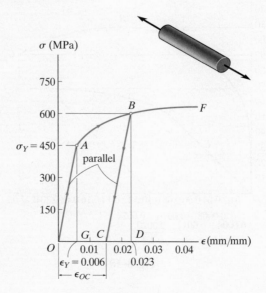

σ (MPa)

Fig. 8–19

SOLUTION

Permanent Strain. When the specimen is subjected to the load, it strain hardens until point B is reached on the σ–ϵ diagram. The strain at this point is approximately 0.023 mm/mm. When the load is released, the material behaves by following the straight line BC, which is parallel to line OA. Since both of these lines have the same slope, the strain at point C can be determined analytically. The slope of line OA is the modulus of elasticity, i.e.,

$$E = \frac{450 \text{ MPa}}{0.006 \text{ mm/mm}} = 75.0 \text{ GPa}$$

From triangle *CBD,* we require

$$E = \frac{BD}{CD}; \qquad 75.0\,(10^9)\ \text{Pa} = \frac{600\,(10^6)\ \text{Pa}}{CD}$$

$$CD = 0.008\ \text{mm/mm}$$

This strain represents the amount of *recovered elastic strain.* The permanent set or strain, ϵ_{OC}, is thus

$$\epsilon_{OC} = 0.023\ \text{mm/mm} - 0.008\ \text{mm/mm}$$

$$= 0.0150\ \text{mm/mm} \qquad\qquad Ans.$$

NOTE: If gage marks on the specimen were originally 50 mm apart, then after the load is *released* these marks will be 50 mm + (0.0150)(50 mm) = 50.75 mm apart.

Modulus of Resilience.   Applying Eq. 8–8, the areas under *OAG* and *CBD* in Fig. 8–19 are*

$$(u_r)_{\text{initial}} = \frac{1}{2}\,\sigma_{pl}\,\epsilon_{pl} = \frac{1}{2}(450\ \text{MPa})\,(0.006\ \text{mm/mm})$$

$$= 1.35\ \text{MJ/m}^3 \qquad\qquad Ans.$$

$$(u_r)_{\text{final}} = \frac{1}{2}\,\sigma_{pl}\,\epsilon_{pl} = \frac{1}{2}(600\ \text{MPa})\,(0.008\ \text{mm/mm})$$

$$= 2.40\ \text{MJ/m}^3 \qquad\qquad Ans.$$

NOTE: By comparison, the effect of strain hardening the material has caused an increase in the modulus of resilience; however, note that the modulus of toughness for the material has decreased, since the area under the original curve, *OABF,* is larger than the area under curve *CBF.*

---

*Work in the SI system of units is measured in joules, where 1 J = 1 N · m.

FUNDAMENTAL PROBLEMS

F8–1. Define a homogeneous material.

F8–2. Indicate the points on the stress–strain diagram which represent the proportional limit and the ultimate stress.

F8–10. The material for the 50-mm-long specimen has the stress–strain diagram shown. If $P = 100$ kN, determine the elongation of the specimen.

F8–11. The material for the 50-mm-long specimen has the stress–strain diagram shown. If $P = 150$ kN is applied and then released, determine the permanent elongation of the specimen.

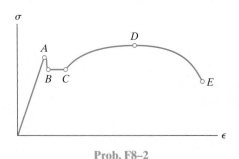

Prob. F8–2

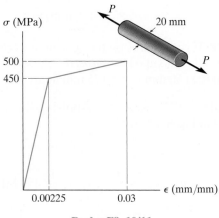

Probs. F8–10/11

F8–3. Define the modulus of elasticity E.

F8–4. At room temperature, mild steel is a ductile material. True or false?

F8–5. Engineering stress and strain are calculated using the *actual* cross-sectional area and length of the specimen. True or false?

F8–6. As the temperature increases the modulus of elasticity will increase. True or false?

F8–7. A 100-mm-long rod has a diameter of 15 mm. If an axial tensile load of 100 kN is applied, determine its change in length. Assume linear elastic behavior with $E = 200$ GPa.

F8–8. A bar has a length of 8 in. and cross-sectional area of 12 in². Determine the modulus of elasticity of the material if it is subjected to an axial tensile load of 10 kip and stretches 0.003 in. The material has linear elastic behavior.

F8–9. A 10-mm-diameter rod has a modulus of elasticity of $E = 100$ GPa. If it is 4 m long and subjected to an axial tensile load of 6 kN, determine its elongation. Assume linear elastic behavior.

F8–12. If the elongation of wire BC is 0.2 mm after the force **P** is applied, determine the magnitude of **P**. The wire is A-36 steel and has a diameter of 3 mm.

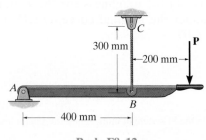

Prob. F8–12

PROBLEMS

8–1. A tension test was performed on a steel specimen having an original diameter of 0.503 in. and gage length of 2.00 in. The data is listed in the table. Plot the stress–strain diagram and determine approximately the modulus of elasticity, the yield stress, the ultimate stress, and the fracture stress. Use a scale of 1 in. = 20 ksi and 1 in. = 0.05 in./in. Redraw the elastic region, using the same stress scale but a strain scale of 1 in. = 0.001 in./in.

Load (kip)	Elongation (in.)
0	0
1.50	0.0005
4.60	0.0015
8.00	0.0025
11.00	0.0035
11.80	0.0050
11.80	0.0080
12.00	0.0200
16.60	0.0400
20.00	0.1000
21.50	0.2800
19.50	0.4000
18.50	0.4600

Prob. 8–1

8–2. Data taken from a stress–strain test for a ceramic are given in the table. The curve is linear between the origin and the first point. Plot the diagram, and determine the modulus of elasticity and the modulus of resilience.

8–3. Data taken from a stress–strain test for a ceramic are given in the table. The curve is linear between the origin and the first point. Plot the diagram, and determine approximately the modulus of toughness. The fracture stress is $\sigma_f = 53.4$ ksi.

σ (ksi)	ϵ (in./in.)
0	0
33.2	0.0006
45.5	0.0010
49.4	0.0014
51.5	0.0018
53.4	0.0022

Probs. 8–2/3

*8–4. The stress–strain diagram for a steel alloy having an original diameter of 0.5 in. and a gage length of 2 in. is given in the figure. Determine approximately the modulus of elasticity for the material, the load on the specimen that causes yielding, and the ultimate load the specimen will support.

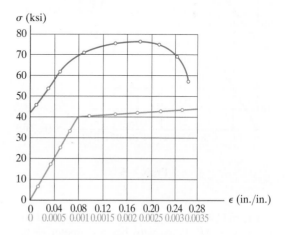

Prob. 8–4

8–5. The stress–strain diagram for a steel alloy having an original diameter of 0.5 in. and a gage length of 2 in. is given in the figure. If the specimen is loaded until it is stressed to 70 ksi, determine the approximate amount of elastic recovery and the increase in the gage length after it is unloaded.

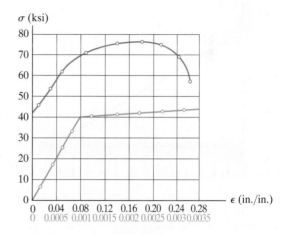

Prob. 8–5

8–6. The stress–strain diagram for a steel alloy having an original diameter of 0.5 in. and a gage length of 2 in. is given in the figure. Determine approximately the modulus of resilience and the modulus of toughness for the material.

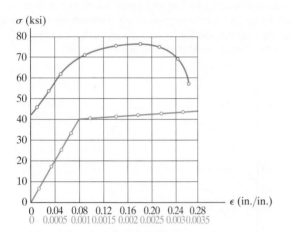

Prob. 8–6

8–7. The rigid beam is supported by a pin at C and an A-36 steel guy wire AB. If the wire has a diameter of 0.2 in., determine how much it stretches when a distributed load of $w = 100$ lb/ft acts on the beam. The material remains elastic.

***8–8.** The rigid beam is supported by a pin at C and an A-36 steel guy wire AB. If the wire has a diameter of 0.2 in., determine the distributed load w if the end B is displaced 0.75 in. downward.

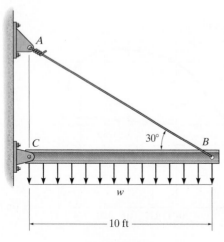

Probs. 8–7/8

8–9. Acetal plastic has a stress–strain diagram as shown. If a bar of this material has a length of 3 ft and cross-sectional area of 0.875 in², and is subjected to an axial load of 2.5 kip, determine its elongation.

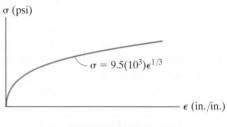

Prob. 8–9

8–10. The stress–strain diagram for an aluminum alloy specimen having an original diameter of 0.5 in. and a gage length of 2 in. is given in the figure. Determine approximately the modulus of elasticity for the material, the load on the specimen that causes yielding, and the ultimate load the specimen will support.

8–11. The stress–strain diagram for an aluminum alloy specimen having an original diameter of 0.5 in. and a gage length of 2 in. is given in the figure. If the specimen is loaded until it is stressed to 60 ksi, determine the approximate amount of elastic recovery and the increase in the gage length after it is unloaded.

***8–12.** The stress–strain diagram for an aluminum alloy specimen having an original diameter of 0.5 in. and a gage length of 2 in. is given in the figure. Determine approximately the modulus of resilience and the modulus of toughness for the material.

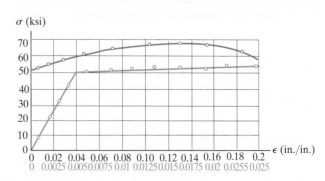

Probs. 8–10/11/12

8–13. A bar having a length of 5 in. and cross-sectional area of 0.7 in.² is subjected to an axial force of 8000 lb. If the bar stretches 0.002 in., determine the modulus of elasticity of the material. The material has linear elastic behavior.

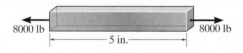

8000 lb   5 in.   8000 lb

Prob. 8–13

8–14. The rigid pipe is supported by a pin at A and an A-36 steel guy wire BD. If the wire has a diameter of 0.25 in., determine how much it stretches when a load of $P = 600$ lb acts on the pipe.

8–15. The rigid pipe is supported by a pin at A and an A-36 guy wire BD. If the wire has a diameter of 0.25 in., determine the load P if the end C is displaced 0.15 in. downward.

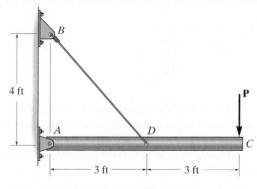

4 ft

3 ft   3 ft

Probs. 8–14/15

*8–16. Direct tension indicators are sometimes used instead of torque wrenches to ensure that a bolt has a prescribed tension when used for connections. If a nut on the bolt is tightened so that the six 3-mm high heads of the indicator are strained 0.1 mm/mm, and leave a contact area on each head of 1.5 mm², determine the tension in the bolt shank. The material has the stress–strain diagram shown.

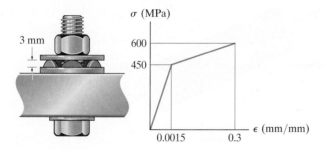

3 mm

σ (MPa)

600
450

0.0015   0.3 ϵ (mm/mm)

Prob. 8–16

8–17. The rigid beam is supported by a pin at C and an A992 steel guy wire AB of length 6 ft. If the wire has a diameter of 0.2 in., determine how much it stretches when a distributed load of $w = 200$ lb/ft acts on the beam. The wire remains elastic.

8–18. The rigid beam is supported by a pin at C and an A992 steel guy wire AB of length 6 ft. If the wire has a diameter of 0.2 in., determine the distributed load w if the end B is displaced 0.12 in. downward. The wire remains elastic.

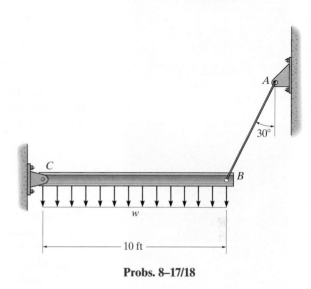

A

30°

C B

w

10 ft

Probs. 8–17/18

8–19. The stress–strain diagram for a bone is shown, and can be described by the equation $\epsilon = 0.45(10^{-6})\,\sigma + 0.36(10^{-12})\,\sigma^3$, where σ is in kPa. Determine the yield strength assuming a 0.3% offset.

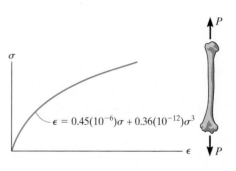

P

σ

$\epsilon = 0.45(10^{-6})\sigma + 0.36(10^{-12})\sigma^3$

ϵ P

Prob. 8–19

8.5   POISSON'S RATIO

When a deformable body is subjected to a force, not only does it elongate but it also contracts laterally. For example, consider the bar in Fig. 8–20 that has an original radius r and length L, and is subjected to the tensile force P. This force elongates the bar by an amount δ, and its radius contracts by an amount δ'. The strains in the longitudinal or axial direction and in the lateral or radial direction become

$$\epsilon_{\text{long}} = \frac{\delta}{L} \quad \text{and} \quad \epsilon_{\text{lat}} = \frac{\delta'}{r}$$

In the early 1800s, the French scientist S. D. Poisson realized that *within the elastic range* the *ratio* of these strains is a *constant*, since the displacements δ and δ' are proportional to the same applied force. This ratio is referred to as **Poisson's ratio**, ν (nu), and it has a numerical value that is unique for any material that is both *homogeneous and isotropic.* Stated mathematically it is

$$\nu = -\frac{\epsilon_{\text{lat}}}{\epsilon_{\text{long}}} \qquad (8\text{–}9)$$

The negative sign is included here since *longitudinal elongation* (positive strain) causes *lateral contraction* (negative strain), and vice versa. Keep in mind that these strains are caused only by the single axial or longitudinal force P; i.e., no force acts in a lateral direction in order to strain the material in this direction.

Poisson's ratio is a *dimensionless* quantity, and it will be shown in Sec. 10.6 that its *maximum* possible value is 0.5, so that $0 \le \nu \le 0.5$. For most nonporous solids it has a value that is generally between 0.25 and 0.355. Typical values for common engineering materials are listed on the inside back cover.

When the rubber block is compressed (negative strain), its sides will expand (positive strain). The ratio of these strains remains constant.

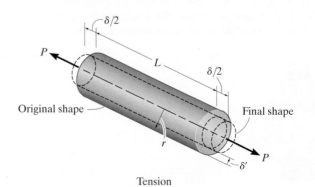

Tension

Fig. 8–20

EXAMPLE 8.3

A bar made of A-36 steel has the dimensions shown in Fig. 8–21. If an axial force of $P = 80$ kN is applied to the bar, determine the change in its length and the change in the dimensions of its cross section. The material behaves elastically.

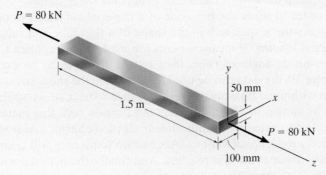

Fig. 8–21

SOLUTION

The normal stress in the bar is

$$\sigma_z = \frac{N}{A} = \frac{80(10^3)\ \text{N}}{(0.1\ \text{m})(0.05\ \text{m})} = 16.0(10^6)\ \text{Pa}$$

From the table on the inside back cover for A-36 steel $E_{st} = 200$ GPa, and so the strain in the z direction is

$$\epsilon_z = \frac{\sigma_z}{E_{st}} = \frac{16.0(10^6)\ \text{Pa}}{200(10^9)\ \text{Pa}} = 80(10^{-6})\ \text{mm/mm}$$

The axial elongation of the bar is therefore

$$\delta_z = \epsilon_z L_z = [80(10^{-6})](1.5\ \text{m}) = 120\ \mu\text{m} \qquad \textit{Ans.}$$

Using Eq. 8–9, where $\nu_{st} = 0.32$ as found from the inside back cover, the lateral contraction strains in *both* the x and y directions are

$$\epsilon_x = \epsilon_y = -\nu_{st}\,\epsilon_z = -0.32[80(10^{-6})] = -25.6\ \mu\text{m/m}$$

Thus the changes in the dimensions of the cross section are

$$\delta_x = \epsilon_x L_x = -[25.6(10^{-6})](0.1\ \text{m}) = -2.56\ \mu\text{m} \qquad \textit{Ans.}$$
$$\delta_y = \epsilon_y L_y = -[25.6(10^{-6})](0.05\ \text{m}) = -1.28\ \mu\text{m} \qquad \textit{Ans.}$$

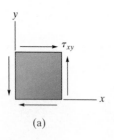

(a)

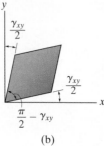

(b)

Fig. 8–22

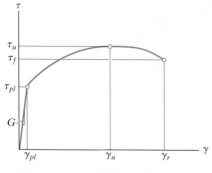

Fig. 8–23

8.6   THE SHEAR STRESS–STRAIN DIAGRAM

In Sec. 7.5 it was shown that when a small element of material is subjected to *pure shear*, equilibrium requires that equal shear stresses must be developed on four faces of the element, Fig. 8–22*a*. Furthermore, if the material is *homogeneous* and *isotropic*, then this shear stress will distort the element *uniformly*, Fig. 8–22*b*, producing shear strain.

In order to study the behavior of a material subjected to pure shear, engineers use a specimen in the shape of a thin tube and subject it to a torsional loading. If measurements are made of the applied torque and the resulting angle of twist, then by the methods to be explained in Chapter 10, the data can be used to determine the shear stress and shear strain within the tube and thereby produce a shear stress–strain diagram such as shown in Fig. 8–23. Like the tension test, this material when subjected to shear will exhibit linear elastic behavior and it will have a defined *proportional limit* τ_{pl}. Also, strain hardening will occur until an *ultimate shear stress* τ_u is reached. And finally, the material will begin to lose its shear strength until it reaches a point where it fractures, τ_f.

For most engineering materials, like the one just described, the elastic behavior is *linear*, and so Hooke's law for shear can be written as

$$\tau = G\gamma \tag{8–10}$$

Here G is called the **shear modulus of elasticity** or the **modulus of rigidity**. Its value represents the slope of the line on the τ–γ diagram, that is, $G = \tau_{pl}/\gamma_{pl}$. Units of measurement for G will be the *same* as those for τ (Pa or psi), since γ is measured in radians, a dimensionless quantity. Typical values for common engineering materials are listed on the inside back cover.

Later it will be shown in Sec. 14.11 that the three material constants, E, ν, and G can all be *related* by the equation

$$G = \frac{E}{2(1 + \nu)} \tag{8–11}$$

Therefore, if E and G are known, the value of ν can then be determined from this equation rather than through experimental measurement.

EXAMPLE 8.4

A specimen of titanium alloy is tested in torsion and the shear stress–strain diagram is shown in Fig. 8–24a. Determine the shear modulus G, the proportional limit, and the ultimate shear stress. Also, determine the maximum distance d that the top of a block of this material, shown in Fig. 8–24b, could be displaced horizontally if the material behaves elastically when acted upon by a shear force $\mathbf{V}$. What is the magnitude of $\mathbf{V}$ necessary to cause this displacement?

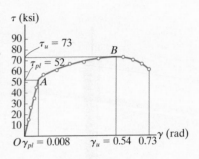

(a)

SOLUTION

Shear Modulus. This value represents the slope of the straight-line portion OA of the $\tau{-}\gamma$ diagram. The coordinates of point A are (0.008 rad, 52 ksi). Thus,

$$G = \frac{52 \text{ ksi}}{0.008 \text{ rad}} = 6500 \text{ ksi} \qquad Ans.$$

The equation of line OA is therefore $\tau = G\gamma = 6500\gamma$, which is Hooke's law for shear.

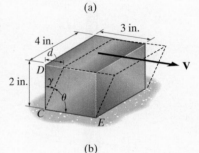

(b)

Fig. 8–24

Proportional Limit. By inspection, the graph ceases to be linear at point A. Thus,

$$\tau_{pl} = 52 \text{ ksi} \qquad Ans.$$

Ultimate Stress. This value represents the maximum shear stress, point B. From the graph,

$$\tau_u = 73 \text{ ksi} \qquad Ans.$$

Maximum Elastic Displacement and Shear Force. The shear strain at the corner C of the block in Fig. 8–24b is determined by finding the difference in the 90° angle DCE and the angle θ. This angle is $\gamma = 90° - \theta$ as shown. From the $\tau{-}\gamma$ diagram the maximum elastic shear strain is 0.008 rad, a very small angle. The top of the block in Fig. 8–24b will therefore be displaced horizontally a distance d given by

$$\tan(0.008 \text{ rad}) \approx 0.008 \text{ rad} = \frac{d}{2 \text{ in.}}$$

$$d = 0.016 \text{ in.}$$

The corresponding *average* shear stress in the block is $\tau_{pl} = 52$ ksi. Thus, the shear force V needed to cause the displacement is

$$\tau_{avg} = \frac{V}{A};\qquad 52 \text{ ksi} = \frac{V}{(3 \text{ in.})(4 \text{ in.})}$$

$$V = 624 \text{ kip} \qquad Ans.$$

EXAMPLE   8.5

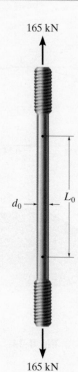

165 kN

d_0 —◄— L_0

165 kN

Fig. 8–25

An aluminum specimen shown in Fig. 8–25 has a diameter of $d_0 = 25$ mm and a gage length of $L_0 = 250$ mm. If a force of 165 kN elongates the gage length 1.20 mm, determine the modulus of elasticity. Also, determine by how much the force causes the diameter of the specimen to contract. Take $G_{al} = 26$ GPa and $\sigma_Y = 440$ MPa.

SOLUTION

Modulus of Elasticity.   The average normal stress in the specimen is

$$\sigma = \frac{P}{A} = \frac{165\,(10^3)\text{ N}}{(\pi/4)\,(0.025\text{ m})^2} = 336.1\text{ MPa}$$

and the average normal strain is

$$\epsilon = \frac{\delta}{L} = \frac{1.20\text{ mm}}{250\text{ mm}} = 0.00480\text{ mm/mm}$$

Since $\sigma < \sigma_Y = 440$ MPa, the material behaves elastically. The modulus of elasticity is therefore

$$E_{al} = \frac{\sigma}{\epsilon} = \frac{336.1\,(10^6)\text{ Pa}}{0.00480} = 70.0\text{ GPa} \qquad \textit{Ans.}$$

Contraction of Diameter.   First we will determine Poisson's ratio for the material using Eq. 8–11.

$$G = \frac{E}{2\,(1 + \nu)}$$

$$26\text{ GPa} = \frac{70.0\text{ GPa}}{2\,(1 + \nu)}$$

$$\nu = 0.347$$

Since $\epsilon_{long} = 0.00480$ mm/mm, then by Eq. 8–9,

$$\nu = -\frac{\epsilon_{lat}}{\epsilon_{long}}$$

$$0.347 = -\frac{\epsilon_{lat}}{0.00480\text{ mm/mm}}$$

$$\epsilon_{lat} = -0.00166\text{ mm/mm}$$

The contraction of the diameter is therefore

$$\delta' = (0.00166)\,(25\text{ mm})$$

$$= 0.0416\text{ mm} \qquad \textit{Ans.}$$

IMPORTANT POINTS

- *Poisson's ratio*, ν, is a ratio of the lateral strain of a homogeneous and isotropic material to its longitudinal strain. Generally these strains are of opposite signs, that is, if one is an elongation, the other will be a contraction.

- The *shear stress–strain diagram* is a plot of the shear stress versus the shear strain. If the material is homogeneous and isotropic, and is also linear elastic, the slope of the straight line within the elastic region is called the modulus of rigidity or the shear modulus, G.

- There is a mathematical relationship between G, E, and ν.

FUNDAMENTAL PROBLEMS

F8–13. A 100 mm long rod has a diameter of 15 mm. If an axial tensile load of 10 kN is applied to it, determine the change in its diameter. $E = 70$ GPa, $\nu = 0.35$.

F8–14. A solid circular rod that is 600 mm long and 20 mm in diameter is subjected to an axial force of $P = 50$ kN. The elongation of the rod is $\delta = 1.40$ mm, and its diameter becomes $d' = 19.9837$ mm. Determine the modulus of elasticity and the modulus of rigidity of the material. Assume that the material does not yield.

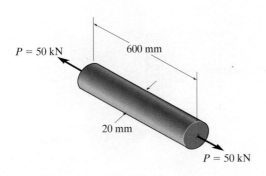

Prob. F8–14

F8–15. A 20-mm-wide block is firmly bonded to rigid plates at its top and bottom. When the force **P** is applied the block deforms into the shape shown by the dashed line. Determine the magnitude of **P**. The block's material has a modulus of rigidity of $G = 26$ GPa. Assume that the material does not yield and use small angle analysis.

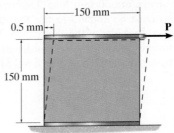

Prob. F8–15

F8–16. A 20-mm-wide block is bonded to rigid plates at its top and bottom. When the force **P** is applied the block deforms into the shape shown by the dashed line. If $a = 3$ mm and **P** is released, determine the permanent shear strain in the block.

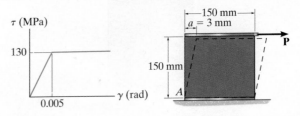

Prob. F8–16

PROBLEMS

*8–20. The acrylic plastic rod is 200 mm long and 15 mm in diameter. If an axial load of 300 N is applied to it, determine the change in its length and the change in its diameter. $E_p = 2.70$ GPa, $\nu_p = 0.4$.

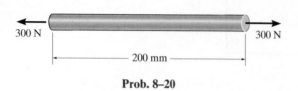

Prob. 8–20

8–21. The plug has a diameter of 30 mm and fits within a rigid sleeve having an inner diameter of 32 mm. Both the plug and the sleeve are 50 mm long. Determine the axial pressure p that must be applied to the top of the plug to cause it to contact the sides of the sleeve. Also, how far must the plug be compressed downward in order to do this? The plug is made from a material for which $E = 5$ MPa, $\nu = 0.45$.

8–22. The elastic portion of the stress–strain diagram for an aluminum alloy is shown in the figure. The specimen from which it was obtained has an original diameter of 12.7 mm and a gage length of 50.8 mm. When the applied load on the specimen is 50 kN, the diameter is 12.67494 mm. Determine Poisson's ratio for the material.

8–23. The elastic portion of the stress–strain diagram for an aluminum alloy is shown in the figure. The specimen from which it was obtained has an original diameter of 12.7 mm and a gage length of 50.8 mm. If a load of $P = 60$ kN is applied to the specimen, determine its new diameter and length. Take $\nu = 0.35$.

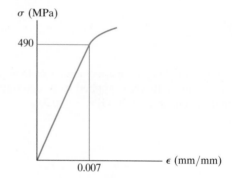

Probs. 8–22/23

*8–24. The brake pads for a bicycle tire are made of rubber. If a frictional force of 50 N is applied to each side of the tires, determine the average shear strain in the rubber. Each pad has cross-sectional dimensions of 20 mm and 50 mm. $G_r = 0.20$ MPa.

Prob. 8–21

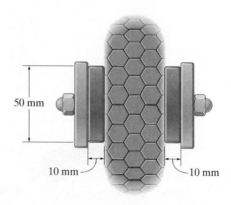

Prob. 8–24

8–25. The lap joint is connected together using a 1.25 in. diameter bolt. If the bolt is made from a material having a shear stress–strain diagram that is approximated as shown, determine the shear strain developed in the shear plane of the bolt when $P = 75$ kip.

8–26. The lap joint is connected together using a 1.25 in. diameter bolt. If the bolt is made from a material having a shear stress–strain diagram that is approximated as shown, determine the permanent shear strain in the shear plane of the bolt when the applied force $P = 150$ kip is removed.

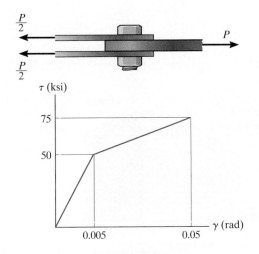

Probs. 8–25/26

8–27. The rubber block is subjected to an elongation of 0.03 in. along the x axis, and its vertical faces are given a tilt so that $\theta = 89.3°$. Determine the strains ϵ_x, ϵ_y and γ_{xy}. Take $\nu_r = 0.5$.

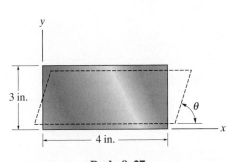

Prob. 8–27

*8–28. The shear stress–strain diagram for an alloy is shown in the figure. If a bolt having a diameter of 0.25 in. is made of this material and used in the lap joint, determine the modulus of elasticity E and the force P required to cause the material to yield. Take $\nu = 0.3$.

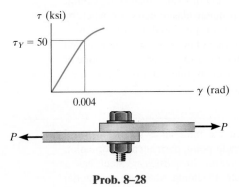

Prob. 8–28

8–29. A shear spring is made from two blocks of rubber, each having a height h, width b, and thickness a. The blocks are bonded to three plates as shown. If the plates are rigid and the shear modulus of the rubber is G, determine the displacement of plate A when the vertical load **P** is applied. Assume that the displacement is small so that $\delta = a \tan \gamma \approx a\gamma$.

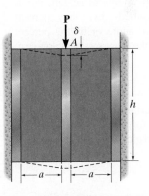

Prob. 8–29

CHAPTER REVIEW

One of the most important tests for material strength is the tension test. The results, found from stretching a specimen of known size, are plotted as normal stress on the vertical axis and normal strain on the horizontal axis.

Many engineering materials exhibit initial linear elastic behavior, whereby stress is proportional to strain, defined by Hooke's law, $\sigma = E\epsilon$. Here E, called the modulus of elasticity, is the slope of this straight line on the stress–strain diagram.	$\sigma = E\epsilon$	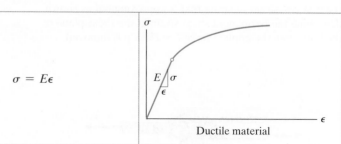 Ductile material

When the material is stressed beyond the yield point, permanent deformation will occur. In particular, steel has a region of yielding, whereby the material will exhibit an increase in strain with no increase in stress. The region of strain hardening causes further yielding of the material with a corresponding increase in stress. Finally, at the ultimate stress, a localized region on the specimen will begin to constrict, forming a neck. It is after this that the fracture occurs.

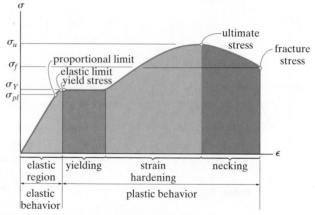

Conventional and true stress–strain diagrams for ductile material (steel) (not to scale)

Ductile materials, such as most metals, exhibit both elastic and plastic behavior. Wood is moderately ductile. Ductility is usually specified by the percent elongation to failure or by the percent reduction in the cross-sectional area.	Percent elongation $= \dfrac{L_f - L_0}{L_0}\,(100\%)$ Percent reduction of area $= \dfrac{A_0 - A_f}{A_0}\,(100\%)$	

Brittle materials exhibit little or no yielding before failure. Cast iron, concrete, and glass are typical examples.		

The yield point of a material at A can be increased by strain hardening. This is accomplished by applying a load that causes the stress to be greater than the yield stress, then releasing the load. The larger stress A' becomes the new yield point for the material.

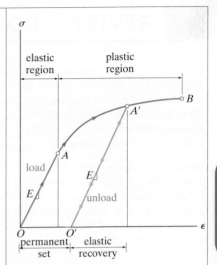

When a load is applied to a member, the deformations cause strain energy to be stored in the material. The strain energy per unit volume or strain energy density is equivalent to the area under the stress–strain curve. This area up to the yield point is called the modulus of resilience. The entire area under the stress–strain diagram is called the modulus of toughness.

Poisson's ratio ν is a dimensionless material property that relates the lateral strain to the longitudinal strain. Its range of values is $0 \leq \nu \leq 0.5$.

$$\nu = -\frac{\epsilon_{\text{lat}}}{\epsilon_{\text{long}}}$$

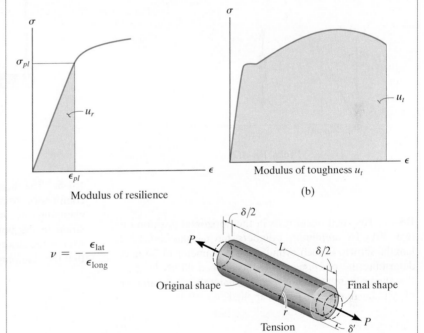

Modulus of resilience

Modulus of toughness u_t

(b)

Shear stress versus shear strain diagrams can also be established for a material. Within the elastic region, $\tau = G\gamma$, where G is the shear modulus found from the slope of the line. The value of ν can be obtained from the relationship that exists between G, E, and ν.

$$G = \frac{E}{2(1 + \nu)}$$

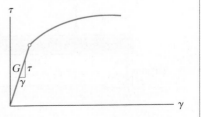

REVIEW PROBLEMS

R8–1. The elastic portion of the tension stress–strain diagram for an aluminum alloy is shown in the figure. The specimen used for the test has a gage length of 2 in. and a diameter of 0.5 in. When the applied load is 9 kip, the new diameter of the specimen is 0.49935 in. Calculate the shear modulus G_{al} for the aluminum.

R8–2. The elastic portion of the tension stress–strain diagram for an aluminum alloy is shown in the figure. The specimen used for the test has a gage length of 2 in. and a diameter of 0.5 in. If the applied load is 10 kip, determine the new diameter of the specimen. The shear modulus is $G_{al} = 3.8(10^3)$ ksi.

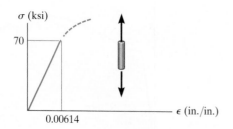

Probs. R8–1/2

R8–3. The rigid beam rests in the horizontal position on two 2014-T6 aluminum cylinders having the *unloaded* lengths shown. If each cylinder has a diameter of 30 mm, determine the placement x of the applied 80-kN load so that the beam remains horizontal. What is the new diameter of cylinder A after the load is applied? $\nu_{al} = 0.35$.

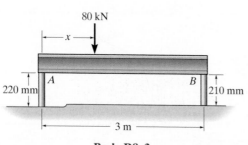

Prob. R8–3

*R8–4. The wires each have a diameter of $\frac{1}{2}$ in., length of 2 ft, and are made from 304 stainless steel. If $P = 6$ kip, determine the angle of tilt of the rigid beam AB.

R8–5. The wires each have a diameter of $\frac{1}{2}$ in., length of 2 ft, and are made from 304 stainless steel. Determine the magnitude of force $\mathbf{P}$ so that the rigid beam tilts 0.015°.

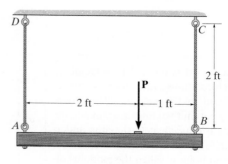

Probs. R8–4/5

R8–6. The head H is connected to the cylinder of a compressor using six $\frac{3}{16}$ in. diameter steel bolts. If the clamping force in each bolt is 800 lb, determine the normal strain in the bolts. If $\sigma_Y = 40$ ksi and $E_{st} = 29(10^3)$ ksi, what is the strain in each bolt when the nut is unscrewed so that the clamping force is released?

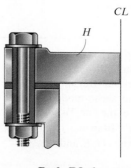

Prob. R8–6

R8–7. The stress–strain diagram for polyethylene, which is used to sheath coaxial cables, is determined from testing a specimen that has a gage length of 10 in. If a load P on the specimen develops a strain of $\epsilon = 0.024$ in./in., determine the approximate length of the specimen, measured between the gage points, when the load is removed. Assume the specimen recovers elastically.

R8–9. The 8-mm-diameter bolt is made of an aluminum alloy. It fits through a magnesium sleeve that has an inner diameter of 12 mm and an outer diameter of 20 mm. If the original lengths of the bolt and sleeve are 80 mm and 50 mm, respectively, determine the strains in the sleeve and the bolt if the nut on the bolt is tightened so that the tension in the bolt is 8 kN. Assume the material at A is rigid. $E_{al} = 70$ GPa, $E_{mg} = 45$ GPa.

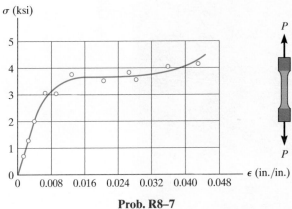

Prob. R8–7

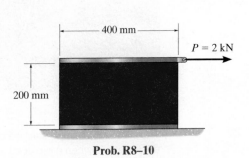

Prob. R8–9

*****R8–8.** The pipe with two rigid caps attached to its ends is subjected to an axial force P. If the pipe is made from a material having a modulus of elasticity E and Poisson's ratio ν, determine the change in volume of the material.

R8–10. An acetal polymer block is fixed to the rigid plates at its top and bottom surfaces. If the top plate displaces 2 mm horizontally when it is subjected to a horizontal force $P = 2$ kN, determine the shear modulus of the polymer. The width of the block is 100 mm. Assume that the polymer is linearly elastic and use small angle analysis.

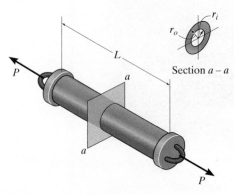

Prob. R8–8

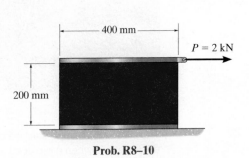

Prob. R8–10

CHAPTER 9

(© Hazlan Abdul Hakim/Getty Images)

The string of drill pipe stacked on this oil rig will be subjected to large axial deformations when it is placed in the hole.

AXIAL LOAD

CHAPTER OBJECTIVES

■ In this chapter we will discuss how to determine the deformation
of an axially loaded member, and we will also develop a method
for finding the support reactions when these reactions cannot be
determined strictly from the equations of equilibrium. An analysis
of the effects of thermal stress, stress concentrations, inelastic
deformations, and residual stress will also be discussed.

9.1 SAINT-VENANT'S PRINCIPLE

In the previous chapters, we have developed the concept of stress as a
means of measuring the force distribution within a body and strain as a
means of measuring a body's deformation. We have also shown that the
mathematical relationship between stress and strain depends on the type
of material from which the body is made. In particular, if the material
behaves in a linear elastic manner, then Hooke's law applies, and there is
a proportional relationship between stress and strain.

Using this idea, consider the manner in which a rectangular bar will
deform elastically when the bar is subjected to the force **P** applied along
its centroidal axis, Fig. 9–1a. The once horizontal and vertical grid lines
drawn on the bar become distorted, and *localized deformation* occurs at
each end. Throughout the midsection of the bar, the lines remain
horizontal and vertical.

P

Load distorts lines
located near load

a——a
b——b

c——c

Lines located away
from the load and support
remain straight

Load distorts lines
located near support

(a)

Fig. 9–1

If the material remains elastic, then the *strains* caused by this deformation are directly related to the *stress* in the bar through Hooke's law, $\sigma = E\epsilon$. As a result, a profile of the variation of the stress distribution acting at sections a–a, b–b, and c–c, will look like that shown in Fig. 9–1b. By comparison, the stress tends to reach a uniform value at section c–c, which is sufficiently removed from the end since the localized deformation caused by **P** *vanishes*. The minimum distance from the bar's end where this occurs can be determined using a mathematical analysis based on the theory of elasticity. It has been found that this distance should at least be equal to the *largest dimension* of the loaded cross section. Hence, section c–c should be located at a distance at least equal to the width (not the thickness) of the bar.*

In the same way, the stress distribution at the support in Fig. 9–1a will also even out and become uniform over the cross section located the same distance away from the support.

The fact that the localized stress and deformation behave in this manner is referred to as **Saint-Venant's principle**, since it was first noticed by the French scientist Barré de Saint-Venant in 1855. Essentially it states that the stress and strain produced at points in a body *sufficiently removed* from the region of external load application will be *the same* as the stress and strain produced by *any other applied external loading* that has the same statically equivalent resultant and is applied to the body within the same region. For example, if two symmetrically applied forces $P/2$ act on the bar, Fig. 9–1c, the stress distribution at section c–c will be uniform and therefore equivalent to $\sigma_{\text{avg}} = P/A$ as in Fig. 9–1c.

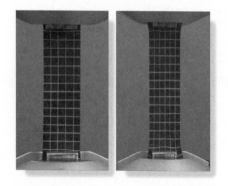

Notice how the lines on this rubber membrane distort after it is stretched. The localized distortions at the grips smooth out as stated by Saint-Venant's principle.

*When section c–c is so located, the theory of elasticity predicts the maximum stress to be $\sigma_{\text{max}} = 1.02\,\sigma_{\text{avg}}$.

section *a–a*   section *b–b*   section *c–c*   section *c–c*

$$\sigma_{avg} = \frac{P}{A}$$

$$\sigma_{avg} = \frac{P}{A}$$

(b)   (c)

Fig. 9–1 (cont.)

9.2 ELASTIC DEFORMATION OF AN AXIALLY LOADED MEMBER

Using Hooke's law and the definitions of stress and strain, we will now develop an equation that can be used to determine the *elastic* displacement of a member subjected to axial loads. To generalize the development, consider the bar shown in Fig. 9–2a, which has a cross-sectional area that gradually varies along its length L, and is made of a material that has a variable stiffness or modulus of elasticity. The bar is subjected to concentrated loads at its ends and a variable external load distributed along its length. This distributed load could, for example, represent the weight of the bar if it is in the vertical position, or friction forces acting on the bar's surface.

Here we wish to find the ***relative displacement*** δ (delta) of one end of the bar with respect to the other end as caused by the loading. We will *neglect* the localized deformations that occur at points of concentrated loading and where the cross section suddenly changes. From Saint-Venant's principle, these effects occur within small regions of the bar's length and will therefore have only a slight effect on the final result. For the most part, the bar will deform uniformly, so the normal stress will be uniformly distributed over the cross section.

The vertical displacement of the rod at the top floor B only depends upon the force in the rod along length AB. However, the displacement at the bottom floor C depends upon the force in the rod along its entire length, ABC.

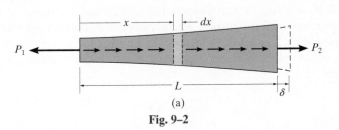

(a)

Fig. 9–2

Using the method of sections, a differential element (or wafer) of length dx and cross-sectional area $A(x)$ is isolated from the bar at the arbitrary position x, where the modulus of elasticity is $E(x)$. The free-body diagram of this element is shown in Fig. 9–2b. The resultant internal axial force will be a function of x since the external distributed loading will cause it to vary along the length of the bar. This load, $N(x)$, will deform the element into the shape indicated by the dashed outline, and therefore the displacement of one end of the element with respect to the other end becomes $d\delta$. The stress and strain in the element are therefore

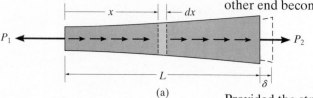

(a)

$$\sigma = \frac{N(x)}{A(x)} \text{ and } \epsilon = \frac{d\delta}{dx}$$

Provided the stress does not exceed the proportional limit, we can apply Hooke's law; i.e., $\sigma = E(x)\epsilon$, and so

$$\frac{N(x)}{A(x)} = E(x)\left(\frac{d\delta}{dx}\right)$$

$$d\delta = \frac{N(x)dx}{A(x)E(x)}$$

(b)

Fig. 9–2 (Repeated)

For the entire length L of the bar, we must integrate this expression to find δ. This yields

$$\delta = \int_0^L \frac{N(x)dx}{A(x)E(x)} \qquad (9\text{–}1)$$

Here
$\quad \delta =$ displacement of one point on the bar relative to the other point
$\quad L =$ original length of bar
$\quad N(x) =$ internal axial force at the section, located a distance x from one end
$\quad A(x) =$ cross-sectional area of the bar expressed as a function of x
$\quad E(x) =$ modulus of elasticity for the material expressed as a function of x

Constant Load and Cross-Sectional Area. In many cases the bar will have a constant cross-sectional area A; and the material will be homogeneous, so E is constant. Furthermore, if a constant external force is applied at each end, Fig. 9–3a, then the internal force N throughout the length of the bar is also constant. As a result, Eq. 9–1 when integrated becomes

$$\delta = \frac{NL}{AE} \qquad (9\text{–}2)$$

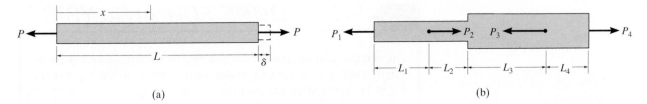

(a)

(b)

Fig. 9–3

If the bar is subjected to several different axial forces along its length, or the cross-sectional area or modulus of elasticity changes abruptly from one region of the bar to the next, as in Fig. 9–3b, then the above equation can be applied to each *segment of the bar* where these quantities remain *constant*. The displacement of one end of the bar with respect to the other is then found from the *algebraic addition* of the relative displacements of the ends of each segment. For this general case,

$$\delta = \sum \frac{NL}{AE}$$
(9–3)

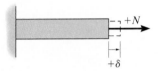

Sign Convention. When applying Eqs. 9–1 through 9–3, it is best to use a consistent sign convention for the internal axial force and the displacement of the bar. To do so, we will consider both the force and displacement to be *positive* if they cause *tension and elongation*, Fig. 9–4; whereas a *negative* force and displacement will cause *compression* and *contraction*.

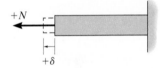

Fig. 9–4

> ## IMPORTANT POINTS

- *Saint-Venant's principle* states that both the localized deformation and stress which occur within the regions of load application or at the supports tend to "even out" at a distance sufficiently removed from these regions.

- The displacement of one end of an axially loaded member relative to the other end is determined by relating the applied *internal* load to the stress using $\sigma = N/A$ and relating the displacement to the strain using $\epsilon = d\delta/dx$. Finally these two equations are combined using Hooke's law, $\sigma = E\epsilon$, which yields Eq. 9–1.

- Since Hooke's law has been used in the development of the displacement equation, it is important that no internal load causes yielding of the material, and that the material behaves in a linear elastic manner.

▶ PROCEDURE FOR ANALYSIS

The relative displacement between any two points A and B on an axially loaded member can be determined by applying Eq. 9–1 (or Eq. 9–2). Application requires the following steps.

Internal Force.

- Use the method of sections to determine the internal axial force N within the member.

- If this force varies along the member's length due to an *external distributed loading*, a section should be made at the arbitrary location x from one end of the member, and the internal force represented as a function of x, i.e., $N(x)$.

- If several *constant external forces* act on the member, the internal force in each *segment* of the member between any two external forces must be determined.

- For any segment, an internal *tensile force* is *positive* and an internal *compressive force* is *negative*. For convenience, the results of the internal loading throughout the member can be shown graphically by constructing the normal-force diagram.

Displacement.

- When the member's cross-sectional area *varies* along its length, the area must be expressed as a function of its position x, i.e., $A(x)$.

- If the cross-sectional area, the modulus of elasticity, or the internal loading *suddenly changes*, then Eq. 9–2 should be applied to each segment for which these quantities are constant.

- When substituting the data into Eqs. 9–1 through 9–3, be sure to account for the proper sign of the internal force N. Tensile forces are positive and compressive forces are negative. Also, use a consistent set of units. For any segment, if the result is a *positive* numerical quantity, it indicates *elongation*; if it is *negative*, it indicates a *contraction*.

EXAMPLE 9.1

The uniform A-36 steel bar in Fig. 9–5*a* has a diameter of 50 mm and is subjected to the loading shown. Determine the displacement at *D*, and the displacement of point *B* relative to *C*.

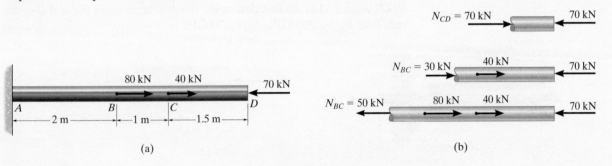

(a)   (b)

Fig. 9–5

SOLUTION

Internal Forces.   The internal forces within the bar are determined using the method of sections and horizontal equilibrium. The results are shown on the free-body diagrams in Fig. 9–5*b*. The normal-force diagram in Fig. 9–5*c* shows the variation of these forces along the bar.

Displacement.   From the table on the inside back cover, for A-36 steel, $E = 200$ GPa. Using the established sign convention, the displacement of the end of the bar is therefore

$$\delta_D = \sum \frac{NL}{AE} = \frac{[-70(10^3)\text{ N}](1.5\text{ m})}{\pi(0.025\text{ m})^2[200(10^9)\text{ N/m}^2]}$$

$$+ \frac{[-30(10^3)\text{ N}](1\text{ m})}{\pi(0.025\text{ m})^2[200(10^9)\text{ N/m}^2]} + \frac{[50(10^3)\text{ N}](2\text{ m})}{\pi(0.025\text{ m})^2[200(10^9)\text{ N/m}^2]}$$

$$\delta_D = -89.1(10^{-3})\text{ mm} \qquad\qquad Ans.$$

This negative result indicates that point *D* moves to the left.

The displacement of *B* relative to *C*, $\delta_{B/C}$, is caused only by the internal load within region *BC*. Thus,

$$\delta_{B/C} = \frac{NL}{AE} = \frac{[-30(10^3)\text{ N}](1\text{ m})}{\pi(0.025\text{ m})^2[200(10^9)\text{ N/m}^2]} = -76.4(10^{-3})\text{ mm} \qquad Ans.$$

Here the negative result indicates that *B* will move towards *C*.

EXAMPLE   9.2

The assembly shown in Fig. 9–6a consists of an aluminum tube AB having a cross-sectional area of 400 mm². A steel rod having a diameter of 10 mm is attached to a rigid collar and passes through the tube. If a tensile load of 80 kN is applied to the rod, determine the displacement of the end C of the rod. Take $E_{st} = 200$ GPa, $E_{al} = 70$ GPa.

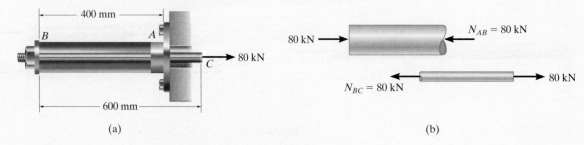

(a)   (b)

Fig. 9–6

SOLUTION

Internal Force.   The free-body diagrams of the tube and rod segments in Fig. 9–6b show that the rod is subjected to a tension of 80 kN, and the tube is subjected to a compression of 80 kN.

Displacement.   We will first determine the displacement of C with respect to B. Working in units of newtons and meters, we have

$$\delta_{C/B} = \frac{NL}{AE} = \frac{[+80(10^3) \text{ N}] (0.6 \text{ m})}{\pi(0.005 \text{ m})^2[200 (10^9) \text{ N/m}^2]} = +0.003056 \text{ m} \rightarrow$$

The positive sign indicates that C moves *to the right* relative to B, since the bar elongates.

The displacement of B with respect to the *fixed* end A is

$$\delta_B = \frac{NL}{AE} = \frac{[-80(10^3) \text{ N}](0.4 \text{ m})}{[400 \text{ mm}^2(10^{-6}) \text{ m}^2/\text{mm}^2][70(10^9) \text{ N/m}^2]}$$

$$= -0.001143 \text{ m} = 0.001143 \text{ m} \rightarrow$$

Here the negative sign indicates that the tube shortens, and so B moves to the *right* relative to A.

Since both displacements are to the right, the displacement of C relative to the fixed end A is therefore

$$(\overset{+}{\rightarrow}) \qquad \delta_C = \delta_B + \delta_{C/B} = 0.001143 \text{ m} + 0.003056 \text{ m}$$

$$= 0.00420 \text{ m} = 4.20 \text{ mm} \rightarrow \qquad\qquad Ans.$$

EXAMPLE 9.3

Rigid beam AB rests on the two short posts shown in Fig. 9–7a. AC is made of steel and has a diameter of 20 mm, and BD is made of aluminum and has a diameter of 40 mm. Determine the displacement of point F on AB if a vertical load of 90 kN is applied over this point. Take $E_{st} = 200$ GPa, $E_{al} = 70$ GPa.

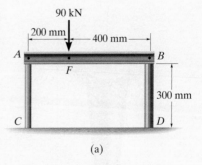

(a)

SOLUTION

Internal Force.   The compressive forces acting at the top of each post are determined from the equilibrium of member AB, Fig. 9–7b. These forces are equal to the internal forces in each post, Fig. 9–7c.

Displacement.   The displacement of the top of each post is

Post AC:

$$\delta_A = \frac{N_{AC}L_{AC}}{A_{AC}E_{st}} = \frac{[-60(10^3)\ \text{N}](0.300\ \text{m})}{\pi(0.010\ \text{m})^2[200(10^9)\ \text{N/m}^2]} = -286(10^{-6})\ \text{m}$$

$$= 0.286\ \text{mm}\ \downarrow$$

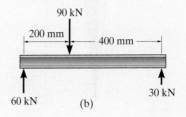

(b)

Post BD:

$$\delta_B = \frac{N_{BD}L_{BD}}{A_{BD}E_{al}} = \frac{[-30(10^3)\ \text{N}](0.300\ \text{m})}{\pi(0.020\ \text{m})^2[70(10^9)\ \text{N/m}^2]} = -102(10^{-6})\ \text{m}$$

$$= 0.102\ \text{mm}\ \downarrow$$

A diagram showing the centerline displacements at A, B, and F on the beam is shown in Fig. 9–7d. By proportion of the blue shaded triangle, the displacement of point F is therefore

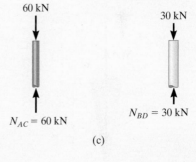

(c)

$$\delta_F = 0.102\ \text{mm} + (0.184\ \text{mm})\left(\frac{400\ \text{mm}}{600\ \text{mm}}\right) = 0.225\ \text{mm}\ \downarrow \qquad Ans.$$

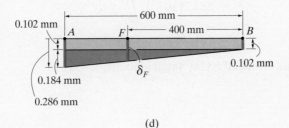

(d)

Fig. 9–7

EXAMPLE 9.4

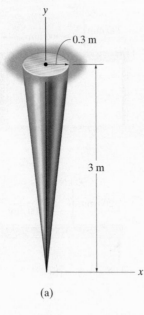

(a)

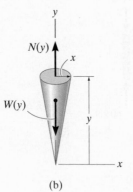

(b)

Fig. 9–8

A member is made of a material that has a specific weight of $\gamma = 6 \text{ kN/m}^3$ and modulus of elasticity of 9 GPa. If it is in the form of a *cone* having the dimensions shown in Fig. 9–8a, determine how far its end is displaced due to gravity when it is suspended in the vertical position.

SOLUTION

Internal Force.   The internal axial force varies along the member, since it is dependent on the weight $W(y)$ of a segment of the member below any section, Fig. 9–8b. Hence, to calculate the displacement, we must use Eq. 9–1. At the section located a distance y from the cone's free end, the radius x of the cone as a function of y is determined by proportion; i.e.,

$$\frac{x}{y} = \frac{0.3 \text{ m}}{3 \text{ m}}; \qquad x = 0.1y$$

The volume of a cone having a base of radius x and height y is

$$V = \frac{1}{3} \pi y x^2 = \frac{\pi(0.01)}{3} y^3 = 0.01047 y^3$$

Since $W = \gamma V$, the internal force at the section becomes

$$+\uparrow \Sigma F_y = 0; \qquad N(y) = 6(10^3)(0.01047 y^3) = 62.83 y^3$$

Displacement.   The area of the cross section is also a function of position y, Fig. 9–8b. We have

$$A(y) = \pi x^2 = 0.03142 y^2$$

Applying Eq. 9–1 between the limits of $y = 0$ and $y = 3$ m yields

$$\delta = \int_0^L \frac{N(y)\, dy}{A(y)\, E} = \int_0^3 \frac{(62.83 y^3)\, dy}{(0.03142 y^2)\, 9(10^9)}$$

$$= 222.2(10^{-9}) \int_0^3 y\, dy$$

$$= 1(10^{-6}) \text{ m} = 1 \ \mu\text{m} \qquad\qquad Ans.$$

NOTE: This is indeed a very small amount.

PRELIMINARY PROBLEMS

P9–1. In each case, determine the internal normal force between lettered points on the bar. Draw all necessary free-body diagrams.

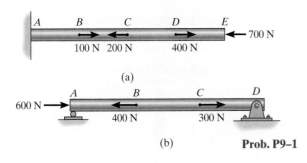

(a)

(b)   **Prob. P9–1**

P9–2. Determine the internal normal force between lettered points on the cable and rod. Draw all necessary free-body diagrams.

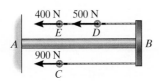

Prob. P9–2

P9–3. The post weighs 8 kN/m. Determine the internal normal force in the post as a function of x.

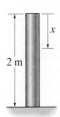

Prob. P9–3

P9–4. The rod is subjected to an external axial force of 800 N and a uniform distributed load of 100 N/m along its length. Determine the internal normal force in the rod as a function of x.

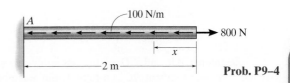

Prob. P9–4

P9–5. The rigid beam supports the load of 60 kN. Determine the displacement at B. Take $E = 60$ GPa, and $A_{BC} = 2 \, (10^{-3}) \, \text{m}^2$.

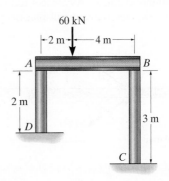

Prob. P9–5

FUNDAMENTAL PROBLEMS

F9–1. The 20-mm-diameter A-36 steel rod is subjected to the axial forces shown. Determine the displacement of end C with respect to the fixed support at A.

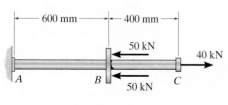

Prob. F9–1

F9–2. Segments AB and CD of the assembly are solid circular rods, and segment BC is a tube. If the assembly is made of 6061-T6 aluminum, determine the displacement of end D with respect to end A.

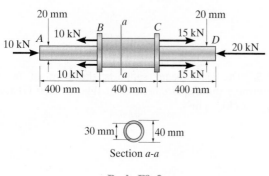

Prob. F9–2

F9–3. The 30-mm-diameter A992 steel rod is subjected to the loading shown. Determine the displacement of end C.

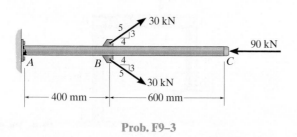

Prob. F9–3

F9–4. If the 20-mm-diameter rod is made of A-36 steel and the stiffness of the spring is $k = 50$ MN/m, determine the displacement of end A when the 60-kN force is applied.

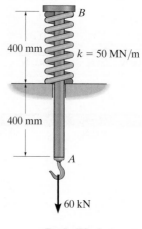

Prob. F9–4

F9–5. The 20-mm-diameter 2014-T6 aluminum rod is subjected to the uniform distributed axial load. Determine the displacement of end A.

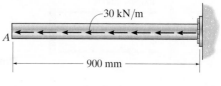

Prob. F9–5

F9–6. The 20-mm-diameter 2014-T6 aluminum rod is subjected to the triangular distributed axial load. Determine the displacement of end A.

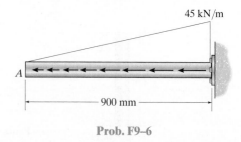

Prob. F9–6

PROBLEMS

9–1. The A992 steel rod is subjected to the loading shown. If the cross-sectional area of the rod is 80 mm², determine the displacement of B and A. Neglect the size of the couplings at B and C.

9–3. The composite shaft, consisting of aluminum, copper, and steel sections, is subjected to the loading shown. Determine the displacement of end A with respect to end D and the normal stress in each section. The cross-sectional area and modulus of elasticity for each section are shown in the figure. Neglect the size of the collars at B and C.

*9–4. The composite shaft, consisting of aluminum, copper, and steel sections, is subjected to the loading shown. Determine the displacement of B with respect to C. The cross-sectional area and modulus of elasticity for each section are shown in the figure. Neglect the size of the collars at B and C.

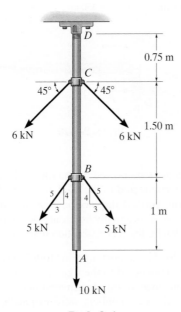

Prob. 9–1

Aluminum	Copper	Steel
$E_{al} = 10(10^3)$ ksi	$E_{cu} = 18(10^3)$ ksi	$E_{st} = 29(10^3)$ ksi
$A_{AB} = 0.09$ in²	$A_{BC} = 0.12$ in²	$A_{CD} = 0.06$ in²

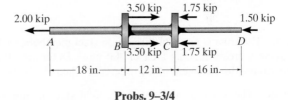

Probs. 9–3/4

9–2. The copper shaft is subjected to the axial loads shown. Determine the displacement of end A with respect to end D if the diameters of each segment are $d_{AB} = 0.75$ in., $d_{BC} = 1$ in., and $d_{CD} = 0.5$ in. Take $E_{cu} = 18(10^3)$ ksi.

9–5. The 2014-T6 aluminum rod has a diameter of 30 mm and supports the load shown. Determine the displacement of end A with respect to end E. Neglect the size of the couplings.

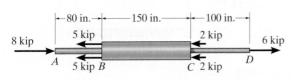

Prob. 9–2

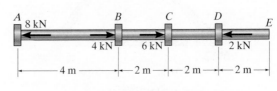

Prob. 9–5

9–6. The A-36 steel drill shaft of an oil well extends 12 000 ft into the ground. Assuming that the pipe used to drill the well is suspended freely from the derrick at *A*, determine the maximum average normal stress in each pipe string and the elongation of its end *D* with respect to the fixed end at *A*. The shaft consists of three different sizes of pipe, *AB*, *BC*, and *CD*, each having the length, weight per unit length, and cross-sectional area indicated.

9–9. The assembly consists of two 10-mm diameter red brass C83400 copper rods *AB* and *CD*, a 15-mm diameter 304 stainless steel rod *EF*, and a rigid bar *G*. If $P = 5$ kN, determine the horizontal displacement of end *F* of rod *EF*.

9–10. The assembly consists of two 10-mm diameter red brass C83400 copper rods *AB* and *CD*, a 15-mm diameter 304 stainless steel rod *EF*, and a rigid bar *G*. If the horizontal displacement of end *F* of rod *EF* is 0.45 mm, determine the magnitude of *P*.

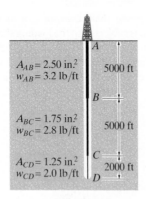

$A_{AB} = 2.50$ in.^{2}
$w_{AB} = 3.2$ lb/ft
5000 ft

$A_{BC} = 1.75$ in.^{2}
$w_{BC} = 2.8$ lb/ft
5000 ft

$A_{CD} = 1.25$ in.^{2}
$w_{CD} = 2.0$ lb/ft
2000 ft

Prob. 9–6

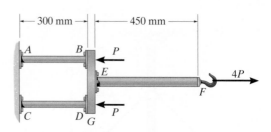

Probs. 9–9/10

9–7. The truss is made of three A-36 steel members, each having a cross-sectional area of 400 mm². Determine the horizontal displacement of the roller at *C* when $P = 8$ kN.

*9–8. The truss is made of three A-36 steel members, each having a cross-sectional area of 400 mm². Determine the magnitude *P* required to displace the roller to the right 0.2 mm.

9–11. The load is supported by the four 304 stainless steel wires that are connected to the rigid members *AB* and *DC*. Determine the vertical displacement of the 500-lb load if the members were originally horizontal when the load was applied. Each wire has a cross-sectional area of 0.025 in².

*9–12. The load is supported by the four 304 stainless steel wires that are connected to the rigid members *AB* and *DC*. Determine the angle of tilt of each member after the 500-lb load is applied. The members were originally horizontal, and each wire has a cross-sectional area of 0.025 in².

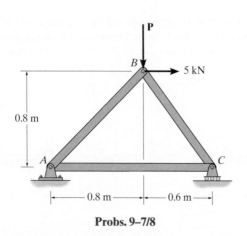

Probs. 9–7/8

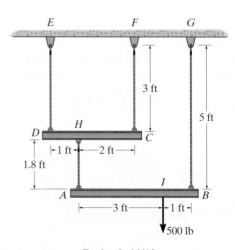

Probs. 9–11/12

9–13. The rigid bar is supported by the pin-connected rod *CB* that has a cross-sectional area of 14 mm² and is made from 6061-T6 aluminum. Determine the vertical deflection of the bar at *D* when the distributed load is applied.

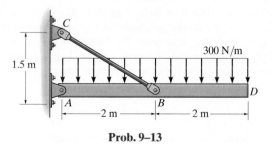

Prob. 9–13

9–14. The post is made of Douglas fir and has a diameter of 100 mm. If it is subjected to the load of 20 kN and the soil provides a frictional resistance distributed around the post that is triangular along its sides; that is, it varies from $w = 0$ at $y = 0$ to $w = 12$ kN/m at $y = 2$ m, determine the force **F** at its bottom needed for equilibrium. Also, what is the displacement of the top of the post *A* with respect to its bottom *B*? Neglect the weight of the post.

9–15. The post is made of Douglas fir and has a diameter of 100 mm. If it is subjected to the load of 20 kN and the soil provides a frictional resistance that is distributed along its length and varies linearly from $w = 4$ kN/m at $y = 0$ to $w = 12$ kN/m at $y = 2$ m, determine the force **F** at its bottom needed for equilibrium. Also, what is the displacement of the top of the post *A* with respect to its bottom *B*? Neglect the weight of the post.

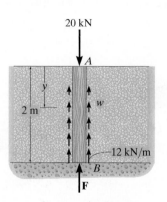

Probs. 9–14/15

*9–16. The coupling rod is subjected to a force of 5 kip. Determine the distance *d* between *C* and *E* accounting for the compression of the spring and the deformation of the bolts. When no load is applied the spring is unstretched and $d = 10$ in. The material is A-36 steel and each bolt has a diameter of 0.25 in. The plates at *A, B,* and *C* are rigid and the spring has a stiffness of $k = 12$ kip/in.

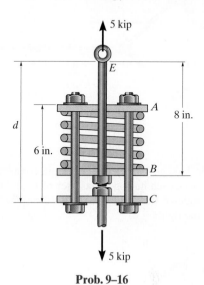

Prob. 9–16

9–17. The pipe is stuck in the ground so that when it is pulled upward the frictional force along its length varies linearly from zero at *B* to f_{max} (force/length) at *C*. Determine the initial force *P* required to pull the pipe out and the pipe's elongation just before it starts to slip. The pipe has a length *L*, cross-sectional area *A*, and the material from which it is made has a modulus of elasticity *E*.

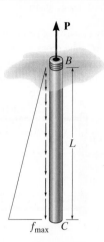

Prob. 9–17

9

9–18. The linkage is made of three pin-connected A992 steel members, each having a diameter of $1\frac{1}{4}$ in. If a horizontal force of $P = 60$ kip is applied to the end B of member AB, determine the displacement of point B.

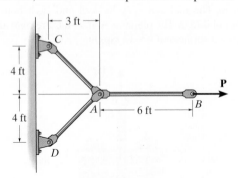

Prob. 9–18

9–19. The linkage is made of three pin-connected A992 steel members, each having a diameter of $1\frac{1}{4}$ in. Determine the magnitude of the force **P** needed to displace point B 0.25 in. to the right.

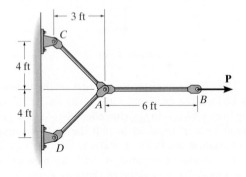

Prob. 9–19

*9–20. The assembly consists of three titanium (Ti-6A1-4V) rods and a rigid bar AC. The cross-sectional area of each rod is given in the figure. If a force of 60 kip is applied to the ring F, determine the horizontal displacement of point F.

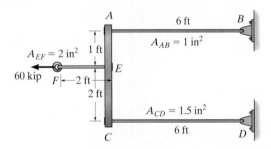

Prob. 9–20

9–21. The rigid beam is supported at its ends by two A-36 steel tie rods. If the allowable stress for the steel is $\sigma_{allow} = 16.2$ ksi, the load $w = 3$ kip/ft, and $x = 4$ ft, determine the smallest diameter of each rod so that the beam remains in the horizontal position when it is loaded.

9–22. The rigid beam is supported at its ends by two A-36 steel tie rods. The rods have diameters $d_{AB} = 0.5$ in. and $d_{CD} = 0.3$ in. If the allowable stress for the steel is $\sigma_{allow} = 16.2$ ksi, determine the largest intensity of the distributed load w and its length x on the beam so that the beam remains in the horizontal position when it is loaded.

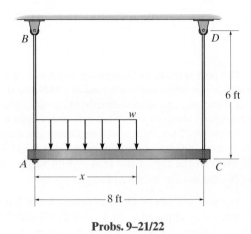

Probs. 9–21/22

9–23. The steel bar has the original dimensions shown in the figure. If it is subjected to an axial load of 50 kN, determine the change in its length and its new cross-sectional dimensions at section a–a. $E_{st} = 200$ GPa, $\nu_{st} = 0.29$.

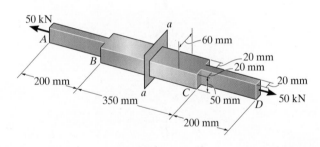

Prob. 9–23

*9–24.   Determine the relative displacement of one end of the tapered plate with respect to the other end when it is subjected to an axial load P.

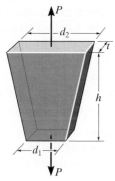

Prob. 9–24

9–25.   The assembly consists of two rigid bars that are originally horizontal. They are supported by pins and 0.25-in.-diameter A-36 steel rods. If the vertical load of 5 kip is applied to the bottom bar AB, determine the displacement at C, B, and E.

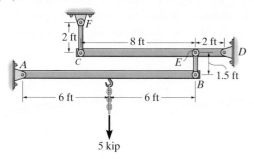

Prob. 9–25

9–26.   The truss consists of three members, each made from A-36 steel and having a cross-sectional area of 0.75 in². Determine the greatest load P that can be applied so that the roller support at B is not displaced more than 0.03 in.

9–27.   Solve Prob. 9–26 when the load **P** acts vertically downward at C.

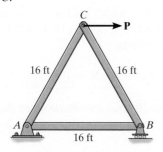

Probs. 9–26/27

*9–28.   The observation cage C has a weight of 250 kip and through a system of gears, travels upward at constant velocity along the A-36 steel column, which has a height of 200 ft. The column has an outer diameter of 3 ft and is made from steel plate having a thickness of 0.25 in. Neglect the weight of the column, and determine the average normal stress in the column at its base, B, as a function of the cage's position y. Also, determine the displacement of end A as a function of y.

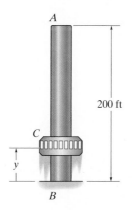

Prob. 9–28

9–29.   Determine the elongation of the aluminum strap when it is subjected to an axial force of 30 kN. $E_{al} = 70$ GPa.

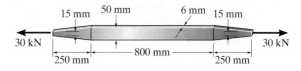

Prob. 9–29

9–30.   The ball is truncated at its ends and is used to support the bearing load **P**. If the modulus of elasticity for the material is E, determine the decrease in the ball's height when the load is applied.

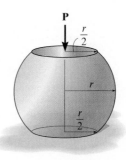

Prob. 9–30

9.3   PRINCIPLE OF SUPERPOSITION

The principle of superposition is often used to determine the stress or displacement at a point in a member when the member is subjected to a complicated loading. By subdividing the loading into components, this principle states that the resultant stress or displacement at the point can be determined by algebraically summing the stress or displacement caused by each load component applied separately to the member.

The following two conditions must be satisfied if the principle of superposition is to be applied.

1. *The loading N must be linearly related to the stress σ or displacement δ that is to be determined*. For example, the equations $\sigma = N/A$ and $\delta = NL/AE$ involve a linear relationship between σ and N, and δ and N.

2. *The loading must not significantly change the original geometry or configuration of the member*. If significant changes do occur, the direction and location of the applied forces and their moment arms will change. For example, consider the slender rod shown in Fig. 9–9a, which is subjected to the load **P**. In Fig. 9–9b, **P** is replaced by two of its components, $\mathbf{P} = \mathbf{P}_1 + \mathbf{P}_2$. If **P** causes the rod to deflect a large amount, as shown, the moment of this load about its support, Pd, will not equal the sum of the moments of its component loads, $Pd \neq P_1d_1 + P_2d_2$, because $d_1 \neq d_2 \neq d$.

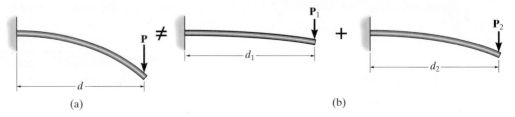

(a)   (b)

Fig. 9–9

9.4   STATICALLY INDETERMINATE AXIALLY LOADED MEMBERS

Consider the bar shown in Fig. 9–10a, which is fixed supported at both of its ends. From its free-body diagram, Fig. 9–10b, there are two unknown support reactions. Equilibrium requires

$$+\uparrow \Sigma F = 0; \qquad\qquad F_B + F_A - 500 \text{ N} = 0$$

This type of problem is called ***statically indeterminate***, since the equilibrium equation is *not sufficient* to determine both reactions on the bar.

In order to establish an additional equation needed for solution, it is necessary to consider how points on the bar are displaced. Specifically, an equation that specifies the conditions for displacement is referred to as a *compatibility* or *kinematic condition*. In this case, a suitable compatibility condition would require the displacement of end A of the bar with respect to end B to equal zero, since the end supports are *fixed*, and so no relative movement can occur between them. Hence, the compatibility condition becomes

$$\delta_{A/B} = 0$$

This equation can be expressed in terms of the internal loads by using a *load–displacement relationship*, which depends on the material behavior. For example, if linear elastic behavior occurs, then $\delta = NL/AE$ can be used. Realizing that the internal force in segment AC is $+F_A$, and in segment CB it is $-F_B$, Fig. 9–10c, then the compatibility equation can be written as

$$\frac{F_A(2 \text{ m})}{AE} - \frac{F_B(3 \text{ m})}{AE} = 0$$

Since AE is constant, then $F_A = 1.5 F_B$. Finally, using the equilibrium equation, the reactions are therefore

$$F_A = 300 \text{ N} \quad \text{and} \quad F_B = 200 \text{ N}$$

Since both of these results are positive, the directions of the reactions are shown correctly on the free-body diagram.

To solve for the reactions on any statically indeterminate problem, we must therefore satisfy both the equilibrium and compatibility equations, and relate the displacements to the loads using the load–displacement relations.

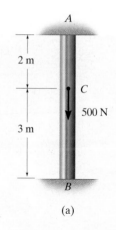

(a)

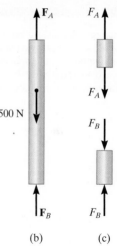

(b)   (c)

Fig. 9–10

▶ IMPORTANT POINTS

- The *principle of superposition* is sometimes used to simplify stress and displacement problems having complicated loadings. This is done by subdividing the loading into components, then algebraically adding the results.

- Superposition requires that the loading be linearly related to the stress or displacement, and the loading must not significantly change the original geometry of the member.

- A problem is *statically indeterminate* if the equations of equilibrium are not sufficient to determine all the reactions on a member.

- *Compatibility conditions* specify the displacement constraints that occur at the supports or other points on a member.

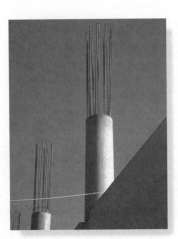

Most concrete columns are reinforced with steel rods; and since these two materials work together in supporting the applied load, the force in each material must be determined using a statically indeterminate analysis.

PROCEDURE FOR ANALYSIS

The support reactions for statically indeterminate problems are determined by satisfying equilibrium, compatibility, and load–displacement requirements for the member.

Equilibrium.

- Draw a free-body diagram of the member in order to identify all the forces that act on it.

- The problem can be classified as statically indeterminate if the number of unknown reactions on the free-body diagram is greater than the number of available equations of equilibrium.

- Write the equations of equilibrium for the member.

Compatibility.

- Consider drawing a displacement diagram in order to investigate the way the member will elongate or contract when subjected to the external loads.

- Express the compatibility conditions in terms of the displacements caused by the loading.

Load–Displacement.

- Use a load–displacement relation, such as $\delta = NL/AE$, to relate the unknown displacements in the compatibility equation to the reactions.

- Solve all the equations for the reactions. If any of the results has a negative numerical value, it indicates that this force acts in the opposite sense of direction to that indicated on the free-body diagram.

EXAMPLE 9.5

The steel rod shown in Fig. 9–11a has a diameter of 10 mm. It is fixed to the wall at A, and before it is loaded, there is a gap of 0.2 mm between the wall at B' and the rod. Determine the reactions on the rod if it is subjected to an axial force of P = 20 kN. Neglect the size of the collar at C. Take E_{st} = 200 GPa.

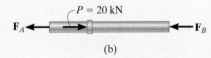

(a)

SOLUTION

Equilibrium. As shown on the free-body diagram, Fig. 9–11b, we will *assume* that force P is large enough to cause the rod's end B to contact the wall at B'. When this occurs, the problem becomes statically indeterminate since there are two unknowns and only one equation of equilibrium.

$$\xrightarrow{+} \ \Sigma F_x = 0; \qquad -F_A - F_B + 20(10^3)\,\text{N} = 0 \qquad (1)$$

Compatibility. The force P causes point B to move to B', with no further displacement. Therefore the compatibility condition for the rod is

$$\delta_{B/A} = 0.0002\,\text{m}$$

Load–Displacement. This displacement can be expressed in terms of the unknown reactions using the load–displacement relationship, Eq. 9–2, applied to segments AC and CB, Fig. 9–11c. Working in units of newtons and meters, we have

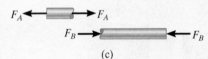

(b)

(c)

Fig. 9–11

$$\delta_{B/A} = \frac{F_A\,L_{AC}}{AE} - \frac{F_B\,L_{CB}}{AE} = 0.0002\,\text{m}$$

$$\frac{F_A(0.4\,\text{m})}{\pi(0.005\,\text{m})^2\,[200(10^9)\,\text{N/m}^2]}$$

$$- \frac{F_B\,(0.8\,\text{m})}{\pi(0.005\,\text{m})^2\,[200(10^9)\,\text{N/m}^2]} = 0.0002\,\text{m}$$

or

$$F_A\,(0.4\,\text{m}) - F_B\,(0.8\,\text{m}) = 3141.59\,\text{N} \cdot \text{m} \qquad (2)$$

Solving Eqs. 1 and 2 yields

$$F_A = 16.0\,\text{kN} \qquad F_B = 4.05\,\text{kN} \qquad\qquad\quad Ans.$$

Since the answer for F_B is *positive*, indeed end B contacts the wall at B' as originally assumed.

NOTE: If F_B were a negative quantity, the problem would be statically determinate, so that $F_B = 0$ and $F_A = 20$ kN.

EXAMPLE 9.6

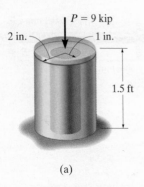

$P = 9$ kip

2 in. — 1 in.

1.5 ft

(a)

The aluminum post shown in Fig. 9–12a is reinforced with a brass core. If this assembly supports an axial compressive load of $P = 9$ kip, applied to the rigid cap, determine the average normal stress in the aluminum and the brass. Take $E_{al} = 10(10^3)$ ksi and $E_{br} = 15(10^3)$ ksi.

SOLUTION

Equilibrium. The free-body diagram of the post is shown in Fig. 9–12b. Here the resultant axial force at the base is represented by the unknown components carried by the aluminum, $\mathbf{F}_{al}$, and brass, $\mathbf{F}_{br}$. The problem is statically indeterminate.

Vertical force equilibrium requires

$$+\uparrow \Sigma F_y = 0; \qquad -9 \text{ kip} + F_{al} + F_{br} = 0 \qquad (1)$$

$P = 9$ kip

$\mathbf{F}_{br}$

$\mathbf{F}_{al}$

(b)

Compatibility. The rigid cap at the top of the post causes both the aluminum and brass to be displaced the *same amount*. Therefore,

$$\delta_{al} = \delta_{br}$$

Load–Displacement. Using the load–displacement relationships,

$$\frac{F_{al} L}{A_{al} E_{al}} = \frac{F_{br} L}{A_{br} E_{br}}$$

$$F_{al} = F_{br} \left(\frac{A_{al}}{A_{br}} \right) \left(\frac{E_{al}}{E_{br}} \right)$$

$$F_{al} = F_{br} \left[\frac{\pi[(2 \text{ in.})^2 - (1 \text{ in.})^2]}{\pi(1 \text{ in.})^2} \right] \left[\frac{10(10^3) \text{ ksi}}{15(10^3) \text{ ksi}} \right]$$

$$F_{al} = 2F_{br} \qquad (2)$$

Solving Eqs. 1 and 2 simultaneously yields

$$F_{al} = 6 \text{ kip} \qquad F_{br} = 3 \text{ kip}$$

The average normal stress in the aluminum and brass is therefore

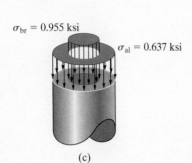

$\sigma_{br} = 0.955$ ksi

$\sigma_{al} = 0.637$ ksi

(c)

Fig. 9–12

$$\sigma_{al} = \frac{6 \text{ kip}}{\pi[(2 \text{ in.})^2 - (1 \text{ in.})^2]} = 0.637 \text{ ksi} \qquad Ans.$$

$$\sigma_{br} = \frac{3 \text{ kip}}{\pi(1 \text{ in.})^2} = 0.955 \text{ ksi} \qquad Ans.$$

NOTE: Using these results, the stress distributions within the materials are shown in Fig. 9–12c. Here the stiffer material, brass, is subjected to the larger stress.

EXAMPLE 9.7

The three A992 steel bars shown in Fig. 9–13a are pin connected to a *rigid* member. If the applied load on the member is 15 kN, determine the force developed in each bar. Bars *AB* and *EF* each have a cross-sectional area of 50 mm², and bar *CD* has a cross-sectional area of 30 mm².

SOLUTION

Equilibrium. The free-body diagram of the rigid member is shown in Fig. 9–13b. This problem is statically indeterminate since there are three unknowns and only two available equilibrium equations.

$$+\uparrow \Sigma F_y = 0; \qquad F_A + F_C + F_E - 15 \text{ kN} = 0 \qquad\qquad (1)$$

$$\zeta + \Sigma M_C = 0; \quad -F_A(0.4 \text{ m}) + 15 \text{ kN}(0.2 \text{ m}) + F_E(0.4 \text{ m}) = 0 \quad (2)$$

Compatibility. The applied load will cause the horizontal line *ACE* shown in Fig. 9–13c to move to the inclined position $A'C'E'$. The red displacements δ_A, δ_C, δ_E can be related by similar triangles. Thus the compatibility equation that relates these displacements is

$$\frac{\delta_A - \delta_E}{0.8 \text{ m}} = \frac{\delta_C - \delta_E}{0.4 \text{ m}}$$

$$\delta_C = \frac{1}{2}\delta_A + \frac{1}{2}\delta_E$$

Load–Displacement. Using the load–displacement relationship, Eq. 9–2, we have

$$\frac{F_C L}{(30 \text{ mm}^2)E_{st}} = \frac{1}{2}\left[\frac{F_A L}{(50 \text{ mm}^2)E_{st}}\right] + \frac{1}{2}\left[\frac{F_E L}{(50 \text{ mm}^2)E_{st}}\right]$$

$$F_C = 0.3F_A + 0.3F_E \qquad\qquad (3)$$

Solving Eqs. 1–3 simultaneously yields

$$F_A = 9.52 \text{ kN} \qquad\qquad Ans.$$

$$F_C = 3.46 \text{ kN} \qquad\qquad Ans.$$

$$F_E = 2.02 \text{ kN} \qquad\qquad Ans.$$

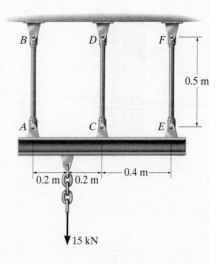

(a)

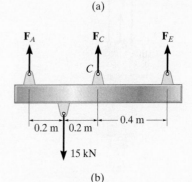

(b)

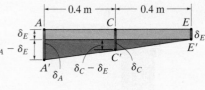

(c)

Fig. 9–13

9

EXAMPLE   9.8

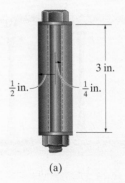

(a)

(b)

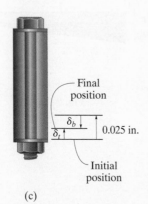

(c)

Fig. 9–14

The bolt shown in Fig. 9–14a is made of 2014-T6 aluminum alloy, and it passes through the cylindrical tube made of Am 1004-T61 magnesium alloy. The tube has an outer radius of $\frac{1}{2}$ in., and it is assumed that both the inner radius of the tube and the radius of the bolt are $\frac{1}{4}$ in. When the bolt is snug against the tube it produces negligible force in the tube. Using a wrench, the nut is then further tightened one-half turn. If the bolt has 20 threads per inch, determine the stress in the bolt.

SOLUTION

Equilibrium. The free-body diagram of a section of the bolt and the tube, Fig. 9–14b, is considered in order to relate the force in the bolt F_b to that in the tube, F_t. Equilibrium requires

$$+\uparrow \Sigma F_y = 0; \qquad\qquad F_b - F_t = 0 \qquad\qquad (1)$$

Compatibility. As noted in Fig. 9–14c, when the nut is tightened one-half turn on the bolt, it advances a distance of $\left(\frac{1}{2}\right)\left(\frac{1}{20}\text{ in.}\right) = 0.025$ in. This will cause the tube to shorten δ_t and the bolt to elongate δ_b. Thus, the compatibility of these displacements requires

$$(+\uparrow) \qquad\qquad \delta_t + \delta_b = 0.025 \text{ in.}$$

Load–Displacement. Taking the moduli of elasticity from the table on the inside back cover, and applying the load–displacement relationship, Eq. 9–2, yields

$$\frac{F_t\,(3\text{ in.})}{\pi[(0.5\text{ in.})^2 - (0.25\text{ in.})^2]\,[6.48(10^3)\text{ ksi}]}$$

$$+ \frac{F_b\,(3\text{ in.})}{\pi(0.25\text{ in.})^2\,[10.6(10^3)\text{ ksi}]} = 0.025 \text{ in.}$$

$$0.78595F_t + 1.4414F_b = 25 \qquad\qquad (2)$$

Solving Eqs. 1 and 2 simultaneously, we get

$$F_b = F_t = 11.22 \text{ kip}$$

The stresses in the bolt and tube are therefore

$$\sigma_b = \frac{F_b}{A_b} = \frac{11.22 \text{ kip}}{\pi(0.25\text{ in.})^2} = 57.2 \text{ ksi} \qquad\qquad Ans.$$

$$\sigma_t = \frac{F_t}{A_t} = \frac{11.22 \text{ kip}}{\pi[(0.5\text{ in.})^2 - (0.25\text{ in.})^2]} = 19.1 \text{ ksi}$$

These stresses are less than the reported yield stress for each material, $(\sigma_Y)_{al} = 60$ ksi and $(\sigma_Y)_{mg} = 22$ ksi (see the inside back cover), and therefore this "elastic" analysis is valid.

9.5 THE FORCE METHOD OF ANALYSIS FOR AXIALLY LOADED MEMBERS

It is also possible to solve statically indeterminate problems by writing the compatibility equation using the principle of superposition. This method of solution is often referred to as the *flexibility or force method of analysis*. To show how it is applied, consider again the bar in Fig. 9–15a. If we choose the support at B as "redundant" and *temporarily* remove it from the bar, then the bar will become statically determinate, as in Fig. 9–15b. Using the principle of superposition, however, we must add back the unknown redundant load $\mathbf{F}_B$, as shown in Fig. 9–15c.

Since the load $\mathbf{P}$ causes B to be displaced *downward* by an amount δ_P, the reaction $\mathbf{F}_B$ must displace end B of the bar *upward* by an amount δ_B, so that no displacement occurs at B when the two loadings are superimposed. Assuming displacements are positive downward, we have

$(+\downarrow)$ $\qquad\qquad 0 = \delta_P - \delta_B$

This condition of $\delta_P = \delta_B$ represents the compatibility equation for displacements at point B.

Applying the load–displacement relationship to each bar, we have $\delta_P = 500\,\text{N}(2\,\text{m})/AE$ and $\delta_B = F_B(5\,\text{m})/AE$. Consequently,

$$0 = \frac{500\,\text{N}(2\,\text{m})}{AE} - \frac{F_B(5\,\text{m})}{AE}$$

$$F_B = 200\,\text{N}$$

From the free-body diagram of the bar, Fig. 9–15d, equilibrium requires

$+\uparrow \Sigma F_y = 0; \qquad\qquad 200\,\text{N} + F_A - 500\,\text{N} = 0$

Then

$$F_A = 300\,\text{N}$$

These results are the same as those obtained in Sec. 9.4.

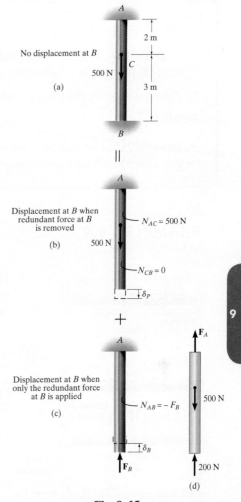

No displacement at B
(a)

Displacement at B when redundant force at B is removed
(b)

Displacement at B when only the redundant force at B is applied
(c)

Fig. 9–15

> ## PROCEDURE FOR ANALYSIS
>
> The force method of analysis requires the following steps.
>
> **Compatibility.**
>
> - Choose one of the supports as redundant and write the equation of compatibility. To do this, the known displacement at the redundant support, which is usually zero, is equated to the displacement at the support caused *only* by the external loads acting on the member *plus* (vectorially) the displacement at this support caused *only* by the redundant reaction acting on the member.

Load–Displacement.

- Express the external load and redundant displacements in terms of the loadings by using a load–displacement relationship, such as $\delta = NL/AE$.

- Once established, the compatibility equation can then be solved for the magnitude of the redundant force.

Equilibrium.

- Draw a free-body diagram and write the appropriate equations of equilibrium for the member using the calculated result for the redundant. Solve these equations for the other reactions.

EXAMPLE 9.9

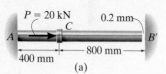

$P = 20$ kN C 0.2 mm

A B'

400 mm   800 mm

(a)

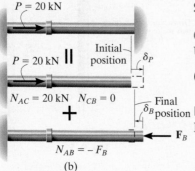

$P = 20$ kN

Initial position δ_P

$P = 20$ kN $\parallel$

$N_{AC} = 20$ kN $N_{CB} = 0$

Final δ_B position

$+$

$N_{AB} = -F_B$ F_B

(b)

F_A 20 kN   4.05 kN

(c)

Fig. 9–16

The A-36 steel rod shown in Fig. 9–16a has a diameter of 10 mm. It is fixed to the wall at A, and before it is loaded there is a gap between the wall and the rod of 0.2 mm. Determine the reactions at A and B'. Neglect the size of the collar at C. Take $E_{st} = 200$ GPa.

SOLUTION

Compatibility.   Here we will consider the support at B' as redundant. Using the principle of superposition, Fig. 9–16b, we have

$$(\overset{+}{\rightarrow}) \qquad\qquad 0.0002 \text{ m} = \delta_P - \delta_B \qquad\qquad (1)$$

Load–Displacement.   The deflections δ_P and δ_B are determined from Eq. 9–2.

$$\delta_P = \frac{N_{AC}L_{AC}}{AE} = \frac{[20(10^3)\text{ N}](0.4\text{ m})}{\pi(0.005\text{ m})^2\,[200(10^9)\text{ N/m}^2]} = 0.5093(10^{-3})\text{ m}$$

$$\delta_B = \frac{N_{AB}L_{AB}}{AE} = \frac{F_B\,(1.20\text{ m})}{\pi(0.005\text{ m})^2\,[200(10^9)\text{ N/m}^2]} = 76.3944(10^{-9})F_B$$

Substituting into Eq. 1, we get

$$0.0002 \text{ m} = 0.5093(10^{-3})\text{ m} - 76.3944(10^{-9})F_B$$

$$F_B = 4.05(10^3)\text{ N} = 4.05 \text{ kN} \qquad\qquad Ans.$$

Equilibrium.   From the free-body diagram, Fig. 9–16c,

$$\overset{+}{\rightarrow}\ \Sigma F_x = 0; \qquad -F_A + 20 \text{ kN} - 4.05 \text{ kN} = 0 \quad F_A = 16.0 \text{ kN} \qquad Ans.$$

PROBLEMS

9–31. The column is constructed from high-strength concrete and eight A992 steel reinforcing rods. If the column is subjected to an axial force of 200 kip, determine the average normal stress in the concrete and in each rod. Each rod has a diameter of 1 in.

*9–32. The column is constructed from high-strength concrete and eight A992 steel reinforcing rods. If the column is subjected to an axial force of 200 kip, determine the required diameter of each rod so that 60% of the axial force is carried by the concrete.

9–34. If column AB is made from high strength precast concrete and reinforced with four $\frac{3}{4}$ in. diameter A-36 steel rods, determine the average normal stress developed in the concrete and in each rod. Set $P = 75$ kip.

9–35. If column AB is made from high strength precast concrete and reinforced with four $\frac{3}{4}$ in. diameter A-36 steel rods, determine the maximum allowable floor loadings **P**. The allowable normal stresses for the concrete and the steel are $(\sigma_{allow})_{con} = 2.5$ ksi and $(\sigma_{allow})_{st} = 24$ ksi, respectively.

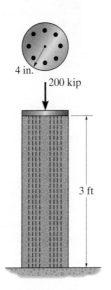

Probs. 9–31/32

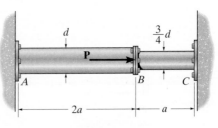

Probs. 9–34/35

*9–36. Determine the support reactions at the rigid supports A and C. The material has a modulus of elasticity of E.

9–37. If the supports at A and C are flexible and have a stiffness k, determine the support reactions at A and C. The material has a modulus of elasticity of E.

9–33. The A-36 steel pipe has a 6061-T6 aluminum core. It is subjected to a tensile force of 200 kN. Determine the average normal stress in the aluminum and the steel due to this loading. The pipe has an outer diameter of 80 mm and an inner diameter of 70 mm.

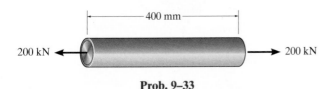

Prob. 9–33

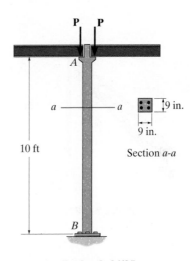

Probs. 9–36/37

9–38. The load of 2000 lb is to be supported by the two vertical steel wires for which $\sigma_Y = 70$ ksi. Originally wire AB is 60 in. long and wire AC is 60.04 in. long. Determine the force developed in each wire after the load is suspended. Each wire has a cross-sectional area of 0.02 in². $E_{st} = 29.0(10^3)$ ksi.

9–39. The load of 2000 lb is to be supported by the two vertical steel wires for which $\sigma_Y = 70$ ksi. Originally wire AB is 60 in. long and wire AC is 60.04 in. long. Determine the cross-sectional area of AB if the load is to be shared equally between both wires. Wire AC has a cross-sectional area of 0.02 in². $E_{st} = 29.0(10^3)$ ksi.

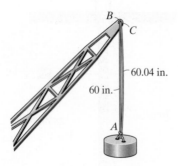

Probs. 9–38/39

9–41. The 10-mm-diameter steel bolt is surrounded by a bronze sleeve. The outer diameter of this sleeve is 20 mm, and its inner diameter is 10 mm. If the yield stress for the steel is $(\sigma_Y)_{st} = 640$ MPa, and for the bronze $(\sigma_Y)_{br} = 520$ MPa, determine the magnitude of the largest elastic load P that can be applied to the assembly. $E_{st} = 200$ GPa, $E_{br} = 100$ GPa.

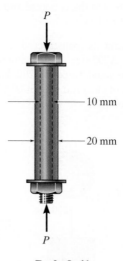

Prob. 9–41

9–42. The 10-mm-diameter steel bolt is surrounded by a bronze sleeve. The outer diameter of this sleeve is 20 mm, and its inner diameter is 10 mm. If the bolt is subjected to a compressive force of $P = 20$ kN, determine the average normal stress in the steel and the bronze. $E_{st} = 200$ GPa, $E_{br} = 100$ GPa.

*9–40. The A-36 steel pipe has an outer radius of 20 mm and an inner radius of 15 mm. If it fits snugly between the fixed walls before it is loaded, determine the reaction at the walls when it is subjected to the load shown.

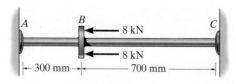

Prob. 9–40

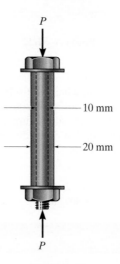

Prob. 9–42

9–43. The assembly consists of two red brass C83400 copper rods *AB* and *CD* of diameter 30 mm, a stainless 304 steel alloy rod *EF* of diameter 40 mm, and a rigid cap *G*. If the supports at *A*, *C*, and *F* are rigid, determine the average normal stress developed in the rods.

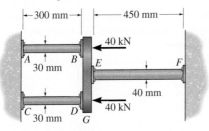

Prob. 9–43

*__9–44.__ The rigid beam is supported by the three suspender bars. Bars *AB* and *EF* are made of aluminum and bar *CD* is made of steel. If each bar has a cross-sectional area of 450 mm², determine the maximum value of *P* if the allowable stress is $(\sigma_{allow})_{st} = 200$ MPa for the steel and $(\sigma_{allow})_{al} = 150$ MPa for the aluminum. $E_{st} = 200$ GPa, $E_{al} = 70$ GPa.

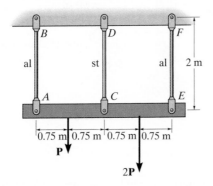

Prob. 9–44

9–45. The bolt *AB* has a diameter of 20 mm and passes through a sleeve that has an inner diameter of 40 mm and an outer diameter of 50 mm. The bolt and sleeve are made of A-36 steel and are secured to the rigid brackets as shown. If the bolt length is 220 mm and the sleeve length is 200 mm, determine the tension in the bolt when a force of 50 kN is applied to the brackets.

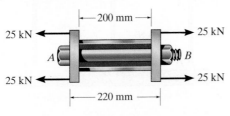

Prob. 9–45

9–46. If the gap between *C* and the rigid wall at *D* is initially 0.15 mm, determine the support reactions at *A* and *D* when the force *P* = 200 kN is applied. The assembly is made of solid A-36 steel cylinders.

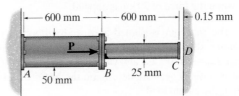

Prob. 9–46

9–47. The support consists of a solid red brass C83400 copper post surrounded by a 304 stainless steel tube. Before the load is applied the gap between these two parts is 1 mm. Given the dimensions shown, determine the greatest axial load that can be applied to the rigid cap *A* without causing yielding of any one of the materials.

Prob. 9–47

*__9–48.__ The specimen represents a filament-reinforced matrix system made from plastic (matrix) and glass (fiber). If there are *n* fibers, each having a cross-sectional area of A_f and modulus of E_f, embedded in a matrix having a cross-sectional area of A_m and modulus of E_m, determine the stress in the matrix and in each fiber when the force *P* is applied on the specimen.

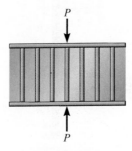

Prob. 9–48

9

9–49. The rigid bar is pinned at A and supported by two aluminum rods, each having a diameter of 1 in., a modulus of elasticity $E_{al} = 10(10^3)$ ksi, and yield stress of $(\sigma_Y)_{al} = 40$ ksi. If the bar is initially vertical, determine the displacement of the end B when the force of 20 kip is applied.

9–50. The rigid bar is pinned at A and supported by two aluminum rods, each having a diameter of 1 in., a modulus of elasticity $E_{al} = 10(10^3)$ ksi, and yield stress of $(\sigma_Y)_{al} = 40$ ksi. If the bar is initially vertical, determine the angle of tilt of the bar when the 20-kip load is applied.

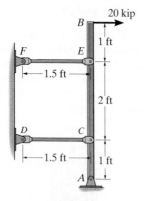

Probs. 9–49/50

9–51. The rigid bar is pinned at A and supported by two aluminum rods, each having a diameter of 1 in. and a modulus of elasticity $E_{al} = 10(10^3)$ ksi. If the bar is initially vertical, determine the displacement of the end B when the force of 2 kip is applied.

*9–52. The rigid bar is pinned at A and supported by two aluminum rods, each having a diameter of 1 in. and a modulus of elasticity $E_{al} = 10(10^3)$ ksi. If the bar is initially vertical, determine the force in each rod when the 2-kip load is applied.

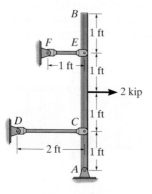

Probs. 9–51/52

9–53. The 2014-T6 aluminum rod AC is reinforced with the firmly bonded A992 steel tube BC. If the assembly fits snugly between the rigid supports so that there is no gap at C, determine the support reactions when the axial force of 400 kN is applied. The assembly is attached at D.

9–54. The 2014-T6 aluminum rod AC is reinforced with the firmly bonded A992 steel tube BC. When no load is applied to the assembly, the gap between end C and the rigid support is 0.5 mm. Determine the support reactions when the axial force of 400 kN is applied.

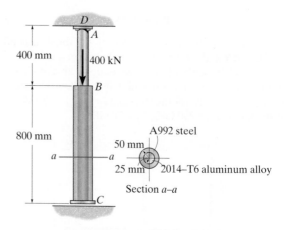

Probs. 9–53/54

9–55. The three suspender bars are made of A992 steel and have equal cross-sectional areas of 450 mm². Determine the average normal stress in each bar if the rigid beam is subjected to the loading shown.

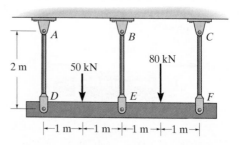

Prob. 9–55

9.6 THERMAL STRESS

A change in temperature can cause a body to change its dimensions. Generally, if the temperature increases, the body will expand, whereas if the temperature decreases, it will contract.* Ordinarily this expansion or contraction is *linearly* related to the temperature increase or decrease that occurs. If this is the case, and the material is homogeneous and isotropic, it has been found from experiment that the displacement of the end of a member having a length L can be calculated using the formula

$$\delta_T = \alpha \Delta T L \qquad (9\text{–}4)$$

Here

Most traffic bridges are designed with expansion joints to accommodate the thermal movement of the deck and thus avoid any thermal stress.

α = a property of the material, referred to as the ***linear coefficient of thermal expansion***. The units measure strain per degree of temperature. They are $1/°\text{F}$ (Fahrenheit) in the FPS system, and $1/°\text{C}$ (Celsius) or $1/\text{K}$ (Kelvin) in the SI system. Typical values are given on the inside back cover.

ΔT = the algebraic change in temperature of the member

L = the original length of the member

δ_T = the algebraic change in the length of the member

The change in length of a *statically determinate* member can easily be calculated using Eq. 9–4, since the member is free to expand or contract when it undergoes a temperature change. However, for a *statically indeterminate* member, these thermal displacements will be constrained by the supports, thereby producing ***thermal stresses*** that must be considered in design. Using the methods outlined in the previous sections, it is possible to determine these thermal stresses, as illustrated in the following examples.

Long extensions of ducts and pipes that carry fluids are subjected to variations in temperature that will cause them to expand and contract. Expansion joints, such as the one shown, are used to mitigate thermal stress in the material.

---

*There are some materials, like Invar, an iron-nickel alloy, and scandium trifluoride, that behave in the opposite way, but we will not consider these here.

EXAMPLE 9.10

0.5 in.

0.5 in.

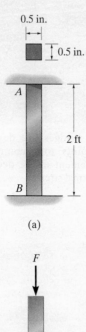

A

2 ft

B

(a)

F

F

(b)

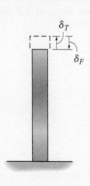

δ_T

δ_F

(c)

Fig. 9–17

The A-36 steel bar shown in Fig. 9–17a is constrained to just fit between two fixed supports when $T_1 = 60°F$. If the temperature is raised to $T_2 = 120°F$, determine the average normal thermal stress developed in the bar.

SOLUTION

Equilibrium. The free-body diagram of the bar is shown in Fig. 9–17b. Since there is no external load, the force at A is equal but opposite to the force at B; that is,

$$+\uparrow\Sigma F_y = 0; \qquad\qquad F_A = F_B = F$$

The problem is statically indeterminate since this force cannot be determined from equilibrium.

Compatibility. Since $\delta_{A/B} = 0$, the thermal displacement δ_T at A that occurs, Fig. 9–17c, is counteracted by the force **F** that is required to push the bar δ_F back to its original position. The compatibility condition at A becomes

$$(+\uparrow) \qquad\qquad \delta_{A/B} = 0 = \delta_T - \delta_F$$

Load–Displacement. Applying the thermal and load–displacement relationships, we have

$$0 = \alpha\Delta TL - \frac{FL}{AE}$$

Using the value of α on the inside back cover yields

$$\begin{aligned} F &= \alpha\Delta TAE \\ &= [6.60(10^{-6})/°F](120°F - 60°F)(0.5 \text{ in.})^2 [29(10^3) \text{ kip/in}^2] \\ &= 2.871 \text{ kip} \end{aligned}$$

Since **F** also represents the internal axial force within the bar, the average normal compressive stress is thus

$$\sigma = \frac{F}{A} = \frac{2.871 \text{ kip}}{(0.5 \text{ in.})^2} = 11.5 \text{ ksi} \qquad\qquad Ans.$$

NOTE: The magnitude of **F** indicates that changes in temperature can cause large reaction forces in statically indeterminate members.

EXAMPLE   9.11

The rigid beam shown in Fig. 9–18a is fixed to the top of the three posts made of A992 steel and 2014-T6 aluminum. The posts each have a length of 250 mm when no load is applied to the beam, and the temperature is $T_1 = 20°C$. Determine the force supported by each post if the bar is subjected to a uniform distributed load of 150 kN/m and the temperature is raised to $T_2 = 80°C$.

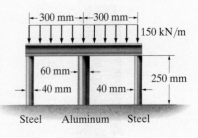

(a)

SOLUTION

Equilibrium.   The free-body diagram of the beam is shown in Fig. 9–18b. Moment equilibrium about the beam's center requires the forces in the steel posts to be equal. Summing forces on the free-body diagram, we have

$$+\uparrow \Sigma F_y = 0; \qquad\qquad 2F_{st} + F_{al} - 90(10^3) \text{ N} = 0 \qquad\qquad (1)$$

Compatibility.   Due to load, geometry, and material symmetry, the top of each post is displaced by an equal amount. Hence,

$$(+\downarrow) \qquad\qquad \delta_{st} = \delta_{al} \qquad\qquad (2)$$

The final position of the top of each post is equal to its displacement caused by the temperature increase, plus its displacement caused by the internal axial compressive force, Fig. 9–18c. Thus, for the steel and aluminum post, we have

$$(+\downarrow) \qquad\qquad \delta_{st} = -(\delta_{st})_T + (\delta_{st})_F$$
$$(+\downarrow) \qquad\qquad \delta_{al} = -(\delta_{al})_T + (\delta_{al})_F$$

Applying Eq. 2 gives

$$-(\delta_{st})_T + (\delta_{st})_F = -(\delta_{al})_T + (\delta_{al})_F$$

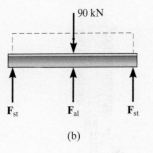

(b)

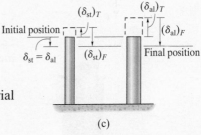

(c)

Fig. 9–18

Load–Displacement.   Using Eqs. 9–2 and 9–4 and the material properties on the inside back cover, we get

$$-[12(10^{-6})/°C](80°C - 20°C)(0.250 \text{ m}) + \frac{F_{st}(0.250 \text{ m})}{\pi(0.020 \text{ m})^2 [200(10^9) \text{ N/m}^2]}$$

$$= -[23(10^{-6})/°C](80°C - 20°C)(0.250 \text{ m}) + \frac{F_{al}(0.250 \text{ m})}{\pi(0.030 \text{ m})^2 [73.1(10^9) \text{ N/m}^2]}$$

$$F_{st} = 1.216F_{al} - 165.9(10^3) \qquad\qquad (3)$$

To be *consistent*, all numerical data has been expressed in terms of newtons, meters, and degrees Celsius. Solving Eqs. 1 and 3 simultaneously yields

$$F_{st} = -16.4 \text{ kN} \quad F_{al} = 123 \text{ kN} \qquad\qquad Ans.$$

The negative value for F_{st} indicates that this force acts opposite to that shown in Fig. 9–18b. In other words, the steel posts are in tension and the aluminum post is in compression.

EXAMPLE 9.12

150 mm

(a)

F_s

F_b

(b)

A 2014-T6 aluminum tube having a cross-sectional area of 600 mm² is used as a sleeve for an A-36 steel bolt having a cross-sectional area of 400 mm², Fig. 9–19a. When the temperature is $T_1 = 15°C$, the nut holds the assembly in a snug position such that the axial force in the bolt is negligible. If the temperature increases to $T_2 = 80°C$, determine the force in the bolt and sleeve.

SOLUTION

Equilibrium. The free-body diagram of a top segment of the assembly is shown in Fig. 9–19b. The forces F_b and F_s are produced since the sleeve has a higher coefficient of thermal expansion than the bolt, and therefore the sleeve will expand more when the temperature is increased. It is required that

$$+\uparrow \Sigma F_y = 0; \qquad\qquad F_s = F_b \qquad\qquad (1)$$

Compatibility. The temperature increase causes the sleeve and bolt to expand $(\delta_s)_T$ and $(\delta_b)_T$, Fig. 9–19c. However, the redundant forces F_b and F_s elongate the bolt and shorten the sleeve. Consequently, the end of the assembly reaches a final position, which is not the same as its initial position. Hence, the compatibility condition becomes

$$(+\downarrow) \qquad\qquad \delta = (\delta_b)_T + (\delta_b)_F = (\delta_s)_T - (\delta_s)_F$$

Load–Displacement. Applying Eqs. 9–2 and 9–4, and using the mechanical properties from the table on the inside back cover, we have

$$[12(10^{-6})/°C](80°C - 15°C)(0.150\ m) +$$

$$\frac{F_b(0.150\ m)}{(400\ mm^2)(10^{-6}\ m^2/mm^2)[200(10^9)\ N/m^2]}$$

$$= [23(10^{-6})/°C](80°C - 15°C)(0.150\ m)$$

$$- \frac{F_s(0.150\ m)}{(600\ mm^2)(10^{-6}\ m^2/mm^2)[73.1(10^9)\ N/m^2]}$$

Using Eq. 1 and solving gives

Initial position

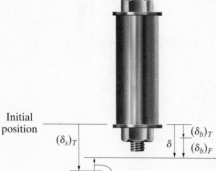

$(\delta_s)_T$

$(\delta_b)_T$

$(\delta_b)_F$

δ

Final position

$(\delta_s)_F$

(c)

Fig. 9–19

$$F_s = F_b = 20.3\ kN \qquad\qquad Ans.$$

NOTE: Since linear elastic material behavior was assumed in this analysis, the average normal stresses should be checked to make sure that they do not exceed the proportional limits for the material.

PROBLEMS

*9–56. The C83400-red-brass rod AB and 2014-T6-aluminum rod BC are joined at the collar B and fixed connected at their ends. If there is no load in the members when $T_1 = 50°F$, determine the average normal stress in each member when $T_2 = 120°F$. Also, how far will the collar be displaced? The cross-sectional area of each member is 1.75 in^2.

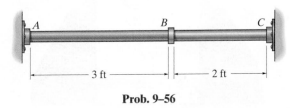

Prob. 9–56

9–57. The assembly has the diameters and material indicated. If it fits securely between its fixed supports when the temperature is $T_1 = 70°F$, determine the average normal stress in each material when the temperature reaches $T_2 = 110°F$.

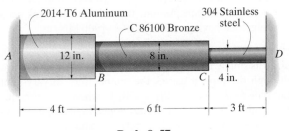

Prob. 9–57

9–58. The rod is made of A992 steel and has a diameter of 0.25 in. If the rod is 4 ft long when the springs are compressed 0.5 in. and the temperature of the rod is $T = 40°F$, determine the force in the rod when its temperature is $T = 160°F$.

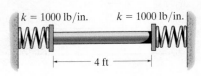

Prob. 9–58

9–59. The two cylindrical rod segments are fixed to the rigid walls such that there is a gap of 0.01 in. between them when $T_1 = 60°F$. What larger temperature T_2 is required in order to just close the gap? Each rod has a diameter of 1.25 in. Determine the average normal stress in each rod if $T_2 = 300°F$. Take $\alpha_{al} = 13(10^{-6})/°F$, $E_{al} = 10(10^3)$ ksi, $(\sigma_Y)_{al} = 40$ ksi, $\alpha_{cu} = 9.4(10^{-6})/°F$, $E_{cu} = 15(10^3)$ ksi, and $(\sigma_Y)_{cu} = 50$ ksi.

*9–60. The two cylindrical rod segments are fixed to the rigid walls such that there is a gap of 0.01 in. between them when $T_1 = 60°F$. Each rod has a diameter of 1.25 in. Determine the average normal stress in each rod if $T_2 = 400°F$, and also calculate the new length of the aluminum segment. Take $\alpha_{al} = 13(10^{-6})/°F$, $E_{al} = 10(10^3)$ ksi, $(\sigma_Y)_{al} = 40$ ksi, $\alpha_{cu} = 9.4(10^{-6})/°F$, $(\sigma_Y)_{cu} = 50$ ksi, and $E_{cu} = 15(10^3)$ ksi.

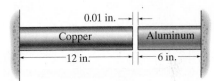

Probs. 9–59/60

9–61. The pipe is made of A992 steel and is connected to the collars at A and B. When the temperature is 60°F, there is no axial load in the pipe. If hot gas traveling through the pipe causes its temperature to rise by $\Delta T = (40 + 15x)°F$, where x is in feet, determine the average normal stress in the pipe. The inner diameter is 2 in., the wall thickness is 0.15 in.

9–62. The bronze C86100 pipe has an inner radius of 0.5 in. and a wall thickness of 0.2 in. If the gas flowing through it changes the temperature of the pipe uniformly from $T_A = 200°F$ at A to $T_B = 60°F$ at B, determine the axial force it exerts on the walls. The pipe was fitted between the walls when $T = 60°F$.

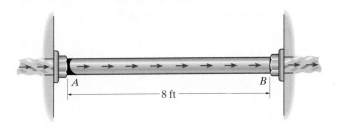

Probs. 9–61/62

9–63. The 40-ft-long A-36 steel rails on a train track are laid with a small gap between them to allow for thermal expansion. Determine the required gap δ so that the rails just touch one another when the temperature is increased from $T_1 = -20°F$ to $T_2 = 90°F$. Using this gap, what would be the axial force in the rails if the temperature rises to $T_3 = 110°F$? The cross-sectional area of each rail is 5.10 in².

Prob. 9–63

*9–64. The device is used to measure a change in temperature. Bars AB and CD are made of A-36 steel and 2014-T6 aluminum alloy, respectively. When the temperature is at 75°F, ACE is in the horizontal position. Determine the vertical displacement of the pointer at E when the temperature rises to 150°F.

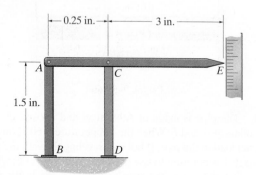

Prob. 9–64

9–65. The bar has a cross-sectional area A, length L, modulus of elasticity E, and coefficient of thermal expansion α. The temperature of the bar changes uniformly along its length from T_A at A to T_B at B so that at any point x along the bar $T = T_A + x(T_B - T_A)/L$. Determine the force the bar exerts on the rigid walls. Initially no axial force is in the bar and the bar has a temperature of T_A.

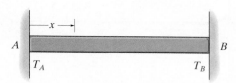

Prob. 9–65

9–66. When the temperature is at 30°C, the A-36 steel pipe fits snugly between the two fuel tanks. When fuel flows through the pipe, the temperatures at ends A and B rise to 130°C and 80°C, respectively. If the temperature drop along the pipe is linear, determine the average normal stress developed in the pipe. Assume each tank provides a rigid support at A and B.

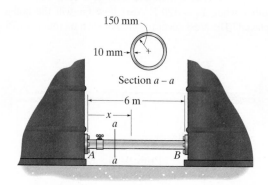

Prob. 9–66

9–67. When the temperature is at 30°C, the A-36 steel pipe fits snugly between the two fuel tanks. When fuel flows through the pipe, the temperatures at ends A and B rise to 130°C and 80°C, respectively. If the temperature drop along the pipe is linear, determine the average normal stress developed in the pipe. Assume the walls of each tank act as a spring, each having a stiffness of $k = 900$ MN/m.

*9–68. When the temperature is at 30°C, the A-36 steel pipe fits snugly between the two fuel tanks. When fuel flows through the pipe, it causes the temperature to vary along the pipe as $T = (\frac{5}{3}x^2 - 20x + 120)°C$, where x is in meters. Determine the normal stress developed in the pipe. Assume each tank provides a rigid support at A and B.

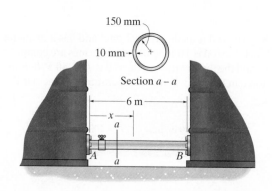

Probs. 9–67/68

9–69. The 50-mm-diameter cylinder is made from Am 1004-T61 magnesium and is placed in the clamp when the temperature is $T_1 = 20°$ C. If the 304-stainless-steel carriage bolts of the clamp each have a diameter of 10 mm, and they hold the cylinder snug with negligible force against the rigid jaws, determine the force in the cylinder when the temperature rises to $T_2 = 130°C$.

9–70. The 50-mm-diameter cylinder is made from Am 1004-T61 magnesium and is placed in the clamp when the temperature is $T_1 = 15°C$. If the two 304-stainless-steel carriage bolts of the clamp each have a diameter of 10 mm, and they hold the cylinder snug with negligible force against the rigid jaws, determine the temperature at which the average normal stress in either the magnesium or the steel first becomes 12 MPa.

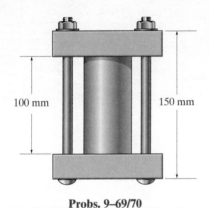

100 mm   150 mm

Probs. 9–69/70

9–71. The wires AB and AC are made of steel, and wire AD is made of copper. Before the 150-lb force is applied, AB and AC are each 60 in. long and AD is 40 in. long. If the temperature is increased by 80°F, determine the force in each wire needed to support the load. Each wire has a cross-sectional area of 0.0123 in². Take $E_{st} = 29(10^3)$ ksi, $E_{cu} = 17(10^3)$ ksi, $\alpha_{st} = 8(10^{-6})/°F$, $\alpha_{cu} = 9.60(10^{-6})/°F$.

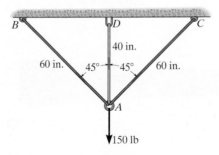

B   D   C
40 in.
60 in.   45°   45°   60 in.
A
150 lb

Prob. 9–71

*9–72. The cylinder CD of the assembly is heated from $T_1 = 30°C$ to $T_2 = 180°C$ using electrical resistance. At the lower temperature T_1 the gap between C and the rigid bar is 0.7 mm. Determine the force in rods AB and EF caused by the increase in temperature. Rods AB and EF are made of steel, and each has a cross-sectional area of 125 mm². CD is made of aluminum and has a cross-sectional area of 375 mm². $E_{st} = 200$ GPa, $E_{al} = 70$ GPa, and $\alpha_{al} = 23(10^{-6})/°C$.

9–73. The cylinder CD of the assembly is heated from $T_1 = 30°C$ to $T_2 = 180°C$ using electrical resistance. Also, the two end rods AB and EF are heated from $T_1 = 30°C$ to $T_2 = 50°C$. At the lower temperature T_1 the gap between C and the rigid bar is 0.7 mm. Determine the force in rods AB and EF caused by the increase in temperature. Rods AB and EF are made of steel, and each has a cross-sectional area of 125 mm². CD is made of aluminum and has a cross-sectional area of 375 mm². $E_{st} = 200$ GPa, $E_{al} = 70$ GPa, $\alpha_{st} = 12(10^{-6})/°C$, and $\alpha_{al} = 23(10^{-6})/°C$.

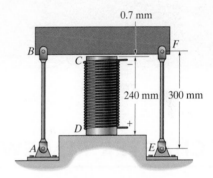

0.7 mm
B   F
C
240 mm   300 mm
D
A   E

Probs. 9–72/73

9–74. The metal strap has a thickness t and width w and is subjected to a temperature gradient T_1 to T_2 ($T_1 < T_2$). This causes the modulus of elasticity for the material to vary linearly from E_1 at the top to a smaller amount E_2 at the bottom. As a result, for any vertical position y, measured from the top surface, $E = [(E_2 - E_1)/w]y + E_1$. Determine the position d where the axial force P must be applied so that the bar stretches uniformly over its cross section.

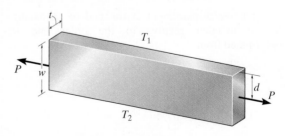

t
T_1
P w
d P
T_2

Prob. 9–74

CONCEPTUAL PROBLEMS

C9–1. The concrete footing A was poured when this column was put in place. Later the rest of the foundation stab was poured. Can you explain why the 45° cracks occured at each corner? Can you think of a better design that would avoid such cracks?

C9–2. The row of bricks, along with mortar and an internal steel reinforcing rod, was intended to serve as a lintel beam to support the bricks above this ventilation opening on an exterior wall of a building. Explain what may have caused the bricks to fail in the manner shown.

Prob. C9–1

Prob. C9–2

CHAPTER REVIEW

When a loading is applied at a point on a body, it tends to create a stress distribution within the body that becomes more uniformly distributed at regions removed from the point of application of the load. This is called Saint-Venant's principle.	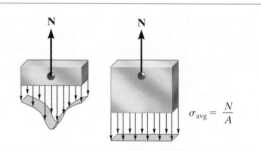 $\sigma_{avg} = \dfrac{N}{A}$
The relative displacement at the end of an axially loaded member relative to the other end is determined from $$\delta = \int_0^L \frac{N(x)dx}{AE}$$	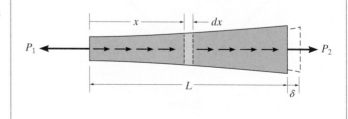

If a series of concentrated external axial forces are applied to a member and AE is also piecewise constant, then $$\delta = \sum \frac{NL}{AE}$$ For application, it is necessary to use a sign convention for the internal load N and displacement δ. We consider tension and elongation as positive values. Also, the material must not yield, but rather it must remain linear elastic.	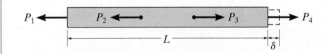
Superposition of load and displacement is possible provided the material remains linear elastic and no significant changes in the geometry of the member occur after loading.	
The reactions on a statically indeterminate bar can be determined using the equilibrium equations and compatibility conditions that specify the displacement at the supports. These displacements are related to the loads using a load–displacement relationship such as $\delta = NL/AE$.	
A change in temperature can cause a member made of homogeneous isotropic material to change its length by $$\delta = \alpha \Delta T L$$ If the member is confined, this change will produce thermal stress in the member.	
Holes and sharp transitions at a cross section will create stress concentrations. For the design of a member made of brittle material one obtains the stress concentration factor K from a graph, which has been determined from experiment. This value is then multiplied by the average stress to obtain the maximum stress at the cross section. $$\sigma_{max} = K \sigma_{avg}$$	

9

REVIEW PROBLEMS

R9–1. The assembly consists of two A992 steel bolts *AB* and *EF* and an 6061-T6 aluminum rod *CD*. When the temperature is at 30° C, the gap between the rod and rigid member *AE* is 0.1 mm. Determine the normal stress developed in the bolts and the rod if the temperature rises to 130° C. Assume *BF* is also rigid.

R9–3. The rods each have the same 25-mm diameter and 600-mm length. If they are made of A992 steel, determine the forces developed in each rod when the temperature increases by 50° C.

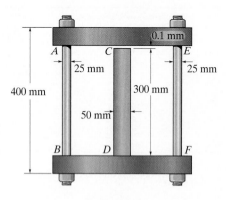

Prob. R9–1

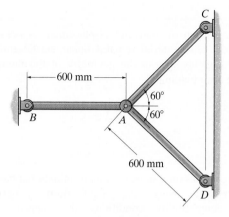

Prob. R9–3

R9–2. The assembly shown consists of two A992 steel bolts *AB* and *EF* and an 6061-T6 aluminum rod *CD*. When the temperature is at 30° C, the gap between the rod and rigid member *AE* is 0.1 mm. Determine the highest temperature to which the assembly can be raised without causing yielding either in the rod or the bolts. Assume *BF* is also rigid.

*R9–4. Two A992 steel pipes, each having a cross-sectional area of 0.32 in^2, are screwed together using a union at *B*. Originally the assembly is adjusted so that no load is on the pipe. If the union is then tightened so that its screw, having a lead of 0.15 in., undergoes two full turns, determine the average normal stress developed in the pipe. Assume that the union and couplings at *A* and *C* are rigid. Neglect the size of the union. *Note:* The lead would cause the pipe, when *unloaded*, to shorten 0.15 in. when the union is rotated one revolution.

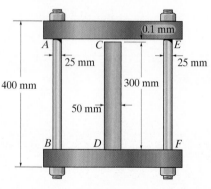

Prob. R9–2

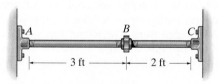

Prob. R9–4

R9–5. The 2014-T6 aluminum rod has a diameter of 0.5 in. and is lightly attached to the rigid supports at A and B when $T_1 = 70°F$. If the temperature becomes $T_2 = -10°F$, and an axial force of $P = 16$ lb is applied to the rigid collar as shown, determine the reactions at the rigid supports A and B.

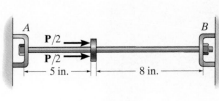

Prob. R9–5

R9–6. The 2014-T6 aluminum rod has a diameter of 0.5 in. and is lightly attached to the rigid supports at A and B when $T_1 = 70°F$. Determine the force P that must be applied to the collar so that, when $T = 0°F$, the reaction at B is zero.

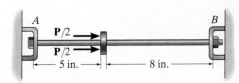

Prob. R9–6

R9–7. The rigid link is supported by a pin at A and two A-36 steel wires, each having an unstretched length of 12 in. and cross-sectional area of 0.0125 in². Determine the force developed in the wires when the link supports the vertical load of 350 lb.

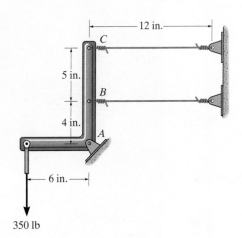

Prob. R9–7

*R9–8. The joint is made from three A992 steel plates that are bonded together at their seams. Determine the displacement of end A with respect to end B when the joint is subjected to the axial loads. Each plate has a thickness of 5 mm.

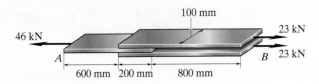

Prob. R9–8

CHAPTER 10

(© Jill Fromer/Getty Images)

The torsional stress and angle of twist of this soil auger depend upon the output of the machine turning the bit as well as the resistance of the soil in contact with the shaft.

TORSION

- To determine the torsional stress and deformation of an elastic circular shaft.

- To determine the support reactions on a statically indeterminate, torsionally loaded shaft when these reactions cannot be determined solely from the moment equilibrium equation.

10.1 TORSIONAL DEFORMATION OF A CIRCULAR SHAFT

Torque is a moment that tends to twist a member about its longitudinal axis. Its effect is of primary concern in the design of drive shafts used in vehicles and machinery, and for this reason it is important to be able to determine the stress and the deformation that occurs in a shaft when it is subjected to torsional loads.

We can physically illustrate what happens when a torque is applied to a circular shaft by considering the shaft to be made of a highly deformable material such as rubber. When the torque is applied, the longitudinal grid lines originally marked on the shaft, Fig. 10–1a, tend to distort into a helix, Fig. 10–1b, that intersects the circles at equal angles. Also, all the cross sections of the shaft will remain flat—that is, they do not warp or bulge in or out—and radial lines remain straight and rotate during this deformation. Provided the angle of twist is *small*, then the length of the shaft and its radius will remain practically unchanged.

10

If the shaft is fixed at one end and a torque is applied to its other end, then the dark green shaded plane in Fig. 10–2a will distort into a skewed form as shown. Here a radial line located on the cross section at a distance x from the fixed end of the shaft will rotate through an angle $\phi(x)$. This angle is called the ***angle of twist***. It depends on the position x and will vary along the shaft as shown.

In order to understand how this distortion strains the material, we will now isolate a small disk element located at x from the end of the shaft, Fig. 10–2b. Due to the deformation, the front and rear faces of the element will undergo rotation—the back face by $\phi(x)$, and the front face by $\phi(x) + d\phi$. As a result, the *difference* in these rotations, $d\phi$, causes the element to be subjected to a *shear strain*, γ (see Fig. 8–24b).

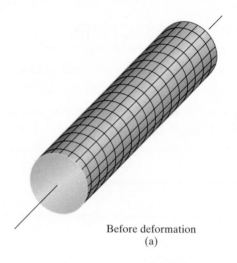

Before deformation
(a)

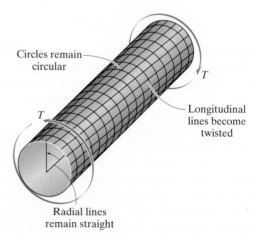

Circles remain circular

T

T

Longitudinal lines become twisted

Radial lines remain straight

After deformation
(b)

Fig. 10–1

Notice the deformation of the rectangular element when this rubber bar is subjected to a torque.

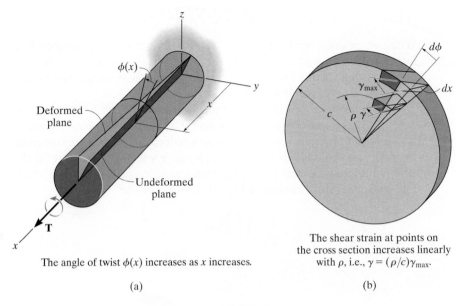

The angle of twist $\phi(x)$ increases as x increases.

(a)

The shear strain at points on the cross section increases linearly with ρ, i.e., $\gamma = (\rho/c)\gamma_{max}$.

(b)

Fig. 10–2

This angle (or shear strain) can be related to the angle $d\phi$ by noting that the length of the red arc in Fig. 10–2b is

$$\rho \, d\phi = dx \, \gamma$$

or

$$\gamma = \rho \, \frac{d\phi}{dx} \qquad\qquad (10\text{–}1)$$

Since dx and $d\phi$ are the same for all elements, then $d\phi/dx$ is constant over the cross section, and Eq. 10–1 states that the magnitude of the shear strain varies only with its radial distance ρ from the axis of the shaft. Since $d\phi/dx = \gamma/\rho = \gamma_{max}/c$, then

$$\gamma = \left(\frac{\rho}{c}\right)\gamma_{max} \qquad\qquad (10\text{–}2)$$

In other words, the shear strain within the shaft *varies linearly* along any radial line, from zero at the axis of the shaft to a maximum γ_{max} at its outer boundary, Fig. 10–2b.

10.2   THE TORSION FORMULA

When an external torque is applied to a shaft, it creates a corresponding internal torque within the shaft. In this section, we will develop an equation that relates this internal torque to the shear stress distribution acting on the cross section of the shaft.

If the material is linear elastic, then Hooke's law applies, $\tau = G\gamma$, or $\tau_{max} = G\gamma_{max}$, and consequently a *linear variation in shear strain*, as noted in the previous section, leads to a corresponding *linear variation in shear stress* along any radial line. Hence, τ will vary from zero at the shaft's longitudinal axis to a maximum value, τ_{max}, at its outer surface, Fig. 10–3. Therefore, similar to Eq. 10–2, we can write

$$\tau = \left(\frac{\rho}{c}\right)\tau_{max} \tag{10–3}$$

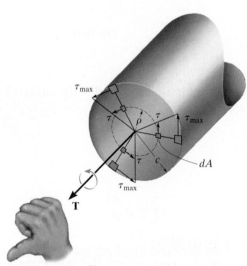

Shear stress varies linearly along
each radial line of the cross section.

Fig. 10–3

Since each element of area dA, located at ρ, is subjected to a force of $dF = \tau \, dA$, Fig. 10–3, the torque produced by this force is then $dT = \rho(\tau \, dA)$. For the entire cross section we have

$$T = \int_A \rho(\tau \, dA) = \int_A \rho\left(\frac{\rho}{c}\right)\tau_{max} \, dA \qquad (10\text{–}4)$$

However, τ_{max}/c is constant, and so

$$T = \frac{\tau_{max}}{c} \int_A \rho^2 \, dA \qquad (10\text{–}5)$$

The integral represents the **polar moment of inertia** of the shaft's cross-sectional area about the shaft's longitudinal axis. On the next page we will calculate its value, but here we will symbolize its value as J. As a result, the above equation can be rearranged and written in a more compact form, namely,

$$\boxed{\tau_{max} = \frac{Tc}{J}} \qquad (10\text{–}6)$$

Here

τ_{max} = the maximum shear stress in the shaft, which occurs at its outer surface

T = the resultant *internal torque* acting at the cross section. Its value is determined from the method of sections and the equation of moment equilibrium applied about the shaft's longitudinal axis

J = the polar moment of inertia of the cross-sectional area

c = the outer radius of the shaft

If Eq. 10–6 is substituted in Eq. 10–3, the shear stress at the intermediate distance ρ on the cross section can be determined.

$$\boxed{\tau = \frac{T\rho}{J}} \qquad (10\text{–}7)$$

Either of the above two equations is often referred to as the **torsion formula**. Recall that it is used only if the shaft has a circular cross section and the material is homogeneous and behaves in a linear elastic manner, since the derivation of Eq. 10–3 is based on Hooke's law.

The shaft attached to the center of this wheel is subjected to a torque, and the maximum stress it creates must be resisted by the shaft to prevent failure.

10

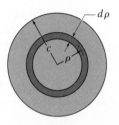

Fig. 10–4

Polar Moment of Inertia. If the shaft has a *solid* circular cross section, the polar moment of inertia J can be determined using an area element in the form of a *differential ring* or annulus having a thickness $d\rho$ and circumference $2\pi\rho$, Fig. 10–4. For this ring, $dA = 2\pi\rho \, d\rho$, and so

$$J = \int_A \rho^2 \, dA = \int_0^c \rho^2 (2\pi\rho \, d\rho)$$

$$= 2\pi \int_0^c \rho^3 \, d\rho = 2\pi \left(\frac{1}{4}\right)\rho^4 \Big|_0^c$$

$$\boxed{J = \frac{\pi}{2} c^4} \qquad (10\text{–}8)$$

Solid Section

Note that J is always positive. Common units used for its measurement are mm^4 or in^4.

If a shaft has a tubular cross section, with inner radius c_i and outer radius c_o, Fig. 10–5, then from Eq. 10–8 we can determine its polar moment of inertia by subtracting J for a shaft of radius c_i from that determined for a shaft of radius c_o. The result is

$$\boxed{J = \frac{\pi}{2} (c_o^4 - c_i^4)} \qquad (10\text{–}9)$$

Tube

Fig. 10–5

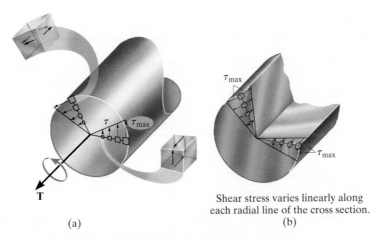

(a)

Shear stress varies linearly along
each radial line of the cross section.

(b)

Fig. 10–6

Shear Stress Distribution. If an element of material on the cross section of the shaft or tube is isolated, then due to the complementary property of shear, equal shear stresses must also act on four of its adjacent faces, as shown in Fig. 10–6a. As a result, *the internal torque T develops a linear distribution of shear stress along each radial line in the plane of the cross-sectional area, and also an associated shear-stress distribution is developed along an axial plane*, Fig. 10–6b. It is interesting to note that because of this axial distribution of shear stress, shafts made of wood tend to *split* along the axial plane when subjected to excessive torque, Fig. 10–7. This is because wood is an anisotropic material, whereby its shear resistance parallel to its grains or fibers, directed along the axis of the shaft, is much less than its resistance perpendicular to the fibers within the plane of the cross section.

The tubular drive shaft for a truck was subjected to an excessive torque, resulting in failure caused by yielding of the material. Engineers deliberately design drive shafts to fail before torsional damage can occur to parts of the engine or transmission.

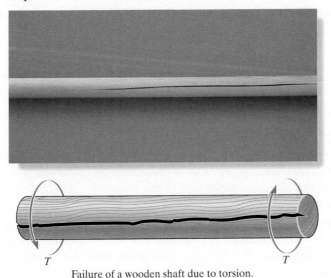

Failure of a wooden shaft due to torsion.

Fig. 10–7

IMPORTANT POINTS

- When a shaft having a *circular cross section* is subjected to a torque, the cross section *remains plane* while radial lines rotate. This causes a *shear strain* within the material that *varies linearly* along any radial line, from zero at the axis of the shaft to a maximum at its outer boundary.

- For linear elastic homogeneous material, the *shear stress* along any radial line of the shaft also *varies linearly*, from zero at its axis to a maximum at its outer boundary. This maximum shear stress *must not* exceed the proportional limit.

- Due to the complementary property of shear, the linear shear stress distribution within the plane of the cross section is also distributed along an adjacent axial plane of the shaft.

- The torsion formula is based on the requirement that the resultant torque on the cross section is equal to the torque produced by the shear stress distribution about the longitudinal axis of the shaft. It is required that the shaft or tube have a *circular* cross section and that it is made of *homogeneous* material which has *linear elastic* behavior.

PROCEDURE FOR ANALYSIS

The torsion formula can be applied using the following procedure.

Internal Torque.

- Section the shaft perpendicular to its axis at the point where the shear stress is to be determined, and use the necessary free-body diagram and equations of equilibrium to obtain the internal torque at the section.

Section Property.

- Calculate the polar moment of inertia of the cross-sectional area. For a solid section of radius c, $J = \pi c^4/2$, and for a tube of outer radius c_o and inner radius c_i, $J = \pi(c_o^4 - c_i^4)/2$.

Shear Stress.

- Specify the radial distance ρ, measured from the center of the cross section to the point where the shear stress is to be found. Then apply the torsion formula $\tau = T\rho/J$, or if the maximum shear stress is to be determined use $\tau_{\max} = Tc/J$. When substituting the data, make sure to use a consistent set of units.

- The shear stress acts on the cross section in a direction that is always perpendicular to ρ. The force it creates must contribute a torque about the axis of the shaft that is in the *same direction* as the internal resultant torque **T** acting on the section. Once this direction is established, a volume element located at the point where τ is determined can be isolated, and the direction of τ acting on the remaining three adjacent faces of the element can be shown.

EXAMPLE 10.1

The solid shaft and tube shown in Fig. 10–8 are made of a material having an allowable shear stress of 75 MPa. Determine the maximum torque that can be applied to each cross section, and show the stress acting on a small element of material at point A of the shaft, and points B and C of the tube.

SOLUTION

Section Properties.   The polar moments of inertia for the solid and tubular shafts are

$$J_s = \frac{\pi}{2} c^4 = \frac{\pi}{2} (0.1 \text{ m})^4 = 0.1571(10^{-3}) \text{ m}^4$$

$$J_t = \frac{\pi}{2} (c_o^4 - c_i^4) = \frac{\pi}{2} \left[(0.1 \text{ m})^4 - (0.075 \text{ m})^4 \right] = 0.1074(10^{-3}) \text{ m}^4$$

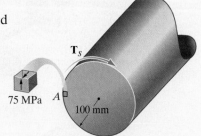

Shear Stress.   The maximum torque in each case is

$$(\tau_{max})_s = \frac{Tc}{J}; \qquad 75(10^6) \text{ N/m}^2 = \frac{T_s(0.1 \text{ m})}{0.1571(10^{-3}) \text{ m}^4}$$

$$T_s = 118 \text{ kN} \cdot \text{m} \qquad\qquad Ans.$$

$$(\tau_{max})_t = \frac{Tc}{J}; \qquad 75(10^6) \text{ N/m}^2 = \frac{T_t(0.1 \text{ m})}{0.1074(10^{-3}) \text{ m}^4}$$

$$T_t = 80.5 \text{ kN} \cdot \text{m} \qquad\qquad Ans.$$

Also, the shear stress at the inner radius of the tube is

$$(\tau_i)_t = \frac{80.5(10^3) \text{ N} \cdot \text{m} (0.075 \text{ m})}{0.1074(10^{-3}) \text{ m}^4} = 56.2 \text{ MPa}$$

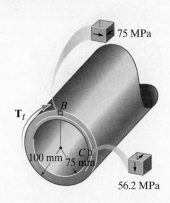

Fig. 10–8

These results are shown acting on small elements in Fig. 10–8. Notice how the shear stress on the front (shaded) face of the element contributes to the torque. As a consequence, shear stress components act on the other three faces. No shear stress acts on the outer surface of the shaft or tube or on the inner surface of the tube because it must be stress free.

EXAMPLE 10.2

42.5 kip·in.

30 kip·in.

12.5 kip·in.

a

a

(a)

The 1.5-in.-diameter shaft shown in Fig. 10–9a is supported by two bearings and is subjected to three torques. Determine the shear stress developed at points A and B, located at section a–a of the shaft, Fig. 10–9c.

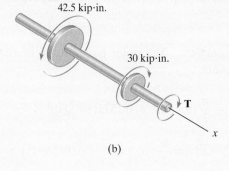

42.5 kip·in.

30 kip·in.

T

x

(b)

A

18.9 ksi

12.5 kip·in.

B

3.77 ksi

0.75 in.   0.15 in.

x

(c)

Fig. 10–9

SOLUTION

Internal Torque. Since the bearing reactions do not offer resistance to shaft rotation, the applied torques satisfy moment equilibrium about the shaft's axis.

The internal torque at section a–a will be determined from the free-body diagram of the left segment, Fig. 10–9b. We have

$$\Sigma M_x = 0; \quad 42.5 \text{ kip·in.} - 30 \text{ kip·in.} - T = 0 \quad T = 12.5 \text{ kip·in.}$$

Section Property. The polar moment of inertia for the shaft is

$$J = \frac{\pi}{2}(0.75 \text{ in.})^4 = 0.497 \text{ in}^4$$

Shear Stress. Since point A is at $\rho = c = 0.75$ in.,

$$\tau_A = \frac{Tc}{J} = \frac{(12.5 \text{ kip·in.})(0.75 \text{ in.})}{(0.497 \text{ in}^4)} = 18.9 \text{ ksi} \qquad Ans.$$

Likewise for point B, at $\rho = 0.15$ in., we have

$$\tau_B = \frac{T\rho}{J} = \frac{(12.5 \text{ kip·in.})(0.15 \text{ in.})}{(0.497 \text{ in}^4)} = 3.77 \text{ ksi} \qquad Ans.$$

NOTE: The directions of these stresses on each element at A and B, Fig. 10–9c, are established on the planes of each of these elements, so that they match the required clockwise torque.

EXAMPLE 10.3

The pipe shown in Fig. 10–10a has an inner radius of 40 mm and an outer radius of 50 mm. If its end is tightened against the support at A using the torque wrench, determine the shear stress developed in the material at the inner and outer walls along the central portion of the pipe.

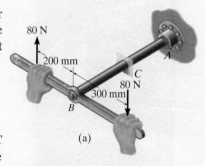

(a)

SOLUTION

Internal Torque. A section is taken at the intermediate location C along the pipe's axis, Fig. 10–10b. The only unknown at the section is the internal torque **T**. We require

$$\Sigma M_x = 0; \qquad 80 \text{ N}(0.3 \text{ m}) + 80 \text{ N}(0.2 \text{ m}) - T = 0$$

$$T = 40 \text{ N} \cdot \text{m}$$

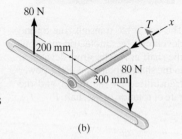

(b)

Section Property. The polar moment of inertia for the pipe's cross-sectional area is

$$J = \frac{\pi}{2} \left[(0.05 \text{ m})^4 - (0.04 \text{ m})^4 \right] = 5.796 (10^{-6}) \text{ m}^4$$

Shear Stress. For any point lying on the outside surface of the pipe, $\rho = c_o = 0.05$ m, we have

$$\tau_o = \frac{Tc_o}{J} = \frac{40 \text{ N} \cdot \text{m}(0.05 \text{ m})}{5.796(10^{-6}) \text{ m}^4} = 0.345 \text{ MPa} \qquad Ans.$$

And for any point located on the inside surface, $\rho = c_i = 0.04$ m, and so

$$\tau_i = \frac{Tc_i}{J} = \frac{40 \text{ N} \cdot \text{m}(0.04 \text{ m})}{5.796(10^{-6}) \text{ m}^4} = 0.276 \text{ MPa} \qquad Ans.$$

The results are shown on two small elements in Fig. 10–10c.

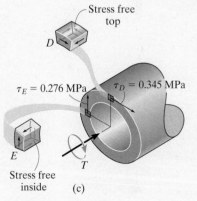

(c)

Fig. 10–10

NOTE: Since the top face of D and the inner face of E are in stress-free regions, no shear stress can exist on these faces or on the other corresponding faces of the elements.

10.3   POWER TRANSMISSION

Shafts and tubes having circular cross sections are often used to transmit power developed by a machine. When used for this purpose, they are subjected to a torque that depends on both the power generated by the machine and the angular speed of the shaft. *Power* is defined as the work performed per unit of time. Also, the work transmitted by a rotating shaft equals the torque applied times the angle of rotation. Therefore, if during an instant of time dt an applied torque $\mathbf{T}$ causes the shaft to rotate $d\theta$, then the work done is $T\,d\theta$ and the instantaneous power is

$$P = \frac{T\,d\theta}{dt}$$

Since the shaft's angular velocity is $\omega = d\theta/dt$, then the power is

$$\boxed{P = T\omega} \tag{10–10}$$

The belt drive transmits the torque developed by an electric motor to the shaft at A. The stress developed in the shaft depends upon the power transmitted by the motor and the rate of rotation of the shaft. $P = T\omega$.

In the SI system, power is expressed in *watts* when torque is measured in newton-meters ($N \cdot m$) and ω is in radians per second (rad/s) ($1\,W = 1\,N \cdot m/s$). In the FPS system, the basic units of power are foot-pounds per second ($ft \cdot lb/s$); however, *horsepower* (hp) is often used in engineering practice, where

$$1\,hp = 550\,ft \cdot lb/s$$

For machinery, the *frequency* of a shaft's rotation, f, is often reported. This is a measure of the number of revolutions or "cycles" the shaft makes per second and is expressed in hertz ($1\,Hz = 1\,cycle/s$). Since $1\,cycle = 2\pi\,rad$, then $\omega = 2\pi f$, and so the above equation for power can also be written as

$$\boxed{P = 2\pi f T} \tag{10–11}$$

Shaft Design.   When the power transmitted by a shaft and its frequency of rotation are known, the torque developed in the shaft can be determined from Eq. 10–11, that is, $T = P/2\pi f$. Knowing T and the allowable shear stress for the material, τ_{allow}, we can then determine the size of the shaft's cross section using the torsion formula. Specifically, the design or geometric parameter J/c becomes

$$\frac{J}{c} = \frac{T}{\tau_{allow}} \tag{10–12}$$

For a *solid shaft*, $J = (\pi/2)c^4$, and thus, upon substitution, a *unique value* for the shaft's radius c can be determined. If the shaft is *tubular*, so that $J = (\pi/2)(c_o^4 - c_i^4)$, design permits a wide range of possibilities for the solution. This is because an *arbitrary choice* can be made for either c_o or c_i and the other radius can then be determined from Eq. 10–12.

EXAMPLE 10.4

A solid steel shaft AB, shown in Fig. 10–11, is to be used to transmit 5 hp from the motor M to which it is attached. If the shaft rotates at $\omega = 175$ rpm and the steel has an allowable shear stress of $\tau_{allow} = 14.5$ ksi, determine the required diameter of the shaft to the nearest $\frac{1}{8}$ in.

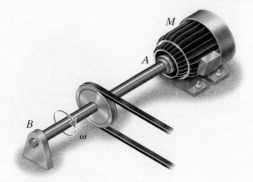

Fig. 10–11

SOLUTION

The torque on the shaft is determined from Eq. 10–10, that is, $P = T\omega$. Expressing P in foot-pounds per second and ω in radians/second, we have

$$P = 5 \text{ hp}\left(\frac{550 \text{ ft} \cdot \text{lb/s}}{1 \text{ hp}}\right) = 2750 \text{ ft} \cdot \text{lb/s}$$

$$\omega = \frac{175 \text{ rev}}{\text{min}}\left(\frac{2\pi \text{ rad}}{1 \text{ rev}}\right)\left(\frac{1 \text{ min}}{60 \text{ s}}\right) = 18.33 \text{ rad/s}$$

Thus,

$$P = T\omega; \qquad 2750 \text{ ft} \cdot \text{lb/s} = T(18.33 \text{ rad/s})$$

$$T = 150.1 \text{ ft} \cdot \text{lb}$$

Applying Eq. 10–12,

$$\frac{J}{c} = \frac{\pi}{2}\frac{c^4}{c} = \frac{T}{\tau_{allow}}$$

$$c = \left(\frac{2T}{\pi\tau_{allow}}\right)^{1/3} = \left(\frac{2(150.1 \text{ ft} \cdot \text{lb})(12 \text{ in./ft})}{\pi(14\,500 \text{ lb/in}^2)}\right)^{1/3}$$

$$c = 0.429 \text{ in.}$$

Since $2c = 0.858$ in., select a shaft having a diameter of

$$d = \frac{7}{8} \text{ in.} = 0.875 \text{ in.} \qquad \qquad Ans.$$

PRELIMINARY PROBLEMS

P10–1. Determine the internal torque at each section and show the shear stress on differential volume elements located at A, B, C, and D.

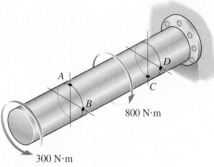

800 N·m

300 N·m

Prob. P10–1

P10–2. Determine the internal torque at each section and show the shear stress on differential volume elements located at A, B, C, and D.

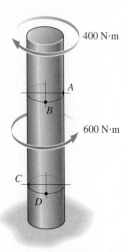

400 N·m

600 N·m

Prob. P10–2

P10–3. The solid and hollow shafts are each subjected to the torque T. In each case, sketch the shear stress distribution along the two radial lines.

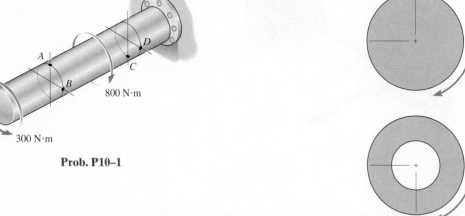

T

T

Prob. P10–3

P10–4. The motor delivers 10 hp to the shaft. If it rotates at 1200 rpm, detemine the torque produced by the motor.

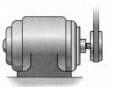

Prob. P10–4

FUNDAMENTAL PROBLEMS

F10–1. The solid circular shaft is subjected to an internal torque of $T = 5\,\text{kN}\cdot\text{m}$. Determine the shear stress at points A and B. Represent each state of stress on a volume element.

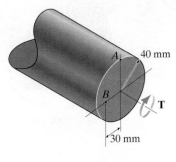

Prob. F10–1

F10–2. The hollow circular shaft is subjected to an internal torque of $T = 10\,\text{kN}\cdot\text{m}$. Determine the shear stress at points A and B. Represent each state of stress on a volume element.

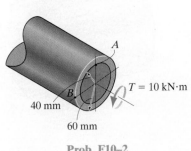

Prob. F10–2

F10–3. The shaft is hollow from A to B and solid from B to C. Determine the maximum shear stress in the shaft. The shaft has an outer diameter of 80 mm, and the thickness of the wall of the hollow segment is 10 mm.

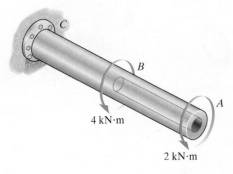

Prob. F10–3

F10–4. Determine the maximum shear stress in the 40-mm-diameter shaft.

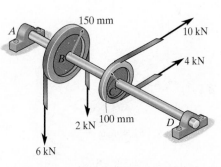

Prob. F10–4

F10–5. Determine the maximum shear stress in the shaft at section a–a.

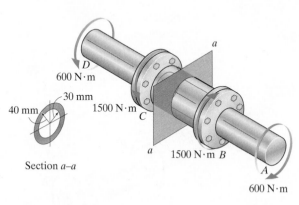

Prob. F10–5

F10–6. Determine the shear stress at point A on the surface of the shaft. Represent the state of stress on a volume element at this point. The shaft has a radius of 40 mm.

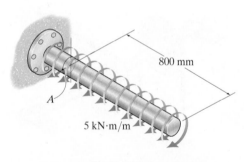

Prob. F10–6

F10–7. The solid 50-mm-diameter shaft is subjected to the torques applied to the gears. Determine the absolute maximum shear stress in the shaft.

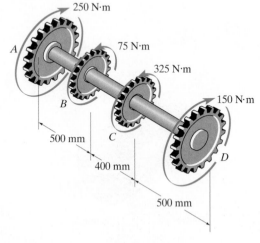

Prob. F10–7

F10–8. The gear motor can develop 3 hp when it turns at 150 rev/min. If the allowable shear stress for the shaft is $\tau_{allow} = 12$ ksi, determine the smallest diameter of the shaft to the nearest $\frac{1}{8}$ in. that can be used.

Prob. F10–8

PROBLEMS

10–1. The solid shaft of radius r is subjected to a torque **T**. Determine the radius r' of the inner core of the shaft that resists one-half of the applied torque $(T/2)$. Solve the problem two ways: (a) by using the torsion formula, (b) by finding the resultant of the shear-stress distribution.

10–2. The solid shaft of radius r is subjected to a torque **T**. Determine the radius r' of the inner core of the shaft that resists one-quarter of the applied torque $(T/4)$. Solve the problem two ways: (a) by using the torsion formula, (b) by finding the resultant of the shear-stress distribution.

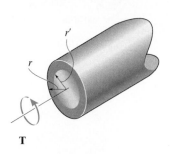

Probs. 10–1/2

10–3. A shaft is made of an aluminum alloy having an allowable shear stress of $\tau_{\text{allow}} = 100$ MPa. If the diameter of the shaft is 100 mm, determine the maximum torque **T** that can be transmitted. What would be the maximum torque **T'** if a 75-mm-diameter hole were bored through the shaft? Sketch the shear-stress distribution along a radial line in each case.

Prob. 10–3

*****10–4.** The copper pipe has an outer diameter of 40 mm and an inner diameter of 37 mm. If it is tightly secured to the wall and three torques are applied to it, determine the absolute maximum shear stress developed in the pipe.

Prob. 10–4

10–5. The copper pipe has an outer diameter of 2.50 in. and an inner diameter of 2.30 in. If it is tightly secured to the wall and three torques are applied to it, determine the shear stress developed at points A and B. These points lie on the pipe's outer surface. Sketch the shear stress on volume elements located at A and B.

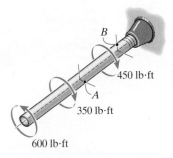

Prob. 10–5

10–6. The solid aluminum shaft has a diameter of 50 mm and an allowable shear stress of $\tau_{allow} = 60$ MPa. Determine the largest torque T_1 that can be applied to the shaft if it is also subjected to the other torsional loadings. It is required that T_1 act in the direction shown. Also, determine the maximum shear stress within regions CD and DE.

10–7. The solid aluminum shaft has a diameter of 50 mm. Determine the absolute maximum shear stress in the shaft and sketch the shear-stress distribution along a radial line of the shaft where the shear stress is maximum. Set $T_1 = 2000$ N $\cdot$ m.

10–9. The solid shaft is fixed to the support at C and subjected to the torsional loadings. Determine the shear stress at points A and B on the surface, and sketch the shear stress on volume elements located at these points.

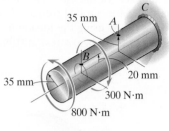

Prob. 10–9

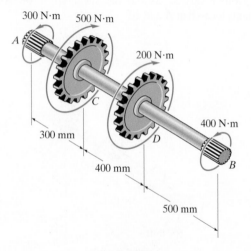

Probs. 10–6/7

*10–8. The solid 30-mm-diameter shaft is used to transmit the torques applied to the gears. Determine the absolute maximum shear stress in the shaft.

10–10. The link acts as part of the elevator control for a small airplane. If the attached aluminum tube has an inner diameter of 25 mm and a wall thickness of 5 mm, determine the maximum shear stress in the tube when the cable force of 600 N is applied to the cables. Also, sketch the shear-stress distribution over the cross section.

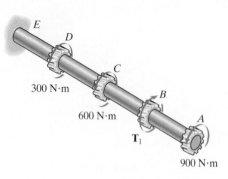

Prob. 10–8

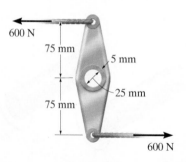

Prob. 10–10

10–11. The assembly consists of two sections of galvanized steel pipe connected together using a reducing coupling at *B*. The smaller pipe has an outer diameter of 0.75 in. and an inner diameter of 0.68 in., whereas the larger pipe has an outer diameter of 1 in. and an inner diameter of 0.86 in. If the pipe is tightly secured to the wall at *C*, determine the maximum shear stress in each section of the pipe when the couple is applied to the handles of the wrench.

10–14. A steel tube having an outer diameter of 2.5 in. is used to transmit 9 hp when turning at 27 rev/min. Determine the inner diameter *d* of the tube to the nearest $\frac{1}{8}$ in. if the allowable shear stress is $\tau_{\text{allow}} = 10$ ksi.

Prob. 10–14

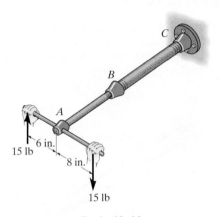

Prob. 10–11

10–12. The shaft has an outer diameter of 100 mm and an inner diameter of 80 mm. If it is subjected to the three torques, determine the absolute maximum shear stress in the shaft. The smooth bearings *A* and *B* do not resist torque.

10–13. The shaft has an outer diameter of 100 mm and an inner diameter of 80 mm. If it is subjected to the three torques, plot the shear stress distribution along a radial line for the cross section within region *CD* of the shaft. The smooth bearings at *A* and *B* do not resist torque.

10–15. If the gears are subjected to the torques shown, determine the maximum shear stress in the segments *AB* and *BC* of the A-36 steel shaft. The shaft has a diameter of 40 mm.

10–16. If the gears are subjected to the torques shown, determine the required diameter of the A-36 steel shaft to the nearest mm if $\tau_{\text{allow}} = 60$ MPa.

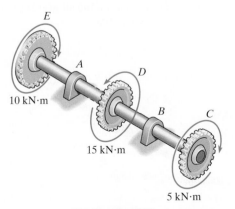

Probs. 10–12/13

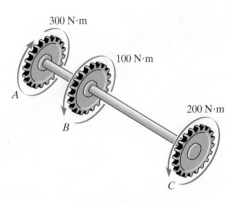

Probs. 10–15/16

10–17. The rod has a diameter of 1 in. and a weight of 10 lb/ft. Determine the maximum torsional stress in the rod at a section located at A due to the rod's weight.

10–18. The rod has a diameter of 1 in. and a weight of 15 lb/ft. Determine the maximum torsional stress in the rod at a section located at B due to the rod's weight.

4.5 ft
A
1.5 ft
B
1.5 ft
4 ft

Probs. 10–17/18

10–19. The copper pipe has an outer diameter of 3 in. and an inner diameter of 2.5 in. If it is tightly secured to the wall at C and a uniformly distributed torque is applied to it as shown, determine the shear stress at points A and B. These points lie on the pipe's outer surface. Sketch the shear stress on volume elements located at A and B.

***10–20.** The copper pipe has an outer diameter of 3 in. and an inner diameter of 2.50 in. If it is tightly secured to the wall at C and it is subjected to the uniformly distributed torque along its entire length, determine the absolute maximum shear stress in the pipe. Discuss the validity of this result.

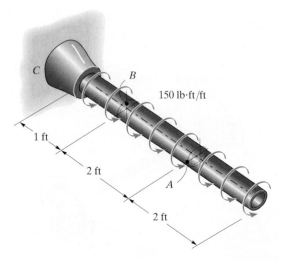

C
B
150 lb·ft/ft
1 ft
2 ft
A
2 ft

Probs. 10–19/20

10–21. The 60-mm-diameter solid shaft is subjected to the distributed and concentrated torsional loadings shown. Determine the shear stress at points A and B, and sketch the shear stress on volume elements located at these points.

10–22. The 60-mm-diameter solid shaft is subjected to the distributed and concentrated torsional loadings shown. Determine the absolute maximum and minimum shear stresses on the shaft's surface, and specify their locations, measured from the fixed end C.

10–23. The solid shaft is subjected to the distributed and concentrated torsional loadings shown. Determine the required diameter d of the shaft if the allowable shear stress for the material is $\tau_{\text{allow}} = 1.6$ MPa.

C
2 kN·m/m
0.4 m
B
600 N·m
0.4 m
A
400 N·m
0.3 m
d
0.3 m

Probs. 10–21/22/23

***10–24.** The 60-mm-diameter solid shaft is subjected to the distributed and concentrated torsional loadings shown. Determine the absolute maximum and minimum shear stresses in the shaft's surface and specify their locations, measured from the free end.

10–25. The solid shaft is subjected to the distributed and concentrated torsional loadings shown. Determine the required diameter d of the shaft if the allowable shear stress for the material is $\tau_{\text{allow}} = 60$ MPa.

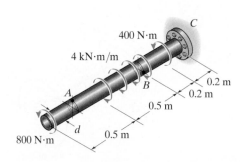

C
400 N·m
4 kN·m/m
A
B
0.2 m
0.2 m
0.5 m
d
0.5 m
800 N·m

Probs. 10–24/25

10–26. The pump operates using the motor that has a power of 85 W. If the impeller at *B* is turning at 150 rev/min, determine the maximum shear stress in the 20-mm-diameter transmission shaft at *A*.

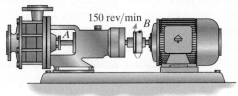

Prob. 10–26

10–27. The gear motor can develop $\frac{1}{10}$ hp when it turns at 300 rev/min. If the shaft has a diameter of $\frac{1}{2}$ in., determine the maximum shear stress in the shaft.

*10–28. The gear motor can develop $\frac{1}{10}$ hp when it turns at 80 rev/min. If the allowable shear stress for the shaft is $\tau_{\text{allow}} = 4$ ksi, determine the smallest diameter of the shaft to the nearest $\frac{1}{8}$ in. that can be used.

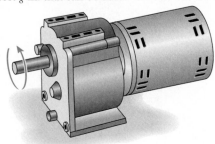

Probs. 10–27/28

10–29. The gear motor can develop $\frac{1}{4}$ hp when it turns at 600 rev/min. If the shaft has a diameter of $\frac{1}{2}$ in., determine the maximum shear stress in the shaft.

10–30. The gear motor can develop 2 hp when it turns at 150 rev/min. If the allowable shear stress for the shaft is $\tau_{\text{allow}} = 8$ ksi, determine the smallest diameter of the shaft to the nearest $\frac{1}{8}$ in. that can be used.

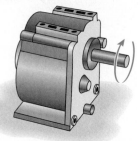

Probs. 10–29/30

10–31. The 6-hp reducer motor can turn at 1200 rev/min. If the allowable shear stress for the shaft is $\tau_{\text{allow}} = 6$ ksi, determine the smallest diameter of the shaft to the nearest $\frac{1}{16}$ in. that can be used.

*10–32. The 6-hp reducer motor can turn at 1200 rev/min. If the shaft has a diameter of $\frac{5}{8}$ in., determine the maximum shear stress in the shaft.

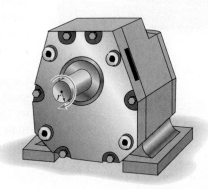

Probs. 10–31/32

10–33. The solid steel shaft *DF* has a diameter of 25 mm and is supported by smooth bearings at *D* and *E*. It is coupled to a motor at *F*, which delivers 12 kW of power to the shaft while it is turning at 50 rev/s. If gears *A*, *B*, and *C* remove 3 kW, 4 kW, and 5 kW respectively, determine the maximum shear stress developed in the shaft within regions *CF* and *BC*. The shaft is free to turn in its support bearings *D* and *E*.

10–34. The solid steel shaft *DF* has a diameter of 25 mm and is supported by smooth bearings at *D* and *E*. It is coupled to a motor at *F*, which delivers 12 kW of power to the shaft while it is turning at 50 rev/s. If gears *A*, *B*, and *C* remove 3 kW, 4 kW, and 5 kW respectively, determine the absolute maximum shear stress in the shaft.

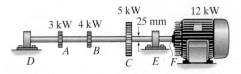

Probs. 10–33/34

10.4   ANGLE OF TWIST

In this section we will develop a formula for determining the *angle of twist* ϕ (phi) of one end of a shaft with respect to its other end. To generalize this development, we will assume the shaft has a circular cross section that can gradually vary along its length, Fig. 10–12a. Also, the material is assumed to be homogeneous and to behave in a linear elastic manner when the torque is applied. As in the case of an axially loaded bar, we will neglect the localized deformations that occur at points of application of the torques and where the cross section changes abruptly. By Saint-Venant's principle, these effects occur within small regions of the shaft's length, and generally they will have only a slight effect on the final result.

Using the method of sections, a differential disk of thickness dx, located at position x, is isolated from the shaft, Fig. 10–12b. At this location, the internal torque is $T(x)$, since the external loading may cause it to change along the shaft. Due to $T(x)$, the disk will twist, such that the *relative rotation* of one of its faces with respect to the other face is $d\phi$. As a result an element of material located at an arbitrary radius ρ within the disk will undergo a shear strain γ. The values of γ and $d\phi$ are related by Eq. 10–1, namely,

$$d\phi = \gamma \frac{dx}{\rho} \tag{10–13}$$

Long shafts subjected to torsion can, in some cases, have a noticeable elastic twist.

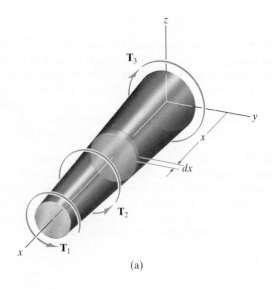

(a)

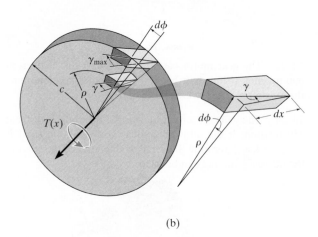

(b)

Fig. 10–12

Since Hooke's law, $\gamma = \tau/G$, applies and the shear stress can be expressed in terms of the applied torque using the torsion formula $\tau = T(x)\rho/J(x)$, then $\gamma = T(x)\rho/J(x)G(x)$. Substituting this into Eq. 10–13, the angle of twist for the disk is therefore

$$d\phi = \frac{T(x)}{J(x)G(x)}\,dx$$

Integrating over the entire length L of the shaft, we can obtain the angle of twist for the entire shaft, namely,

$$\phi = \int_0^L \frac{T(x)\,dx}{J(x)G(x)} \qquad (10\text{–}14)$$

Here

ϕ = the angle of twist of one end of the shaft with respect to the other end, measured in radians

$T(x)$ = the *internal torque* at the arbitrary position x, found from the method of sections and the equation of moment equilibrium applied about the shaft's axis

$J(x)$ = the shaft's polar moment of inertia expressed as a function of x

$G(x)$ = the shear modulus of elasticity for the material expressed as a function of x

Constant Torque and Cross-Sectional Area.

Usually in engineering practice the material is homogeneous so that G is constant. Also, the cross-sectional area and the external torque are constant along the length of the shaft, Fig. 10–13. When this is the case, the internal torque $T(x) = T$, the polar moment of inertia $J(x) = J$, and Eq. 10–14 can be integrated, which gives

$$\phi = \frac{TL}{JG} \qquad (10\text{–}15)$$

Note the similarities between the above two equations and those for an axially loaded bar.

When calculating both the stress and the angle of twist of this soil auger, it is necessary to consider the variable torsional loading which acts along its length.

Fig. 10–13

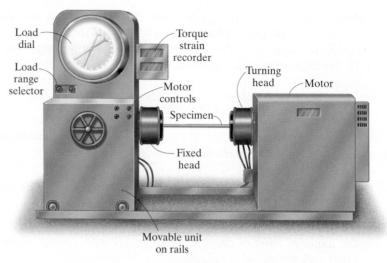

Fig. 10–14

Equation 10–15 is often used to determine the shear modulus of elasticity, G, of a material. To do so, a specimen of known length and diameter is placed in a torsion testing machine like the one shown in Fig. 10–14. The applied torque T and angle of twist ϕ are then measured along the length L. From Eq. 10–15, we get $G = TL/J\phi$. To obtain a more reliable value of G, several of these tests are performed and the average value is used.

Multiple Torques.   If the shaft is subjected to several different torques, or the cross-sectional area or shear modulus changes abruptly from one region of the shaft to the next, as in Fig. 10–12, then Eq. 10–15 should be applied to each segment of the shaft where these quantities are all constant. The angle of twist of one end of the shaft with respect to the other is found from the algebraic addition of the angles of twist of each segment. For this case,

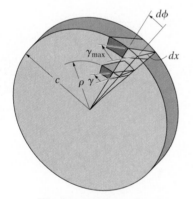

The shear strain at points on the cross section increases linearly with ρ, i.e., $\gamma = (\rho/c)\gamma_{max}$.

(b)

Fig. 10–12 (Repeated)

$$\phi = \sum \frac{TL}{JG} \qquad (10\text{–}16)$$

Sign Convention.   The best way to apply this equation is to use a sign convention for both the internal torque and the angle of twist of one end of the shaft with respect to the other end. To do this, we will apply the right-hand rule, whereby both the torque and angle will be *positive*, provided the *thumb* is directed *outward* from the shaft while the fingers curl in the direction of the torque, Fig. 10–15.

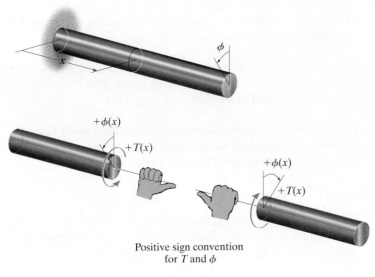

Positive sign convention
for T and ϕ

Fig. 10–15

IMPORTANT POINT

- When applying Eq. 10–14 to determine the angle of twist, it is important that the applied torques do not cause yielding of the material and that the material is homogeneous and behaves in a linear elastic manner.

PROCEDURE FOR ANALYSIS

The angle of twist of one end of a shaft or tube with respect to the other end can be determined using the following procedure.

Internal Torque.

- The *internal torque* is found at a point on the axis of the shaft by using the method of sections and the equation of moment equilibrium, applied along the shaft's axis.

- If the torque varies along the shaft's length, a section should be made at the arbitrary position x along the shaft and the *internal torque* represented as a function of x, i.e., $T(x)$.

- If several constant external torques act on the shaft between its ends, the internal torque in each *segment* of the shaft, between any two external torques, must be determined.

Angle of Twist.

- When the circular cross-sectional area of the shaft varies along the shaft's axis, the polar moment of inertia must be expressed as a function of its position x along the axis, $J(x)$.

- If the polar moment of inertia or the internal torque *suddenly changes* between the ends of the shaft, then $\phi = \int (T(x)/J(x)G(x)) \, dx$ or $\phi = TL/JG$ must be applied to *each segment* for which J, G, and T are continuous or constant.

- When the internal torque in each segment is determined, be sure to use a consistent sign convention for the shaft or its segments, such as the one shown in Fig. 10–15. Also make sure that a consistent set of units is used when substituting numerical data into the equations.

EXAMPLE 10.5

Determine the angle of twist of the end A of the A-36 steel shaft shown in Fig. 10–16a. Also, what is the angle of twist of A relative to C? The shaft has a diameter of 200 mm.

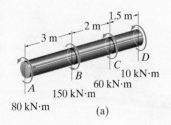

(a)

SOLUTION

Internal Torque. Using the method of sections, the *internal torques* are found in each segment as shown in Fig. 10–16b. By the right-hand rule, with positive torques directed away from the *sectioned end* of the shaft, we have $T_{AB} = +80$ kN · m, $T_{BC} = -70$ kN · m, and $T_{CD} = -10$ kN · m. These results are also shown on the *torque diagram*, which indicates how the internal torque varies along the axis of the shaft, Fig. 10–16c.

Angle of Twist. The polar moment of inertia for the shaft is

$$J = \frac{\pi}{2}(0.1 \text{ m})^4 = 0.1571(10^{-3}) \text{ m}^4$$

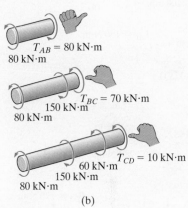

For A-36 steel, the table on the back cover gives $G = 75$ GPa. Therefore, the end A of the shaft has a rotation of

$$\phi_A = \Sigma\frac{TL}{JG} = \frac{80(10^3) \text{ N} \cdot \text{m} (3 \text{ m})}{(0.1571(10^{-3}) \text{ m}^4)(75(10^9) \text{ N/m}^2)}$$

$$+ \frac{-70(10^3) \text{ N} \cdot \text{m} (2 \text{ m})}{(0.1571(10^{-3}) \text{ m}^4)(75(10^9) \text{ N/m}^2)} + \frac{-10(10^3) \text{ N} \cdot \text{m} (1.5 \text{ m})}{(0.1571(10^{-3}) \text{ m}^4)(75(10^9) \text{ N/m}^2)}$$

$$\phi_A = 7.22(10^{-3}) \text{ rad} \qquad\qquad Ans.$$

The relative angle of twist of A with respect to C involves only two segments of the shaft.

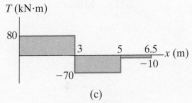

(c)

Fig. 10–16

$$\phi_{A/C} = \Sigma\frac{TL}{JG} = \frac{80(10^3) \text{ N} \cdot \text{m} (3 \text{ m})}{(0.1571(10^{-3}) \text{ m}^4)(75(10^9) \text{ N/m}^2)}$$

$$+ \frac{-70(10^3) \text{ N} \cdot \text{m} (2 \text{ m})}{(0.1571(10^{-3}) \text{ m}^4)(75(10^9) \text{ N/m}^2)}$$

$$\phi_{A/C} = 8.49(10^{-3}) \text{ rad} \qquad\qquad Ans.$$

Both results are positive, which means that end A will rotate as indicated by the curl of the right-hand fingers when the thumb is directed away from the shaft.

10

EXAMPLE 10.6

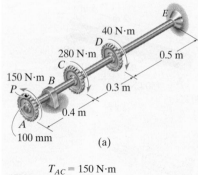

(a)

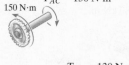

$T_{AC} = 150 \text{ N·m}$

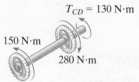

$T_{CD} = 130 \text{ N·m}$

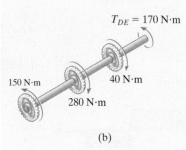

$T_{DE} = 170 \text{ N·m}$

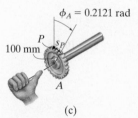

(b)

$\phi_A = 0.2121 \text{ rad}$

(c)

Fig. 10–17

The gears attached to the fixed-end steel shaft are subjected to the torques shown in Fig. 10–17a. If the shaft has a diameter of 14 mm, determine the displacement of the tooth P on gear A. $G = 80$ GPa.

SOLUTION

Internal Torque. By inspection, the torques in segments AC, CD, and DE are different yet *constant* throughout each segment. Free-body diagrams of these segments along with the calculated internal torques are shown in Fig. 10–17b. Using the right-hand rule and the established sign convention that positive torque is directed away from the sectioned end of the shaft, we have

$$T_{AC} = +150 \text{ N·m} \qquad T_{CD} = -130 \text{ N·m} \qquad T_{DE} = -170 \text{ N·m}$$

Angle of Twist. The polar moment of inertia for the shaft is

$$J = \frac{\pi}{2}(0.007 \text{ m})^4 = 3.771(10^{-9}) \text{ m}^4$$

Applying Eq. 10–16 to each segment and adding the results algebraically, we have

$$\phi_A = \Sigma \frac{TL}{JG} = \frac{(+150 \text{ N·m})(0.4 \text{ m})}{3.771(10^{-9})\text{m}^4\,[80(10^9)\text{N/m}^2]}$$

$$+ \frac{(-130 \text{ N·m})(0.3 \text{ m})}{3.771(10^{-9})\text{m}^4\,[80(10^9)\text{N/m}^2]} + \frac{(-170 \text{ N·m})(0.5 \text{ m})}{3.771(10^{-9})\text{m}^4\,[80(10^9)\text{ N/m}^2]}$$

$$\phi_A = -0.2121 \text{ rad}$$

Since the answer is negative, by the right-hand rule the thumb is directed *toward* the support E of the shaft, and therefore gear A will rotate as shown in Fig. 10–17c.

The displacement of tooth P on gear A is

$$s_P = \phi_A r = (0.2121 \text{ rad})(100 \text{ mm}) = 21.2 \text{ mm} \qquad \text{Ans.}$$

EXAMPLE 10.7

The 2-in.-diameter solid cast-iron post shown in Fig. 10–18a is buried 24 in. in soil. If a torque is applied to its top using a rigid wrench, determine the maximum shear stress in the post and the angle of twist of the wrench. Assume that the torque is about to turn the bottom of the post, and the soil exerts a uniform torsional resistance of t lb · in./in. along its 24-in. buried length. $G = 5.5(10^3)$ ksi.

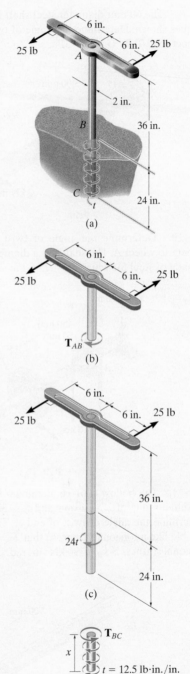

SOLUTION

Internal Torque. The internal torque in segment AB of the post is constant. From the free-body diagram, Fig. 10–18b, we have

$$\Sigma M_z = 0; \qquad T_{AB} = 25 \text{ lb } (12 \text{ in.}) = 300 \text{ lb} \cdot \text{in.}$$

The magnitude of the uniform distribution of torque along the buried segment BC can be determined from equilibrium of the entire post, Fig. 10–18c. Here

$$\Sigma M_z = 0 \qquad 25 \text{ lb } (12 \text{ in.}) - t(24 \text{ in.}) = 0$$
$$t = 12.5 \text{ lb} \cdot \text{in./in.}$$

Hence, from a free-body diagram of the bottom segment of the post, located at the position x, Fig. 10–18d, we have

$$\Sigma M_z = 0; \qquad T_{BC} - 12.5x = 0$$
$$T_{BC} = 12.5x$$

Maximum Shear Stress. The largest shear stress occurs in region AB, since the torque is largest there and J is constant for the post. Applying the torsion formula, we have

$$\tau_{max} = \frac{T_{AB} c}{J} = \frac{(300 \text{ lb} \cdot \text{in.})(1 \text{ in.})}{(\pi/2)(1 \text{ in.})^4} = 191 \text{ psi} \qquad Ans.$$

Angle of Twist. The angle of twist at the top can be determined relative to the bottom of the post, since it is fixed and yet is about to turn. Both segments AB and BC twist, and so in this case we have

$$\phi_A = \frac{T_{AB} L_{AB}}{JG} + \int_0^{L_{BC}} \frac{T_{BC}\, dx}{JG}$$

$$= \frac{(300 \text{ lb} \cdot \text{in.})\, 36 \text{ in.}}{JG} + \int_0^{24 \text{ in.}} \frac{12.5x\, dx}{JG}$$

$$= \frac{10\,800 \text{ lb} \cdot \text{in}^2}{JG} + \frac{12.5[(24)^2/2]\, \text{lb} \cdot \text{in}^2}{JG}$$

$$= \frac{14\,400 \text{ lb} \cdot \text{in}^2}{(\pi/2)(1 \text{ in.})^4\, 5500(10^3)\, \text{lb/in}^2} = 0.00167 \text{ rad} \qquad Ans.$$

Fig. 10–18

FUNDAMENTAL PROBLEMS

F10–9. The 60-mm-diameter steel shaft is subjected to the torques shown. Determine the angle of twist of end A with respect to C. Take $G = 75$ GPa.

Prob. F10–9

F10–10. Determine the angle of twist of wheel B with respect to wheel A. The shaft has a diameter of 40 mm and is made of steel for which $G = 75$ GPa.

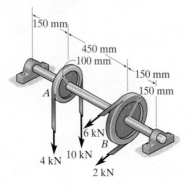

Prob. F10–10

F10–11. The hollow 6061-T6 aluminum shaft has an outer and inner radius of $c_o = 40$ mm and $c_i = 30$ mm, respectively. Determine the angle of twist of end A. The support at B is flexible like a torsional spring, so that $T_B = k_B \phi_B$, where the torsional stiffness is $k_B = 90$ kN $\cdot$ m/rad.

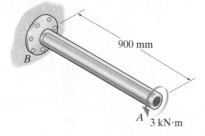

Prob. F10–11

F10–12. A series of gears are mounted on the 40-mm-diameter steel shaft. Determine the angle of twist of gear E relative to gear A. Take $G = 75$ GPa.

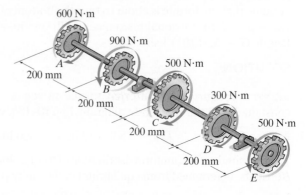

Prob. F10–12

F10–13. The 80-mm-diameter shaft is made of steel. If it is subjected to the uniform distributed torque, determine the angle of twist of end A. Take $G = 75$ GPa.

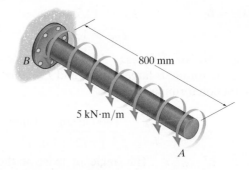

Prob. F10–13

F10–14. The 80-mm-diameter shaft is made of steel. If it is subjected to the triangular distributed load, determine the angle of twist of end A. Take $G = 75$ GPa.

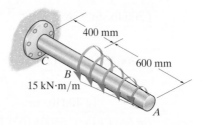

Prob. F10–14

PROBLEMS

10–35. The propellers of a ship are connected to an A-36 steel shaft that is 60 m long and has an outer diameter of 340 mm and inner diameter of 260 mm. If the power output is 4.5 MW when the shaft rotates at 20 rad/s, determine the maximum torsional stress in the shaft and its angle of twist.

*10–36. The solid shaft of radius c is subjected to a torque **T** at its ends. Show that the maximum shear strain in the shaft is $\gamma_{max} = Tc/JG$. What is the shear strain on an element located at point A, $c/2$ from the center of the shaft? Sketch the shear strain distortion of this element.

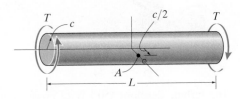

Prob. 10–36

10–37. The splined ends and gears attached to the A992 steel shaft are subjected to the torques shown. Determine the angle of twist of end B with respect to end A. The shaft has a diameter of 40 mm.

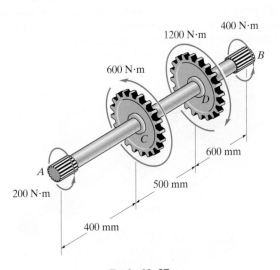

Prob. 10–37

10–38. The A-36 steel shaft has a diameter of 50 mm and is subjected to the distributed and concentrated loadings shown. Determine the absolute maximum shear stress in the shaft and plot a graph of the angle of twist of the shaft in radians versus x.

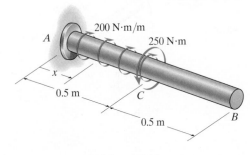

Prob. 10–38

10–39. The 60-mm-diameter shaft is made of 6061-T6 aluminum having an allowable shear stress of $\tau_{allow} = 80$ MPa. Determine the maximum allowable torque **T**. Also, find the corresponding angle of twist of disk A relative to disk C.

*10–40. The 60-mm-diameter shaft is made of 6061-T6 aluminum. If the allowable shear stress is $\tau_{allow} = 80$ MPa, and the angle of twist of disk A relative to disk C is limited so that it does not exceed 0.06 rad, determine the maximum allowable torque **T**.

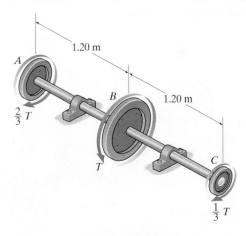

Probs. 10–39/40

10

10–41. The 50-mm-diameter A992 steel shaft is subjected to the torques shown. Determine the angle of twist of the end A.

Prob. 10–41

10–42. The shaft is made of A992 steel with the allowable shear stress of $\tau_{allow} = 75$ MPa. If gear B supplies 15 kW of power, while gears A, C and D withdraw 6 kW, 4 kW and 5 kW, respectively, determine the required minimum diameter d of the shaft to the nearest millimeter. Also, find the corresponding angle of twist of gear A relative to gear D. The shaft is rotating at 600 rpm.

10–43. Gear B supplies 15 kW of power, while gears A, C, and D withdraw 6 kW, 4 kW and 5 kW, respectively. If the shaft is made of steel with the allowable shear stress of $\tau_{allow} = 75$ MPa, and the relative angle of twist between any two gears cannot exceed 0.05 rad, determine the required minimum diameter d of the shaft to the nearest millimeter. The shaft is rotating at 600 rpm.

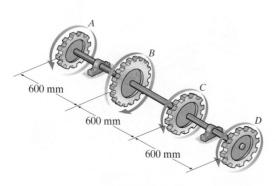

Probs. 10–42/43

*10–44.** The rotating flywheel-and-shaft, when brought to a sudden stop at D, begins to oscillate clockwise-counterclockwise such that a point A on the outer edge of the fly-wheel is displaced through a 6-mm arc. Determine the maximum shear stress developed in the tubular A-36 steel shaft due to this oscillation. The shaft has an inner diameter of 24 mm and an outer diameter of 32 mm. The bearings at B and C allow the shaft to rotate freely, whereas the support at D holds the shaft fixed.

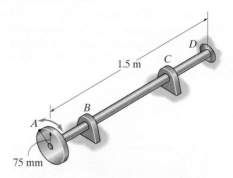

Prob. 10–44

10–45. The turbine develops 150 kW of power, which is transmitted to the gears such that C receives 70% and D receives 30%. If the rotation of the 100-mm-diameter A-36 steel shaft is $\omega = 800$ rev/min., determine the absolute maximum shear stress in the shaft and the angle of twist of end E of the shaft relative to B. The journal bearing at E allows the shaft to turn freely about its axis.

10–46. The turbine develops 150 kW of power, which is transmitted to the gears such that both C and D receive an equal amount. If the rotation of the 100-mm-diameter A-36 steel shaft is $\omega = 500$ rev/min., determine the absolute maximum shear stress in the shaft and the rotation of end B of the shaft relative to E. The journal bearing at E allows the shaft to turn freely about its axis.

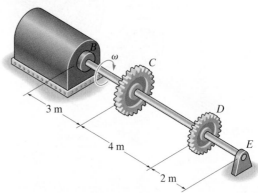

Probs. 10–45/46

10–47. The shaft is made of A992 steel. It has a diameter of 1 in. and is supported by bearings at A and D, which allow free rotation. Determine the angle of twist of B with respect to D.

***10–48.** The shaft is made of A-36 steel. It has a diameter of 1 in. and is supported by bearings at A and D, which allow free rotation. Determine the angle of twist of gear C with respect to B.

10–50. The turbine develops 300 kW of power, which is transmitted to the gears such that both B and C receive an equal amount. If the rotation of the 100-mm-diameter A992 steel shaft is $\omega = 600$ rev/min., determine the absolute maximum shear stress in the shaft and the rotation of end D of the shaft relative to A. The journal bearing at D allows the shaft to turn freely about its axis.

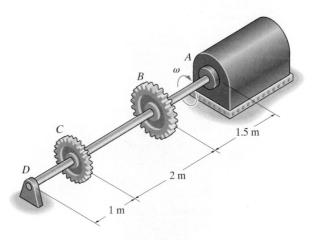

Prob. 10–50

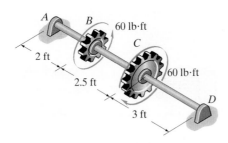

Probs. 10–47/48

10–51. The device shown is used to mix soils in order to provide in-situ stabilization. If the mixer is connected to an A-36 steel tubular shaft that has an inner diameter of 3 in. and an outer diameter of 4.5 in., determine the angle of twist of the shaft at A relative to C if each mixing blade is subjected to the torques shown.

10–49. The A992 steel shaft has a diameter of 50 mm and is subjected to the distributed loadings shown. Determine the absolute maximum shear stress in the shaft and plot a graph of the angle of twist of the shaft in radians versus x.

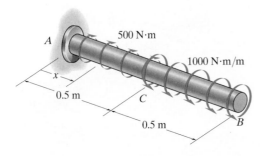

Prob. 10–49

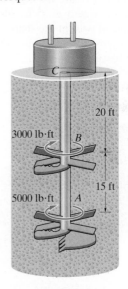

Prob. 10–51

*10–52. The device shown is used to mix soils in order to provide in-situ stabilization. If the mixer is connected to an A-36 steel tubular shaft that has an inner diameter of 3 in. and an outer diameter of 4.5 in, determine the angle of twist of the shaft at A relative to B and the absolute maximum shear stress in the shaft if each mixing blade is subjected to the torques shown.

10–54. The A-36 hollow steel shaft is 2 m long and has an outer diameter of 40 mm. When it is rotating at 80 rad/s, it transmits 32 kW of power from the engine E to the generator G. Determine the smallest thickness of the shaft if the allowable shear stress is $\tau_{allow} = 140$ MPa and the shaft is restricted not to twist more than 0.05 rad.

10–55. The A-36 solid steel shaft is 3 m long and has a diameter of 50 mm. It is required to transmit 35 kW of power from the engine E to the generator G. Determine the smallest angular velocity of the shaft if it is restricted not to twist more than 1°.

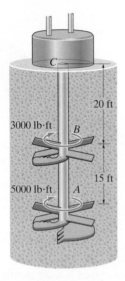

Probs. 10–54/55

Prob. 10–52

10–53. The 6-in.-diameter L-2 steel shaft on the turbine is supported on journal bearings at A and B. If C is held fixed and the turbine blades create a torque on the shaft that increases linearly from zero at C to 2000 lb · ft at D, determine the angle of twist of the shaft at D relative to C. Also, calculate the absolute maximum shear stress in the shaft. Neglect the size of the blades.

*10–56. The shaft of radius c is subjected to a distributed torque t, measured as torque/length of shaft. Determine the angle of twist at end A. The shear modulus is G.

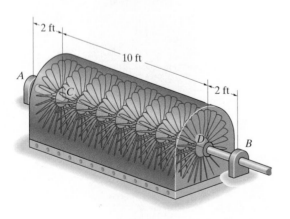

Prob. 10–53

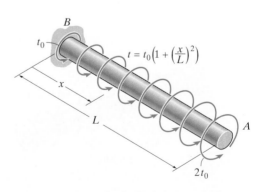

Prob. 10–56

10–57. The A-36 steel bolt is tightened within a hole so that the reactive torque on the shank AB can be expressed by the equation $t = (kx^2)\,\text{N} \cdot \text{m/m}$, where x is in meters. If a torque of $T = 50\,\text{N} \cdot \text{m}$ is applied to the bolt head, determine the constant k and the amount of twist in the 50-mm length of the shank. Assume the shank has a constant radius of 4 mm.

10–58. Solve Prob. 10–57 if the distributed torque is $t = (kx^{2/3})\,\text{N} \cdot \text{m/m}$.

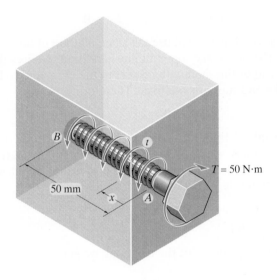

Probs. 10–57/58

10–59. The tubular drive shaft for the propeller of a hovercraft is 6 m long. If the motor delivers 4 MW of power to the shaft when the propellers rotate at 25 rad/s, determine the required inner diameter of the shaft if the outer diameter is 250 mm. What is the angle of twist of the shaft when it is operating? Take $\tau_{\text{allow}} = 90\,\text{MPa}$ and $G = 75\,\text{GPa}$.

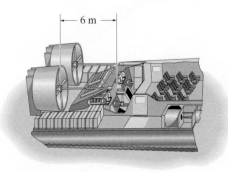

Prob. 10–59

*10–60.** The 60-mm diameter solid shaft is made of 2014-T6 aluminum and is subjected to the distributed and concentrated torsional loadings shown. Determine the angle of twist at the free end A of the shaft.

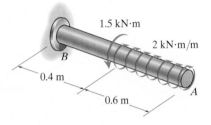

Prob. 10–60

10–61. The motor produces a torque of $T = 20\,\text{N} \cdot \text{m}$ on gear A. If gear C is suddenly locked so it does not turn, yet B can freely turn, determine the angle of twist of F with respect to E and F with respect to D of the L2-steel shaft, which has an inner diameter of 30 mm and an outer diameter of 50 mm. Also, calculate the absolute maximum shear stress in the shaft. The shaft is supported on journal bearings at G at H.

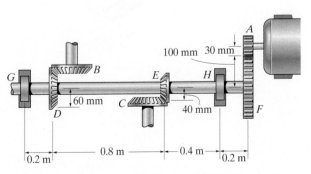

Prob. 10–61

10.5 STATICALLY INDETERMINATE TORQUE-LOADED MEMBERS

A torsionally loaded shaft will be statically indeterminate if the moment equation of equilibrium, applied about the axis of the shaft, is not adequate to determine the unknown torques acting on the shaft. An example of this situation is shown in Fig. 10–19a. As shown on the free-body diagram, Fig. 10–19b, the reactive torques at the supports A and B are unknown. Along the axis of the shaft, we require

$$\Sigma M = 0; \qquad\qquad 500 \text{ N} \cdot \text{m} - T_A - T_B = 0$$

In order to obtain a solution, we will use the same method of analysis discussed in Sec. 9.4. The necessary compatibility condition requires the angle of twist of one end of the shaft with respect to the other end to be equal to zero, since the end supports are fixed. Therefore,

$$\phi_{A/B} = 0$$

Provided the material is linear elastic, we can then apply the load–displacement relation $\phi = TL/JG$ to express this equation in terms of the unknown torques. Realizing that the internal torque in segment AC is $+T_A$ and in segment CB it is $-T_B$, Fig. 10–19c, we have

$$\frac{T_A(3 \text{ m})}{JG} - \frac{T_B(2 \text{ m})}{JG} = 0$$

Solving the above two equations for the reactions, we get

$$T_A = 200 \text{ N} \cdot \text{m} \quad \text{and} \quad T_B = 300 \text{ N} \cdot \text{m}$$

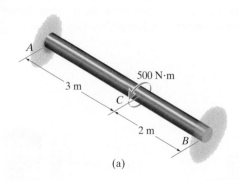

(a)

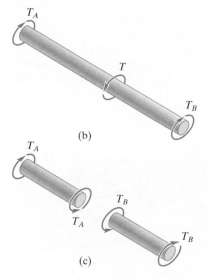

(b)

(c)

Fig. 10–19

▶ PROCEDURE FOR ANALYSIS

The unknown torques in statically indeterminate shafts are determined by satisfying equilibrium, compatibility, and load–displacement requirements for the shaft.

Equilibrium.

- Draw a free-body diagram of the shaft in order to identify all the external torques that act on it. Then write the equation of moment equilibrium about the axis of the shaft.

Compatibility.

- Write the compatibility equation. Give consideration as to how the supports constrain the shaft when it is twisted.

Load-Displacement.

- Express the angles of twist in the compatibility condition in terms of the torques, using a load–displacement relation, such as $\phi = TL/JG$.

- Solve the equations for the unknown reactive torques. If any of the magnitudes have a negative numerical value, it indicates that this torque acts in the opposite sense of direction to that shown on the free-body diagram.

The shaft of this cutting machine is fixed at its ends and subjected to a torque at its center, allowing it to act as a torsional spring.

EXAMPLE   10.8

The solid steel shaft shown in Fig. 10–20a has a diameter of 20 mm. If it is subjected to the two torques, determine the reactions at the fixed supports A and B.

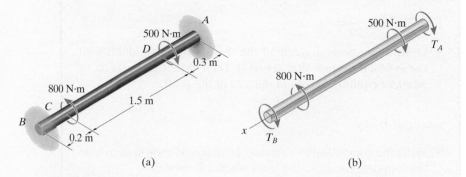

(a)   (b)

SOLUTION

Equilibrium. By inspection of the free-body diagram, Fig. 10–20b, it is seen that the problem is statically indeterminate, since there is only *one* available equation of equilibrium and there are two unknowns. We require

$$\Sigma M_x = 0; \qquad -T_B + 800 \text{ N} \cdot \text{m} - 500 \text{ N} \cdot \text{m} - T_A = 0 \qquad (1)$$

Compatibility. Since the ends of the shaft are fixed, the angle of twist of one end of the shaft with respect to the other must be zero. Hence, the compatibility equation becomes

$$\phi_{A/B} = 0$$

Load–Displacement. This condition can be expressed in terms of the unknown torques by using the load-displacement relationship, $\phi = TL/JG$. Here there are three regions of the shaft where the internal torque is constant. On the free-body diagrams in Fig. 10–20c we have shown the internal torques acting on the left segments of the shaft. This way the internal torque is only a function of T_B. Using the sign convention established in Sec. 10.4, we have

$$\frac{-T_B(0.2 \text{ m})}{JG} + \frac{(800 - T_B)(1.5 \text{ m})}{JG} + \frac{(300 - T_B)(0.3 \text{ m})}{JG} = 0$$

so that

$$T_B = 645 \text{ N} \cdot \text{m} \qquad \qquad Ans.$$

Using Eq. 1,

$$T_A = -345 \text{ N} \cdot \text{m} \qquad \qquad Ans.$$

The negative sign indicates that $\mathbf{T}_A$ acts in the opposite direction of that shown in Fig. 10–20b.

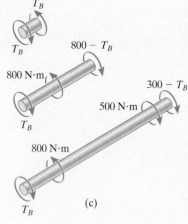

(c)

Fig. 10–20

EXAMPLE 10.9

The shaft shown in Fig. 10–21a is made from a steel tube, which is bonded to a brass core. If a torque of $T = 250$ lb · ft is applied at its end, plot the shear stress distribution along a radial line on its cross section. Take $G_{st} = 11.4(10^3)$ ksi, $G_{br} = 5.20(10^3)$ ksi.

SOLUTION

Equilibrium. A free-body diagram of the shaft is shown in Fig. 10–21b. The reaction at the wall has been represented by the amount of torque resisted by the steel, T_{st}, and by the brass, T_{br}. Working in units of pounds and inches, equilibrium requires

$$-T_{st} - T_{br} + (250 \text{ lb} \cdot \text{ft})(12 \text{ in./ft}) = 0 \qquad (1)$$

Compatibility. We require the angle of twist of end A to be the same for both the steel and brass since they are bonded together. Thus,

$$\phi = \phi_{st} = \phi_{br}$$

Load–Displacement. Applying the load–displacement relationship, $\phi = TL/JG$,

$$\frac{T_{st}L}{(\pi/2)[(1 \text{ in.})^4 - (0.5 \text{ in.})^4]\,11.4(10^3) \text{ kip/in}^2}$$

$$= \frac{T_{br}L}{(\pi/2)(0.5 \text{ in.})^4\,5.20(10^3) \text{ kip/in}^2}$$

$$T_{st} = 32.88\,T_{br} \qquad (2)$$

Solving Eqs. 1 and 2, we get

$$T_{st} = 2911.5 \text{ lb} \cdot \text{in.} = 242.6 \text{ lb} \cdot \text{ft}$$
$$T_{br} = 88.5 \text{ lb} \cdot \text{in.} = 7.38 \text{ lb} \cdot \text{ft}$$

The shear stress in the brass core varies from zero at its center to a maximum at the interface where it contacts the steel tube. Using the torsion formula,

$$(\tau_{br})_{max} = \frac{(88.5 \text{ lb} \cdot \text{in.})(0.5 \text{ in.})}{(\pi/2)(0.5 \text{ in.})^4} = 451 \text{ psi}$$

For the steel, the minimum and maximum shear stresses are

$$(\tau_{st})_{min} = \frac{(2911.5 \text{ lb} \cdot \text{in.})(0.5 \text{ in.})}{(\pi/2)[(1 \text{ in.})^4 - (0.5 \text{ in.})^4]} = 989 \text{ psi}$$

$$(\tau_{st})_{max} = \frac{(2911.5 \text{ lb} \cdot \text{in.})(1 \text{ in.})}{(\pi/2)[(1 \text{ in.})^4 - (0.5 \text{ in.})^4]} = 1977 \text{ psi}$$

The results are plotted in Fig. 10–21c. Note the discontinuity of *shear stress* at the brass and steel interface. This is to be expected, since the materials have different moduli of rigidity; i.e., steel is stiffer than brass ($G_{st} > G_{br}$), and thus it carries more shear stress at the interface. Although the shear stress is discontinuous here, the *shear strain* is not. Rather, it is the *same* on either side of the brass–steel interface, Fig. 10–21d.

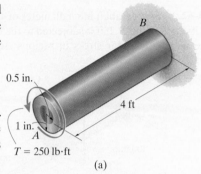

0.5 in.

1 in.

A

$T = 250$ lb·ft

B

4 ft

(a)

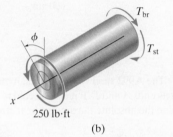

ϕ

T_{br}

T_{st}

x

250 lb·ft

(b)

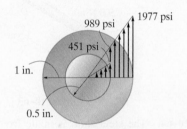

989 psi

451 psi

1977 psi

1 in.

0.5 in.

Shear–stress distribution

(c)

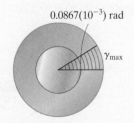

$0.0867(10^{-3})$ rad

γ_{max}

Shear–strain distribution

(d)

Fig. 10–21

PROBLEMS

10–62. The steel shaft has a diameter of 40 mm and is fixed at its ends A and B. If it is subjected to the couple, determine the maximum shear stress in regions AC and CB of the shaft. $G_{st} = 75$ GPa.

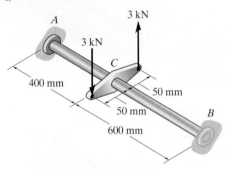

Prob. 10–62

10–63. The A992 steel shaft has a diameter of 60 mm and is fixed at its ends A and B. If it is subjected to the torques shown, determine the absolute maximum shear stress in the shaft.

Prob. 10–63

*10–64. The steel shaft is made from two segments: AC has a diameter of 0.5 in., and CB has a diameter of 1 in. If the shaft is fixed at its ends A and B and subjected to a torque of 500 lb · ft, determine the maximum shear stress in the shaft. $G_{st} = 10.8(10^3)$ ksi.

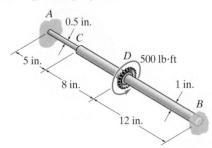

Prob. 10–64

10–65. The bronze C86100 pipe has an outer diameter of 1.5 in. and a thickness of 0.125 in. The coupling on it at C is being tightened using a wrench. If the torque developed at A is 125 lb · in., determine the magnitude F of the couple forces. The pipe is fixed supported at end B.

10–66. The bronze C86100 pipe has an outer diameter of 1.5 in. and a thickness of 0.125 in. The coupling on it at C is being tightened using a wrench. If the applied force is $F = 20$ lb, determine the maximum shear stress in the pipe.

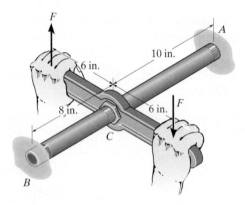

Probs. 10–65/66

10–67. The shaft is made of L2 tool steel, has a diameter of 40 mm, and is fixed at its ends A and B. If it is subjected to the torque, determine the maximum shear stress in regions AC and CB.

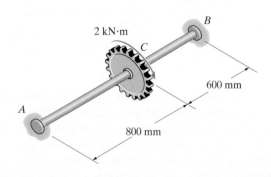

Prob. 10–67

***10–68.** The shaft is made of L2 tool steel, has a diameter of 40 mm, and is fixed at its ends A and B. If it is subjected to the couple, determine the maximum shear stress in regions AC and CB.

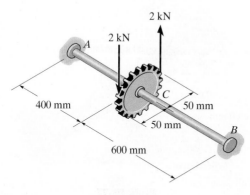

Prob. 10–68

10–69. The Am1004-T61 magnesium tube is bonded to the A-36 steel rod. If the allowable shear stresses for the magnesium and steel are $(\tau_{\text{allow}})_{\text{mg}} = 45$ MPa and $(\tau_{\text{allow}})_{\text{st}} = 75$ MPa, respectively, determine the maximum allowable torque that can be applied at A. Also, find the corresponding angle of twist of end A.

10–70. The Am1004-T61 magnesium tube is bonded to the A-36 steel rod. If a torque of $T = 5$ kN $\cdot$ m is applied to end A, determine the maximum shear stress in each material. Sketch the shear stress distribution.

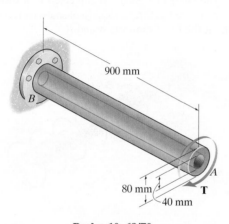

Probs. 10–69/70

10–71. The two shafts are made of A-36 steel. Each has a diameter of 25 mm and they are connected using the gears fixed to their ends. Their other ends are attached to fixed supports at A and B. They are also supported by journal bearings at C and D, which allow free rotation of the shafts along their axes. If a torque of 500 N $\cdot$ m is applied to the gear at E, determine the reactions at A and B.

10–72. The two shafts are made of A-36 steel. Each has a diameter of 25 mm and they are connected using the gears fixed to their ends. Their other ends are attached to fixed supports at A and B. They are also supported by journal bearings at C and D, which allow free rotation of the shafts along their axes. If a torque of 500 N $\cdot$ m is applied to the gear at E, determine the rotation of this gear.

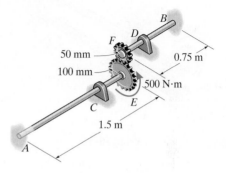

Probs. 10–71/72

10–73. A rod is made from two segments: AB is steel and BC is brass. It is fixed at its ends and subjected to a torque of $T = 680$ N $\cdot$ m. If the steel portion has a diameter of 30 mm, determine the required diameter of the brass portion so the reactions at the walls will be the same. $G_{\text{st}} = 75$ GPa, $G_{\text{br}} = 39$ GPa.

10–74. Determine the absolute maximum shear stress in the shaft of Prob. 10–73.

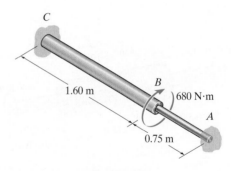

Probs. 10–73/74

10–75. The two 3-ft-long shafts are made of 2014-16 aluminum. Each has a diameter of 1.5 in. and they are connected using the gears fixed to their ends. Their other ends are attached to fixed supports at A and B. They are also supported by bearings at C and D, which allow free rotation of the shafts along their axes. If a torque of 600 lb·ft is applied to the top gear as shown, determine the maximum shear stress in each shaft.

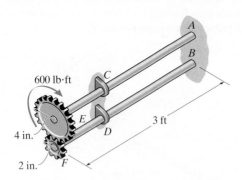

Prob. 10–75

*10–76. The composite shaft consists of a mid-section that includes the 1-in.-diameter solid shaft and a tube that is welded to the rigid flanges at A and B. Neglect the thickness of the flanges and determine the angle of twist of end C of the shaft relative to end D. The shaft is subjected to a torque of 800 lb · ft. The material is A-36 steel.

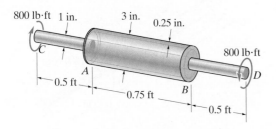

Prob. 10–76

10–77. If the shaft is subjected to a uniform distributed torque of $t = 20$ kN · m/m, determine the maximum shear stress developed in the shaft. The shaft is made of 2014-T6 aluminum alloy and is fixed at A and C.

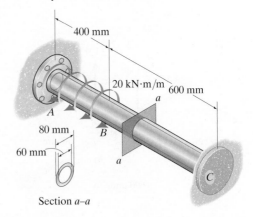

Section a–a

Prob. 10–77

10–78. The tapered shaft is confined by the fixed supports at A and B. If a torque **T** is applied at its mid-point, determine the reactions at the supports.

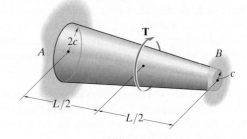

Prob. 10–78

10–79. The shaft of radius c is subjected to a distributed torque t, measured as torque/length of shaft. Determine the reactions at the fixed supports A and B.

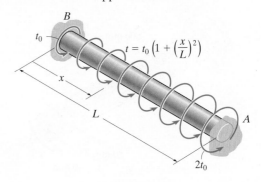

$$t = t_0 \left(1 + \left(\frac{x}{L}\right)^2\right)$$

Prob. 10–79

CHAPTER 11

(© Construction Photography/Corbis)

The girders of this bridge have been designed on the basis of their ability to resist bending stress.

R10–6. Segments *AB* and *BC* of the assembly are made from 6061-T6 aluminum and A992 steel, respectively. If couple forces *P* = 3 kip are applied to the lever arm, determine the maximum shear stress developed in each segment. The assembly is fixed at *A* and *C*.

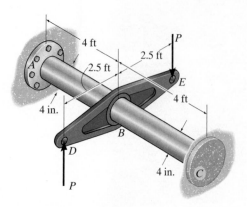

Prob. R10–6

R10–7. Segments *AB* and *BC* of the assembly are made from 6061-T6 aluminum and A992 steel, respectively. If the allowable shear stress for the aluminum is $(\tau_{allow})_{al} = 12$ ksi and for the steel $(\tau_{allow})_{st} = 10$ ksi, determine the maximum allowable couple forces *P* that can be applied to the lever arm. The assembly is fixed at *A* and *C*.

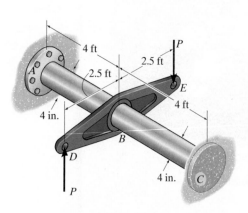

Prob. R10–7

***R10–8.** The tapered shaft is made from 2014-T6 aluminum alloy, and has a radius which can be described by the equation $r = 0.02(1 + x^{3/2})$ m, where *x* is in meters. Determine the angle of twist of its end *A* if it is subjected to a torque of 450 N · m.

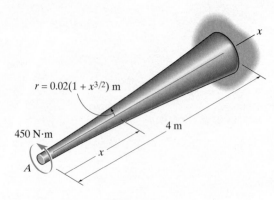

Prob. R10–8

R10–9. The 60-mm-diameter shaft rotates at 300 rev/min. This motion is caused by the unequal belt tensions on the pulley of 800 N and 450 N. Determine the power transmitted and the maximum shear stress developed in the shaft.

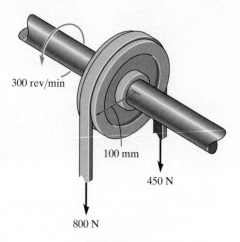

Prob. R10–9

REVIEW PROBLEMS

R10–1.   The shaft is made of A992 steel and has an allowable shear stress of $\tau_{allow} = 75$ MPa. when the shaft is rotating at 300 rpm, the motor supplies 8 kW of power, while gears A and B withdraw 5 kW and 3 kW, respectively. Determine the required minimum diameter of the shaft to the nearest millimeter. Also, find the rotation of gear A relative to C.

R10–2.   The shaft is made of A992 steel and has an allowable shear stress of $\tau_{allow} = 75$ MPa. when the shaft is rotating at 300 rpm, the motor supplies 8 kW of power, while gears A and B withdraw 5 kW and 3 kW, respectively. If the angle of twist of gear A relative to C is not allowed to exceed 0.03 rad, determine the required minimum diameter of the shaft to the nearest millimeter.

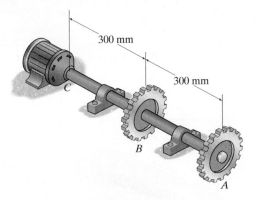

Probs. R10–1/2

R10–3.   The A-36 steel circular tube is subjected to a torque of 10 kN · m. Determine the shear stress at the mean radius $\rho = 60$ mm and calculate the angle of twist of the tube if it is 4 m long and fixed at its far end. Solve the problem using Eqs. 10–7 and 10–15 and using Eqs. 10–18 and 10–20.

Prob. R10–3

*R10–4.   The shaft has a radius c and is subjected to a torque per unit length of t_0, which is distributed uniformly over the shaft's entire length L. If it is fixed at its far end A, determine the angle of twist ϕ of end B. The shear modulus is G.

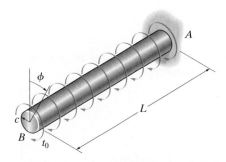

Prob. R10–4

R10–5.   The motor delivers 50 hp while turning at a constant rate of 1350 rpm at A. Using the belt and pulley system this loading is delivered to the steel blower shaft BC. Determine to the nearest $\frac{1}{8}$ in. the smallest diameter of this shaft if the allowable shear stress for steel is $\tau_{allow} = 12$ ksi.

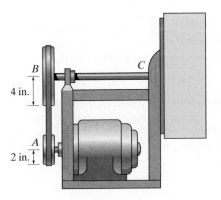

Prob. R10–5

CHAPTER REVIEW

Torque causes a shaft having a circular cross section to twist, such that whatever the torque, the shear strain in the shaft is always proportional to its radial distance from the center of the shaft.

Provided the material is homogeneous and linear elastic, then the shear stress is determined from the torsion formula,

$$\tau = \frac{T\rho}{J}$$

The design of a shaft requires finding the geometric parameter,

$$\frac{J}{c} = \frac{T}{\tau_{\text{allow}}}$$

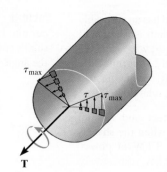
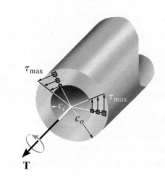

Often the power P supplied to a shaft rotating at ω is reported, in which case the torque is determined from $P = T\omega$.

The angle of twist of a circular shaft is determined from

$$\phi = \int_0^L \frac{T(x)\, dx}{J(x)G(x)}$$

If the internal torque and JG are constant within each segment of the shaft then

$$\phi = \sum \frac{TL}{JG}$$

For application, it is necessary to use a sign convention for the internal torque and to be sure the material remains linear elastic.

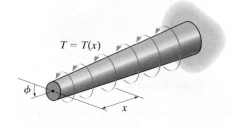
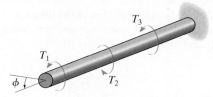

If the shaft is statically indeterminate, then the reactive torques are determined from equilibrium, compatibility of twist, and a load–displacement relationship, such as $\phi = TL/JG$.

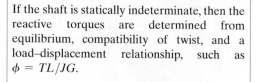

10–75. The two 3-ft-long shafts are made of 2014-16 aluminum. Each has a diameter of 1.5 in. and they are connected using the gears fixed to their ends. Their other ends are attached to fixed supports at A and B. They are also supported by bearings at C and D, which allow free rotation of the shafts along their axes. If a torque of 600 lb·ft is applied to the top gear as shown, determine the maximum shear stress in each shaft.

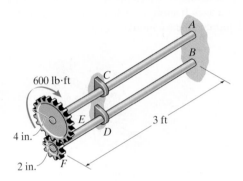

Prob. 10–75

*10–76. The composite shaft consists of a mid-section that includes the 1-in.-diameter solid shaft and a tube that is welded to the rigid flanges at A and B. Neglect the thickness of the flanges and determine the angle of twist of end C of the shaft relative to end D. The shaft is subjected to a torque of 800 lb · ft. The material is A-36 steel.

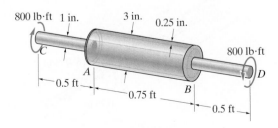

Prob. 10–76

10–77. If the shaft is subjected to a uniform distributed torque of $t = 20$ kN · m/m, determine the maximum shear stress developed in the shaft. The shaft is made of 2014-T6 aluminum alloy and is fixed at A and C.

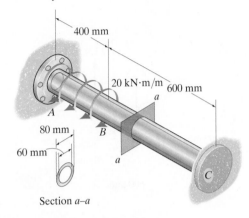

Prob. 10–77

10–78. The tapered shaft is confined by the fixed supports at A and B. If a torque **T** is applied at its mid-point, determine the reactions at the supports.

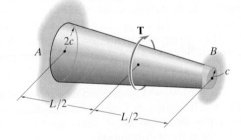

Prob. 10–78

10–79. The shaft of radius c is subjected to a distributed torque t, measured as torque/length of shaft. Determine the reactions at the fixed supports A and B.

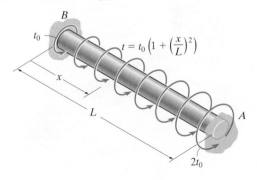

Prob. 10–79

*10–68. The shaft is made of L2 tool steel, has a diameter of 40 mm, and is fixed at its ends A and B. If it is subjected to the couple, determine the maximum shear stress in regions AC and CB.

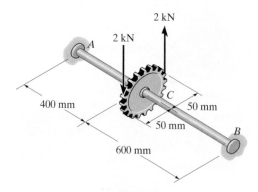

Prob. 10–68

10–69. The Am1004-T61 magnesium tube is bonded to the A-36 steel rod. If the allowable shear stresses for the magnesium and steel are $(\tau_{allow})_{mg} = 45$ MPa and $(\tau_{allow})_{st} = 75$ MPa, respectively, determine the maximum allowable torque that can be applied at A. Also, find the corresponding angle of twist of end A.

10–70. The Am1004-T61 magnesium tube is bonded to the A-36 steel rod. If a torque of $T = 5$ kN $\cdot$ m is applied to end A, determine the maximum shear stress in each material. Sketch the shear stress distribution.

Probs. 10–69/70

10–71. The two shafts are made of A-36 steel. Each has a diameter of 25 mm and they are connected using the gears fixed to their ends. Their other ends are attached to fixed supports at A and B. They are also supported by journal bearings at C and D, which allow free rotation of the shafts along their axes. If a torque of 500 N $\cdot$ m is applied to the gear at E, determine the reactions at A and B.

10–72. The two shafts are made of A-36 steel. Each has a diameter of 25 mm and they are connected using the gears fixed to their ends. Their other ends are attached to fixed supports at A and B. They are also supported by journal bearings at C and D, which allow free rotation of the shafts along their axes. If a torque of 500 N $\cdot$ m is applied to the gear at E, determine the rotation of this gear.

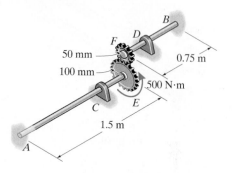

Probs. 10–71/72

10–73. A rod is made from two segments: AB is steel and BC is brass. It is fixed at its ends and subjected to a torque of $T = 680$ N $\cdot$ m. If the steel portion has a diameter of 30 mm, determine the required diameter of the brass portion so the reactions at the walls will be the same. $G_{st} = 75$ GPa, $G_{br} = 39$ GPa.

10–74. Determine the absolute maximum shear stress in the shaft of Prob. 10–73.

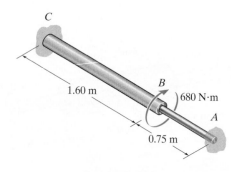

Probs. 10–73/74

BENDING

11.1 SHEAR AND MOMENT DIAGRAMS

Members that are slender and support loadings that are applied perpendicular to their longitudinal axis are called *beams*. In general, beams are long, straight bars having a constant cross-sectional area. Often they are classified as to how they are supported. For example, a *simply supported beam* is pinned at one end and roller supported at the other, Fig. 11–1, a *cantilevered beam* is fixed at one end and free at the other, and an *overhanging beam* has one or both of its ends freely extended over the

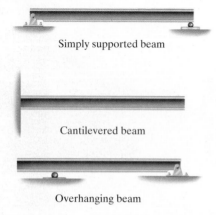

Simply supported beam

Cantilevered beam

Overhanging beam

Fig. 11–1

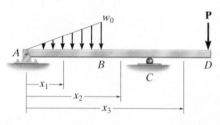

Fig. 11–2

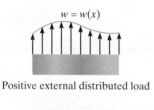

$w = w(x)$

Positive external distributed load

Positive internal shear

Positive internal moment

Beam sign convention

Fig. 11–3

supports. Beams are considered among the most important of all structural elements. They are used to support the floor of a building, the deck of a bridge, or the wing of an aircraft. Also, the axle of an automobile, the boom of a crane, even many of the bones of the body act as beams.

Because of the applied loadings, beams develop an internal shear force and bending moment that, in general, vary from point to point along the axis of the beam. In order to properly design a beam it therefore becomes important to determine the *maximum* shear and moment in the beam. One way to do this is to express V and M as functions of their arbitrary position x along the beam's axis, and then plot these functions. They represent the ***shear and moment diagrams***, respectively. The maximum values of V and M can then be obtained directly from these graphs. Also, since the shear and moment diagrams provide detailed information about the *variation* of the shear and moment along the beam's axis, they are often used by engineers to decide where to place reinforcement materials within the beam or how to proportion the size of the beam at various points along its length.

In order to formulate V and M in terms of x we must choose the origin and the positive direction for x. Although the choice is arbitrary, most often the origin is located at the left end of the beam and the positive x direction is to the right.

Since beams can support portions of a distributed load and concentrated forces and couple moments, the internal shear and moment functions of x will be *discontinuous*, or their slopes will be discontinuous, at points where the loads are applied. Because of this, these functions must be determined for each region of the beam *between* any two discontinuities of loading. For example, coordinates x_1, x_2, and x_3 will have to be used to describe the variation of V and M throughout the length of the beam in Fig. 11–2. Here the coordinates are valid *only* within the regions from A to B for x_1, from B to C for x_2, and from C to D for x_3.

Beam Sign Convention. Before presenting a method for determining the shear and moment as functions of x, and later plotting these functions (shear and moment diagrams), it is first necessary to establish a *sign convention* in order to define "positive" and "negative" values for V and M. Although the choice of a sign convention is arbitrary, here we will use the one often used in engineering practice. It is shown in Fig. 11–3. The *positive directions* are as follows: the *distributed load* acts *upward* on the beam, the internal *shear force* causes a *clockwise* rotation of the beam segment on which it acts, and the internal *moment* causes *compression* in the *top fibers* of the segment such that it bends the segment so that it "holds water." Loadings that are opposite to these are considered negative.

▶ IMPORTANT POINTS

- *Beams* are long straight members that are subjected to loads perpendicular to their longitudinal axis. They are classified according to the way they are supported, e.g., simply supported, cantilevered, or overhanging.

- In order to properly design a beam, it is important to know the *variation* of the internal shear and moment along its axis in order to find the points where these values are a maximum.

- Using an established sign convention for positive shear and moment, the shear and moment in the beam can be determined as a function of their position x on the beam, and then these functions can be plotted to form the shear and moment diagrams.

▶ PROCEDURE FOR ANALYSIS

The shear and moment diagrams for a beam can be constructed using the following procedure.

Support Reactions.

- Determine all the reactive forces and couple moments acting on the beam, and resolve all the forces into components acting perpendicular and parallel to the beam's axis.

Shear and Moment Functions.

- Specify separate coordinates x having an origin at the beam's *left end* and extending to regions of the beam between concentrated forces and/or couple moments, or where there is no discontinuity of distributed loading.

- Section the beam at each distance x, and draw the free-body diagram of one of the segments. Be sure **V** and **M** are shown acting in their positive sense, in accordance with the sign convention given in Fig. 11–3.

- The shear is obtained by summing forces perpendicular to the beam's axis.

- To eliminate V, the moment is obtained directly by summing moments about the sectioned end of the segment.

Shear and Moment Diagrams.

- Plot the shear diagram (V versus x) and the moment diagram (M versus x). If numerical values of the functions describing V and M are *positive*, the values are plotted above the x axis, whereas negative values are plotted below the axis.

- Generally it is convenient to show the shear and moment diagrams below the free-body diagram of the beam.

11

EXAMPLE 11.1

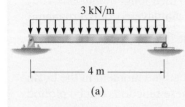

3 kN/m

4 m

(a)

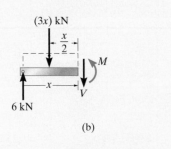

(3x) kN

$\frac{x}{2}$

M

x

V

6 kN

(b)

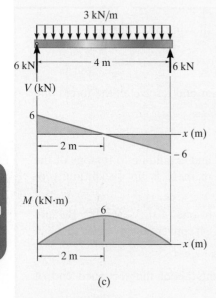

3 kN/m

6 kN   4 m   6 kN

V (kN)

6

2 m

x (m)

−6

M (kN·m)

6

2 m

x (m)

(c)

Fig. 11–4

Draw the shear and moment diagrams for the beam shown in Fig. 11–4a.

SOLUTION

Support Reactions.   The support reactions are shown in Fig. 11–4c.

Shear and Moment Functions.   A free-body diagram of the left segment of the beam is shown in Fig. 11–4b. The distributed loading on this segment is represented by its resultant force $(3x)$ kN, which is found only *after* the segment is isolated as a free-body diagram. This force acts through the centroid of the area under the distributed loading, a distance of $x/2$ from the right end. Applying the two equations of equilibrium yields

$$+\uparrow \Sigma F_y = 0; \qquad 6 \text{ kN} - (3x) \text{ kN} - V = 0$$

$$V = (6 - 3x) \text{ kN} \qquad (1)$$

$$\zeta + \Sigma M = 0; \qquad -6 \text{ kN}(x) + (3x) \text{ kN} \left(\tfrac{1}{2}x\right) + M = 0$$

$$M = (6x - 1.5x^2) \text{ kN} \cdot \text{m} \qquad (2)$$

Shear and Moment Diagrams.   The shear and moment diagrams shown in Fig. 11–4c are obtained by plotting Eqs. 1 and 2. The point of *zero shear* can be found from Eq. 1:

$$V = (6 - 3x) \text{ kN} = 0$$

$$x = 2 \text{ m}$$

NOTE: From the moment diagram, this value of x represents the point on the beam where the *maximum moment* occurs, since by Eq. 11–2 (see Sec. 11.2) the *slope* $V = dM/dx = 0$. From Eq. 2, we have

$$M_{\max} = [6\,(2) - 1.5\,(2)^2] \text{ kN} \cdot \text{m}$$

$$= 6 \text{ kN} \cdot \text{m}$$

EXAMPLE 11.2

Draw the shear and moment diagrams for the beam shown in Fig. 11–5a.

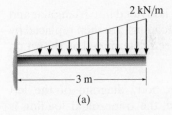

2 kN/m

3 m

(a)

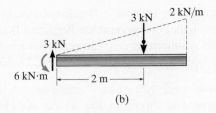

3 kN   2 kN/m

3 kN

6 kN·m   2 m

(b)

SOLUTION

Support Reactions. The distributed load is replaced by its resultant force, and the reactions have been determined, as shown in Fig. 11–5b.

Shear and Moment Functions. A free-body diagram of a beam segment of length x is shown in Fig. 11–5c. The intensity of the triangular load at the section is found by proportion, that is, $w/x = (2 \text{ kN/m})/3 \text{ m}$ or $w = \left(\frac{2}{3}x\right)$ kN/m. The resultant of the distributed loading is found from the area under the diagram. Thus,

$$+\uparrow \Sigma F_y = 0; \qquad 3 \text{ kN} - \frac{1}{2}\left(\frac{2}{3}x\right)x - V = 0$$

$$V = \left(3 - \frac{1}{3}x^2\right) \text{ kN} \qquad (1)$$

$$\zeta + \Sigma M = 0; \qquad 6 \text{ kN} \cdot \text{m} - (3 \text{ kN})(x) + \frac{1}{2}\left(\frac{2}{3}x\right)x\left(\frac{1}{3}x\right) + M = 0$$

$$M = \left(-6 + 3x - \frac{1}{9}x^3\right) \text{ kN} \cdot \text{m} \qquad (2)$$

Shear and Moment Diagrams. The graphs of Eqs. 1 and 2 are shown in Fig. 11–5d.

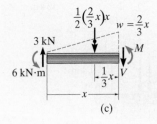

(c)

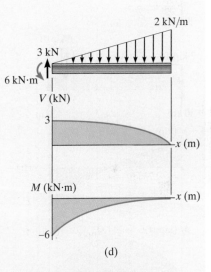

(d)

Fig. 11–5

11

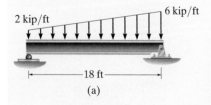

2 kip/ft   6 kip/ft

├────18 ft────┤

(a)

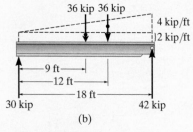

36 kip 36 kip

4 kip/ft
2 kip/ft

├─9 ft─┤
├──12 ft──┤
├────18 ft────┤

30 kip   42 kip

(b)

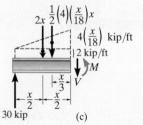

$2x \frac{1}{2}(4)\left(\frac{x}{18}\right)x$

$4\left(\frac{x}{18}\right)$ kip/ft

2 kip/ft

M

$\frac{x}{3}$ V

$\frac{x}{2}$ $\frac{x}{2}$

30 kip   (c)

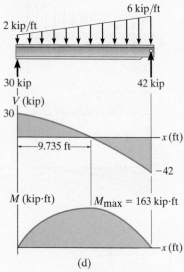

6 kip/ft

2 kip/ft

30 kip   42 kip

V (kip)

30

├─9.735 ft─┤ x (ft)

−42

M (kip·ft) M_{max} = 163 kip·ft

x (ft)

(d)

Fig. 11–6

Draw the shear and moment diagrams for the beam shown in Fig. 11–6a.

SOLUTION

Support Reactions.   The distributed load is divided into triangular and rectangular component loadings, and these loadings are then replaced by their resultant forces. The reactions have been determined as shown on the beam's free-body diagram, Fig. 11–6b.

Shear and Moment Functions.   A free-body diagram of the left segment is shown in Fig. 11–6c. As above, the trapezoidal loading is replaced by rectangular and triangular distributions. Here the intensity of the triangular load at the section is found by proportion. The resultant force and the location of each distributed loading are also shown. Applying the equilibrium equations, we have

$$+\uparrow \Sigma F_y = 0; \qquad 30 \text{ kip} - (2 \text{ kip/ft})x - \frac{1}{2}(4 \text{ kip/ft})\left(\frac{x}{18 \text{ ft}}\right)x - V = 0$$

$$V = \left(30 - 2x - \frac{x^2}{9}\right) \text{kip} \qquad (1)$$

$$\zeta + \Sigma M = 0;$$

$$-30 \text{ kip}(x) + (2 \text{ kip/ft})x\left(\frac{x}{2}\right) + \frac{1}{2}(4 \text{ kip/ft})\left(\frac{x}{18 \text{ ft}}\right)x\left(\frac{x}{3}\right) + M = 0$$

$$M = \left(30x - x^2 - \frac{x^3}{27}\right) \text{kip} \cdot \text{ft} \qquad (2)$$

Shear and Moment Diagrams.   Equations 1 and 2 are plotted in Fig. 11–6d. Since the point of maximum moment occurs when $dM/dx = V = 0$ (Eq. 11–2), then, from Eq. 1,

$$V = 0 = 30 - 2x - \frac{x^2}{9}$$

Choosing the positive root,

$$x = 9.735 \text{ ft}$$

Thus, from Eq. 2,

$$M_{max} = 30(9.735) - (9.735)^2 - \frac{(9.735)^3}{27} = 163 \text{ kip} \cdot \text{ft}$$

EXAMPLE 11.4

Draw the shear and moment diagrams for the beam shown in Fig. 11–7a.

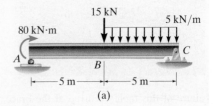

(a)

SOLUTION

Support Reactions. The reactions at the supports are shown on the free-body diagram of the beam, Fig. 11–7d.

Shear and Moment Functions. Since there is a discontinuity of distributed load and also a concentrated load at the beam's center, two regions of x must be considered in order to describe the shear and moment functions for the entire beam.

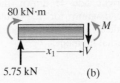

(b)

$0 \le x_1 < 5$ m, Fig. 11–7b:

$$+\uparrow \Sigma F_y = 0; \qquad 5.75 \text{ kN} - V = 0$$

$$V = 5.75 \text{ kN} \qquad (1)$$

$$\zeta + \Sigma M = 0; \qquad -80 \text{ kN} \cdot \text{m} - 5.75 \text{ kN } x_1 + M = 0$$

$$M = (5.75x_1 + 80) \text{ kN} \cdot \text{m} \qquad (2)$$

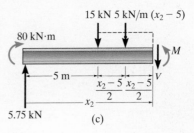

(c)

5 m $< x_2 \le 10$ m, Fig. 11–7c:

$$+\uparrow \Sigma F_y = 0; \quad 5.75 \text{ kN} - 15 \text{ kN} - 5 \text{ kN/m}(x_2 - 5 \text{ m}) - V = 0$$

$$V = (15.75 - 5x_2) \text{ kN} \qquad (3)$$

$$\zeta + \Sigma M = 0; \quad -80 \text{ kN} \cdot \text{m} - 5.75 \text{ kN } x_2 + 15 \text{ kN}(x_2 - 5 \text{ m})$$

$$+ 5 \text{ kN/m}(x_2 - 5 \text{ m})\left(\frac{x_2 - 5 \text{ m}}{2}\right) + M = 0$$

$$M = (-2.5x_2^2 + 15.75x_2 + 92.5) \text{ kN} \cdot \text{m} \qquad (4)$$

Shear and Moment Diagrams. Equations 1 through 4 are plotted in Fig. 11–7d.

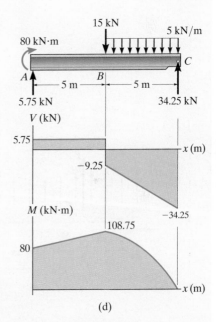

(d)

Fig. 11–7

11

Failure of this table occurred at the brace support on its right side. If drawn, the bending-moment diagram for the table loading would indicate this to be the point of maximum internal moment.

11.2 GRAPHICAL METHOD FOR CONSTRUCTING SHEAR AND MOMENT DIAGRAMS

In cases where a beam is subjected to *several* different loadings, determining V and M as functions of x and then plotting these equations can become quite tedious. In this section a simpler method for constructing the shear and moment diagrams is discussed—a method based on two differential relations, one that exists between the distributed load and shear, and the other between the shear and moment.

Regions of Distributed Load. For purposes of generality, consider the beam shown in Fig. 11–8a, which is subjected to an arbitrary loading. A free-body diagram for a very small segment Δx of the beam is shown in Fig. 11–8b. Since this segment has been chosen at a position x where there is no concentrated force or couple moment, the results to be obtained will *not* apply at these points.

Notice that all the loadings shown on the segment act in their positive directions according to the established sign convention, Fig. 11–3. Also, both the internal resultant shear and moment, acting on the right face of the segment, must be changed by a small amount in order to keep the segment in equilibrium. The distributed load, which is approximately constant over Δx, has been replaced by a resultant force $w\Delta x$ that acts at $\frac{1}{2}(\Delta x)$ from the right side. Applying the equations of equilibrium to the segment, we have

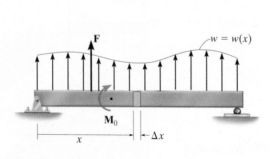

(a)

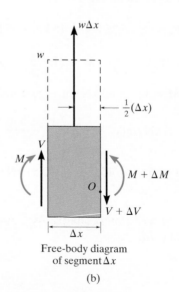

Free-body diagram
of segment Δx

(b)

Fig. 11–8

$$+\uparrow \Sigma F_y = 0; \qquad V + w\,\Delta x - (V + \Delta V) = 0$$

$$\Delta V = w\,\Delta x$$

$$\zeta + \Sigma M_O = 0; \qquad -V\,\Delta x - M - w\,\Delta x\left[\tfrac{1}{2}(\Delta x)\right] + (M + \Delta M) = 0$$

$$\Delta M = V\,\Delta x + w\,\tfrac{1}{2}(\Delta x)^2$$

Dividing by Δx and taking the limit as $\Delta x \to 0$, the above two equations become

$$\boxed{\dfrac{dV}{dx} = w} \qquad (11\text{–}1)$$

slope of	distributed
shear diagram	load intensity
at each point	at each point

$$\boxed{\dfrac{dM}{dx} = V} \qquad (11\text{–}2)$$

slope of	shear
moment diagram	at each
at each point	point

Equation 11–1 states that at any point the *slope* of the shear diagram equals the intensity of the distributed loading. For example, consider the beam in Fig. 11–9a. The distributed loading is negative and increases from zero to w_B. Knowing this provides a quick means for drawing the shape of the shear diagram. It must be a curve that has a *negative slope*, increasing from zero to $-w_B$. Specific slopes $w_A = 0$, $-w_C$, $-w_D$, and $-w_B$ are shown in Fig. 11–9b.

In a similar manner, Eq. 11–2 states that at any point the *slope* of the moment diagram is equal to the shear. Since the shear diagram in Fig. 11–9b starts at $+V_A$, decreases to zero, and then becomes negative and decreases to $-V_B$, the moment diagram (or curve) will then have an initial slope of $+V_A$ which decreases to zero, then the slope becomes negative and decreases to $-V_B$. Specific slopes V_A, V_C, V_D, 0, and $-V_B$ are shown in Fig. 11–9c.

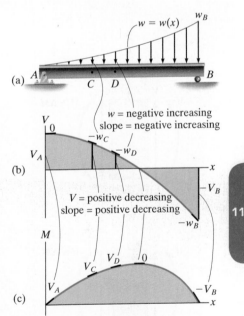

Fig. 11–9

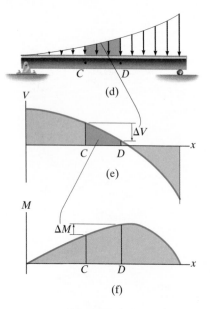

(d)

(e)

(f)

Fig. 11–9 (cont.)

Equations 11–1 and 11–2 may also be rewritten in the form $dV = w\,dx$ and $dM = V\,dx$. Since $w\,dx$ and $V\,dx$ represent differential areas under the distributed loading and the shear diagram, we can then integrate these areas between any two points C and D on the beam, Fig. 11–9d, and write

$$\Delta V = \int w\,dx \qquad (11\text{–}3)$$

change in   area under
shear   distributed loading

$$\Delta M = \int V\,dx \qquad (11\text{–}4)$$

change in   area under
moment   shear diagram

Equation 11–3 states that the *change in shear* between C and D is equal to the *area* under the distributed-loading curve between these two points, Fig. 11–9d. In this case the change is negative since the distributed load acts downward. Similarly, from Eq. 11–4, the change in moment between C and D, Fig. 11–9f, is equal to the area under the shear diagram within the region from C to D. Here the change is positive.

Regions of Concentrated Force and Moment. A free-body diagram of a small segment of the beam in Fig. 11–8a taken from under the force is shown in Fig. 11–10a. Here force equilibrium requires

$$+\uparrow \Sigma F_y = 0; \qquad V + F - (V + \Delta V) = 0$$

$$\Delta V = F \qquad (11\text{–}5)$$

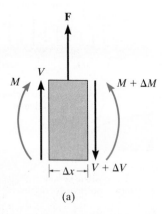

(a)

Thus, when $\mathbf{F}$ acts *upward* on the beam, then the change in shear, ΔV, is *positive* so the values of the shear on the shear diagram will "jump" *upward*. Likewise, if $\mathbf{F}$ acts *downward*, the jump (ΔV) will be *downward*.

When the beam segment includes the couple moment M_0, Fig. 11–10b, then moment equilibrium requires the change in moment to be

$$\zeta + \Sigma M_O = 0; \quad M + \Delta M - M_0 - V\,\Delta x - M = 0$$

Letting $\Delta x \approx 0$, we get

$$\Delta M = M_0 \qquad (11\text{–}6)$$

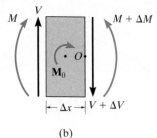

(b)

Fig. 11–10

In this case, if $\mathbf{M}_0$ is applied *clockwise,* the change in moment, ΔM, is *positive* so the moment diagram will "jump" *upward*. Likewise, when $\mathbf{M}_0$ acts *counterclockwise,* the jump (ΔM) will be *downward*.

PROCEDURE FOR ANALYSIS

The following procedure provides a method for constructing the shear and moment diagrams for a beam based on the relations among distributed load, shear, and moment.

Support Reactions.

- Determine the support reactions and resolve the forces acting on the beam into components that are perpendicular and parallel to the beam's axis.

Shear Diagram.

- Establish the V and x axes and plot the known values of the shear at the two *ends* of the beam.
- Notice how the values of the distributed load vary along the beam, such as positive increasing, negative increasing, etc., and realize that each of these successive values indicates the way the shear diagram will slope $(dV/dx = w)$. Here w is positive when it acts upward. Begin by sketching the slope at the end points.
- If a numerical value of the shear is to be determined at a point, one can find this value either by using the method of sections and the equation of force equilibrium, or by using $\Delta V = \int w\, dx$, which states that the *change in the shear* between any two points is equal to the *area under the load diagram* between the two points.

Moment Diagram.

- Establish the M and x axes and plot the known values of the moment at the *ends* of the beam.
- Notice how the values of the shear diagram vary along the beam, such as positive increasing, negative increasing, etc., and realize that each of these successive values indicates the way the moment diagram will slope $(dM/dx = V)$. Begin by sketching the slope at the end points.
- At the point where the shear is zero, $dM/dx = 0$, and therefore this will be a point of maximum or minimum moment.
- If a numerical value of the moment is to be determined at the point, one can find this value either by using the method of sections and the equation of moment equilibrium, or by using $\Delta M = \int V\, dx$, which states that the *change in moment* between any two points is equal to the *area under the shear diagram* between the two points.
- Since w must be *integrated* to obtain ΔV, and V is integrated to obtain M, then if w is a curve of degree n, V will be a curve of degree $n + 1$ and M will be a curve of degree $n + 2$. For example, if w is uniform, V will be linear and M will be parabolic.

11

EXAMPLE 11.5

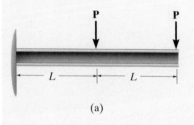

(a)

Draw the shear and moment diagrams for the beam shown in Fig. 11–11a.

SOLUTION

Support Reactions. The reaction at the fixed support is shown on the free-body diagram, Fig. 11–11b.

Shear Diagram. The shear at each end of the beam is plotted first, Fig. 11–11c. Since there is no distributed loading on the beam, the slope of the shear diagram is zero as indicated. Note how the force P at the center of the beam causes the shear diagram to jump downward an amount P, since this force acts downward.

Moment Diagram. The moments at the ends of the beam are plotted, Fig. 11–11d. Here the moment diagram consists of two sloping lines, one with a slope of $+2P$ and the other with a slope of $+P$.

The value of the moment in the center of the beam can be determined by the method of sections, or from the area under the shear diagram. If we choose the left half of the shear diagram,

$$M|_{x=L} = M|_{x=0} + \Delta M$$
$$M|_{x=L} = -3PL + (2P)(L) = -PL$$

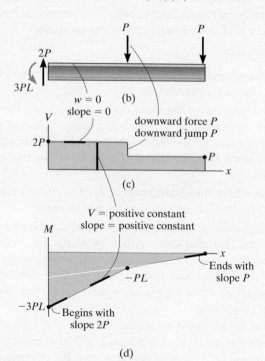

Fig. 11–11

EXAMPLE   11.6

Draw the shear and moment diagrams for the beam shown in Fig. 11–12a.

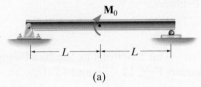

(a)

SOLUTION

Support Reactions.   The reactions are shown on the free-body diagram in Fig. 11–12b.

Shear Diagram.   The shear at each end is plotted first, Fig. 11–12c. Since there is no distributed load on the beam, the shear diagram has zero slope and is therefore a horizontal line.

Moment Diagram.   The moment is zero at each end, Fig. 11–12d. The moment diagram has a constant negative slope of $-M_0/2L$ since this is the shear in the beam at each point. However, here the couple moment M_0 causes a jump in the moment diagram at the beam's center.

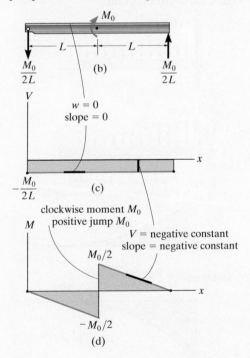

Fig. 11–12

EXAMPLE 11.7

Draw the shear and moment diagrams for each of the beams shown in Figs. 11–13a and 11–14a.

SOLUTION

Support Reactions. The reactions at the fixed support are shown on each free-body diagram, Figs. 11–13b and 11–14b.

Shear Diagram. The shear at each end point is plotted first, Figs. 11–13c and 11–14c. The distributed loading on each beam indicates the slope of the shear diagram and thus produces the shapes shown.

Moment Diagram. The moment at each end point is plotted first, Figs. 11–13d and 11–14d. Various values of the shear at each point on the beam indicate the slope of the moment diagram at the point. Notice how this variation produces the curves shown.

NOTE: Observe how the degree of the curves from w to V to M increases by one due to the integration of $dV = w\,dx$ and $dM = V\,dx$. For example, in Fig. 11–14, the linear distributed load produces a parabolic shear diagram and cubic moment diagram.

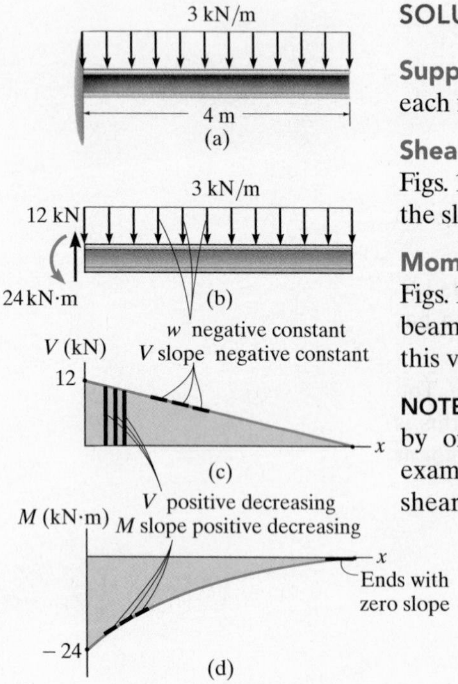

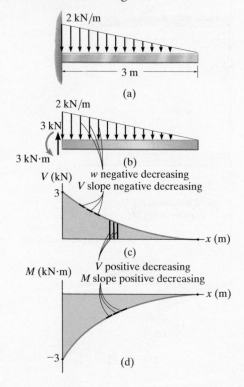

Fig. 11–13

Fig. 11–14

EXAMPLE   11.8

Draw the shear and moment diagrams for the cantilever beam in Fig. 11–15a.

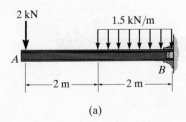

(a)

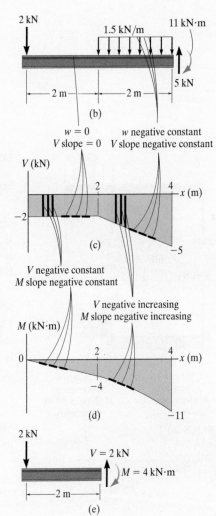

(b)

$w = 0$ w negative constant
V slope $= 0$ V slope negative constant

(c)

V negative constant
M slope negative constant

V negative increasing
M slope negative increasing

(d)

2 kN $V = 2$ kN

$M = 4$ kN·m

(e)

Fig. 11–15

SOLUTION

Support Reactions. The support reactions at the fixed support B are shown in Fig. 11–15b.

Shear Diagram. The shear at the ends is plotted first, Fig. 11–15c. Notice how the shear diagram is constructed by following the slopes defined by the loading w.

Moment Diagram. The moments at the ends of the beam are plotted first, Fig. 11–15d. Notice how the moment diagram is constructed based on knowing its slope, which is equal to the shear at each point. The moment at $x = 2$ m can be found from the area under the shear diagram. We have

$$M|_{x=2\,m} = M|_{x=0} + \Delta M = 0 + [-2\,\text{kN}(2\,\text{m})] = -4\,\text{kN} \cdot \text{m}$$

Of course, this same value can be determined from the method of sections, Fig. 11–15e.

11

EXAMPLE 11.9

Draw the shear and moment diagrams for the overhang beam in Fig. 11–16a.

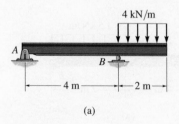

4 kN/m

A

B

4 m   2 m

(a)

SOLUTION

Support Reactions.   The support reactions are shown in Fig. 11–16b.

Shear Diagram.   The shear at the ends is plotted first, Fig. 11–16c. The slopes are determined from the loading and from this the shear diagram is constructed. Notice the positive jump of 10 kN at $x = 4$ m due to the force reaction.

Moment Diagram.   The moments at the ends are plotted first, Fig. 11–16d. Then following the behavior of the slope found from the shear diagram, the moment diagram is constructed. The moment at $x = 4$ m is found from the area under the shear diagram.

$$M|_{x=4\,m} = M|_{x=0} + \Delta M = 0 + [-2\text{ kN}(4\text{ m})] = -8\text{ kN} \cdot \text{m}$$

We can also obtain this value by using the method of sections, as shown in Fig. 11–16e.

4 kN/m

4 m   2 m

(b)

2 kN $w = 0$ 10 kN
V slope = 0
w negative constant
V (kN) V slope negative constant

8

0 ──────── 4 ──→ x (m)
6
−2

V negative   (c) V positive
constant   decreasing
M slope negative M slope positive
constant   decreasing

M (kN·m)

slope = 0
0 ──── 4 ──→ x (m)
6
−8
(d)

Fig. 11–16

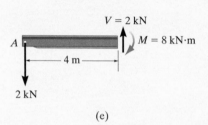

$V = 2$ kN

A $M = 8$ kN·m

4 m

2 kN

(e)

EXAMPLE 11.10

The shaft in Fig. 11–17a is supported by a thrust bearing at A and a journal bearing at B. Draw the shear and moment diagrams.

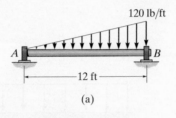

120 lb/ft

A B

— 12 ft —

(a)

SOLUTION

Support Reactions.   The support reactions are shown in Fig. 11–17b.

Shear Diagram.   As shown in Fig. 11–17c, the shears at the ends of the beam are +240 lb and –480 lb. The point where $V = 0$ must be located. To do this we will use the method of sections. The free-body diagram of the left segment of the shaft, sectioned at an arbitrary position x, is shown in Fig. 11–17e. Here the intensity of the distributed load at x is $w = 10x$, which has been found by proportional triangles, i.e., $120/12 = w/x$. Thus, for $V = 0$,

$$+\uparrow \Sigma F_y = 0; \qquad 240 \text{ lb} - \tfrac{1}{2}(10x)x = 0$$

$$x = 6.93 \text{ ft}$$

Moment Diagram.   The moment diagram starts and ends at 0. The maximum moment occurs at $x = 6.93$ ft, where the shear is equal to zero, since $dM/dx = V = 0$, Fig. 11–17d. From Fig. 11–17e, we have

$$\zeta + \Sigma M = 0; \quad M_{max} + \tfrac{1}{2}[(10)(6.93)]\, 6.93 \left(\tfrac{1}{3}(6.93)\right) - 240(6.93) = 0$$

$$M_{max} = 1109 \text{ lb} \cdot \text{ft}$$

Finally, notice how integration, first of the loading w, which is linear, produces a shear diagram which is parabolic, and then a moment diagram which is cubic.

NOTE: Now test yourself by covering over the shear and moment diagrams in Examples 11.1 through 11.4, and see if you can construct them based on the concepts discussed here.

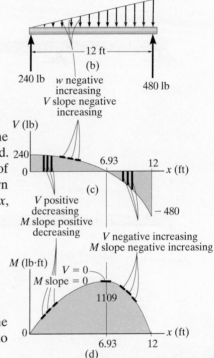

120 lb/ft

— 12 ft —

(b)

240 lb w negative
increasing
V slope negative
increasing

480 lb

V (lb)

240

0   6.93   12
                                    x (ft)

(c)

V positive
decreasing
M slope positive
decreasing V negative increasing
                     M slope negative increasing

– 480

M (lb·ft) $V = 0$
              M slope $= 0$

                   1109

0
        6.93   12
               x (ft)
(d)

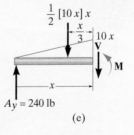

$\tfrac{1}{2}[10x]\,x$

$\dfrac{x}{3}$ $10x$

V

M

x

$A_y = 240$ lb

(e)

Fig. 11–17

PRELIMINARY PROBLEM

P11–1. In each case, the beam is subjected to the loadings shown. Draw the free-body diagram of the beam, and sketch the general shape of the shear and moment diagrams. The loads and geometry are assumed to be known.

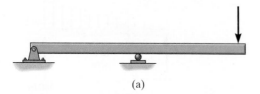

(a)

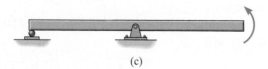

(b)

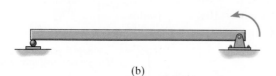

(c)

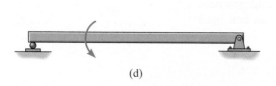

(d)

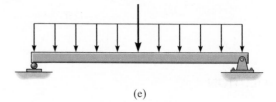

(e)

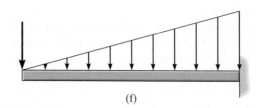

(f)

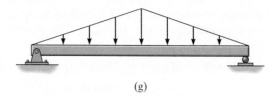

(g)

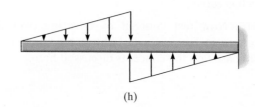

(h)

Prob. P11–1

FUNDAMENTAL PROBLEMS

In each case, express the shear and moment functions in terms of x, and then draw the shear and moment diagrams for the beam.

In each case, draw the shear and moment diagrams for the beam.

F11–1.

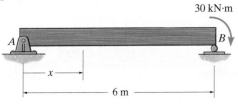

30 kN·m

A

B

x

6 m

Prob. F11–1

F11–2.

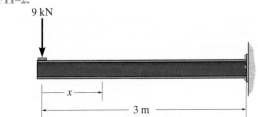

9 kN

x

3 m

Prob. F11–2

F11–3.

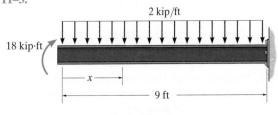

2 kip/ft

18 kip·ft

x

9 ft

Prob. F11–3

F11–4.

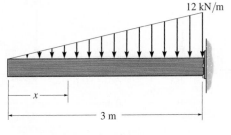

12 kN/m

x

3 m

Prob. F11–4

F11–5.

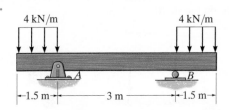

4 kN/m   4 kN/m

A B

1.5 m   3 m   1.5 m

Prob. F11–5

F11–6.

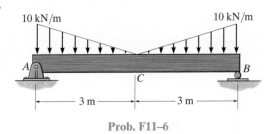

10 kN/m   10 kN/m

A B

C

3 m   3 m

Prob. F11–6

F11–7.

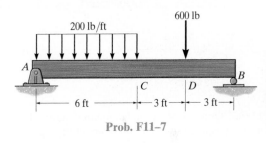

600 lb

200 lb/ft

A B

C D

6 ft   3 ft   3 ft

Prob. F11–7

F11–8.

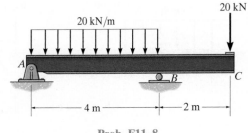

20 kN

20 kN/m

A B C

4 m   2 m

Prob. F11–8

11

PROBLEMS

11–1. Draw the shear and moment diagrams for the shaft and determine the shear and moment throughout the shaft as a function of x for $0 \leq x < 3$ ft, 3 ft $< x < 5$ ft, and 5 ft $< x < 6$ ft. The bearings at A and B exert only vertical reactions on the shaft.

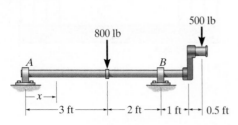

Prob. 11–1

11–2. Draw the shear and moment diagrams for the beam, and determine the shear and moment in the beam as functions of x for $0 \leq x < 4$ ft, 4 ft $< x < 10$ ft, and 10 ft $< x < 14$ ft.

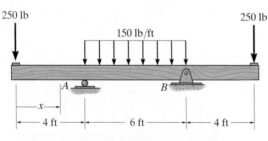

Prob. 11–2

11–3. Draw the shear and moment diagrams for the beam, and determine the shear and moment throughout the beam as functions of x for $0 \leq x \leq 6$ ft and 6 ft $\leq x \leq 10$ ft.

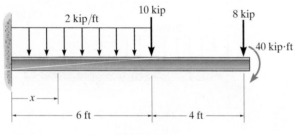

Prob. 11–3

*11–4. Express the shear and moment in terms of x for $0 < x < 3$ m and 3 m $< x < 4.5$ m, and then draw the shear and moment diagrams for the simply supported beam.

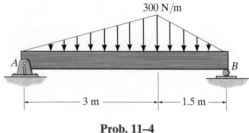

Prob. 11–4

11–5. Express the internal shear and moment in the cantilevered beam as a function of x and then draw the shear and moment diagrams.

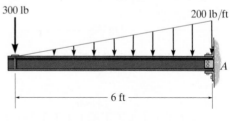

Prob. 11–5

11–6. Draw the shear and moment diagrams for the shaft. The bearings at A and B exert only vertical reactions on the shaft. Also, express the shear and moment in the shaft as a function of x within the region 125 mm $< x < 725$ mm.

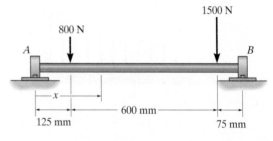

Prob. 11–6

11–7. Express the internal shear and moment in terms of x for $0 \leq x < L/2$, and $L/2 < x < L$, and then draw the shear and moment diagrams.

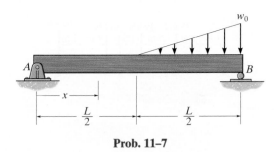

Prob. 11–7

*11–8. Draw the shear and moment diagrams for the beam, and determine the shear and moment throughout the beam as functions of x for $0 \leq x \leq 6$ ft and 6 ft $\leq x \leq 9$ ft.

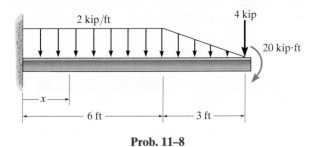

Prob. 11–8

11–9. If the force applied to the handle of the load binder is 50 lb, determine the tensions T_1 and T_2 in each end of the chain and then draw the shear and moment diagrams for the arm ABC.

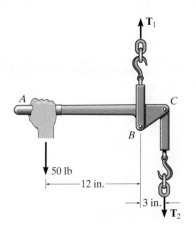

Prob. 11–9

11–10. Draw the shear and moment diagrams for the shaft. The bearings at A and D exert only vertical reactions on the shaft.

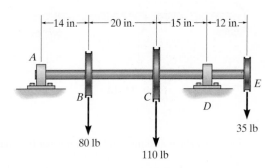

Prob. 11–10

11–11. The crane is used to support the engine, which has a weight of 1200 lb. Draw the shear and moment diagrams of the boom ABC when it is in the horizontal position.

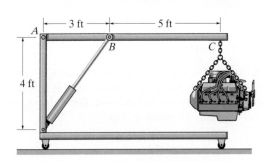

Prob. 11–11

*11–12. Draw the shear and moment diagrams for the beam.

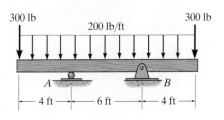

Prob. 11–12

11–13. Draw the shear and moment diagrams for the beam.

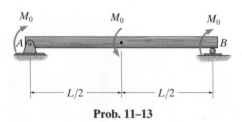

Prob. 11–13

11–14. Draw the shear and moment diagrams for the beam.

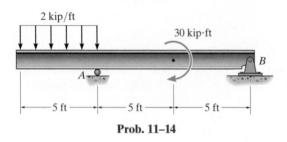

Prob. 11–14

11–15. Members ABC and BD of the counter chair are rigidly connected at B and the smooth collar at D is allowed to move freely along the vertical post. Draw the shear and moment diagrams for member ABC.

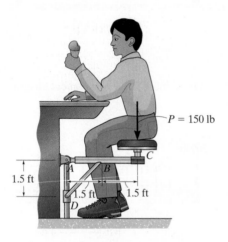

Prob. 11–15

*11–16. A reinforced concrete pier is used to support the stringers for a bridge deck. Draw the shear and moment diagrams for the pier. Assume the columns at A and B exert only vertical reactions on the pier.

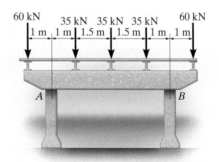

Prob. 11–16

11–17. Draw the shear and moment diagrams for the beam and determine the shear and moment in the beam as functions of x, where $4 \text{ ft} < x < 10 \text{ ft}$.

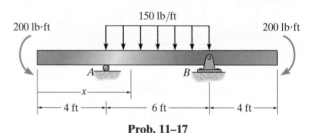

Prob. 11–17

11–18. The industrial robot is held in the stationary position shown. Draw the shear and moment diagrams of the arm ABC if it is pin connected at A and connected to a hydraulic cylinder (two-force member) BD. Assume the arm and grip have a uniform weight of 1.5 lb/in. and support the load of 40 lb at C.

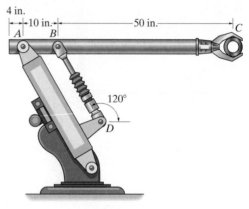

Prob. 11–18

11

11–19. Determine the placement distance a of the roller support so that the largest absolute value of the moment is a minimum. Draw the shear and moment diagrams for this condition.

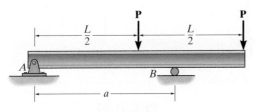

Prob. 11–19

*11–20. Draw the shear and moment diagrams for the beam.

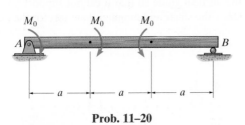

Prob. 11–20

11–21. Draw the shear and moment diagrams for the beam.

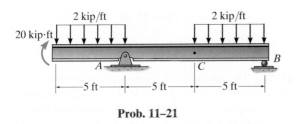

Prob. 11–21

11–22. Draw the shear and moment diagrams for the overhanging beam.

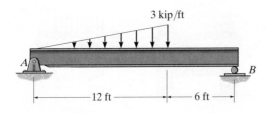

Prob. 11–22

11–23. The 150-lb man sits in the center of the boat, which has a uniform width and a weight per linear foot of 3 lb/ft. Determine the maximum internal bending moment. Assume that the water exerts a uniform distributed load upward on the bottom of the boat.

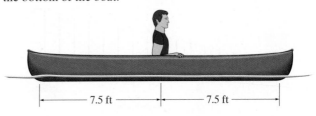

Prob. 11–23

*11–24. Draw the shear and moment diagrams for the beam.

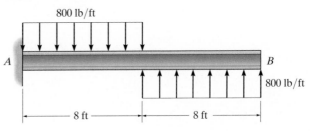

Prob. 11–24

11–25. The footing supports the load transmitted by the two columns. Draw the shear and moment diagrams for the footing if the soil pressure on the footing is assumed to be uniform.

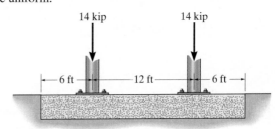

Prob. 11–25

11–26. Draw the shear and moment diagrams for the beam.

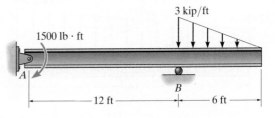

Prob. 11–26

11–27. Draw the shear and moment diagrams for the beam.

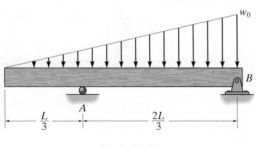

Prob. 11–27

*11–28.** Draw the shear and moment diagrams for the beam.

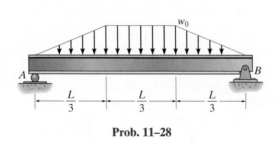

Prob. 11–28

11–29. Draw the shear and moment diagrams for the beam.

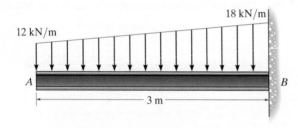

Prob. 11–29

11–30. Draw the shear and moment diagrams for the beam.

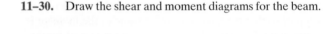
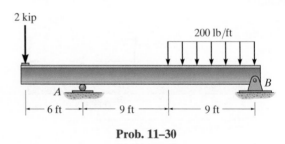

Prob. 11–30

11–31. The support at A allows the beam to slide freely along the vertical guide so that it cannot support a vertical force. Draw the shear and moment diagrams for the beam.

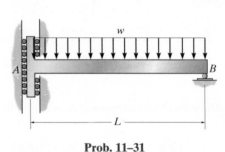

Prob. 11–31

*11–32.** The smooth pin is supported by two leaves A and B and subjected to a compressive load of 0.4 kN/m caused by bar C. Determine the intensity of the distributed load w_0 of the leaves on the pin and draw the shear and moment diagram for the pin.

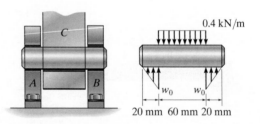

Prob. 11–32

11–33. The shaft is supported by a smooth thrust bearing at *A* and smooth journal bearing at *B*. Draw the shear and moment diagrams for the shaft.

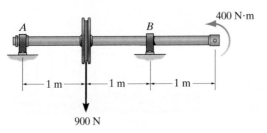

900 N

Prob. 11–33

11–34. Draw the shear and moment diagrams for the cantilever beam.

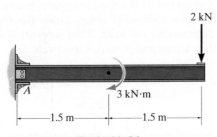

Prob. 11–34

11–35. Draw the shear and moment diagrams for the beam.

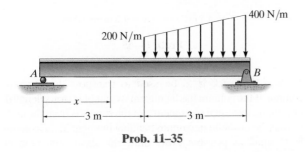

Prob. 11–35

*11–36. Draw the shear and moment diagrams for the rod. Only vertical reactions occur at its ends *A* and *B*.

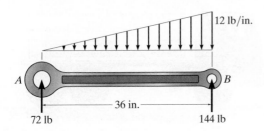

Prob. 11–36

11–37. Draw the shear and moment diagrams for the beam.

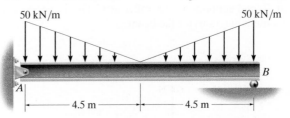

Prob. 11–37

11–38. The beam is used to support a uniform load along *CD* due to the 6-kN weight of the crate. Also, the reaction at the bearing support *B* can be assumed uniformly distributed along its width. Draw the shear and moment diagrams for the beam.

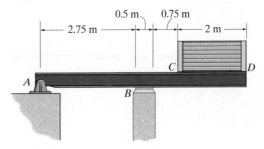

Prob. 11–38

11–39. Draw the shear and moment diagrams for the double overhanging beam.

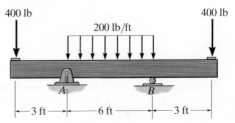

Prob. 11–39

*11–40. Draw the shear and moment diagrams for the simply supported beam.

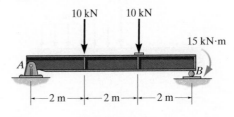

Prob. 11–40

11

11–41. The compound beam is fixed at A, pin connected at B, and supported by a roller at C. Draw the shear and moment diagrams for the beam.

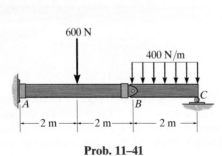

600 N

400 N/m

A B C

—2 m—|—2 m—|— 2 m —

Prob. 11–41

11–42. Draw the shear and moment diagrams for the compound beam.

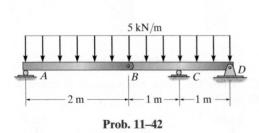

5 kN/m

A B C D

|— 2 m —|— 1 m —|—1 m —|

Prob. 11–42

11–43. The compound beam is fixed at A, pin connected at B, and supported by a roller at C. Draw the shear and moment diagrams for the beam.

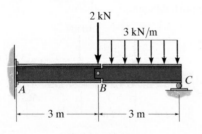

2 kN

3 kN/m

A B C

|— 3 m —|— 3 m —|

Prob. 11–43

*11–44. Draw the shear and moment diagrams for the beam.

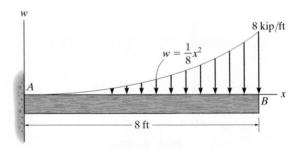

w

8 kip/ft

$w = \dfrac{1}{8}x^2$

A B x

|— 8 ft —|

Prob. 11–44

11–45. A short link at B is used to connect beams AB and BC to form the compound beam. Draw the shear and moment diagrams for the beam if the supports at A and C are considered fixed and pinned, respectively.

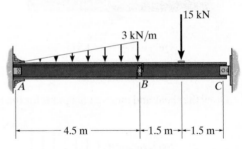

15 kN

3 kN/m

A B C

|— 4.5 m —|—1.5 m—|—1.5 m—|

Prob. 11–45

11–46. The truck is to be used to transport the concrete column. If the column has a uniform weight of w (force/length), determine the equal placement a of the supports from the ends so that the absolute maximum bending moment in the column is as small as possible. Also, draw the shear and moment diagrams for the column.

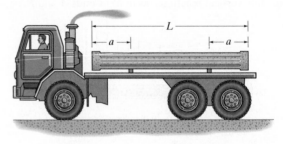

L

a a

Prob. 11–46

11.3 BENDING DEFORMATION OF A STRAIGHT MEMBER

In this section, we will discuss the deformations that occur when a straight prismatic beam, made of homogeneous material, is subjected to bending. The discussion will be limited to beams having a cross-sectional area that is symmetrical with respect to an axis, and the bending moment is applied about an axis perpendicular to this axis of symmetry, as shown in Fig. 11–18. The behavior of members that have unsymmetrical cross sections, or are made of several different materials, is based on similar observations and will be discussed separately in later sections of this chapter.

Consider the undeformed bar in Fig. 11–19*a*, which has a square cross section and is marked with horizontal and vertical grid lines. When a bending moment is applied, it tends to distort these lines into the pattern shown in Fig. 11–19*b*. Here the horizontal lines become *curved*, while the vertical lines *remain straight* but undergo a *rotation*. The bending moment causes the material within the *bottom* portion of the bar to *stretch* and the material within the *top* portion to *compress*. Consequently, between these two regions there must be a surface, called the **neutral surface**, in which horizontal fibers of the material will not undergo a change in length, Fig. 11–18. As noted, we will refer to the z axis that lies along the neutral surface as the **neutral axis**.

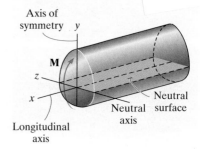

Fig. 11–18

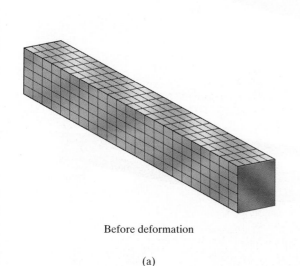

Before deformation

(a)

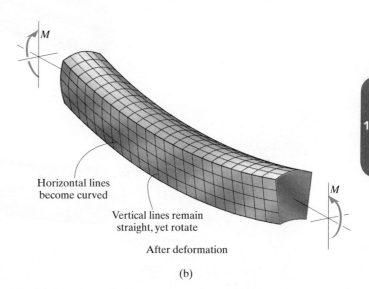

Horizontal lines become curved

Vertical lines remain straight, yet rotate

After deformation

(b)

Fig. 11–19

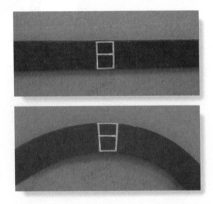

Note the distortion of the lines due to bending of this rubber bar. The top line stretches, the bottom line compresses, and the center line remains the same length. Furthermore the vertical lines rotate and yet remain straight.

From these observations we will make the following three assumptions regarding the way the moment deforms the material. First, the longitudinal axis, which lies within the neutral surface, Fig. 11–20a, does not experience any change in length. Rather the moment will tend to deform the beam so that this line becomes a curve that lies in the vertical plane of symmetry, Fig. 11–20b. Second, all cross sections of the beam remain plane and perpendicular to the longitudinal axis during the deformation. And third, the small lateral strains due to the Poisson effect discussed in Sec. 3.6 will be neglected. In other words, the cross section in Fig. 11–19 retains its shape.

With the above assumptions, we will now consider how the bending moment distorts a small element of the beam located a distance x along the beam's length, Fig. 11–20. This element is shown in profile view in the undeformed and deformed positions in Fig. 11–21. Here the line segment

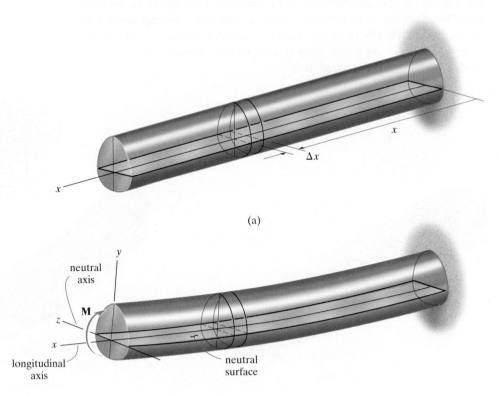

(a)

(b)

Fig. 11–20

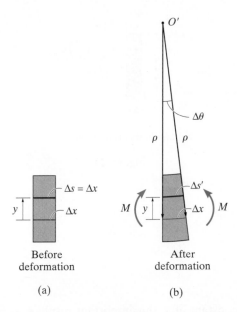

Before
deformation

After
deformation

(a)

(b)

Fig. 11–21

Δx, located on the neutral surface, does not change its length, whereas any line segment Δs, located at the arbitrary distance y above the neutral surface, will contract and become $\Delta s'$ after deformation. By definition, the normal strain along Δs is determined from Eq. 7–11, namely,

$$\epsilon = \lim_{\Delta s \to 0} \frac{\Delta s' - \Delta s}{\Delta s}$$

Now let's represent this strain in terms of the location y of the segment and the radius of curvature ρ of the longitudinal axis of the element. Before deformation, $\Delta s = \Delta x$, Fig. 11–21a. After deformation, Δx has a radius of curvature ρ, with center of curvature at point O', Fig. 11–21b, so that $\Delta x = \Delta s = \rho \Delta \theta$. Also, since $\Delta s'$ has a radius of curvature of $\rho - y$, then $\Delta s' = (\rho - y)\Delta \theta$. Substituting these results into the above equation, we get

$$\epsilon = \lim_{\Delta \theta \to 0} \frac{(\rho - y)\Delta \theta - \rho \Delta \theta}{\rho \Delta \theta}$$

or

$$\epsilon = -\frac{y}{\rho} \qquad (11\text{–}7)$$

11

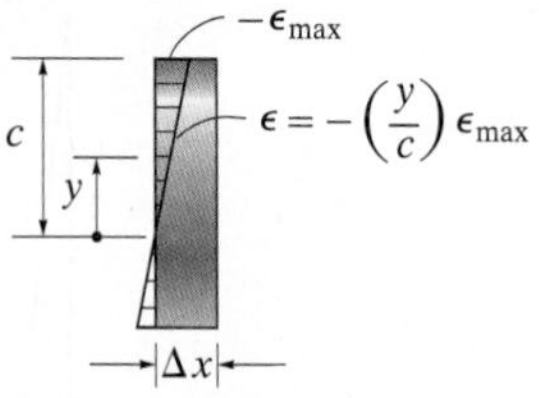

Normal strain distribution

Fig. 11–22

Since $1/\rho$ is constant at x, this important result, $\epsilon = -y/\rho$, indicates that the *longitudinal normal strain will vary linearly* with y measured from the neutral axis. A contraction $(-\epsilon)$ will occur in fibers located above the neutral axis $(+y)$, whereas elongation $(+\epsilon)$ will occur in fibers located below the axis $(-y)$. This variation in strain over the cross section is shown in Fig. 11–22. Here the maximum strain occurs at the outermost fiber, located a distance of $y = c$ from the neutral axis. Using Eq. 11–7, since $\epsilon_{max} = c/\rho$, then by division,

$$\frac{\epsilon}{\epsilon_{max}} = -\left(\frac{y/\rho}{c/\rho}\right)$$

So that

$$\epsilon = -\left(\frac{y}{c}\right)\epsilon_{max} \tag{11–8}$$

This normal strain depends only on the assumptions made with regard to the deformation.

11.4 THE FLEXURE FORMULA

In this section, we will develop an equation that relates the stress distribution within a straight beam to the bending moment acting on its cross section. To do this we will assume that the material behaves in a linear elastic manner, so that by Hooke's law, a linear variation of normal strain, Fig. 11–23a, must result in a linear variation in normal stress, Fig. 11–23b. Hence, like the normal strain variation, σ will vary from zero at the member's neutral axis to a maximum value, σ_{max}, a distance c farthest from the neutral axis. Because of the proportionality of triangles, Fig. 11–23b, or by using Hooke's law, $\sigma = E\epsilon$, and Eq. 11–8, we can write

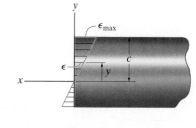

Normal strain variation
(profile view)

(a)

$$\sigma = -\left(\frac{y}{c}\right)\sigma_{max} \qquad (11\text{–}9)$$

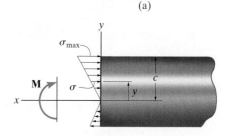

Bending stress variation
(profile view)

(b)

Fig. 11–23

This equation describes the stress distribution over the cross-sectional area. The sign convention established here is significant. For positive **M**, which acts in the $+z$ direction, positive values of y give negative values for σ, that is, a compressive stress, since it acts in the negative x direction. Similarly, negative y values will give positive or tensile values for σ.

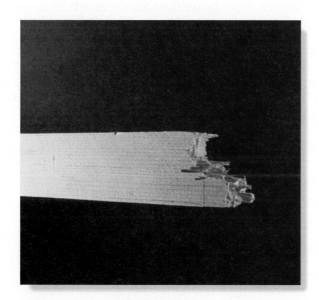

This wood specimen failed in bending due to its fibers being crushed at its top and torn apart at its bottom.

Location of Neutral Axis. To locate the position of the neutral axis, we require the *resultant force* produced by the stress distribution acting over the cross-sectional area to be equal to *zero*. Noting that the force $dF = \sigma\, dA$ acts on the arbitrary element dA in Fig. 11–24, we have

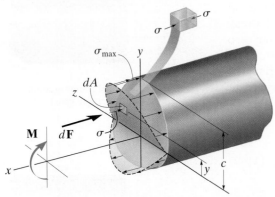

Bending stress variation

Fig. 11–24

$$F_R = \Sigma F_x; \qquad 0 = \int_A dF = \int_A \sigma\, dA$$

$$= \int_A -\left(\frac{y}{c}\right)\sigma_{max}\, dA$$

$$= \frac{-\sigma_{max}}{c}\int_A y\, dA$$

Since σ_{max}/c is not equal to zero, then

$$\int_A y\, dA = 0 \qquad (11\text{–}10)$$

In other words, the first moment of the member's cross-sectional area about the neutral axis must be zero. This condition can only be satisfied if the neutral axis is also the horizontal centroidal axis for the cross section.* Therefore, once the centroid for the member's cross-sectional area is determined, the location of the neutral axis is known.

Bending Moment. We can determine the stress in the beam if we require the moment M to be equal to the moment produced by the stress distribution about the neutral axis. The moment of $d\mathbf{F}$ in Fig. 11–24 is $dM = y\, dF$. Since $dF = \sigma\, dA$, using Eq. 11–9, we have for the entire cross section,

$$(M_R)_z = \Sigma M_z; \qquad M = \int_A y\, dF = \int_A y\, (\sigma\, dA) = \int_A y\left(\frac{y}{c}\sigma_{max}\right) dA$$

or

$$M = \frac{\sigma_{max}}{c}\int_A y^2\, dA \qquad (11\text{–}11)$$

*Recall that the location $\bar{y}$ for the centroid of an area is defined from the equation $\bar{y} = \int y\, dA/\int dA$. If $\int y\, dA = 0$, then $\bar{y} = 0$, and so the centroid lies on the reference (neutral) axis. See Appendix A.

The integral represents the **moment of inertia** of the cross-sectional area about the neutral axis.* We will symbolize its value as I. Hence, Eq. 11–11 can be solved for σ_{max} and written as

$$\sigma_{max} = \frac{Mc}{I} \qquad (11\text{–}12)$$

Here

σ_{max} = the maximum normal stress in the member, which occurs at a point on the cross-sectional area *farthest away* from the neutral axis

M = the resultant internal moment, determined from the method of sections and the equations of equilibrium, and calculated about the neutral axis of the cross section

c = perpendicular distance from the neutral axis to a point farthest away from the neutral axis. This is where σ_{max} acts.

I = moment of inertia of the cross-sectional area about the neutral axis

Since $\sigma_{max}/c = -\sigma/y$, Eq. 11–9, the normal stress at any distance y can be determined from an equation similar to Eq. 11–12. We have

$$\sigma = -\frac{My}{I} \qquad (11\text{–}13)$$

Either of the above two equations is often referred to as the **flexure formula**. Although we have assumed that the member is prismatic, we can conservatively also use the flexure formula to determine the normal stress in members that have a *slight taper*. For example, using a mathematical analysis based on the theory of elasticity, a member having a rectangular cross section and a length that is tapered 15° will have an actual maximum normal stress that is about 5.4% *less* than that calculated using the flexure formula.

---

*See Appendix A for a discussion on how to determine the moment of inertia for various shapes.

IMPORTANT POINTS

- The cross section of a straight beam *remains plane* when the beam deforms due to bending. This causes tensile stress on one portion of the cross section and compressive stress on the other portion. In between these portions, there exists the *neutral axis* which is subjected to *zero stress*.

- Due to the deformation, the *longitudinal strain* varies *linearly* from zero at the neutral axis to a maximum at the outer fibers of the beam. Provided the material is homogeneous and linear elastic, then the *stress* also varies in a *linear* fashion over the cross section.

- Since there is no resultant normal force on the cross section, then the neutral axis must pass through the *centroid* of the cross-sectional area.

- The flexure formula is based on the requirement that the internal moment on the cross section is equal to the moment produced by the normal stress distribution about the neutral axis.

PROCEDURE FOR ANALYSIS

In order to apply the flexure formula, the following procedure is suggested.

Internal Moment.

- Section the member at the point where the bending or normal stress is to be determined, and obtain the internal moment M at the section. The centroidal or neutral axis for the cross section must be known, since M *must* be calculated about this axis.

- If the absolute maximum bending stress is to be determined, then draw the moment diagram in order to determine the maximum moment in the member.

Section Property.

- Determine the moment of inertia of the cross-sectional area about the neutral axis. Methods used for its calculation are discussed in Appendix A, and a table listing values of I for several common shapes is given on the inside front cover.

Normal Stress.

- Specify the location y, measured perpendicular to the neutral axis to the point where the normal stress is to be determined. Then apply the equation $\sigma = -My/I$, or if the maximum bending stress is to be calculated, use $\sigma_{max} = Mc/I$. When substituting the data, make sure the units are consistent.

- The stress acts in a direction such that the force it creates at the point contributes a moment about the neutral axis that is in the same direction as the internal moment $\mathbf{M}$. In this manner the stress distribution acting over the entire cross section can be sketched, or a volume element of the material can be isolated and used to graphically represent the normal stress acting at the point, see Fig. 11–24.

EXAMPLE 11.11

The simply supported beam in Fig. 11–25a has the cross-sectional area shown in Fig. 11–25b. Determine the absolute maximum bending stress in the beam and draw the stress distribution over the cross section at this location. Also, what is the stress at point B?

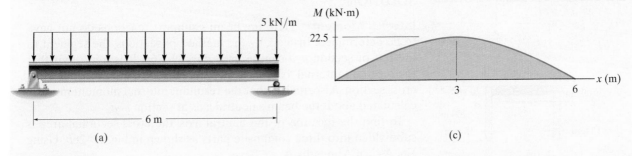

(a)

(c)

SOLUTION

Maximum Internal Moment.   The maximum internal moment in the beam, $M = 22.5$ kN · m, occurs at the center, as indicated on the moment diagram, Fig. 11–25c.

Section Property.   By reasons of symmetry, the neutral axis passes through the centroid C at the midheight of the beam, Fig. 11–25b. The area is subdivided into the three parts shown, and the moment of inertia of each part is calculated about the neutral axis using the parallel-axis theorem. (See Eq. A–5 of Appendix A.) Choosing to work in meters, we have

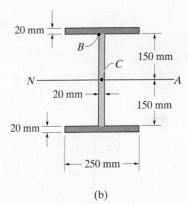

(b)

$$I = \Sigma(\bar{I} + Ad^2)$$

$$= 2\left[\frac{1}{12}(0.25 \text{ m})(0.020 \text{ m})^3 + (0.25 \text{ m})(0.020 \text{ m})(0.160 \text{ m})^2\right]$$

$$+ \left[\frac{1}{12}(0.020 \text{ m})(0.300 \text{ m})^3\right]$$

$$= 301.3(10^{-6}) \text{ m}^4$$

$$\sigma_{max} = \frac{Mc}{I}; \quad \sigma_{max} = \frac{22.5(10^3) \text{ N} \cdot \text{m}(0.170 \text{ m})}{301.3(10^{-6}) \text{ m}^4} = 12.7 \text{ MPa} \quad Ans.$$

A three-dimensional view of the stress distribution is shown in Fig. 11–25d. Specifically, at point B, $y_B = 150$ mm, and so as shown in Fig. 11–25d,

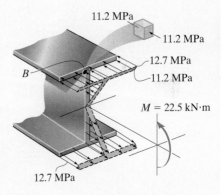

(d)

Fig. 11–25

$$\sigma_B = -\frac{My_B}{I}; \quad \sigma_B = -\frac{22.5(10^3) \text{ N} \cdot \text{m}(0.150 \text{ m})}{301.3(10^{-6}) \text{ m}^4} = -11.2 \text{ MPa} \quad Ans.$$

EXAMPLE   11.12

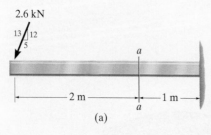

(a)

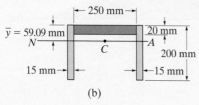

(b)

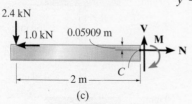

(c)

Fig. 11–26

The beam shown in Fig. 11–26a has a cross-sectional area in the shape of a channel, Fig. 11–26b. Determine the maximum bending stress that occurs in the beam at section a–a.

SOLUTION

Internal Moment.   Here the beam's support reactions do not have to be determined. Instead, by the method of sections, the segment to the left of section a–a can be used, Fig. 11–26c. It is important that the resultant internal axial force **N** passes through the centroid of the cross section. Also, realize that the resultant internal moment must be calculated about the beam's neutral axis at section a–a.

To find the location of the neutral axis, the cross-sectional area is subdivided into three composite parts as shown in Fig. 11–26b. Using Eq. A–2 of Appendix A, we have

$$\bar{y} = \frac{\Sigma \bar{y} A}{\Sigma A} = \frac{2[0.100 \text{ m}](0.200 \text{ m})(0.015 \text{ m}) + [0.010 \text{ m}](0.02 \text{ m})(0.250 \text{ m})}{2(0.200 \text{ m})(0.015 \text{ m}) + 0.020 \text{ m}(0.250 \text{ m})}$$

$$= 0.05909 \text{ m} = 59.09 \text{ mm}$$

This dimension is shown in Fig. 11–26c.

Applying the moment equation of equilibrium about the neutral axis, we have

$$\zeta + \Sigma M_{NA} = 0; \quad 2.4 \text{ kN}(2 \text{ m}) + 1.0 \text{ kN}(0.05909 \text{ m}) - M = 0$$

$$M = 4.859 \text{ kN} \cdot \text{m}$$

Section Property.   The moment of inertia of the cross-sectional area about the neutral axis is determined using $I = \Sigma (\bar{I} + Ad^2)$ applied to each of the three composite parts of the area. Working in meters, we have

$$I = \left[\frac{1}{12}(0.250 \text{ m})(0.020 \text{ m})^3 + (0.250 \text{ m})(0.020 \text{ m})(0.05909 \text{ m} - 0.010 \text{ m})^2 \right]$$

$$+ 2 \left[\frac{1}{12}(0.015 \text{ m})(0.200 \text{ m})^3 + (0.015 \text{ m})(0.200 \text{ m})(0.100 \text{ m} - 0.05909 \text{ m})^2 \right]$$

$$= 42.26(10^{-6}) \text{ m}^4$$

Maximum Bending Stress.   The maximum bending stress occurs at points farthest away from the neutral axis. This is at the bottom of the beam, $c = 0.200 \text{ m} - 0.05909 \text{ m} = 0.1409 \text{ m}$. Here the stress is compressive. Thus,

$$\sigma_{max} = \frac{Mc}{I} = \frac{4.859(10^3) \text{ N} \cdot \text{m}(0.1409 \text{ m})}{42.26(10^{-6}) \text{ m}^4} = 16.2 \text{ MPa} \text{ (C)} \qquad Ans.$$

Show that at the top of the beam the bending stress is $\sigma' = 6.79 \text{ MPa}$.

NOTE: The normal force of $N = 1 \text{ kN}$ and shear force $V = 2.4 \text{ kN}$ will also contribute additional stress on the cross section. The superposition of all these effects will be discussed in Chapter 13.

EXAMPLE 11.13

The member having a rectangular cross section, Fig. 11–27a, is designed to resist a moment of 40 N·m. In order to increase its strength and rigidity, it is proposed that two small ribs be added at its bottom, Fig. 11–27b. Determine the maximum normal stress in the member for both cases.

SOLUTION

Without Ribs. Clearly the neutral axis is at the center of the cross section, Fig. 11–27a, so $\bar{y} = c = 15$ mm $= 0.015$ m. Thus,

$$I = \frac{1}{12}bh^3 = \frac{1}{12}(0.060 \text{ m})(0.030 \text{ m})^3 = 0.135(10^{-6}) \text{ m}^4$$

Therefore the maximum normal stress is

$$\sigma_{max} = \frac{Mc}{I} = \frac{(40 \text{ N} \cdot \text{m})(0.015 \text{ m})}{0.135(10^{-6}) \text{ m}^4} = 4.44 \text{ MPa} \qquad Ans.$$

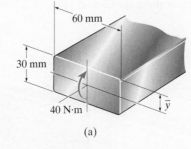

(a)

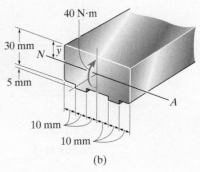

(b)

Fig. 11–27

With Ribs. From Fig. 11–27b, segmenting the area into the large main rectangle and the bottom two rectangles (ribs), the location $\bar{y}$ of the centroid and the neutral axis is determined as follows:

$$\bar{y} = \frac{\Sigma \tilde{y} A}{\Sigma A}$$

$$= \frac{[0.015 \text{ m}](0.030 \text{ m})(0.060 \text{ m}) + 2[0.0325 \text{ m}](0.005 \text{ m})(0.010 \text{ m})}{(0.03 \text{ m})(0.060 \text{ m}) + 2(0.005 \text{ m})(0.010 \text{ m})}$$

$$= 0.01592 \text{ m}$$

This value does not represent c. Instead

$$c = 0.035 \text{ m} - 0.01592 \text{ m} = 0.01908 \text{ m}$$

Using the parallel-axis theorem, the moment of inertia about the neutral axis is

$$I = \left[\frac{1}{12}(0.060 \text{ m})(0.030 \text{ m})^3 + (0.060 \text{ m})(0.030 \text{ m})(0.01592 \text{ m} - 0.015 \text{ m})^2 \right]$$

$$+ 2\left[\frac{1}{12}(0.010 \text{ m})(0.005 \text{ m})^3 + (0.010 \text{ m})(0.005 \text{ m})(0.0325 \text{ m} - 0.01592 \text{ m})^2 \right]$$

$$= 0.1642(10^{-6}) \text{ m}^4$$

Therefore, the maximum normal stress is

$$\sigma_{max} = \frac{Mc}{I} = \frac{40 \text{ N} \cdot \text{m}(0.01908 \text{ m})}{0.1642(10^{-6}) \text{ m}^4} = 4.65 \text{ MPa} \qquad Ans.$$

NOTE: This surprising result indicates that the addition of the ribs to the cross section will *increase* the maximum normal stress rather than decrease it, and for this reason, the ribs should be omitted.

PRELIMINARY PROBLEMS

P11–2. Determine the moment of inertia of the cross section about the neutral axis.

P11–4. In each case, show how the bending stress acts on a differential volume element located at point A and point B.

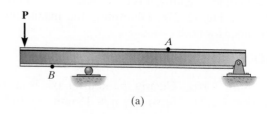

(a)

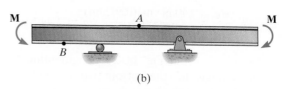

(b)

Prob. P11–4

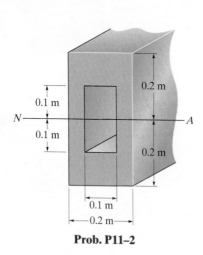

0.2 m

0.1 m

N ———— A

0.1 m

0.2 m

0.1 m

0.2 m

Prob. P11–2

P11–3. Determine the location of the centroid, y, and the moment of inertia of the cross section about the neutral axis.

P11–5. Sketch the bending stress distribution over each cross section.

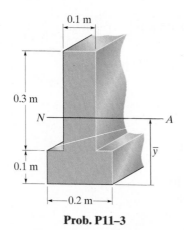

0.1 m

0.3 m

N ———— A

$\bar{y}$

0.1 m

0.2 m

Prob. P11–3

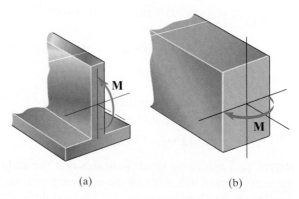

(a)   (b)

Prob. P11–5

FUNDAMENTAL PROBLEMS

F11–9. If the beam is subjected to a bending moment of $M = 20 \text{ kN} \cdot \text{m}$, determine the maximum bending stress in the beam.

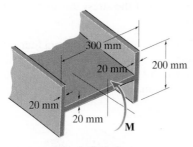

Prob. F11–9

F11–10. If the beam is subjected to a bending moment of $M = 50 \text{ kN} \cdot \text{m}$, sketch the bending stress distribution over the beam's cross section.

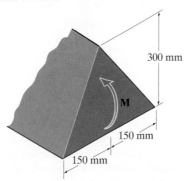

Prob. F11–10

F11–11. If the beam is subjected to a bending moment of $M = 50 \text{ kN} \cdot \text{m}$, determine the maximum bending stress in the beam.

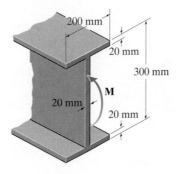

Prob. F11–11

F11–12. If the beam is subjected to a bending moment of $M = 10 \text{ kN} \cdot \text{m}$, determine the bending stress in the beam at points A and B, and sketch the results on a differential element at each of these points.

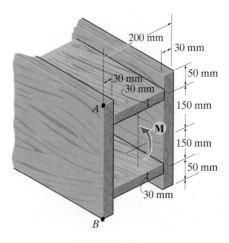

Prob. F11–12

F11–13. If the beam is subjected to a bending moment of $M = 5 \text{ kN} \cdot \text{m}$, determine the bending stress developed at point A and sketch the result on a differential element at this point.

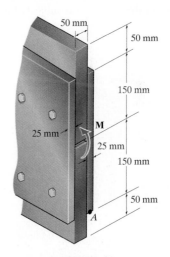

Prob. F11–13

PROBLEMS

11–47. An A-36 steel strip has an allowable bending stress of 165 MPa. If it is rolled up, determine the smallest radius r of the spool if the strip has a width of 10 mm and a thickness of 1.5 mm. Also, find the corresponding maximum internal moment developed in the strip.

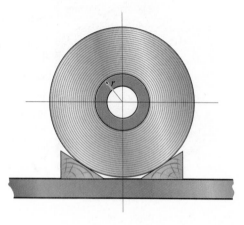

Prob. 11–47

*11–48. Determine the moment M that will produce a maximum stress of 10 ksi on the cross section.

11–49. Determine the maximum tensile and compressive bending stress in the beam if it is subjected to a moment of $M = 4$ kip · ft.

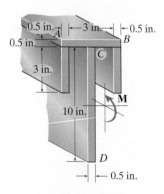

Probs. 11–48/49

11–50. The beam is constructed from four pieces of wood, glued together as shown. If $M = 10$ kip · ft, determine the maximum bending stress in the beam. Sketch a three-dimensional view of the stress distribution acting over the cross section.

11–51. The beam is constructed from four pieces of wood, glued together as shown. If $M = 10$ kip · ft, determine the resultant force this moment exerts on the top and bottom boards of the beam.

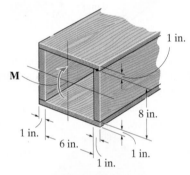

Probs. 11–50/51

*11–52. The beam is made from three boards nailed together as shown. If the moment acting on the cross section is $M = 600$ N · m, determine the maximum bending stress in the beam. Sketch a three-dimensional view of the stress distribution and cover the cross section.

11–53. The beam is made from three boards nailed together as shown. If the moment acting on the cross section is $M = 600$ N · m, determine the resultant force the bending stress produces on the top board.

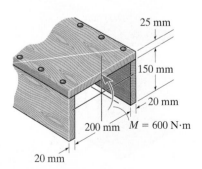

Probs. 11–52/53

11–54. If the built-up beam is subjected to an internal moment of $M = 75$ kN · m, determine the maximum tensile and compressive stress acting in the beam.

11–55. If the built-up beam is subjected to an internal moment of $M = 75$ kN · m, determine the amount of this internal moment resisted by plate A.

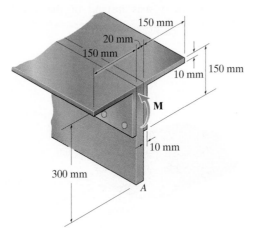

Probs. 11–54/55

11–58. The beam is made from three boards nailed together as shown. If the moment acting on the cross section is $M = 1$ kip · ft, determine the maximum bending stress in the beam. Sketch a three-dimensional view of the stress distribution acting over the cross section.

11–59. If $M = 1$ kip · ft, determine the resultant force the bending stresses produce on the top board A of the beam.

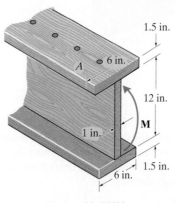

Probs. 11–58/59

*11–56. The beam is subjected to a moment M. Determine the percentage of this moment that is resisted by the stresses acting on both the top and bottom boards of the beam.

11–57. Determine the moment M that should be applied to the beam in order to create a compressive stress at point D of $\sigma_D = 10$ MPa. Also sketch the stress distribution acting over the cross section and calculate the maximum stress developed in the beam.

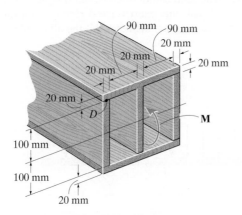

Probs. 11–56/57

*11–60. The beam is subjected to a moment of 15 kip · ft. Determine the resultant force the bending stress produces on the top flange A and bottom flange B. Also calculate the maximum bending stress developed in the beam.

11–61. The beam is subjected to a moment of 15 kip · ft. Determine the percentage of this moment that is resisted by the web D of the beam.

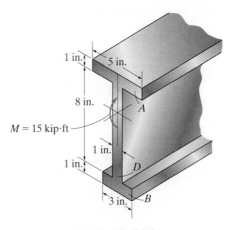

Probs. 11–60/61

11–62. The beam is subjected to a moment of $M = 40$ kN $\cdot$ m. Determine the bending stress at points A and B. Sketch the results on a volume element acting at each of these points.

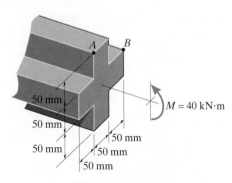

50 mm
50 mm
50 mm
50 mm
50 mm
50 mm

$M = 40$ kN·m

Prob. 11–62

11–63. The steel shaft has a diameter of 2 in. It is supported on smooth journal bearings A and B, which exert only vertical reactions on the shaft. Determine the absolute maximum bending stress in the shaft if it is subjected to the pulley loadings shown.

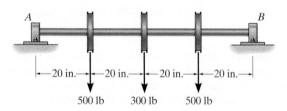

A

B

20 in. — 20 in. — 20 in. — 20 in.

500 lb   300 lb   500 lb

Prob. 11–63

*11–64. The beam is made of steel that has an allowable stress of $\sigma_{allow} = 24$ ksi. Determine the largest internal moment the beam can resist if the moment is applied (a) about the z axis, (b) about the y axis.

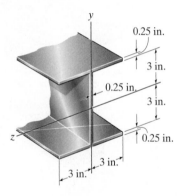

y

0.25 in.

3 in.

0.25 in.

3 in.

0.25 in.

3 in.

3 in.

z

Prob. 11–64

11–65. A shaft is made of a polymer having an elliptical cross section. If it resists an internal moment of $M = 50$ N $\cdot$ m, determine the maximum bending stress in the material (a) using the flexure formula, where $I_z = \frac{1}{4}\pi(0.08 \text{ m})(0.04 \text{ m})^3$, (b) using integration. Sketch a three-dimensional view of the stress distribution acting over the cross-sectional area. Here $I_x = \frac{1}{4}\pi(0.08 \text{ m})(0.04 \text{ m})^3$.

11–66. Solve Prob. 11–65 if the moment $M = 50$ N $\cdot$ m is applied about the y axis instead of the x axis. Here $I_y = \frac{1}{4}\pi(0.04 \text{ m})(0.08 \text{ m})^3$.

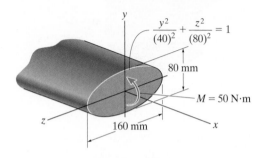

y

$\dfrac{y^2}{(40)^2} + \dfrac{z^2}{(80)^2} = 1$

80 mm

$M = 50$ N·m

z   160 mm   x

Probs. 11–65/66

11–67. The shaft is supported by smooth journal bearings at A and B that only exert vertical reactions on the shaft. If $d = 90$ mm, determine the absolute maximum bending stress in the beam, and sketch the stress distribution acting over the cross section.

*11–68. The shaft is supported by smooth journal bearings at A and B that only exert vertical reactions on the shaft. Determine its smallest diameter d if the allowable bending stress is $\sigma_{allow} = 180$ MPa.

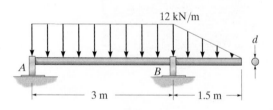

12 kN/m

d

A

B

3 m   1.5 m

Probs. 11–67/68

11–69. The axle of the freight car is subjected to a wheel loading of 20 kip. If it is supported by two journal bearings at C and D, determine the maximum bending stress developed at the center of the axle, where the diameter is 5.5 in.

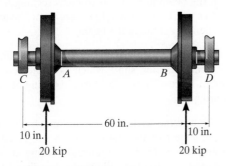

60 in.

10 in.

10 in.

20 kip

20 kip

A   B   D   C

Prob. 11–69

11–70. The strut on the utility pole supports the cable having a weight of 600 lb. Determine the absolute maximum bending stress in the strut if A, B, and C are assumed to be pinned.

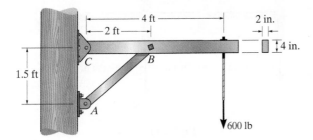

4 ft

2 ft

2 in.

4 in.

1.5 ft

C   B

A

600 lb

Prob. 11–70

11–71. The boat has a weight of 2300 lb and a center of gravity at G. If it rests on the trailer at the smooth contact A and can be considered pinned at B, determine the absolute maximum bending stress developed in the main strut of the trailer which is pinned at C. Consider the strut to be a box-beam having the dimensions shown.

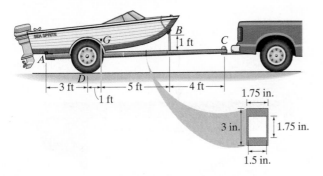

B
1 ft
G
C
A
D
3 ft   5 ft   4 ft
1 ft

1.75 in.

3 in.   1.75 in.

1.5 in.

Prob. 11–71

*11–72. Determine the absolute maximum bending stress in the 1.5-in.-diameter shaft. The shaft is supported by a thrust bearing at A and a journal bearing at B.

11–73. Determine the smallest allowable diameter of the shaft. The shaft is supported by a thrust bearing at A and a journal bearing at B. The allowable bending stress is $\sigma_{\text{allow}} = 22$ ksi.

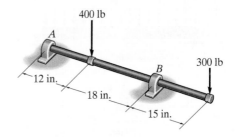

400 lb

A

B

300 lb

12 in.

18 in.

15 in.

Probs. 11–72/73

11–74. The pin is used to connect the three links together. Due to wear, the load is distributed over the top and bottom of the pin as shown on the free-body diagram. If the diameter of the pin is 0.40 in., determine the maximum bending stress on the cross-sectional area at the center section a–a. For the solution it is first necessary to determine the load intensities w_1 and w_2.

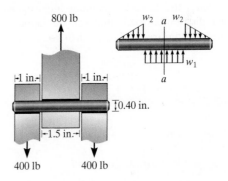

800 lb

w_2 a w_2

w_1

1 in.   1 in.

0.40 in.

1.5 in.

a

400 lb   400 lb

Prob. 11–74

11–75. The shaft is supported by a thrust bearing at A and journal bearing at D. If the shaft has the cross section shown, determine the absolute maximum bending stress in the shaft.

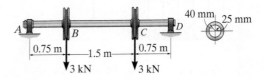

40 mm   25 mm

A   B   C   D

0.75 m   1.5 m   0.75 m

3 kN   3 kN

Prob. 11–75

*11–76. If the intensity of the load $w = 15$ kN/m, determine the absolute maximum tensile and compressive stress in the beam.

11–77. If the allowable bending stress is $\sigma_{allow} = 150$ MPa, determine the maximum intensity w of the uniform distributed load.

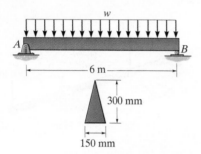

Probs. 11–76/77

11–78. The beam is subjected to the triangular distributed load with a maximum intensity of $w_0 = 300$ lb/ft. If the allowable bending stress is $\sigma_{allow} = 1.40$ ksi, determine the required dimension b of its cross section to the nearest $\frac{1}{8}$ in. Assume the support at A is a pin and B is a roller.

11–79. The beam has a rectangular cross section with $b = 4$ in. Determine the largest maximum intensity w_0 of the triangular distributed loads that can be supported if the allowable bending stress is $\sigma_{allow} = 1.40$ ksi.

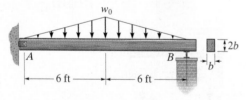

Probs. 11–78/79

*11–80. Determine the absolute maximum bending stress in the beam. Each segment has a rectangular cross section with a base of 4 in. and height of 12 in.

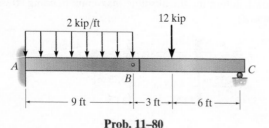

Prob. 11–80

11–81. If the compound beam in Prob. 11–42 has a square cross section of side length a, determine the minimum value of a if the allowable bending stress is $\sigma_{allow} = 150$ MPa.

11–82. If the beam in Prob. 11–28 has a rectangular cross section with a width b and a height h, determine the absolute maximum bending stress in the beam.

11–83. Determine the absolute maximum bending stress in the 80-mm-diameter shaft which is subjected to the concentrated forces. There is a journal bearing at A and a thrust bearing at B.

*11–84. Determine, to the nearest millimeter, the smallest allowable diameter of the shaft which is subjected to the concentrated forces. There is a journal bearing at A and a thrust bearing at B. The allowable bending stress is $\sigma_{allow} = 150$ MPa.

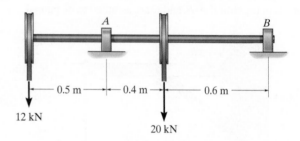

Probs. 11–83/84

11–85. Determine the absolute maximum bending stress in the beam, assuming that the support at B exerts a uniformly distributed reaction on the beam. The cross section is rectangular with a base of 3 in. and height of 6 in.

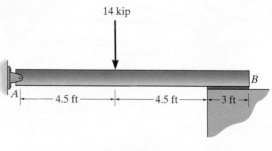

Prob. 11–85

11–86. Determine the absolute maximum bending stress in the 2-in.-diameter shaft. There is a journal bearing at A and a thrust bearing at B.

11–87. Determine the smallest diameter of the shaft to the nearest $\frac{1}{8}$ in. There is a journal bearing at A and a thrust bearing at B. The allowable bending stress is $\sigma_{allow} = 22$ ksi.

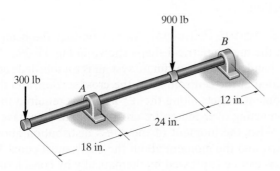

Probs. 11–86/87

*11–88. A log that is 2 ft in diameter is to be cut into a rectangular section for use as a simply supported beam. If the allowable bending stress is $\sigma_{allow} = 8$ ksi, determine the required width b and height h of the beam that will support the largest load possible. What is this load?

11–89. A log that is 2 ft in diameter is to be cut into a rectangular section for use as a simply supported beam. If the allowable bending stress is $\sigma_{allow} = 8$ ksi, determine the largest load P that can be supported if the width of the beam is $b = 8$ in.

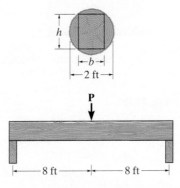

Probs. 11–88/89

11–90. If the beam in Prob. 11–19 has a rectangular cross section with a width of 8 in. and a height of 16 in., determine the absolute maximum bending stress in the beam.

11–91. The simply supported truss is subjected to the central distributed load. Neglect the effect of the diagonal lacing and determine the absolute maximum bending stress in the truss. The top member is a pipe having an outer diameter of 1 in. and thickness of $\frac{3}{16}$ in., and the bottom member is a solid rod having a diameter of $\frac{1}{2}$ in.

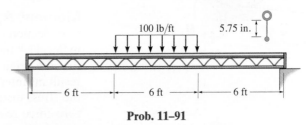

Prob. 11–91

*11–92. If $d = 450$ mm, determine the absolute maximum bending stress in the overhanging beam.

11–93. If the allowable bending stress is $\sigma_{allow} = 6$ MPa, determine the minimum dimension d of the beam's cross-sectional area to the nearest mm.

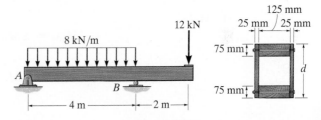

Probs. 11–92/93

11–94. The beam has a rectangular cross section as shown. Determine the largest intensity w of the uniform distributed load so that the bending stress in the beam does not exceed $\sigma_{max} = 10$ MPa.

11–95. The beam has the rectangular cross section shown. If $w = 1$ kN/m, determine the maximum bending stress in the beam. Sketch the stress distribution acting over the cross section.

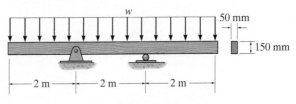

Probs. 11–94/95

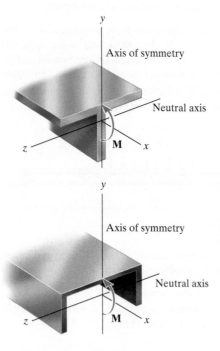

Fig. 11–28

11.5 UNSYMMETRIC BENDING

When developing the flexure formula, we required the cross-sectional area to be *symmetric* about an axis perpendicular to the neutral axis and the resultant moment **M** act along the neutral axis. Such is the case for the "T" sections shown in Fig. 11–28. In this section we will show how to apply the flexure formula either to a beam having a cross-sectional area of any shape or to a beam supporting a moment that acts in any direction.

Moment Applied About Principal Axis. Consider the beam's cross section to have the unsymmetrical shape shown in Fig. 11–29a. As in Sec. 11.4, the right-handed x, y, z coordinate system is established such that the origin is located at the centroid C on the cross section, and the resultant internal moment **M** acts along the $+z$ axis. It is required that the stress distribution acting over the entire cross-sectional area have a zero force resultant. Also, the moment of the stress distribution about the y axis must be zero, and the moment about the z axis must equal **M**. These three conditions can be expressed mathematically by considering the force acting on the differential element dA located at $(0, y, z)$, Fig. 11–29a. Since this force is $dF = \sigma\, dA$, we have

$$F_R = \Sigma F_x; \qquad\qquad 0 = -\int_A \sigma\, dA \qquad\qquad (11\text{–}14)$$

$$(M_R)_y = \Sigma M_y; \qquad\qquad 0 = -\int_A z\sigma\, dA \qquad\qquad (11\text{–}15)$$

$$(M_R)_z = \Sigma M_z; \qquad\qquad M = \int_A y\sigma\, dA \qquad\qquad (11\text{–}16)$$

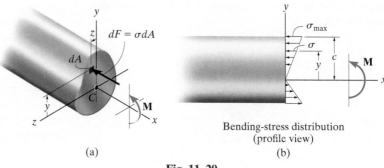

Bending-stress distribution
(profile view)

(a)   (b)

Fig. 11–29

As shown in Sec. 11.4, Eq. 11–14 is satisfied since the z axis passes through the *centroid* of the area. Also, since the z axis represents the *neutral axis* for the cross section, the normal stress will vary linearly from zero at the neutral axis to a maximum at $|y| = c$, Fig. 11–29b. Hence the stress distribution is defined by $\sigma = -(y/c)\sigma_{max}$. When this equation is substituted into Eq. 11–16 and integrated, it leads to the flexure formula $\sigma_{max} = Mc/I$. When it is substituted into Eq. 11–15, we get

$$0 = \frac{-\sigma_{max}}{c} \int_A yz \, dA$$

which requires

$$\int_A yz \, dA = 0$$

Z- sectioned members are often used in light-gage metal building construction to support roofs. To design them to support bending loads, it is necessary to determine their principal axes of inertia.

This integral is called the **product of inertia** for the area. As indicated in Appendix A, it will indeed be zero provided the y and z axes are chosen as **principal axes of inertia** for the area. For an arbitrarily shaped area, such as the one in Fig. 11–29a, the orientation of the principal axes can always be determined, using the inertia transformation equations as explained in Appendix A, Sec. A.4. If the area has an axis of symmetry, however, the **principal axes** can easily be established **since they will always be oriented along the axis of symmetry and perpendicular to it**.

For example, consider the members shown in Fig. 11–30. In each of these cases, y and z represent the principal axes of inertia for the cross section. In Fig. 11–30a the principal axes are located by symmetry, and in Figs. 11–30b and 11–30c their orientation is determined using the methods of Appendix A. Since **M** is applied only about one of the principal axes (the z axis), the stress distribution has a linear variation, and is determined from the flexure formula, $\sigma = -My/I_z$, as shown for each case.

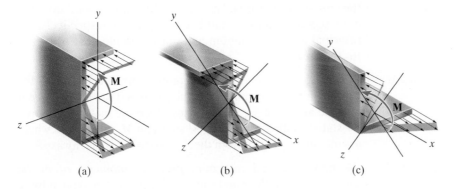

(a)   (b)   (c)

Fig. 11–30

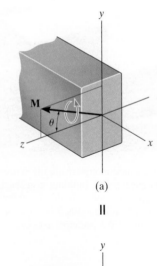

(a)

||

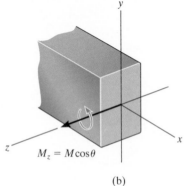

$M_z = M\cos\theta$

(b)

+

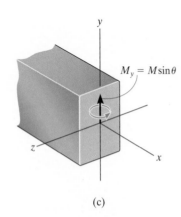

$M_y = M\sin\theta$

(c)

Fig. 11–31

Moment Arbitrarily Applied. Sometimes a member may be loaded such that **M** does not act about one of the principal axes of the cross section. When this occurs, the moment should first be resolved into components directed along the principal axes, then the flexure formula can be used to determine the normal stress caused by *each* moment component. Finally, using the principle of superposition, the resultant normal stress at the point can be determined.

To formalize this procedure, consider the beam to have a rectangular cross section and to be subjected to the moment **M**, Fig. 11–31*a*, where **M** makes an angle θ with the maximum principal *z* axis, i.e., the axis of maximum moment of inertia for the cross section. We will assume θ is positive when it is directed from the +*z* axis towards the +*y* axis. Resolving **M** into components, we have $M_z = M\cos\theta$ and $M_y = M\sin\theta$, Figs. 11–31*b* and 11–31*c*. The normal-stress distributions that produce **M** and its components **M**$_z$ and **M**$_y$ are shown in Figs. 11–31*d*, 11–31*e*, and 11–31*f*, where it is assumed that $(\sigma_x)_{max} > (\sigma'_x)_{max}$. By inspection, the maximum tensile and compressive stresses $[(\sigma_x)_{max} + (\sigma'_x)_{max}]$ occur at two opposite corners of the cross section, Fig. 11–31*d*.

Applying the flexure formula to each moment component in Figs. 11–31*b* and 11–31*c*, and adding the results algebraically, the resultant normal stress at any point on the cross section, Fig. 11–31*d*, is therefore

$$\sigma = -\frac{M_z y}{I_z} + \frac{M_y z}{I_y} \qquad (11\text{–}17)$$

Here,

σ = the normal stress at the point. Tensile stress is positive and compressive stress is negative.

y, z = the coordinates of the point measured from a *right-handed coordinate system, x, y, z,* having their origin at the centroid of the cross-sectional area. The *x* axis is directed outward from the cross section and the *y* and *z* axes represent, respectively, the principal axes of minimum and maximum moment of inertia for the area.

M_z, M_y = the resultant internal moment components directed along the maximum *z* and minimum *y* principal axes. They are positive if directed along the +*z* and +*y* axes, otherwise they are negative. Or, stated another way, $M_y = M\sin\theta$ and $M_z = M\cos\theta$, where θ is measured positive from the +*z* axis towards the +*y* axis.

I_z, I_y = the maximum and minimum *principal moments of inertia* calculated about the *z* and *y* axes, respectively. See Appendix A.

Orientation of the Neutral Axis. The equation defining the neutral axis, and its inclination α, Fig. 11–31d, can be determined by applying Eq. 11–17 to a point y, z where $\sigma = 0$, since by definition no normal stress acts on the neutral axis. We have

$$y = \frac{M_y I_z}{M_z I_y} z$$

Since $M_z = M \cos \theta$ and $M_y = M \sin \theta$, then

$$y = \left(\frac{I_z}{I_y} \tan \theta \right) z \qquad (11\text{–}18)$$

Since the slope of this line is $\tan \alpha = y/z$, then

$$\boxed{\tan \alpha = \frac{I_z}{I_y} \tan \theta} \qquad (11\text{–}19)$$

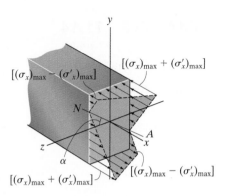

(d)

$\|$

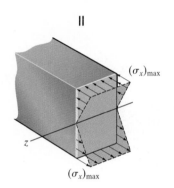

(e)

$+$

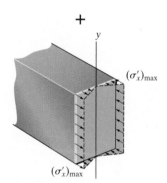

(f)

Fig. 11–31 (cont.)

▶ *IMPORTANT POINTS*

- The flexure formula can be applied only when bending occurs about axes that represent the *principal axes of inertia* for the cross section. These axes have their origin at the centroid and are oriented along an axis of symmetry, if there is one, and perpendicular to it.

- If the moment is applied about some arbitrary axis, then the moment must be resolved into components along each of the principal axes, and the stress at a point is determined by superposition of the stress caused by each of the moment components.

EXAMPLE 11.14

The rectangular cross section shown in Fig. 11–32a is subjected to a bending moment of $M = 12$ kN $\cdot$ m. Determine the normal stress developed at each corner of the section, and specify the orientation of the neutral axis.

SOLUTION

Internal Moment Components.   By inspection it is seen that the y and z axes represent the principal axes of inertia since they are axes of symmetry for the cross section. As required we have established the z axis as the principal axis for *maximum* moment of inertia. The moment is resolved into its y and z components, where

$$M_y = -\frac{4}{5}(12 \text{ kN} \cdot \text{m}) = -9.60 \text{ kN} \cdot \text{m}$$

$$M_z = \frac{3}{5}(12 \text{ kN} \cdot \text{m}) = 7.20 \text{ kN} \cdot \text{m}$$

Section Properties.   The moments of inertia about the y and z axes are

$$I_y = \frac{1}{12}(0.4 \text{ m})(0.2 \text{ m})^3 = 0.2667(10^{-3}) \text{ m}^4$$

$$I_z = \frac{1}{12}(0.2 \text{ m})(0.4 \text{ m})^3 = 1.067(10^{-3}) \text{ m}^4$$

Bending Stress.   Thus,

$$\sigma = -\frac{M_z y}{I_z} + \frac{M_y z}{I_y}$$

$$\sigma_B = -\frac{7.20(10^3) \text{ N} \cdot \text{m}(0.2 \text{ m})}{1.067(10^{-3}) \text{ m}^4} + \frac{-9.60(10^3) \text{ N} \cdot \text{m}(-0.1 \text{ m})}{0.2667(10^{-3}) \text{ m}^4} = 2.25 \text{ MPa} \qquad \textit{Ans.}$$

$$\sigma_C = -\frac{7.20(10^3) \text{ N} \cdot \text{m}(0.2 \text{ m})}{1.067(10^{-3}) \text{ m}^4} + \frac{-9.60(10^3) \text{ N} \cdot \text{m}(0.1 \text{ m})}{0.2667(10^{-3}) \text{ m}^4} = -4.95 \text{ MPa} \qquad \textit{Ans.}$$

$$\sigma_D = -\frac{7.20(10^3) \text{ N} \cdot \text{m}(-0.2 \text{ m})}{1.067(10^{-3}) \text{ m}^4} + \frac{-9.60(10^3) \text{ N} \cdot \text{m}(0.1 \text{ m})}{0.2667(10^{-3}) \text{ m}^4} = -2.25 \text{ MPa} \quad \textit{Ans.}$$

$$\sigma_E = -\frac{7.20(10^3) \text{ N} \cdot \text{m}(-0.2 \text{ m})}{1.067(10^{-3}) \text{ m}^4} + \frac{-9.60(10^3) \text{ N} \cdot \text{m}(-0.1 \text{ m})}{0.2667(10^{-3}) \text{ m}^4} = 4.95 \text{ MPa} \qquad \textit{Ans.}$$

The resultant normal-stress distribution has been sketched using these values, Fig. 11–32b. Since superposition applies, the distribution is linear as shown.

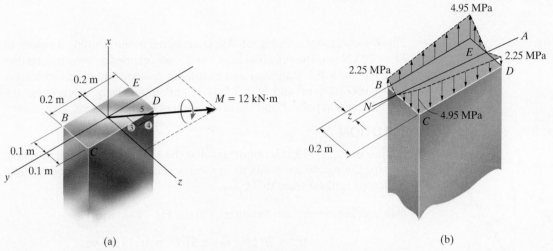

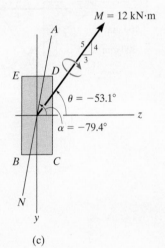

(a)

(b)

Fig. 11–32

Orientation of Neutral Axis. The location z of the neutral axis (NA), Fig. 11–32b, can be established by proportion. Along the edge BC, we require

$$\frac{2.25 \text{ MPa}}{z} = \frac{4.95 \text{ MPa}}{(0.2 \text{ m} - z)}$$

$$0.450 - 2.25z = 4.95z$$

$$z = 0.0625 \text{ m}$$

In the same manner this is also the distance from D to the neutral axis.

We can also establish the orientation of the NA using Eq. 11–19, which is used to specify the angle α that the axis makes with the z or *maximum* principal axis. According to our sign convention, θ must be measured from the $+z$ axis toward the $+y$ axis. By comparison, in Fig. 11–32c, $\theta = -\tan^{-1}\frac{14}{3} = -53.1°$ (or $\theta = +306.9°$). Thus,

$$\tan \alpha = \frac{I_z}{I_y} \tan \theta$$

$$\tan \alpha = \frac{1.067(10^{-3}) \text{ m}^4}{0.2667(10^{-3}) \text{ m}^4} \tan(-53.1°)$$

$$\alpha = -79.4° \qquad\qquad Ans.$$

This result is shown in Fig. 11–32c. Using the value of z calculated above, verify, using the geometry of the cross section, that one obtains the same answer.

(c)

11

EXAMPLE   11.15

The Z-section shown in Fig. 11–33a is subjected to the bending moment of $M = 20$ kN · m. The principal axes y and z are oriented as shown, such that they represent the minimum and maximum principal moments of inertia, $I_y = 0.960(10^{-3})$ m^4 and $I_z = 7.54(10^{-3})$ m^4, respectively.* Determine the normal stress at point P and the orientation of the neutral axis.

SOLUTION

For use of Eq. 11–19, it is important that the z axis represent the principal axis for the *maximum* moment of inertia. (For this case most of the area is located farthest from this axis.)

Internal Moment Components.   From Fig. 11–33a,

$$M_y = 20 \text{ kN · m} \sin 57.1° = 16.79 \text{ kN · m}$$

$$M_z = 20 \text{ kN · m} \cos 57.1° = 10.86 \text{ kN · m}$$

Bending Stress.   The y and z coordinates of point P must be determined first. Note that the y', z' coordinates of P are $(-0.2$ m, 0.35 m$)$. Using the colored triangles from the construction shown in Fig. 11–33b, we have

$$y_P = -0.35 \sin 32.9° - 0.2 \cos 32.9° = -0.3580 \text{ m}$$

$$z_P = 0.35 \cos 32.9° - 0.2 \sin 32.9° = 0.1852 \text{ m}$$

Applying Eq. 11–17,

$$\sigma_P = -\frac{M_z y_P}{I_z} + \frac{M_y z_P}{I_y}$$

$$= -\frac{(10.86(10^3) \text{ N · m})(-0.3580 \text{ m})}{7.54(10^{-3}) \text{ m}^4} + \frac{(16.79(10^3) \text{ N · m})(0.1852 \text{ m})}{0.960(10^{-3}) \text{ m}^4}$$

$$= 3.76 \text{ MPa} \qquad\qquad\qquad\qquad\qquad Ans.$$

Orientation of Neutral Axis.   Using the angle $\theta = 57.1°$ between **M** and the z axis, Fig. 11–33a, we have

$$\tan \alpha = \left[\frac{7.54(10^{-3}) \text{ m}^4}{0.960(10^{-3}) \text{ m}^4} \right] \tan 57.1°$$

$$\alpha = 85.3° \qquad\qquad\qquad\qquad\qquad Ans.$$

The neutral axis is oriented as shown in Fig. 11–33b.

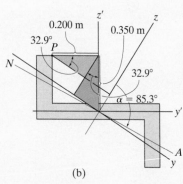

(a)

(b)

Fig. 11–33

---

* These values are obtained using the methods of Appendix A. (See Example A.4 or A.5.)

FUNDAMENTAL PROBLEMS

F11–14. Determine the bending stress developed at corners A and B. What is the orientation of the neutral axis?

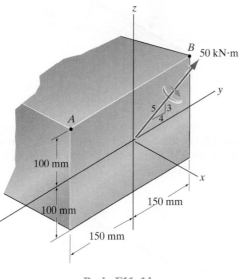

F11–15. Determine the maximum stress in the beam's cross section.

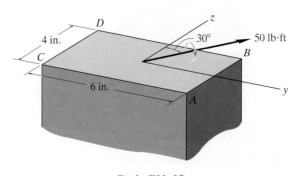

Prob. F11–14   Prob. F11–15

PROBLEMS

***11–96.** The member has a square cross section and is subjected to the moment $M = 850$ N·m. Determine the stress at each corner and sketch the stress distribution. Set $\theta = 45°$.

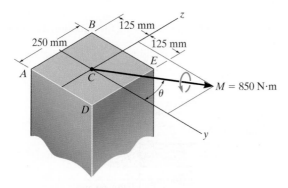

11–97. The member has a square cross section and is subjected to the moment $M = 850$ N·m as shown. Determine the stress at each corner and sketch the stress distribution. Set $\theta = 30°$.

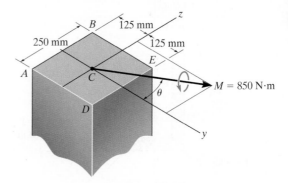

Prob. 11–96   Prob. 11–97

11

11–98. Consider the general case of a prismatic beam subjected to bending-moment components $\mathbf{M}_y$ and $\mathbf{M}_z$ as shown, when the x, y, z axes pass through the centroid of the cross section. If the material is linear elastic, the normal stress in the beam is a linear function of position such that $\sigma = a + by + cz$. Using the equilibrium conditions $0 = \int_A \sigma \, dA$, $M_y = \int_A z\sigma \, dA$, $M_z = \int_A -y\sigma \, dA$, determine the constants a, b, and c, and show that the normal stress can be determined from the equation $\sigma = [-(M_z I_y + M_y I_{yz})y + (M_y I_z + M_z I_{yz})z]/(I_y I_z - I_{yz}^2)$, where the moments and products of inertia are defined in Appendix A.

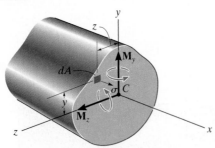

Prob. 11–98

11–99. Determine the bending stress at point A of the beam, and the orientation of the neutral axis. Using the method in Appendix A, the principal moments of inertia of the cross section are $I_{z'} = 8.828 \text{ in}^4$ and $I_{y'} = 2.295 \text{ in}^4$, where z' and y' are the principal axes. Solve the problem using Eq. 11–17.

***11–100.** Determine the bending stress at point A of the beam using the result obtained in Prob. 11–98. The moments of inertia of the cross-sectional area about the z and y axes are $I_z = I_y = 5.561 \text{ in}^4$ and the product of inertia of the cross sectional area with respect to the z and y axes is $I_{yz} = -3.267 \text{ in}^4$. (See Appendix A.)

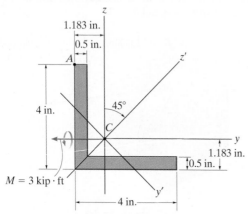

Probs. 11–99/100

11–101. The steel shaft is subjected to the two loads. If the journal bearings at A and B do not exert an axial force on the shaft, determine the required diameter of the shaft if the allowable bending stress is $\sigma_{\text{allow}} = 180 \text{ MPa}$.

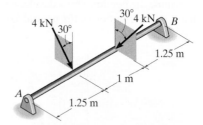

Prob. 11–101

11–102. The 65-mm-diameter steel shaft is subjected to the two loads If the journal bearings at A and B do not exert an axial force on the shaft, determine the absolute maximum bending stress developed in the shaft.

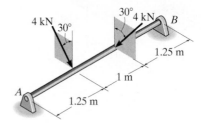

Prob. 11–102

11–103. For the section, $I_{z'} = 31.7(10^{-6}) \text{ m}^4$, $I_{y'} = 114(10^{-6}) \text{ m}^4$, $I_{y'z'} = -15.1(10^{-6}) \text{ m}^4$. Using the techniques outlined in Appendix A, the member's cross-sectional area has principal moments of inertia of $I_z = 29.0(10^{-6}) \text{ m}^4$ and $I_y = 117(10^{-6}) \text{ m}^4$, calculated about the principal axes of inertia y and z, respectively. If the section is subjected to a moment $M = 15 \text{ kN} \cdot \text{m}$, determine the stress at point A using Eq. 11–17.

***11–104.** Solve Prob. 11–103 using the equation developed in Prob. 11–98.

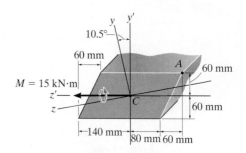

Probs. 11–103/104

CONCEPTUAL PROBLEMS

C11–1. The steel saw blade passes over the drive wheel of the band saw. Using appropriate measurements and data, explain how to determine the bending stress in the blade.

Prob. C11–1

C11–2. The crane boom has a noticeable taper along its length. Explain why. To do so, assume the boom is in the horizontal position and in the process of hoisting a load at its end, so that the reaction on the support *A* becomes zero. Use realistic dimensions and a load, to justify your reasoning.

Prob. C11–2

C11–3. Use reasonable dimensions for this hammer and a loading to show through an analysis why this hammer failed in the manner shown.

Prob. C11–3

C11–4. These garden shears were manufactured using an inferior material. Using a loading of 50 lb applied normal to the blades, and appropriate dimensions for the shears, determine the absolute maximum bending stress in the material and show why the failure occurred at the critical location on the handle.

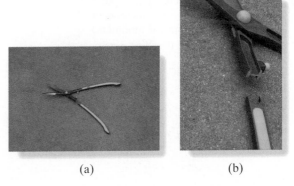

(a)   (b)

Prob. C11–4

11

CHAPTER REVIEW

Shear and moment diagrams are graphical representations of the internal shear and moment within a beam. They can be constructed by sectioning the beam an arbitrary distance x from the left end, using the equilibrium equations to find V and M as functions of x, and then plotting the results. A sign convention for positive distributed load, shear, and moment must be followed.

Positive external distributed load

Positive internal shear

Positive internal moment

Beam sign convention

It is also possible to plot the shear and moment diagrams by realizing that at each point the slope of the shear diagram is equal to the intensity of the distributed loading at the point.

$$w = \frac{dV}{dx}$$

Likewise, the slope of the moment diagram is equal to the shear at the point.

$$V = \frac{dM}{dx}$$

The area under the distributed-loading diagram between the points represents the change in shear.

$$\Delta V = \int w \, dx$$

And the area under the shear diagram represents the change in moment.

$$\Delta M = \int V \, dx$$

The shear and moment at any point can be obtained using the method of sections. The maximum (or minimum) moment occurs where the shear is zero.

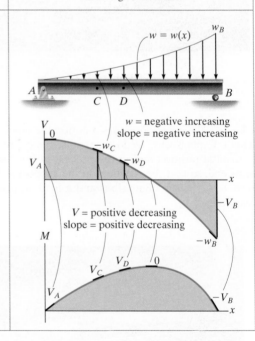

A bending moment tends to produce a linear variation of normal strain within a straight beam. Provided the material is homogeneous and linear elastic, then equilibrium can be used to relate the internal moment in the beam to the stress distribution. The result is the flexure formula,

$$\sigma_{max} = \frac{Mc}{I}$$

where I and c are determined from the neutral axis that passes through the centroid of the cross section.

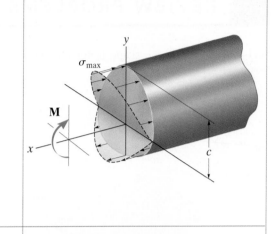

If the cross-sectional area of the beam is not symmetric about an axis that is perpendicular to the neutral axis, then unsymmetrical bending will occur. The maximum stress can be determined from formulas, or the problem can be solved by considering the superposition of bending caused by the moment components $\mathbf{M}_y$ and $\mathbf{M}_z$ about the principal axes of inertia for the area.

$$\sigma = -\frac{M_z y}{I_z} + \frac{M_y z}{I_y}$$

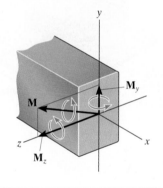

11

REVIEW PROBLEMS

R11–1. Determine the shape factor for the wide-flange beam.

R11–3. A shaft is made of a polymer having a parabolic upper and lower cross section. If it resists a moment of $M = 125\ \mathrm{N \cdot m}$, determine the maximum bending stress in the material (a) using the flexure formula and (b) using integration. Sketch a three-dimensional view of the stress distribution acting over the cross-sectional area. *Hint*: The moment of inertia is determined using Eq. A–3 of Appendix A.

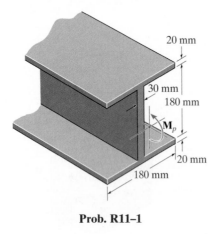

Prob. R11–1

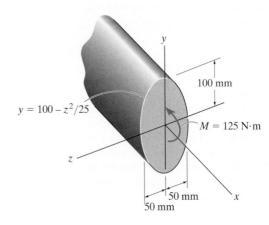

Prob. R11–3

R11–2. The compound beam consists of two segments that are pinned together at B. Draw the shear and moment diagrams if it supports the distributed loading shown.

*R11–4. Determine the maximum bending stress in the handle of the cable cutter at section a–a. A force of 45 lb is applied to the handles.

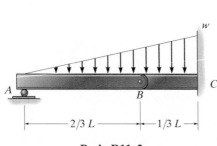

Prob. R11–2

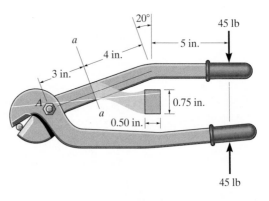

Prob. R11–4

R11–5. Determine the shear and moment in the beam as functions of x, where $0 \leq x < 6$ ft, then draw the shear and moment diagrams for the beam.

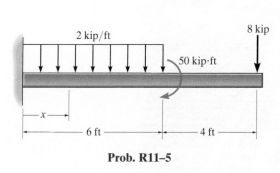

Prob. R11–5

R11–6. A wooden beam has a square cross section as shown. Determine which orientation of the beam provides the greatest strength at resisting the moment **M**. What is the difference in the resulting maximum stress in both cases?

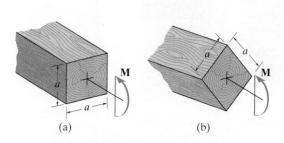

(a)   (b)

Prob. R11–6

R11–7. Draw the shear and moment diagrams for the shaft if it is subjected to the vertical loadings. The bearings at A and B exert only vertical reactions on the shaft.

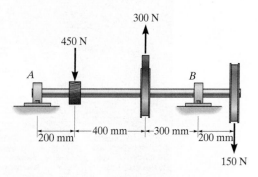

Prob. R11–7

*R11–8. The strut has a square cross section a by a and is subjected to the bending moment **M** applied at an angle θ as shown. Determine the maximum bending stress in terms of a, M, and θ. What angle θ will give the largest bending stress in the strut? Specify the orientation of the neutral axis for this case.

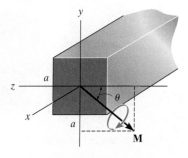

Prob. R11–8

CHAPTER 12

(© Bert Folsom/Alamy)

Railroad ties act as beams that support very large transverse shear loadings. As a result, if they are made of wood they will tend to split at their ends, where the shear loads are the largest.

TRANSVERSE SHEAR

CHAPTER OBJECTIVES

- To determine the shear stress in beams subjected to a transverse loading.
- To calculate the shear in fasteners used to construct beams made from several members.

12.1 SHEAR IN STRAIGHT MEMBERS

In general, a beam will support both an internal shear and a moment. The shear **V** is the result of a *transverse* shear-stress distribution that acts over the beam's cross section, Fig. 12–1. Due to the complementary property of shear, this stress will also create corresponding *longitudinal* shear stress that acts along the length of the beam.

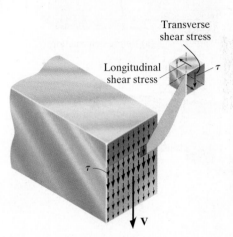

Fig. 12–1

559

Boards not bonded together
(a)

Boards bonded together
(b)

Fig. 12–2

Shear connectors are "tack welded" to this corrugated metal floor liner so that when the concrete floor is poured, the connectors will prevent the concrete slab from slipping on the liner surface. The two materials will thus act as a composite slab.

To illustrate the effect caused by the longitudinal shear stress, consider the beam made from three boards shown in Fig. 12–2a. If the top and bottom surfaces of each board are smooth, and the boards are *not* bonded together, then application of the load **P** will cause the boards to *slide* relative to one another when the beam deflects. However, if the boards are bonded together, then the longitudinal shear stress acting between the boards will prevent their relative sliding, and consequently the beam will act as a single unit, Fig. 12–2b.

As a result of the shear stress, shear strains will be developed and these will tend to distort the cross section in a rather complex manner. For example, consider the short bar in Fig. 12–3a made of a highly deformable material and marked with horizontal and vertical grid lines. When the shear force **V** is applied, it tends to deform these lines into the pattern shown in Fig. 12–3b. This nonuniform shear-strain distribution will cause the cross section to *warp*; and as a result, when a beam is subjected to *both* bending and shear, the cross section will not remain plane as assumed in the development of the flexure formula.

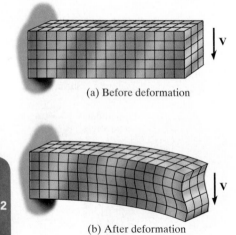

(a) Before deformation

(b) After deformation

Fig. 12–3

12.2 THE SHEAR FORMULA

Because the strain distribution for shear is not easily defined, as in the case of axial load, torsion, and bending, we will obtain the shear-stress distribution in an indirect manner. To do this we will consider the horizontal force equilibrium of a portion of an element taken from the beam in Fig. 12–4a. A free-body diagram of the entire element is shown in Fig. 12–4b. The normal-stress distribution acting on it is caused by the bending moments M and $M + dM$. Here we have excluded the effects of V, $V + dV$, and $w(x)$, since these loadings are vertical and will therefore not be involved in a horizontal force summation. Notice that $\Sigma F_x = 0$ is satisfied since the stress distribution on each side of the element forms only a couple moment, and therefore a zero force resultant.

12

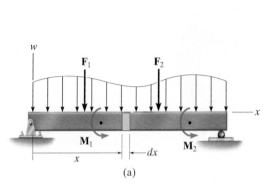

$\Sigma F_x = 0$ satisfied

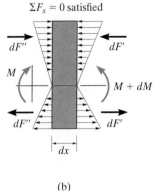

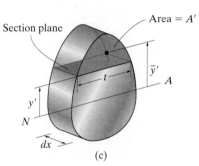

(a)   (b)   (c)

Fig. 12–4

Now let's consider the shaded *top portion* of the element that has been sectioned at y' from the neutral axis, Fig. 12–4c. It is on this sectioned plane that we want to find the shear stress. This top segment has a width t at the section, and the two cross-sectional sides each have an area A'. The segment's free-body diagram is shown in Fig. 12–4d. The resultant moments on each side of the element differ by dM, so that $\Sigma F_x = 0$ will not be satisfied unless a longitudinal shear stress τ acts over the bottom sectioned plane. To simplify the analysis, we will assume that this shear stress is *constant* across the width t of the bottom face. To find the horizontal force created by the bending moments, we will assume that the effect of warping due to shear is small, so that it can generally be *neglected*. This assumption is particularly true for the most common case of a *slender beam*, that is, one that has a small depth compared to its length. Therefore, using the flexure formula, Eq. 11–13, we have

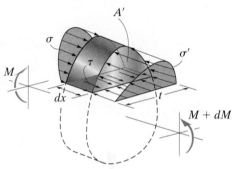

Three-dimensional view

$$\overset{+}{\leftarrow}\Sigma F_x = 0; \qquad \int_{A'} \sigma' \, dA' - \int_{A'} \sigma \, dA' - \tau(t \, dx) = 0$$

$$\int_{A'} \left(\frac{M + dM}{I}\right) y \, dA' - \int_{A'} \left(\frac{M}{I}\right) y \, dA' - \tau(t \, dx) = 0$$

$$\left(\frac{dM}{I}\right) \int_{A'} y \, dA' = \tau(t \, dx) \qquad (12\text{–}1)$$

Solving for τ, we get

$$\tau = \frac{1}{It}\left(\frac{dM}{dx}\right) \int_{A'} y \, dA'$$

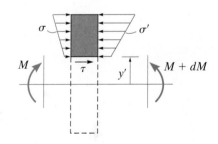

Profile view

(d)

Here $V = dM/dx$ (Eq. 11–2). Also, the integral represents the *moment of the area A'* about the neutral axis, which we will denote by the symbol Q. Since the location of the centroid of A' is determined from $\bar{y}' = \int_{A'} y \, dA'/A'$, we can also write

$$Q = \int_{A'} y \, dA' = \bar{y}'A' \qquad (12\text{–}2)$$

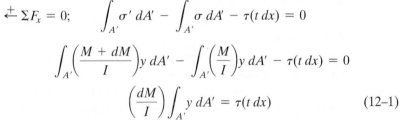

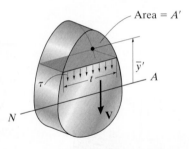

Fig. 12–5

The final result is called the *shear formula*, namely

$$\tau = \frac{VQ}{It} \qquad (12\text{--}3)$$

With reference to Fig. 12–5,

τ = the shear stress in the member at the point located a distance y from the neutral axis. This stress is assumed to be constant and therefore *averaged* across the width t of the member

V = the shear force, determined from the method of sections and the equations of equilibrium

I = the moment of inertia of the *entire* cross-sectional area calculated about the neutral axis

t = the width of the member's cross section, measured at the point where τ is to be determined

Q = $\bar{y}'A'$, where A' is the area of the top (or bottom) portion of the member's cross section, above (or below) the section plane where t is measured, and $\bar{y}'$ is the distance from the neutral axis to the centroid of A'

Although for the derivation we considered only the shear stress acting on the beam's longitudinal plane, the formula applies as well for finding the transverse shear stress on the beam's cross section, because these stresses are complementary and numerically equal.

Calculating Q. Of all the variables in the shear formula, Q is usually the most difficult to define properly. Try to remember that it represents *the moment of the cross-sectional area that is above or below the point where the shear stress is to be determined*. It is this area A' that is "held onto" the rest of the beam by the longitudinal shear stress as the beam undergoes bending, Fig. 12–4d. The examples shown in Fig. 12–6 will help to illustrate this point. Here the stress at point P is to be determined, and so A' represents the dark shaded region. The value of Q for each case is reported under each figure. These same results can *also* be obtained for Q by considering A' to be the light shaded area below P, although here y' is a negative quantity when a portion of A' is below the neutral axis.

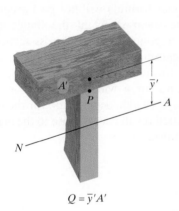

$$Q = \bar{y}'A'$$

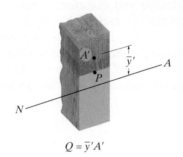

$$Q = \bar{y}'A'$$

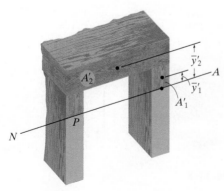

$$Q = 2\bar{y}'_1 A'_1 + \bar{y}'_2 A'_2$$

Fig. 12–6

12

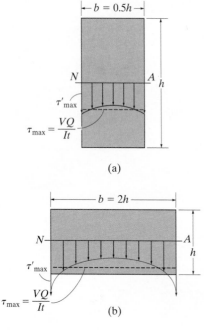

$$\tau_{max} = \frac{VQ}{It}$$

(a)

$$\tau_{max} = \frac{VQ}{It}$$

(b)

Fig. 12–7

Limitations on the Use of the Shear Formula. One of the major assumptions used in the development of the shear formula is that the shear stress is *uniformly* distributed over the *width t* at the section. In other words, the *average shear stress* is calculated across the width. We can test the accuracy of this assumption by comparing it with a more exact mathematical analysis based on the theory of elasticity. For example, if the beam's cross section is rectangular, the shear-stress distribution across the neutral axis actually varies as shown in Fig. 12–7. The maximum value, τ'_{max}, occurs at the *sides* of the cross section, and its magnitude depends on the ratio b/h (width/depth). For sections having a $b/h = 0.5$, τ'_{max} is only about 3% greater than the shear stress calculated from the shear formula, Fig. 12–7a. However, for *flat sections*, say $b/h = 2$, τ'_{max} is about 40% greater than τ_{max}, Fig. 12–7b. The error becomes even greater as the section becomes flatter, that is, as the b/h ratio increases. Errors of this magnitude are certainly intolerable if one attempts to use the shear formula to determine the shear stress in the *flange* of the wide-flange beam shown in Fig. 12–8.

It should also be noted that the shear formula will not give accurate results when used to determine the shear stress at the flange–web junction of this beam, since this is a point of sudden cross-sectional change and therefore a *stress concentration* occurs here. Fortunately, engineers must only use the shear formula to calculate the average maximum shear stress in a beam, and for a wide-flange section this occurs at the neutral axis, where the b/h (width/depth) ratio for the web is *very small*, and therefore the calculated result is very close to the *actual* maximum shear stress as explained above.

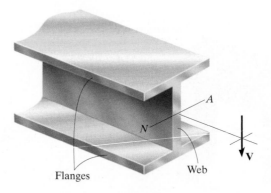

Fig. 12–8

Another important limitation on the use of the shear formula can be illustrated with reference to Fig. 12–9a, which shows a member having a cross section with an irregular boundary. If we apply the shear formula to determine the (average) shear stress τ along the line AB, it will be directed downward across this line as shown in Fig. 12–9b. However, an element of material taken from the boundary point B, Fig. 12–9c, must not have any shear stress on its outer surface. In other words, the shear stress acting on this element must *be directed tangent to the boundary*, and so the shear-stress distribution across line AB is actually directed as shown in Fig. 12–9d. As a result, the shear formula can only be applied at sections shown by the blue lines in Fig. 12–9a, because these lines intersect the tangents to the boundary at *right angles*, Fig. 12–9e.

To summarize the above points, the shear formula does not give accurate results when applied to members having cross sections that are *short or flat*, or at points where the cross section suddenly changes. Nor should it be applied across a section that intersects the boundary of the member at an angle other than 90°.

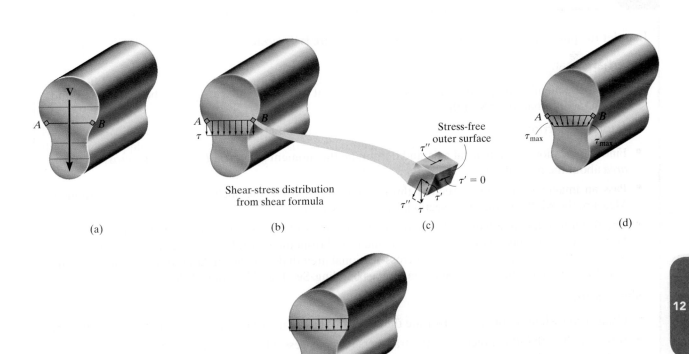

(a)

(b)

Shear-stress distribution from shear formula

(c)

Stress-free outer surface

$\tau' = 0$

(d)

(e)

Fig. 12–9

EXAMPLE 12.2

Determine the distribution of the shear stress over the cross section of the beam shown in Fig. 12–11a.

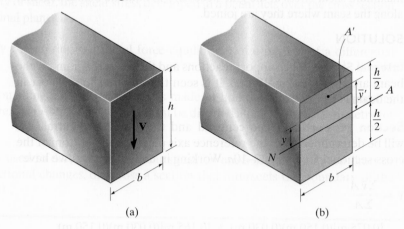

(a)   (b)

SOLUTION

The distribution can be determined by finding the shear stress at an *arbitrary height y* from the neutral axis, Fig. 12–11b, and then plotting this function. Here, the dark colored area A' will be used for Q.* Hence

$$Q = \bar{y}'A' = \left[y + \frac{1}{2}\left(\frac{h}{2} - y\right)\right]\left(\frac{h}{2} - y\right)b = \frac{1}{2}\left(\frac{h^2}{4} - y^2\right)b$$

Applying the shear formula, we have

$$\tau = \frac{VQ}{It} = \frac{V\left(\frac{1}{2}\right)\left[(h^2/4) - y^2\right]b}{\left(\frac{1}{12}bh^3\right)b} = \frac{6V}{bh^3}\left(\frac{h^2}{4} - y^2\right) \tag{1}$$

This result indicates that the shear-stress distribution over the cross section is *parabolic*. As shown in Fig. 12–11c, the intensity varies from zero at the top and bottom, $y = \pm h/2$, to a maximum value at the neutral axis, $y = 0$. Specifically, since the area of the cross section is $A = bh$, then at $y = 0$ Eq. 1 becomes

$$\boxed{\tau_{max} = 1.5\frac{V}{A}} \tag{2}$$

Rectangular cross section

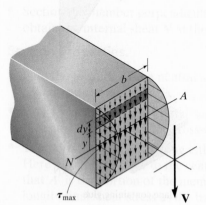

Shear–stress distribution

(c)

Fig. 12–11

*The area below y can also be used $[A' = b(h/2 + y)]$, but doing so involves a bit more algebraic manipulation.

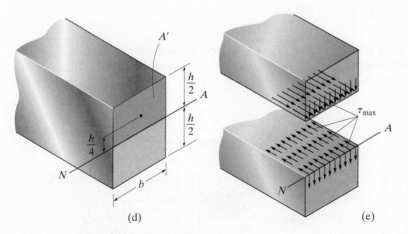

(d)

(e)

Typical shear failure of this wooden beam occurred at the support and through the approximate center of its cross section.

Fig. 12–11 (cont.)

This same value for τ_{max} can be obtained directly from the shear formula, $\tau = VQ/It$, by realizing that τ_{max} occurs where Q is *largest*, since V, I, and t are *constant*. By inspection, Q will be a maximum when the entire area above (or below) the neutral axis is considered; that is, $A' = bh/2$ and $\bar{y}' = h/4$, Fig. 12–11d. Thus,

$$\tau_{max} = \frac{VQ}{It} = \frac{V(h/4)(bh/2)}{\left[\frac{1}{12}bh^3\right]b} = 1.5\frac{V}{A}$$

By comparison, τ_{max} is 50% greater than the *average* shear stress determined from Eq. 7–4; that is, $\tau_{avg} = V/A$.

It is important to realize that τ_{max} also acts in the longitudinal direction of the beam, Fig. 12–11e. It is this stress that can cause a timber beam to fail at its supports, as shown Fig. 12–11f. Here horizontal splitting of the wood starts to occur through the neutral axis at the beam's ends, since there the vertical reactions subject the beam to large shear stress, and wood has a low resistance to shear along its grains, which are oriented in the longitudinal direction.

It is instructive to show that when the shear-stress distribution, Eq. 1, is integrated over the cross section it produces the resultant shear V. To do this, a differential strip of area $dA = b\,dy$ is chosen, Fig. 12–11c, and since τ acts uniformly over this strip, we have

$$\int_A \tau\,dA = \int_{-h/2}^{h/2} \frac{6V}{bh^3}\left(\frac{h^2}{4} - y^2\right)b\,dy$$

$$= \frac{6V}{h^3}\left[\frac{h^2}{4}y - \frac{1}{3}y^3\right]_{-h/2}^{h/2}$$

$$= \frac{6V}{h^3}\left[\frac{h^2}{4}\left(\frac{h}{2} + \frac{h}{2}\right) - \frac{1}{3}\left(\frac{h^3}{8} + \frac{h^3}{8}\right)\right] = V$$

(f)

12

EXAMPLE 12.3

A steel wide-flange beam has the dimensions shown in Fig. 12–12a. If it is subjected to a shear of $V = 80$ kN, plot the shear-stress distribution acting over the beam's cross section.

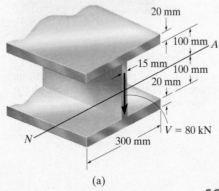

(a)

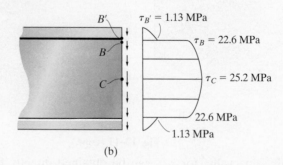

(b)

SOLUTION

Since the flange and web are rectangular elements, then like the previous example, the shear-stress distribution will be parabolic and in this case it will vary in the manner shown in Fig. 12–12b. Due to symmetry, only the shear stresses at points B', B, and C have to be determined. To show how these values are obtained, we must first determine the moment of inertia of the cross-sectional area about the neutral axis. Working in meters, we have

$$I = \left[\frac{1}{12}(0.015 \text{ m})(0.200 \text{ m})^3\right]$$

$$+ 2\left[\frac{1}{12}(0.300 \text{ m})(0.02 \text{ m})^3 + (0.300 \text{ m})(0.02 \text{ m})(0.110 \text{ m})^2\right]$$

$$= 155.6(10^{-6}) \text{ m}^4$$

For point B', $t_{B'} = 0.300$ m, and A' is the dark shaded area shown in Fig. 12–12c. Thus,

$$Q_{B'} = \bar{y}'A' = [0.110 \text{ m}](0.300 \text{ m})(0.02 \text{ m}) = 0.660(10^{-3}) \text{ m}^3$$

so that

$$\tau_{B'} = \frac{VQ_{B'}}{It_{B'}} = \frac{80(10^3) \text{ N}(0.660(10^{-3}) \text{ m}^3)}{155.6(10^{-6}) \text{ m}^4(0.300 \text{ m})} = 1.13 \text{ MPa}$$

For point B, $t_B = 0.015$ m and $Q_B = Q_{B'}$, Fig. 12–12c. Hence

$$\tau_B = \frac{VQ_B}{It_B} = \frac{80(10^3)\text{N}(0.660(10^{-3}) \text{ m}^3)}{155.6(10^{-6}) \text{ m}^4(0.015 \text{ m})} = 22.6 \text{ MPa}$$

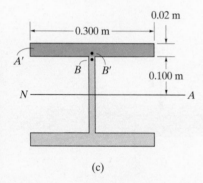

(c)

Fig. 12–12

12

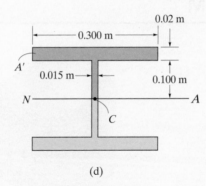

(d)

Fig. 12–12 (cont.)

Note from our discussion of the limitations on the use of the shear formula that the calculated values for both $\tau_{B'}$ and τ_B are actually very misleading. Why?

For point C, $t_C = 0.015$ m and A' is the dark shaded area shown in Fig. 12–12d. Considering this area to be composed of two rectangles, we have

$$Q_C = \Sigma \bar{y}'A' = [0.110 \text{ m}](0.300 \text{ m})(0.02 \text{ m})$$

$$+ |0.05 \text{ m}|(0.015 \text{ m})(0.100 \text{ m})$$

$$= 0.735(10^{-3}) \text{ m}^3$$

Thus,

$$\tau_C = \tau_{max} = \frac{VQ_C}{It_C} = \frac{80(10^3) \text{ N}[0.735(10^{-3}) \text{ m}^3]}{155.6(10^{-6}) \text{ m}^4(0.015 \text{ m})} = 25.2 \text{ MPa}$$

NOTE: From Fig. 12–12b, the largest shear stress occurs in the web and is almost uniform throughout its depth, varying from 22.6 MPa to 25.2 MPa. It is for this reason that for design, some codes permit the use of calculating the *average* shear stress on the cross section of the web, rather than using the shear formula; that is,

$$\tau_{avg} = \frac{V}{A_w} = \frac{80(10^3) \text{ N}}{(0.015 \text{ m})(0.2 \text{ m})} = 26.7 \text{ MPa}$$

This will be discussed further in Chapter 15.

PRELIMINARY PROBLEM

P12–1. In each case, calculate the value of Q and t that are used in the shear formula for finding the shear stress at A. Also, show how the shear stress acts on a differential volume element located at point A.

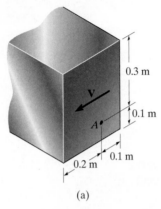

(a)

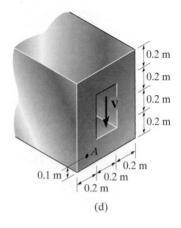

(d)

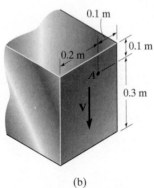

(b)

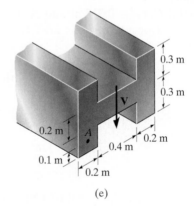

(e)

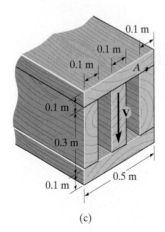

(c)

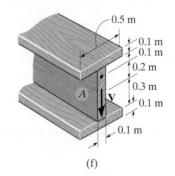

(f)

Prob. P12–1

FUNDAMENTAL PROBLEMS

F12–1. If the beam is subjected to a shear force of $V = 100$ kN, determine the shear stress at point A. Represent the state of stress on a volume element at this point.

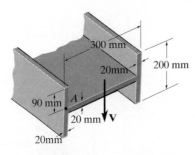

Prob. F12–1

F12–2. Determine the shear stress at points A and B if the beam is subjected to a shear force of $V = 600$ kN. Represent the state of stress on a volume element of these points.

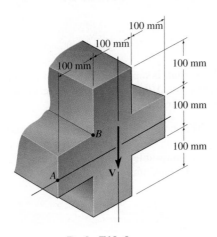

Prob. F12–2

F12–3. Determine the absolute maximum shear stress in the beam.

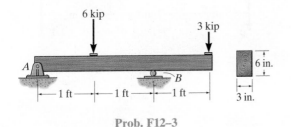

Prob. F12–3

F12–4. If the beam is subjected to a shear force of $V = 20$ kN, determine the maximum shear stress in the beam.

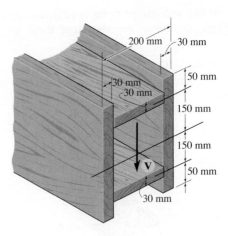

Prob. F12–4

F12–5. If the beam is made from four plates and subjected to a shear force of $V = 20$ kN, determine the shear stress at point A. Represent the state of stress on a volume element at this point.

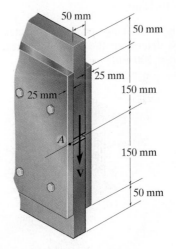

Prob. F12–5

12

PROBLEMS

12–1. If the wide-flange beam is subjected to a shear of $V = 20$ kN, determine the shear stress on the web at A. Indicate the shear-stress components on a volume element located at this point.

12–2. If the wide-flange beam is subjected to a shear of $V = 20$ kN, determine the maximum shear stress in the beam.

12–3. If the wide-flange beam is subjected to a shear of $V = 20$ kN, determine the shear force resisted by the web of the beam.

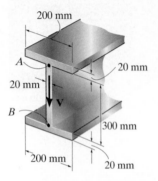

Probs. 12–1/2/3

*12–4. If the beam is subjected to a shear of $V = 30$ kN, determine the web's shear stress at A and B. Indicate the shear-stress components on a volume element located at these points. Set $w = 200$ mm. Show that the neutral axis is located at $\bar{y} = 0.2433$ m from the bottom and $I = 0.5382(10^{-3})$ m^4.

12–5. If the wide-flange beam is subjected to a shear of $V = 30$ kN, determine the maximum shear stress in the beam. Set $w = 300$ mm.

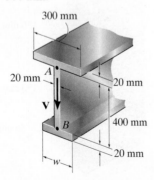

Probs. 12–4/5

12–6. The wood beam has an allowable shear stress of $\tau_{allow} = 7$ MPa. Determine the maximum shear force V that can be applied to the cross section.

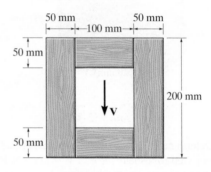

Prob. 12–6

12–7. The shaft is supported by a thrust bearing at A and a journal bearing at B. If $P = 20$ kN, determine the absolute maximum shear stress in the shaft.

*12–8. The shaft is supported by a thrust bearing at A and a journal bearing at B. If the shaft is made from a material having an allowable shear stress of $\tau_{allow} = 75$ MPa, determine the maximum value for P.

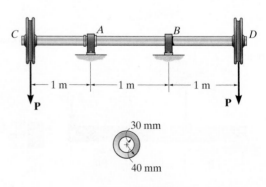

Probs. 12–7/8

12–9. Determine the largest shear force V that the member can sustain if the allowable shear stress is $\tau_{\text{allow}} = 8$ ksi.

12–10. If the applied shear force $V = 18$ kip, determine the maximum shear stress in the member.

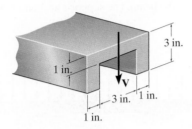

Probs. 12–9/10

12–11. The overhang beam is subjected to the uniform distributed load having an intensity of $w = 50$ kN/m. Determine the maximum shear stress in the beam.

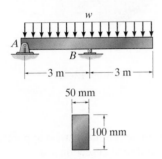

Prob. 12–11

*12–12. The beam is made from a polymer and is subjected to a shear of $V = 7$ kip. Determine the maximum shear stress in the beam and plot the shear-stress distribution over the cross section. Report the values of the shear stress every 0.5 in. of beam depth.

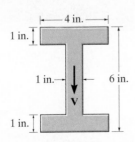

Prob. 12–12

12–13. Determine the maximum shear stress in the strut if it is subjected to a shear force of $V = 20$ kN.

12–14. Determine the maximum shear force V that the strut can support if the allowable shear stress for the material is $\tau_{\text{allow}} = 40$ MPa.

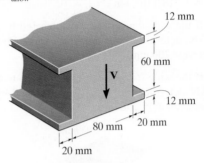

Probs. 12–13/14

12–15. Sketch the intensity of the shear-stress distribution acting over the beam's cross-sectional area, and determine the resultant shear force acting on the segment AB. The shear force acting at the section is $V = 35$ kip. Show that $I_{NA} = 872.49$ in⁴.

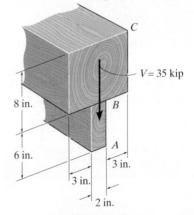

Prob. 12–15

*12–16. Plot the shear-stress distribution over the cross section of a rod that has a radius c. By what factor is the maximum shear stress greater than the average shear stress acting over the cross section?

Prob. 12–16

12

12–17. If the beam is subjected to a shear of $V = 15$ kN, determine the web's shear stress at A and B. Indicate the shear-stress components on a volume element located at these points. Set $w = 125$ mm. Show that the neutral axis is located at $\bar{y} = 0.1747$ m from the bottom and $I_{NA} = 0.2182(10^{-3})$ m^4.

12–18. If the wide-flange beam is subjected to a shear of $V = 30$ kN, determine the maximum shear stress in the beam. Set $w = 200$ mm.

12–19. If the wide-flange beam is subjected to a shear of $V = 30$ kN, determine the shear force resisted by the web of the beam. Set $w = 200$ mm.

200 mm

30 mm

A

25 mm

V

250 mm B

30 mm

w

Probs. 12–17/18/19

*12–20. Determine the length of the cantilevered beam so that the maximum bending stress in the beam is equivalent to the maximum shear stress.

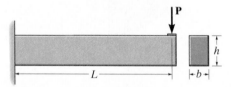

P

L $|{-}b{-}|$ h

Prob. 12–20

12–21. If the beam is made from wood having an allowable shear stress $\tau_{allow} = 400$ psi, determine the maximum magnitude of **P**. Set $d = 4$ in.

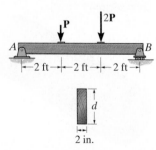

P   **2P**

A B

$|{-}2\text{ ft}{-}|{-}2\text{ ft}{-}|{-}2\text{ ft}{-}|$

d

2 in.

Prob. 12–21

12–22. Determine the largest intensity w of the distributed load that the member can support if the allowable shear stress is $\tau_{allow} = 800$ psi. The supports at A and B are smooth.

12–23. If $w = 800$ lb/ft, determine the absolute maximum shear stress in the beam. The supports at A and B are smooth.

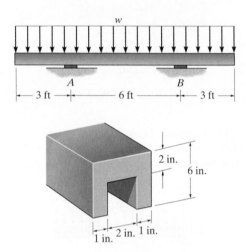

w

A B

$|{-}$ 3 ft $-|-$ 6 ft $-|-$ 3 ft $-|$

2 in.

6 in.

1 in.   2 in.   1 in.

Probs. 12–22/23

*12–24. Determine the shear stress at point B on the web of the cantilevered strut at section a–a.

12–25. Determine the maximum shear stress acting at section a–a of the cantilevered strut.

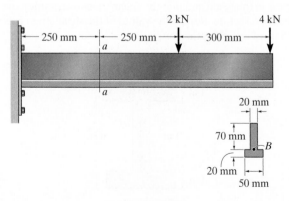

2 kN   4 kN

$|{-}$ 250 mm $-|-$ 250 mm $-|-$ 300 mm $-|$

a

a

20 mm

70 mm

B

20 mm

50 mm

Probs. 12–24/25

12

12–26.   Railroad ties must be designed to resist large shear loadings. If the tie is subjected to the 34-kip rail loadings and an assumed uniformly distributed ground reaction, determine the intensity w for equilibrium, and calculate the maximum shear stress in the tie at section a–a, which is located just to the left of the rail.

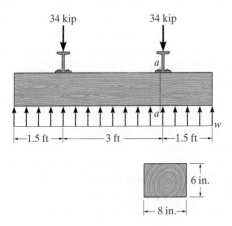

Prob. 12–26

12–27.   The beam is slit longitudinally along both sides. If it is subjected to a shear of $V = 250$ kN, compare the maximum shear stress in the beam before and after the cuts were made.

***12–28.**   The beam is to be cut longitudinally along both sides as shown. If it is made from a material having an allowable shear stress of $\tau_{allow} = 75$ MPa, determine the maximum allowable shear force **V** that can be applied before and after the cut is made.

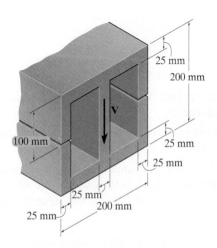

Probs. 12–27/28

12–29.   Determine the maximum shear stress in the T-beam at the critical section where the internal shear force is maximum.

12–30.   Determine the maximum shear stress in the T-beam at section C. Show the result on a volume element at this point.

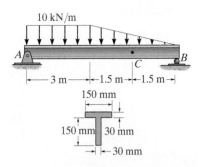

Probs. 12–29/30

12–31.   The beam has a square cross section and is made of wood having an allowable shear stress of $\tau_{allow} = 1.4$ ksi. If it is subjected to a shear of $V = 1.5$ kip, determine the smallest dimension a of its sides.

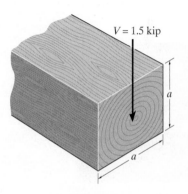

Prob. 12–31

12.3 SHEAR FLOW IN BUILT-UP MEMBERS

Occasionally in engineering practice, members are "built up" from several composite parts in order to achieve a greater resistance to loads. An example is shown in Fig. 12–13. If the loads cause the members to bend, fasteners such as nails, bolts, welding material, or glue will be needed to keep the component parts from sliding relative to one another, Fig. 12–2. In order to design these fasteners or determine their spacing, it is necessary to know the shear force that they must resist. This loading, when measured as a force per unit length of beam, is referred to as *shear flow, q.**

Fig. 12–13

The magnitude of the shear flow is obtained using a procedure similar to that for finding the shear stress in a beam. To illustrate, consider finding the shear flow along the juncture where the segment in Fig. 12–14a is connected to the flange of the beam. Three horizontal forces must act on this segment, Fig. 12–14b. Two of these forces, F and $F + dF$, are the result of the normal stresses caused by the moments M and $M + dM$, respectively. The third force, which for equilibrium equals dF, acts at the juncture. Realizing that dF is the result of dM, then, like Eq. 12–1, we have

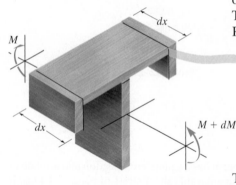

(a)

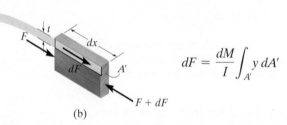

(b)

$$dF = \frac{dM}{I} \int_{A'} y \, dA'$$

The integral represents Q, that is, the moment of the segment's area A' about the neutral axis. Since the segment has a length dx, the shear flow, or *force per unit length along the beam*, is $q = dF/dx$. Hence dividing both sides by dx and noting that $V = dM/dx$, Eq. 11–2, we have

$$\boxed{q = \frac{VQ}{I}} \qquad (12\text{–}4)$$

Here

q = the shear flow, measured as a force per unit length along the beam

V = the shear force, determined from the method of sections and the equations of equilibrium

I = the moment of inertia of the *entire* cross-sectional area calculated about the neutral axis

$Q = \bar{y}'A'$, where A' is the cross-sectional area of the segment that is *connected to the beam* at the juncture where the shear flow is calculated, and $\bar{y}'$ is the distance from the neutral axis to the centroid of A'

Fig. 12–14

*The use of the word "flow" in this terminology will become meaningful as it pertains to the discussion in Sec. 12.4.

Fastener Spacing. When segments of a beam are connected by fasteners, such as nails or bolts, their spacing *s* along the beam can be determined. For example, let's say that a fastener, such as a nail, can support a maximum shear force of *F* (N) before it fails, Fig. 12–15a. If these nails are used to construct the beam made from two boards, as shown in Fig. 12–15b, then the nails must resist the shear flow *q* (N/m) between the boards. In other words, the nails are used to "hold" the top board to the bottom board so that no slipping occurs during bending. (See Fig. 12–2a). As shown in Fig. 12–15c, the nail spacing is therefore determined from

$$F \text{ (N)} = q \text{ (N/m)} \, s \text{ (m)}$$

The examples that follow illustrate application of this equation.

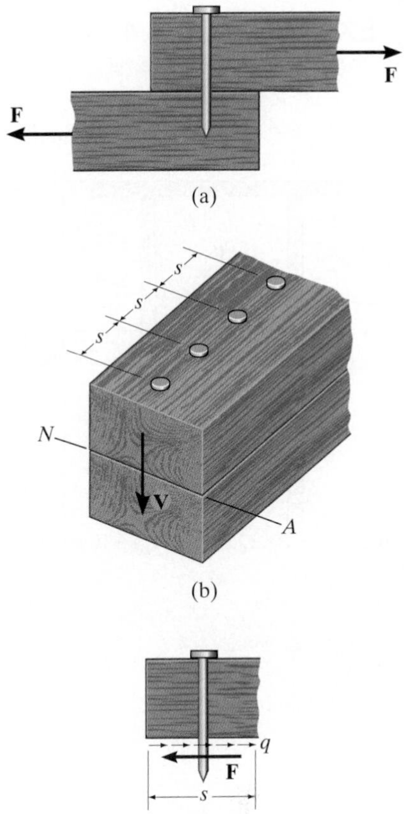

Other examples of shaded segments connected to built-up beams by fasteners are shown in Fig. 12–16. The shear flow here must be found at the thick black line, and is determined by using a value of *Q* calculated from *A'* and $\bar{y}'$ indicated in each figure. This value of *q* will be resisted by a *single* fastener in Fig. 12–16a, by *two* fasteners in Fig. 12–16b, and by *three* fasteners in Fig. 12–16c. In other words, the fastener in Fig. 12–16a supports the calculated value of *q*, and in Figs. 12–16b and 12–16c each fastener supports *q*/2 and *q*/3, respectively.

> ### IMPORTANT POINT
>
> - *Shear flow* is a measure of the force per unit length along the axis of a beam. This value is found from the shear formula and is used to determine the shear force developed in fasteners and glue that holds the various segments of a composite beam together.

Fig. 12–15

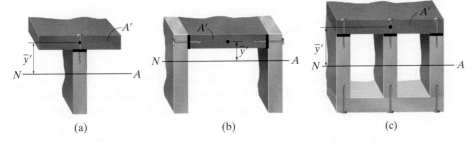

(a)   (b)   (c)

Fig. 12–16

EXAMPLE 12.4

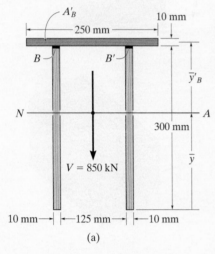

(a)

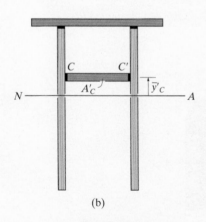

(b)

Fig. 12–17

The beam is constructed from three boards glued together as shown in Fig. 12–17a. If it is subjected to a shear of $V = 850$ kN, determine the shear flow at B and B' that must be resisted by the glue.

SOLUTION

Section Properties. The neutral axis (centroid) will be located from the bottom of the beam, Fig. 12–17a. Working in units of meters, we have

$$\bar{y} = \frac{\Sigma \tilde{y} A}{\Sigma A} = \frac{2[0.15 \text{ m}](0.3 \text{ m})(0.01 \text{ m}) + [0.305 \text{ m}](0.250 \text{ m})(0.01 \text{ m})}{2(0.3 \text{ m})(0.01 \text{ m}) + 0.250 \text{ m}(0.01 \text{ m})}$$

$$= 0.1956 \text{ m}$$

The moment of inertia of the cross section about the neutral axis is thus

$$I = 2\left[\frac{1}{12}(0.01 \text{ m})(0.3 \text{ m})^3 + (0.01 \text{ m})(0.3 \text{ m})(0.1956 \text{ m} - 0.150 \text{ m})^2\right]$$

$$+ \left[\frac{1}{12}(0.250 \text{ m})(0.01 \text{ m})^3 + (0.250 \text{ m})(0.01 \text{ m})(0.305 \text{ m} - 0.1956 \text{ m})^2\right]$$

$$= 87.42(10^{-6}) \text{ m}^4$$

The glue at both B and B' in Fig. 12–17a "holds" the top board to the beam. Here

$$Q_B = \bar{y}'_B A'_B = [0.305 \text{ m} - 0.1956 \text{ m}](0.250 \text{ m})(0.01 \text{ m})$$

$$= 0.2735(10^{-3}) \text{ m}^3$$

Shear Flow.

$$q = \frac{VQ_B}{I} = \frac{850(10^3) \text{ N}(0.2735(10^{-3}) \text{ m}^3)}{87.42(10^{-6}) \text{ m}^4} = 2.66 \text{ MN/m}$$

Since *two seams* are used to secure the board, the glue per meter length of beam at each seam must be strong enough to resist *one-half* of this shear flow. Thus,

$$q_B = q_{B'} = \frac{q}{2} = 1.33 \text{ MN/m} \qquad Ans.$$

NOTE: If the board CC' is added to the beam, Fig. 12–17b, then $\bar{y}$ and I have to be recalculated, and the shear flow at C and C' determined from $q = V \, y'_C \, A'_C / I$. Finally, this value is divided by one-half to obtain q_C and $q_{C'}$.

12

EXAMPLE 12.5

A box beam is constructed from four boards nailed together as shown in Fig. 12–18a. If each nail can support a maximum shear force of 30 lb, determine the maximum spacing s of the nails at B and at C so that the beam can support the force of 80 lb.

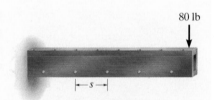

SOLUTION

Internal Shear. If the beam is sectioned at an *arbitrary point* along its length, the internal shear required for equilibrium is always $V = 80$ lb.

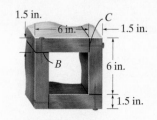

(a)

Section Properties. The moment of inertia of the cross-sectional area about the neutral axis can be determined by considering a 7.5 in. × 7.5 in. square minus a 4.5 in. × 4.5 in. square.

$$I = \frac{1}{12}(7.5 \text{ in.})(7.5 \text{ in.})^3 - \frac{1}{12}(4.5 \text{ in.})(4.5 \text{ in.})^3 = 229.5 \text{ in}^4$$

The shear flow at B is determined using Q_B found from the darker shaded area shown in Fig. 12–18b. It is this "symmetric" portion of the beam that is to be "held onto" the rest of the beam by nails on the left side and by the fibers of the board on the right side, B'.
Thus,

$$Q_B = \bar{y}'A' = [3 \text{ in.}](7.5 \text{ in.})(1.5 \text{ in.}) = 33.75 \text{ in}^3$$

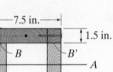

(b)

Likewise, the shear flow at C can be determined using the "symmetric" shaded area shown in Fig. 12–18c. We have

$$Q_C = \bar{y}'A' = [3 \text{ in.}](4.5 \text{ in.})(1.5 \text{ in.}) = 20.25 \text{ in}^3$$

Shear Flow.

$$q_B = \frac{VQ_B}{I} = \frac{80 \text{ lb}(33.75 \text{ in}^3)}{229.5 \text{ in}^4} = 11.76 \text{ lb/in.}$$

$$q_C = \frac{VQ_C}{I} = \frac{80 \text{ lb}(20.25 \text{ in}^3)}{229.5 \text{ in}^4} = 7.059 \text{ lb/in.}$$

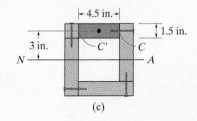

(c)

Fig. 12–18

These values represent the shear force per unit length of the beam that must be resisted by the nails at B and the fibers at B', Fig. 12–18b, and the nails at C and the fibers at C', Fig. 12–18c, respectively. Since in each case the shear flow is resisted at *two* surfaces and each nail can resist 30 lb, for B the spacing is

$$s_B = \frac{30 \text{ lb}}{(11.76/2) \text{ lb/in.}} = 5.10 \text{ in.} \quad \text{Use } s_B = 5 \text{ in.} \quad Ans.$$

And for C,

$$s_C = \frac{30 \text{ lb}}{(7.059/2) \text{ lb/in.}} = 8.50 \text{ in.} \quad \text{Use } s_C = 8.5 \text{ in.} \quad Ans.$$

12

EXAMPLE 12.6

Nails, each having a total shear strength of 40 lb, are used in a beam that can be constructed either as in Case I or as in Case II, Fig. 12–19. If the nails are spaced at 9 in., determine the largest vertical shear that can be supported in each case so that the fasteners will not fail.

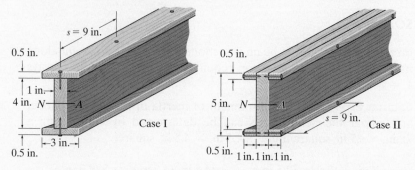

Fig. 12–19

SOLUTION

Since the cross section is the same in both cases, the moment of inertia about the neutral axis is calculated using one large rectangle and two smaller side rectangles.

$$I = \frac{1}{12}(3 \text{ in.})(5 \text{ in.})^3 - 2\left[\frac{1}{12}(1 \text{ in.})(4 \text{ in.})^3\right] = 20.58 \text{ in}^4$$

Case I. For this design a single row of nails holds the top or bottom flange onto the web. For one of these flanges,

$$Q = \bar{y}'A' = [2.25 \text{ in.}](3 \text{ in.}(0.5 \text{ in.})) = 3.375 \text{ in}^3$$

so that

$$q = \frac{VQ}{I}; \qquad \frac{40 \text{ lb}}{9 \text{ in.}} = \frac{V(3.375 \text{ in}^3)}{20.58 \text{ in}^4}$$
$$V = 27.1 \text{ lb} \qquad\qquad Ans.$$

Case II. Here a single row of nails holds *one* of the side boards onto the web. Thus,

$$Q = \bar{y}'A' = [2.25 \text{ in.}](1 \text{ in.}(0.5 \text{ in.})) = 1.125 \text{ in}^3$$

$$q = \frac{F}{s} = \frac{VQ}{I}; \qquad \frac{40 \text{ lb}}{9 \text{ in.}} = \frac{V(1.125 \text{ in}^3)}{20.58 \text{ in}^4}$$
$$V = 81.3 \text{ lb} \qquad\qquad Ans.$$

Or, we can also say two rows of nails hold two side boards onto the web, so that

$$q = \frac{F}{s} = \frac{VQ}{I}; \qquad \frac{2(40 \text{ lb})}{9 \text{ in.}} = \frac{V[2(1.125 \text{ in}^3)]}{20.58 \text{ in}^4}$$
$$V = 81.3 \text{ lb} \qquad\qquad Ans.$$

FUNDAMENTAL PROBLEMS

F12–6. The two identical boards are bolted together to form the beam. Determine the maximum spacing s of the bolts to the nearest mm if each bolt has a shear strength of 15 kN. The beam is subjected to a shear force of $V = 50$ kN.

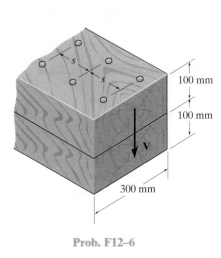

Prob. F12–6

F12–7. Two identical 20-mm-thick plates are bolted to the top and bottom flange to form the built-up beam. If the beam is subjected to a shear force of $V = 300$ kN, determine the maximum spacing s of the bolts to the nearest mm if each bolt has a shear strength of 30 kN.

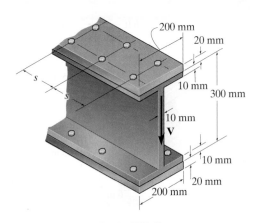

Prob. F12–7

F12–8. The boards are bolted together to form the built-up beam. If the beam is subjected to a shear force of $V = 20$ kN, determine the maximum spacing s of the bolts to the nearest mm if each bolt has a shear strength of 8 kN.

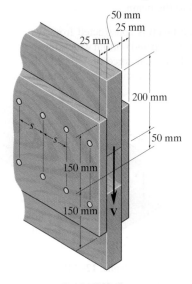

Prob. F12–8

F12–9. The boards are bolted together to form the built-up beam. If the beam is subjected to a shear force of $V = 15$ kip, determine the maximum spacing s of the bolts to the nearest $\frac{1}{8}$ in. if each bolt has a shear strength of 6 kip.

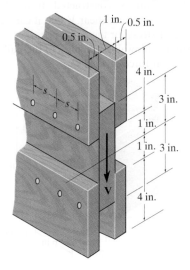

Prob. F12–9

12

12–45. The member consists of two plastic channel strips 0.5 in. thick, glued together at A and B. If the distributed load has a maximum intensity of $w_0 = 3$ kip/ft, determine the maximum shear stress resisted by the glue.

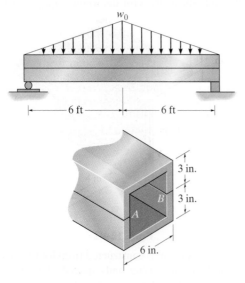

Prob. 12–45

12–46. The member consists of two plastic channel strips 0.5 in. thick, glued together at A and B. If the glue can support an allowable shear stress of $\tau_{allow} = 600$ psi, determine the maximum intensity w_0 of the triangular distributed loading that can be applied to the member based on the strength of the glue.

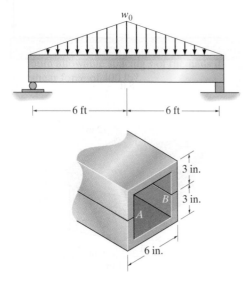

Prob. 12–46

12–47. The beam is made from four boards nailed together as shown. If the nails can each support a shear force of 100 lb., determine their required spacing s' and s if the beam is subjected to a shear of $V = 700$ lb.

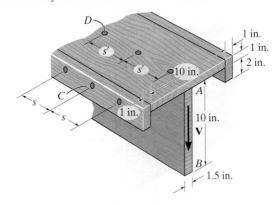

Prob. 12–47

*12–48. The beam is made from three polystyrene strips that are glued together as shown. If the glue has a shear strength of 80 kPa, determine the maximum load P that can be applied without causing the glue to lose its bond.

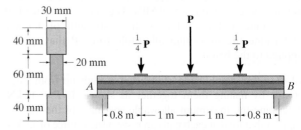

Prob. 12–48

12–49. The timber T-beam is subjected to a load consisting of n concentrated forces, P_n. If the allowable shear V_{nail} for each of the nails is known, write a computer program that will specify the nail spacing between each load. Show an application of the program using the values $L = 15$ ft, $a_1 = 4$ ft, $P_1 = 600$ lb, $a_2 = 8$ ft, $P_2 = 1500$ lb, $b_1 = 1.5$ in., $h_1 = 10$ in., $b_2 = 8$ in., $h_2 = 1$ in., and $V_{nail} = 200$ lb.

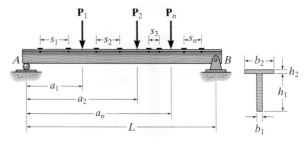

Prob. 12–49

CHAPTER REVIEW

Transverse shear stress in beams is determined indirectly by using the flexure formula and the relationship between moment and shear ($V = dM/dx$). The result is the shear formula

$$\tau = \frac{VQ}{It}$$

In particular, the value for Q is the moment of the area A' about the neutral axis, $Q = \bar{y}'A'$. This area is the portion of the cross-sectional area that is "held onto" the beam above (or below) the thickness t where τ is to be determined.

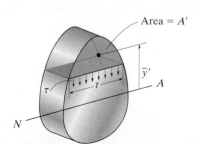

If the beam has a rectangular cross section, then the shear-stress distribution will be parabolic, having a maximum value at the neutral axis. For this special case, the maximum shear stress can be determined using

$$\tau_{max} = 1.5\,\frac{V}{A}$$

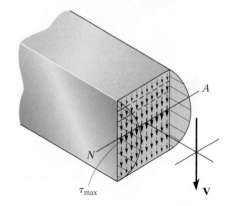

Shear–stress distribution

Fasteners, such as nails, bolts, glue, or weld, are used to connect the composite parts of a "built-up" section. The shear force resisted by these fasteners is determined from the shear flow, q, or force per unit length, that must be supported by the beam. The shear flow is

$$q = \frac{VQ}{I}$$

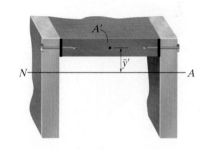

12

REVIEW PROBLEMS

R12–1. The beam is fabricated from four boards nailed together as shown. Determine the shear force each nail along the sides C and the top D must resist if the nails are uniformly spaced at $s = 3$ in. The beam is subjected to a shear of $V = 4.5$ kip.

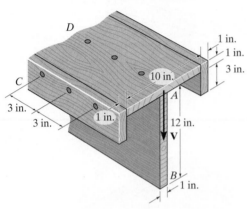

Prob. R12–1

R12–2. The T-beam is subjected to a shear of $V = 150$ kN. Determine the amount of this force that is supported by the web B.

Prob. R12–2

R12–3. The member is subjected to a shear force of $V = 2$ kN. Determine the shear flow at points A, B, and C. The thickness of each thin-walled segment is 15 mm.

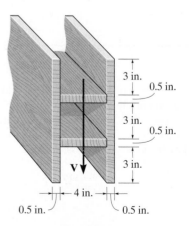

Prob. R12–3

*R12–4. The beam is constructed from four boards glued together at their seams. If the glue can withstand 75 lb/in., what is the maximum vertical shear V that the beam can support? What is the maximum vertical shear V that the beam can support if it is rotated 90° from the position shown?

Prob. R12–4

R12–5. Determine the shear stress at points B and C on the web of the beam located at section a–a.

R12–6. Determine the maximum shear stress acting at section a–a in the beam.

*R12–8. The member consists of two triangular plastic strips bonded together along AB. If the glue can support an allowable shear stress of $\tau_{allow} = 600$ psi, determine the maximum vertical shear V that can be applied to the member based on the strength of the glue.

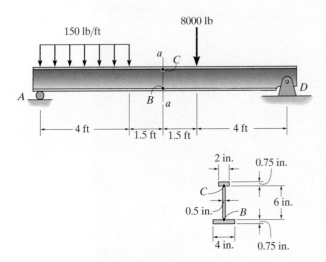

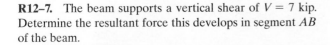

Probs. R12–5/6

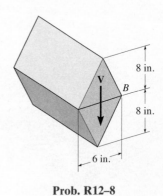

Prob. R12–8

R12–9. If the pipe is subjected to a shear of $V = 15$ kip, determine the maximum shear stress in the pipe.

R12–7. The beam supports a vertical shear of $V = 7$ kip. Determine the resultant force this develops in segment AB of the beam.

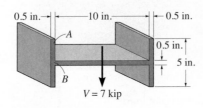

Prob. R12–7

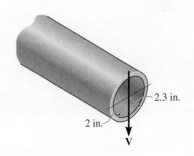

Prob. R12–9

CHAPTER 13

(© ImageBroker/Alamy)

The offset hanger supporting this ski gondola is subjected to the combined loadings of axial force and bending moment.

COMBINED LOADINGS

13.1 THIN-WALLED PRESSURE VESSELS

Cylindrical or spherical pressure vessels are commonly used in industry to serve as boilers or storage tanks. The stresses acting in the wall of these vessels can be analyzed in a simple manner provided it has a *thin wall*, that is, the inner-radius-to-wall-thickness ratio is 10 or more ($r/t \geq 10$). Specifically, when $r/t = 10$ the results of a thin-wall analysis will predict a stress that is approximately 4% *less* than the actual maximum stress in the vessel. For larger r/t ratios this error will be even smaller.

In the following analysis, we will assume the gas pressure in the vessel is the *gage pressure,* that is, it is the pressure *above* atmospheric pressure, since atmospheric pressure is assumed to exist both inside and outside the vessel's wall before the vessel is pressurized.

Cylindrical pressure vessels, such as this gas tank, have semispherical end caps rather than flat ones in order to reduce the stress in the tank.

591

13

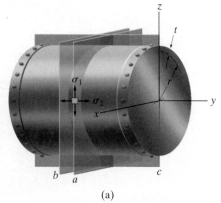

(a)

(b)

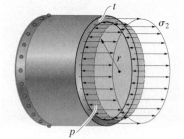

(c)

Fig. 13–1

Cylindrical Vessels. The cylindrical vessel in Fig. 13–1a has a wall thickness t, inner radius r, and is subjected to an internal gas pressure p. To find the *circumferential* or *hoop stress*, we can section the vessel by planes a, b, and c. A free-body diagram of the back segment along with its contained gas is then shown in Fig. 13–1b. Here only the loadings in the x direction are shown. They are caused by the uniform hoop stress σ_1, acting on the vessel's wall, and the pressure acting on the vertical face of the gas. For equilibrium in the x direction, we require

$$\Sigma F_x = 0; \qquad 2[\sigma_1(t\,dy)] - p(2r\,dy) = 0$$

$$\boxed{\sigma_1 = \frac{pr}{t}} \qquad (13\text{–}1)$$

The longitudinal stress can be determined by considering the left portion of section b, Fig. 13–1a. As shown on its free-body diagram, Fig. 13–1c, σ_2 acts uniformly throughout the wall, and p acts on the section of the contained gas. Since the mean radius is approximately equal to the vessel's inner radius, equilibrium in the y direction requires

$$\Sigma F_y = 0; \qquad \sigma_2(2\pi rt) - p(\pi r^2) = 0$$

$$\boxed{\sigma_2 = \frac{pr}{2t}} \qquad (13\text{–}2)$$

For these two equations,

σ_1, σ_2 = the normal stress in the hoop and longitudinal directions, respectively. Each is assumed to be *constant* throughout the wall of the cylinder, and each subjects the material to tension.
p = the internal gage pressure developed by the contained gas
r = the inner radius of the cylinder
t = the thickness of the wall ($r/t \geq 10$)

By comparison, note that the hoop or circumferential stress is *twice as large* as the longitudinal or axial stress. Consequently, when fabricating cylindrical pressure vessels from rolled-formed plates, it is important that the longitudinal joints be designed to carry twice as much stress as the circumferential joints.

Spherical Vessels. We can analyze a spherical pressure vessel in a similar manner. If the vessel in Fig. 13–2a is sectioned in half, the resulting free-body diagram is shown in Fig. 13–2b. Like the cylinder, equilibrium in the y direction requires

$$\Sigma F_y = 0; \qquad \sigma_2(2\pi rt) - p(\pi r^2) = 0$$

This thin-walled pipe was subjected to an excessive gas pressure that caused it to rupture in the circumferential or hoop direction. The stress in this direction is twice that in the axial direction as noted by Eqs. 13–1 and 13–2.

$$\boxed{\sigma_2 = \frac{pr}{2t}} \qquad (13\text{–}3)$$

This is the same result as that obtained for the longitudinal stress in the cylindrical pressure vessel, although this stress will be the same regardless of the orientation of the hemispheric free-body diagram.

Limitations. The above analysis indicates that an element of material taken from either a cylindrical or a spherical pressure vessel is subjected to **biaxial stress**, i.e., normal stress existing in only two directions. Actually, however, the pressure also subjects the material to a **radial stress**, σ_3, which acts along a radial line. This stress has a maximum value equal to the pressure p at the interior wall and it decreases through the wall to zero at the exterior surface of the vessel, since the pressure there is zero. For thin-walled vessels, however, we will *ignore* this stress component, since our limiting assumption of $r/t = 10$ results in σ_2 and σ_1 being, respectively, 5 and 10 times *higher* than the maximum radial stress, $(\sigma_3)_{\text{max}} = p$. Finally, note that if the vessel is subjected to an *external pressure,* the resulting compressive stresses within the wall may cause the wall to suddenly collapse inward or buckle rather than causing the material to fracture.

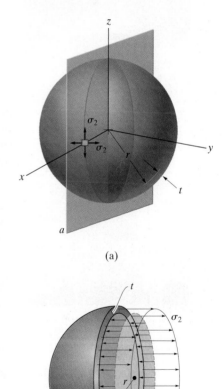

(a)

(b)

Fig. 13–2

EXAMPLE 13.1

13

A cylindrical pressure vessel has an inner diameter of 4 ft and a thickness of $\frac{1}{2}$ in. Determine the maximum internal pressure it can sustain so that neither its circumferential nor its longitudinal stress component exceeds 20 ksi. Under the same conditions, what is the maximum internal pressure that a similar-size spherical vessel can sustain?

SOLUTION

Cylindrical Pressure Vessel. The maximum stress occurs in the circumferential direction. From Eq. 13–1 we have

$$\sigma_1 = \frac{pr}{t}; \qquad\qquad 20 \text{ kip/in}^2 = \frac{p(24 \text{ in.})}{\frac{1}{2} \text{ in.}}$$

$$p = 417 \text{ psi} \qquad\qquad Ans.$$

   Note that when this pressure is reached, from Eq. 13–2, the stress in the longitudinal direction will be $\sigma_2 = \frac{1}{2}(20 \text{ ksi}) = 10 \text{ ksi}$, and the *maximum stress* in the *radial direction* is at the inner wall of the vessel, $(\sigma_3)_{max} = p = 417$ psi. This value is 48 times smaller than the circumferential stress (20 ksi), and as stated earlier, its effect will be neglected.

Spherical Vessel. Here the maximum stress occurs in any two perpendicular directions on an element of the vessel, Fig. 13–2a. From Eq. 13–3, we have

$$\sigma_2 = \frac{pr}{2t}; \qquad\qquad 20 \text{ kip/in}^2 = \frac{p(24 \text{ in.})}{2\left(\frac{1}{2} \text{ in.}\right)}$$

$$p = 833 \text{ psi} \qquad\qquad Ans.$$

NOTE: Although it is more difficult to fabricate, the spherical pressure vessel will carry twice as much internal pressure as a cylindrical vessel.

PROBLEMS

13–1.   A spherical gas tank has an inner radius of $r = 1.5$ m. If it is subjected to an internal pressure of $p = 300$ kPa, determine its required thickness if the maximum normal stress is not to exceed 12 MPa.

13–2.   A pressurized spherical tank is made of 0.5-in.-thick steel. If it is subjected to an internal pressure of $p = 200$ psi, determine its outer radius if the maximum normal stress is not to exceed 15 ksi.

13–3.   The thin-walled cylinder can be supported in one of two ways as shown. Determine the state of stress in the wall of the cylinder for both cases if the piston P causes the internal pressure to be 65 psi. The wall has a thickness of 0.25 in., and the inner diameter of the cylinder is 8 in.

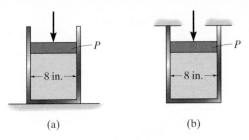

(a)   (b)

Prob. 13–3

*13–4.   The tank of the air compressor is subjected to an internal pressure of 90 psi. If the inner diameter of the tank is 22 in., and the wall thickness is 0.25 in., determine the stress components acting at point A. Draw a volume element of the material at this point, and show the results on the element.

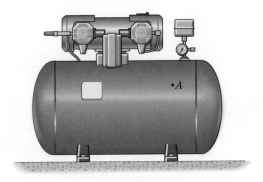

Prob. 13–4

13–5.   Air pressure in the cylinder is increased by exerting forces $P = 2$ kN on the two pistons, each having a radius of 45 mm. If the cylinder has a wall thickness of 2 mm, determine the state of stress in the wall of the cylinder.

13–6.   Determine the maximum force P that can be exerted on each of the two pistons so that the circumferential stress in the cylinder does not exceed 3 MPa. Each piston has a radius of 45 mm and the cylinder has a wall thickness of 2 mm.

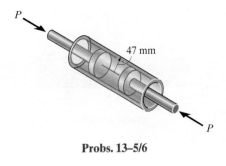

47 mm

Probs. 13–5/6

13–7.   A boiler is constructed of 8-mm-thick steel plates that are fastened together at their ends using a butt joint consisting of two 8-mm cover plates and rivets having a diameter of 10 mm and spaced 50 mm apart as shown. If the steam pressure in the boiler is 1.35 MPa, determine (a) the circumferential stress in the boiler's plate away from the seam, (b) the circumferential stress in the outer cover plate along the rivet line a–a, and (c) the shear stress in the rivets.

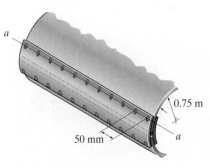

0.75 m

50 mm

Prob. 13–7

13

*13–8. The steel water pipe has an inner diameter of 12 in. and a wall thickness of 0.25 in. If the valve A is opened and the flowing water has a pressure of 250 psi as it passes point B, determine the longitudinal and hoop stress developed in the wall of the pipe at point B.

13–9. The steel water pipe has an inner diameter of 12 in. and a wall thickness of 0.25 in. If the valve A is closed and the water pressure is 300 psi, determine the longitudinal and hoop stress developed in the wall of the pipe at point B. Draw the state of stress on a volume element located on the wall.

13–11. The gas pipe line is supported every 20 ft by concrete piers and also lays on the ground. If there are rigid retainers at the piers that hold the pipe fixed, determine the longitudinal and hoop stress in the pipe if the temperature rises 60° F from the temperature at which it was installed. The gas within the pipe is at a pressure of 600 lb/in². The pipe has an inner diameter of 20 in. and thickness of 0.25 in. The material is A-36 steel.

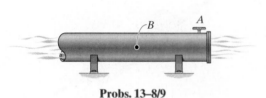

Probs. 13–8/9

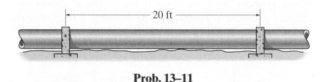

Prob. 13–11

13–10. The A-36-steel band is 2 in. wide and is secured around the smooth rigid cylinder. If the bolts are tightened so that the tension in them is 400 lb, determine the normal stress in the band, the pressure exerted on the cylinder, and the distance half the band stretches.

*13–12. A pressure-vessel head is fabricated by welding the circular plate to the end of the vessel as shown. If the vessel sustains an internal pressure of 450 kPa, determine the average shear stress in the weld and the state of stress in the wall of the vessel.

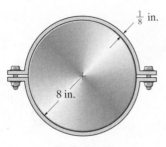

Prob. 13–10

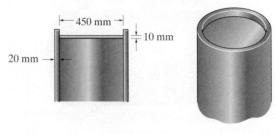

Prob. 13–12

13–13. An A-36-steel hoop has an inner diameter of 23.99 in., thickness of 0.25 in., and width of 1 in. If it and the 24-in.-diameter rigid cylinder have a temperature of 65° F, determine the temperature to which the hoop should be heated in order for it to just slip over the cylinder. What is the pressure the hoop exerts on the cylinder, and the tensile stress in the ring when it cools back down to 65° F?

24 in.

Prob. 13–13

13–14. The ring, having the dimensions shown, is placed over a flexible membrane which is pumped up with a pressure p. Determine the change in the inner radius of the ring after this pressure is applied. The modulus of elasticity for the ring is E.

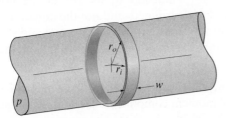

r_o
r_i
w
p

Prob. 13–14

13–15. The inner ring A has an inner radius r_1 and outer radius r_2. The outer ring B has an inner radius r_3 and an outer radius r_4, and $r_2 > r_3$. If the outer ring is heated and then fitted over the inner ring, determine the pressure between the two rings when ring B reaches the temperature of the inner ring. The material has a modulus of elasticity of E and a coefficient of thermal expansion of α.

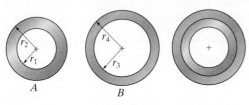

r_2
r_1
A

r_4
r_3
B

Prob. 13–15

*13–16. Two hemispheres having an inner radius of 2 ft and wall thickness of 0.25 in. are fitted together, and the inside pressure is reduced to -10 psi. If the coefficient of static friction is $\mu_s = 0.5$ between the hemispheres, determine (a) the torque T needed to initiate the rotation of the top hemisphere relative to the bottom one, (b) the vertical force needed to pull the top hemisphere off the bottom one, and (c) the horizontal force needed to slide the top hemisphere off the bottom one.

Prob. 13–16

13–17. In order to increase the strength of the pressure vessel, filament winding of the same material is wrapped around the circumference of the vessel as shown. If the pretension in the filament is T and the vessel is subjected to an internal pressure p, determine the hoop stresses in the filament and in the wall of the vessel. Use the free-body diagram shown, and assume the filament winding has a thickness t' and width w for a corresponding length L of the vessel.

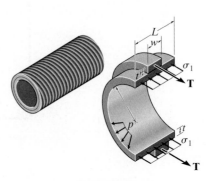

L
w
σ_1
T
r
p
t'
σ_1
T

Prob. 13–17

This chimney is subjected to the combined internal loading caused by the wind and the chimney's weight.

13.2   STATE OF STRESS CAUSED BY COMBINED LOADINGS

In the previous chapters we showed how to determine the stress in a member subjected to either an internal axial force, a shear force, a bending moment, or a torsional moment. Most often, however, the cross section of a member will be subjected to several of these loadings simultaneously, and when this occurs, then the method of superposition should be used to determine the resultant stress. The following procedure for analysis provides a method for doing this.

▶ PROCEDURE FOR ANALYSIS

Here it is required that the material be homogeneous and behave in a linear elastic manner. Also, Saint-Venant's principle requires that the stress be determined at a point far removed from any discontinuities in the cross section or points of applied load.

Internal Loading.

- Section the member perpendicular to its axis at the point where the stress is to be determined; and use the equations of equilibrium to obtain the resultant internal normal and shear force components, and the bending and torsional moment components.

- The force components should act through the *centroid* of the cross section, and the moment components should be calculated about *centroidal axes*, which represent the principal axes of inertia for the cross section.

Stress Components.

- Determine the stress component associated with *each* internal loading.

   Normal Force.

   - The normal force is related to a uniform normal-stress distribution determined from $\sigma = N/A$.

Shear Force.

- The shear force is related to a shear-stress distribution determined from the shear formula, $\tau = VQ/It$.

Bending Moment.

- For *straight members* the bending moment is related to a normal-stress distribution that varies linearly from zero at the neutral axis to a maximum at the outer boundary of the member. This stress distribution is determined from the flexure formula, $\sigma = -My/I$. If the member is *curved*, the stress distribution is nonlinear and is determined from $\sigma = My/[Ae(R - y)]$.

Torsional Moment.

- For circular shafts and tubes the torsional moment is related to a shear-stress distribution that varies linearly from zero at the center of the shaft to a maximum at the shaft's outer boundary. This stress distribution is determined from the torsion formula, $\tau = T\rho/J$.

Thin-Walled Pressure Vessels.

- If the vessel is a thin-walled cylinder, the internal pressure p will cause a biaxial state of stress in the material such that the hoop or circumferential stress component is $\sigma_1 = pr/t$, and the longitudinal stress component is $\sigma_2 = pr/2t$. If the vessel is a thin-walled sphere, then the biaxial state of stress is represented by two equivalent components, each having a magnitude of $\sigma_2 = pr/2t$.

Superposition.

- Once the normal and shear stress components for each loading have been calculated, use the principle of superposition and determine the resultant normal and shear stress components.

- Represent the results on an element of material located at a point, or show the results as a distribution of stress acting over the member's cross-sectional area.

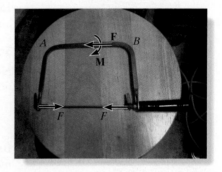

When a pretension force F is developed in the blade of this coping saw, it will produce both a compressive force F and bending moment M at the section AB of the frame. The material must therefore resist the normal stress produced by both of these loadings.

Problems in this section, which involve combined loadings, serve as a basic *review* of the application of the stress equations mentioned above. A thorough understanding of how these equations are applied, as indicated in the previous chapters, is necessary if one is to successfully solve the problems at the end of this section. The following examples should be carefully studied before proceeding to solve the problems.

13

EXAMPLE 13.2

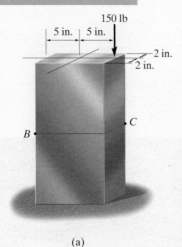

(a)

Fig. 13–3

A force of 150 lb is applied to the edge of the member shown in Fig. 13–3a. Neglect the weight of the member and determine the state of stress at points B and C.

SOLUTION

Internal Loadings. The member is sectioned through B and C, Fig. 13–3b. For equilibrium at the section there must be an axial force of 150 lb acting through the *centroid* and a bending moment of 750 lb · in. about the centroidal principal axis, Fig. 13–3b.

Stress Components.

Normal Force. The uniform normal-stress distribution due to the normal force is shown in Fig. 13–3c. Here

$$\sigma = \frac{N}{A} = \frac{150 \text{ lb}}{(10 \text{ in.})(4 \text{ in.})} = 3.75 \text{ psi}$$

Bending Moment. The normal-stress distribution due to the bending moment is shown in Fig. 13–3d. The maximum stress is

$$\sigma_{\max} = \frac{Mc}{I} = \frac{750 \text{ lb} \cdot \text{in. } (5 \text{ in.})}{\frac{1}{12} (4 \text{ in.}) (10 \text{ in.})^3} = 11.25 \text{ psi}$$

Superposition. Algebraically adding the stresses at B and C, we get

$$\sigma_B = -\frac{N}{A} + \frac{Mc}{I} = -3.75 \text{ psi} + 11.25 \text{ psi} = 7.5 \text{ psi} \quad \text{(tension)} \qquad Ans.$$

$$\sigma_C = -\frac{N}{A} - \frac{Mc}{I} = -3.75 \text{ psi} - 11.25 \text{ psi} = -15 \text{ psi (compression)}$$

Ans.

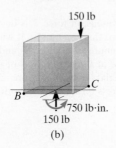

(b)

These results are shown in Figs. 13–3f and 13–3g.

NOTE: The resultant stress distribution over the cross section is shown in Fig. 13–3e, where the location of the line of zero stress can be determined by proportional triangles; i.e.,

$$\frac{7.5 \text{ psi}}{x} = \frac{15 \text{ psi}}{(10 \text{ in.} - x)}; \quad x = 3.33 \text{ in.}$$

Normal force

(c)

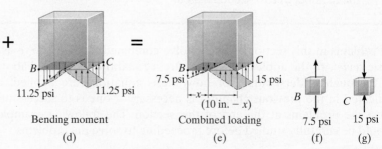

+   =

Bending moment   Combined loading

(d)   (e)   (f)   (g)

EXAMPLE 13.3

The gas tank in Fig. 13–4a has an inner radius of 24 in. and a thickness of 0.5 in. If it supports the 1500-lb load at its top, and the gas pressure within it is 2 lb/in², determine the state of stress at point A.

SOLUTION

Internal Loadings. The free-body diagram of the section of the tank above point A is shown in Fig. 13–4b.

Stress Components.

Circumferential Stress. Since $r/t = 24$ in./0.5 in. $= 48 > 10$, the tank is a thin-walled vessel. Applying Eq. 13–1, using the inner radius $r = 24$ in., we have

$$\sigma_1 = \frac{pr}{t} = \frac{2 \text{ lb/in}^2 \,(24 \text{ in.})}{0.5 \text{ in.}} = 96 \text{ psi} \qquad \textit{Ans.}$$

Longitudinal Stress. Here the wall of the tank uniformly supports the load of 1500 lb (compression) and the pressure stress (tensile). Thus, we have

$$\sigma_2 = -\frac{N}{A} + \frac{pr}{2t} = -\frac{1500 \text{ lb}}{\pi[(24.5 \text{ in.})^2 - (24 \text{ in.})^2]} + \frac{2 \text{ lb/in}^2 \,(24 \text{ in.})}{2\,(0.5 \text{ in.})}$$

$$= 28.3 \text{ psi} \qquad\qquad\qquad\qquad\qquad\qquad \textit{Ans.}$$

Point A is therefore subjected to the biaxial stress shown in Fig. 13–4c.

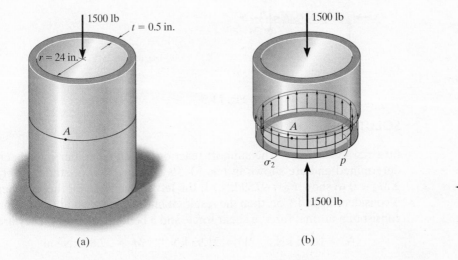

(a)

(b)

(c)

Fig. 13–4

EXAMPLE 13.4

The member shown in Fig. 13–5a has a rectangular cross section. Determine the state of stress that the loading produces at point C and point D.

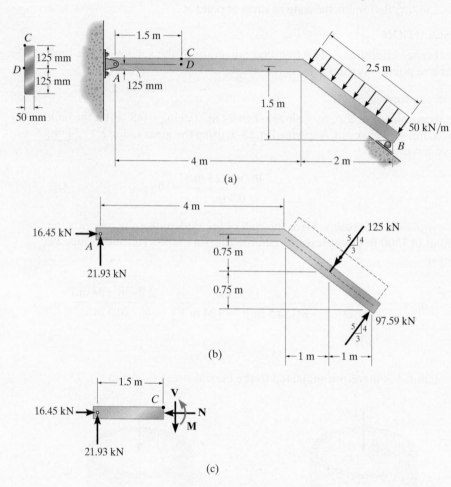

Fig. 13–5

SOLUTION

Internal Loadings. The support reactions on the member have been determined and are shown in Fig. 13–5b. (As a review of statics, apply $\Sigma M_A = 0$ to show $F_B = 97.59$ kN.) If the left segment AC of the member is considered, Fig. 13–5c, then the resultant internal loadings at the section consist of a normal force, a shear force, and a bending moment. They are

$$N = 16.45 \text{ kN} \qquad V = 21.93 \text{ kN} \qquad M = 32.89 \text{ kN} \cdot \text{m}$$

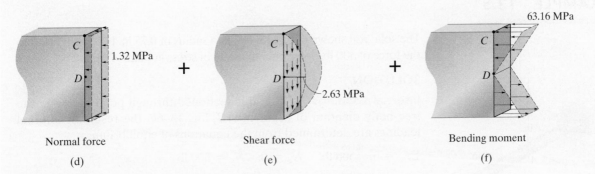

Normal force	Shear force	Bending moment
(d)	(e)	(f)

Fig. 13–5 (cont.)

Stress Components at C.

Normal Force.   The uniform normal-stress distribution acting over the cross section is produced by the normal force, Fig. 13–5d. At point C,

$$\sigma_C = \frac{N}{A} = \frac{16.45(10^3)\ \text{N}}{(0.050\ \text{m})\ (0.250\ \text{m})} = 1.32\ \text{MPa}$$

Shear Force.   Here the area $A' = 0$, since point C is located at the top of the member. Thus $Q = \bar{y}'A' = 0$, Fig. 13–5e. The shear stress is therefore

$$\tau_C = 0$$

Bending Moment.   Point C is located at $y = c = 0.125$ m from the neutral axis, so the bending stress at C, Fig. 13–5f, is

$$\sigma_C = \frac{Mc}{I} = \frac{(32.89(10^3)\ \text{N}\cdot\text{m})(0.125\ \text{m})}{\left[\frac{1}{12}(0.050\ \text{m})(0.250\ \text{m})^3\right]} = 63.16\ \text{MPa}$$

Superposition.   There is no shear-stress component. Adding the normal stresses gives a compressive stress at C having a value of

$$\sigma_C = 1.32\ \text{MPa} + 63.16\ \text{MPa} = 64.5\ \text{MPa} \qquad \textit{Ans.}$$

This result, acting on an element at C, is shown in Fig. 13–5g.

64.5 MPa

(g)

Stress Components at D.

Normal Force.   This is the same as at C, $\sigma_D = 1.32$ MPa, Fig. 13–5d.

Shear Force.   Since D is at the neutral axis, and the cross section is rectangular, we can use the special form of the shear formula, Fig. 13–5e.

$$\tau_D = 1.5\frac{V}{A} = 1.5\frac{21.93(10^3)\ \text{N}}{(0.25\ \text{m})(0.05\ \text{m})} = 2.63\ \text{MPa} \qquad \textit{Ans.}$$

Bending Moment.   Here D is on the neutral axis and so $\sigma_D = 0$.

Superposition.   The resultant stress on the element is shown in Fig. 13–5h.

2.63 MPa

1.32 MPa

(h)

EXAMPLE 13.5

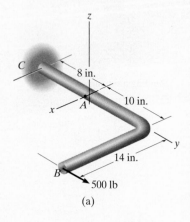

(a)

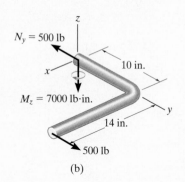

(b)

The solid rod shown in Fig. 13–6a has a radius of 0.75 in. If it is subjected to the force of 500 lb, determine the state of stress at point A.

SOLUTION

Internal Loadings. The rod is sectioned through point A. Using the free-body diagram of segment AB, Fig. 13–6b, the resultant internal loadings are determined from the equations of equilibrium.

$$\Sigma F_y = 0; \quad 500 \text{ lb} - N_y = 0; \quad N_y = 500 \text{ lb}$$

$$\Sigma M_z = 0; \quad 500 \text{ lb}(14 \text{ in.}) - M_z = 0; \quad M_z = 7000 \text{ lb} \cdot \text{in.}$$

In order to better "visualize" the stress distributions due to these loadings, we can consider the *equal but opposite resultants* acting on segment AC, Fig. 13–6c.

Stress Components.

Normal Force. The normal-stress distribution is shown in Fig. 13–6d. For point A, we have

$$(\sigma_A)_y = \frac{N}{A} = \frac{500 \text{ lb}}{\pi(0.75 \text{ in.})^2} = 283 \text{ psi} = 0.283 \text{ ksi}$$

Bending Moment. For the moment, $c = 0.75$ in., so the bending stress at point A, Fig. 13–6e, is

$$(\sigma_A)_y = \frac{Mc}{I} = \frac{7000 \text{ lb} \cdot \text{in.}(0.75 \text{ in.})}{[\frac{1}{4}\pi(0.75 \text{ in.})^4]}$$

$$= 21\,126 \text{ psi} = 21.13 \text{ ksi}$$

Superposition. When the above results are superimposed, it is seen that an element at A, Fig. 13–6f, is subjected to the normal stress

$$(\sigma_A)_y = 0.283 \text{ ksi} + 21.13 \text{ ksi} = 21.4 \text{ ksi} \qquad Ans.$$

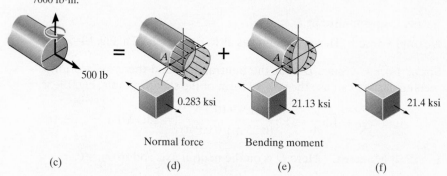

(c)   (d)   (e)   (f)

Fig. 13–6

EXAMPLE 13.6

The solid rod shown in Fig. 13–7a has a radius of 0.75 in. If it is subjected to the force of 800 lb, determine the state of stress at point A.

SOLUTION

Internal Loadings. The rod is sectioned through point A. Using the free-body diagram of segment AB, Fig. 13–7b, the resultant internal loadings are determined from the equations of equilibrium. Take a moment to verify these results. The *equal but opposite resultants* are shown acting on segment AC, Fig. 13–7c.

$$\Sigma F_z = 0; \quad V_z - 800 \text{ lb} = 0; \quad V_z = 800 \text{ lb}$$

$$\Sigma M_x = 0; \quad M_x - 800 \text{ lb}(10 \text{ in.}) = 0; \quad M_x = 8000 \text{ lb} \cdot \text{in.}$$

$$\Sigma M_y = 0; \quad -M_y + 800 \text{ lb}(14 \text{ in.}) = 0; \quad M_y = 11\,200 \text{ lb} \cdot \text{in.}$$

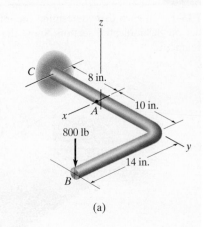

(a)

Stress Components.

Shear Force. The shear-stress distribution is shown in Fig. 13–7d. For point A, Q is determined from the gray shaded *semicircular* area. Using the table in Appendix B, we have

$$Q = \bar{y}'A' = \frac{4(0.75 \text{ in.})}{3\pi}\left[\frac{1}{2}\pi(0.75 \text{ in.})^2\right] = 0.28125 \text{ in}^3$$

so that

$$(\tau_{yz})_A = \frac{VQ}{It} = \frac{800 \text{ lb}(0.28125 \text{ in}^3)}{\left[\frac{1}{4}\pi(0.75 \text{ in.})^4\right]2(0.75 \text{ in.})}$$

$$= 604 \text{ psi} = 0.604 \text{ ksi}$$

Bending Moment. Since point A lies on the neutral axis, Fig. 13–7e, the bending stress is

$$\sigma_A = 0$$

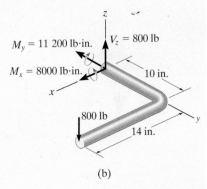

(b)

Fig. 13–7

Torque. At point A, $\rho_A = c = 0.75$ in., Fig. 13–7f. Thus the shear stress is

$$(\tau_{yz})_A = \frac{Tc}{J} = \frac{11\,200 \text{ lb} \cdot \text{in.}(0.75 \text{ in.})}{\left[\frac{1}{2}\pi(0.75 \text{ in.})^4\right]} = 16\,901 \text{ psi} = 16.90 \text{ ksi}$$

Superposition. Here the element of material at A is subjected only to a shear stress component, Fig. 13–7g, where

$$(\tau_{yz})_A = 0.604 \text{ ksi} + 16.90 \text{ ksi} = 17.5 \text{ ksi} \qquad \textit{Ans.}$$

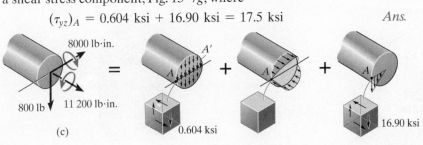

(c)

Shear force   Bending moment   Torsional moment

(d)   (e)   (f)   (g)

13

PRELIMINARY PROBLEMS

P13–1. In each case, determine the internal loadings that act on the indicated section. Show the results on the left segment.

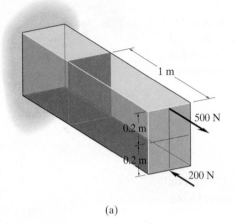

(a)

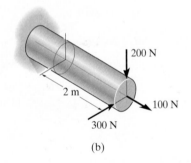

(b)

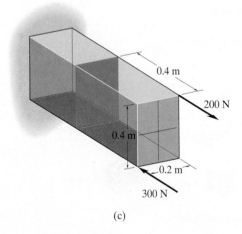

(c)

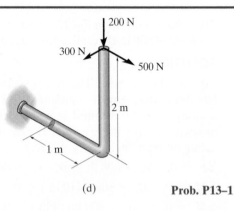

(d)   **Prob. P13–1**

P13–2. The internal loadings act on the section. Show the stress that each of these loads produce on differential elements located at point A and point B.

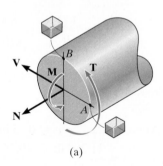

(a)

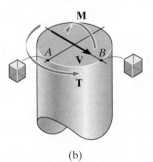

(b)   **Prob. P13–2**

FUNDAMENTAL PROBLEMS

F13–1. Determine the normal stress at corners A and B of the column.

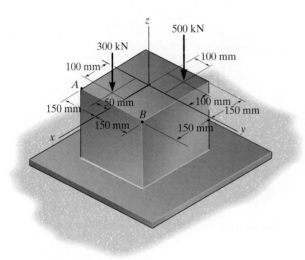

Prob. F13–1

F13–2. Determine the state of stress at point A on the cross section at section a–a of the cantilever beam. Show the results in a differential element at the point.

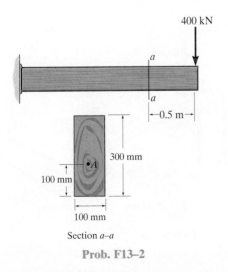

Section a–a

Prob. F13–2

F13–3. Determine the state of stress at point A on the cross section of the beam at section a–a. Show the results in a differential element at the point.

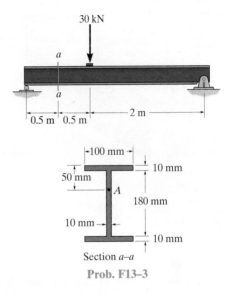

Section a–a

Prob. F13–3

F13–4. Determine the magnitude of the load P that will cause a maximum normal stress of $\sigma_{max} = 30$ ksi in the link along section a–a.

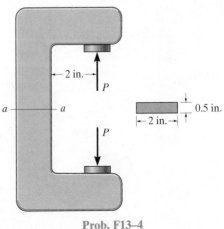

Prob. F13–4

13

F13–5. The beam has a rectangular cross section and is subjected to the loading shown. Determine the state of stress at point B. Show the results in a differential element at the point.

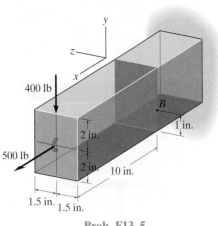

Prob. F13–5

F13–7. Determine the state of stress at point A on the cross section of the pipe at section a–a. Show the results in a differential element at the point.

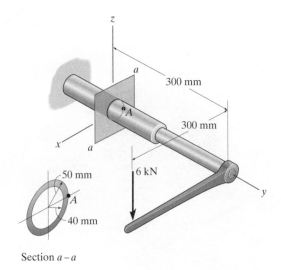

Section a–a

Prob. F13–7

F13–6. Determine the state of stress at point A on the cross section of the pipe assembly at section a–a. Show the results in a differential element at the point.

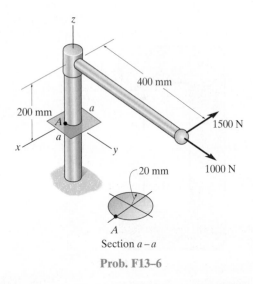

Section a–a

Prob. F13–6

F13–8. Determine the state of stress at point A on the cross section of the shaft at section a–a. Show the results in a differential element at the point.

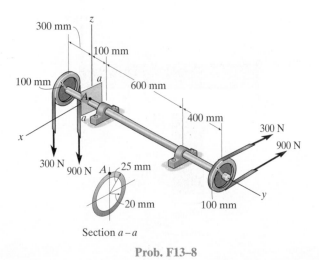

Section a–a

Prob. F13–8

PROBLEMS

13–18. Determine the shortest distance d to the edge of the plate at which the force **P** can be applied so that it produces no compressive stresses in the plate at section a–a. The plate has a thickness of 10 mm and **P** acts along the centerline of this thickness.

13–21. If the load has a weight of 600 lb, determine the maximum normal stress on the cross section of the supporting member at section a–a. Also, plot the normal-stress distribution over the cross section.

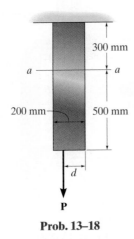

300 mm

a ——— a

200 mm —— | 500 mm

d

P

Prob. 13–18

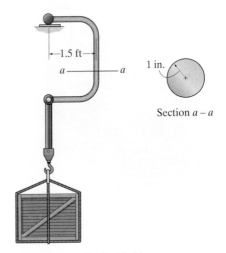

1.5 ft

a ——— a

1 in.

Section a – a

Prob. 13–21

13–19. Determine the maximum distance d to the edge of the plate at which the force **P** can be applied so that it produces no compressive stresses on the plate at section a–a. The plate has a thickness of 20 mm and **P** acts along the centerline of this thickness.

*13–20. The plate has a thickness of 20 mm and the force $P = 3$ kN acts along the centerline of this thickness such that $d = 150$ mm. Plot the distribution of normal stress acting along section a–a.

13–22. The steel bracket is used to connect the ends of two cables. If the allowable normal stress for the steel is $\sigma_{\text{allow}} = 30$ ksi, determine the largest tensile force P that can be applied to the cables. Assume the bracket is a rod having a diameter of 1.5 in.

13–23. The steel bracket is used to connect the ends of two cables. If the applied force $P = 1.50$ kip, determine the maximum normal stress in the bracket. Assume the bracket is a rod having a diameter of 1.5 in.

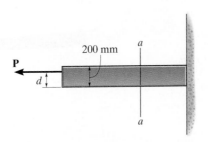

200 mm

a

P

d

a

Probs. 13–19/20

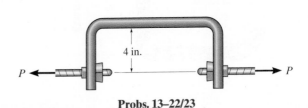

4 in.

P ◀—————————————▶ P

Probs. 13–22/23

*13–24. The column is built up by gluing the two boards together. Determine the maximum normal stress on the cross section when the eccentric force of $P = 50$ kN is applied.

13–25. The column is built up by gluing the two boards together. If the wood has an allowable normal stress of $\sigma_{allow} = 6$ MPa, determine the maximum allowable eccentric force **P** that can be applied to the column.

*13–28. The joint is subjected to the force system shown. Sketch the normal-stress distribution acting over section *a–a* if the member has a rectangular cross section of width 0.5 in. and thickness 1 in.

13–29. The joint is subjected to the force system shown. Determine the state of stress at points *A* and *B*, and sketch the results on differential elements located at these points. The member has a rectangular cross-sectional area of width 0.5 in. and thickness 1 in.

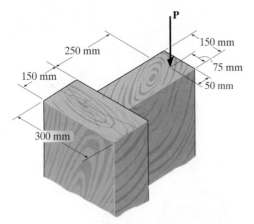

Probs. 13–24/25

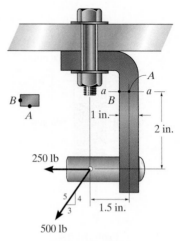

Probs. 13–28/29

13–26. The screw of the clamp exerts a compressive force of 500 lb on the wood blocks. Determine the maximum normal stress along section *a–a*. The cross section is rectangular, 0.75 in. by 0.50 in.

13–27. The screw of the clamp exerts a compressive force of 500 lb on the wood blocks. Sketch the stress distribution along section *a–a* of the clamp. The cross section is rectangular, 0.75 in. by 0.50 in.

13–30. The rib-joint pliers are used to grip the smooth pipe *C*. If the force of 100 N is applied to the handles, determine the state of stress at points *A* and *B* on the cross section of the jaw at section *a–a*. Indicate the results on an element at each point.

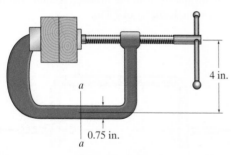

Probs. 13–26/27

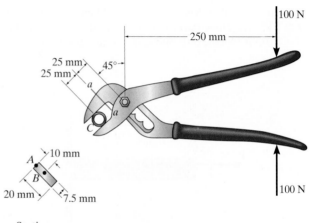

Section *a – a*

Prob. 13–30

13–31. The $\frac{1}{2}$-in.-diameter bolt hook is subjected to the load of $F = 150$ lb. Determine the stress components at point A on the shank. Show the result on a volume element located at this point.

*13–32. The $\frac{1}{2}$-in.-diameter bolt hook is subjected to the load of $F = 150$ lb. Determine the stress components at point B on the shank. Show the result on a volume element located at this point.

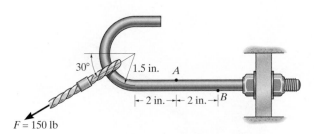

F = 150 lb

Probs. 13–31/32

13–33. The block is subjected to the eccentric load shown. Determine the normal stress developed at points A and B. Neglect the weight of the block.

13–34. The block is subjected to the eccentric load shown. Sketch the normal-stress distribution acting over the cross section at section a–a. Neglect the weight of the block.

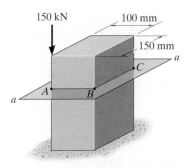

Probs. 13–33/34

13–35. The spreader bar is used to lift the 2000-lb tank. Determine the state of stress at points A and B, and indicate the results on a differential volume element.

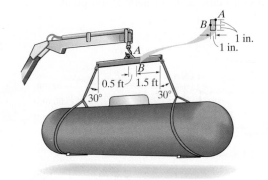

Prob. 13–35

*13–36. The drill is jammed in the wall and is subjected to the torque and force shown. Determine the state of stress at point A on the cross section of the drill bit at section a–a.

13–37. The drill is jammed in the wall and is subjected to the torque and force shown. Determine the state of stress at point B on the cross section of the drill bit at section a–a.

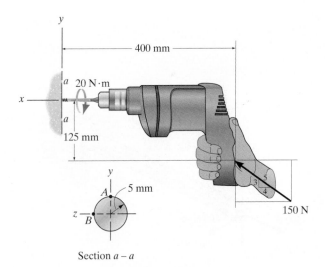

Section a – a

Probs. 13–36/37

13–38. The frame supports the distributed load shown. Determine the state of stress acting at point D. Show the results on a differential element at this point.

13–39. The frame supports the distributed load shown. Determine the state of stress acting at point E. Show the results on a differential element at this point.

13–42. The beveled gear is subjected to the loads shown. Determine the stress components acting on the shaft at point A, and show the results on a volume element located at this point. The shaft has a diameter of 1 in. and is fixed to the wall at C.

13–43. The beveled gear is subjected to the loads shown. Determine the stress components acting on the shaft at point B, and show the results on a volume element located at this point. The shaft has a diameter of 1 in. and is fixed to the wall at C.

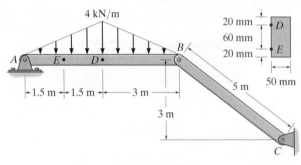

Probs. 13–38/39

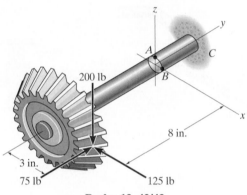

Probs. 13–42/43

*13–40.** The rod has a diameter of 40 mm. If it is subjected to the force system shown, determine the stress components that act at point A, and show the results on a volume element located at this point.

13–41. Solve Prob. 13–40 for point B.

*13–44.** Determine the normal-stress developed at points A and B. Neglect the weight of the block.

13–45. Sketch the normal-stress distribution acting over the cross section at section a–a. Neglect the weight of the block.

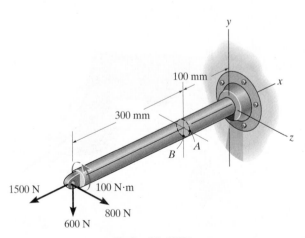

Probs. 13–40/41

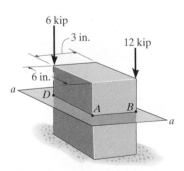

Probs. 13–44/45

13–46. The vertebra of the spinal column can support a maximum compressive stress of σ_{max}, before undergoing a compression fracture. Determine the smallest force P that can be applied to a vertebra if we assume this load is applied at an eccentric distance e from the centerline of the bone, and the bone remains elastic. Model the vertebra as a hollow cylinder with an inner radius r_i and outer radius r_o.

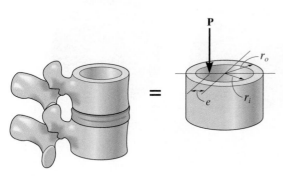

Prob. 13–46

13–47. The solid rod is subjected to the loading shown. Determine the state of stress at point A, and show the results on a differential volume element located at this point.

*****13–48.** The solid rod is subjected to the loading shown. Determine the state of stress at point B, and show the results on a differential volume element at this point.

13–49. The solid rod is subjected to the loading shown. Determine the state of stress at point C, and show the results on a differential volume element at this point.

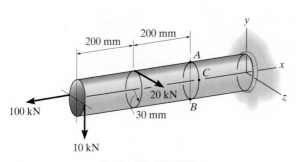

Probs. 13–47/48/49

13–50. The C-frame is used in a riveting machine. If the force at the ram on the clamp at D is $P = 8$ kN, sketch the stress distribution acting over the section a–a.

13–51. Determine the maximum ram force P that can be applied to the clamp at D if the allowable normal stress for the material is $\sigma_{allow} = 180$ MPa.

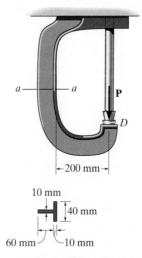

Probs. 13–50/51

*****13–52.** The uniform sign has a weight of 1500 lb and is supported by the pipe AB, which has an inner radius of 2.75 in. and an outer radius of 3.00 in. If the face of the sign is subjected to a uniform wind pressure of $p = 150$ lb/ft², determine the state of stress at points C and D. Show the results on a differential volume element located at each of these points. Neglect the thickness of the sign, and assume that it is supported along the outside edge of the pipe.

13–53. Solve Prob. 13–52 for points E and F.

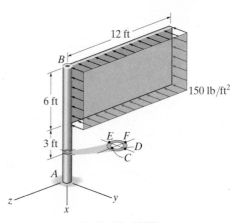

Probs. 13–52/53

CHAPTER REVIEW

A pressure vessel is considered to have a thin wall provided $r/t \geq 10$. If the vessel contains gas having a gage pressure p, then for a cylindrical vessel, the circumferential or hoop stress is

$$\sigma_1 = \frac{pr}{t}$$

This stress is twice as great as the longitudinal stress,

$$\sigma_2 = \frac{pr}{2t}$$

Thin-walled spherical vessels have the same stress within their walls in all directions. It is

$$\sigma_1 = \sigma_2 = \frac{pr}{2t}$$

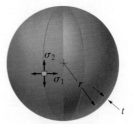

Superposition of stress components can be used to determine the normal and shear stress at a point in a member subjected to a combined loading. To do this, it is first necessary to determine the resultant axial and shear forces and the resultant torsional and bending moments at the section where the point is located. Then the normal and shear stress resultant components at the point are determined by algebraically adding the normal and shear stress components of each loading.

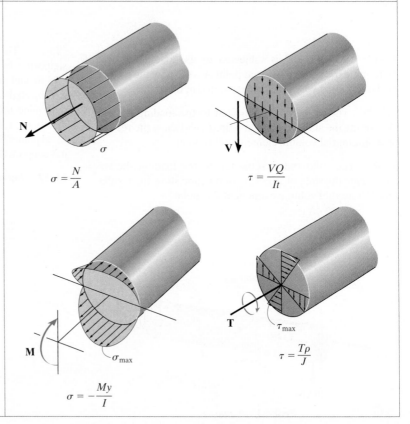

$$\sigma = \frac{N}{A}$$

$$\tau = \frac{VQ}{It}$$

$$\sigma = -\frac{My}{I}$$

$$\tau = \frac{T\rho}{J}$$

CONCEPTUAL PROBLEMS

C13–1. Explain why failure of this garden hose occurred near its end and why the tear occurred along its length. Use numerical values to explain your result. Assume the water pressure is 30 psi.

Prob. C13–1

C13–2. This open-ended silo contains granular material. It is constructed from wood slats and held together with steel bands. Explain, using numerical values, why the bands are not spaced evenly along the height of the cylinder. Also, how would you find this spacing if each band is to be subjected to the same stress?

Prob. C13–2

C13–3. Unlike the turnbuckle at B, which is connected along the axis of the rod, the one at A has been welded to the edges of the rod, and so it will be subjected to additional stress. Use the same numerical values for the tensile load in each rod and the rod's diameter, and compare the stress in each rod.

Prob. C13–3

C13–4. A constant wind blowing against the side of this chimney has caused creeping strains in the mortar joints, such that the chimney has a noticeable deformation. Explain how to obtain the stress distribution over a section at the base of the chimney, and sketch this distribution over the section.

Prob. C13–4

REVIEW PROBLEMS

R13–1. The post has a circular cross section of radius c. Determine the maximum radius e at which the load **P** can be applied so that no part of the post experiences a tensile stress. Neglect the weight of the post.

Prob. R13–1

R13–2. The 20-kg drum is suspended from the hook mounted on the wooden frame. Determine the state of stress at point E on the cross section of the frame at section a–a. Indicate the results on an element.

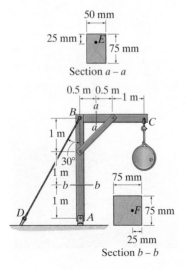

Prob. R13–2

R13–3. The 20-kg drum is suspended from the hook mounted on the wooden frame. Determine the state of stress at point F on the cross section of the frame at section b–b. Indicate the results on an element.

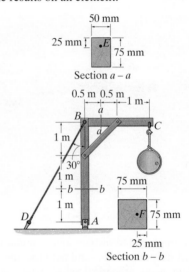

Prob. R13–3

*R13–4. The gondola and passengers have a weight of 1500 lb and center of gravity at G. The suspender arm AE has a square cross-sectional area of 1.5 in. by 1.5 in., and is pin connected at its ends A and E. Determine the largest tensile stress developed in regions AB and DC of the arm.

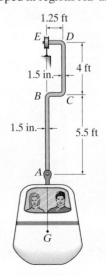

Prob. R13–4

R13–5. If the cross section of the femur at section *a–a* can be approximated as a circular tube as shown, determine the maximum normal stress developed on the cross section at section *a–a* due to the load of 75 lb.

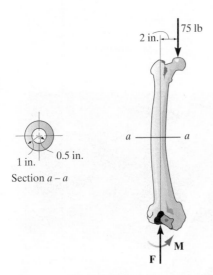

Section *a – a*

Prob. R13–5

R13–6. A bar having a square cross section of 30 mm by 30 mm is 2 m long and is held upward. If it has a mass of 5 kg/m, determine the largest angle θ, measured from the vertical, at which it can be supported before it is subjected to a tensile stress along its axis near the grip.

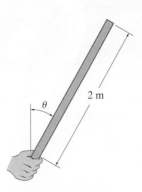

Prob. R13–6

R13–7. The wall hanger has a thickness of 0.25 in. and is used to support the vertical reactions of the beam that is loaded as shown. If the load is transferred uniformly to each strap of the hanger, determine the state of stress at points *C* and *D* on the strap at *A*. Assume the vertical reaction **F** at this end acts in the center and on the edge of the bracket as shown.

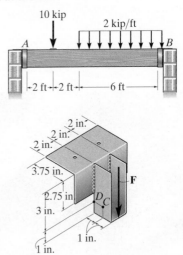

Prob. R13–7

*R13–8. The wall hanger has a thickness of 0.25 in. and is used to support the vertical reactions of the beam that is loaded as shown. If the load is transferred uniformly to each strap of the hanger, determine the state of stress at points *C* and *D* on the strap at *B*. Assume the vertical reaction **F** at this end acts in the center and on the edge of the bracket as shown.

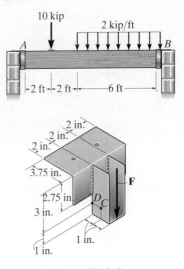

Prob. R13–8

CHAPTER 14

(© R.G. Henry/Fotolia)

These turbine blades are subjected to a complex pattern of stress. For design it is necessary to determine where and in what direction the maximum stress occurs.

STRESS AND STRAIN TRANSFORMATION

14.1 PLANE-STRESS TRANSFORMATION

It was shown in Sec. 7.3 that the general state of stress at a point is characterized by *six* normal and shear-stress components, shown in Fig. 14–1a. This state of stress, however, is not often encountered in engineering practice. Instead, most loadings are coplanar, and so the stress these loadings produce can be analyzed in a *single plane*. When this is the case, the material is then said to be subjected to *plane stress*.

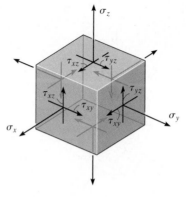

General state of stress

(a)

Fig. 14–1

619

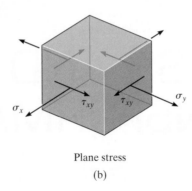

Plane stress

(b)

Fig. 14–1 (cont.)

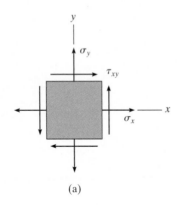

(a)

||

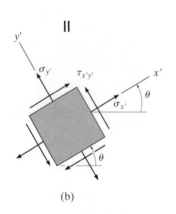

(b)

Fig. 14–2

The general state of plane stress at a point, shown in Fig. 14–1b, is therefore represented by a combination of two normal-stress components, σ_x, σ_y, and one shear-stress component, τ_{xy}, which act on only four faces of the element. For convenience, in this book we will view this state of stress in the x–y plane, as shown in Fig. 14–2a. Realize, however, that if this state of stress is produced on an element having a *different orientation* θ, as in Fig. 14–2b, then it will be subjected to three *different* stress components, $\sigma_{x'}$, $\sigma_{y'}$, $\tau_{x'y'}$, measured relative to the x', y' axes. In other words, ***the state of plane stress at the point is uniquely represented by two normal-stress components and one shear-stress component acting on an element. To be equivalent, these three components will be different for each specific orientation θ of the element at the point.***

If these three stress components act on the element in Fig. 14–2a, we will now show what their values will have to be when they act on the element in Fig. 14–2b. This is similar to knowing the two force components $\mathbf{F}_x$ and $\mathbf{F}_y$ directed along the x, y axes, and then finding the force components $\mathbf{F}_{x'}$ and $\mathbf{F}_{y'}$ directed along the x', y' axes, so they produce the *same* resultant force. The transformation of force must only account for the force component's magnitude and direction. The transformation of stress components, however, is more difficult since it must account for the magnitude and direction of each stress *and* the orientation of the area upon which it acts.

PROCEDURE FOR ANALYSIS

If the state of stress at a point is known for a given orientation of an element, Fig. 14–3a, then the state of stress on an element having some other orientation θ, Fig. 14–3b, can be determined as follows.

- The normal and shear stress components $\sigma_{x'}$, $\tau_{x'y'}$ acting on the $+x'$ face of the element, Fig. 14–3b, can be determined from an arbitrary section of the element in Fig. 14–3a as shown in Fig. 14–3c. If the sectioned area is ΔA, then the adjacent areas of the segment will be $\Delta A \sin \theta$ and $\Delta A \cos \theta$.

- Draw the *free-body diagram* of the segment, which requires showing the *forces* that act on the segment, Fig. 14–3d. This is done by multiplying the stress components on each face by the area upon which they act.

- When $\Sigma F_{x'} = 0$ is applied to the free-body diagram, the area ΔA will cancel out of each term and a *direct* solution for $\sigma_{x'}$ will be possible. Likewise, $\Sigma F_{y'} = 0$ will yield $\tau_{x'y'}$.

- If $\sigma_{y'}$, acting on the $+y'$ face of the element in Fig. 14–3b, is to be determined, then it is necessary to consider an arbitrary segment of the element as shown in Fig. 14–3e. Applying $\Sigma F_{y'} = 0$ to its free-body diagram will give $\sigma_{y'}$.

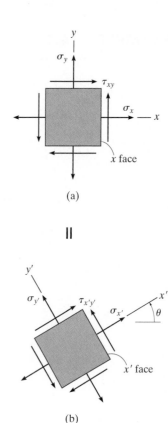

(a)

(b)

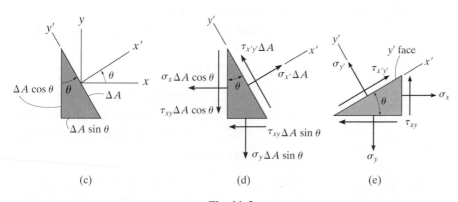

(c)

(d)

(e)

Fig. 14–3

EXAMPLE 14.1

The state of plane stress at a point on the surface of the airplane fuselage is represented on the element oriented as shown in Fig. 14–4a. Represent the state of stress at the point on an element that is oriented 30° clockwise from this position.

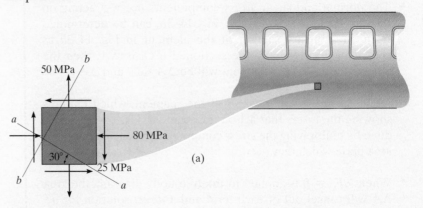

(a)

SOLUTION

The rotated element is shown in Fig. 14–4d. To obtain the stress components on this element we will first section the element in Fig. 14–4a by the line a–a. The bottom segment is removed, and assuming the sectioned (inclined) plane has an area ΔA, the horizontal and vertical planes have the areas shown in Fig. 14–4b. The free-body diagram of this segment is shown in Fig. 14–4c. Notice that the sectioned x' face is defined by the *outward normal* x' axis, and the y' axis is *along* the face.

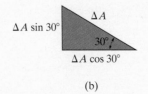

(b)

Equilibrium. If we apply the equations of force equilibrium in the x' and y' directions, not the x and y directions, we will be able to obtain *direct solutions* for $\sigma_{x'}$ and $\tau_{x'y'}$.

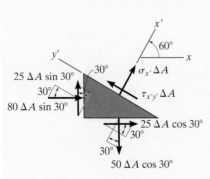

(c)

Fig. 14–4

$$+\nearrow\Sigma F_{x'} = 0; \qquad \sigma_{x'}\Delta A - (50\ \Delta A \cos 30°)\cos 30°$$
$$+ (25\ \Delta A \cos 30°)\sin 30° + (80\ \Delta A \sin 30°)\sin 30°$$
$$+ (25\ \Delta A \sin 30°)\cos 30° = 0$$
$$\sigma_{x'} = -4.15\ \text{MPa} \qquad\qquad Ans.$$

$$+\nwarrow\Sigma F_{y'} = 0; \qquad \tau_{x'y'}\Delta A - (50\ \Delta A \cos 30°)\sin 30°$$
$$- (25\ \Delta A \cos 30°)\cos 30° - (80\ \Delta A \sin 30°)\cos 30°$$
$$+ (25\ \Delta A \sin 30°)\sin 30° = 0$$
$$\tau_{x'y'} = 68.8\ \text{MPa} \qquad\qquad Ans.$$

Since $\sigma_{x'}$ is negative, it acts in the opposite direction of that shown in Fig. 14–4c. The results are shown on the *top* of the element in Fig. 14–4d, since this surface is the one considered in Fig. 14–4c.

We must now repeat the procedure to obtain the stress on the *perpendicular* plane b–b. Sectioning the element in Fig. 14–4a along b–b results in a segment having sides with areas shown in Fig. 14–4e. Orienting the +x' axis outward, perpendicular to the sectioned face, the associated free-body diagram is shown in Fig. 14–4f. Thus,

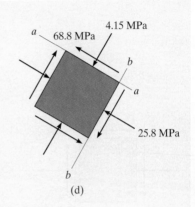

$$+\searrow\Sigma F_{x'} = 0; \quad \sigma_{x'}\Delta A - (25\,\Delta A \cos 30°)\sin 30°$$
$$+ (80\,\Delta A \cos 30°)\cos 30° - (25\,\Delta A \sin 30°)\cos 30°$$
$$- (50\,\Delta A \sin 30°)\sin 30° = 0$$
$$\sigma_{x'} = -25.8 \text{ MPa} \qquad Ans.$$

(d)

$$+\nearrow\Sigma F_{y'} = 0; \quad \tau_{x'y'}\,\Delta A + (25\,\Delta A \cos 30°)\cos 30°$$
$$+ (80\,\Delta A \cos 30°)\sin 30° - (25\,\Delta A \sin 30°)\sin 30°$$
$$+ (50\,\Delta A \sin 30°)\cos 30° = 0$$
$$\tau_{x'y'} = -68.8 \text{ MPa} \qquad Ans.$$

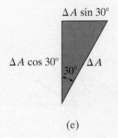

(e)

Since both $\sigma_{x'}$ and $\tau_{x'y'}$ are negative quantities, they act opposite to their direction shown in Fig. 14–4f. The stress components are shown acting on the *right side* of the element in Fig. 14–4d.

From this analysis we may therefore conclude that the state of stress at the point can be represented by a stress component acting on an element removed from the fuselage and oriented as shown in Fig. 14–4a, or by choosing one removed and oriented as shown in Fig. 14–4d. In other words, these states of stress are equivalent.

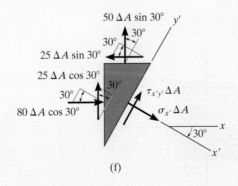

(f)

Fig. 14–4 (cont.)

14.2 GENERAL EQUATIONS OF PLANE-STRESS TRANSFORMATION

The method of transforming the normal and shear stress components from the x, y to the x', y' coordinate axes, as discussed in the previous section, can be developed in a general manner and expressed as a set of stress-transformation equations.

Sign Convention. To apply these equations we must first establish a sign convention for the stress components. As shown in Fig. 14–5, the $+x$ and $+x'$ axes are used to define the outward normal on the right-hand face of the element, so that σ_x and $\sigma_{x'}$ are positive when they act in the positive x and x' directions, and τ_{xy} and $\tau_{x'y'}$ are positive when they act in the positive y and y' directions.

The orientation of the face upon which the normal and shear stress components are to be determined will be defined by the angle θ, which is measured from the $+x$ axis to the $+x'$ axis, Fig. 14–5b. Notice that the unprimed and primed sets of axes in this figure both form right-handed coordinate systems; that is, the positive z (or z') axis always points out of the page. The *angle* θ will be *positive* when it follows the curl of the right-hand fingers, i.e., counterclockwise as shown in Fig. 14–5b.

Normal and Shear Stress Components. Using this established sign convention, the element in Fig. 14–6a is sectioned along the inclined plane and the segment shown in Fig. 14–6b is isolated. Assuming the sectioned area is ΔA, then the horizontal and vertical faces of the segment have an area of $\Delta A \sin \theta$ and $\Delta A \cos \theta$, respectively.

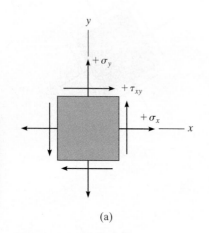

(a)

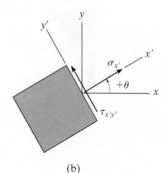

(b)

Positive sign convention

Fig. 14–5

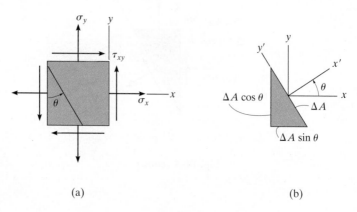

(a)   (b)

Fig. 14–6

The resulting *free-body diagram* of the segment is shown in Fig. 14–6c. If we apply the equations of equilibrium along the x' and y' axes, we can obtain a direct solution for $\sigma_{x'}$ and $\tau_{x'y'}$. We have

$$+\nearrow\Sigma F_{x'} = 0; \quad \sigma_{x'}\Delta A - (\tau_{xy}\Delta A \sin \theta) \cos \theta - (\sigma_y\Delta A \sin \theta) \sin \theta$$
$$- (\tau_{xy}\Delta A \cos \theta) \sin \theta - (\sigma_x\Delta A \cos \theta) \cos \theta = 0$$
$$\sigma_{x'} = \sigma_x \cos^2 \theta + \sigma_y \sin^2 \theta + \tau_{xy}(2 \sin \theta \cos \theta)$$

$$+\nwarrow\Sigma F_{y'} = 0; \quad \tau_{x'y'}\Delta A + (\tau_{xy}\Delta A \sin \theta) \sin \theta - (\sigma_y\Delta A \sin \theta) \cos \theta$$
$$- (\tau_{xy}\Delta A \cos \theta) \cos \theta + (\sigma_x \Delta A \cos \theta) \sin \theta = 0$$
$$\tau_{x'y'} = (\sigma_y - \sigma_x) \sin\theta \cos \theta + \tau_{xy} (\cos^2 \theta - \sin^2 \theta)$$

To simplify these two equations, use the trigonometric identities $\sin 2\theta = 2 \sin \theta \cos \theta$, $\sin^2 \theta = (1 - \cos 2\theta)/2$, and $\cos^2 \theta = (1 + \cos 2\theta)/2$. Therefore,

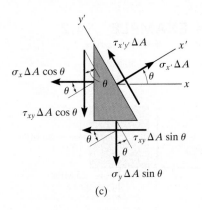

(c)

$$\sigma_{x'} = \frac{\sigma_x + \sigma_y}{2} + \frac{\sigma_x - \sigma_y}{2} \cos 2\theta + \tau_{xy} \sin 2\theta \qquad (14\text{–}1)$$

$$\tau_{x'y'} = -\frac{\sigma_x - \sigma_y}{2} \sin 2\theta + \tau_{xy} \cos 2\theta \qquad (14\text{–}2)$$

Stress Components Acting along x', y' Axes

If the normal stress acting in the y' direction is needed, it can be obtained by simply substituting $\theta + 90°$ for θ into Eq. 14–1, Fig. 14–6d. This yields

$$\sigma_{y'} = \frac{\sigma_x + \sigma_y}{2} - \frac{\sigma_x - \sigma_y}{2} \cos 2\theta - \tau_{xy} \sin 2\theta \qquad (14\text{–}3)$$

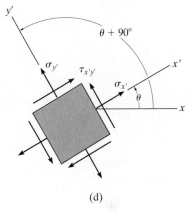

(d)

Fig. 14–6

▶ PROCEDURE FOR ANALYSIS

To apply the stress transformation Eqs. 14–1 and 14–2, it is simply necessary to substitute in the known data for σ_x, σ_y, τ_{xy}, and θ in accordance with the established sign convention, Fig. 14–5. Remember that the x' axis is *always* directed *positive outward* from the plane upon which the normal stress is to be determined. The angle θ is *positive counterclockwise*, from the x to the x' axis. If $\sigma_{x'}$ and $\tau_{x'y'}$ are calculated as positive quantities, then these stresses act in the positive direction of the x' and y' axes.

For convenience, these equations can easily be programmed on a pocket calculator.

EXAMPLE 14.2

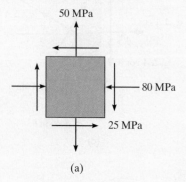

50 MPa

80 MPa

25 MPa

(a)

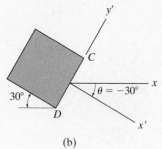

y'

C

$\theta = -30°$

x

30°

D

x'

(b)

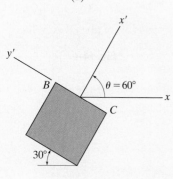

x'

y'

B

$\theta = 60°$

C

x

30°

(c)

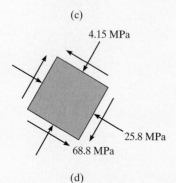

4.15 MPa

25.8 MPa

68.8 MPa

(d)

Fig. 14–7

The state of plane stress at a point is represented on the element shown in Fig. 14–7a. Determine the state of stress at this point on another element oriented 30° clockwise from the position shown.

SOLUTION

This problem was solved in Example 14.1 using basic principles. Here we will apply Eqs. 14–1 and 14–2. From the established sign convention, Fig. 14–5, it is seen that

$$\sigma_x = -80 \text{ MPa} \qquad \sigma_y = 50 \text{ MPa} \qquad \tau_{xy} = -25 \text{ MPa}$$

Plane CD. To obtain the stress components on plane CD, Fig. 14–7b, the positive x' axis must be directed outward, perpendicular to CD, and the associated y' axis is directed along CD. The angle measured from the x to the x' axis is $\theta = -30°$ (clockwise). Applying Eqs. 14–1 and 14–2 yields

$$\sigma_{x'} = \frac{\sigma_x + \sigma_y}{2} + \frac{\sigma_x - \sigma_y}{2} \cos 2\theta + \tau_{xy} \sin 2\theta$$

$$= \frac{-80 + 50}{2} + \frac{-80 - 50}{2} \cos 2(-30°) + (-25) \sin 2(-30°)$$

$$= -25.8 \text{ MPa} \qquad\qquad Ans.$$

$$\tau_{x'y'} = -\frac{\sigma_x - \sigma_y}{2} \sin 2\theta + \tau_{xy} \cos 2\theta$$

$$= -\frac{-80 - 50}{2} \sin 2(-30°) + (-25) \cos 2(-30°)$$

$$= -68.8 \text{ MPa} \qquad\qquad Ans.$$

The negative signs indicate that $\sigma_{x'}$ and $\tau_{x'y'}$ act in the negative x' and y' directions, respectively. The results are shown acting on the element in Fig. 14–7d.

Plane BC. Establishing the x' axis outward from plane BC, Fig. 14–7c, then between the x and x' axes, $\theta = 60°$ (counterclockwise). Applying Eqs. 14–1 and 14–2,* we get

$$\sigma_{x'} = \frac{-80 + 50}{2} + \frac{-80 - 50}{2} \cos 2(60°) + (-25) \sin 2(60°)$$

$$= -4.15 \text{ MPa} \qquad\qquad Ans.$$

$$\tau_{x'y'} = -\frac{-80 - 50}{2} \sin 2(60°) + (-25) \cos 2(60°)$$

$$= 68.8 \text{ MPa} \qquad\qquad Ans.$$

Here $\tau_{x'y'}$ has been calculated twice in order to provide a check. The negative sign for $\sigma_{x'}$ indicates that this stress acts in the negative x' direction, Fig. 14–7c. The results are shown on the element in Fig. 14–7d.

*Alternatively, we could apply Eq. 14–3 with $\theta = -30°$ rather than Eq. 14–1.

14.3   PRINCIPAL STRESSES AND MAXIMUM IN-PLANE SHEAR STRESS

Since σ_x, σ_y, τ_{xy} are all constant, then from Eqs. 14–1 and 14–2 it can be seen that the magnitudes of $\sigma_{x'}$ and $\tau_{x'y'}$ only depend on the angle of inclination θ of the planes on which these stresses act. In engineering practice it is often important to determine the orientation that causes the normal stress to be a maximum, and the orientation that causes the shear stress to be a maximum. We will now consider each of these cases.

In-Plane Principal Stresses. To determine the maximum and minimum *normal stress*, we must differentiate Eq. 14–1 with respect to θ and set the result equal to zero. This gives

$$\frac{d\sigma_{x'}}{d\theta} = -\frac{\sigma_x - \sigma_y}{2}(2 \sin 2\theta) + 2\tau_{xy} \cos 2\theta = 0$$

Solving we obtain the orientation $\theta = \theta_p$ of the planes of maximum and minimum normal stress.

$$\tan 2\theta_p = \frac{\tau_{xy}}{(\sigma_x - \sigma_y)/2} \qquad (14\text{–}4)$$

Orientation of Principal Planes

The solution has two roots, θ_{p_1} and θ_{p_2}. Specifically, the values of $2\theta_{p_1}$ and $2\theta_{p_2}$ are 180° apart, so θ_{p_1} and θ_{p_2} will be 90° apart.

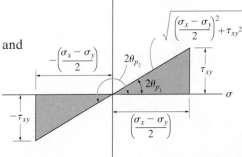

Fig. 14–8

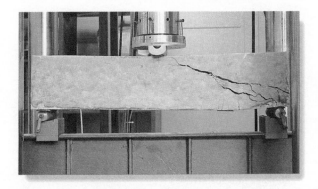

The cracks in this concrete beam were caused by tension stress, even though the beam was subjected to both an internal moment and shear. The stress transformation equations can be used to predict the direction of the cracks, and the principal normal stresses that caused them.

To obtain the maximum and minimum normal stress, we must substitute these angles into Eq. 14–1. Here the necessary sine and cosine of $2\theta_{p_1}$ and $2\theta_{p_2}$ can be found from the shaded triangles shown in Fig. 14–8, which are constructed based on Eq. 14–4, assuming that τ_{xy} and $(\sigma_x - \sigma_y)$ are both positive or both negative quantities.

After substituting and simplifying, we obtain two roots, σ_1 and σ_2. They are

$$\sigma_{1,2} = \frac{\sigma_x + \sigma_y}{2} \pm \sqrt{\left(\frac{\sigma_x - \sigma_y}{2}\right)^2 + \tau_{xy}^2} \qquad (14\text{–}5)$$

Principal Stresses

These two values, with $\sigma_1 \geq \sigma_2$, are called the in-plane **principal stresses**, and the corresponding planes on which they act are called the **principal planes** of stress, Fig. 14–9. Finally, if the trigonometric relations for θ_{p_1} or θ_{p_2} are substituted into Eq. 14–2, it will be seen that $\tau_{x'y'} = 0$; in other words, **no shear stress acts on the principal planes**, Fig. 14–9.

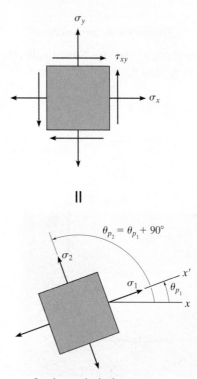

In-plane principal stresses

Fig. 14–9

Maximum In-Plane Shear Stress. The orientation of the element that is subjected to maximum shear stress can be determined by taking the derivative of Eq. 14–2 with respect to θ, and setting the result equal to zero. This gives

$$\tan 2\theta_s = \frac{-(\sigma_x - \sigma_y)/2}{\tau_{xy}} \qquad (14\text{–}6)$$

Orientation of Maximum In-Plane Shear Stress

The two roots of this equation, θ_{s_1} and θ_{s_2}, can be determined from the shaded triangles shown in Fig. 14–10a. Since $\tan 2\theta_s$, Eq. 14–6, is the negative reciprocal of $\tan 2\theta_p$, Eq. 14–4, then each root $2\theta_s$ is 90° from $2\theta_p$, and the roots θ_s and θ_p are 45° apart. Therefore, an element subjected to *maximum shear stress must be oriented 45° from the position of an element that is subjected to the principal stress*.

The maximum shear stress can be found by taking the trigonometric values of $\sin 2\theta_s$ and $\cos 2\theta_s$ from Fig. 14–10 and substituting them into Eq. 14–2. The result is

$$\tau_{\substack{max \\ in\text{-}plane}} = \sqrt{\left(\frac{\sigma_x - \sigma_y}{2}\right)^2 + \tau_{xy}{}^2} \qquad (14\text{–}7)$$

Maximum In-Plane Shear Stress

Here $\tau_{\substack{max \\ in\text{-}plane}}$ is referred to as the *maximum in-plane shear stress*, because it acts on the element in the x–y plane.

Finally, when the values for $\sin 2\theta_s$ and $\cos 2\theta_s$ are substituted into Eq. 14–1, we see that there is *also* an *average normal stress* on the planes of maximum in-plane shear stress. It is

$$\sigma_{avg} = \frac{\sigma_x + \sigma_y}{2} \qquad (14\text{–}8)$$

Average Normal Stress

For numerical applications, it is suggested that Eqs. 14–1 through 14–8 be programmed for use on a pocket calculator.

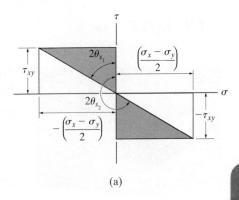

(a)

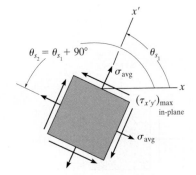

Maximum in-plane shear stresses

(b)

Fig. 14–10

> ### IMPORTANT POINTS
>
> - The *principal stresses* represent the maximum and minimum normal stress at the point.
> - When the state of stress is represented by the principal stresses, *no shear stress* will act on the element.
> - The state of stress at the point can also be represented in terms of the *maximum in-plane shear stress*. In this case an *average normal stress* will also act on the element.
> - The element representing the maximum in-plane shear stress with the associated average normal stresses is oriented 45° from the element representing the principal stresses.

EXAMPLE 14.3

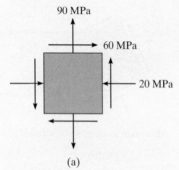

90 MPa

60 MPa

20 MPa

(a)

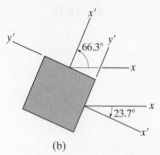

(b)

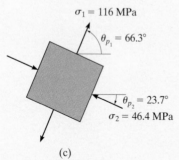

(c)

Fig. 14–11

The state of stress at a point just before failure of this shaft is shown in Fig. 14–11a. Represent this state of stress in terms of its principal stresses.

SOLUTION

From the established sign convention,

$$\sigma_x = -20 \text{ MPa} \qquad \sigma_y = 90 \text{ MPa} \qquad \tau_{xy} = 60 \text{ MPa}$$

Orientation of Element. Applying Eq. 14–4,

$$\tan 2\theta_p = \frac{\tau_{xy}}{(\sigma_x - \sigma_y)/2} = \frac{60}{(-20 - 90)/2}$$

Solving, and referring to this first angle as θ_{p_2}, we have

$$2\theta_{p_2} = -47.49° \qquad \theta_{p_2} = -23.7°$$

Since the difference between $2\theta_{p_1}$ and $2\theta_{p_2}$ is 180°, the second angle is

$$2\theta_{p_1} = 180° + 2\theta_{p_2} = 132.51° \qquad \theta_{p_1} = 66.3°$$

In both cases, θ must be measured positive *counterclockwise* from the x axis to the outward normal (x' axis) on the face of the element, and so the element showing the principal stresses will be oriented as shown in Fig. 14–11b.

Principal Stress. We have

$$\sigma_{1,2} = \frac{\sigma_x + \sigma_y}{2} \pm \sqrt{\left(\frac{\sigma_x - \sigma_y}{2}\right)^2 + \tau_{xy}^2}$$

$$= \frac{-20 + 90}{2} \pm \sqrt{\left(\frac{-20 - 90}{2}\right)^2 + (60)^2}$$

$$= 35.0 \pm 81.4$$

$$\sigma_1 = 116 \text{ MPa} \qquad\qquad\qquad Ans.$$

$$\sigma_2 = -46.4 \text{ MPa} \qquad\qquad\qquad Ans.$$

The principal plane on which each normal stress acts can be determined by applying Eq. 14–1 with, say, $\theta = \theta_{p_2} = -23.7°$. We have

$$\sigma_{x'} = \frac{\sigma_x + \sigma_y}{2} + \frac{\sigma_x - \sigma_y}{2} \cos 2\theta + \tau_{xy} \sin 2\theta$$

$$= \frac{-20 + 90}{2} + \frac{-20 - 90}{2} \cos 2(-23.7°) + 60 \sin 2(-23.7°)$$

$$= -46.4 \text{ MPa}$$

Hence, $\sigma_2 = -46.4 \text{ MPa}$ acts on the plane defined by $\theta_{p_2} = -23.7°$, whereas $\sigma_1 = 116 \text{ MPa}$ acts on the plane defined by $\theta_{p_1} = 66.3°$, Fig. 14–11c. Recall that no shear stress acts on this element.

EXAMPLE 14.4

The state of plane stress at a point on a body is represented on the element shown in Fig. 14–12a. Represent this state of stress in terms of its maximum in-plane shear stress and associated average normal stress.

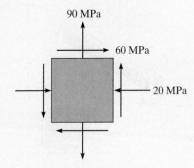

(a)

SOLUTION

Orientation of Element. Since $\sigma_x = -20$ MPa, $\sigma_y = 90$ MPa, and $\tau_{xy} = 60$ MPa, applying Eq. 14–6, the two angles are

$$\tan 2\theta_s = \frac{-(\sigma_x - \sigma_y)/2}{\tau_{xy}} = \frac{-(-20 - 90)/2}{60}$$

$$2\theta_{s_2} = 42.5° \qquad\qquad \theta_{s_2} = 21.3°$$

$$2\theta_{s_1} = 180° + 2\theta_{s_2} \qquad\qquad \theta_{s_1} = 111.3°$$

Note how these angles are formed between the x and x' axes, Fig. 14–12b. They happen to be 45° away from the principal planes of stress, which were determined in Example 14.3.

Maximum In-Plane Shear Stress. Applying Eq. 14–7,

$$\tau_{\substack{max \\ in\text{-}plane}} = \sqrt{\left(\frac{\sigma_x - \sigma_y}{2}\right)^2 + \tau_{xy}^2} = \sqrt{\left(\frac{-20 - 90}{2}\right)^2 + (60)^2}$$

$$= \pm 81.4 \text{ MPa} \qquad\qquad\qquad Ans.$$

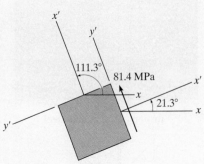

(b)

The proper direction of $\tau_{\substack{max \\ in\text{-}plane}}$ on the element can be determined by substituting $\theta = \theta_{s_2} = 21.3°$ into Eq. 14–2. We have

$$\tau_{x'y'} = -\left(\frac{\sigma_x - \sigma_y}{2}\right)\sin 2\theta + \tau_{xy}\cos 2\theta$$

$$= -\left(\frac{-20 - 90}{2}\right)\sin 2(21.3°) + 60 \cos 2(21.3°)$$

$$= 81.4 \text{ MPa}$$

This positive result indicates that $\tau_{\substack{max \\ in\text{-}plane}} = \tau_{x'y'}$ acts in the *positive y'* direction on this face ($\theta = 21.3°$), Fig. 14–12b. The shear stresses on the other three faces are directed as shown in Fig. 14–12c.

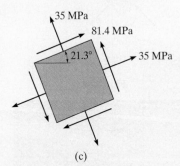

(c)

Fig. 14–12

Average Normal Stress. Besides the maximum shear stress, the element is also subjected to an average normal stress determined from Eq. 14–8; that is,

$$\sigma_{avg} = \frac{\sigma_x + \sigma_y}{2} = \frac{-20 + 90}{2} = 35 \text{ MPa} \qquad\qquad Ans.$$

This is a tensile stress. The results are shown in Fig. 14–12c.

EXAMPLE 14.5

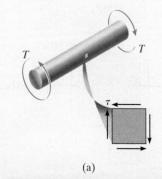

(a)

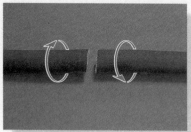

Torsion failure of mild steel.

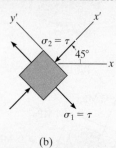

(b)

Fig. 14–13

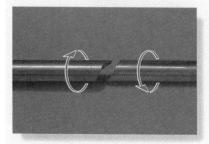

Torsion failure of cast iron.

When the torsional loading T is applied to the bar in Fig. 14–13a, it produces a state of pure shear stress in the material. Determine (a) the maximum in-plane shear stress and the associated average normal stress, and (b) the principal stress.

SOLUTION

From the established sign convention,

$$\sigma_x = 0 \qquad \sigma_y = 0 \qquad \tau_{xy} = -\tau$$

Maximum In-Plane Shear Stress.   Applying Eqs. 14–7 and 14–8, we have

$$\tau_{\substack{max \\ \text{in-plane}}} = \sqrt{\left(\frac{\sigma_x - \sigma_y}{2}\right)^2 + \tau_{xy}^2} = \sqrt{(0)^2 + (-\tau)^2} = \pm\tau \quad Ans.$$

$$\sigma_{avg} = \frac{\sigma_x + \sigma_y}{2} = \frac{0 + 0}{2} = 0 \qquad\qquad Ans.$$

Thus, as expected, the maximum in-plane shear stress is represented by the element in Fig. 14–13a.

NOTE: Through experiment it has been found that materials that are *ductile* actually fail due to *shear stress*. As a result, if the bar in Fig. 14–13a is made of mild steel, the maximum in-plane shear stress will cause it to fail as shown in the adjacent photo.

Principal Stress.   Applying Eqs. 14–4 and 14–5 yields

$$\tan 2\theta_p = \frac{\tau_{xy}}{(\sigma_x - \sigma_y)/2} = \frac{-\tau}{(0 - 0)/2}, \theta_{p_2} = 45°, \theta_{p_1} = -45°$$

$$\sigma_{1,2} = \frac{\sigma_x + \sigma_y}{2} \pm \sqrt{\left(\frac{\sigma_x - \sigma_y}{2}\right)^2 + \tau_{xy}^2} = 0 \pm \sqrt{(0)^2 + \tau^2} = \pm\tau \ Ans.$$

If we now apply Eq. 14–1 with $\theta_{p_2} = 45°$, then

$$\sigma_{x'} = \frac{\sigma_x + \sigma_y}{2} + \frac{\sigma_x - \sigma_y}{2}\cos 2\theta + \tau_{xy}\sin 2\theta$$

$$= 0 + 0 + (-\tau)\sin 90° = -\tau$$

Thus, $\sigma_2 = -\tau$ acts at $\theta_{p_2} = 45°$ as shown in Fig. 14–13b, and $\sigma_1 = \tau$ acts on the other face, $\theta_{p_1} = -45°$.

NOTE: Materials that are *brittle* fail due to *normal stress*. Therefore, if the bar in Fig. 14–13a is made of cast iron it will fail in tension at a 45° inclination as seen in the adjacent photo.

EXAMPLE 14.6

When the axial loading P is applied to the bar in Fig. 14–14a, it produces a tensile stress in the material. Determine (a) the principal stress and (b) the maximum in-plane shear stress and associated average normal stress.

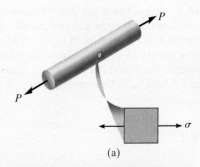

(a)

SOLUTION

From the established sign convention,

$$\sigma_x = \sigma \qquad \sigma_y = 0 \qquad \tau_{xy} = 0$$

Principal Stress. By observation, the element oriented as shown in Fig. 14–14a illustrates a condition of principal stress since no shear stress acts on this element. This can also be shown by direct substitution of the above values into Eqs. 14–4 and 14–5. Thus,

$$\sigma_1 = \sigma \qquad \sigma_2 = 0 \qquad Ans.$$

NOTE: *Brittle materials* will fail due to normal stress, and therefore, if the bar in Fig. 14–14a is made of cast iron, it will fail as shown in the adjacent photo.

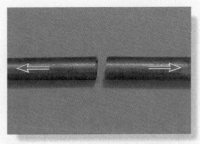

Axial failure of cast iron.

Maximum In-Plane Shear Stress. Applying Eqs. 14–6, 14–7, and 14–8, we have

$$\tan 2\theta_s = \frac{-(\sigma_x - \sigma_y)/2}{\tau_{xy}} = \frac{-(\sigma - 0)/2}{0}; \theta_{s_1} = 45°, \theta_{s_2} = -45°$$

$$\tau_{\substack{max \\ in\text{-}plane}} = \sqrt{\left(\frac{\sigma_x - \sigma_y}{2}\right)^2 + \tau_{xy}^2} = \sqrt{\left(\frac{\sigma - 0}{2}\right)^2 + (0)^2} = \pm\frac{\sigma}{2} \qquad Ans.$$

$$\sigma_{avg} = \frac{\sigma_x + \sigma_y}{2} = \frac{\sigma + 0}{2} = \frac{\sigma}{2} \qquad Ans.$$

To determine the proper orientation of the element, apply Eq. 14–2.

$$\tau_{x'y'} = -\frac{\sigma_x - \sigma_y}{2}\sin 2\theta + \tau_{xy}\cos 2\theta = -\frac{\sigma - 0}{2}\sin 90° + 0 = -\frac{\sigma}{2}$$

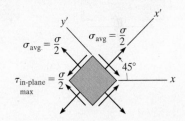

(b)

Fig. 14–14

This negative shear stress acts on the x' face in the negative y' direction, as shown in Fig. 14–14b.

NOTE: If the bar in Fig. 14–14a is made of a *ductile material* such as mild steel then *shear stress* will cause it to fail. This can be noted in the adjacent photo, where within the region of necking, shear stress has caused "slipping" along the steel's crystalline boundaries, resulting in a plane of failure that has formed a *cone* around the bar oriented at approximately 45° as calculated above.

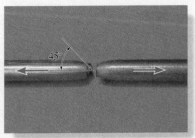

Axial failure of mild steel.

PRELIMINARY PROBLEMS

P14–1. In each case, the state of stress σ_x, σ_y, τ_{xy} produces normal and shear stress components along section AB of the element that have values of $\sigma_{x'} = -5$ kPa and $\tau_{x'y'} = 8$ kPa when calculated using the stress transformation equations. Establish the x' and y' axes for each segment and specify the angle θ, then show these results acting on each segment.

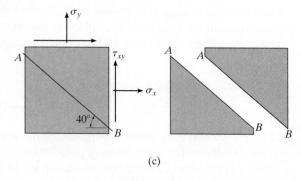

(c)

Prob. P14–1

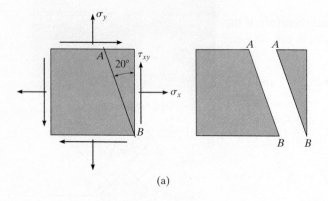

(a)

P14–2. Given the state of stress shown on the element, find σ_{avg} and $\tau_{\substack{max \\ in\text{-}plane}}$ and show the results on a properly oriented element.

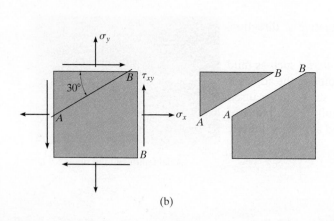

(b)

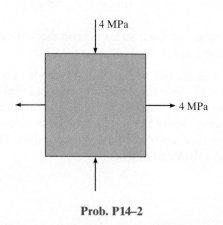

Prob. P14–2

FUNDAMENTAL PROBLEMS

F14–1. Determine the normal stress and shear stress acting on the inclined plane AB.

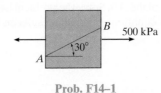

Prob. F14–1

F14–2. Determine the equivalent state of stress on an element at the same point oriented 45° clockwise with respect to the element shown.

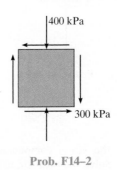

Prob. F14–2

F14–3. Determine the equivalent state of stress on an element at the same point that represents the principal stresses at the point 1. Also, find the corresponding orientation of the element with respect to the element shown.

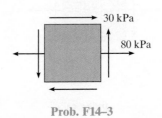

Prob. F14–3

F14–4. Determine the equivalent state of stress on an element at the same point that represents the maximum in-plane shear stress at the point.

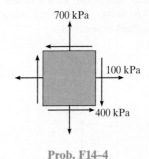

Prob. F14–4

F14–5. The beam is subjected to the load at its end. Determine the maximum principal stress at point B.

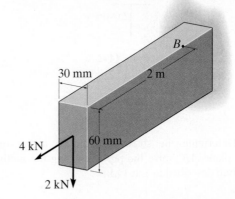

Prob. F14–5

F14–6. The beam is subjected to the loading shown. Determine the principal stress at point C.

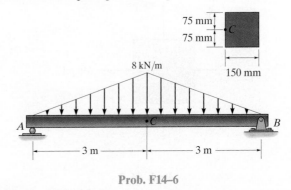

Prob. F14–6

PROBLEMS

14–1. Prove that the sum of the normal stresses $\sigma_x + \sigma_y = \sigma_{x'} + \sigma_{y'}$ is constant. See Figs. 14–2a and 14–2b.

14–2. Determine the stress components acting on the inclined plane AB. Solve the problem using the method of equilibrium described in Sec. 14.1.

*14–4. Determine the normal stress and shear stress acting on the inclined plane AB. Solve the problem using the method of equilibrium described in Sec. 14.1.

14–5. Determine the normal stress and shear stress acting on the inclined plane AB. Solve the problem using the stress transformation equations. Show the results on the sectional element.

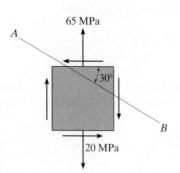

Prob. 14–2

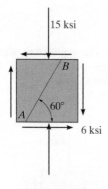

Probs. 14–4/5

14–3. Determine the stress components acting on the inclined plane AB. Solve the problem using the method of equilibrium described in Sec. 14.1.

14–6. Determine the stress components acting on the inclined plane AB. Solve the problem using the method of equilibrium described in Sec. 14.1.

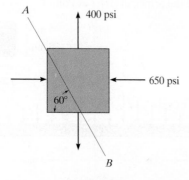

Prob. 14–3

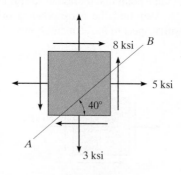

Prob. 14–6

14–7. Determine the stress components acting on the inclined plane AB. Solve the problem using the method of equilibrium described in Sec. 14.1.

*14–8. Solve Prob. 14–7 using the stress-transformation equations developed in Sec. 14.2.

14–11. Determine the equivalent state of stress on an element at the same point oriented 60° clockwise with respect to the element shown. Sketch the results on the element.

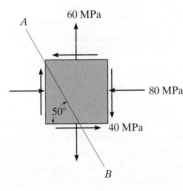

Probs. 14–7/8

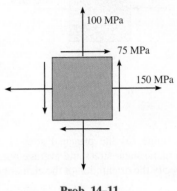

Prob. 14–11

14–9. Determine the stress components acting on the inclined plane AB. Solve the problem using the method of equilibrium described in Sec. 14.1.

14–10. Solve Prob. 14–9 using the stress-transformation equation developed in Sec. 14.2.

*14–12. Determine the equivalent state of stress on an element at the same point oriented 60° counterclockwise with respect to the element shown. Sketch the results on the element.

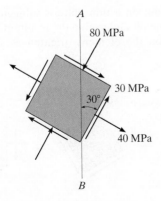

Probs. 14–9/10

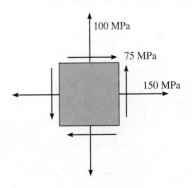

Prob. 14–12

14–13. Determine the stress components acting on the inclined plane *AB*. Solve the problem using the method of equilibrium described in Sec. 14.1.

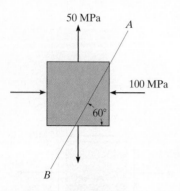

Prob. 14–13

14–14. Determine (a) the principal stresses and (b) the maximum in-plane shear stress and average normal stress at the point. Specify the orientation of the element in each case.

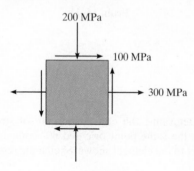

Prob. 14–14

14–15. The state of stress at a point is shown on the element. Determine (a) the principal stresses and (b) the maximum in-plane shear stress and average normal stress at the point. Specify the orientation of the element in each case.

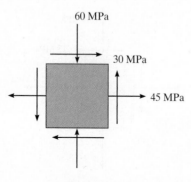

Prob. 14–15

*14–16. Determine the equivalent state of stress on an element at the point which represents (a) the principal stresses and (b) the maximum in-plane shear stress and the associated average normal stress. Also, for each case, determine the corresponding orientation of the element with respect to the element shown and sketch the results on the element.

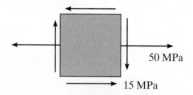

Prob. 14–16

14–17. Determine the equivalent state of stress on an element at the same point which represents (a) the principal stress, and (b) the maximum in-plane shear stress and the associated average normal stress. Also, for each case, determine the corresponding orientation of the element with respect to the element shown and sketch the results on each element.

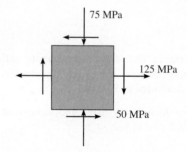

Prob. 14–17

14–18. A point on a thin plate is subjected to the two stress components. Determine the resultant state of stress represented on the element oriented as shown on the right.

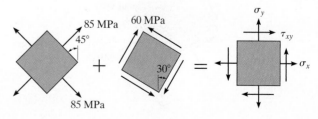

Prob. 14–18

14–19. Determine the equivalent state of stress on an element at the same point which represents (a) the principal stress, and (b) the maximum in-plane shear stress and the associated average normal stress. Also, for each case, determine the corresponding orientation of the element with respect to the element shown and sketch the results on the element.

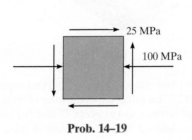

Prob. 14–19

14–21. The stress acting on two planes at a point is indicated. Determine the shear stress on plane a–a and the principal stresses at the point.

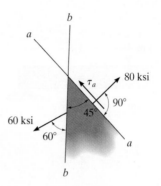

Prob. 14–21

*14–20. The stress along two planes at a point is indicated. Determine the normal stresses on plane b–b and the principal stresses.

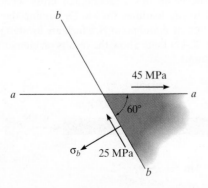

Prob. 14–20

14–22. The state of stress at a point in a member is shown on the element. Determine the stress components acting on the plane AB.

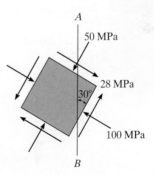

Prob. 14–22

The following problems involve material covered in Chapter 13.

14–23. The wood beam is subjected to a load of 12 kN. If grains of wood in the beam at point A make an angle of 25° with the horizontal as shown, determine the normal and shear stress that act perpendicular to the grains due to the loading.

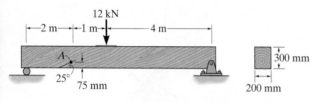

Prob. 14–23

*14–24. The internal loadings at a section of the beam are shown. Determine the in-plane principal stresses at point A. Also compute the maximum in-plane shear stress at this point.

14–25. Solve Prob. 14–24 for point B.

14–26. Solve Prob. 14–24 for point C.

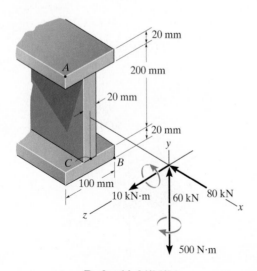

Probs. 14–24/25/26

14–27. A rod has a circular cross section with a diameter of 2 in. It is subjected to a torque of 12 kip · in. and a bending moment M. The greater principal stress at the point of maximum flexural stress is 15 ksi. Determine the magnitude of the bending moment.

Prob. 14–27

*14–28. The bell crank is pinned at A and supported by a short link BC. If it is subjected to the force of 80 N, determine the principal stresses at (a) point D and (b) point E. The crank is constructed from an aluminum plate having a thickness of 20 mm.

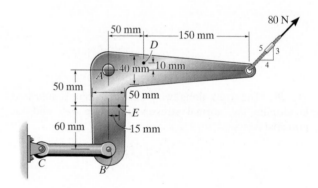

Prob. 14–28

14–29. The beam has a rectangular cross section and is subjected to the loadings shown. Determine the principal stresses at point A and point B, which are located just to the left of the 20-kN load. Show the results on elements located at these points.

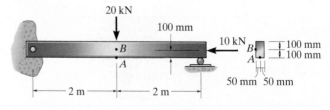

Prob. 14–29

14-30.   A paper tube is formed by rolling a cardboard strip in a spiral and then gluing the edges together as shown. Determine the shear stress acting along the seam, which is at 50° from the horizontal, when the tube is subjected to an axial compressive force of 200 N. The paper is 2 mm thick and the tube has an outer diameter of 100 mm.

14-31.   Solve Prob. 14–30 for the normal stress acting perpendicular to the seam.

14-33.   Determine the principal stresses in the cantilevered beam at points A and B.

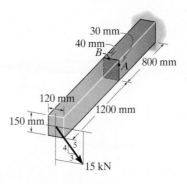

Prob. 14–33

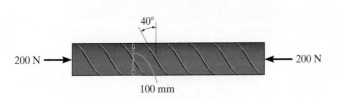

Probs. 14–30/31

*14-32.   The 2-in.-diameter drive shaft AB on the helicopter is subjected to an axial tension of 10 000 lb and a torque of 300 lb · ft. Determine the principal stresses and the maximum in-plane shear stress that act at a point on the surface of the shaft.

14-34.   The internal loadings at a cross section through the 6-in.-diameter drive shaft of a turbine consist of an axial force of 2500 lb, a bending moment of 800 lb · ft, and a torsional moment of 1500 lb · ft. Determine the principal stresses at point A. Also calculate the maximum in-plane shear stress at this point.

14-35.   The internal loadings at a cross section through the 6-in.-diameter drive shaft of a turbine consist of an axial force of 2500 lb, a bending moment of 800 lb · ft, and a torsional moment of 1500 lb · ft. Determine the principal stresses at point B. Also calculate the maximum in-plane shear stress at this point.

Prob. 14–32

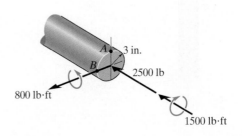

Probs. 14–34/35

14

***14–36.** The shaft has a diameter d and is subjected to the loadings shown. Determine the principal stresses and the maximum in-plane shear stress at point A. The bearings only support vertical reactions.

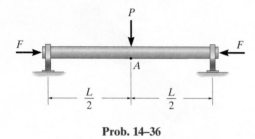

Prob. 14–36

14–37. The steel pipe has an inner diameter of 2.75 in. and an outer diameter of 3 in. If it is fixed at C and subjected to the horizontal 60-lb force acting on the handle of the pipe wrench at its end, determine the principal stresses in the pipe at point A, which is located on the outer surface of the pipe.

14–38. Solve Prob. 14–37 for point B, which is located on the outer surface of the pipe.

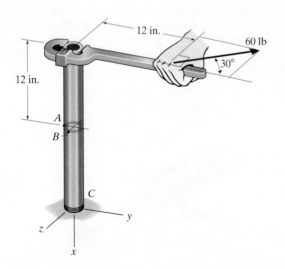

Probs. 14–37/38

14–39. The wide-flange beam is subjected to the 50-kN force. Determine the principal stresses in the beam at point A located on the *web* at the bottom of the upper flange. Although it is not very accurate, use the shear formula to calculate the shear stress.

***14–40.** Solve Prob. 14–39 for point B located on the *web* at the top of the bottom flange.

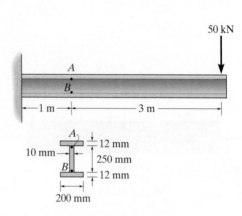

Probs. 14–39/40

14–41. The box beam is subjected to the 26-kN force that is applied at the center of its width, 75 mm from each side. Determine the principal stresses at point A and show the results in an element located at this point. Use the shear formula to calculate the shear stress.

14–42. Solve Prob. 14–41 for point B.

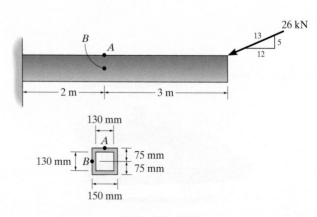

Probs. 14–41/42

14.4 MOHR'S CIRCLE—PLANE STRESS

In this section, we will show how to apply the equations for plane-stress transformation using a *graphical* procedure that is often convenient to use and easy to remember. Furthermore, this approach will allow us to "visualize" how the normal and shear stress components $\sigma_{x'}$ and $\tau_{x'y'}$ vary as the plane on which they act changes its direction, Fig. 14–15a.

If we write Eqs. 14–1 and 14–2 in the form

$$\sigma_{x'} - \left(\frac{\sigma_x + \sigma_y}{2}\right) = \left(\frac{\sigma_x - \sigma_y}{2}\right)\cos 2\theta + \tau_{xy}\sin 2\theta \qquad (14\text{–}9)$$

$$\tau_{x'y'} = -\left(\frac{\sigma_x - \sigma_y}{2}\right)\sin 2\theta + \tau_{xy}\cos 2\theta \qquad (14\text{–}10)$$

then the parameter θ can be eliminated by squaring each equation and adding them together. The result is

$$\left[\sigma_{x'} - \left(\frac{\sigma_x + \sigma_y}{2}\right)\right]^2 + \tau_{x'y'}^{\,2} = \left(\frac{\sigma_x - \sigma_y}{2}\right)^2 + \tau_{xy}^{\,2}$$

Finally, since σ_x, σ_y, τ_{xy} are *known constants*, then the above equation can be written in a more compact form as

$$(\sigma_{x'} - \sigma_{\text{avg}})^2 + \tau_{x'y'}^{\,2} = R^2 \qquad (14\text{–}11)$$

where

$$\sigma_{\text{avg}} = \frac{\sigma_x + \sigma_y}{2}$$

$$R = \sqrt{\left(\frac{\sigma_x - \sigma_y}{2}\right)^2 + \tau_{xy}^{\,2}} \qquad (14\text{–}12)$$

If we establish coordinate axes, σ *positive to the right* and τ *positive downward*, and then plot Eq. 14–11, it will be seen that this equation represents a *circle* having a radius R and center on the σ axis at point $C(\sigma_{\text{avg}}, 0)$, Fig. 14–15b. This circle is called **Mohr's circle**, because it was developed by the German engineer Otto Mohr.

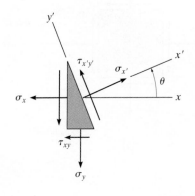

(a)

Fig. 14–15

14

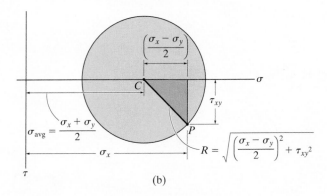

(b)

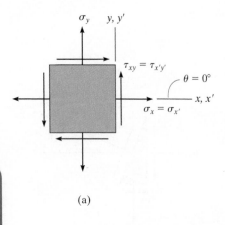

(a)

Each point on Mohr's circle represents the two stress components $\sigma_{x'}$ and $\tau_{x'y'}$ acting on the side of the element defined by the outward x' axis, when this axis is in a specific direction θ. For example, when x' is coincident with the x axis as shown in Fig. 14–16a, then $\theta = 0°$ and $\sigma_{x'} = \sigma_x$, $\tau_{x'y'} = \tau_{xy}$. We will refer to this as the "reference point" A and plot its coordinates $A(\sigma_x, \tau_{xy})$, Fig. 14–16c.

Now consider rotating the x' axis 90° counterclockwise, Fig. 14–16b. Then $\sigma_{x'} = \sigma_y$, $\tau_{x'y'} = -\tau_{xy}$. These values are the coordinates of point $G(\sigma_y, -\tau_{xy})$ on the circle, Fig. 14–16c. Hence, the radial line CG is 180° counterclockwise from the radial "reference line" CA. In other words, a rotation θ of the x' axis on the element will correspond to a rotation 2θ on the circle in the *same direction*.

As discussed in the following procedure, Mohr's circle can be used to determine the principal stresses, the maximum in-plane shear stress, or the stress on any arbitrary plane.

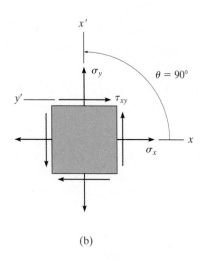

(b)

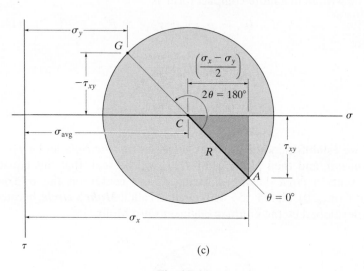

(c)

Fig. 14–16

PROCEDURE FOR ANALYSIS

The following steps are required to draw and use Mohr's circle.

Construction of the Circle.

- Establish a coordinate system such that the horizontal axis represents the normal stress σ, with *positive to the right*, and the vertical axis represents the shear stress τ, with *positive downwards*, Fig. 14–17a.*
- Using the positive sign convention for σ_x, σ_y, τ_{xy}, Fig. 14–17a, plot the center of the circle C, which is located on the σ axis at a distance $\sigma_{avg} = (\sigma_x + \sigma_y)/2$ from the origin, Fig. 14–17a.
- Plot the "reference point" A having coordinates $A(\sigma_x, \tau_{xy})$. This point represents the normal and shear stress components on the element's right-hand vertical face, and since the x' axis coincides with the x axis, this represents $\theta = 0°$, Fig. 14–17a.
- Connect point A with the center C of the circle and determine CA by trigonometry. This represents the radius R of the circle, Fig. 14–17a.
- Once R has been determined, sketch the circle.

Principal Stress.

- The principal stresses σ_1 and σ_2 ($\sigma_1 \geq \sigma_2$) are the coordinates of points B and D, where the circle intersects the σ axis, i.e., where $\tau = 0$, Fig. 14–17a.
- These stresses act on planes defined by angles θ_{p_1} and θ_{p_2}, Fig. 14–17b. One of these angles is represented on the circle as $2\theta_{p_1}$. It is measured *from* the radial reference line CA to line CB.
- Using trigonometry, determine θ_{p_1} from the circle. Remember that the direction of rotation $2\theta_p$ on the circle (here it happens to be counterclockwise) represents the *same* direction of rotation θ_p from the reference axis ($+x$) to the principal plane ($+x'$), Fig. 14–17b.*

Maximum In-Plane Shear Stress.

- The average normal stress and maximum in-plane shear stress components are determined from the circle as the coordinates of either point E or F, Fig. 14–17a.
- In this case the angles θ_{s_1} and θ_{s_2} give the orientation of the planes that contain these components, Fig. 14–17c. The angle $2\theta_{s_1}$ is shown in Fig. 14–17a and can be determined using trigonometry. Here the rotation happens to be clockwise, from CA to CE, and so θ_{s_1} must be clockwise on the element, Fig. 14–17c.*

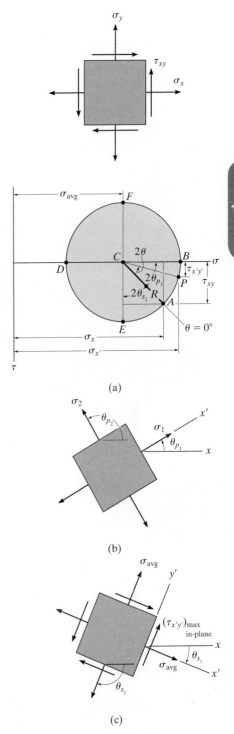

(a)

(b)

(c)

Fig. 14–17

Stresses on Arbitrary Plane.

- The normal and shear stress components $\sigma_{x'}$ and $\tau_{x'y'}$ acting on a specified plane or x' axis, defined by the angle θ, Fig. 14–17d, can be obtained by finding the coordinates of point P on the circle using trigonometry, Fig. 14–17a.

- To locate P, the known angle θ (in this case counterclockwise), Fig. 14–17d, must be measured on the circle in the *same direction* 2θ (counterclockwise) *from* the radial reference line *CA to* the radial line *CP*, Fig. 14–17a.*

   *If the τ axis were constructed *positive upwards,* then the angle 2θ on the circle would be measured in the *opposite direction* to the orientation θ of the x' axis.

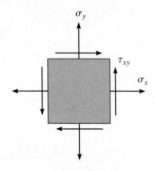

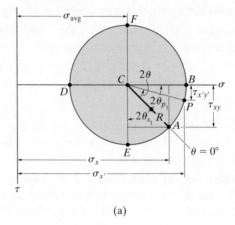

(a)

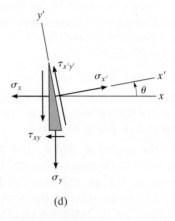

(d)

Fig. 14–17 (cont.)

EXAMPLE   14.7

Due to the applied loading, the element at point A on the solid shaft in Fig. 14–18a is subjected to the state of stress shown. Determine the principal stresses acting at this point.

SOLUTION

Construction of the Circle.   From Fig. 14–18a,

$$\sigma_x = -12 \text{ ksi} \qquad \sigma_y = 0 \qquad \tau_{xy} = -6 \text{ ksi}$$

The center of the circle is located on the σ axis at the point

$$\sigma_{avg} = \frac{-12 + 0}{2} = -6 \text{ ksi}$$

The reference point $A(-12, -6)$ and the center $C(-6, 0)$ are plotted in Fig. 14–18b. From the shaded triangle, the circle is constructed having a radius of

$$R = \sqrt{(12 - 6)^2 + (6)^2} = 8.49 \text{ ksi}$$

Principal Stress.   The principal stresses are indicated by the coordinates of points B and D. We have, for $\sigma_1 > \sigma_2$,

$$\sigma_1 = 8.49 - 6 = 2.49 \text{ ksi} \qquad\qquad Ans.$$

$$\sigma_2 = -6 - 8.49 = -14.5 \text{ ksi} \qquad\qquad Ans.$$

The orientation of the element can be determined by calculating the angle $2\theta_{p_2}$ in Fig. 14–18b, which here is measured *counterclockwise* from CA to CD. It defines the direction θ_{p_2} of σ_2 and its associated principal plane. We have

$$2\theta_{p_2} = \tan^{-1}\frac{6}{12 - 6} = 45.0°$$

$$\theta_{p_2} = 22.5°$$

The element is oriented such that the x' axis or σ_2 is directed 22.5° *counterclockwise* from the horizontal (x axis), as shown in Fig. 14–18c.

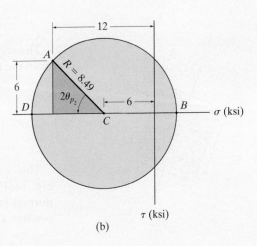

(a)

(b)

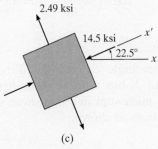

(c)

Fig. 14–18

EXAMPLE 14.8

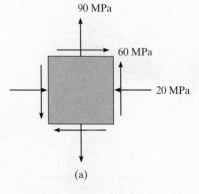

(a)

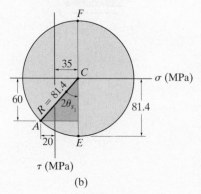

(b)

The state of plane stress at a point is shown on the element in Fig. 14–19a. Determine the maximum in-plane shear stress at this point.

SOLUTION

Construction of the Circle. From the problem data,

$$\sigma_x = -20 \text{ MPa} \qquad \sigma_y = 90 \text{ MPa} \qquad \tau_{xy} = 60 \text{ MPa}$$

The σ, τ axes are established in Fig. 14–19b. The center of the circle C is located on the σ axis, at the point

$$\sigma_{avg} = \frac{-20 + 90}{2} = 35 \text{ MPa}$$

Point C and the reference point $A(-20, 60)$ are plotted. Applying the Pythagorean theorem to the shaded triangle to determine the circle's radius CA, we have

$$R = \sqrt{(60)^2 + (55)^2} = 81.4 \text{ MPa}$$

Maximum In-Plane Shear Stress. The maximum in-plane shear stress and the average normal stress are identified by point E (or F) on the circle. The coordinates of point $E(35, 81.4)$ give

$$\sigma_{avg} = 35 \text{ MPa} \hspace{3cm} Ans.$$

$$\tau_{\substack{max \\ in\text{-}plane}} = 81.4 \text{ MPa} \hspace{2cm} Ans.$$

The angle θ_{s_1}, measured *counterclockwise* from CA to CE, can be found from the circle, identified as $2\theta_{s_1}$. We have

$$2\theta_{s_1} = \tan^{-1}\left(\frac{20 + 35}{60}\right) = 42.5°$$

$$\theta_{s_1} = 21.3° \hspace{3.5cm} Ans.$$

This *counterclockwise* angle defines the direction of the x' axis, Fig. 14–19c. Since point E has *positive* coordinates, then the average normal stress and the maximum in-plane shear stress both act in the *positive x'* and y' directions as shown.

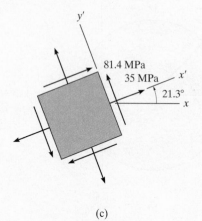

(c)

Fig. 14–19

EXAMPLE   14.9

The state of plane stress at a point is shown on the element in Fig. 14–20*a*. Represent this state of stress on an element oriented 30° counterclockwise from the position shown.

SOLUTION

Construction of the Circle.   From the problem data,

$$\sigma_x = -8 \text{ ksi} \qquad \sigma_y = 12 \text{ ksi} \qquad \tau_{xy} = -6 \text{ ksi}$$

The σ and τ axes are established in Fig. 14–20*b*. The center of the circle C is on the σ axis at the point

$$\sigma_{avg} = \frac{-8 + 12}{2} = 2 \text{ ksi}$$

The reference point for $\theta = 0°$ has coordinates $A(-8, -6)$. Hence from the shaded triangle the radius CA is

$$R = \sqrt{(10)^2 + (6)^2} = 11.66$$

Stresses on 30° Element.   Since the element is to be rotated 30° *counterclockwise*, we must construct a radial line CP, $2(30°) = 60°$ *counterclockwise*, measured from CA ($\theta = 0°$), Fig. 14–20*b*. The coordinates of point $P(\sigma_{x'}, \tau_{x'y'})$ must then be obtained. From the geometry of the circle,

$$\phi = \tan^{-1}\frac{6}{10} = 30.96° \qquad \psi = 60° - 30.96° = 29.04°$$

$$\sigma_{x'} = 2 - 11.66 \cos 29.04° = -8.20 \text{ ksi} \qquad \textit{Ans.}$$

$$\tau_{x'y'} = 11.66 \sin 29.04° = 5.66 \text{ ksi} \qquad \textit{Ans.}$$

These two stress components act on face BD of the element shown in Fig. 14–20*c*, since the x' axis for this face is oriented 30° *counterclockwise* from the x axis.

The stress components acting on the adjacent face DE of the element, which is 60° *clockwise* from the positive x axis, Fig. 14–20*c*, are represented by the coordinates of point Q on the circle. This point lies on the radial line CQ, which is 180° from CP, or 120° *clockwise* from CA. The coordinates of point Q are

$$\sigma_{x'} = 2 + 11.66 \cos 29.04° = 12.2 \text{ ksi} \qquad \textit{Ans.}$$

$$\tau_{x'y'} = -(11.66 \sin 29.04) = -5.66 \text{ ksi} \quad \text{(check)} \qquad \textit{Ans.}$$

NOTE: Here $\tau_{x'y'}$ acts in the $-y'$ direction, Fig. 14–20*c*.

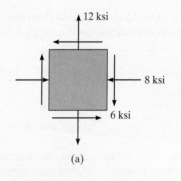

(a)

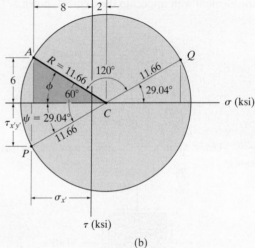

(b)

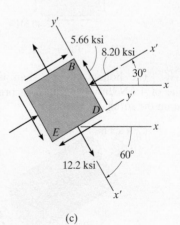

(c)

Fig. 14–20

FUNDAMENTAL PROBLEMS

F14–7. Use Mohr's circle to determine the normal stress and shear stress acting on the inclined plane AB.

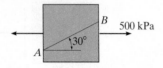

Prob. F14–7

14

F14–8. Use Mohr's circle to determine the principal stresses at the point. Also, find the corresponding orientation of the element with respect to the element shown.

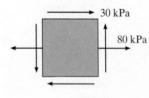

Prob. F14–8

F14–9. Draw Mohr's circle and determine the principal stresses.

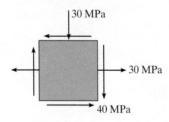

Prob. F14–9

F14–10. The hollow circular shaft is subjected to the torque of $4\ kN \cdot m$. Determine the principal stresses at a point on the surface of the shaft.

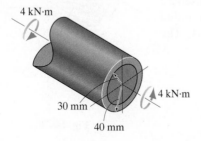

Prob. F14–10

F14–11. Determine the principal stresses at point A on the cross section of the beam at section a–a.

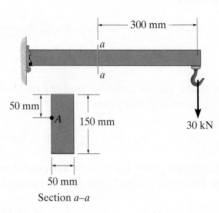

Prob. F14–11

F14–12. Determine the maximum in-plane shear stress at point A on the cross section of the beam at section a–a, which is located just to the left of the 60-kN force. Point A is just below the flange.

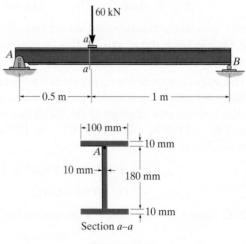

Prob. F14–12

PROBLEMS

14–43.   Solve Prob. 14–2 using Mohr's circle.

*14–44.   Solve Prob. 14–3 using Mohr's circle.

14–45.   Solve Prob. 14–6 using Mohr's circle.

14–46.   Solve Prob. 14–10 using Mohr's circle.

14–47.   Solve Prob. 14–15 using Mohr's circle.

*14–48.   Solve Prob. 14–16 using Mohr's circle.

14–49.   Mohr's circle for the state of stress is shown in Fig. 14–17a. Show that finding the coordinates of point $P(\sigma_{x'}, \tau_{x'y'})$ on the circle gives the same value as the stress transformation Eqs. 14–1 and 14–2.

14–50.   Determine (a) the principal stresses and (b) the maximum in-plane shear stress and average normal stress. Specify the orientation of the element in each case.

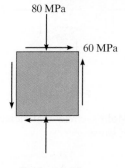

Prob. 14–50

14–51.   Determine (a) the principal stresses and (b) the maximum in-plane shear stress and average normal stress. Specify the orientation of the element in each case.

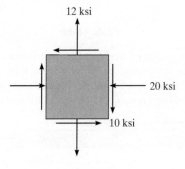

Prob. 14–51

*14–52.   Determine the equivalent state of stress if an element is oriented 60° clockwise from the element shown.

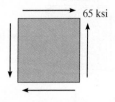

Prob. 14–52

14–53.   Draw Mohr's circle that describes each of the following states of stress.

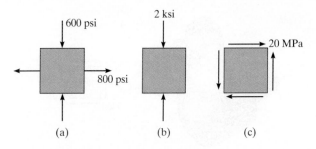

(a)   (b)   (c)

Prob. 14–53

14–54.   Draw Mohr's circle that describes each of the following states of stress.

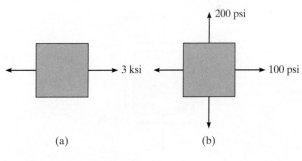

(a)   (b)

Prob. 14–54

14–55. Determine (a) the principal stresses and (b) the maximum in-plane shear stress and average normal stress. Specify the orientation of the element in each case.

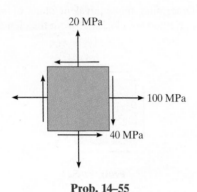

Prob. 14–55

*14–56.** Determine (a) the principal stress and (b) the maximum in-plane shear stress and average normal stress. Specify the orientation of the element in each case.

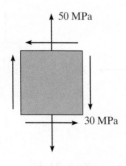

Prob. 14–56

14–57. Determine (a) the principal stresses and (b) the maximum in-plane shear stress and average normal stress. Specify the orientation of the element in each case.

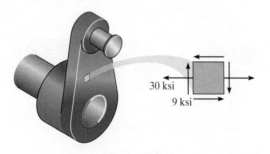

Prob. 14–57

14–58. Determine (a) the principal stresses and (b) the maximum in-plane shear stress and average normal stress. Specify the orientation of the element in each case.

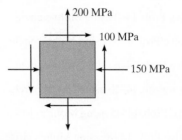

Prob. 14–58

14–59. Determine (a) the principal stresses and (b) the maximum in-plane shear stress and average normal stress. Specify the orientation of the element in each case.

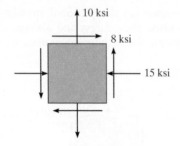

Prob. 14–59

*14–60.** Draw Mohr's circle that describes each of the following states of stress.

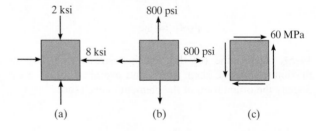

Prob. 14–60

14–61. The grains of wood in the board make an angle of 20° with the horizontal as shown. Determine the normal and shear stresses that act perpendicular and parallel to the grains if the board is subjected to an axial load of 250 N.

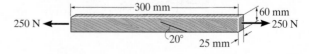

Prob. 14–61

14–62. The post is fixed supported at its base and a horizontal force is applied at its end as shown, determine (a) the maximum in-plane shear stress developed at A and (b) the principal stresses at A.

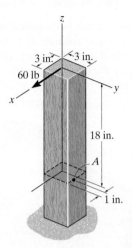

Prob. 14–62

14–63. Determine the principal stresses, the maximum in-plane shear stress, and average normal stress. Specify the orientation of the element in each case.

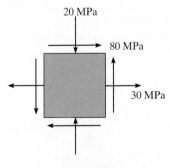

Prob. 14–63

*14–64. The thin-walled pipe has an inner diameter of 0.5 in. and a thickness of 0.025 in. If it is subjected to an internal pressure of 500 psi and the axial tension and torsional loadings shown, determine the principal stress at a point on the surface of the pipe.

Prob. 14–64

14–65. The frame supports the triangular distributed load shown. Determine the normal and shear stresses at point D that act perpendicular and parallel, respectively, to the grains. The grains at this point make an angle of 35° with the horizontal as shown.

14–66. The frame supports the triangular distributed load shown. Determine the normal and shear stresses at point E that act perpendicular and parallel, respectively, to the grains. The grains at this point make an angle of 45° with the horizontal as shown.

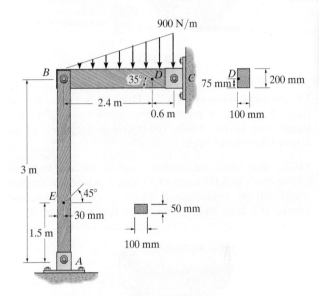

Probs. 14–65/66

14–67. The rotor shaft of the helicopter is subjected to the tensile force and torque shown when the rotor blades provide the lifting force to suspend the helicopter at midair. If the shaft has a diameter of 6 in., determine the principal stresses and maximum in-plane shear stress at a point located on the surface of the shaft.

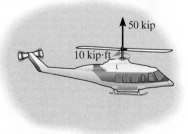

Prob. 14–67

*14–68. The pedal crank for a bicycle has the cross section shown. If it is fixed to the gear at B and does not rotate while subjected to a force of 75 lb, determine the principal stresses on the cross section at point C.

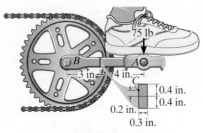

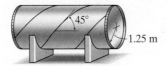

Prob. 14–68

14–69. A spherical pressure vessel has an inner radius of 5 ft and a wall thickness of 0.5 in. Draw Mohr's circle for the state of stress at a point on the vessel and explain the significance of the result. The vessel is subjected to an internal pressure of 80 psi.

14–70. The cylindrical pressure vessel has an inner radius of 1.25 m and a wall thickness of 15 mm. It is made from steel plates that are welded along the 45° seam. Determine the normal and shear stress components along this seam if the vessel is subjected to an internal pressure of 8 MPa.

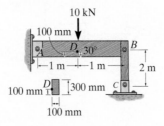

Prob. 14–70

14–71. Determine the normal and shear stresses at point D that act perpendicular and parallel, respectively, to the grains. The grains at this point make an angle of 30° with the horizontal as shown. Point D is located just to the left of the 10-kN force.

*14–72. Determine the principal stress at point D, which is located just to the left of the 10-kN force.

14–73. If the box wrench is subjected to the 50 lb force, determine the principal stresses and maximum in-plane shear stress at point A on the cross section of the wrench at section a–a. Specify the orientation of these states of stress and indicate the results on elements at the point.

14–74. If the box wrench is subjected to the 50-lb force, determine the principal stresses and maximum in-plane shear stress at point B on the cross section of the wrench at section a–a. Specify the orientation of these states of stress and indicate the results on elements at the point.

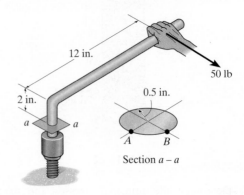

Probs. 14–73/74

14–75. The post is fixed supported at its base and the loadings are applied at its end as shown. Determine (a) the maximum in-plane shear stress developed at A and (b) the principal stresses at A.

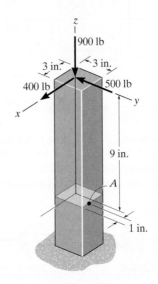

Prob. 14–75

Probs. 14–71/72

14.5 ABSOLUTE MAXIMUM SHEAR STRESS

Since the strength of a ductile material depends upon its ability to resist shear stress, it becomes important to find the ***absolute maximum shear stress*** in the material when it is subjected to a loading. To show how this can be done, we will confine our attention only to the most common case of plane stress,* as shown in Fig. 14–21a. Here *both* σ_1 and σ_2 are tensile. If we view the element in two dimensions at a time, that is, in the y–z, x–z, and x–y planes, Figs. 14–21b, 14–21c, and 14–21d, then we can use Mohr's circle to determine the maximum in-plane shear stress for each case. For example, Mohr's circle extends between 0 and σ_2 for the case shown in Fig. 14–21b. From this circle, Fig. 14–21e, the maximum in-plane shear stress is $\tau_{\substack{max \\ in\text{-}plane}} = \sigma_2/2$. Mohr's circles for the other two cases are also shown in Fig. 14–21e. Comparing all three circles, the absolute maximum shear stress is

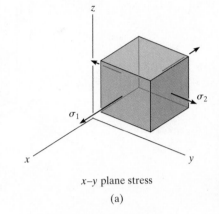

x–y plane stress

(a)

$$\tau_{\substack{abs \\ max}} = \frac{\sigma_1}{2} \qquad (14\text{--}13)$$

σ_1 and σ_2 have
the same sign

It occurs on an element that is rotated 45° about the y axis from the element shown in Fig. 14–21a or Fig. 14–21c. It is this out of plane shear stress that will cause the material to fail, not $\tau_{\substack{max \\ in\text{-}plane}}$.

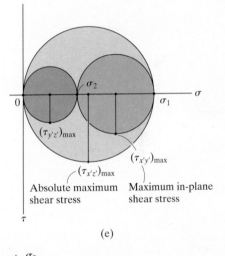

(e)

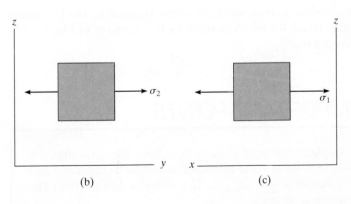

(b)   (c)   (d)

Fig. 14–21

---

*The case for three-dimensional stress is discussed in books related to advanced mechanics of materials and the theory of elasticity.

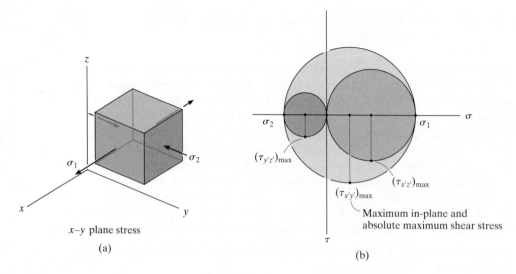

x–y plane stress

(a)

Maximum in-plane and
absolute maximum shear stress

(b)

Fig. 14–22

In a similar manner, if one of the in-plane principal stresses has the *opposite sign* of the other, Fig. 14–22a, then the three Mohr's circles that describe the state of stress for the element when viewed from each plane are shown in Fig. 14–22b. Clearly, in this case

$$\tau_{\substack{abs \\ max}} = \frac{\sigma_1 - \sigma_2}{2} \qquad (14\text{–}14)$$

σ_1 and σ_2 have
opposite signs

Here the absolute maximum shear stress is equal to the maximum in-plane shear stress found from rotating the element in Fig. 14–22a, 45° about the z axis.

IMPORTANT POINTS

- If the in-plane principal stresses both have the same sign, the absolute maximum shear stress will occur out of the plane and has a value of $\tau_{\substack{abs \\ max}} = \sigma_{max}/2$. This value is greater than the in-plane shear stress.

- If the in-plane principal stresses are of opposite signs, then the absolute maximum shear stress will equal the maximum in-plane shear stress; that is, $\tau_{\substack{abs \\ max}} = (\sigma_{max} - \sigma_{min})/2$.

EXAMPLE 14.10

The point on the surface of the pressure vessel in Fig. 14–23a is subjected to the state of plane stress. Determine the absolute maximum shear stress at this point.

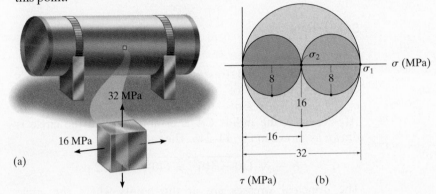

(a)

τ (MPa)   (b)

Fig. 14–23

SOLUTION

The principal stresses are $\sigma_1 = 32$ MPa, $\sigma_2 = 16$ MPa. If these stresses are plotted along the σ axis, the three Mohr's circles can be constructed that describe the state of stress viewed in each of the three perpendicular planes, Fig. 14–23b. The largest circle has a radius of 16 MPa and describes the state of stress in the plane only containing $\sigma_1 = 32$ MPa, shown shaded in Fig. 14–23a. An orientation of an element 45° within this plane yields the state of absolute maximum shear stress and the associated average normal stress, namely,

$$\tau_{\substack{abs \\ max}} = 16 \text{ MPa} \qquad\qquad \textit{Ans.}$$

$$\sigma_{avg} = 16 \text{ MPa}$$

This same result for $\tau_{\substack{abs \\ max}}$ can be obtained from direct application of Eq. 14–13.

$$\tau_{\substack{abs \\ max}} = \frac{\sigma_1}{2} = \frac{32}{2} = 16 \text{ MPa} \qquad\qquad \textit{Ans.}$$

$$\sigma_{avg} = \frac{32 + 0}{2} = 16 \text{ MPa}$$

By comparison, the maximum in-plane shear stress can be determined from the Mohr's circle drawn between $\sigma_1 = 32$ MPa and $\sigma_2 = 16$ MPa, Fig. 14–23b. This gives a value of

$$\tau_{\substack{max \\ in\text{-}plane}} = \frac{32 - 16}{2} = 8 \text{ MPa}$$

$$\sigma_{avg} = \frac{32 + 16}{2} = 24 \text{ MPa}$$

EXAMPLE 14.11

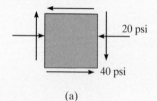

(a)

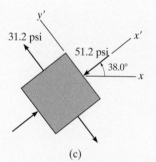

(b)

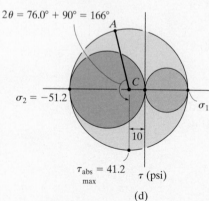

(c)

$2\theta = 76.0° + 90° = 166°$

(d)

Fig. 14-24

Due to an applied loading, an element at a point on a machine shaft is subjected to the state of plane stress shown in Fig. 14–24a. Determine the principal stresses and the absolute maximum shear stress at the point.

SOLUTION

Principal Stresses.
The in-plane principal stresses can be determined from Mohr's circle. The center of the circle is on the σ axis at $\sigma_{avg} = (-20 + 0)/2 = -10$ psi. Plotting the reference point $A(-20, -40)$, the radius CA is established and the circle is drawn as shown in Fig. 14–24b. The radius is

$$R = \sqrt{(20 - 10)^2 + (40)^2} = 41.2 \text{ psi}$$

The principal stresses are at the points where the circle intersects the σ axis; i.e.,

$$\sigma_1 = -10 + 41.2 = 31.2 \text{ psi}$$
$$\sigma_2 = -10 - 41.2 = -51.2 \text{ psi}$$

From the circle, the *counterclockwise* angle 2θ, measured from CA to the $-\sigma$ axis, is

$$2\theta = \tan^{-1}\left(\frac{40}{20 - 10}\right) = 76.0°$$

Thus,

$$\theta = 38.0°$$

This *counterclockwise* rotation defines the direction of the x' axis and σ_2, Fig. 14–24c. We have

$$\sigma_1 = 31.2 \text{ psi} \quad \sigma_2 = -51.2 \text{ psi} \qquad Ans.$$

Absolute Maximum Shear Stress.   Since these stresses have opposite signs, applying Eq. 14–14 we have

$$\tau_{\substack{abs\\max}} = \frac{\sigma_1 - \sigma_2}{2} = \frac{31.2 - (-51.2)}{2} = 41.2 \text{ psi} \qquad Ans.$$

$$\sigma_{avg} = \frac{31.2 - 51.2}{2} = -10 \text{ psi}$$

These same results can also be obtained by drawing Mohr's circle for each orientation of an element about the x, y, and z axes, Fig. 14–24d. Since σ_1 and σ_2 are of *opposite signs*, then the absolute maximum shear stress as noted equals the maximum in-plane shear stress.

PROBLEMS

*14–76. Draw the three Mohr's circles that describe each of the following states of stress.

14–78. Draw the three Mohr's circles that describe the following state of stress.

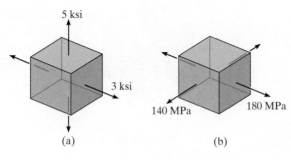

(a)   (b)

Prob. 14–76

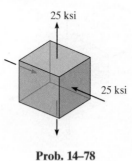

Prob. 14–78

14–79. Determine the principal stresses and the absolute maximum shear stress.

14–77. Draw the three Mohr's circles that describe the following state of stress.

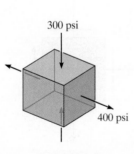

Prob. 14–77

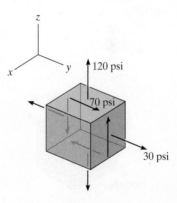

Prob. 14–79

*14–80. Determine the principal stresses and the absolute maximum shear stress.

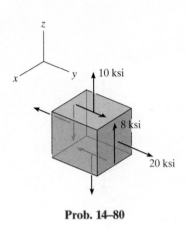

Prob. 14–80

14–81. Determine the principal stresses and the absolute maximum shear stress.

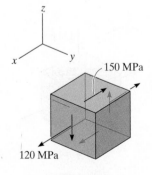

Prob. 14–81

14–82. Determine the principal stresses and the absolute maximum shear stress.

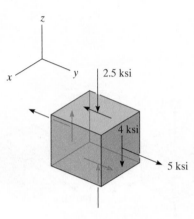

Prob. 14–82

14–83. Consider the general case of plane stress as shown. Write a computer program that will show a plot of the three Mohr's circles for the element, and will also determine the maximum in-plane shear stress and the absolute maximum shear stress.

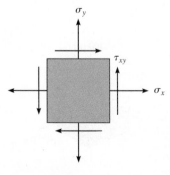

Prob. 14–83

14.6 PLANE STRAIN

As outlined in Sec. 7.9, the general state of strain at a point in a body is represented by a combination of three components of normal strain, ϵ_x, ϵ_y, ϵ_z, and three components of shear strain, γ_{xy}, γ_{xz}, γ_{yz}. The normal strains cause a change in the volume of the element, and the shear strains cause a change in its shape. Like stress, these six components depend upon the orientation of the element, and in many situations, engineers must transform the strains in order to obtain their values in other directions.

To understand how this is done, we will direct our attention to a study of *plane strain*, whereby the element is subjected to two components of normal strain, ϵ_x, ϵ_y, and one component of shear strain, γ_{xy}. Although plane strain and plane stress each have three components lying in the same plane, realize that plane stress *does not* necessarily cause plane strain or vice versa. The reason for this has to do with the Poisson effect discussed in Sec. 8.5. For example, the element in Fig. 14–25 is subjected to *plane stress* caused by σ_x and σ_y. Not only are normal strains ϵ_x and ϵ_y produced, but there is *also* an associated normal strain, ϵ_z, and so this is *not* a case of plane strain.

Actually, a case of plane strain rarely occurs in practice, because few materials are constrained between rigid surfaces so as not to permit any distortion in, say, the z direction. In spite of this, the analysis of plane strain, as outlined in the following section, is still of great importance, because it will allow us to convert strain-gage data, measured at a point on the surface of a body, into plane stress at the point.

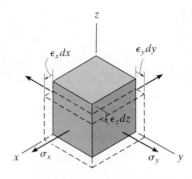

Plane stress, σ_x, σ_y, does not cause plane
strain in the x–y plane since $\epsilon_z \neq 0$.

Fig. 14–25

14.7 GENERAL EQUATIONS OF PLANE-STRAIN TRANSFORMATION

For plane-strain analysis it is important to establish strain transformation equations that can be used to determine the components of normal and shear strain at a point, $\epsilon_{x'}, \epsilon_{y'}, \gamma_{x'y'}$, Fig. 14–26c, provided the components $\epsilon_x, \epsilon_y, \gamma_{xy}$ are known, Fig. 14–26a. So in other words, if we know how the element of material in Fig. 14–26a deforms, we want to know how the tipped element of material in Fig. 14–26b will deform. To do this requires relating the deformations and rotations of line segments which represent the sides of differential elements that are parallel to the x, y and x', y' axes.

Sign Convention. To begin, we must first establish a sign convention for strain. The *normal strains* ϵ_x and ϵ_y in Fig. 14–26a are *positive* if they cause *elongation* along the x and y axes, respectively, and the *shear strain* γ_{xy} is *positive* if the interior angle *AOB* becomes *smaller* than 90°. This sign convention also follows the corresponding one used for plane stress, Fig. 14–5a, that is, positive $\sigma_x, \sigma_y, \tau_{xy}$ will cause the element to *deform* in the positive $\epsilon_x, \epsilon_y, \gamma_{xy}$ directions, respectively. Finally, if the angle between the x and x' axes is θ, then, like the case of plane stress, θ will be *positive* provided it follows the curl of the right-hand fingers, i.e., counterclockwise, as shown in Fig. 14–26c.

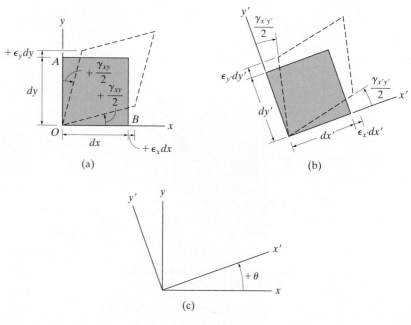

(a)

(b)

(c)

Fig. 14–26

Normal and Shear Strains. To determine $\epsilon_{x'}$, we must find the elongation of a line segment dx' that lies along the x' axis and is subjected to strain components $\epsilon_x, \epsilon_y, \gamma_{xy}$. As shown in Fig. 14–27a, the components of line dx' along the x and y axes are

$$dx = dx' \cos \theta$$
$$dy = dx' \sin \theta \qquad (14\text{–}15)$$

When the positive normal strain ϵ_x occurs, dx is elongated $\epsilon_x\, dx$, Fig. 14–27b, which causes dx' to elongate $\epsilon_x\, dx \cos \theta$. Likewise, when ϵ_y occurs, dy elongates $\epsilon_y\, dy$, Fig. 14–27c, which causes dx' to elongate $\epsilon_y\, dy \sin \theta$. Finally, assuming that dx remains fixed in position, the shear strain γ_{xy} in Fig. 14–27d, which is the change in angle between dx and dy, causes the top of line dy to be displaced $\gamma_{xy}\, dy$ to the right. This causes dx' to elongate $\gamma_{xy}\, dy \cos \theta$. If all three of these (red) elongations are added together, the resultant elongation of dx' is then

$$\delta x' = \epsilon_x\, dx \cos \theta + \epsilon_y\, dy \sin \theta + \gamma_{xy}\, dy \cos \theta$$

Since the normal strain along line dx' is $\epsilon_{x'} = \delta x'/dx'$, then using Eqs. 14–15, we have

$$\epsilon_{x'} = \epsilon_x \cos^2 \theta + \epsilon_y \sin^2 \theta + \gamma_{xy} \sin \theta \cos \theta \qquad (14\text{–}16)$$

This normal strain is shown in Fig. 14–26b.

The rubber specimen is constrained between the two fixed supports, and so it will undergo plane strain when loads are applied to it in the horizontal plane.

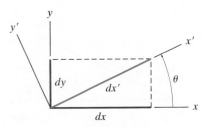

Before deformation

(a)

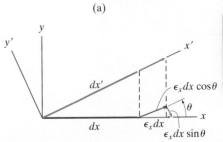

Normal strain ϵ_x

(b)

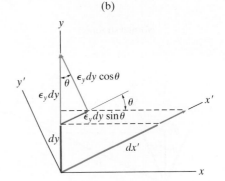

Normal strain ϵ_y

(c)

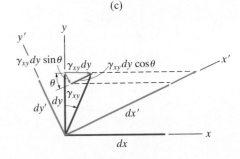

Shear strain γ_{xy}

(d)

Fig. 14–27

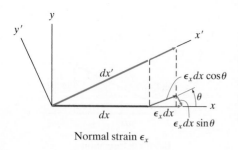

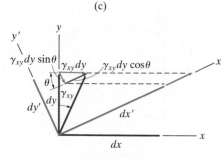

Normal strain ϵ_x

(b)

Normal strain ϵ_y

(c)

Shear strain γ_{xy}

(d)

To determine $\gamma_{x'y'}$, we must find the rotation of each of the line segments dx' and dy' when they are subjected to the strain components ϵ_x, ϵ_y, γ_{xy}. First we will consider the counterclockwise rotation α of dx', Fig. 14–27e. Here $\alpha = \delta y'/dx'$. The displacement $\delta y'$ consists of three displacement components: one from ϵ_x, giving $-\epsilon_x\, dx \sin \theta$, Fig. 14–27b; another from ϵ_y, giving $\epsilon_y\, dy \cos \theta$, Fig. 14–27c; and the last from γ_{xy}, giving $-\gamma_{xy}\, dy \sin \theta$, Fig. 14–27d. Thus, $\delta y'$ is

$$\delta y' = -\epsilon_x\, dx \sin \theta + \epsilon_y\, dy \cos \theta - \gamma_{xy}\, dy \sin \theta$$

Using Eq. 14–15, we therefore have

$$\alpha = \frac{\delta y'}{dx'} = (-\epsilon_x + \epsilon_y) \sin \theta \cos \theta - \gamma_{xy} \sin^2 \theta \qquad (14\text{–}17)$$

Finally, line dy' rotates by an amount β, Fig. 14–27e. We can determine this angle by a similar analysis, or by simply substituting $\theta + 90°$ for θ into Eq. 14–17. Using the identities $\sin(\theta + 90°) = \cos \theta$, $\cos(\theta + 90°) = -\sin \theta$, we have

$$\beta = (-\epsilon_x + \epsilon_y) \sin(\theta + 90°) \cos(\theta + 90°) - \gamma_{xy} \sin^2(\theta + 90°)$$

$$= -(-\epsilon_x + \epsilon_y) \cos \theta \sin \theta - \gamma_{xy} \cos^2 \theta$$

Since α and β must represent the rotation of the sides dx' and dy' in the manner shown in Fig. 14–27e, then the element is subjected to a shear strain of

$$\gamma_{x'y'} = \alpha - \beta = -2(\epsilon_x - \epsilon_y) \sin \theta \cos \theta + \gamma_{xy}(\cos^2 \theta - \sin^2 \theta) \quad (14\text{–}18)$$

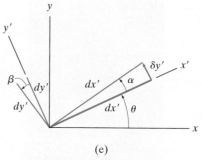

(e)

Fig. 14–27 (cont.)

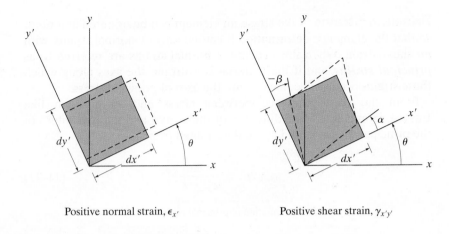

Positive normal strain, $\epsilon_{x'}$

Positive shear strain, $\gamma_{x'y'}$

(a)

(b)

Fig. 14–28

Using the trigonometric identities $\sin 2\theta = 2 \sin \theta \cos \theta$, $\cos^2 \theta = (1 + \cos 2\theta)/2$, and $\sin^2 \theta + \cos^2 \theta = 1$, Eqs. 14–16 and 14–18 can be written in the final form

$$\epsilon_{x'} = \frac{\epsilon_x + \epsilon_y}{2} + \frac{\epsilon_x - \epsilon_y}{2} \cos 2\theta + \frac{\gamma_{xy}}{2} \sin 2\theta \qquad (14\text{–}19)$$

$$\frac{\gamma_{x'y'}}{2} = -\left(\frac{\epsilon_x - \epsilon_y}{2}\right) \sin 2\theta + \frac{\gamma_{xy}}{2} \cos 2\theta \qquad (14\text{–}20)$$

Normal and Shear Strain Components

According to our sign convention, if $\epsilon_{x'}$ is *positive*, the element *elongates* in the positive x' direction, Fig. 14–28a, and if $\gamma_{x'y'}$ is positive, the element deforms as shown in Fig. 14–28b.

If the normal strain in the y' direction is required, it can be obtained from Eq. 14–19 by simply substituting $(\theta + 90°)$ for θ. The result is

$$\epsilon_{y'} = \frac{\epsilon_x + \epsilon_y}{2} - \frac{\epsilon_x - \epsilon_y}{2} \cos 2\theta - \frac{\gamma_{xy}}{2} \sin 2\theta \qquad (14\text{–}21)$$

The similarity between the above three equations and those for plane-stress transformation, Eqs. 14–1, 14–2, and 14–3, should be noted. Making the comparison, σ_x, σ_y, $\sigma_{x'}$, $\sigma_{y'}$ correspond to ϵ_x, ϵ_y, $\epsilon_{x'}$, $\epsilon_{y'}$; and τ_{xy}, $\tau_{x'y'}$ correspond to $\gamma_{xy}/2$, $\gamma_{x'y'}/2$.

Complex stresses are often developed at the joints where the cylindrical and hemispherical vessels are joined together. The stresses are determined by making measurements of strain.

Principal Strains. Like stress, an element can be oriented at a point so that the element's deformation is caused only by normal strains, with *no* shear strain. When this occurs the normal strains are referred to as *principal strains*, and if the material is isotropic, the axes along which these strains occur will coincide with the axes of principal stress.

From the correspondence between stress and strain, then like Eqs. 14–18 and 14–19, the direction of the x' axis and the two values of the principal strains ϵ_1 and ϵ_2 are determined from

$$\tan 2\theta_p = \frac{\gamma_{xy}}{\epsilon_x - \epsilon_y}$$

(14–22)

Orientation of principal planes

$$\epsilon_{1,2} = \frac{\epsilon_x + \epsilon_y}{2} \pm \sqrt{\left(\frac{\epsilon_x - \epsilon_y}{2}\right)^2 + \left(\frac{\gamma_{xy}}{2}\right)^2}$$

(14–23)

Principal strains

Maximum In-Plane Shear Strain. Similar to Eqs. 14–20, 14–21, and 14–22, the direction of the x' axis and the maximum in-plane shear strain and associated average normal strain are determined from the following equations:

$$\tan 2\theta_s = -\left(\frac{\epsilon_x - \epsilon_y}{\gamma_{xy}}\right)$$

(14–24)

Orientation of maximum in-plane shear strain

$$\frac{\gamma_{\text{in-plane}}^{\text{max}}}{2} = \sqrt{\left(\frac{\epsilon_x - \epsilon_y}{2}\right)^2 + \left(\frac{\gamma_{xy}}{2}\right)^2}$$

(14–25)

Maximum in-plane shear strain

$$\epsilon_{\text{avg}} = \frac{\epsilon_x + \epsilon_y}{2}$$

(14–26)

Average normal strain

▶ IMPORTANT POINTS

- In the case of plane stress, plane-strain analysis may be used within the plane of the stresses to analyze the data from strain gages. Remember, though, there will be a normal strain that is perpendicular to the gages due to the Poisson effect.

- When the state of strain is represented by the principal strains, no shear strain will act on the element.

- When the state of strain is represented by the maximum in-plane shear strain, an associated average normal strain will also act on the element.

EXAMPLE   14.12

The state of plane strain at a point has components of $\epsilon_x = 500(10^{-6})$, $\epsilon_y = -300(10^{-6})$, $\gamma_{xy} = 200(10^{-6})$, which tends to distort the element as shown in Fig. 14–29a. Determine the equivalent strains acting on an element of the material oriented *clockwise* 30°.

SOLUTION
The strain transformation Eqs. 14–19 and 14–20 will be used to solve the problem. Since θ is *positive counterclockwise*, then for this problem $\theta = -30°$. Thus,

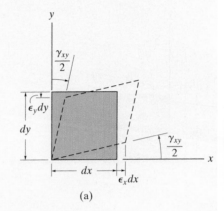

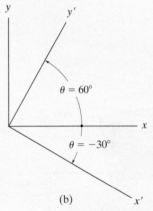

$$\epsilon_{x'} = \frac{\epsilon_x + \epsilon_y}{2} + \frac{\epsilon_x - \epsilon_y}{2}\cos 2\theta + \frac{\gamma_{xy}}{2}\sin 2\theta$$

$$= \left[\frac{500 + (-300)}{2}\right](10^{-6}) + \left[\frac{500 - (-300)}{2}\right](10^{-6})\cos(2(-30°))$$

$$+ \left[\frac{200(10^{-6})}{2}\right]\sin(2(-30°))$$

$$\epsilon_{x'} = 213(10^{-6}) \qquad\qquad Ans.$$

$$\frac{\gamma_{x'y'}}{2} = -\left(\frac{\epsilon_x - \epsilon_y}{2}\right)\sin 2\theta + \frac{\gamma_{xy}}{2}\cos 2\theta$$

$$= -\left[\frac{500 - (-300)}{2}\right](10^{-6})\sin(2(-30°)) + \frac{200(10^{-6})}{2}\cos(2(-30°))$$

$$\gamma_{x'y'} = 793(10^{-6}) \qquad\qquad Ans.$$

The strain in the y' direction can be obtained from Eq. 14–21 with $\theta = -30°$. However, we can also obtain $\epsilon_{y'}$ using Eq. 14–19 with $\theta = 60°$ ($\theta = -30° + 90°$), Fig. 14–29b. We have with $\epsilon_{y'}$ replacing $\epsilon_{x'}$,

$$\epsilon_{y'} = \frac{\epsilon_x + \epsilon_y}{2} + \frac{\epsilon_x - \epsilon_y}{2}\cos 2\theta + \frac{\gamma_{xy}}{2}\sin 2\theta$$

$$= \left[\frac{500 + (-300)}{2}\right](10^{-6}) + \left[\frac{500 - (-300)}{2}\right](10^{-6})\cos(2(60°))$$

$$+ \frac{200(10^{-6})}{2}\sin(2(60°))$$

$$\epsilon_{y'} = -13.4(10^{-6}) \qquad\qquad Ans.$$

These results tend to distort the element as shown in Fig. 14–29c.

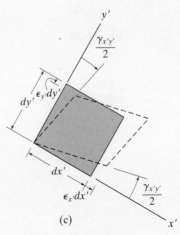

Fig. 14–29

EXAMPLE 14.13

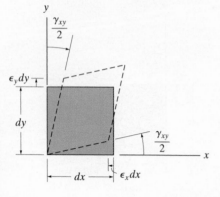

(a)

(b)

Fig. 14–30

The state of plane strain at a point has components of $\epsilon_x = -350(10^{-6})$, $\epsilon_y = 200(10^{-6})$, $\gamma_{xy} = 80(10^{-6})$, Fig. 14–30a. Determine the principal strains at the point and the orientation of the element upon which they act.

SOLUTION

Orientation of the Element. From Eq. 14–22 we have

$$\tan 2\theta_p = \frac{\gamma_{xy}}{\epsilon_x - \epsilon_y}$$

$$= \frac{80(10^{-6})}{(-350 - 200)(10^{-6})}$$

Thus, $2\theta_p = -8.28°$ and $-8.28° + 180° = 171.72°$, so that

$$\theta_p = -4.14° \text{ and } 85.9° \qquad Ans.$$

Each of these angles is measured *positive counterclockwise*, from the x axis to the outward normals on each face of the element. The angle of $-4.14°$ is shown in Fig. 14–30b.

Principal Strains. The principal strains are determined from Eq. 14–23. We have

$$\epsilon_{1,2} = \frac{\epsilon_x + \epsilon_y}{2} \pm \sqrt{\left(\frac{\epsilon_x - \epsilon_y}{2}\right)^2 + \left(\frac{\gamma_{xy}}{2}\right)^2}$$

$$= \frac{(-350 + 200)(10^{-6})}{2} \pm \left[\sqrt{\left(\frac{-350 - 200}{2}\right)^2 + \left(\frac{80}{2}\right)^2}\right](10^{-6})$$

$$= -75.0(10^{-6}) \pm 277.9(10^{-6})$$

$$\epsilon_1 = 203(10^{-6}) \qquad \epsilon_2 = -353(10^{-6}) \qquad Ans.$$

To determine the direction of each of these strains we will apply Eq. 14–19 with $\theta = -4.14°$, Fig. 14–30b. Thus,

$$\epsilon_{x'} = \frac{\epsilon_x + \epsilon_y}{2} + \frac{\epsilon_x - \epsilon_y}{2} \cos 2\theta + \frac{\gamma_{xy}}{2} \sin 2\theta$$

$$= \left(\frac{-350 + 200}{2}\right)(10^{-6}) + \left(\frac{-350 - 200}{2}\right)(10^{-6}) \cos 2(-4.14°)$$

$$+ \frac{80(10^{-6})}{2} \sin 2(-4.14°)$$

$$\epsilon_{x'} = -353(10^{-6})$$

Hence $\epsilon_{x'} = \epsilon_2$. When subjected to the principal strains, the element is distorted as shown in Fig. 14–30b.

EXAMPLE   14.14

The state of plane strain at a point has components of $\epsilon_x = -350(10^{-6})$, $\epsilon_y = 200(10^{-6})$, $\gamma_{xy} = 80(10^{-6})$, Fig. 14–31a. Determine the maximum in-plane shear strain at the point and the orientation of the element upon which it acts.

SOLUTION

Orientation of the Element.   From Eq. 14–24 we have

$$\tan 2\theta_s = -\left(\frac{\epsilon_x - \epsilon_y}{\gamma_{xy}}\right) = -\frac{(-350 - 200)(10^{-6})}{80(10^{-6})}$$

Thus, $2\theta_s = 81.72°$ and $81.72° + 180° = 261.72°$, so that

$$\theta_s = 40.9° \text{ and } 131°$$

Notice that this orientation is 45° from that shown in Fig. 14–31b.

Maximum In-Plane Shear Strain.   Applying Eq. 14–25 gives

$$\frac{\gamma_{\substack{max \\ \text{in-plane}}}}{2} = \sqrt{\left(\frac{\epsilon_x - \epsilon_y}{2}\right)^2 + \left(\frac{\gamma_{xy}}{2}\right)^2}$$

$$= \left[\sqrt{\left(\frac{-350 - 200}{2}\right)^2 + \left(\frac{80}{2}\right)^2}\right](10^{-6})$$

$$\gamma_{\substack{max \\ \text{in-plane}}} = 556(10^{-6}) \qquad\qquad Ans.$$

The square root gives two signs for $\gamma_{\substack{max \\ \text{in-plane}}}$. The proper one for each angle can be obtained by applying Eq. 14–20. When $\theta_s = 40.9°$, we have

$$\frac{\gamma_{x'y'}}{2} = -\frac{\epsilon_x - \epsilon_y}{2}\sin 2\theta + \frac{\gamma_{xy}}{2}\cos 2\theta$$

$$= -\left(\frac{-350 - 200}{2}\right)(10^{-6})\sin 2(40.9°) + \frac{80(10^{-6})}{2}\cos 2(40.9°)$$

$$\gamma_{x'y'} = 556(10^{-6})$$

This result is positive and so $\gamma_{\substack{max \\ \text{in-plane}}}$ tends to distort the element so that the right angle between dx' and dy' is *decreased* (positive sign convention), Fig. 14–31b.

Also, there are associated average normal strains imposed on the element that are determined from Eq. 14–26.

$$\epsilon_{avg} = \frac{\epsilon_x + \epsilon_y}{2} = \frac{-350 + 200}{2}(10^{-6}) = -75(10^{-6})$$

These strains tend to cause the element to contract, Fig. 14–31b.

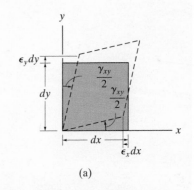

(a)

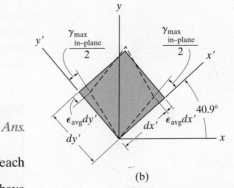

(b)

Fig. 14–31

14

*14.8 MOHR'S CIRCLE—PLANE STRAIN

Since the equations of plane-strain transformation are mathematically similar to the equations of plane-stress transformation, we can also solve problems involving the transformation of strain using Mohr's circle.

Like the case for stress, the parameter θ in Eqs. 14–19 and 14–20 can be eliminated and the result rewritten in the form

$$(\epsilon_{x'} - \epsilon_{avg})^2 + \left(\frac{\gamma_{x'y'}}{2}\right)^2 = R^2 \qquad (14\text{–}27)$$

where

$$\epsilon_{avg} = \frac{\epsilon_x + \epsilon_y}{2}$$

$$R = \sqrt{\left(\frac{\epsilon_x - \epsilon_y}{2}\right)^2 + \left(\frac{\gamma_{xy}}{2}\right)^2}$$

Equation 14–27 represents the equation of Mohr's circle for strain. It has a center on the ϵ axis at point $C(\epsilon_{avg}, 0)$ and a radius R. As described in the following procedure, Mohr's circle can be used to determine the principal strains, the maximum in-plane strain, or the strains on an arbitrary plane.

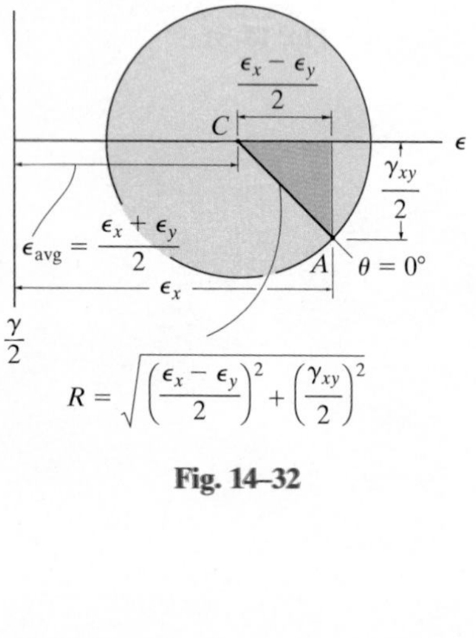

Fig. 14–32

PROCEDURE FOR ANALYSIS

The procedure for drawing Mohr's circle for strain follows the same one established for stress.

Construction of the Circle.

- Establish a coordinate system such that the horizontal axis represents the normal strain ϵ, with *positive to the right*, and the vertical axis represents *half* the value of the shear strain, $\gamma/2$, with *positive downward*, Fig. 14–32.
- Using the positive sign convention for ϵ_x, ϵ_y, γ_{xy}, Fig. 14–26, determine the center of the circle C, located $\epsilon_{avg} = (\epsilon_x + \epsilon_y)/2$ from the origin, Fig. 14–32.
- Plot the reference point A having coordinates $A(\epsilon_x, \gamma_{xy}/2)$. This point represents the case when the x' axis coincides with the x axis. Hence $\theta = 0°$, Fig. 14–32.
- Connect point A with C and from the shaded triangle determine the radius R of the circle, Fig. 14–32.
- Once R has been determined, sketch the circle.

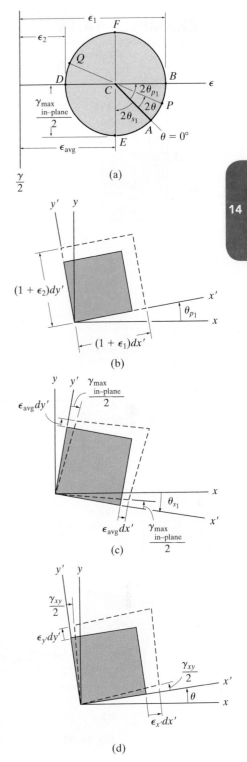

Principal Strains.

- The principal strains ϵ_1 and ϵ_2 are determined from the circle as the coordinates of points B and D, that is, where $\gamma/2 = 0$, Fig. 14–33a.

- The orientation of the plane on which ϵ_1 acts can be determined from the circle by calculating $2\theta_{p_1}$ using trigonometry. Here this angle happens to be counterclockwise, measured *from CA to CB*, Fig. 14–33a. Remember that the *rotation* of θ_{p_1} must be in this *same direction*, from the element's reference axis x to the x' axis, Fig. 14–33b.*

- When ϵ_1 and ϵ_2 are positive as in Fig. 14–33a, the element in Fig. 14–33b will elongate in the x' and y' directions as shown by the dashed outline.

Maximum In-Plane Shear Strain.

- The average normal strain and half the maximum in-plane shear strain are determined from the circle as the coordinates of point E or F, Fig. 14–33a.

- The orientation of the plane on which $\gamma_{\substack{max \\ in\text{-}plane}}$ and ϵ_{avg} act can be determined from the circle in Fig. 14–33a, by calculating $2\theta_{s_1}$ using trigonometry. Here this angle happens to be clockwise *from CA to CE*. Remember that the *rotation* of θ_{s_1} must be in this *same direction*, from the element's reference axis x to the x' axis, Fig. 14–33c.*

Strains on Arbitrary Plane.

- The normal and shear strain components $\epsilon_{x'}$ and $\gamma_{x'y'}$ for an element oriented at an angle θ, Fig. 14–33d, can be obtained from the circle using trigonometry to determine the coordinates of point P, Fig. 14–33a.

- To locate P, the known counterclockwise angle θ of the x' axis, Fig. 14–33d, is measured counterclockwise on the circle as 2θ. This measurement is made *from CA to CP*.

- If the value of $\epsilon_{y'}$ is required, it can be determined by calculating the ϵ coordinate of point Q in Fig. 14–33a. The line CQ lies $180°$ away from CP and thus represents a $90°$ rotation of the x' axis.

---

*If the $\gamma/2$ axis were constructed *positive upwards*, then the angle 2θ on the circle would be measured in the *opposite direction* to the orientation θ of the plane.

Fig. 14–33

EXAMPLE 14.15

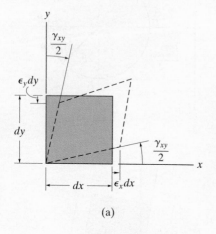

(a)

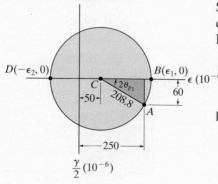

(b)

(c)

Fig. 14–34

The state of plane strain at a point has components of $\epsilon_x = 250(10^{-6})$, $\epsilon_y = -150(10^{-6})$, $\gamma_{xy} = 120(10^{-6})$, Fig. 14–34a. Determine the principal strains and the orientation of the element upon which they act.

SOLUTION

Construction of the Circle. The ϵ and $\gamma/2$ axes are established in Fig. 14–34b. Remember that the *positive* $\gamma/2$ axis must be directed *downward* so that *counterclockwise* rotations of the element correspond to *counterclockwise* rotation around the circle, and vice versa. The center of the circle C is located at

$$\epsilon_{avg} = \frac{250 + (-150)}{2} (10^{-6}) = 50(10^{-6})$$

Since $\gamma_{xy}/2 = 60(10^{-6})$, the reference point A ($\theta = 0°$) has coordinates $A(250(10^{-6}), 60(10^{-6}))$. From the shaded triangle in Fig. 14–34b, the radius of the circle is

$$R = \left[\sqrt{(250 - 50)^2 + (60)^2} \right](10^{-6}) = 208.8(10^{-6})$$

Principal Strains. The ϵ coordinates of points B and D are therefore

$$\epsilon_1 = (50 + 208.8)(10^{-6}) = 259(10^{-6}) \qquad Ans.$$
$$\epsilon_2 = (50 - 208.8)(10^{-6}) = -159(10^{-6}) \qquad Ans.$$

The direction of the positive principal strain ϵ_1 in Fig. 14–34b is defined by the *counterclockwise* angle $2\theta_{p_1}$, measured from CA ($\theta = 0°$) to CB. We have

$$\tan 2\theta_{p_1} = \frac{60}{(250 - 50)}$$

$$\theta_{p_1} = 8.35° \qquad Ans.$$

Hence, the side dx' of the element is inclined *counterclockwise* 8.35° as shown in Fig. 14–34c. This also defines the direction of ϵ_1. The deformation of the element is also shown in the figure.

EXAMPLE 14.16

The state of plane strain at a point has components of $\epsilon_x = 250(10^{-6})$, $\epsilon_y = -150(10^{-6})$, $\gamma_{xy} = 120(10^{-6})$, Fig. 14–35a. Determine the maximum in-plane shear strains and the orientation of the element upon which they act.

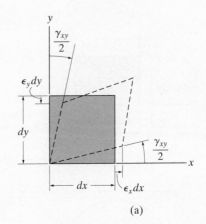

(a)

SOLUTION
The circle has been established in the previous example and is shown in Fig. 14–35b.

Maximum In-Plane Shear Strain. Half the maximum in-plane shear strain and average normal strain are represented by the coordinates of point E or F on the circle. From the coordinates of point E,

$$\frac{(\gamma_{x'y'})^{max}_{in\text{-}plane}}{2} = 208.8(10^{-6})$$

$$(\gamma_{x'y'})^{max}_{in\text{-}plane} = 418(10^{-6}) \qquad Ans.$$

$$\epsilon_{avg} = 50(10^{-6})$$

To orient the element, we will determine the clockwise angle $2\theta_{s_1}$, measured from CA ($\theta = 0°$) to CE.

$$2\theta_{s_1} = 90° - 2(8.35°)$$

$$\theta_{s_1} = 36.7° \qquad Ans.$$

This angle is shown in Fig. 14–35c. Since the shear strain defined from point E on the circle has a positive value and the average normal strain is also positive, these strains deform the element into the dashed shape shown in the figure.

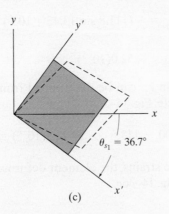

(c)

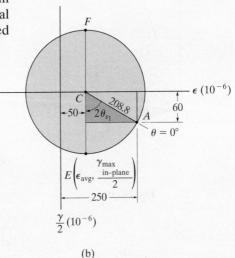

(b)

Fig. 14–35

EXAMPLE 14.17

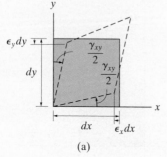

(a)

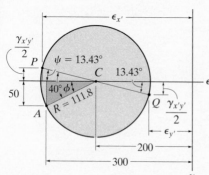

(b)

(c)

Fig. 14–36

The state of plane strain at a point has components of $\epsilon_x = -300(10^{-6})$, $\epsilon_y = -100(10^{-6})$, $\gamma_{xy} = 100(10^{-6})$, Fig. 14–36a. Determine the state of strain on an element oriented 20° clockwise from this position.

SOLUTION

Construction of the Circle.   The ϵ and $\gamma/2$ axes are established in Fig. 14–36b. The center of the circle is at

$$\epsilon_{avg} = \left(\frac{-300 - 100}{2} \right)(10^{-6}) = -200(10^{-6})$$

The reference point A has coordinates $A(-300(10^{-6}), 50(10^{-6}))$, and so the radius CA, determined from the shaded triangle, is

$$R = \left[\sqrt{(300 - 200)^2 + (50)^2} \right](10^{-6}) = 111.8(10^{-6})$$

Strains on Inclined Element.   Since the element is to be oriented 20° *clockwise*, we must consider the radial line CP, $2(20°) = 40°$ *clockwise*, measured from CA ($\theta = 0°$), Fig. 14–36a. The coordinates of point P are obtained from the geometry of the circle. Note that

$$\phi = \tan^{-1}\left(\frac{50}{(300 - 200)} \right) = 26.57°, \qquad \psi = 40° - 26.57° = 13.43°$$

Thus,

$$\epsilon_{x'} = -(200 + 111.8 \cos 13.43°)(10^{-6})$$

$$= -309(10^{-6}) \qquad\qquad Ans.$$

$$\frac{\gamma_{x'y'}}{2} = -(111.8 \sin 13.43°)(10^{-6})$$

$$\gamma_{x'y'} = -52.0(10^{-6}) \qquad\qquad Ans.$$

The normal strain $\epsilon_{y'}$ can be determined from the ϵ coordinate of point Q on the circle, Fig. 14–36b.

$$\epsilon_{y'} = -(200 - 111.8 \cos 13.43°)(10^{-6}) = -91.3(10^{-6}) \qquad Ans.$$

As a result of these strains, the element deforms relative to the x', y' axes as shown in Fig. 14–36c.

PROBLEMS

*14–84. Prove that the sum of the normal strains in perpendicular directions is constant, i.e., $\epsilon_x + \epsilon_y = \epsilon_{x'} + \epsilon_{y'}$.

14–85. The state of strain at the point on the arm has components of $\epsilon_x = 200(10^{-6})$, $\epsilon_y = -300(10^{-6})$, and $\gamma_{xy} = 400(10^{-6})$. Use the strain transformation equations to determine the equivalent in-plane strains on an element oriented at an angle of 30° counterclockwise from the original position. Sketch the deformed element due to these strains within the x–y plane.

*14–88. The state of strain at the point on the leaf of the caster assembly has components of $\epsilon_x = -400(10^{-6})$, $\epsilon_y = 860(10^{-6})$, and $\gamma_{xy} = 375(10^{-6})$. Use the strain transformation equations to determine the equivalent in-plane strains on an element oriented at an angle of $\theta = 30°$ counterclockwise from the original position. Sketch the deformed element due to these strains within the x–y plane.

Prob. 14–88

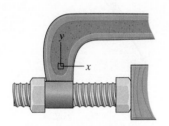

Prob. 14–85

14–89. The state of strain at a point on the bracket has component: $\epsilon_x = 150(10^{-6})$, $\epsilon_y = 200(10^{-6})$, $\gamma_{xy} = -700(10^{-6})$. Use the strain transformation equations and determine the equivalent in plane strains on an element oriented at an angle of $\theta = 60°$ counterclockwise from the original position. Sketch the deformed element within the x–y plane due to these strains.

14–86. The state of strain at the point on the pin leaf has components of $\epsilon_x = 200(10^{-6})$, $\epsilon_y = 180(10^{-6})$, and $\gamma_{xy} = -300(10^{-6})$. Use the strain transformation equations and determine the equivalent in-plane strains on an element oriented at an angle of $\theta = 60°$ counterclockwise from the original position. Sketch the deformed element due to these strains within the x–y plane.

14–90. Solve Prob. 14–89 for an element oriented $\theta = 30°$ clockwise.

14–87. Solve Prob. 14–86 for an element oriented $\theta = 30°$ clockwise.

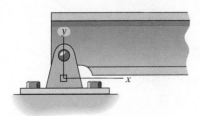

Probs. 14–86/87

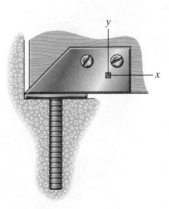

Probs. 14–89/90

14–91.   The state of strain at the point on the spanner wrench has components of $\epsilon_x = 260(10^{-6})$, $\epsilon_y = 320(10^{-6})$, and $\gamma_{xy} = 180(10^{-6})$. Use the strain transformation equations to determine (a) the in-plane principal strains and (b) the maximum in-plane shear strain and average normal strain. In each case specify the orientation of the element and show how the strains deform the element within the x–y plane.

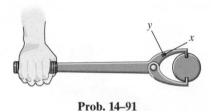

Prob. 14–91

14–93.   The state of strain at the point on the support has components of $\epsilon_x = 350(10^{-6})$, $\epsilon_y = 400(10^{-6})$, $\gamma_{xy} = -675(10^{-6})$. Use the strain transformation equations to determine (a) the in-plane principal strains and (b) the maximum in-plane shear strain and average normal strain. In each case specify the orientation of the element and show how the strains deform the element within the x–y plane.

Prob. 14–93

*14–92.   The state of strain at the point on the member has components of $\epsilon_x = 180(10^{-6})$, $\epsilon_y = -120(10^{-6})$, and $\gamma_{xy} = -100(10^{-6})$. Use the strain transformation equations to determine (a) the in-plane principal strains and (b) the maximum in-plane shear strain and average normal strain. In each case specify the orientation of the element and show how the strains deform the element within the x–y plane.

14–94.   Due to the load **P**, the state of strain at the point on the bracket has components of $\epsilon_x = 500(10^{-6})$, $\epsilon_y = 350(10^{-6})$, and $\gamma_{xy} = -430(10^{-6})$. Use the strain transformation equations to determine the equivalent in-plane strains on an element oriented at an angle of $\theta = 30°$ clockwise from the original position. Sketch the deformed element due to these strains within the x–y plane.

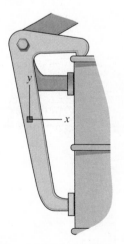

Prob. 14–92

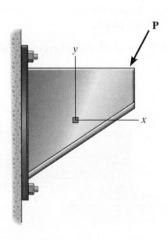

Prob. 14–94

14–95. The state of strain on an element has components $\epsilon_x = -400(10^{-6})$, $\epsilon_y = 0$, $\gamma_{xy} = 150(10^{-6})$. Determine the equivalent state of strain on an element at the same point oriented 30° clockwise with respect to the original element. Sketch the results on this element.

*14–96. The state of plane strain on the element is $\epsilon_x = -300(10^{-6})$, $\epsilon_y = 0$, and $\gamma_{xy} = 150(10^{-6})$. Determine the equivalent state of strain which represents (a) the principal strains, and (b) the maximum in-plane shear strain and the associated average normal strain. Specify the orientation of the corresponding elements for these states of strain with respect to the original element.

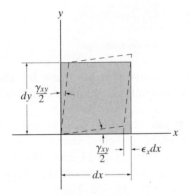

Probs. 14–95/96

14–97. The state of strain at the point on a boom of a shop crane has components of $\epsilon_x = 250(10^{-6})$, $\epsilon_y = 300(10^{-6})$, $\gamma_{xy} = -180(10^{-6})$. Use the strain transformation equations to determine (a) the in-plane principal strains and (b) the maximum in-plane shear strain and average normal strain. In each case, specify the orientation of the element and show how the strains deform the element within the x–y plane.

Prob. 14–97

14–98. Consider the general case of plane strain where ϵ_x, ϵ_y, and γ_{xy} are known. Write a computer program that can be used to determine the normal and shear strain, $\epsilon_{x'}$ and $\gamma_{x'y'}$, on the plane of an element oriented θ from the horizontal. Also, include the principal strains and the element's orientation, and the maximum in-plane shear strain, the average normal strain, and the element's orientation.

14–99. The state of strain on the element has components $\epsilon_x = -300(10^{-6})$, $\epsilon_y = 100(10^{-6})$, $\gamma_{xy} = 150(10^{-6})$. Determine the equivalent state of strain, which represents (a) the principal strains, and (b) the maximum in-plane shear strain and the associated average normal strain. Specify the orientation of the corresponding elements for these states of strain with respect to the original element.

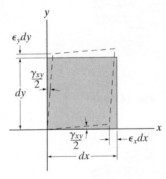

Prob. 14–99

*14–100. Solve Prob. 14–86 using Mohr's circle.

14–101. Solve Prob. 14–87 using Mohr's circle.

14–102. Solve Prob. 14–88 using Mohr's circle.

14–103. Solve Prob. 14–91 using Mohr's circle.

*14–104. Solve Prob. 14–90 using Mohr's circle.

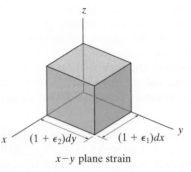

$x-y$ plane strain

(a)

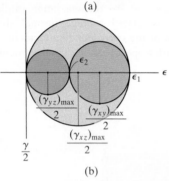

(b)

Fig. 14–37

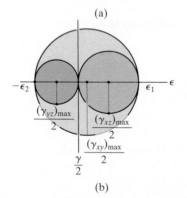

$x-y$ plane strain

(a)

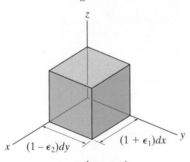

(b)

Fig. 14–38

*14.9   ABSOLUTE MAXIMUM SHEAR STRAIN

In Sec. 14.5 it was pointed out that in the case of plane stress, the absolute maximum shear stress in an element of material will occur *out of the plane* when the principal stresses have the *same sign*, i.e., both are tensile or both are compressive. A similar result occurs for plane strain. For example, if the principal in-plane strains cause elongations, Fig. 14–37a, then the three Mohr's circles describing the normal and shear strain components for the element rotations about the x, y, and z axes are shown in Fig. 14–37b. By inspection, the largest circle has a radius $R = (\gamma_{xz})_{max}/2$, and so

$$\gamma_{\substack{abs \\ max}} = (\gamma_{xz})_{max} = \epsilon_1 \qquad (14\text{–}28)$$

ϵ_1 and ϵ_2 have the same sign

This value gives the *absolute maximum shear strain* for the material. Note that it is *larger* than the maximum in-plane shear strain, which is $(\gamma_{xy})_{max} = \epsilon_1 - \epsilon_2$.

Now consider the case where one of the in-plane principal strains is of *opposite sign* to the other in-plane principal strain, so that ϵ_1 causes elongation and ϵ_2 causes contraction, Fig. 14–38a. The three Mohr's circles, which describe the strain components on the element rotated about the x, y, z axes, are shown in Fig. 14–38b. Here

$$\gamma_{\substack{abs \\ max}} = (\gamma_{xy})_{\substack{max \\ in\text{-}plane}} = \epsilon_1 - \epsilon_2 \qquad (14\text{–}29)$$

ϵ_1 and ϵ_2 have opposite signs

▶ IMPORTANT POINTS

- If the in-plane principal strains both have the same sign, the absolute maximum shear strain will occur out of plane and has a value of $\gamma_{\substack{abs \\ max}} = \epsilon_{max}$. This value is greater than the maximum in-plane shear strain.

- If the in-plane principal strains are of opposite signs, then the absolute maximum shear strain equals the maximum in-plane shear strain, $\gamma_{\substack{abs \\ max}} = \epsilon_1 - \epsilon_2$.

EXAMPLE 14.18

The state of plane strain at a point has strain components of $\epsilon_x = -400(10^{-6})$, $\epsilon_y = 200(10^{-6})$, and $\gamma_{xy} = 150(10^{-6})$, Fig. 14–39a. Determine the maximum in-plane shear strain and the absolute maximum shear strain.

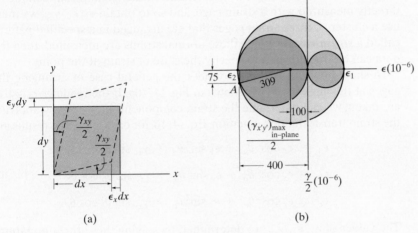

(a)   (b)

Fig. 14–39

SOLUTION

Maximum In-Plane Shear Strain.   We will solve this problem using Mohr's circle. The center of the circle is at

$$\epsilon_{avg} = \frac{-400 + 200}{2}(10^{-6}) = -100(10^{-6})$$

Since $\gamma_{xy}/2 = 75(10^{-6})$, the reference point A has coordinates $(-400(10^{-6}), 75(10^{-6}))$, Fig. 14–39b. The radius of the circle is therefore

$$R = \left[\sqrt{(400 - 100)^2 + (75)^2}\right](10^{-6}) = 309(10^{-6})$$

From the circle, the in-plane principal strains are

$$\epsilon_1 = (-100 + 309)(10^{-6}) = 209(10^{-6})$$
$$\epsilon_2 = (-100 - 309)(10^{-6}) = -409(10^{-6})$$

Also, the maximum in-plane shear strain is

$$\gamma_{\substack{max \\ \text{in-plane}}} = \epsilon_1 - \epsilon_2 = [209 - (-409)](10^{-6}) = 618(10^{-6}) \qquad Ans.$$

Absolute Maximum Shear Strain.   Since the *principal in-plane strains have opposite signs*, the maximum in-plane shear strain is *also* the absolute maximum shear strain; i.e.,

$$\gamma_{\substack{abs \\ max}} = 618(10^{-6}) \qquad Ans.$$

The three Mohr's circles, plotted for element orientations about each of the x, y, z axes, are also shown in Fig. 14–39b.

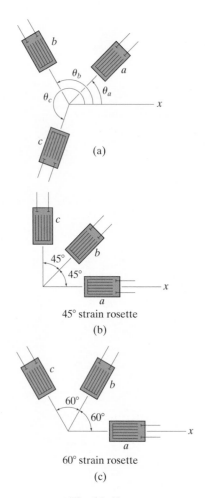

45° strain rosette

(b)

60° strain rosette

(c)

Fig. 14–40

Typical electrical resistance 45° strain rosette.

14.10 STRAIN ROSETTES

The normal strain on the free surface of a body can be measured in a particular direction using an electrical resistance strain gage. For example, in Sec. 8.1 we showed how this type of gage is used to find the axial strain in a specimen when performing a tension test. When the body is subjected to several loads, however, then the strains $\epsilon_x, \epsilon_y, \gamma_{xy}$ at a point on its surface may have to be determined. Unfortunately, the shear strain cannot be directly measured with a strain gage, and so to obtain $\epsilon_x, \epsilon_y, \gamma_{xy}$, we must use a cluster of three strain gages that are arranged in a specified pattern called a ***strain rosette***. Once these normal strains are measured, then the data can be transformed to specify the state of strain at the point.

To show how this is done, consider the general case of arranging the gages at the angles $\theta_a, \theta_b, \theta_c$ shown in Fig. 14–40a. If the readings $\epsilon_a, \epsilon_b, \epsilon_c$ are taken, we can determine the strain components $\epsilon_x, \epsilon_y, \gamma_{xy}$ by applying the strain transformation equation, Eq. 14–16, for each gage. The results are

$$\epsilon_a = \epsilon_x \cos^2 \theta_a + \epsilon_y \sin^2 \theta_a + \gamma_{xy} \sin \theta_a \cos \theta_a$$

$$\epsilon_b = \epsilon_x \cos^2 \theta_b + \epsilon_y \sin^2 \theta_b + \gamma_{xy} \sin \theta_b \cos \theta_b \qquad (14\text{–}30)$$

$$\epsilon_c = \epsilon_x \cos^2 \theta_c + \epsilon_y \sin^2 \theta_c + \gamma_{xy} \sin \theta_c \cos \theta_c$$

The values of $\epsilon_x, \epsilon_y, \gamma_{xy}$ are determined by solving these three equations simultaneously.

Normally, strain rosettes are arranged in 45° or 60° patterns. In the case of the 45° or "rectangular" strain rosette, Fig. 14–40b, $\theta_a = 0°, \theta_b = 45°$, $\theta_c = 90°$, so that Eq. 14–30 gives

$$\epsilon_x = \epsilon_a$$

$$\epsilon_y = \epsilon_c$$

$$\gamma_{xy} = 2\epsilon_b - (\epsilon_a + \epsilon_c)$$

And for the 60° strain rosette, Fig. 14–40c, $\theta_a = 0°, \theta_b = 60°, \theta_c = 120°$. Here Eq. 14–30 gives

$$\epsilon_x = \epsilon_a$$

$$\epsilon_y = \frac{1}{3}(2\epsilon_b + 2\epsilon_c - \epsilon_a) \qquad (14\text{–}31)$$

$$\gamma_{xy} = \frac{2}{\sqrt{3}}(\epsilon_b - \epsilon_c)$$

Once $\epsilon_x, \epsilon_y, \gamma_{xy}$ are determined, then the strain transformation equations or Mohr's circle can be used to determine the principal in-plane strains ϵ_1 and ϵ_2, or the maximum in-plane shear strain $\gamma_{\text{in-plane}}^{\max}$. The stress in the material that causes these strains can then be determined using Hooke's law, which is discussed in the next section.

EXAMPLE   14.19

The state of strain at point A on the bracket in Fig. 14–41a is measured using the strain rosette shown in Fig. 14–41b. The readings from the gages give $\epsilon_a = 60(10^{-6})$, $\epsilon_b = 135(10^{-6})$, and $\epsilon_c = 264(10^{-6})$. Determine the in-plane principal strains at the point and the directions in which they act.

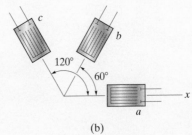

(a)

SOLUTION
We will use Eqs. 14–30 for the solution. Establishing an x axis, Fig. 14–41b, and measuring the angles counterclockwise from this axis to the centerlines of each gage, we have $\theta_a = 0°$, $\theta_b = 60°$, and $\theta_c = 120°$. Substituting these results, along with the problem data, into the equations gives

$$60(10^{-6}) = \epsilon_x \cos^2 0° + \epsilon_y \sin^2 0° + \gamma_{xy} \sin 0° \cos 0°$$
$$= \epsilon_x \qquad\qquad (1)$$
$$135(10^{-6}) = \epsilon_x \cos^2 60° + \epsilon_y \sin^2 60° + \gamma_{xy} \sin 60° \cos 60°$$
$$= 0.25\epsilon_x + 0.75\epsilon_y + 0.433\gamma_{xy} \qquad (2)$$
$$264(10^{-6}) = \epsilon_x \cos^2 120° + \epsilon_y \sin^2 120° + \gamma_{xy} \sin 120° \cos 120°$$
$$= 0.25\epsilon_x + 0.75\epsilon_y - 0.433\gamma_{xy} \qquad (3)$$

Using Eq. 1 and solving Eqs. 2 and 3 simultaneously, we get

$$\epsilon_x = 60(10^{-6}) \quad \epsilon_y = 246(10^{-6}) \quad \gamma_{xy} = -149(10^{-6})$$

These same results can also be obtained in a more direct manner from Eq. 14–31.

The in-plane principal strains will be determined using Mohr's circle. The center, C, is at $\epsilon_{avg} = 153(10^{-6})$, and the reference point on the circle is at $A[60(10^{-6}), -74.5(10^{-6})]$, Fig. 14–41$c$. From the shaded triangle, the radius is

$$R = \left[\sqrt{(153 - 60)^2 + (74.5)^2} \right](10^{-6}) = 119.1(10^{-6})$$

The in-plane principal strains are therefore

$$\epsilon_1 = 153(10^{-6}) + 119.1(10^{-6}) = 272(10^{-6}) \qquad Ans.$$
$$\epsilon_2 = 153(10^{-6}) - 119.1(10^{-6}) = 33.9(10^{-6}) \qquad Ans.$$
$$2\theta_{p_2} = \tan^{-1} \frac{74.5}{(153 - 60)} = 38.7°$$
$$\theta_{p_2} = 19.3° \qquad\qquad Ans.$$

NOTE: The deformed element is shown in the dashed position in Fig. 14–41d. Realize that, due to the Poisson effect, the element is *also* subjected to an out-of-plane strain, i.e., in the z direction, although this value will not influence the calculated results.

(b)

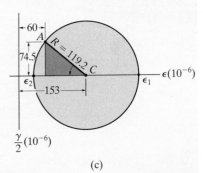

(c)

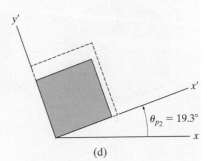

(d)

Fig. 14–41

14.11 MATERIAL PROPERTY RELATIONSHIPS

In this section we will present some important material property relationships that are used when the material is subjected to multiaxial stress and strain. In all cases, we will assume that the material is homogeneous and isotropic, and behaves in a linear elastic manner.

Generalized Hooke's Law. When the material at a point is subjected to a state of ***triaxial stress***, σ_x, σ_y, σ_z, Fig. 14–42a, then these stresses can be related to the normal strains ϵ_x, ϵ_y, ϵ_z by using the principle of superposition, Poisson's ratio, $\epsilon_{lat} = -\nu\epsilon_{long}$, and Hooke's law as it applies in the uniaxial direction, $\epsilon = \sigma/E$. For example, consider the normal strain of the element in the x direction, caused by separate application of each normal stress. When σ_x is applied, Fig. 14–42b, the element elongates with a strain ϵ_x', where

$$\epsilon_x' = \frac{\sigma_x}{E}$$

Application of σ_y causes the element to contract with a strain ϵ_x'', Fig. 14–42c. Here

$$\epsilon_x'' = -\nu\frac{\sigma_y}{E}$$

Finally, application of σ_z, Fig. 14–42d, causes a contraction strain ϵ_x''', so that

$$\epsilon_x''' = -\nu\frac{\sigma_z}{E}$$

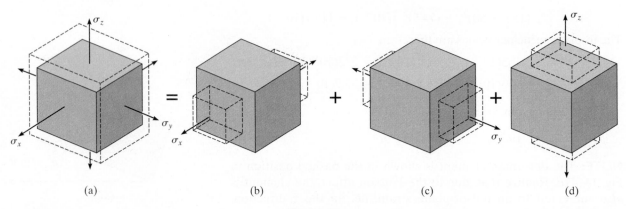

(a) = (b) + (c) + (d)

Fig. 14–42

We can obtain the resultant strain ϵ_x by adding these three strains algebraically. Similar equations can be developed for the normal strains in the y and z directions, and so the final results can be written as

$$
\begin{aligned}
\epsilon_x &= \frac{1}{E}[\sigma_x - \nu(\sigma_y + \sigma_z)] \\[4pt]
\epsilon_y &= \frac{1}{E}[\sigma_y - \nu(\sigma_x + \sigma_z)] \\[4pt]
\epsilon_z &= \frac{1}{E}[\sigma_z - \nu(\sigma_x + \sigma_y)]
\end{aligned}
\qquad (14\text{–}32)
$$

These three equations represent the general form of Hooke's law for a triaxial state of stress. For application, tensile stress is considered a positive quantity, and a compressive stress is negative. If a resulting normal strain is *positive*, it indicates that the material *elongates*, whereas a *negative* normal strain indicates the material *contracts*.

If we only apply a shear stress τ_{xy} to the element, Fig. 14–43a, experimental observations indicate that the material will change its shape, but it will not change its volume. In other words, τ_{xy} will only cause the shear strain γ_{xy} in the material. Likewise, τ_{yz} and τ_{xz} will only cause shear strains γ_{yz} and γ_{xz}, Figs. 14–43b and 14–43c. Therefore, Hooke's law for shear stress and shear strain becomes

$$
\gamma_{xy} = \frac{1}{G}\tau_{xy} \qquad \gamma_{yz} = \frac{1}{G}\tau_{yz} \qquad \gamma_{xz} = \frac{1}{G}\tau_{xz} \qquad (14\text{–}33)
$$

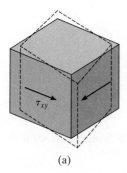 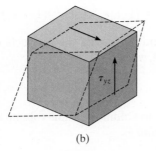 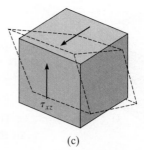

(a)   (b)   (c)

Fig. 14–43

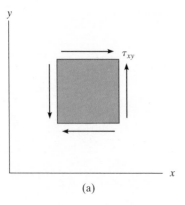

(a)

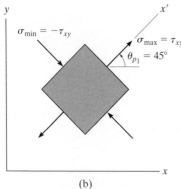

(b)

Fig. 14–44

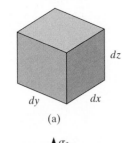

(a)

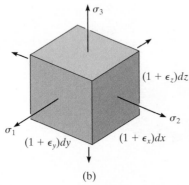

(b)

Fig. 14–45

Relationship Involving E, ν, and G. In Sec. 8.6 it was stated that the modulus of elasticity E is related to the shear modulus G by Eq. 8–11, namely,

$$G = \frac{E}{2(1 + \nu)} \qquad (14\text{–}34)$$

One way to derive this relationship is to consider an element of the material to be subjected only to shear, Fig. 14–44a. Applying Eq. 14–5 (see Example 14.5) the principal stresses at the point are $\sigma_{max} = \tau_{xy}$ and $\sigma_{min} = -\tau_{xy}$, where this element must be oriented $\theta_{p_1} = 45°$ counterclockwise from the x axis, as shown in Fig. 14–44b. If the three principal stresses $\sigma_{max} = \tau_{xy}$, $\sigma_{int} = 0$, and $\sigma_{min} = -\tau_{xy}$ are then substituted into the first of Eqs. 14–32, the principal strain ϵ_{max} can be related to the shear stress τ_{xy}. The result is

$$\epsilon_{max} = \frac{\tau_{xy}}{E}(1 + \nu) \qquad (14\text{–}35)$$

This strain, which deforms the element along the x' axis, can also be related to the shear strain γ_{xy}. From Fig. 14–44a, $\sigma_x = \sigma_y = \sigma_z = 0$. Substituting these results into the first and second Eqs. 14–32 gives $\epsilon_x = \epsilon_y = 0$. Now apply the strain transformation Eq. 14–23, which gives

$$\epsilon_1 = \epsilon_{max} = \frac{\gamma_{xy}}{2}$$

By Hooke's law, $\gamma_{xy} = \tau_{xy}/G$, so that $\epsilon_{max} = \tau_{xy}/2G$. Finally, substituting this into Eq. 14–35 and rearranging the terms gives our result, namely, Eq. 14–34.

Dilatation. When an elastic material is subjected to normal stress, the strains that are produced will cause its volume to change. For example, if the volume element in Fig. 14–45a is subjected to the principal stresses $\sigma_1, \sigma_2, \sigma_3$, Fig. 14–45b, then the lengths of the sides of the element become $(1 + \epsilon_x)\,dx$, $(1 + \epsilon_y)\,dy$, $(1 + \epsilon_z)\,dz$. The change in volume of the element is therefore

$$\delta V = (1 + \epsilon_x)(1 + \epsilon_y)(1 + \epsilon_z)\,dx\,dy\,dz - dx\,dy\,dz$$

Expanding, and neglecting the products of the strains, since the strains are very small, we get

$$\delta V = (\epsilon_x + \epsilon_y + \epsilon_z)\,dx\,dy\,dz$$

The change in volume per unit volume is called the "volumetric strain" or the **dilatation** e.

$$e = \frac{\delta V}{dV} = \epsilon_x + \epsilon_y + \epsilon_z \qquad (14\text{–}36)$$

If we use Hooke's law, Eq. 14–32, we can also express the dilatation in terms of the applied stress. We have

$$e = \frac{1 - 2\nu}{E}(\sigma_1 + \sigma_2 + \sigma_3) \qquad (14\text{--}37)$$

Bulk Modulus. According to Pascal's law, when a volume element of material is subjected to a uniform pressure p caused by a static fluid, the pressure will be the same in all directions. Shear stresses will not be present, since the fluid does not flow around the element. This state of "hydrostatic" loading therefore requires $\sigma_1 = \sigma_2 = \sigma_3 = -p$, Fig. 14–46. Substituting into Eq. 14–37 and rearranging terms yields

$$\frac{p}{e} = -\frac{E}{3(1 - 2\nu)} \qquad (14\text{--}38)$$

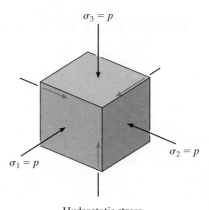

Hydrostatic stress

Fig. 14–46

The term on the right is called the **volume modulus of elasticity** or the **bulk modulus**, since this ratio, p/e, is *similar* to the ratio of one-dimensional linear elastic stress to strain, which defines E, i.e., $\sigma/\epsilon = E$. The bulk modulus has the same units as stress and is symbolized by the letter k, so that

$$k = \frac{E}{3(1 - 2\nu)} \qquad (14\text{--}39)$$

For most metals $\nu \approx \frac{1}{3}$ and so $k \approx E$. However, if we assume the material did not change its volume when loaded, then $\delta V = e = 0$, and k would be infinite. As a result, Eq. 14–39 would then indicate the theoretical *maximum* value for Poisson's ratio to be $\nu = 0.5$.

> ## IMPORTANT POINTS
>
> - When a homogeneous isotropic material is subjected to a state of triaxial stress, the strain in each direction is influenced by the strains produced by *all* the stresses. This is the result of the Poisson effect, and the stress is then related to the strain in the form of a generalized Hooke's law.
> - When a shear stress is applied to homogeneous isotropic material, it will only produce shear strain in the same plane.
> - The material constants E, G, and ν are all related by Eq. 14–34.
> - *Dilatation*, or *volumetric strain*, is caused only by normal strain, not shear strain.
> - The *bulk modulus* is a measure of the stiffness of a volume of material. This material property provides an upper limit to Poisson's ratio of $\nu = 0.5$.

EXAMPLE   14.20

The bracket in Example 14.19, Fig. 14–47a, is made of steel for which $E_{st} = 200$ GPa, $\nu_{st} = 0.3$. Determine the principal stresses at point A.

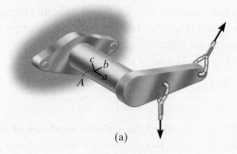

(a)

Fig. 14–47

SOLUTION I
From Example 14.19 the principal strains have been determined as

$$\epsilon_1 = 272(10^{-6})$$

$$\epsilon_2 = 33.9(10^{-6})$$

Since point A is on the *surface* of the bracket, for which there is no loading, the stress on the surface is zero, and so point A is subjected to plane stress (not plane strain). Applying Hooke's law with $\sigma_3 = 0$, we have

$$\epsilon_1 = \frac{\sigma_1}{E} - \frac{\nu}{E}\sigma_2 ; \quad 272(10^{-6}) = \frac{\sigma_1}{200(10^9)} - \frac{0.3}{200(10^9)}\sigma_2$$

$$54.4(10^6) = \sigma_1 - 0.3\sigma_2 \tag{1}$$

$$\epsilon_2 = \frac{\sigma_2}{E} - \frac{\nu}{E}\sigma_1 ; \quad 33.9(10^{-6}) = \frac{\sigma_2}{200(10^9)} - \frac{0.3}{200(10^9)}\sigma_1$$

$$6.78(10^6) = \sigma_2 - 0.3\sigma_1 \tag{2}$$

Solving Eqs. 1 and 2 simultaneously yields

$$\sigma_1 = 62.0 \text{ MPa} \qquad\qquad Ans.$$

$$\sigma_2 = 25.4 \text{ MPa} \qquad\qquad Ans.$$

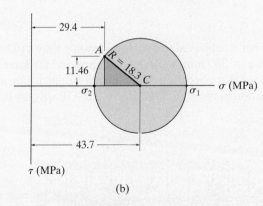

Fig. 14–47 (cont.)

SOLUTION II

It is also possible to solve this problem using the given state of strain as specified in Example 14.19.

$$\epsilon_x = 60(10^{-6}) \qquad \epsilon_y = 246(10^{-6}) \qquad \gamma_{xy} = -149(10^{-6})$$

Applying Hooke's law in the x–y plane, we have

$$\epsilon_x = \frac{\sigma_x}{E} - \frac{\nu}{E}\sigma_y; \quad 60(10^{-6}) = \frac{\sigma_x}{200(10^9)\ \text{Pa}} - \frac{0.3\sigma_y}{200(10^9)\ \text{Pa}}$$

$$\epsilon_y = \frac{\sigma_y}{E} - \frac{\nu}{E}\sigma_x; \quad 246(10^{-6}) = \frac{\sigma_y}{200(10^9)\ \text{Pa}} - \frac{0.3\sigma_x}{200(10^9)\ \text{Pa}}$$

$$\sigma_x = 29.4\ \text{MPa} \qquad \sigma_y = 58.0\ \text{MPa}$$

The shear stress is determined using Hooke's law for shear. First, however, we must calculate G.

$$G = \frac{E}{2(1 + \nu)} = \frac{200\ \text{GPa}}{2(1 + 0.3)} = 76.9\ \text{GPa}$$

Thus,

$$\tau_{xy} = G\gamma_{xy}; \quad \tau_{xy} = 76.9(10^9)[-149(10^{-6})] = -11.46\ \text{MPa}$$

The Mohr's circle for this state of plane stress has a center at $\sigma_{\text{avg}} = 43.7\ \text{MPa}$ and a reference point $A(29.4\ \text{MPa}, -11.46\ \text{MPa})$, Fig. 14–47b. The radius is determined from the shaded triangle.

$$R = \sqrt{(43.7 - 29.4)^2 + (11.46)^2} = 18.3\ \text{MPa}$$

Therefore,

$$\sigma_1 = 43.7\ \text{MPa} + 18.3\ \text{MPa} = 62.0\ \text{MPa} \qquad \textit{Ans.}$$

$$\sigma_2 = 43.7\ \text{MPa} - 18.3\ \text{MPa} = 25.4\ \text{MPa} \qquad \textit{Ans.}$$

NOTE: Each of these solutions is valid provided the material is both linear elastic and isotropic, since only then will the directions of the principal stress and strain coincide.

EXAMPLE   14.21

The copper bar is subjected to a uniform loading shown in Fig. 14–48. If it has a length $a = 300$ mm, width $b = 50$ mm, and thickness $t = 20$ mm before the load is applied, determine its new length, width, and thickness after application of the load. Take $E_{\text{cu}} = 120$ GPa, $\nu_{\text{cu}} = 0.34$.

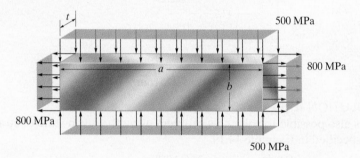

Fig. 14–48

SOLUTION

By inspection, the bar is subjected to a state of plane stress. From the loading we have

$$\sigma_x = 800 \text{ MPa} \qquad \sigma_y = -500 \text{ MPa} \qquad \tau_{xy} = 0, \qquad \sigma_z = 0$$

The associated normal strains are determined from Hooke's law, Eq. 14–32; that is,

$$\epsilon_x = \frac{\sigma_x}{E} - \frac{\nu}{E}(\sigma_y + \sigma_z)$$

$$= \frac{800 \text{ MPa}}{120(10^3) \text{ MPa}} - \frac{0.34}{120(10^3) \text{ MPa}}(-500 \text{ MPa} + 0) = 0.00808$$

$$\epsilon_y = \frac{\sigma_y}{E} - \frac{\nu}{E}(\sigma_x + \sigma_z)$$

$$= \frac{-500 \text{ MPa}}{120(10^3) \text{ MPa}} - \frac{0.34}{120(10^3) \text{ MPa}}(800 \text{ MPa} + 0) = -0.00643$$

$$\epsilon_z = \frac{\sigma_z}{E} - \frac{\nu}{E}(\sigma_x + \sigma_y)$$

$$= 0 - \frac{0.34}{120(10^3) \text{ MPa}}(800 \text{ MPa} - 500 \text{ MPa}) = -0.000850$$

The new bar length, width, and thickness are therefore

$$a' = 300 \text{ mm} + 0.00808(300 \text{ mm}) = 302.4 \text{ mm} \qquad Ans.$$

$$b' = 50 \text{ mm} + (-0.00643)(50 \text{ mm}) = 49.68 \text{ mm} \qquad Ans.$$

$$t' = 20 \text{ mm} + (-0.000850)(20 \text{ mm}) = 19.98 \text{ mm} \qquad Ans.$$

EXAMPLE 14.22

If the rectangular block shown in Fig. 14–49 is subjected to a uniform pressure of $p = 20$ psi, determine the dilatation and the change in length of each side. Take $E = 600$ psi, $\nu = 0.45$.

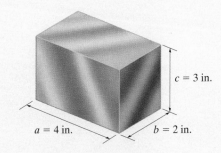

$c = 3$ in.

$a = 4$ in. $b = 2$ in.

Fig. 14–49

SOLUTION

Dilatation.   The dilatation can be determined using Eq. 14–37 with $\sigma_x = \sigma_y = \sigma_z = -20$ psi. We have

$$e = \frac{1 - 2\nu}{E}(\sigma_x + \sigma_y + \sigma_z)$$

$$= \frac{1 - 2(0.45)}{600 \text{ psi}}[3(-20 \text{ psi})]$$

$$= -0.01 \text{ in}^3/\text{in}^3 \qquad\qquad Ans.$$

Change in Length.   The normal strain on each side is determined from Hooke's law, Eq. 14–32; that is,

$$\epsilon = \frac{1}{E}[\sigma_x - \nu(\sigma_y + \sigma_z)]$$

$$= \frac{1}{600 \text{ psi}}[-20 \text{ psi} - (0.45)(-20 \text{ psi} - 20 \text{ psi})] = -0.00333 \text{ in./in.}$$

Thus, the change in length of each side is

$$\delta a = -0.00333(4 \text{ in.}) = -0.0133 \text{ in.} \qquad Ans.$$

$$\delta b = -0.00333(2 \text{ in.}) = -0.00667 \text{ in.} \qquad Ans.$$

$$\delta c = -0.00333(3 \text{ in.}) = -0.0100 \text{ in.} \qquad Ans.$$

The negative signs indicate that each dimension is decreased.

PROBLEMS

14–105. The strain at point A on the bracket has components $\epsilon_x = 300(10^{-6})$, $\epsilon_y = 550(10^{-6})$, $\gamma_{xy} = -650(10^{-6})$, $\epsilon_z = 0$. Determine (a) the principal strains at A in the x–y plane, (b) the maximum shear strain in the x–y plane, and (c) the absolute maximum shear strain.

14–107. The strain at point A on the pressure-vessel wall has components $\epsilon_x = 480(10^{-6})$, $\epsilon_y = 720(10^{-6})$, $\gamma_{xy} = 650(10^{-6})$. Determine (a) the principal strains at A, in the x–y plane, (b) the maximum shear strain in the x–y plane, and (c) the absolute maximum shear strain.

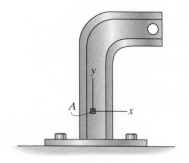

Prob. 14–105

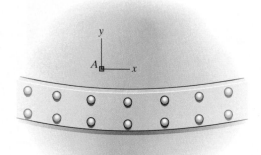

Prob. 14–107

14–106. The strain at point A on a beam has components $\epsilon_x = 450(10^{-6})$, $\epsilon_y = 825(10^{-6})$, $\gamma_{xy} = 275(10^{-6})$, $\epsilon_z = 0$. Determine (a) the principal strains at A, (b) the maximum shear strain in the x–y plane, and (c) the absolute maximum shear strain.

*14–108. The 45° strain rosette is mounted on the surface of a shell. The following readings are obtained for each gage: $\epsilon_a = -200(10^{-6})$, $\epsilon_b = 300(10^{-6})$, and $\epsilon_c = 250(10^{-6})$. Determine the in-plane principal strains.

Prob. 14–106

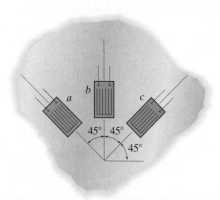

Prob. 14–108

14–109. For the case of plane stress, show that Hooke's law can be written as

$$\sigma_x = \frac{E}{(1 - \nu^2)}(\epsilon_x + \nu\epsilon_y), \quad \sigma_y = \frac{E}{(1 - \nu^2)}(\epsilon_y + \nu\epsilon_x)$$

14–110. Use Hooke's law, Eq. 14–32, to develop the strain tranformation equations, Eqs. 14–19 and 14–20, from the stress tranformation equations, Eqs. 14–1 and 14–2.

14–111. The principal plane stresses and associated strains in a plane at a point are $\sigma_1 = 36$ ksi, $\sigma_2 = 16$ ksi, $\epsilon_1 = 1.02(10^{-3})$, $\epsilon_2 = 0.180(10^{-3})$. Determine the modulus of elasticity and Poisson's ratio.

*14–112. A rod has a radius of 10 mm. If it is subjected to an axial load of 15 N such that the axial strain in the rod is $\epsilon_x = 2.75(10^{-6})$, determine the modulus of elasticity E and the change in the rod's diameter. $\nu = 0.23$.

14–113. The polyvinyl chloride bar is subjected to an axial force of 900 lb. If it has the original dimensions shown, determine the change in the angle θ after the load is applied. $E_{pvc} = 800(10^3)$ psi, $\nu_{pvc} = 0.20$.

14–114. The polyvinyl chloride bar is subjected to an axial force of 900 lb. If it has the original dimensions shown, determine the value of Poisson's ratio if the angle θ decreases by $\Delta\theta = 0.01°$ after the load is applied. $E_{pvc} = 800(10^3)$ psi.

14–115. The spherical pressure vessel has an inner diameter of 2 m and a thickness of 10 mm. A strain gage having a length of 20 mm is attached to it, and it is observed to increase in length by 0.012 mm when the vessel is pressurized. Determine the pressure causing this deformation, and find the maximum in-plane shear stress, and the absolute maximum shear stress at a point on the outer surface of the vessel. The material is steel, for which $E_{st} = 200$ GPa and $\nu_{st} = 0.3$.

20 mm

Prob. 14–115

*14–116. Determine the bulk modulus for each of the following materials: (a) rubber, $E_r = 0.4$ ksi, $\nu_r = 0.48$, and (b) glass, $E_g = 8(10^3)$ ksi, $\nu_g = 0.24$.

14–117. The strain gage is placed on the surface of the steel boiler as shown. If it is 0.5 in. long, determine the pressure in the boiler when the gage elongates $0.2(10^{-3})$ in. The boiler has a thickness of 0.5 in. and inner diameter of 60 in. Also, determine the maximum x, y in-plane shear strain in the material. $E_{st} = 29(10^3)$ ksi, $\nu_{st} = 0.3$.

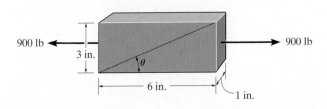

900 lb   900 lb

3 in.

θ

6 in.

1 in.

Probs. 14–113/114

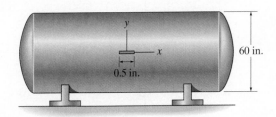

y

x

0.5 in.

60 in.

Prob. 14–117

14–118. The principal strains at a point on the aluminum fuselage of a jet aircraft are $\epsilon_1 = 780(10^{-6})$ and $\epsilon_2 = 400(10^{-6})$. Determine the associated principal stresses at the point in the same plane. $E_{al} = 10(10^3)$ ksi, $\nu_{al} = 0.33$. *Hint:* See Prob. 14–109.

14–119. The strain in the x direction at point A on the A-36 structural-steel beam is measured and found to be $\epsilon_x = 200(10^{-6})$. Determine the applied load P. What is the shear strain γ_{xy} at point A?

*14–120. If a load of $P = 3$ kip is applied to the A-36 structural-steel beam, determine the strain ϵ_x and γ_{xy} at point A.

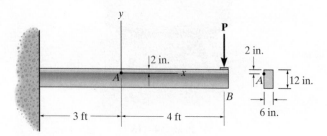

Probs. 14–119/120

14–121. The cube of aluminum is subjected to the three stresses shown. Determine the principal strains. Take $E_{al} = 10(10^3)$ ksi and $\nu_{al} = 0.33$.

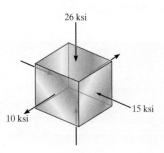

Prob. 14–121

14–122. The principal strains at a point on the aluminum surface of a tank are $\epsilon_1 = 630(10^{-6})$ and $\epsilon_2 = 350(10^{-6})$. If this is a case of plane stress, determine the associated principal stresses at the point in the same plane. $E_{al} = 10(10^3)$ ksi, $\nu_{al} = 0.33$. *Hint:* See Prob. 14–109.

14–123. A uniform edge load of 500 lb/in. and 350 lb/in. is applied to the polystyrene specimen. If the specimen is originally square and has dimensions of $a = 2$ in., $b = 2$ in., and a thickness of $t = 0.25$ in., determine its new dimensions $a', b',$ and t' after the load is applied. $E_p = 597(10^3)$ psi and $\nu_p = 0.25$.

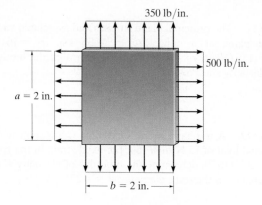

Prob. 14–123

*14–124. A material is subjected to principal stresses σ_x and σ_y. Determine the orientation θ of the strain gage so that its reading of normal strain responds only to σ_y and not σ_x. The material constants are E and ν.

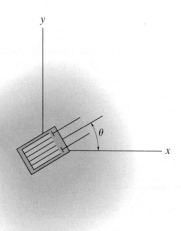

Prob. 14–124

CHAPTER REVIEW

Plane stress occurs when the material at a point is subjected to two normal stress components σ_x and σ_y and a shear stress τ_{xy}. Provided these components are known, then the stress components acting on an element having a different orientation θ can be determined using the two force equations of equilibrium or the equations of stress transformation.

$$\sigma_{x'} = \frac{\sigma_x + \sigma_y}{2} + \frac{\sigma_x - \sigma_y}{2}\cos 2\theta + \tau_{xy}\sin 2\theta$$

$$\tau_{x'y'} = -\frac{\sigma_x - \sigma_y}{2}\sin 2\theta + \tau_{xy}\cos 2\theta$$

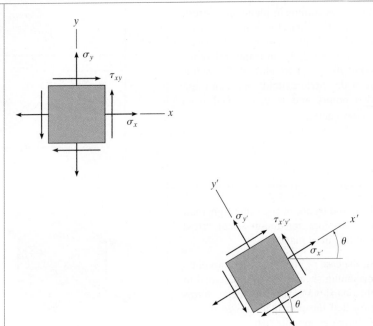

For design, it is important to determine the orientation of the element that produces the maximum principal normal stresses and the maximum in-plane shear stress. Using the stress transformation equations, it is found that no shear stress acts on the planes of principal stress. The principal stresses are

$$\sigma_{1,2} = \frac{\sigma_x + \sigma_y}{2} \pm \sqrt{\left(\frac{\sigma_x - \sigma_y}{2}\right)^2 + \tau_{xy}^2}$$

The planes of maximum in-plane shear stress are oriented $45°$ from this orientation, and on these shear planes there is an associated average normal stress.

$$\tau_{\substack{\max \\ \text{in-plane}}} = \sqrt{\left(\frac{\sigma_x - \sigma_y}{2}\right)^2 + \tau_{xy}^2}$$

$$\sigma_{\text{avg}} = \frac{\sigma_x + \sigma_y}{2}$$

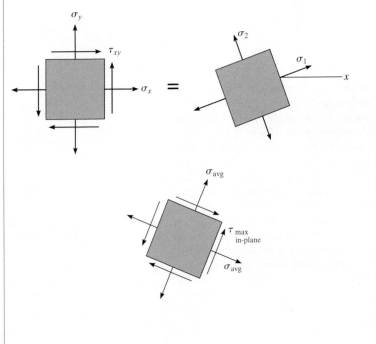

14

Mohr's circle provides a semi-graphical method for finding the stress on any plane, the principal normal stresses, and the maximum in-plane shear stress. To draw the circle, the σ and τ axes are established, the center of the circle $C[(\sigma_x + \sigma_y)/2, 0]$ and the reference point $A(\sigma_x, \tau_{xy})$ are plotted. The radius R of the circle extends between these two points and is determined from trigonometry.

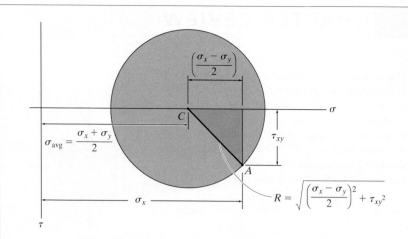

$$\sigma_{avg} = \frac{\sigma_x + \sigma_y}{2}$$

$$R = \sqrt{\left(\frac{\sigma_x - \sigma_y}{2}\right)^2 + \tau_{xy}^2}$$

If σ_1 and σ_2 are of the same sign, then the absolute maximum shear stress will lie out of plane.

In the case of plane stress, the absolute maximum shear stress will be equal to the maximum in-plane shear stress provided the principal stresses σ_1 and σ_2 have the opposite sign.

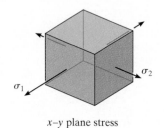

x–y plane stress

$$\tau_{\substack{abs \\ max}} = \frac{\sigma_1}{2}$$

x–y plane stress

$$\tau_{\substack{abs \\ max}} = \frac{\sigma_1 - \sigma_2}{2}$$

When an element of material is subjected to deformations that only occur in a single plane, then it undergoes plane strain. If the strain components ϵ_x, ϵ_y, and γ_{xy} are known for a specified orientation of the element, then the strains acting for some other orientation of the element can be determined using the plane-strain transformation equations. Likewise, the principal normal strains and maximum in-plane shear strain can be determined using transformation equations.

$$\epsilon_{x'} = \frac{\epsilon_x + \epsilon_y}{2} + \frac{\epsilon_x - \epsilon_y}{2}\cos 2\theta + \frac{\gamma_{xy}}{2}\sin 2\theta$$

$$\epsilon_{y'} = \frac{\epsilon_x + \epsilon_y}{2} - \frac{\epsilon_x - \epsilon_y}{2}\cos 2\theta - \frac{\gamma_{xy}}{2}\sin 2\theta$$

$$\frac{\gamma_{x'y'}}{2} = -\left(\frac{\epsilon_x - \epsilon_y}{2}\right)\sin 2\theta + \frac{\gamma_{xy}}{2}\cos 2\theta$$

$$\epsilon_{1,2} = \frac{\epsilon_x + \epsilon_y}{2} \pm \sqrt{\left(\frac{\epsilon_x - \epsilon_y}{2}\right)^2 + \left(\frac{\gamma_{xy}}{2}\right)^2}$$

$$\frac{\gamma_{\substack{max \\ in\text{-}plane}}}{2} = \sqrt{\left(\frac{\epsilon_x - \epsilon_y}{2}\right)^2 + \left(\frac{\gamma_{xy}}{2}\right)^2}$$

$$\epsilon_{avg} = \frac{\epsilon_x + \epsilon_y}{2}$$

Strain transformation problems can also be solved in a semi-graphical manner using Mohr's circle. To draw the circle, the ϵ and $\gamma/2$ axes are established and the center of the circle $C\,[(\epsilon_x + \epsilon_y)/2, 0]$ and the "reference point" $A\,(\epsilon_x, \gamma_{xy}/2)$ are plotted. The radius of the circle extends between these two points and is determined from trigonometry.

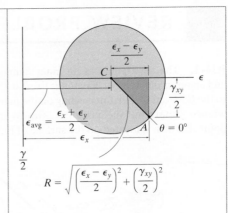

$$R = \sqrt{\left(\frac{\epsilon_x - \epsilon_y}{2}\right)^2 + \left(\frac{\gamma_{xy}}{2}\right)^2}$$

If the material is subjected to triaxial stress, then the strain in each direction is influenced by the strain produced by all three stresses. Hooke's law then involves the material properties E and ν.

$$\epsilon_x = \frac{1}{E}[\sigma_x - \nu(\sigma_y + \sigma_z)]$$

$$\epsilon_y = \frac{1}{E}[\sigma_y - \nu(\sigma_x + \sigma_z)]$$

$$\epsilon_z = \frac{1}{E}[\sigma_z - \nu(\sigma_x + \sigma_y)]$$

If E and ν are known, then G can be determined.

$$G = \frac{E}{2(1 + \nu)}$$

The dilatation is a measure of volumetric strain.

$$e = \frac{1 - 2\nu}{E}(\sigma_x + \sigma_y + \sigma_z)$$

The bulk modulus is used to measure the stiffness of a volume of material.

$$k = \frac{E}{3(1 - 2\nu)}$$

REVIEW PROBLEMS

R14–1. The steel pipe has an inner diameter of 2.75 in. and an outer diameter of 3 in. If it is fixed at C and subjected to the horizontal 20-lb force acting on the handle of the pipe wrench, determine the principal stresses in the pipe at point A, which is located on the surface of the pipe.

Prob. R14–1

R14–2. The steel pipe has an inner diameter of 2.75 in. and an outer diameter of 3 in. If it is fixed at C and subjected to the horizontal 20-lb force acting on the handle of the pipe wrench, determine the principal stresses in the pipe at point B, which is located on the surface of the pipe.

Prob. R14–2

R14–3. Determine the equivalent state of stress if an element is oriented 40° clockwise from the element shown. Use Mohr's circle.

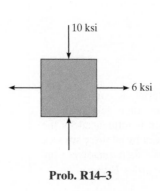

Prob. R14–3

*R14–4. The crane is used to support the 350-lb load. Determine the principal stresses acting in the boom at points A and B. The cross section is rectangular and has a width of 6 in. and a thickness of 3 in. Use Mohr's circle.

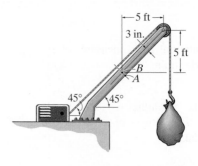

Prob. R14–4

R14–5. In the case of plane stress, where the in-plane principal strains are given by ϵ_1 and ϵ_2, show that the third principal strain can be obtained from

$$\epsilon_3 = \frac{-\nu(\epsilon_1 + \epsilon_2)}{(1 - \nu)}$$

where ν is Poisson's ratio for the material.

R14–6. The plate is made of material having a modulus of elasticity $E = 200$ GPa and Poisson's ratio $\nu = \frac{1}{3}$. Determine the change in width a, height b, and thickness t when it is subjected to the uniform distributed loading shown.

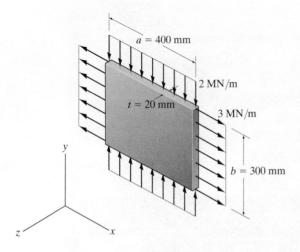

Prob. R14–6

R14–7. If the material is graphite for which $E_g = 800$ ksi and $\nu_g = 0.23$, determine the principal strains.

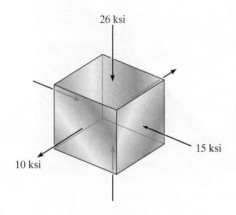

Prob. R14–7

*R14–8. A single strain gage, placed in the vertical plane on the outer surface and at an angle 60° to the axis of the pipe, gives a reading at point A of $\epsilon_A = -250(10^{-6})$. Determine the principal strains in the pipe at this point. The pipe has an outer diameter of 1 in. and an inner diameter of 0.6 in. and is made of C86100 bronze.

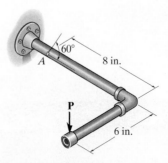

Prob. R14–8

R14–9. The 60° strain rosette is mounted on a beam. The following readings are obtained for each gage: $\epsilon_a = 600(10^{-6})$, $\epsilon_b = -700(10^{-6})$, and $\epsilon_c = 350(10^{-6})$. Determine (a) the in-plane principal strains and (b) the maximum in-plane shear strain and average normal strain. In each case show the deformed element due to these strains.

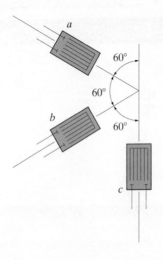

Prob. R14–9

CHAPTER 15

(© Olaf Speier/Alamy)

Beams are important structural members used to support roof and floor loadings.

DESIGN OF BEAMS AND SHAFTS

CHAPTER OBJECTIVES

■ To develop methods for designing beams to resist both bending and shear loads.

15.1 BASIS FOR BEAM DESIGN

Beams are said to be designed on the *basis of strength* when they can resist the internal shear and moment developed along their length. To design a beam in this way requires application of the shear and flexure formulas provided the material is homogeneous and has linear elastic behavior. Although some beams may also be subjected to an axial force, the effects of this force are often neglected in design since the axial stress is generally much smaller than the stress developed by shear and bending.

As shown in Fig. 15–1, the external loadings on a beam will create additional stresses in the beam *directly under the load*. Notably, a compressive stress σ_y will be developed, in addition to the bending stress σ_x and shear stress τ_{xy} discussed previously in Chapters 11 and 12. Using advanced methods of analysis, as treated in the theory of elasticity, it can be shown that σ_y diminishes rapidly throughout the beam's depth, and for *most* beam span-to-depth ratios used in engineering practice, the maximum value of σ_y remains small compared to the bending stress σ_x, that is, $\sigma_x \gg \sigma_y$. Furthermore, the direct application of concentrated loads is generally avoided in beam design. Instead, **bearing plates** are used to spread these loads more evenly onto the surface of the beam, thereby further reducing σ_y.

Beams must also be braced properly along their sides so that they do not sidesway or suddenly become unstable. In some cases they must also be designed to resist *deflection*, as when they support ceilings made of brittle materials such as plaster. Methods for finding beam deflections will be discussed in Chapter 16, and limitations placed on beam sidesway are often discussed in codes related to structural or mechanical design.

Knowing how the magnitude and direction of the principal stress change from point to point within a beam is important if the beam is made of a brittle material, because brittle materials, such as concrete, fail in tension. To give some idea as to how to determine this variation, let's consider the cantilever beam shown in Fig. 15–2a, which has a rectangular cross section and supports a load **P** at its end.

In general, at an arbitrary section *a–a* along the beam, Fig. 15–2b, the internal shear **V** and moment **M** create a *parabolic* shear-stress distribution and a *linear* normal-stress distribution, Fig. 15–2c. As a result, the stresses acting on elements located at points 1 through 5 along the section are shown in Fig. 15–2d. Note that elements 1 and 5 are subjected only to a maximum normal stress, whereas element 3, which is on the neutral axis, is subjected only to a maximum in-plane shear stress. The intermediate elements 2 and 4 must resist *both* normal and shear stress.

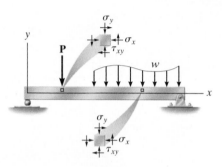

Fig. 15–1

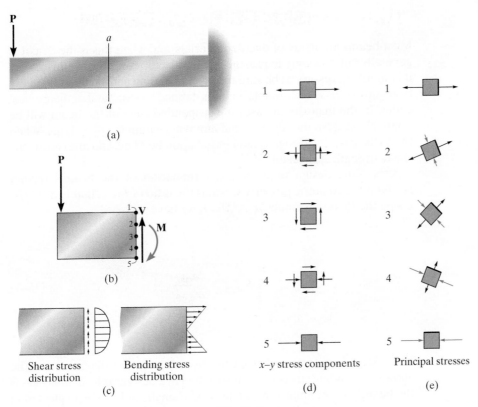

(a)

(b)

Shear stress
distribution

Bending stress
distribution

(c)

1

2

3

4

5

x–y stress components

(d)

1

2

3

4

5

Principal stresses

(e)

Fig. 15–2

15

When these states of stress are transformed into *principal stresses*, using either the stress transformation equations or Mohr's circle, the results will look like those shown in Fig. 15–2e. If this analysis is extended to many vertical sections along the beam other than *a–a*, a profile of the results can be represented by curves called ***stress trajectories***. Each of these curves indicates the *direction* of a principal stress having a constant magnitude. Some of these trajectories are shown in Fig. 15–3. Here the solid lines represent the direction of the tensile principal stresses and the dashed lines represent the direction of the compressive principal stresses. As expected, the lines intersect the neutral axis at 45° angles (like element 3), and the solid and dashed lines will intersect at 90° because the principal stresses are always 90° apart. Once the directions of these lines are established, it can help engineers decide where and how to place reinforcement in a beam if it is made of brittle material, so that it does not fail.

Stress trajectories for
cantilevered beam

Fig. 15–3

15

15.2 PRISMATIC BEAM DESIGN

Most beams are made of ductile materials, and when this is the case it is generally not necessary to plot the stress trajectories for the beam. Instead, it is simply necessary to be sure the *actual* bending and shear stress in the beam do not exceed allowable limits as defined by structural or mechanical codes. In the majority of cases the suspended span of the beam will be relatively long, so that the internal moments within it will be large. When this is the case, the design is then based upon bending, and afterwards the shear strength is checked.

A bending design requires a determination of the beam's **section modulus**, a geometric property which is the ratio of I to c, that is, $S = I/c$. Using the flexure formula, $\sigma = Mc/I$, we have

$$S_{\text{req'd}} = \frac{M_{\text{max}}}{\sigma_{\text{allow}}} \qquad (15\text{–}1)$$

Here M_{max} is determined from the beam's moment diagram, and the allowable bending stress, σ_{allow}, is specified in a design code. In many cases the beam's as yet unknown weight will be small, and can be neglected in comparison with the loads the beam must carry. However, if the additional moment caused by the weight is to be included in the design, a selection for S is made so that it slightly *exceeds* $S_{\text{req'd}}$.

Once $S_{\text{req'd}}$ is known, if the beam has a simple cross-sectional shape, such as a square, a circle, or a rectangle of known width-to-height proportions, its *dimensions* can be determined directly from $S_{\text{req'd}}$, since $S_{\text{req'd}} = I/c$. However, if the cross section is made from several elements, such as a wide-flange section, then an infinite number of web and flange dimensions can be determined that satisfy the value of $S_{\text{req'd}}$. In practice, however, engineers choose a particular beam meeting the requirement that $S > S_{\text{req'd}}$ from a table that lists the standard sizes available from manufacturers. Often several beams that have the same section modulus can be selected, and if deflections are not restricted, usually the beam having the smallest cross-sectional area is chosen, since it is made of less material, and is therefore both lighter and more economical than the others.

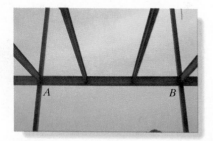

The two floor beams are connected to the beam AB, which transmits the load to the columns of this building frame. For design, all the connections can be considered to act as pins.

Once the beam has been selected, the shear formula can then be used to be sure the allowable shear stress is not exceeded, $\tau_{allow} \geq VQ/It$. Often this requirement will not present a problem; however, if the beam is "short" and supports large concentrated loads, the shear-stress limitation may dictate the size of the beam.

Steel Sections. Most manufactured steel beams are produced by rolling a hot ingot of steel until the desired shape is formed. These so-called **rolled shapes** have properties that are tabulated in the American Institute of Steel Construction (AISC) manual. A representative listing of different cross sections taken from this manual is given in Appendix B. Here the wide-flange shapes are designated by their depth and weight per unit length; for example, W18 × 46 indicates a wide-flange cross section (W) having a depth of 18 in. and a weight of 46 lb/ft, Fig. 15–4. For any given selection, the weight per length, dimensions, cross-sectional area, moment of inertia, and section modulus are reported. Also included is the radius of gyration, r, which is a geometric property related to the section's buckling strength. This will be discussed in Chapter 17.

Typical profile view of a steel wide-flange beam

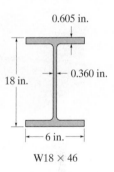

W18 × 46

Fig. 15–4

The large shear force that occurs at the support of this steel beam can cause localized buckling of the beam's flanges or web. To avoid this, a "stiffener" A is placed along the web to maintain stability.

Wood Sections. Most beams made of wood have rectangular cross sections because such beams are easy to manufacture and handle. Manuals, such as that of the National Forest Products Association, list the dimensions of lumber often used in the design of wood beams. Lumber is identified by its **nominal dimensions,** such as 2 × 4 (2 in. by 4 in.); however, its *actual* or "dressed" dimensions are smaller, being 1.5 in. by 3.5 in. The reduction in the dimensions occurs in order to obtain a smooth surface from lumber that is rough sawn. Obviously, the *actual dimensions* must be used whenever stress calculations are performed on wood beams.

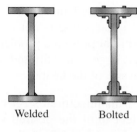

Welded   Bolted

Steel plate girders

Fig. 15–5

15

Wooden box beam

(a)

Glulam beam

(b)

Fig. 15–6

Built-up Sections. A ***built-up section*** is constructed from two or more parts joined together to form a single unit. The capacity of this section to resist a moment will vary directly with its section modulus, since $S_{\text{req'd}} = M/\sigma_{\text{allow}}$. If $S_{\text{req'd}}$ is *increased*, then so is I because by definition $S_{\text{req'd}} = I/c$. For this reason, *most of the material* for a built-up section should be placed as far away from the neutral axis as practical. This, of course, is what makes a deep wide-flange beam so efficient in resisting a moment. For a very large load, however, an available rolled-steel section may not have a section modulus great enough to support the load. When this is the case, engineers will usually "build up" a beam made from plates and angles. A deep I-shaped section having this form is called a ***plate girder***. For example, the steel plate girder in Fig. 15–5 has two flange plates that are either welded or, using angles, bolted to the web plate.

Wood beams are also "built up," usually in the form of a box beam, Fig. 15–6a. They may also be made having plywood webs and larger boards for the flanges. For very large spans, ***glulam beams*** are used. These members are made from several boards glue-laminated together to form a single unit, Fig. 15–6b.

Just as in the case of rolled sections or beams made from a single piece, the design of built-up sections requires that the bending and shear stresses be checked. In addition, the shear stress in the fasteners, such as weld, glue, nails, etc., must be checked to be certain the beam performs as a single unit.

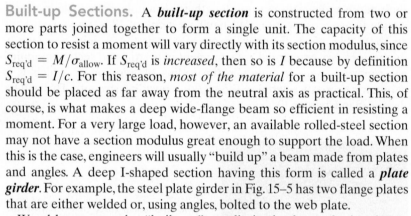

IMPORTANT POINTS

- Beams support loadings that are applied perpendicular to their axes. If they are designed on the basis of strength, they must resist their allowable shear and bending stresses.

- The maximum bending stress in the beam is assumed to be much greater than the localized stresses caused by the application of loadings on the surface of the beam.

PROCEDURE FOR ANALYSIS

Based on the previous discussion, the following procedure provides a rational method for the design of a beam on the basis of strength.

Shear and Moment Diagrams.

- Determine the maximum shear and moment in the beam. Often this is done by constructing the beam's shear and moment diagrams.

Bending Stress.

- If the beam is relatively long, it is designed by finding its section modulus using the flexure formula, $S_{req'd} = M_{max}/\sigma_{allow}$.
- Once $S_{req'd}$ is determined, the cross-sectional dimensions for simple shapes can then be calculated, since $S_{req'd} = I/c$.
- If rolled-steel sections are to be used, several possible beams can be selected from the tables in Appendix B that meet the requirement that $S \geq S_{req'd}$. Of these, choose the one having the smallest cross-sectional area, since this beam has the least weight and is therefore the most economical.
- Make sure that the selected section modulus, S, is *slightly greater* than $S_{req'd}$, so that the additional moment created by the beam's weight is considered.

Shear Stress.

- Normally beams that are short and carry large loads, especially those made of wood, are first designed to resist shear and then later checked against the allowable bending stress requirement.
- Using the shear formula, check to see that the allowable shear stress is not exceeded; that is, use $\tau_{allow} \geq V_{max}Q/It$.
- If the beam has a solid *rectangular* cross section, the shear formula becomes $\tau_{allow} \geq 1.5\,(V_{max}/A)$ (see Eq. 2 of Example 12.2.), and if the cross section is a *wide flange*, it is generally appropriate to assume that the shear stress is *constant* over the cross-sectional area of the beam's web so that $\tau_{allow} \geq V_{max}/A_{web}$, where A_{web} is determined from the product of the web's depth and its thickness. (See the note at the end of Example 12.3.)

Adequacy of Fasteners.

- The adequacy of fasteners used on built-up beams depends upon the shear stress the fasteners can resist. Specifically, the required spacing of nails or bolts of a particular size is determined from the allowable shear flow, $q_{allow} = VQ/I$, calculated at points on the cross section where the fasteners are located. (See Sec. 12.3.)

15

EXAMPLE 15.1

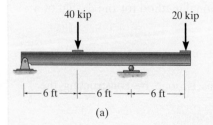

40 kip   20 kip

├── 6 ft ──┼── 6 ft ──┼── 6 ft ──┤

(a)

40 kip   20 kip

├── 6 ft ──┼── 6 ft ──┼── 6 ft ──┤

10 kip

V (kip)

10 |‾‾‾| 20
                          |‾‾‾‾| x (ft)
         |−30 |
50 kip

M (kip·ft)   60

├──── 8 ft ────┤   x (ft)

−120

(b)

Fig. 15–7

A beam is to be made of steel that has an allowable bending stress of $\sigma_{\text{allow}} = 24$ ksi and an allowable shear stress of $\tau_{\text{allow}} = 14.5$ ksi. Select an appropriate W shape that will carry the loading shown in Fig. 15–7a.

SOLUTION

Shear and Moment Diagrams. The support reactions have been calculated, and the shear and moment diagrams are shown in Fig. 15–7b. From these diagrams, $V_{\text{max}} = 30$ kip and $M_{\text{max}} = 120$ kip · ft.

Bending Stress. The required section modulus for the beam is determined from the flexure formula,

$$S_{\text{req'd}} = \frac{M_{\text{max}}}{\sigma_{\text{allow}}} = \frac{120 \text{ kip} \cdot \text{ft} \,(12 \text{ in./ft})}{24 \text{ kip/in}^2} = 60 \text{ in}^3$$

Using the table in Appendix B, the following beams are adequate:

W18 × 40	$S = 68.4 \text{ in}^3$
W16 × 45	$S = 72.7 \text{ in}^3$
W14 × 43	$S = 62.7 \text{ in}^3$
W12 × 50	$S = 64.7 \text{ in}^3$
W10 × 54	$S = 60.0 \text{ in}^3$
W8 × 67	$S = 60.4 \text{ in}^3$

The beam having the least weight per foot is chosen,* i.e.,

$$\text{W18} \times 40$$

Shear Stress. Since the beam is a *wide-flange section*, the *average shear stress* within the web will be considered. (See Example 12.3.) Here the web is assumed to extend from the very top to the very bottom of the beam. From Appendix B, for a W18 × 40, $d = 17.90$ in., $t_w = 0.315$ in. Thus,

$$\tau_{\text{avg}} = \frac{V_{\text{max}}}{A_w} = \frac{30 \text{ kip}}{(17.90 \text{ in.})(0.315 \text{ in.})} = 5.32 \text{ ksi} < 14.5 \text{ ksi} \quad \text{OK}$$

Use a W18 × 40.   *Ans.*

---

*The additional moment caused by the weight of the beam, (0.040 kip/ft) (18 ft) = 0.720 kip, will only slightly increase $S_{\text{req'd}}$.

EXAMPLE 15.2

The laminated wooden beam shown in Fig. 15–8a supports a uniform distributed loading of 12 kN/m. If the beam is to have a height-to-width ratio of 1.5, determine its smallest width. Take $\sigma_{allow} = 9$ MPa, and $\tau_{allow} = 0.6$ MPa. Neglect the weight of the beam.

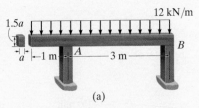

(a)

SOLUTION

Shear and Moment Diagrams. The support reactions at A and B have been calculated, and the shear and moment diagrams are shown in Fig. 15–8b. Here $V_{max} = 20$ kN, $M_{max} = 10.67$ kN·m.

Bending Stress. Applying the flexure formula,

$$S_{req'd} = \frac{M_{max}}{\sigma_{allow}} = \frac{10.67(10^3) \text{ N} \cdot \text{m}}{9(10^6) \text{ N/m}^2} = 0.00119 \text{ m}^3$$

Assuming that the width is a, then the height is $1.5a$, Fig. 15–8a. Thus,

$$S_{req'd} = \frac{I}{c} = 0.00119 \text{ m}^3 = \frac{\frac{1}{12}(a)(1.5a)^3}{(0.75a)}$$

$$a^3 = 0.003160 \text{ m}^3$$

$$a = 0.147 \text{ m}$$

Shear Stress. Applying the shear formula for rectangular sections (which is a special case of $\tau_{max} = VQ/It$, as shown in Example 12.2), we have

$$\tau_{max} = 1.5\frac{V_{max}}{A} = (1.5)\frac{20(10^3) \text{ N}}{(0.147 \text{ m})(1.5)(0.147 \text{ m})}$$

$$= 0.929 \text{ MPa} > 0.6 \text{ MPa}$$

Since the design based on bending fails the shear criterion, the beam must be redesigned on the basis of shear.

$$\tau_{allow} = 1.5\frac{V_{max}}{A}$$

$$600 \text{ kN/m}^2 = 1.5\frac{20(10^3) \text{ N}}{(a)(1.5a)}$$

$$a = 0.183 \text{ m} = 183 \text{ mm} \qquad \text{Ans.}$$

This larger section will also adequately resist the bending stress.

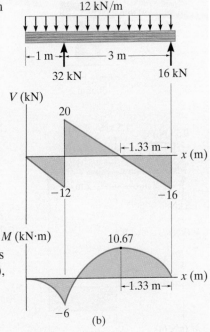

Fig. 15–8

EXAMPLE   15.3

The wooden T-beam shown in Fig. 15–9a is made from two 200 mm × 30 mm boards. If $\sigma_{\text{allow}} = 12$ MPa and $\tau_{\text{allow}} = 0.8$ MPa, determine if the beam can safely support the loading shown. Also, specify the maximum spacing of nails needed to hold the two boards together if each nail can safely resist 1.50 kN in shear.

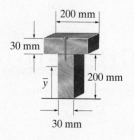

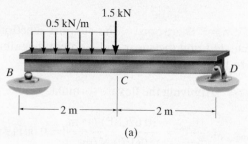

(a)

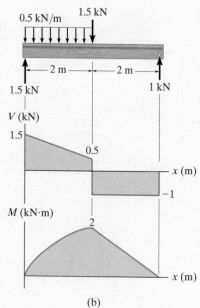

(b)

Fig. 15–9

SOLUTION

Shear and Moment Diagrams. The reactions on the beam are shown, and the shear and moment diagrams are drawn in Fig. 15–9b. Here $V_{\text{max}} = 1.5$ kN, $M_{\text{max}} = 2$ kN · m.

Bending Stress. The neutral axis (centroid) will be located from the bottom of the beam. Working in units of meters, we have

$$\bar{y} = \frac{\Sigma \bar{y} A}{\Sigma A}$$

$$= \frac{(0.1 \text{ m})(0.03 \text{ m})(0.2 \text{ m}) + 0.215 \text{ m}(0.03 \text{ m})(0.2 \text{ m})}{0.03 \text{ m}(0.2 \text{ m}) + 0.03 \text{ m}(0.2 \text{ m})} = 0.1575 \text{ m}$$

Thus,

$$I = \left[\frac{1}{12}(0.03 \text{ m})(0.2 \text{ m})^3 + (0.03 \text{ m})(0.2 \text{ m})(0.1575 \text{ m} - 0.1 \text{ m})^2 \right]$$

$$+ \left[\frac{1}{12}(0.2 \text{ m})(0.03 \text{ m})^3 + (0.03 \text{ m})(0.2 \text{ m})(0.215 \text{ m} - 0.1575 \text{ m})^2 \right]$$

$$= 60.125(10^{-6}) \text{ m}^4$$

Since $c = 0.1575$ m (not 0.230 m − 0.1575 m = 0.0725 m), we require

$$\sigma_{\text{allow}} \geq \frac{M_{\text{max}} c}{I}$$

$$12(10^6) \text{ Pa} \geq \frac{2(10^3) \text{ N} \cdot \text{m}(0.1575 \text{ m})}{60.125(10^{-6}) \text{ m}^4} = 5.24(10^6) \text{ Pa} \qquad \text{OK}$$

Shear Stress. Maximum shear stress in the beam depends upon the magnitude of Q and t. It occurs at the neutral axis, since Q is a maximum there and at the neutral axis the thickness $t = 0.03$ m is the smallest for the cross section. For simplicity, we will use the rectangular area below the neutral axis to calculate Q, rather than a two-part composite area above this axis, Fig. 15–9c. We have

$$Q = \bar{y}'A' = \left(\frac{0.1575 \text{ m}}{2}\right)[(0.1575 \text{ m})(0.03 \text{ m})] = 0.372(10^{-3}) \text{ m}^3$$

so that

$$\tau_{\text{allow}} \geq \frac{V_{\text{max}}Q}{It}$$

$$800(10^3) \text{ Pa} \geq \frac{1.5(10^3) \text{ N}[0.372(10^{-3})] \text{ m}^3}{60.125(10^{-6}) \text{ m}^4(0.03 \text{ m})} = 309(10^3) \text{ Pa} \quad \text{OK}$$

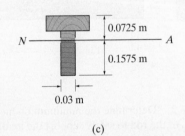

(c)

Nail Spacing. From the shear diagram it is seen that the shear varies over the entire span. Since the nail spacing depends on the magnitude of shear in the beam, for simplicity (and to be conservative), we will design the spacing on the basis of $V = 1.5$ kN for region BC, and $V = 1$ kN for region CD. Since the nails join the flange to the web, Fig. 15–9d, we have

$$Q = \bar{y}'A' = (0.0725 \text{ m} - 0.015 \text{ m})[(0.2 \text{ m})(0.03 \text{ m})] = 0.345(10^{-3}) \text{ m}^3$$

The shear flow for each region is therefore

$$q_{BC} = \frac{V_{BC}Q}{I} = \frac{1.5(10^3) \text{ N}[0.345(10^{-3}) \text{ m}^3]}{60.125(10^{-6}) \text{ m}^4} = 8.61 \text{ kN/m}$$

$$q_{CD} = \frac{V_{CD}Q}{I} = \frac{1(10^3) \text{ N}[0.345(10^{-3}) \text{ m}^3]}{60.125(10^{-6}) \text{ m}^4} = 5.74 \text{ kN/m}$$

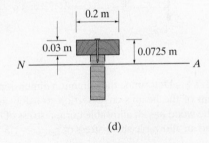

(d)

Fig. 15–9 (cont.)

One nail can resist 1.50 kN in shear, so the maximum spacing becomes

$$s_{BC} = \frac{1.50 \text{ kN}}{8.61 \text{ kN/m}} = 0.174 \text{ m}$$

$$s_{CD} = \frac{1.50 \text{ kN}}{5.74 \text{ kN/m}} = 0.261 \text{ m}$$

For ease of measuring, use

$$s_{BC} = 150 \text{ mm} \qquad \text{Ans.}$$

$$s_{CD} = 250 \text{ mm} \qquad \text{Ans.}$$

FUNDAMENTAL PROBLEMS

F15–1. Determine the minimum dimension a to the nearest mm of the beam's cross section to safely support the load. The wood has an allowable normal stress of $\sigma_{\text{allow}} = 10$ MPa and an allowable shear stress of $\tau_{\text{allow}} = 1$ MPa.

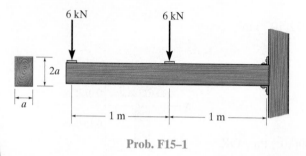

Prob. F15–1

F15–2. Determine the minimum diameter d to the nearest $\frac{1}{8}$ in. of the rod to safely support the load. The rod is made of a material having an allowable normal stress of $\sigma_{\text{allow}} = 20$ ksi and an allowable shear stress of $\tau_{\text{allow}} = 10$ ksi.

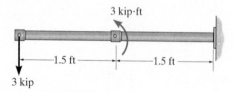

Prob. F15–2

F15–3. Determine the minimum dimension a to the nearest mm of the beam's cross section to safely support the load. The wood has an allowable normal stress of $\sigma_{\text{allow}} = 12$ MPa and an allowable shear stress of $\tau_{\text{allow}} = 1.5$ MPa.

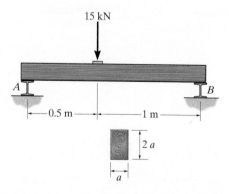

Prob. F15–3

F15–4. Determine the minimum dimension h to the nearest $\frac{1}{8}$ in. of the beam's cross section to safely support the load. The wood has an allowable normal stress of $\sigma_{\text{allow}} = 2$ ksi and an allowable shear stress of $\tau_{\text{allow}} = 200$ psi.

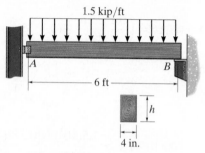

Prob. F15–4

F15–5. Determine the minimum dimension b to the nearest mm of the beam's cross section to safely support the load. The wood has an allowable normal stress of $\sigma_{\text{allow}} = 12$ MPa and an allowable shear stress of $\tau_{\text{allow}} = 1.5$ MPa.

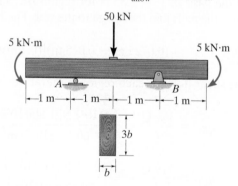

Prob. F15–5

F15–6. Select the lightest W410-shaped section that can safely support the load. The beam is made of steel having an allowable normal stress of $\sigma_{\text{allow}} = 150$ MPa and an allowable shear stress of $\tau_{\text{allow}} = 75$ MPa. Assume the beam is pinned at A and roller supported at B.

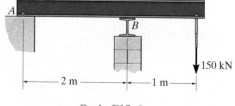

Prob. F15–6

PROBLEMS

15–1. The beam is made of timber that has an allowable bending stress of $\sigma_{allow} = 6.5$ MPa and an allowable shear stress of $\tau_{allow} = 500$ kPa. Determine its dimensions if it is to be rectangular and have a height-to-width ratio of 1.25. Assume the beam rests on smooth supports.

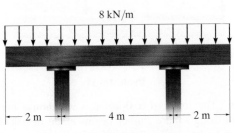

Prob. 15–1

15–2. Determine the minimum width of the beam to the nearest $\frac{1}{4}$ in. that will safely support the loading of $P = 8$ kip. The allowable bending stress is $\sigma_{allow} = 24$ ksi and the allowable shear stress is $\tau_{allow} = 15$ ksi.

15–3. Solve Prob. 15–2 if $P = 10$ kip.

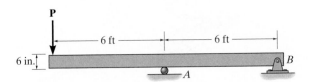

Probs. 15–2/3

*15–4. The brick wall exerts a uniform distributed load of 1.20 kip/ft on the beam. If the allowable bending stress is $\sigma_{allow} = 22$ ksi and the allowable shear stress is $\tau_{allow} = 12$ ksi, select the lightest wide-flange section with the shortest depth from Appendix B that will safely support the load. If there are several choices of equal weight, choose the one with the shortest height.

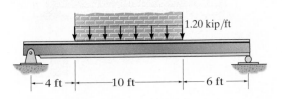

Prob. 15–4

15–5. Select the lightest-weight wide-flange beam from Appendix B that will safely support the machine loading shown. The allowable bending stress is $\sigma_{allow} = 24$ ksi and the allowable shear stress is $\tau_{allow} = 14$ ksi.

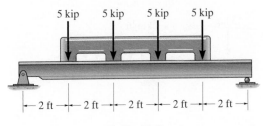

Prob. 15–5

15–6. The spreader beam AB is used to slowly lift the 3000-lb pipe that is centrally located on the straps at C and D. If the beam is a W12 × 45, determine if it can safely support the load. The allowable bending stress is $\sigma_{allow} = 22$ ksi and the allowable shear stress is $\tau_{allow} = 12$ ksi.

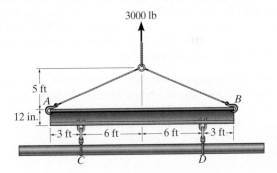

Prob. 15–6

15–7. Select the lightest-weight wide-flange beam with the shortest depth from Appendix B that will safely support the loading shown. The allowable bending stress is $\sigma_{allow} = 24$ ksi and the allowable shear stress of $\tau_{allow} = 14$ ksi.

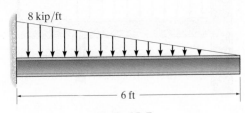

Prob. 15–7

15

*15–8. Select the lightest-weight wide-flange beam from Appendix B that will safely support the loading shown. The allowable bending stress σ_{allow} = 24 ksi and the allowable shear stress of τ_{allow} = 14 ksi.

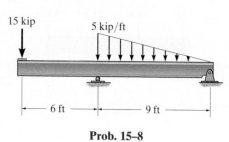

Prob. 15–8

15–9. Select the lightest W360 wide-flange beam from Appendix B that can safely support the loading. The beam has an allowable normal stress of σ_{allow} = 150 MPa and an allowable shear stress of τ_{allow} = 80 MPa. Assume there is a pin at A and a roller support at B.

15–10. Investigate if the W250 × 58 beam can safely support the loading. The beam has an allowable normal stress of σ_{allow} = 150 MPa and an allowable shear stress of τ_{allow} = 80 MPa. Assume there is a pin at A and a roller support at B.

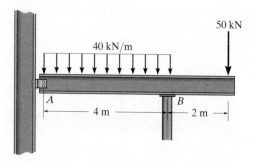

Probs. 15–9/10

15–11. The beam is constructed from two boards. If each nail can support a shear force of 200 lb, determine the maximum spacing of the nails, s, s', and s'', to the nearest $\frac{1}{8}$ inch for regions AB, BC, and CD, respectively.

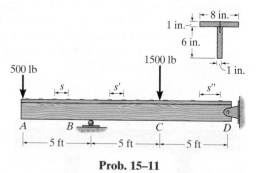

Prob. 15–11

*15–12. The joists of a floor in a warehouse are to be selected using square timber beams made of oak. If each beam is to be designed to carry 90 lb/ft over a simply supported span of 25 ft, determine the dimension a of its square cross section to the nearest $\frac{1}{4}$ in. The allowable bending stress is σ_{allow} = 4.5 ksi and the allowable shear stress is τ_{allow} = 125 psi.

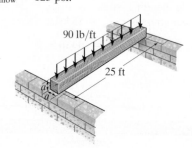

Prob. 15–12

15–13. The timber beam has a width of 6 in. Determine its height h so that it simultaneously reaches its allowable bending stress σ_{allow} = 1.50 ksi and an allowable shear stress of τ_{allow} = 50 psi. Also, what is the maximum load P that the beam can then support?

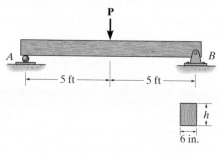

Prob. 15–13

15–14. The beam is constructed from four boards. If each nail can support a shear force of 300 lb, determine the maximum spacing of the nails, s, s' and s'', for regions AB, BC, and CD, respectively.

*15–16. If the cable is subjected to a maximum force of $P = 50$ kN, select the lightest W310 wide-flange beam that can safely support the load. The beam has an allowable normal stress of $\sigma_{allow} = 150$ MPa and an allowable shear stress of $\tau_{allow} = 85$ MPa.

15–17. If the W360 × 45 wide-flange beam has an allowable normal stress of $\sigma_{allow} = 150$ MPa and an allowable shear stress of $\tau_{allow} = 85$ MPa, determine the maximum cable force P that can safely be supported by the beam.

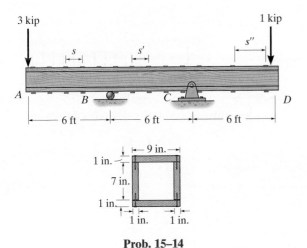

Prob. 15–14

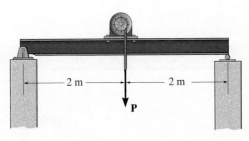

Probs. 15–16/17

15–15. The beam is constructed from two boards. If each nail can support a shear force of 200 lb, determine the maximum spacing of the nails, s, s', and s'', to the nearest $\frac{1}{8}$ in. for regions AB, BC, and CD, respectively.

15–18. If $P = 800$ lb, determine the minimum dimension a of the beam's cross section to the nearest $\frac{1}{8}$ in. to safely support the load. The wood has an allowable normal stress of $\sigma_{allow} = 1.5$ ksi and an allowable shear stress of $\tau_{allow} = 150$ psi.

15–19. If $a = 3$ in. and the wood has an allowable normal stress of $\sigma_{allow} = 1.5$ ksi, and an allowable shear stress of $\tau_{allow} = 150$ psi, determine the maximum allowable value of P that can act on the beam.

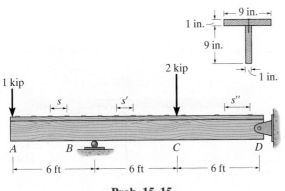

Prob. 15–15

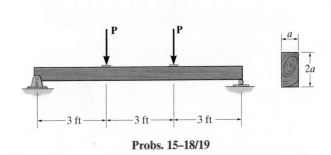

Probs. 15–18/19

*15–20. The beam is constructed from three plastic strips. If the glue can support a shear stress of $\tau_{allow} = 8$ kPa, determine the largest magnitude of the loads **P** that the beam can support.

15–21. If the allowable bending stress is $\sigma_{allow} = 6$ MPa, and the glue can support a shear stress of $\tau_{allow} = 8$ kPa, determine the largest magnitude of the loads **P** that can be applied to the beam.

15–23. Select the lightest-weight wide-flange beam from Appendix B that will safely support the loading. The allowable bending stress is $\sigma_{allow} = 24$ ksi and the allowable shear stress is $\tau_{allow} = 14$ ksi.

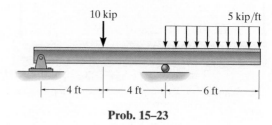

Prob. 15–23

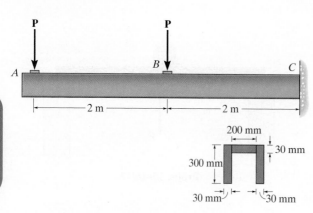

Probs. 15–20/21

*15–24. Draw the shear and moment diagrams for the shaft, and determine its required diameter to the nearest $\frac{1}{8}$ in. if $\sigma_{allow} = 30$ ksi and $\tau_{allow} = 15$ ksi. The journal bearings at *A* and *C* exert only vertical reactions on the shaft. Take $P = 6$ kip.

15–25. Draw the shear and moment diagrams for the shaft, and determine its required diameter to the nearest $\frac{1}{4}$ in. if $\sigma_{allow} = 30$ ksi and $\tau_{allow} = 15$ ksi. The journal bearings at *A* and *C* exert only vertical reactions on the shaft. Take $P = 12$ kip.

15–22. The beam is made of Douglas fir having an allowable bending stress of $\sigma_{allow} = 1.1$ ksi and an allowable shear stress of $\tau_{allow} = 0.70$ ksi. Determine the width *b* if the height $h = 2b$.

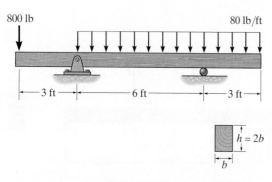

Prob. 15–22

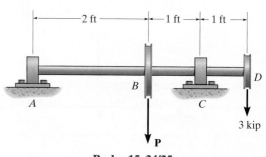

Probs. 15–24/25

CHAPTER REVIEW

Failure of a beam will occur where the internal shear or moment in the beam is a maximum. To resist these loadings, it is therefore important that the maximum shear and bending stress not exceed allowable values as stated in codes. Normally, the cross section of a beam is first designed to resist the allowable bending stress, $$\sigma_{\text{allow}} = \frac{M_{\max}c}{I}$$	
Then the allowable shear stress is checked. For rectangular sections, $\tau_{\text{allow}} \geq 1.5(V_{\max}/A)$, and for wide-flange sections it is appropriate to use $\tau_{\text{allow}} \geq V_{\max}/A_{\text{web}}$. In general, use $$\tau_{\text{allow}} = \frac{V_{\max}Q}{It}$$	
For built-up beams, the spacing of fasteners or the strength of glue or weld is determined using an allowable shear flow $$q_{\text{allow}} = \frac{VQ}{I}$$	

CHAPTER 16

(© Michael Blann/Getty Images)

If the curvature of this pole is measured, it is then possible to determine the bending stress developed within it.

DEFLECTION OF BEAMS AND SHAFTS

CHAPTER OBJECTIVES

- To determine the deflection and slope at specfic points on beams and shafts using the integration method, discontinuity functions, and the method of superposition.

- To use the method of superposition to solve for the support reactions on a beam or shaft that is statically indeterminate.

16.1 THE ELASTIC CURVE

The deflection of a beam or shaft must often be limited in order to provide stability, and for beams, to prevent the cracking of any attached brittle materials such as concrete or plaster. Most importantly, though, slopes and displacements must be determined in order to find the reactions if the beam is statically indeterminate. In this chapter we will find these slopes and displacements caused by the effects of bending.

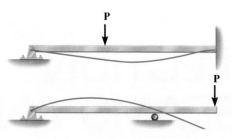

Fig. 16–1

Before finding the slope or displacement, it is often helpful to sketch the deflected shape of the beam, which is represented by its ***elastic curve***. This curve passes through the centroid of each cross section of the beam, and for most cases it can be sketched without much difficulty. When doing so, just remember that supports that resist a *force*, such as a pin, restrict *displacement*, and those that resist a *moment*, such as a fixed wall, restrict *rotation* or *slope* as well as displacement. Two examples of the elastic curves for loaded beams are shown in Fig. 16–1.

If the elastic curve for a beam seems difficult to establish, it is suggested that the moment diagram for the beam be drawn first. Using the beam sign convention established in Sec. 11.1, a positive internal moment tends to bend the beam concave upwards, Fig. 16–2*a*. Likewise, a negative moment tends to bend the beam concave downwards, Fig. 16–2*b*. Therefore, if the moment diagram is *known*, it will be easy to construct the elastic curve. For example, consider the beam in Fig. 16–3*a* with its associated moment diagram shown in Fig. 16–3*b*. Due to the roller and pin supports, the displacement at *B* and *D* must be zero. Within the region of negative moment, *AC*, Fig. 16–3*b*, the elastic curve must be concave downwards, and within the region of positive moment, *CD*, the elastic curve must be concave upwards. There is an *inflection point* at *C*, where the curve changes from concave up to concave down, since this is a point of zero moment. It should also be noted that the displacements Δ_A and Δ_E are especially critical. At point *E* the *slope* of the elastic curve is *zero*, and there the beam's *deflection* may be a *maximum*. Whether Δ_E is actually greater than Δ_A depends on the relative magnitudes of $\mathbf{P}_1$ and $\mathbf{P}_2$ and the location of the roller at *B*.

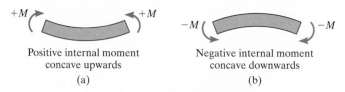

$+M$	$+M$	$-M$	$-M$

Positive internal moment
concave upwards

(a)

Negative internal moment
concave downwards

(b)

Fig. 16–2

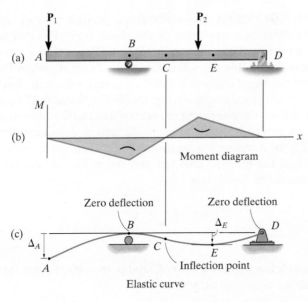

Fig. 16–3

Following these same principles, note how the elastic curve in Fig. 16–4 was constructed. Here the beam is cantilevered from a fixed support at A, and therefore the elastic curve must have both zero displacement and zero slope at this point. Also, the largest displacement will occur either at D, where the slope is zero, or at C.

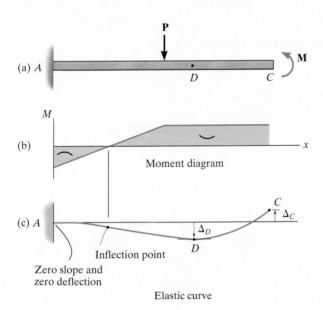

Fig. 16–4

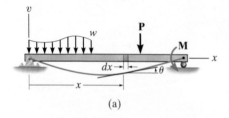

(a)

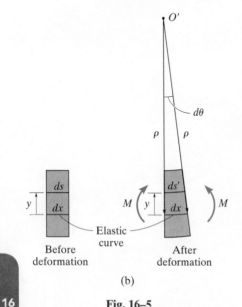

Before
deformation

Elastic
curve

After
deformation

(b)

Fig. 16–5

Moment–Curvature Relationship. Before we can obtain the slope and deflection at any point on the elastic curve, it is first necessary to relate the internal moment to the radius of curvature ρ (rho) of the elastic curve. To do this, we will consider the beam shown in Fig. 16–5a, and remove the small element located a distance x from the left end and having an undeformed length dx, Fig. 16–5b. The "localized" y coordinate is measured from the elastic curve (neutral axis) to the fiber in the beam that has an original length of $ds = dx$ and a deformed length ds'. In Sec. 11.3 we developed a relationship between the normal strain in this fiber and the internal moment and the radius of curvature of the beam element, Fig. 16–5b. It is

$$\frac{1}{\rho} = -\frac{\epsilon}{y} \qquad (16\text{–}1)$$

Since Hooke's law applies, $\epsilon = \sigma/E$, and $\sigma = -My/I$, after substituting into the above equation, we get

$$\boxed{\frac{1}{\rho} = \frac{M}{EI}} \qquad (16\text{–}2)$$

Here

$\rho =$ the radius of curvature at the point on the elastic curve ($1/\rho$ is referred to as the *curvature*)

$M =$ the internal moment in the beam at the point

$E =$ the material's modulus of elasticity

$I =$ the beam's moment of inertia about the neutral axis

The sign for ρ therefore depends on the direction of the moment. As shown in Fig. 16–6, when M is *positive*, ρ extends *above* the beam, and when M is *negative*, ρ extends *below* the beam.

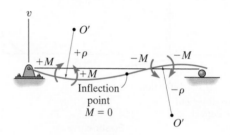

Fig. 16–6

16.2 SLOPE AND DISPLACEMENT BY INTEGRATION

The equation of the elastic curve in Fig. 16–5a will be defined by the coordinates v and x. And so to find the deflection $v = f(x)$ we must be able to represent the curvature $(1/\rho)$ in terms of v and x. In most calculus books it is shown that this relationship is

$$\frac{1}{\rho} = \frac{d^2v/dx^2}{[1 + (dv/dx)^2]^{3/2}} \qquad (16\text{–}3)$$

Substituting into Eq. 16–2, we have

$$\frac{d^2v/dx^2}{[1 + (dv/dx)^2]^{3/2}} = \frac{M}{EI} \qquad (16\text{–}4)$$

Apart from a few cases of simple beam geometry and loading, this equation is difficult to solve, because it represents a nonlinear second-order differential equation. Fortunately it can be modified, because most engineering design codes will restrict the maximum deflection of a beam or shaft. Consequently, the *slope* of the elastic curve, which is determined from dv/dx, will be *very small*, and its square will be negligible compared with unity.* Therefore the curvature, as defined in Eq. 16–3, can be *approximated* by $1/\rho = d^2v/dx^2$. With this simplification, Eq. 16–4 can now be written as

$$\frac{d^2v}{dx^2} = \frac{M}{EI} \qquad (16\text{–}5)$$

It is also possible to write this equation in two alternative forms. If we differentiate each side with respect to x and substitute $V = dM/dx$ (Eq. 11–2), we get

$$\frac{d}{dx}\left(EI\frac{d^2v}{dx^2}\right) = V(x) \qquad (16\text{–}6)$$

Differentiating again, using $w = dV/dx$ (Eq. 11–1), yields

$$\frac{d^2}{dx^2}\left(EI\frac{d^2v}{dx^2}\right) = w(x) \qquad (16\text{–}7)$$

---

*See Example 16.1.

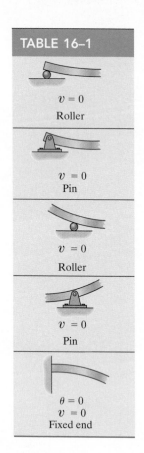

TABLE 16–1

$v = 0$
Roller

$v = 0$
Pin

$v = 0$
Roller

$v = 0$
Pin

$\theta = 0$
$v = 0$
Fixed end

16

For most problems the *flexural rigidity* (EI) will be constant along the length of the beam. Assuming this to be the case, the above results may be reordered into the following set of three equations:

$$EI\frac{d^4v}{dx^4} = w(x) \qquad (16\text{–}8)$$

$$EI\frac{d^3v}{dx^3} = V(x) \qquad (16\text{–}9)$$

$$EI\frac{d^2v}{dx^2} = M(x) \qquad (16\text{–}10)$$

Boundary Conditions. Solution of any of these equations requires successive integrations to obtain v. For each integration, it is necessary to introduce a "constant of integration" and then solve for all the constants to obtain a unique solution for a particular problem. For example, if the distributed load w is expressed as a function of x and Eq. 16–8 is used, then four constants of integration must be evaluated; however, it is generally easier to determine the internal moment M as a function of x and use Eq. 16–10, so that only two constants of integration must be found.

Most often, the integration constants are determined from *boundary conditions* for the beam, Table 16–1. As noted, if the beam is supported by a roller or pin, then it is required that the displacement be *zero* at these points. At the fixed support, the slope and displacement are both zero.

Continuity Conditions. Recall from Sec. 11.1 that if the loading on a beam is discontinuous, that is, it consists of a series of several distributed and concentrated loads, Fig. 16–7a, then several functions must be written for the internal moment, each valid within the region between two discontinuities. For example, the internal moment in regions AB, BC, and CD can be written in terms of the x_1, x_2, and x_3 coordinates selected as shown in Fig. 16–7b.

When each of these functions is integrated twice, it will produce two constants of integration, and since not all of these constants can be determined from the boundary conditions, some must be determined using *continuity conditions*. For example, consider the beam in Fig. 16–8. Here two x coordinates are chosen with origin at A. Once the functions for the slope and deflection are obtained, they must give the *same values* for the slope and deflection at point B so the elastic curve is physically *continuous*. Expressed mathematically, these continuity conditions are $\theta_1(a) = \theta_2(a)$ and $v_1(a) = v_2(a)$. They are used to evaluate the two constants of integration. Once these functions and the constants of integration are determined, they will then give the slope and deflection (elastic curve) for each region of the beam for which they are valid.

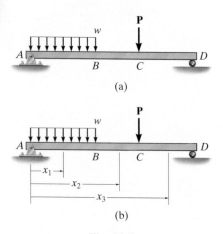

(a)

(b)

Fig. 16–7

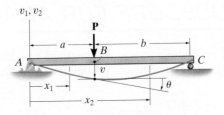

Fig. 16–8

Sign Convention and Coordinates.

When applying Eqs. 16–8 through 16–10, it is important to use the proper signs for *w, V,* or *M* as established for the derivation of these equations, Fig. 16–9a. Also, since *positive deflection, v,* is *upwards* then *positive slope θ* will be measured *counterclockwise* from the *x* axis when *x* is *positive to the right*, Fig. 16–9b. This is because a positive *increase dx* and *dv* creates an increased *θ* that is counterclockwise. By the same reason, if *positive x* is directed to the *left*, then *θ* will be *positive clockwise*, Fig. 16–9c.

Since we have considered $dv/dx \approx 0$, the original horizontal length of the beam's axis and the length of the arc of its elastic curve will almost be the same. In other words, *ds* in Figs. 16–9b and 16–9c is approximately equal to *dx,* since $ds = \sqrt{(dx)^2 + (dv)^2} = \sqrt{1 + (dv/dx)^2}\, dx \approx dx.$ As a result, points on the elastic curve will only be *displaced vertically,* and not horizontally. Also, since the *slope θ* will be *very small,* its value in radians can be determined *directly* from $\theta \approx \tan \theta = dv/dx.$

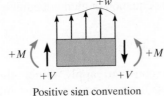

Positive sign convention

(a)

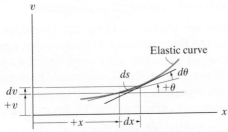

Positive sign convention

(b)

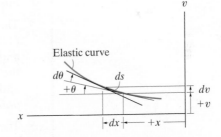

Positive sign convention

(c)

Fig. 16–9

PROCEDURE FOR ANALYSIS

The following procedure provides a method for determining the slope and deflection of a beam (or shaft) using the method of integration.

Elastic Curve.

- Draw an exaggerated view of the beam's elastic curve. Recall that zero slope and zero displacement occur at all fixed supports, and zero displacement occurs at all pin and roller supports.

- Establish the x and v coordinate axes. The x axis must be parallel to the undeflected beam and can have an origin at any point along the beam, with a positive direction either to the right or to the left. The positive v axis should be directed upwards.

- If several discontinuous loads are present, establish x coordinates that are valid for each region of the beam between the discontinuities. Choose these coordinates so that they will simplify subsequent algebraic work.

Load or Moment Function.

- For each region in which there is an x coordinate, express the loading w or the internal moment M as a function of x. In particular, *always* assume that M acts in the *positive direction* when applying the equation of moment equilibrium to determine $M = f(x)$.

Slope and Elastic Curve.

- Provided EI is constant, apply either the load equation $EI\ d^4v/dx^4 = w(x)$, which requires four integrations to get $v = v(x)$, or the moment equation $EI\ d^2v/dx^2 = M(x)$, which requires only two integrations. For each integration it is important to include a constant of integration.

- The constants are evaluated using the boundary conditions (Table 16–1) and the continuity conditions that apply to slope and displacement at points where two functions meet. Once the constants are evaluated and substituted back into the slope and deflection equations, the slope and displacement at *specific points* on the elastic curve can then be determined.

- The numerical values obtained can be checked graphically by comparing them with the sketch of the elastic curve. *Positive* values for *slope* are *counterclockwise* if the x axis extends *positive* to the *right*, and *clockwise* if the x axis extends *positive* to the *left*. In either of these cases, *positive displacement* is *upwards*.

EXAMPLE   16.1

The beam shown in Fig. 16–10a supports the triangular distributed loading. Determine its maximum deflection. EI is constant.

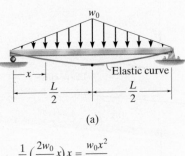

(a)

SOLUTION

Elastic Curve.   Due to symmetry, only one x coordinate is needed for the solution, in this case $0 \le x \le L/2$. The beam deflects as shown in Fig. 16–10a. The maximum deflection occurs at the center since the slope is zero at this point.

Moment Function.   A free-body diagram of the segment on the left is shown in Fig. 16–10b. The equation for the distributed loading is

$$w = \frac{2w_0}{L}x \qquad (1)$$

Hence,

$$\zeta + \Sigma M_{NA} = 0; \qquad M + \frac{w_0 x^2}{L}\left(\frac{x}{3}\right) - \frac{w_0 L}{4}(x) = 0$$

$$M = -\frac{w_0 x^3}{3L} + \frac{w_0 L}{4}x$$

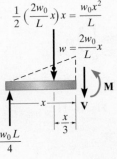

(b)

Fig. 16–10

Slope and Elastic Curve.   Using Eq. 16–10 and integrating twice, we have

$$EI\frac{d^2 v}{dx^2} = M = -\frac{w_0}{3L}x^3 + \frac{w_0 L}{4}x \qquad (2)$$

$$EI\frac{dv}{dx} = -\frac{w_0}{12L}x^4 + \frac{w_0 L}{8}x^2 + C_1$$

$$EIv = -\frac{w_0}{60L}x^5 + \frac{w_0 L}{24}x^3 + C_1 x + C_2$$

The constants of integration are obtained by applying the boundary condition $v = 0$ at $x = 0$ and the symmetry condition that $dv/dx = 0$ at $x = L/2$. This leads to

$$C_1 = -\frac{5w_0 L^3}{192} \qquad C_2 = 0$$

Hence,

$$EI\frac{dv}{dx} = -\frac{w_0}{12L}x^4 + \frac{w_0 L}{8}x^2 - \frac{5w_0 L^3}{192}$$

$$EIv = -\frac{w_0}{60L}x^5 + \frac{w_0 L}{24}x^3 - \frac{5w_0 L^3}{192}x$$

Determining the maximum deflection at $x = L/2$, we get

$$v_{max} = -\frac{w_0 L^4}{120EI} \qquad \qquad Ans.$$

EXAMPLE 16.2

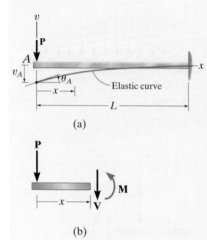

(a)

(b)

Fig. 16–11

The cantilevered beam shown in Fig. 16–11a is subjected to a vertical load **P** at its end. Determine the equation of the elastic curve. EI is constant.

SOLUTION I

Elastic Curve. The load tends to deflect the beam as shown in Fig. 16–11a. By inspection, the internal moment can be represented throughout the beam using a single x coordinate.

Moment Function. From the free-body diagram, with **M** acting in the *positive direction*, Fig. 16–11b, we have

$$M = -Px$$

Slope and Elastic Curve. Applying Eq. 16–10 and integrating twice yields

$$EI\frac{d^2v}{dx^2} = -Px \tag{1}$$

$$EI\frac{dv}{dx} = -\frac{Px^2}{2} + C_1 \tag{2}$$

$$EIv = -\frac{Px^3}{6} + C_1x + C_2 \tag{3}$$

Using the boundary conditions $dv/dx = 0$ at $x = L$ and $v = 0$ at $x = L$, Eqs. 2 and 3 become

$$0 = -\frac{PL^2}{2} + C_1$$

$$0 = -\frac{PL^3}{6} + C_1L + C_2$$

Thus, $C_1 = PL^2/2$ and $C_2 = -PL^3/3$. Substituting these results into Eqs. 2 and 3 with $\theta = dv/dx$, we get

$$\theta = \frac{P}{2EI}(L^2 - x^2)$$

$$v = \frac{P}{6EI}(-x^3 + 3L^2x - 2L^3) \qquad Ans.$$

Maximum slope and displacement occur at $A(x = 0)$, for which

$$\theta_A = \frac{PL^2}{2EI} \tag{4}$$

$$v_A = -\frac{PL^3}{3EI} \tag{5}$$

The *positive* result for θ_A indicates *counterclockwise* rotation and the *negative* result for v_A indicates that v_A is *downward*. This agrees with the results sketched in Fig. 16–11a.

In order to obtain some idea as to the actual *magnitude* of the slope and displacement at the end A, consider the beam in Fig. 16–11a to have a length of 15 ft, support a load of $P = 6$ kip, and be made of A-36 steel having $E_{st} = 29(10^3)$ ksi. Using the methods of Sec. 15.2, if this beam was designed without a factor of safety by assuming the allowable normal stress is equal to the yield stress $\sigma_{allow} = 36$ ksi, then a W12 × 26 would be found to be adequate ($I = 204$ in^4). From Eqs. 4 and 5 we get

$$\theta_A = \frac{6 \text{ kip}(15 \text{ ft})^2(12 \text{ in.}/\text{ft})^2}{2[29(10^3) \text{ kip}/\text{in}^2](204 \text{ in}^4)} = 0.0164 \text{ rad}$$

$$v_A = -\frac{6 \text{ kip}(15 \text{ ft})^3(12 \text{ in.}/\text{ft})^3}{3[29(10^3) \text{ kip}/\text{in}^2](204 \text{ in}^4)} = -1.97 \text{ in.}$$

Since $\theta_A^2 = (dv/dx)^2 = 0.000270 \text{ rad}^2 \ll 1$, this justifies the use of Eq. 16–10, rather than applying the more exact Eq. 16–4. Also, since this numerical application is for a *cantilevered beam*, we have obtained *larger* values for θ and v than would have been obtained if the beam were supported using pins, rollers, or other fixed supports.

SOLUTION II

This problem can also be solved using Eq. 16–8, $EI\, d^4v/dx^4 = w(x)$. Here $w(x) = 0$ for $0 \le x \le L$, Fig. 16–11a, so that upon integrating once we get the form of Eq. 16–9, i.e.,

$$EI\frac{d^4v}{dx^4} = 0$$

$$EI\frac{d^3v}{dx^3} = C_1' = V$$

The shear constant C_1' can be evaluated at $x = 0$, since $V_A = -P$ (negative according to the beam sign convention, Fig. 16–9a). Thus, $C_1' = -P$. Integrating again yields the form of Eq. 16–10, i.e.,

$$EI\frac{d^3v}{dx^3} = -P$$

$$EI\frac{d^2v}{dx^2} = -Px + C_2' = M$$

Here $M = 0$ at $x = 0$, so $C_2' = 0$, and as a result one obtains Eq. 1, and the solution proceeds as before.

16

EXAMPLE 16.3

The simply supported beam shown in Fig. 16–12a is subjected to the concentrated force. Determine the maximum deflection of the beam. EI is constant.

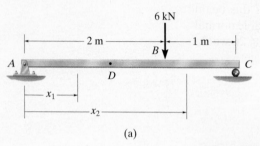

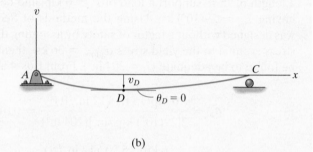

(a)   (b)

SOLUTION

Elastic Curve. The beam deflects as shown in Fig. 16–12b. Two coordinates must be used, since the moment function will change at B. Here we will take x_1 and x_2, having the *same origin* at A.

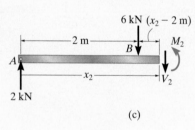

(c)

Fig. 16–12

Moment Function. From the free-body diagrams shown in Fig. 16–12c,

$$M_1 = 2x_1$$

$$M_2 = 2x_2 - 6(x_2 - 2) = 4(3 - x_2)$$

Slope and Elastic Curve. Applying Eq. 16–10 for M_1, for $0 \le x_1 < 2$ m, and integrating twice yields

$$EI\frac{d^2v_1}{dx_1{}^2} = 2x_1$$

$$EI\frac{dv_1}{dx_1} = x_1^2 + C_1 \tag{1}$$

$$EIv_1 = \frac{1}{3}x_1^3 + C_1x_1 + C_2 \tag{2}$$

Likewise for M_2, for 2 m $< x_2 \le 3$ m,

$$EI\frac{d^2v_2}{dx_2{}^2} = 4(3 - x_2)$$

$$EI\frac{dv_2}{dx_2} = 4\left(3x_2 - \frac{x_2^2}{2}\right) + C_3 \tag{3}$$

$$EIv_2 = 4\left(\frac{3}{2}x_2^2 - \frac{x_2^3}{6}\right) + C_3 x_2 + C_4 \tag{4}$$

The four constants are evaluated using *two* boundary conditions, namely, $x_1 = 0$, $v_1 = 0$ and $x_2 = 3$ m, $v_2 = 0$. Also, *two* continuity conditions must be applied at B, that is, $dv_1/dx_1 = dv_2/dx_2$ at $x_1 = x_2 = 2$ m and $v_1 = v_2$ at $x_1 = x_2 = 2$ m. Therefore

$v_1 = 0$ at $x_1 = 0$; $0 = 0 + 0 + C_2$

$v_2 = 0$ at $x_2 = 3$ m; $0 = 4\left(\dfrac{3}{2}(3)^2 - \dfrac{(3)^3}{6}\right) + C_3(3) + C_4$

$\dfrac{dv_1}{dx_1}\bigg|_{x=2m} = \dfrac{dv_2}{dx_2}\bigg|_{x=2m}$; $(2)^2 + C_1 = 4\left(3(2) - \dfrac{(2)^2}{2}\right) + C_3$

$v_1(2 \text{ m}) = v_2(2 \text{ m})$;

$\dfrac{1}{3}(2)^3 + C_1(2) + C_2 = 4\left(\dfrac{3}{2}(2)^2 - \dfrac{(2)^3}{6}\right) + C_3(2) + C_4$

Solving, we get

$$C_1 = -\frac{8}{3} \qquad C_2 = 0$$

$$C_3 = -\frac{44}{3} \qquad C_4 = 8$$

Thus Eqs. 1–4 become

$$EI\frac{dv_1}{dx_1} = x_1^2 - \frac{8}{3} \qquad (5)$$

$$EIv_1 = \frac{1}{3}x_1^3 - \frac{8}{3}x_1 \qquad (6)$$

$$EI\frac{dv_2}{dx_2} = 12x_2 - 2x_2^2 - \frac{44}{3} \qquad (7)$$

$$EIv_2 = 6x_2^2 - \frac{2}{3}x_2^3 - \frac{44}{3}x_2 + 8 \qquad (8)$$

By inspection of the elastic curve, Fig. 16–12b, the maximum deflection occurs at D, somewhere within region AB. Here the slope must be zero. From Eq. 5,

$$x_1{}^2 - \frac{8}{3} = 0$$

$$x_1 = 1.633$$

Substituting into Eq. 6,

$$v_{max} = -\frac{2.90 \text{ kN} \cdot \text{m}^3}{EI} \qquad\qquad Ans.$$

The negative sign indicates that the deflection is downwards.

EXAMPLE   16.4

The beam in Fig. 16–13a is subjected to a load at its end. Determine the displacement at C. EI is constant.

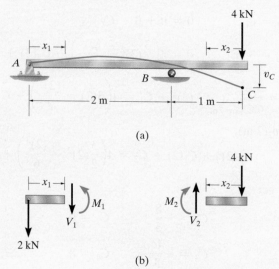

(a)

(b)

Fig. 16–13

SOLUTION

Elastic Curve.   The beam deflects into the shape shown in Fig. 16–13a. Due to the loading, two x coordinates will be considered, namely, $0 \le x_1 < 2$ m and $0 \le x_2 < 1$ m, where x_2 is directed to the left from C, since the internal moment is easy to formulate.

Moment Functions.   Using the free-body diagrams shown in Fig. 16–13b, we have

$$M_1 = -2x_1 \qquad M_2 = -4x_2$$

Slope and Elastic Curve.   Applying Eq. 16–10,

For $0 \le x_1 \le 2$: $$EI\frac{d^2v_1}{dx_1^{\,2}} = -2x_1$$

$$EI\frac{dv_1}{dx_1} = -x_1^2 + C_1 \tag{1}$$

$$EIv_1 = -\frac{1}{3}x_1^3 + C_1x_1 + C_2 \tag{2}$$

For $0 \le x_2 \le 1$ m: $EI\dfrac{d^2v_2}{dx_2^2} = -4x_2$

$$EI\dfrac{dv_2}{dx_2} = -2x_2^2 + C_3 \qquad (3)$$

$$EIv_2 = -\dfrac{2}{3}x_2^3 + C_3x_2 + C_4 \qquad (4)$$

The *four* constants of integration are determined using *three* boundary conditions, namely, $v_1 = 0$ at $x_1 = 0$, $v_1 = 0$ at $x_1 = 2$ m, and $v_2 = 0$ at $x_2 = 1$ m, and *one* continuity equation. Here the continuity of slope at the roller requires $dv_1/dx_1 = -dv_2/dx_2$ at $x_1 = 2$ m and $x_2 = 1$ m. There is a negative sign in this equation because the slope is measured positive counterclockwise from the right, and positive clockwise from the left, Fig. 16–9. (Continuity of displacement at B has been indirectly considered in the boundary conditions, since $v_1 = v_2 = 0$ at $x_1 = 2$ m and $x_2 = 1$ m.) Applying these four conditions yields

$v_1 = 0$ at $x_1 = 0$; $0 = 0 + 0 + C_2$

$v_1 = 0$ at $x_1 = 2$ m; $0 = -\dfrac{1}{3}(2)^3 + C_1(2) + C_2$

$v_2 = 0$ at $x_2 = 1$ m; $0 = -\dfrac{2}{3}(1)^3 + C_3(1) + C_4$

$\left.\dfrac{dv_1}{dx_1}\right|_{x=2\text{m}} = \left.\dfrac{dv_2}{dx_2}\right|_{x=1\text{m}}$; $-(2)^2 + C_1 = -(-2(1)^2 + C_3)$

Solving, we obtain

$$C_1 = \dfrac{4}{3} \qquad C_2 = 0 \qquad C_3 = \dfrac{14}{3} \qquad C_4 = -4$$

Substituting C_3 and C_4 into Eq. 4 gives

$$EIv_2 = -\dfrac{2}{3}x_2^3 + \dfrac{14}{3}x_2 - 4$$

The displacement at C is determined by setting $x_2 = 0$. We get

$$v_C = -\dfrac{4\,\text{kN}\cdot\text{m}^3}{EI} \qquad\qquad Ans.$$

PRELIMINARY PROBLEM

P16–1. In each case, determine the internal bending moment as a function of x, and state the necessary boundary and/or continuity conditions used to determine the elastic curve for the beam.

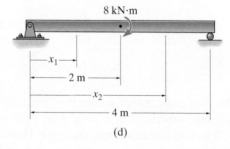

(d)

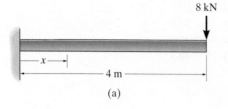

(a)

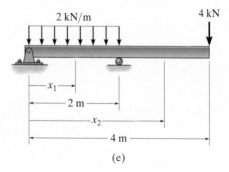

(e)

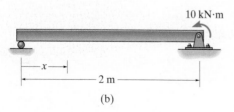

(b)

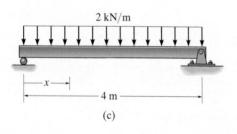

(c)

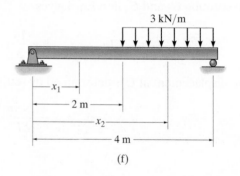

(f)

Prob. P16–1

FUNDAMENTAL PROBLEMS

F16–1. Determine the slope and deflection of end A of the cantilevered beam. $E = 200$ GPa and $I = 65.0(10^6)$ mm^4.

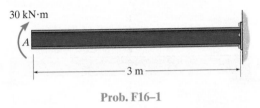

30 kN·m

A

3 m

Prob. F16–1

F16–2. Determine the slope and deflection of end A of the cantilevered beam. $E = 200$ GPa and $I = 65.0(10^6)$ mm^4.

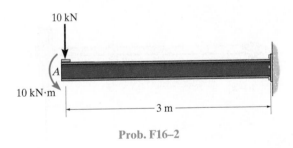

10 kN

A

10 kN·m

3 m

Prob. F16–2

F16–3. Determine the slope of end A of the cantilevered beam. $E = 200$ GPa and $I = 65.0(10^6)$ mm^4.

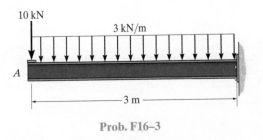

10 kN

3 kN/m

A

3 m

Prob. F16–3

F16–4. Determine the maximum deflection of the simply supported beam. The beam is made of wood having a modulus of elasticity of $E_w = 1.5(10^3)$ ksi and a rectangular cross section of width $b = 3$ in. and height $h = 6$ in.

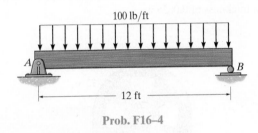

100 lb/ft

A

B

12 ft

Prob. F16–4

F16–5. Determine the maximum deflection of the simply supported beam. $E = 200$ GPa and $I = 39.9(10^{-6})$ m^4.

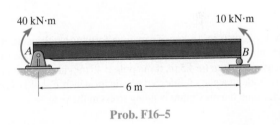

40 kN·m

10 kN·m

A

B

6 m

Prob. F16–5

F16–6. Determine the slope of the simply supported beam at A. $E = 200$ GPa and $I = 39.9(10^{-6})$ m^4.

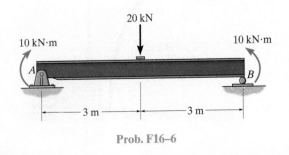

20 kN

10 kN·m

10 kN·m

A

B

3 m

3 m

Prob. F16–6

16

PROBLEMS

16–1.   An L2 steel strap having a thickness of 0.125 in. and a width of 2 in. is bent into a circular arc of radius 600 in. Determine the maximum bending stress in the strap.

16–2.   The L2 steel blade of the band saw wraps around the pulley having a radius of 12 in. Determine the maximum normal stress in the blade. The blade has a width of 0.75 in. and a thickness of 0.0625 in.

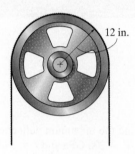

12 in.

Prob. 16–2

16–3.   A picture is taken of a man performing a pole vault, and the minimum radius of curvature of the pole is estimated by measurement to be 4.5 m. If the pole is 40 mm in diameter and it is made of a glass-reinforced plastic for which $E_g = 131$ GPa, determine the maximum bending stress in the pole.

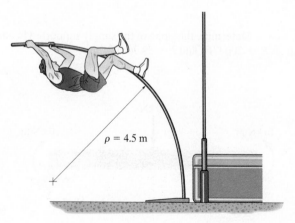

$\rho = 4.5$ m

Prob. 16–3

***16–4.**   Determine the equation of the elastic curve for the beam using the x coordinate that is valid for $0 \le x < L/2$. Specify the slope at A and the beam's maximum deflection. EI is constant.

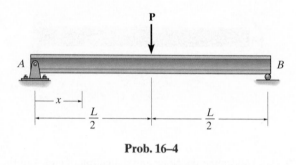

P

A B

x

$\dfrac{L}{2}$ $\dfrac{L}{2}$

Prob. 16–4

16–5.   Determine the deflection of end C of the 100-mm-diameter solid circular shaft. Take $E = 200$ GPa.

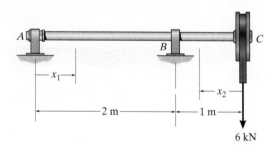

A
B C

x_1

x_2

2 m   1 m

6 kN

Prob. 16–5

16–6.   Determine the elastic curve for the cantilevered beam, which is subjected to the couple moment $\mathbf{M}_0$. Also calculate the maximum slope and maximum deflection of the beam. EI is constant.

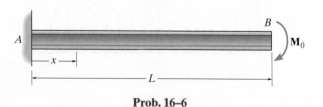

B

A $\mathbf{M}_0$

x

L

Prob. 16–6

16–7. The A-36 steel beam has a depth of 10 in. and is subjected to a constant moment M_0, which causes the stress at the outer fibers to become $\sigma_Y = 36$ ksi. Determine the radius of curvature of the beam and the beam's maximum slope and deflection.

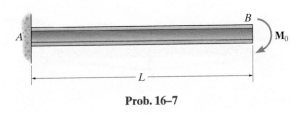

Prob. 16–7

***16–8.** Determine the equations of the elastic curve using the coordinates x_1 and x_2. EI is constant.

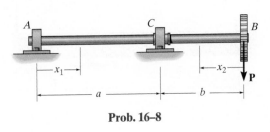

Prob. 16–8

16–9. Determine the equations of the elastic curve for the beam using the x_1 and x_2 coordinates. EI is constant.

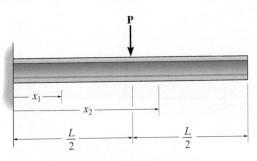

Prob. 16–9

16–10. Determine the equations of the elastic curve using the coordinates x_1 and x_2. What is the slope at C and displacement at B? EI is constant.

16–11. Determine the equations of the elastic curve using the coordinates x_1 and x_3. What is the slope at B and deflection at C? EI is constant.

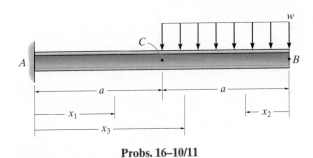

Probs. 16–10/11

***16–12.** Draw the bending-moment diagram for the shaft and then, from this diagram, sketch the deflection or elastic curve for the shaft's centerline. Determine the equations of the elastic curve using the coordinates x_1 and x_2. EI is constant.

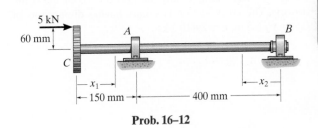

Prob. 16–12

16–13. Determine the maximum deflection of the beam and the slope at A. EI is constant.

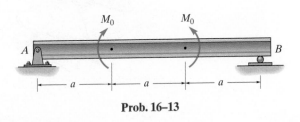

Prob. 16–13

16

16–14. The simply supported shaft has a moment of inertia of $2I$ for region BC and a moment of inertia I for regions AB and CD. Determine the maximum deflection of the shaft due to the load **P**.

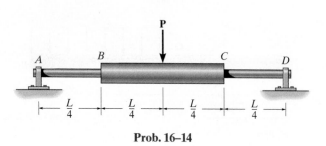

Prob. 16–14

16–15. A torque wrench is used to tighten the nut on a bolt. If the dial indicates that a torque of 60 lb · ft is applied when the bolt is fully tightened, determine the force P acting at the handle and the distance s the needle moves along the scale. Assume only the portion AB of the beam distorts. The cross section is square having dimensions of 0.5 in. by 0.5 in. $E = 29(10^3)$ ksi.

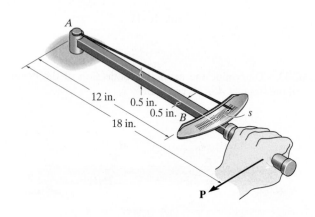

Prob. 16–15

*16–16.** The pipe can be assumed roller supported at its ends and by a rigid saddle C at its center. The saddle rests on a cable that is connected to the supports. Determine the force that should be developed in the cable if the saddle keeps the pipe from sagging or deflecting at its center. The pipe and fluid within it have a combined weight of 12.5 lb/ft. EI is constant.

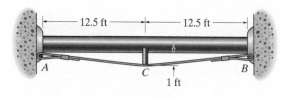

Prob. 16–16

16–17. Determine the equations of the elastic curve for the beam using the x_1 and x_2 coordinates. Specify the beam's maximum deflection. EI is constant.

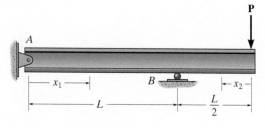

Prob. 16–17

16–18. The bar is supported by a roller constraint at B, which allows vertical displacement but resists axial load and moment. If the bar is subjected to the loading shown, determine the slope at A and the deflection at C. EI is constant.

16–19. Determine the deflection at B of the bar in Prob. 16–18.

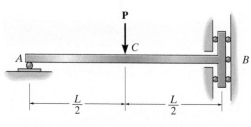

Probs. 16–18/19

*16–20. Determine the equations of the elastic curve using the x_1 and x_2 coordinates. What is the slope at A and the deflection at C? EI is constant.

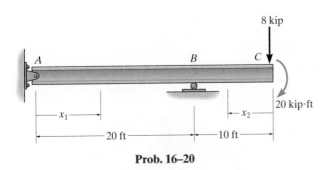

Prob. 16–20

16–21. Determine the maximum deflection of the solid circular shaft. The shaft is made of steel having $E = 200$ GPa. It has a diameter of 100 mm.

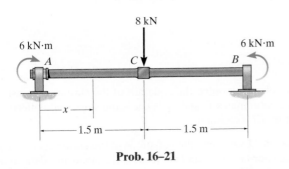

Prob. 16–21

16–22. Determine the elastic curve for the cantilevered W14 × 30 beam using the x coordinate. Specify the maximum slope and maximum deflection. $E = 29(10^3)$ ksi.

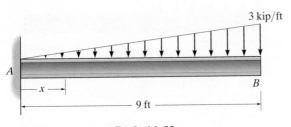

Prob. 16–22

16–23. Determine the equations of the elastic curve using the coordinates x_1 and x_2. What is the deflection and slope at C? EI is constant.

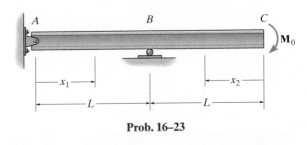

Prob. 16–23

*16–24. Determine the equations of the elastic curve using the coordinates x_1 and x_2. What is the slope at A? EI is constant.

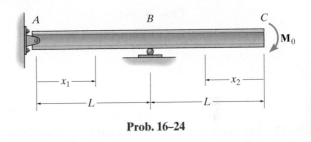

Prob. 16–24

16–25. The floor beam of the airplane is subjected to the loading shown. Assuming that the fuselage exerts only vertical reactions on the ends of the beam, determine the maximum deflection of the beam. EI is constant.

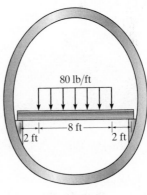

Prob. 16–25

16–26. Determine the maximum deflection of the simply supported beam. The beam is made of wood having a modulus of elasticity of $E = 1.5\,(10^3)$ ksi.

*16–28.** Determine the slope at end B and the maximum deflection of the cantilever triangular plate of constant thickness t. The plate is made of material having a modulus of elasticity of E.

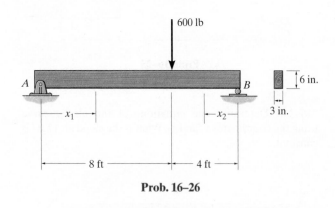

Prob. 16–26

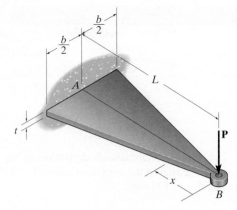

Prob. 16–28

16–27. The beam is made of a material having a specific weight γ. Determine the displacement and slope at its end A due to its weight. The modulus of elasticity for the material is E.

16–29. Determine the equation of the elastic curve using the coordinates x_1 and x_2. What is the slope and deflection at B? EI is constant.

16–30. Determine the equations of the elastic curve using the coordinates x_1 and x_3. What is the slope and deflection at point B? EI is constant.

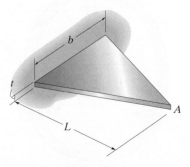

Prob. 16–27

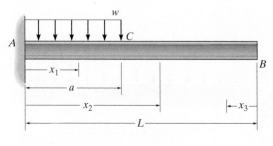

Probs. 16–29/30

*16.3 DISCONTINUITY FUNCTIONS

The method of integration, used to find the equation of the elastic curve for a beam or shaft, is convenient if the load or internal moment can be expressed as a continuous function throughout the beam's entire length. If several different loadings act on the beam, however, this method can become tedious to apply, because separate loading or moment functions must be written for each region of the beam. Furthermore, as noted in Examples 16.3 and 16.4, integration of these functions requires the evaluation of integration constants using both the boundary and continuity conditions.

In this section, we will discuss a method for finding the equation of the elastic curve using a *single expression*, either formulated directly from the loading on the beam, $w = w(x)$, or from the beam's internal moment, $M = M(x)$. Then when this expression for w is substituted into $EI\ d^4v/dx^4 = w(x)$ and integrated four times, or if the expression for M is substituted into $EI\ d^2v/dx^2 = M(x)$ and integrated twice, the constants of integration will only have to be determined from the boundary conditions.

Discontinuity Functions. In order to express the load on the beam or the internal moment within it using a single expression, we will use two types of mathematical operators known as ***discontinuity functions***.

For safety, the beams supporting these bags of cement must be designed for strength and a restricted amount of deflection.

16

Macaulay Functions.

For purposes of beam or shaft deflection, Macaulay functions, named after the mathematician W. H. Macaulay, can be used to describe *distributed loadings*. These functions can be written in general form as

$$\langle x - a \rangle^n = \begin{cases} 0 & \text{for } x < a \\ (x - a)^n & \text{for } x \geq a \end{cases}$$

$$n \geq 0 \qquad (16\text{--}11)$$

Here x represents the location of a point on the beam, and a is the location where the distributed loading *begins*. The Macaulay function $\langle x - a \rangle^n$ is written with angle or Macaulay brackets to distinguish it from an ordinary function $(x - a)^n$ written with parentheses. As stated by the equation, only when $x \geq a$ is $\langle x - a \rangle^n = (x - a)^n$; otherwise it is zero. Furthermore, this function is valid only for exponential values $n \geq 0$. Integration of the Macaulay function follows the same rules as for ordinary functions, i.e.,

$$\int \langle x - a \rangle^n dx = \frac{\langle x - a \rangle^{n+1}}{n + 1} + C \qquad (16\text{--}12)$$

Macaulay functions for a uniform and triangular load are shown in Table 16–2. Using integration, the Macaulay functions for shear, $V = \int w(x)\, dx$, and moment, $M = \int V\, dx$, are also shown in the table.

TABLE 16–2

Loading	Loading Function $w = w(x)$	Shear $V = \int w(x)dx$	Moment $M = \int V dx$
M_0	$w = M_0 \langle x-a \rangle^{-2}$	$V = M_0 \langle x-a \rangle^{-1}$	$M = M_0 \langle x-a \rangle^0$
P	$w = P \langle x-a \rangle^{-1}$	$V = P \langle x-a \rangle^0$	$M = P \langle x-a \rangle^1$
w_0	$w = w_0 \langle x-a \rangle^0$	$V = w_0 \langle x-a \rangle^1$	$M = \frac{w_0}{2} \langle x-a \rangle^2$
slope $= m$	$w = m \langle x-a \rangle^1$	$V = \frac{m}{2} \langle x-a \rangle^2$	$M = \frac{m}{6} \langle x-a \rangle^3$

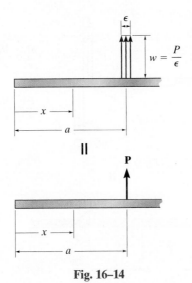

Fig. 16–14

Singularity Functions. These functions are used to describe concentrated forces or couple moments acting on a beam or shaft. Specifically, a concentrated force $\mathbf{P}$ can be considered a special case of a distributed loading having an intensity of $w = P/\epsilon$ when its length $\epsilon \to 0$, Fig. 16–14. The area under this loading diagram is equivalent to P, *positive upwards*, and has this value only when $x = a$. We will use a symbolic representation to express this result, namely

$$w = P\langle x - a \rangle^{-1} = \begin{cases} 0 & \text{for } x \neq a \\ P & \text{for } x = a \end{cases} \tag{16–13}$$

This expression is referred to as a ***singularity function***, since it takes on the value P only at the point $x = a$ where the load acts, otherwise it is zero.*

---

*It is also referred to as a unit impulse function or the Dirac delta.

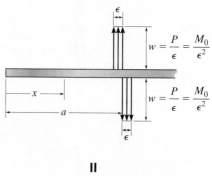

$$w = \frac{P}{\epsilon} = \frac{M_0}{\epsilon^2}$$

$$w = \frac{P}{\epsilon} = \frac{M_0}{\epsilon^2}$$

||

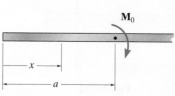

Fig. 16–15

In a similar manner, a couple moment $\mathbf{M}_0$, considered *positive clockwise*, is a limit as $\epsilon \to 0$ of two distributed loadings, as shown in Fig. 16–15. Here the following function describes its value.

$$w = M_0\langle x - a \rangle^{-2} = \begin{cases} 0 & \text{for } x \neq a \\ M_0 & \text{for } x = a \end{cases} \qquad (16\text{--}14)$$

The exponent $n = -2$ in order to ensure that the units of w, force per length, are maintained.

Integration of the above two functions follows the rules of calculus and yields results that are *different* from those of the Macaulay function. Specifically,

$$\int \langle x - a \rangle^n dx = \langle x - a \rangle^{n+1}, n = -1, -2 \qquad (16\text{--}15)$$

Using this formula, notice how M_0 and P, described in Table 16–2, are integrated once, then twice, to obtain the internal shear and moment in the beam.

Application of Eqs. 16–11 through 16–15 provides a direct means for expressing the loading or the internal moment in a beam as a function of x. Close attention, however, must be paid to the signs of the external loadings. As stated above, and as shown in Table 16–2, *concentrated forces and distributed loads are positive upwards, and couple moments are positive clockwise*. If this sign convention is followed, then the internal shear and moment will be in accordance with the beam sign convention established in Sec. 11.1.

Application. As an example of how to apply discontinuity functions to describe the loading or internal moment, we will consider the beam in Fig. 16–16a. Here the reactive 2.75-kN force created by the roller, Fig. 16–16b, is positive since it acts upwards, and the 1.5-kN · m couple moment is also positive since it acts clockwise. Finally, the trapezoidal loading is negative and by superposition has been separated into triangular and uniform loadings. From Table 16–2, the loading at any point x on the beam is therefore

$$w = 2.75 \text{ kN}\langle x - 0\rangle^{-1} + 1.5 \text{ kN} \cdot \text{m}\langle x - 3 \text{ m}\rangle^{-2} - 3 \text{ kN/m}\langle x - 3 \text{ m}\rangle^{0} - 1 \text{ kN/m}^{2}\langle x - 3 \text{ m}\rangle^{1}$$

The reactive force at B is not included here since x is never greater than 6 m. In the same manner, we can determine the moment expression directly from Table 16–2. It is

$$M = 2.75 \text{ kN}\langle x - 0\rangle^{1} + 1.5 \text{ kN} \cdot \text{m}\langle x - 3 \text{ m}\rangle^{0} - \frac{3 \text{ kN/m}}{2}\langle x - 3 \text{ m}\rangle^{2} - \frac{1 \text{ kN/m}^{2}}{6}\langle x - 3 \text{ m}\rangle^{3}$$

$$= 2.75x + 1.5\langle x - 3\rangle^{0} - 1.5\langle x - 3\rangle^{2} - \frac{1}{6}\langle x - 3\rangle^{3}$$

The deflection of the beam can now be determined after this equation is integrated two successive times, and the constants of integration are evaluated using the boundary conditions of zero displacement at A and B.

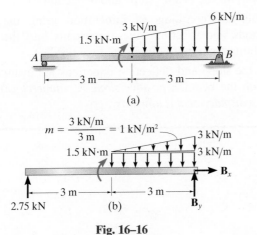

(a)

(b)

2.75 kN

Fig. 16–16

PROCEDURE FOR ANALYSIS

The following procedure provides a method for using discontinuity functions to determine a beam's elastic curve. This method is particularly advantageous for solving problems involving beams or shafts subjected to *several loadings*, since the constants of integration can be evaluated by using *only* the boundary conditions, while the compatibility conditions are automatically satisfied.

Elastic Curve.

- Sketch the beam's elastic curve and identify the boundary conditions at the supports.

- Zero displacement occurs at all pin and roller supports, and zero slope and zero displacement occur at fixed supports.

- Establish the x axis so that it extends to the right and has its origin at the beam's left end.

Load or Moment Function.

- Calculate the support reactions and then use the discontinuity functions in Table 16–2 to express either the loading w or the internal moment M as a function of x. Make sure to follow the sign convention for each loading.

- Note that the distributed loadings must extend all the way to the beam's right end to be valid. If this does not occur, use the method of superposition, which is illustrated in Example 16.5.

Slope and Elastic Curve.

- Substitute w into $EI\, d^4v/dx^4 = w(x)$, or M into the moment curvature relation $EI\, d^2v/dx^2 = M$, and integrate to obtain the equations for the beam's slope and deflection.

- Evaluate the constants of integration using the boundary conditions, and substitute these constants into the slope and deflection equations to obtain the final results.

- When the slope and deflection equations are evaluated at any point on the beam, a *positive slope* is *counterclockwise*, and a *positive displacement* is *upwards*.

EXAMPLE 16.5

Determine the equation of the elastic curve for the cantilevered beam shown in Fig. 16–17a. EI is constant.

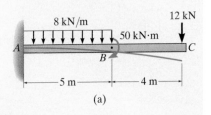

(a)

SOLUTION

Elastic Curve. The loads cause the beam to deflect as shown in Fig. 16–17a. The boundary conditions require zero slope and displacement at A.

Loading Function. The support reactions at A have been calculated and are shown on the free-body diagram in Fig. 16–17b. Since the distributed loading in Fig. 16–17a does not extend to C as required, we will use the superposition of loadings shown in Fig. 16–17b to represent the same effect. By our sign convention, the beam's loading is therefore

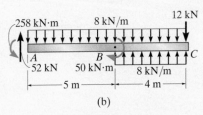

(b)

Fig. 16–17

$$w = 52\,\text{kN}\langle x - 0\rangle^{-1} - 258\,\text{kN} \cdot \text{m}\langle x - 0\rangle^{-2} - 8\,\text{kN/m}\langle x - 0\rangle^{0}$$
$$+ 50\,\text{kN} \cdot \text{m}\langle x - 5\,\text{m}\rangle^{-2} + 8\,\text{kN/m}\langle x - 5\,\text{m}\rangle^{0}$$

The 12-kN load is *not included* here, since x cannot be greater than 9 m. Because $dV/dx = w(x)$, then by integrating, and neglecting the constant of integration since the reactions at A are included in the load function, we get

$$V = 52\langle x - 0\rangle^{0} - 258\langle x - 0\rangle^{-1} - 8\langle x - 0\rangle^{1} + 50\langle x - 5\rangle^{-1} + 8\langle x - 5\rangle^{1}$$

Furthermore, $dM/dx = V$, so that integrating again yields

$$M = -258\langle x - 0\rangle^{0} + 52\langle x - 0\rangle^{1} - \frac{1}{2}(8)\langle x - 0\rangle^{2} + 50\langle x - 5\rangle^{0} + \frac{1}{2}(8)\langle x - 5\rangle^{2}$$

$$= (-258 + 52x - 4x^{2} + 50\langle x - 5\rangle^{0} + 4\langle x - 5\rangle^{2})\,\text{kN} \cdot \text{m}$$

This same result can be obtained *directly* from Table 16–2.

Slope and Elastic Curve. Applying Eq. 16–10 and integrating twice, we have

$$EI\frac{d^{2}v}{dx^{2}} = -258 + 52x - 4x^{2} + 50\langle x - 5\rangle^{0} + 4\langle x - 5\rangle^{2}$$

$$EI\frac{dv}{dx} = -258x + 26x^{2} - \frac{4}{3}x^{3} + 50\langle x - 5\rangle^{1} + \frac{4}{3}\langle x - 5\rangle^{3} + C_{1}$$

$$EIv = -129x^{2} + \frac{26}{3}x^{3} - \frac{1}{3}x^{4} + 25\langle x - 5\rangle^{2} + \frac{1}{3}\langle x - 5\rangle^{4} + C_{1}x + C_{2}$$

Since $dv/dx = 0$ at $x = 0$, $C_{1} = 0$; and $v = 0$ at $x = 0$, so $C_{2} = 0$. Thus,

$$v = \frac{1}{EI}\left(-129x^{2} + \frac{26}{3}x^{3} - \frac{1}{3}x^{4} + 25\langle x - 5\rangle^{2} + \frac{1}{3}\langle x - 5\rangle^{4}\right)\text{m}\qquad Ans.$$

EXAMPLE 16.6

Determine the maximum deflection of the beam shown in Fig. 16–18a. *EI* is constant.

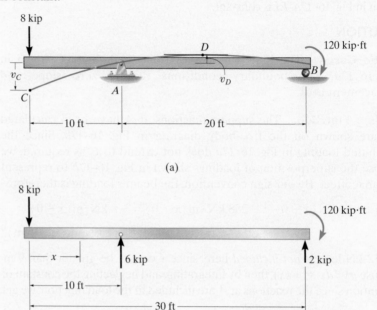

(a)

(b)

Fig. 16–18

SOLUTION

Elastic Curve. The beam deflects as shown in Fig. 16–18a. The boundary conditions require zero displacement at *A* and *B*.

Loading Function. The reactions have been calculated and are shown on the free-body diagram in Fig. 16–18b. The loading function for the beam is

$$w = -8 \text{ kip } \langle x - 0 \rangle^{-1} + 6 \text{ kip } \langle x - 10 \text{ ft} \rangle^{-1}$$

The couple moment and force at *B* are not included here, since they are located at the right end of the beam, and *x* cannot be greater than 30 ft. Integrating $dV/dx = w(x)$, we get

$$V = -8 \langle x - 0 \rangle^0 + 6 \langle x - 10 \rangle^0$$

In a similar manner, $dM/dx = V$ yields

$$M = -8 \langle x - 0 \rangle^1 + 6 \langle x - 10 \rangle^1$$
$$= (-8x + 6 \langle x - 10 \rangle^1) \text{ kip} \cdot \text{ft}$$

Notice how this equation can also be established *directly* using the results of Table 16–2 for moment.

Slope and Elastic Curve.   Integrating twice yields

$$EI\frac{d^2v}{dx^2} = -8x + 6\langle x - 10 \rangle^1$$

$$EI\frac{dv}{dx} = -4x^2 + 3\langle x - 10 \rangle^2 + C_1$$

$$EIv = -\frac{4}{3}x^3 + \langle x - 10 \rangle^3 + C_1x + C_2 \qquad (1)$$

From Eq. 1, the boundary condition $v = 0$ at $x = 10$ ft and $v = 0$ at $x = 30$ ft gives

$$0 = -1333 + (10 - 10)^3 + C_1(10) + C_2$$

$$0 = -36\,000 + (30 - 10)^3 + C_1(30) + C_2$$

Solving these equations simultaneously for C_1 and C_2, we get $C_1 = 1333$ and $C_2 = -12\,000$. Thus,

$$EI\frac{dv}{dx} = -4x^2 + 3\langle x - 10 \rangle^2 + 1333 \qquad (2)$$

$$EIv = -\frac{4}{3}x^3 + \langle x - 10 \rangle^3 + 1333x - 12\,000 \qquad (3)$$

From Fig. 16–18a, maximum displacement can occur either at C, or at D where the slope $dv/dx = 0$. To obtain the displacement of C, set $x = 0$ in Eq. 3. We get

$$v_C = -\frac{12\,000 \text{ kip} \cdot \text{ft}^3}{EI} \qquad\qquad \textit{Ans.}$$

The *negative* sign indicates that the displacement is *downwards* as shown in Fig. 16–18a. To locate point D, use Eq. 2 with $x > 10$ ft and $dv/dx = 0$. This gives

$$0 = -4x_D^2 + 3(x_D - 10)^2 + 1333$$

$$x_D^2 + 60x_D - 1633 = 0$$

Solving for the positive root,

$$x_D = 20.3 \text{ ft}$$

Hence, from Eq. 3,

$$EIv_D = -\frac{4}{3}(20.3)^3 + (20.3 - 10)^3 + 1333(20.3) - 12\,000$$

$$v_D = \frac{5006 \text{ kip} \cdot \text{ft}^3}{EI}$$

Comparing this value with v_C, we see that $v_{\text{max}} = v_C$.

PROBLEMS

16–31. The shaft is supported at A by a journal bearing and at C by a thrust bearing. Determine the equation of the elastic curve. EI is constant.

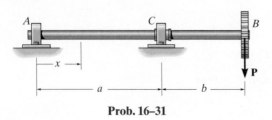

Prob. 16–31

*16–32. The shaft supports the two pulley loads shown. Determine the equation of the elastic curve. EI is constant.

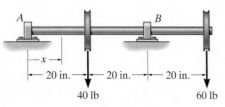

Prob. 16–32

16–33. The beam is made of a ceramic material. If it is subjected to the elastic loading shown, and the moment of inertia is I and the beam has a measured maximum deflection Δ at its center, determine the modulus of elasticity, E. The supports at A and D exert only vertical reactions on the beam.

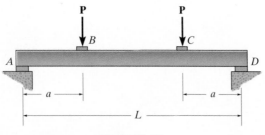

Prob. 16–33

16–34. Determine the equation of the elastic curve, the maximum deflection in region AB, and the deflection of end C. EI is constant.

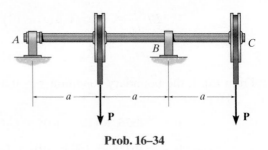

Prob. 16–34

16–35. The beam is subjected to the load shown. Determine the equation of the elastic curve. EI is constant.

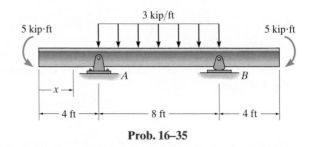

Prob. 16–35

*16–36. Determine the equation of the elastic curve, the slope at A, and the deflection at B. EI is constant.

16–37. Determine the equation of the elastic curve and the maximum deflection of the simply supported beam. EI is constant.

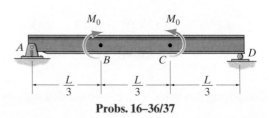

Probs. 16–36/37

16–38. The shaft supports the two pulley loads. Determine the equation of the elastic curve. EI is constant.

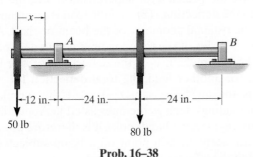

Prob. 16–38

16–39. Determine the maximum deflection of the cantilevered beam. Take $E = 200$ GPa and $I = 65.0(10^6)$ mm⁴.

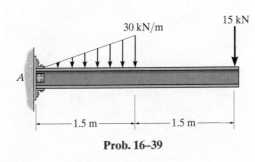

Prob. 16–39

***16–40.** Determine the slope at A and the deflection of end C of the overhang beam. Take $E = 29(10^3)$ ksi and $I = 204$ in⁴.

16–41. Determine the maximum deflection in region AB of the overhang beam. Take $E = 29(10^3)$ ksi and $I = 204$ in⁴.

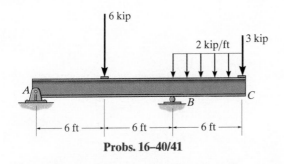

Probs. 16–40/41

16–42. The shaft supports the two pulley loads shown. Determine the slope of the shaft at A and B. The bearings exert only vertical reactions on the shaft. EI is constant.

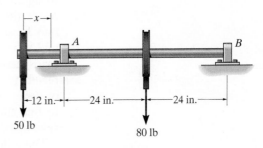

Prob. 16–42

16–43. Determine the equation of the elastic curve. EI is constant.

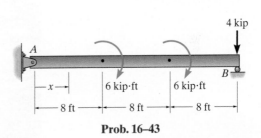

Prob. 16–43

***16–44.** Determine the equation of the elastic curve. EI is constant.

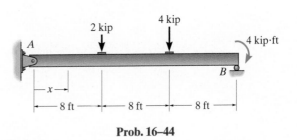

Prob. 16–44

16

16.4   METHOD OF SUPERPOSITION

The differential equation $EI\, d^4v/dx^4 = w(x)$ satisfies the two necessary requirements for applying the principle of superposition; i.e., the load $w(x)$ is linearly related to the deflection $v(x)$, and the load is assumed not to significantly change the original geometry of the beam or shaft. As a result, the deflections for a series of separate loadings acting on a beam may be superimposed. For example, if v_1 is the deflection for one load and v_2 is the deflection for another load, the total deflection for both loads acting together is the algebraic sum $v_1 + v_2$. Using tabulated results for various beam loadings, such as the ones listed in Appendix C, or those found in various engineering handbooks, it is therefore possible to find the slope and displacement at a point on a beam subjected to several loadings by adding the effects of each loading.

The following examples numerically illustrate how to do this.

The resultant deflection at any point on this beam can be determined from the superposition of the deflections caused by each of the separate loadings acting on the beam.

EXAMPLE   16.7

Determine the displacement at point C and the slope at the support A of the beam shown in Fig. 16–19a. EI is constant.

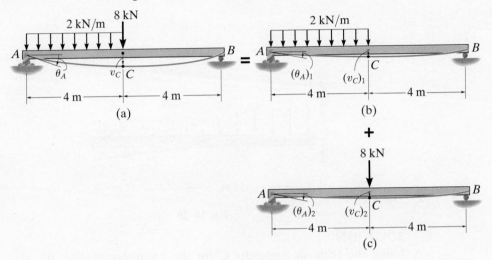

Fig. 16–19

SOLUTION

The loading can be separated into two component parts as shown in Figs. 16–19b and 16–19c. The displacement at C and slope at A are found using the table in Appendix C for each part.

For the distributed loading,

$$(\theta_A)_1 = \frac{3wL^3}{128EI} = \frac{3(2\ \text{kN/m})(8\ \text{m})^3}{128EI} = \frac{24\ \text{kN} \cdot \text{m}^2}{EI} \circlearrowright$$

$$(v_C)_1 = \frac{5wL^4}{768EI} = \frac{5(2\ \text{kN/m})(8\ \text{m})^4}{768EI} = \frac{53.33\ \text{kN} \cdot \text{m}^3}{EI} \downarrow$$

For the 8-kN concentrated force,

$$(\theta_A)_2 = \frac{PL^2}{16EI} = \frac{8\ \text{kN}(8\ \text{m})^2}{16EI} = \frac{32\ \text{kN} \cdot \text{m}^2}{EI} \circlearrowright$$

$$(v_C)_2 = \frac{PL^3}{48EI} = \frac{8\ \text{kN}(8\ \text{m})^3}{48EI} = \frac{85.33\ \text{kN} \cdot \text{m}^3}{EI} \downarrow$$

The displacement at C and the slope at A are the algebraic sums of these components. Hence,

$(+\circlearrowright)$ $\qquad \theta_A = (\theta_A)_1 + (\theta_A)_2 = \dfrac{56\ \text{kN} \cdot \text{m}^2}{EI} \circlearrowright$ $\qquad$ *Ans.*

$(+\downarrow)$ $\qquad v_C = (v_C)_1 + (v_C)_2 = \dfrac{139\ \text{kN} \cdot \text{m}^3}{EI} \downarrow$ $\qquad$ *Ans.*

EXAMPLE   16.8

Determine the displacement at the end C of the cantilever beam shown in Fig. 16–20. EI is constant.

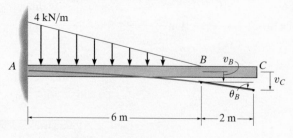

Fig. 16–20

SOLUTION

Using the table in Appendix C for the triangular loading, the slope and displacement at point B are

$$\theta_B = \frac{w_0 L^3}{24EI} = \frac{4 \text{ kN/m}(6 \text{ m})^3}{24EI} = \frac{36 \text{ kN} \cdot \text{m}^2}{EI}$$

$$v_B = \frac{w_0 L^4}{30EI} = \frac{4 \text{ kN/m}(6 \text{ m})^4}{30EI} = \frac{172.8 \text{ kN} \cdot \text{m}^3}{EI}$$

The unloaded region BC of the beam remains straight, as shown in Fig. 16–20. Since θ_B is small, the displacement at C becomes

$$(+\downarrow) \qquad v_C = v_B + \theta_B(L_{BC})$$

$$= \frac{172.8 \text{ kN} \cdot \text{m}^3}{EI} + \frac{36 \text{ kN} \cdot \text{m}^2}{EI}(2 \text{ m})$$

$$= \frac{244.8 \text{ kN} \cdot \text{m}^3}{EI} \downarrow \qquad\qquad\qquad\qquad Ans.$$

EXAMPLE 16.9

Determine the displacement at the end C of the overhanging beam shown in Fig. 16–21a. EI is constant.

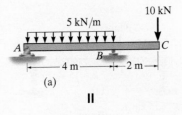

(a)

SOLUTION

Since the table in Appendix C *does not* include beams with overhangs, the beam will be separated into a simply supported and a cantilevered portion. First we will calculate the slope at B, as caused by the distributed load acting on the simply supported span, Fig. 16–21b.

$$(\theta_B)_1 = \frac{wL^3}{24EI} = \frac{5 \text{ kN/m}(4 \text{ m})^3}{24EI} = \frac{13.33 \text{ kN} \cdot \text{m}^2}{EI} \;\circlearrowright$$

Since this angle is *small*, the vertical displacement at point C is

$$(v_C)_1 = (2 \text{ m})\left(\frac{13.33 \text{ kN} \cdot \text{m}^2}{EI}\right) = \frac{26.67 \text{ kN} \cdot \text{m}^3}{EI}\uparrow$$

Next, the 10-kN load on the overhang causes a statically equivalent force of 10 kN and couple moment of 20 kN $\cdot$ m at the support B of the simply supported span, Fig. 16–21c. The 10-kN force does not cause a slope at B; however, the 20-kN $\cdot$ m couple moment does cause a slope. This slope is

$$(\theta_B)_2 = \frac{M_0L}{3EI} = \frac{20 \text{ kN} \cdot \text{m}(4 \text{ m})}{3EI} = \frac{26.67 \text{ kN} \cdot \text{m}^2}{EI}\circlearrowleft$$

so that the displacement of point C is

$$(v_C)_2 = (2 \text{ m})\left(\frac{26.7 \text{ kN} \cdot \text{m}^2}{EI}\right) = \frac{53.33 \text{ kN} \cdot \text{m}^3}{EI}\downarrow$$

Finally, the cantilevered portion BC is displaced by the 10-kN force, Fig. 16–21d. We have

$$(v_C)_3 = \frac{PL^3}{3EI} = \frac{10 \text{ kN}(2 \text{ m})^3}{3EI} = \frac{26.67 \text{ kN} \cdot \text{m}^3}{EI}\downarrow$$

Summing these results algebraically, we get

$$(+\downarrow) \quad v_C = -\frac{26.7}{EI} + \frac{53.3}{EI} + \frac{26.7}{EI} = \frac{53.3 \text{ kN} \cdot \text{m}^3}{EI}\downarrow \qquad Ans.$$

||

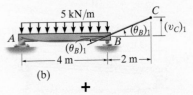

(b)

+

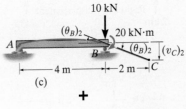

(c)

+

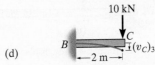

(d)

Fig. 16–21

16

EXAMPLE 16.10

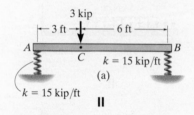

(a)

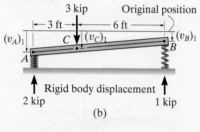

Original position

(b)

+

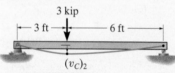

Deformable body displacement

(c)

Fig. 16–22

The steel bar shown in Fig. 16–22a is supported by two springs at its ends A and B. Each spring has a stiffness of $k = 15$ kip/ft and is originally unstretched. If the bar is loaded with a force of 3 kip at point C, determine the vertical displacement of the force. Neglect the weight of the bar and take $E_{st} = 29(10^3)$ ksi, $I = 12$ in^4.

SOLUTION

The end reactions at A and B are calculated and shown in Fig. 16–22b. Each spring deflects by an amount

$$(v_A)_1 = \frac{2 \text{ kip}}{15 \text{ kip/ft}} = 0.1333 \text{ ft}$$

$$(v_B)_1 = \frac{1 \text{ kip}}{15 \text{ kip/ft}} = 0.0667 \text{ ft}$$

If the bar is considered to be *rigid*, then the vertical displacement at C is

$$(v_C)_1 = (v_B)_1 + \frac{6 \text{ ft}}{9 \text{ ft}}[(v_A)_1 - (v_B)_1]$$

$$= 0.0667 \text{ ft} + \frac{2}{3}[0.1333 \text{ ft} - 0.0667 \text{ ft}] = 0.1111 \text{ ft} \downarrow$$

We can find the displacement at C caused by the *deformation* of the bar, Fig. 16–22c, by using the table in Appendix C. We have

$$(v_C)_2 = \frac{Pab}{6EIL}(L^2 - b^2 - a^2)$$

$$= \frac{3 \text{ kip}(3 \text{ ft})(6 \text{ ft})[(9 \text{ ft})^2 - (6 \text{ ft})^2 - (3 \text{ ft})^2]}{6[29(10^3)\text{kip/in}^2](144 \text{ in}^2/1 \text{ ft}^2)(12 \text{ in}^4)(1 \text{ ft}^4/20 \text{ } 736 \text{ in}^4)(9 \text{ ft})}$$

$$= 0.0149 \text{ ft} \downarrow$$

Adding the two displacement components, we get

$$(+\downarrow) \qquad v_C = 0.1111 \text{ ft} + 0.0149 \text{ ft} = 0.126 \text{ ft} = 1.51 \text{ in.} \downarrow \qquad \textit{Ans.}$$

PROBLEMS

16–45. The W10 × 15 cantilevered beam is made of A-36 steel and is subjected to the loading shown. Determine the slope and displacement at its end B.

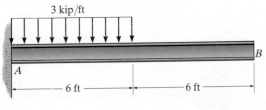

3 kip/ft

A

6 ft   6 ft

B

Prob. 16–45

16–46. The W10 × 15 cantilevered beam is made of A-36 steel and is subjected to the loading shown. Determine the displacement at B and the slope at B.

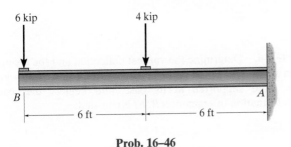

6 kip   4 kip

B A

6 ft   6 ft

Prob. 16–46

16–47. The W14 × 43 simply supported beam is made of A992 steel and is subjected to the loading shown. Determine the deflection at its center C.

*16–48. The W14 × 43 simply supported beam is made of A992 steel and is subjected to the loading shown. Determine the slope at A and B.

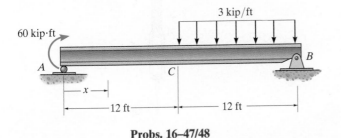

60 kip·ft   3 kip/ft

A C B

x

12 ft   12 ft

Probs. 16–47/48

16–49. The W14 × 43 simply supported beam is made of A-36 steel and is subjected to the loading shown. Determine the deflection at its center C.

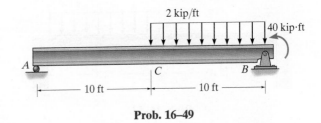

2 kip/ft   40 kip·ft

A C B

10 ft   10 ft

Prob. 16–49

16–50. The W14 × 43 simply supported beam is made of A-36 steel and is subjected to the loading shown. Determine the slope at A and B.

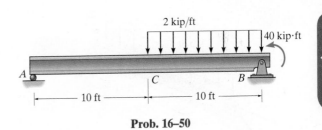

2 kip/ft   40 kip·ft

A C B

10 ft   10 ft

Prob. 16–50

16–51. The W8 × 48 cantilevered beam is made of A-36 steel and is subjected to the loading shown. Determine the displacement at C and the slope at A.

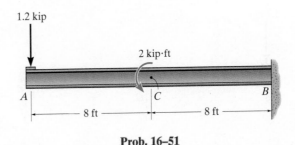

1.2 kip   2 kip·ft

A C B

8 ft   8 ft

Prob. 16–51

16

*16–52. The beam supports the loading shown. Code restrictions, due to a plaster ceiling, require the maximum deflection not to exceed 1/360 of the span length. Select the lightest-weight A-36 steel wide-flange beam from Appendix B that will satisfy this requirement and safely support the load. The allowable bending stress is $\sigma_{allow} = 24$ ksi and the allowable shear stress is $\tau_{allow} = 14$ ksi. Assume A is a roller and B is a pin.

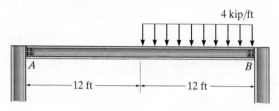

Prob. 16–52

16–53. The W24 × 104 A-36 steel beam is used to support the uniform distributed load and a concentrated force which is applied at its end. If the force acts at an angle with the vertical as shown, determine the horizontal and vertical displacement at A.

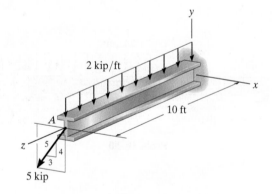

Prob. 16–53

16–54. The W8 × 48 cantilevered beam is made of A-36 steel and is subjected to the loading shown. Determine the displacement at its end A.

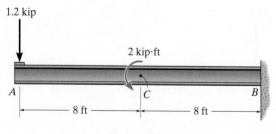

Prob. 16–54

16–55. The rod is pinned at its end A and attached to a torsional spring having a stiffness k, which measures the torque per radian of rotation of the spring. If a force **P** is always applied perpendicular to the end of the rod, determine the displacement of the force. EI is constant.

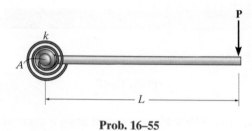

Prob. 16–55

*16–56. Determine the vertical deflection and the change in angle at the end A of the bracket. Assume that the bracket is fixed supported at its base, and neglect the axial deformation of segment AB. EI is constant.

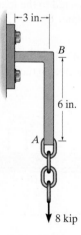

Prob. 16–56

16–57. The pipe assembly consists of three equal-sized pipes with flexibility stiffness EI and torsional stiffness GJ. Determine the vertical deflection at A.

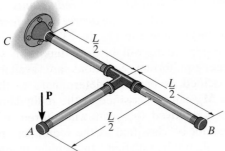

Prob. 16–57

16–58. The assembly consists of a cantilevered beam CB and a simply supported beam AB. If each beam is made of A-36 steel and has a moment of inertia about its principal axis of $I_x = 118$ in⁴, determine the displacement at the center D of beam BA.

16–59. The relay switch consists of a thin metal strip or armature AB that is made of red brass C83400 and is attracted to the solenoid S by a magnetic field. Determine the smallest force F required to attract the armature at C in order that contact is made at the free end B. Also, what should the distance a be for this to occur? The armature is fixed at A and has a moment of inertia of $I = 0.18(10^{-12})$ m⁴.

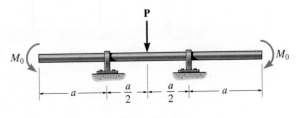

Prob. 16–59

***16–60.** Determine the moment M_0 in terms of the load P and dimension a so that the deflection at the center of the shaft is zero. EI is constant.

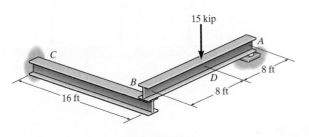

Prob. 16–58

Prob. 16–60

16.5   STATICALLY INDETERMINATE BEAMS AND SHAFTS—METHOD OF SUPERPOSITION

In this section we will illustrate a general method for determining the reactions on a statically indeterminate beam or shaft. Specifically, a member is ***statically indeterminate*** if the number of unknown reactions *exceeds* the available number of equilibrium equations.

The additional support reactions on a beam (or shaft) that are *not needed* to keep it in stable equilibrium are called ***redundants***, and the number of these redundants is referred to as the ***degree of indeterminacy***. For example, consider the beam shown in Fig. 16–23*a*. If its free-body diagram is drawn, Fig. 16–23*b*, there will be four unknown support reactions, and since three equilibrium equations are available for solution, the beam is classified as being "indeterminate to the first degree." Either $\mathbf{A}_y$, $\mathbf{B}_y$, or $\mathbf{M}_A$ can be classified as the redundant, for if any one of these reactions is removed, the beam will still remain stable and in equilibrium ($\mathbf{A}_x$ cannot be classified as the redundant, for if it were removed, $\Sigma F_x = 0$ would not be satisfied.) In a similar manner, the *continuous beam* in Fig. 16–24*a* is "indeterminate to the second degree," since there are five unknown reactions and only three available equilibrium equations, Fig. 16–24*b*. Here any two redundant support reactions can be chosen among $\mathbf{A}_y$, $\mathbf{B}_y$, $\mathbf{C}_y$, and $\mathbf{D}_y$.

The reactions on a beam that is statically indeterminate must satisfy both the equations of equilibrium and the compatibility requirements at the supports. We will now illustrate how this is done using the method of superposition.

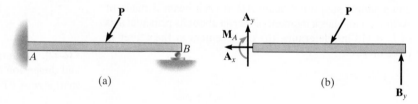

(a)   (b)

Fig. 16–23

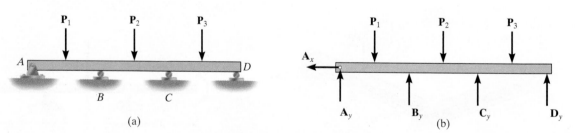

(a)

Fig. 16–24

To do this it is first necessary to identify the redundants and remove them from the beam. This will produce the **primary beam**, which will then be statically determinate and stable. Using superposition, we add to this beam a succession of similarly supported beams, each loaded only with a *separate* redundant. The redundants are determined from the *conditions of compatibility* that exist at each support where a redundant acts. Since the redundant forces are determined directly in this manner, this method of analysis is sometimes called the **force method**.

To clarify these concepts, consider the beam shown in Fig. 16–25a. If we choose the reaction $\mathbf{B}_y$ at the roller as the redundant, then the primary beam is shown in Fig. 16–25b, and the beam with the redundant $\mathbf{B}_y$ acting on it is shown in Fig. 16–25c. The displacement at the roller is to be zero, and since the displacement of B on the primary beam is v_B, and $\mathbf{B}_y$ causes B to be displaced upward v'_B, we can write the compatibility equation at B as

$$(+\uparrow) \qquad\qquad 0 = -v_B + v'_B$$

These displacements can be expressed in terms of the loads using the table in Appendix C. These **load–displacement relations** are

$$v_B = \frac{5PL^3}{48EI} \quad \text{and} \quad v'_B = \frac{B_y L^3}{3EI}$$

Substituting into the compatibility equation, we get

$$0 = -\frac{5PL^3}{48EI} + \frac{B_y L^3}{3EI}$$

$$B_y = \frac{5}{16}P$$

Now that $\mathbf{B}_y$ is known, the reactions at the wall are determined from the three equations of equilibrium applied to the free-body diagram of the beam, Fig. 16–25d. The results are

$$A_x = 0 \quad A_y = \frac{11}{16}P$$

$$M_A = \frac{3}{16}PL$$

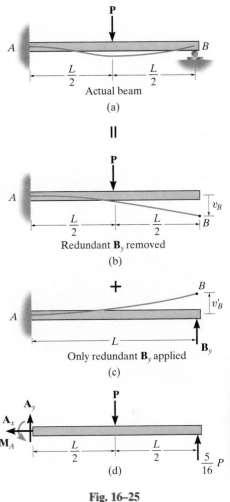

Actual beam
(a)

‖

Redundant $\mathbf{B}_y$ removed
(b)

+

Only redundant $\mathbf{B}_y$ applied
(c)

(d)

Fig. 16–25

16

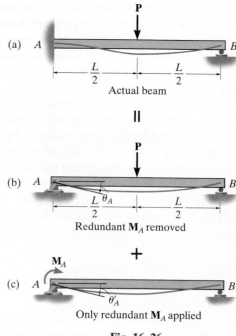

Fig. 16–26

Actually, choice of a redundant is *arbitrary*, provided the primary beam remains stable. For example, the moment at A for the beam in Fig. 16–26a can also be chosen as the redundant. In this case the capacity of the beam to resist M_A is removed, and so the primary beam is then pin supported at A, Fig. 16–26b. To it we add the beam subjected only to the redundant, Fig. 16–26c. Referring to the slope at A caused by the load $\mathbf{P}$ as θ_A, and the slope at A caused by the redundant $\mathbf{M}_A$ as θ'_A, the compatibility equation for the slope at A requires

$$(\circlearrowleft+) \qquad\qquad\qquad 0 = \theta_A + \theta'_A$$

Again using the table in Appendix C to relate these rotations to the loads, we have

$$\theta_A = \frac{PL^2}{16EI} \quad \text{and} \quad \theta'_A = \frac{M_A L}{3EI}$$

Thus,

$$0 = \frac{PL^2}{16EI} + \frac{M_A L}{3EI}$$

$$M_A = -\frac{3}{16}PL$$

which is the same result determined previously. Here, however, the negative sign for M_A simply means that $\mathbf{M}_A$ acts in the opposite sense of direction to that shown in Fig. 16–26c.

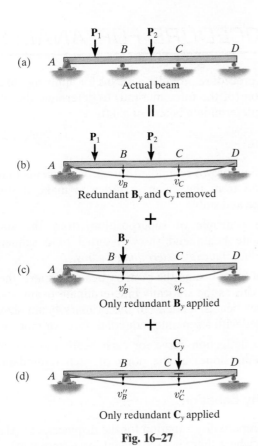

Fig. 16–27

A final example that illustrates this method is shown in Fig. 16–27a. In this case the beam is indeterminate to the second degree, and therefore two redundant reactions must be removed from the beam. We will choose the forces at the roller supports B and C as redundants. The primary (statically determinate) beam deforms as shown in Fig. 16–27b, and each redundant force deforms the beam as shown in Figs. 16–27c and 16–27d. By superposition, the compatibility equations for the displacements at B and C are therefore

$(+\downarrow)$ $\qquad\qquad\qquad 0 = v_B + v_B' + v_B''$

$(+\downarrow)$ $\qquad\qquad\qquad 0 = v_C + v_C' + v_C''$ $\qquad\qquad$ (16–16)

Using the table in Appendix C, all these displacement components can be expressed in terms of the known and unknown loads. Once this is done, the equations can then be solved simultaneously for the two unknowns B_y and C_y.

PROCEDURE FOR ANALYSIS

The following procedure provides a means for applying the method of superposition (or the force method) to determine the reactions on statically indeterminate beams or shafts.

Elastic Curve.

- Specify the unknown redundant forces or moments that must be removed from the beam in order to make it statically determinate and stable.

- Using the principle of superposition, draw the statically indeterminate beam and show it equal to a sequence of corresponding *statically determinate beams*.

- The first of these beams, the primary beam, supports the same external loads as the statically indeterminate beam, and each of the other beams "added" to the primary beam shows the beam loaded with a separate redundant force or moment.

- Sketch the deflection curve for each beam and indicate the displacement (slope) at the point of each redundant force (moment).

Compatibility Equations.

- Write a compatibility equation for the displacement (slope) at each point where there is a redundant force (moment).

Load–Displacement Equations.

- Relate all the displacements or slopes to the forces or moments using the formulas in Appendix C.

- Substitute the results into the compatibility equations and solve for the unknown redundants.

- If a numerical value for a redundant is *positive*, it has the *same sense of direction* as originally assumed. A *negative* numerical value indicates the redundant acts *opposite* to its assumed *sense of direction*.

Equilibrium Equations.

- Once the redundant forces and/or moments have been determined, the remaining unknown reactions can be found from the equations of equilibrium applied to the loadings shown on the beam's free-body diagram.

EXAMPLE 16.11

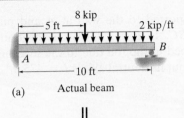

(a)   Actual beam

$$\parallel$$

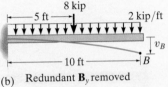

(b)   Redundant $\mathbf{B}_y$ removed

$$+$$

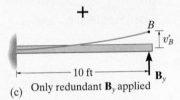

(c)   Only redundant $\mathbf{B}_y$ applied

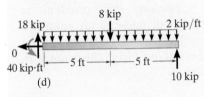

(d)

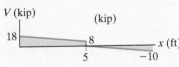

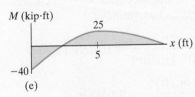

(e)

Fig. 16–28

Determine the reactions at the roller support B of the beam shown in Fig. 16–28a, then draw the shear and moment diagrams. EI is constant.

SOLUTION

Principle of Superposition.   By inspection, the beam is statically indeterminate to the first degree. The roller support at B will be chosen as the redundant so that $\mathbf{B}_y$ will be determined *directly*. Figures 16–28b and 16–28c show application of the principle of superposition. Here we have assumed that $\mathbf{B}_y$ acts upward on the beam.

Compatibility Equation.   Taking positive displacement as downward, the compatibility equation at B is

$$(+\downarrow) \qquad\qquad 0 = v_B - v'_B \qquad\qquad (1)$$

Load–Displacement Equations.   These displacements are related to the loads using the table in Appendix C.

$$v_B = \frac{wL^4}{8EI} + \frac{5PL^3}{48EI}$$

$$= \frac{2 \text{ kip/ft}(10 \text{ ft})^4}{8EI} + \frac{5(8 \text{ kip})(10 \text{ ft})^3}{48EI} = \frac{3333 \text{ kip} \cdot \text{ft}^3}{EI} \downarrow$$

$$v'_B = \frac{PL^3}{3EI} = \frac{B_y\,(10 \text{ ft})^3}{3EI} = \frac{333.3 \text{ ft}^3\,B_y}{EI} \uparrow$$

Substituting into Eq. 1 and solving yields

$$0 = \frac{3333}{EI} - \frac{333.3 B_y}{EI}$$

$$B_y = 10 \text{ kip} \qquad\qquad\qquad Ans.$$

Equilibrium Equations.   Using this result and applying the three equations of equilibrium, we obtain the results shown on the beam's free-body diagram in Fig. 16–28d. The shear and moment diagrams are shown in Fig. 16–28e.

16

EXAMPLE 16.12

The beam in Fig. 16–29a is fixed supported to the wall at A and pin connected to a $\frac{1}{2}$-in.-diameter rod BC. If $E = 29(10^3)$ ksi for both members, determine the force developed in the rod due to the loading. The moment of inertia of the beam about its neutral axis is $I = 475$ in^4.

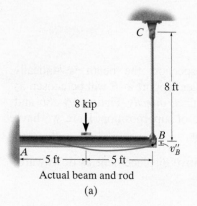

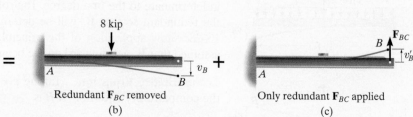

Fig. 16–29

SOLUTION I

Principle of Superposition. By inspection, this problem is indeterminate to the first degree. Here B will undergo an unknown displacement v_B'', since the rod will stretch. The rod will be treated as the redundant and hence the force of the rod is removed from the beam at B, Fig. 16–29b, and then reapplied, Fig. 16–29c.

Compatibility Equation. At point B we require

$$(+\downarrow) \qquad\qquad v_B'' = v_B - v_B' \qquad\qquad (1)$$

Load–Displacement Equations. The displacements v_B and v_B' are related to the loads using Appendix C. The displacement v_B'' is calculated from Eq. 9–2. Working in kilopounds and inches, we have

$$v_B'' = \frac{PL}{AE} = \frac{F_{BC}\,(8\text{ ft})(12\text{ in./ft})}{(\pi/4)\left(\frac{1}{2}\text{ in.}\right)^2[29(10^3)\text{ kip/in}^2]} = 0.01686 F_{BC} \downarrow$$

$$v_B = \frac{5PL^3}{48EI} = \frac{5(8\text{ kip})(10\text{ ft})^3\,(12\text{ in./ft})^3}{48[29(10^3)\text{ kip/in}^2](475\text{ in}^4)} = 0.1045 \text{ in.} \downarrow$$

$$v_B' = \frac{PL^3}{3EI} = \frac{F_{BC}\,(10\text{ ft})^3\,(12\text{ in./ft})^3}{3[29(10)^3\text{ kip/in}^2](475\text{ in}^4)} = 0.04181 F_{BC} \uparrow$$

Thus, Eq. 1 becomes

$$(+\downarrow) \qquad\qquad 0.01686 F_{BC} = 0.1045 - 0.04181 F_{BC}$$

$$F_{BC} = 1.78 \text{ kip} \qquad\qquad\qquad Ans.$$

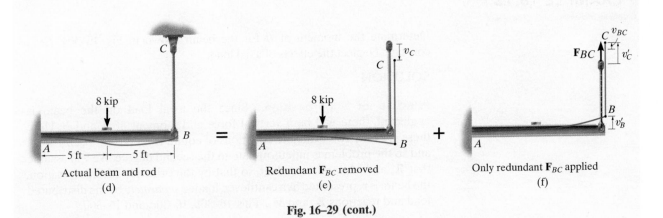

Actual beam and rod	Redundant $\mathbf{F}_{BC}$ removed	Only redundant $\mathbf{F}_{BC}$ applied
(d)	(e)	(f)

Fig. 16–29 (cont.)

SOLUTION II

Principle of Superposition. We can also solve this problem by removing the pin support at C and keeping the rod attached to the beam. In this case the 8-kip load will cause points B and C to be displaced downward the *same amount* v_C, Fig. 16–29e, since no force exists in rod BC. When the redundant force $\mathbf{F}_{BC}$ is applied at point C, it causes the end C of the rod to be displaced upward v'_C and the end B of the beam to be displaced upward v'_B, Fig. 16–29f. The difference in these two displacements, v_{BC}, represents the stretch of the rod due to $\mathbf{F}_{BC}$, so that $v'_C = v_{BC} + v'_B$. Hence, from Figs. 16–29d, 16–29e, and 16–29f, the compatibility of displacement at point C is

$$(+\downarrow) \qquad\qquad 0 = v_C - (v_{BC} + v'_B) \qquad\qquad (2)$$

From Solution I, we have

$$v_C = v_B = 0.1045 \text{ in. } \downarrow$$
$$v_{BC} = v''_B = 0.01686 F_{BC} \uparrow$$
$$v'_B = 0.04181 F_{BC} \uparrow$$

Therefore, Eq. 2 becomes

$$(+\downarrow)\, 0 = 0.1045 - (0.01686 F_{BC} + 0.04181 F_{BC})$$
$$F_{BC} = 1.78 \text{ kip} \qquad\qquad\qquad Ans.$$

EXAMPLE 16.13

Determine the moment at B for the beam shown in Fig. 16–30a. EI is constant. Neglect the effects of axial load.

SOLUTION

Principle of Superposition. Since the axial load on the beam is neglected, there will be a vertical force and moment at A and B. Here there are only two available equations of equilibrium ($\Sigma M = 0, \Sigma F_y = 0$), and so the problem is indeterminate to the second degree. We will assume that $\mathbf{B}_y$ and $\mathbf{M}_B$ are redundant, so that by the principle of superposition, the beam is represented as a cantilever, loaded *separately* by the distributed load and reactions $\mathbf{B}_y$ and $\mathbf{M}_B$, Figs. 16–30b, 16–30c, and 16–30d.

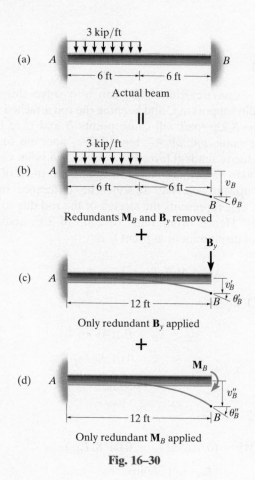

Fig. 16–30

Compatibility Equations. Referring to the displacement and slope at B, we require

$(\circlearrowleft+)$ $$0 = \theta_B + \theta'_B + \theta''_B \tag{1}$$

$(+\downarrow)$ $$0 = v_B + v'_B + v''_B \tag{2}$$

Load–Displacement Equations. Using the table in Appendix C to relate the slopes and displacements to the loads, we have

$$\theta_B = \frac{wL^3}{48EI} = \frac{3\text{ kip/ft }(12\text{ ft})^3}{48EI} = \frac{108\text{ kip}\cdot\text{ft}^2}{EI}\ \circlearrowright$$

$$v_B = \frac{7wL^4}{384EI} = \frac{7(3\text{ kip/ft})(12\text{ ft})^4}{384EI} = \frac{1134\text{ kip}\cdot\text{ft}^3}{EI}\ \downarrow$$

$$\theta'_B = \frac{PL^2}{2EI} = \frac{B_y(12\text{ ft})^2}{2EI} = \frac{72B_y}{EI}\ \circlearrowright$$

$$v'_B = \frac{PL^3}{3EI} = \frac{B_y(12\text{ ft})^3}{3EI} = \frac{576B_y}{EI}\ \downarrow$$

$$\theta''_B = \frac{ML}{EI} = \frac{M_B(12\text{ ft})}{EI} = \frac{12M_B}{EI}\ \circlearrowright$$

$$v''_B = \frac{ML^2}{2EI} = \frac{M_B(12\text{ ft})^2}{2EI} = \frac{72M_B}{EI}\ \downarrow$$

Substituting these values into Eqs. 1 and 2 and canceling out the common factor EI, we get

$(\circlearrowleft+)$ $$0 = 108 + 72B_y + 12M_B$$

$(+\downarrow)$ $$0 = 1134 + 576B_y + 72M_B$$

Solving these equations simultaneously gives

$$B_y = -3.375\text{ kip}$$

$$M_B = 11.25\text{ kip}\cdot\text{ft} \qquad\qquad Ans.$$

NOTE: The reactions at A can now be determined from the equilibrium equations.

FUNDAMENTAL PROBLEMS

F16–7. Determine the reactions at the fixed support A and the roller B. EI is constant.

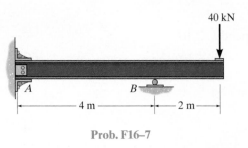

Prob. F16–7

F16–8. Determine the reactions at the fixed support A and the roller B. EI is constant.

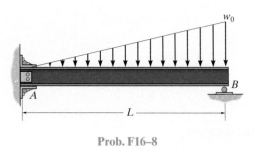

Prob. F16–8

F16–9. Determine the reactions at the fixed support A and the roller B. Support B settles 2 mm. $E = 200$ GPa, $I = 65.0(10^{-6})$ m^4.

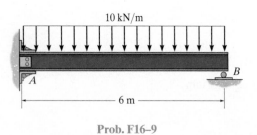

Prob. F16–9

F16–10. Determine the reaction at the roller B. EI is constant.

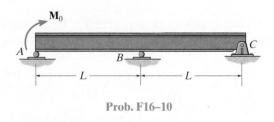

Prob. F16–10

F16–11. Determine the reaction at the roller B. EI is constant.

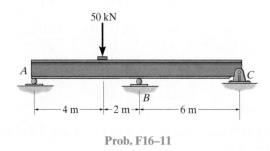

Prob. F16–11

F16–12. Determine the reaction at the roller support B if it settles 5 mm. $E = 200$ GPa and $I = 65.0(10^{-6})$ m^4.

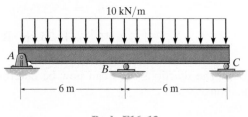

Prob. F16–12

PROBLEMS

16–61. Determine the reactions at the journal bearing supports *A*, *B*, and *C* of the shaft, then draw the shear and moment diagrams. *EI* is constant.

16–63. Determine the reactions at the supports, then draw the shear and moment diagrams. *EI* is constant.

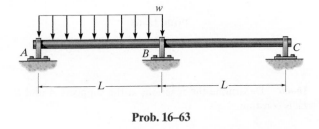

Prob. 16–63

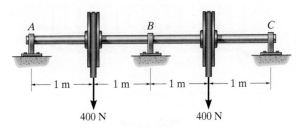

400 N   400 N

Prob. 16–61

16–62.   Determine the reactions at the supports, then draw the shear and moment diagrams. *EI* is constant.

*16–64.** Determine the reactions at the supports *A* and *B*. *EI* is constant.

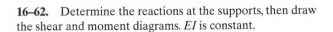

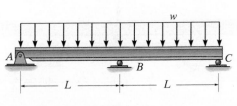

Prob. 16–62

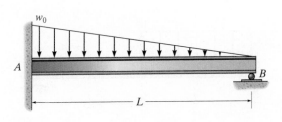

Prob. 16–64

16–65. The beam is used to support the 20-kip load. Determine the reactions at the supports. Assume A is fixed and B is a roller.

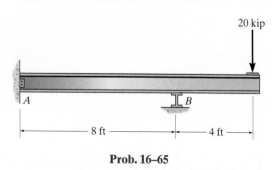

20 kip

|← 8 ft →|← 4 ft →|

Prob. 16–65

16–66. Determine the reactions at the supports A and B. EI is constant.

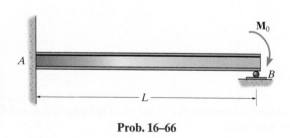

M_0

|← L →|

Prob. 16–66

16–67. Determine the reactions at the supports A and B. EI is constant.

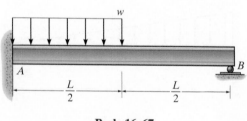

w

|← $\frac{L}{2}$ →|← $\frac{L}{2}$ →|

Prob. 16–67

*16–68. Before the uniform distributed load is applied to the beam, there is a small gap of 0.2 mm between the beam and the post at B. Determine the support reactions at A, B, and C. The post at B has a diameter of 40 mm, and the moment of inertia of the beam is $I = 875(10^6)$ mm^4. The post and the beam are made of material having a modulus of elasticity of $E = 200$ GPa.

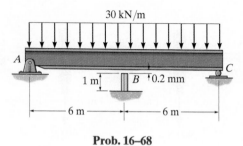

30 kN/m

1 m B 0.2 mm

|← 6 m →|← 6 m →|

Prob. 16–68

16–69. The fixed supported beam AB is strengthened using the simply supported beam CD and the roller at F which is set in place just before application of the load **P**. Determine the reactions at the supports if EI is constant.

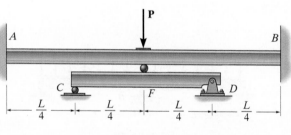

P

|← $\frac{L}{4}$ →|← $\frac{L}{4}$ →|← $\frac{L}{4}$ →|← $\frac{L}{4}$ →|

Prob. 16–69

16–70. The beam has a constant E_1I_1 and is supported by the fixed wall at B and the rod AC. If the rod has a cross-sectional area A_2 and the material has a modulus of elasticity E_2, determine the force in the rod.

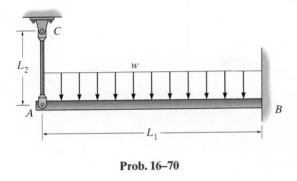

Prob. 16–70

16–71. The beam is supported by the bolted supports at its ends. When loaded these supports initially do not provide an actual fixed connection, but instead allow a slight rotation α before becoming fixed after the load is fully applied. Determine the moment at the support and the maximum deflection of the beam.

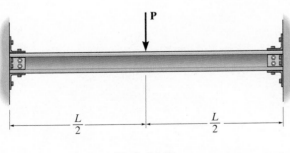

Prob. 16–71

*16–72. Each of the two members is made from 6061-T6 aluminum and has a square cross section 1 in. $\times$ 1 in. They are pin connected at their ends and a jack is placed between them and opened until the force it exerts on each member is 50 lb. Determine the greatest force P that can be applied to the center of the top member without causing either of the two members to yield. For the analysis neglect the axial force in each member. Assume the jack is rigid.

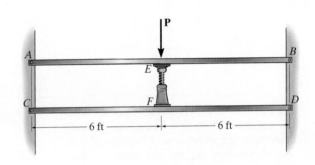

Prob. 16–72

16–73. The beam is made from a soft linear elastic material having a constant EI. If it is originally a distance Δ from the surface of its end support, determine the length a that rests on this support when it is subjected to the uniform load w_0, which is great enough to cause this to happen.

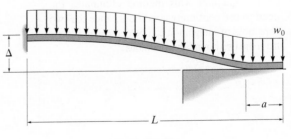

Prob. 16–73

CHAPTER REVIEW

The elastic curve represents the centerline deflection of a beam or shaft. Its shape can be determined using the moment diagram. Positive moments cause the elastic curve to be concave upwards and negative moments cause it to be concave downwards. The radius of curvature at any point is determined from

$$\frac{1}{\rho} = \frac{M}{EI}$$

Moment diagram

Inflection point

Elastic curve

The equation of the elastic curve and its slope can be obtained by first finding the internal moment in the member as a function of x. If several loadings act on the member, then separate moment functions must be determined between each of the loadings. Integrating these functions once using $EI(d^2v/dx^2) = M(x)$ gives the equation for the slope of the elastic curve, and integrating again gives the equation for the deflection. The constants of integration are determined from the boundary conditions at the supports, or in cases where several moment functions are involved, continuity of slope and deflection at points where these functions join must be satisfied.

$\theta = 0$
$v = 0$

$v = 0$

Boundary conditions

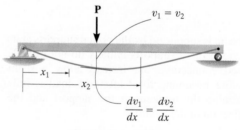

$v_1 = v_2$

$\dfrac{dv_1}{dx} = \dfrac{dv_2}{dx}$

Continuity conditions

Discontinuity functions allow one to express the equation of the elastic curve as a continuous function, regardless of the number of loadings on the member. This method eliminates the need to use continuity conditions, since the two constants of integration can be determined solely from the two boundary conditions.

The deflection or slope at a point on a member subjected to combinations of loadings can be determined using the method of superposition. The table in Appendix D is available for this purpose.

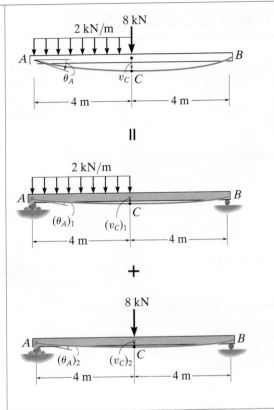

Statically indeterminate beams and shafts have more unknown support reactions than available equations of equilibrium. To solve, one first identifies the redundant reactions. The method of integration can then be used to solve for the unknown redundants. It is also possible to determine the redundants by using the method of superposition, where one considers the conditions of continuity at the redundant. Here the displacement due to the external loading is determined with the redundant removed, and again with the redundant applied and the external loading removed. The tables in Appendix D can be used to determine these necessary displacements.

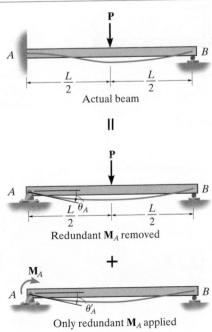

REVIEW PROBLEMS

R16–1. Determine the equation of the elastic curve. Use discontinuity functions EI is constant.

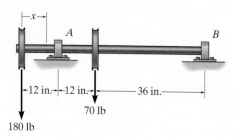

Prob. R16–1

R16–2. Draw the bending-moment diagram for the shaft and then, from this diagram, sketch the deflection or elastic curve for the shaft's centerline. Determine the equations of the elastic curve using the coordinates x_1 and x_2. Use the method of integration. EI is constant.

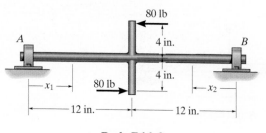

Prob. R16–2

R16–3. Determine the moment reactions at the supports A and B. Use the method of integration. EI is constant.

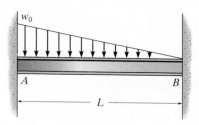

Prob. R16–3

*R16–4. Determine the equations of the elastic curve for the beam using the x_1 and x_2 coordinates. Specify the slope at A and the maximum deflection. Use the method of integration. EI is constant.

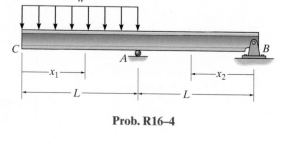

Prob. R16–4

R16–5. Determine the maximum deflection between the supports A and B. Use the method of integration. EI is constant.

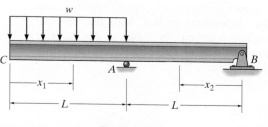

Prob. R16–5

R16–6. If the cantilever beam has a constant thickness t, determine the deflection at end A. The beam is made of material having a modulus of elasticity E.

*R16–8. Using the method of superposition, determine the magnitude of $\mathbf{M}_0$ in terms of the distributed load w and dimension a so that the deflection at the center of the beam is zero. EI is constant.

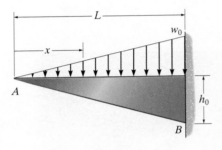

Prob. R16–6

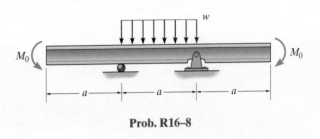

Prob. R16–8

R16–9. Using the method of superposition, determine the deflection at C of beam AB. The beams are made of wood having a modulus of elasticity of $E = 1.5(10^3)$ ksi.

R16–7. The framework consists of two A-36 steel cantilevered beams CD and BA and a simply supported beam CB. If each beam is made of steel and has a moment of inertia about its principal axis of $I_x = 118$ in⁴, determine the deflection at the center G of beam CB. Use the method of superposition.

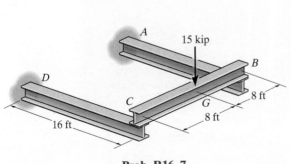

Prob. R16–7

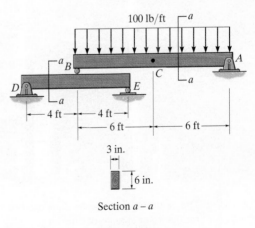

Prob. R16–9

16

CHAPTER 17

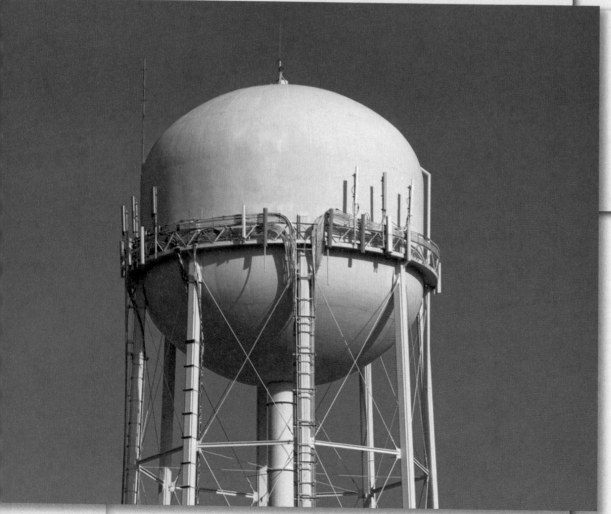

The columns of this water tank are braced at points along their length in order to reduce their chance of buckling.

BUCKLING OF COLUMNS

CHAPTER OBJECTIVES

- To develop methods to calculate the critical loads causing columns to buckle in the elastic region of the stress-strain curve.

- To consider the effect of different supports on the critical load for buckling.

17.1 CRITICAL LOAD

Not only must a member satisfy specific strength and deflection requirements but it must also be stable. Stability is particularly important if the member is long and slender, and it supports a compressive loading that becomes large enough to cause the member to suddenly deflect laterally or sidesway. These members are called *columns*, and the lateral deflection that occurs is called *buckling*. Quite often the buckling of a column can lead to a sudden and dramatic failure of a structure or mechanism, and as a result, special attention must be given to the design of columns so that they can safely support their intended loadings without buckling.

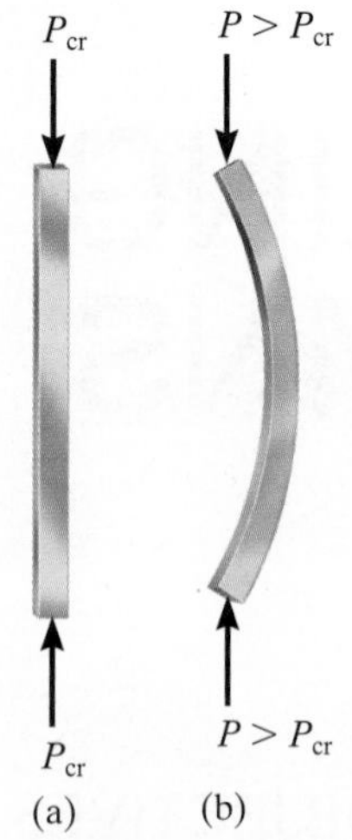

Fig. 17–1

The maximum axial load that a column can support when it is on the *verge of buckling* is called the ***critical load***, P_{cr}, Fig. 17–1a. Any additional loading will cause the column to buckle and therefore deflect laterally as shown in Fig. 17–1b.

We can study the nature of this instability by considering the two-bar mechanism consisting of weightless rigid bars that are pin connected as shown in Fig. 17–2a. When the bars are in the vertical position, the spring, having a stiffness k, is unstretched, and a *small* vertical force **P** is applied at the top of one of the bars. To upset this equilibrium position the pin at A is displaced by a small amount Δ, Fig. 17–2b. As shown on the free-body diagram of the pin, Fig. 17–2c, the spring will produce a restoring force $F = k\Delta$ in order to resist the two horizontal components, $P_x = P \tan \theta$, which tend to push the pin (and the bars) further out of equilibrium. Since θ is small, $\Delta \approx \theta(L/2)$ and $\tan \theta \approx \theta$. Thus the *restoring* spring force becomes $F = k\theta(L/2)$, and the *disturbing* force is $2P_x = 2P\theta$.

If the restoring force is greater than the disturbing force, that is, $k\theta L/2 > 2P\theta$, then, noticing that θ cancels out, we can solve for P, which gives

$$P < \frac{kL}{4} \qquad \text{stable equilibrium}$$

This is a condition for ***stable equilibrium***, since the force developed by the spring would be adequate to restore the bars back to their vertical position. However, if $k\theta(L/2) < 2P\theta$, or

$$P > \frac{kL}{4} \qquad \text{unstable equilibrium}$$

then the bars will be in ***unstable equilibrium***. In other words, if this load is applied, and a slight displacement occurs at A, the bars will tend to move out of equilibrium and not be restored to their original position.

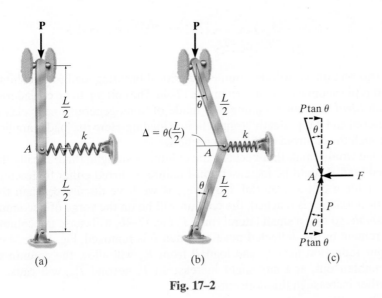

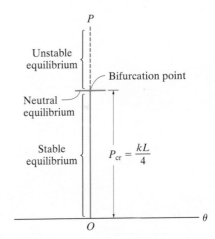

Fig. 17–2

The intermediate value of P, which requires $kL\theta/2 = 2P\theta$, is the *critical load.* Here

$$P_{cr} = \frac{kL}{4} \qquad \text{neutral equilibrium}$$

This loading represents a case of the bars being in **neutral equilibrium**. Since P_{cr} is *independent* of the (small) displacement θ of the bars, any slight disturbance given to the mechanism will not cause it to move further out of equilibrium, nor will it be restored to its original position. Instead, the bars will simply *remain* in the deflected position.

These three different states of equilibrium are represented graphically in Fig. 17–3. The transition point where the load is equal to its critical value $P = P_{cr}$ is called the **bifurcation point**. Here the bars will be in neutral equilibrium for any *small value* of θ. If a larger load P is placed on the bars, then they will undergo a larger deflection, so that the spring is compressed or elongated enough to hold them in equilibrium.

In a similar manner, if the load on an actual column exceeds its critical loading, then this loading will also require the column to undergo a *large* deflection; however, this is generally not tolerated in engineering structures or machines.

Fig. 17–3

17.2   IDEAL COLUMN WITH PIN SUPPORTS

In this section we will determine the critical buckling load for a column that is pin supported as shown in Fig. 17–4*a*. The column to be considered is an ***ideal column***, meaning it is made of homogeneous linear elastic material and it is perfectly straight before loading. Here the load is applied through the centroid of the cross section.

One would think that because the column is straight, theoretically the axial load *P* could be increased until failure occurred either by fracture or yielding of the material. However, as we have discussed, when the critical load P_{cr} is reached, the column will be on the verge of becoming *unstable*, so that a small lateral force *F*, Fig. 17–4*b*, will cause the column to remain in the deflected position when *F* is removed, Fig. 17–4*c*. Any slight reduction in the axial load *P* from P_{cr} will allow the column to straighten out, and any slight increase in *P*, beyond P_{cr}, will cause a further increase in this deflection.

The tendency of a column to remain stable or become unstable when subjected to an axial load actually depends upon its ability to resist bending. Hence, in order to determine the critical load and the buckled shape of the column, we will apply Eq. 16–10, which relates the internal moment in the column to its deflected shape, i.e.,

$$EI\frac{d^2v}{dx^2} = M \qquad (17\text{–}1)$$

The dramatic failure of this off-shore oil platform was caused by the horizontal forces of hurricane winds, which led to buckling of its supporting columns.

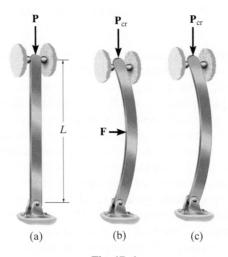

(a)   (b)   (c)

Fig. 17–4

A free-body diagram of a segment of the column in the deflected position is shown in Fig. 17–5a. Here both the displacement v and the internal moment M are shown in the *positive direction.* Since moment equilibrium requires $M = -Pv$, then Eq. 17–1 becomes

$$EI\frac{d^2v}{dx^2} = -Pv$$

$$\frac{d^2v}{dx^2} + \left(\frac{P}{EI}\right)v = 0 \qquad (17\text{–}2)$$

This is a homogeneous, second-order, linear differential equation with constant coefficients. It can be shown by using the methods of differential equations, or by direct substitution into Eq. 17–2, that the general solution is

$$v = C_1 \sin\left(\sqrt{\frac{P}{EI}}x\right) + C_2 \cos\left(\sqrt{\frac{P}{EI}}x\right) \qquad (17\text{–}3)$$

The two constants of integration can be determined from the boundary conditions at the ends of the column. Since $v = 0$ at $x = 0$, then $C_2 = 0$. And since $v = 0$ at $x = L$, then

$$C_1 \sin\left(\sqrt{\frac{P}{EI}}L\right) = 0$$

This equation is satisfied if $C_1 = 0$; however, then $v = 0$, which is a *trivial solution* that requires the column to always remain straight, even though the load may cause the column to become unstable. The other possibility requires

$$\sin\left(\sqrt{\frac{P}{EI}}L\right) = 0$$

which is satisfied if

$$\sqrt{\frac{P}{EI}}L = n\pi$$

or

$$P = \frac{n^2\pi^2 EI}{L^2} \qquad n = 1, 2, 3, \ldots \qquad (17\text{–}4)$$

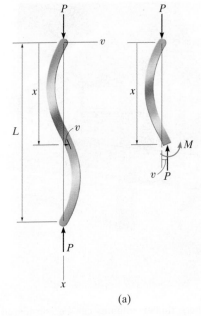

(a)

Fig. 17–5

These timber columns can be considered pinned at their bottom and fixed connected to the beams at their top.

The *smallest value* of P is obtained when $n = 1$, so the *critical load* for the column is therefore*

$$P_{cr} = \frac{\pi^2 EI}{L^2}$$

This load is sometimes referred to as the **Euler load**, named after the Swiss mathematician Leonhard Euler, who originally solved this problem in 1757. From Eq. 17–3, the corresponding buckled shape, shown in Fig. 17–5b, is therefore

$$v = C_1 \sin \frac{\pi x}{L}$$

The constant C_1 represents the maximum deflection, v_{max}, which occurs at the midpoint of the column, Fig. 17–5b. Unfortunately, a specific value for C_1 cannot be obtained once it has buckled. It is assumed, however, that this deflection is small.

As noted above, the critical load depends on the material's stiffness or modulus of elasticity E and not its yield stress. Therefore, a column made of high-strength steel offers no advantage over one made of lower-strength steel, since the modulus of elasticity for both materials is the same. Also note that P_{cr} will increase as the moment of inertia of the cross section increases. Thus, efficient columns are designed so that most of the column's cross-sectional area is located as far away as possible from the center of the section. This is why hollow sections such as tubes are more economical than solid sections. Furthermore, wide-flange sections, and columns that are "built up" from channels, angles, plates, etc., are better than sections that are solid and rectangular.

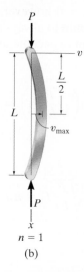

Fig. 17–5 (cont.)

*n represents the number of curves in the deflected shape of the column. For example, if $n = 1$, then one curve appears as in Fig. 17–5b; if $n = 2$, then *two* curves appear as in Fig. 17–5a, etc.

Since P_{cr} is directly related to I, a column will buckle about the principal axis of the cross section having the **least moment of inertia** (the weakest axis), provided it is supported the same way about each axis. For example, a column having a rectangular cross section, like a meter stick, Fig. 17–6, will buckle about the a–a axis, not the b–b axis. Because of this, engineers usually try to achieve a balance, keeping the moments of inertia the same in all directions. Geometrically, then, circular tubes make excellent columns. Square tubes or those shapes having $I_x \approx I_y$ are also often selected for columns.

To summarize, the buckling equation for a pin-supported long slender column is

$$P_{cr} = \frac{\pi^2 EI}{L^2} \qquad (17\text{–}5)$$

where

P_{cr} = critical or maximum axial load on the column just before it begins to buckle. This load must *not* cause the stress in the column to exceed the proportional limit.

E = modulus of elasticity for the material

I = *least* moment of inertia for the column's cross-sectional area

L = unsupported length of the column, whose ends are pinned

For design purposes, the above equation can also be written in terms of stress, by using $I = Ar^2$, where A is the cross-sectional area and r is the **radius of gyration** of the cross-sectional area. We have,

$$P_{cr} = \frac{\pi^2 E(Ar^2)}{L^2}$$

$$\left(\frac{P}{A}\right)_{cr} = \frac{\pi^2 E}{(L/r)^2}$$

or

$$\sigma_{cr} = \frac{\pi^2 E}{(L/r)^2} \qquad (17\text{–}6)$$

Here

σ_{cr} = critical stress, which is an average normal stress in the column just before the column buckles. It is required that $\sigma_{cr} \leq \sigma_Y$.

E = modulus of elasticity for the material

L = unsupported length of the column, whose ends are pinned

r = *smallest* radius of gyration of the column, determined from $r = \sqrt{I/A}$, where I is the *least* moment of inertia of the column's cross-sectional area A

The geometric ratio L/r in Eq. 17–6 is known as the **slenderness ratio**. It is a measure of the column's flexibility, and as we will discuss later, it serves to classify columns as long, intermediate, or short.

P
a   b
b   a

Fig. 17–6

Failure of this crane boom was caused by the localized buckling of one of its tubular struts.

17

A graph of this equation for columns made of structural steel and an aluminum alloy is shown in Fig. 17–7. The curves are hyperbolic and are valid only for critical stresses that are below the material's yield point (proportional limit). Notice that for the steel the yield stress is $(\sigma_Y)_{st} = 36$ ksi $[E_{st} = 29(10^3)$ ksi], and for the aluminum it is $(\sigma_Y)_{al} = 27$ ksi $[E_{al} = 10(10^3)$ ksi]. If we substitute $\sigma_{cr} = \sigma_Y$ into Eq. 17–6, the *smallest* allowable slenderness ratios for the steel and aluminum columns then become $(L/r)_{st} = 89$ and $(L/r)_{al} = 60.5$, Fig. 17–7. Thus, for a steel column, if $(L/r)_{st} < 89$, the column's stress will exceed the yield point before buckling can occur, and so the Euler formula cannot be used.

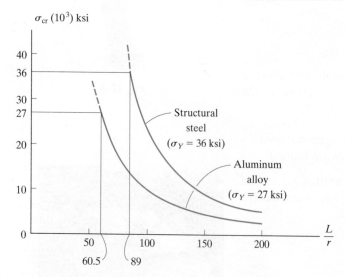

Fig. 17–7

IMPORTANT POINTS

- *Columns* are long slender members that are subjected to axial compressive loads.

- The *critical load* is the maximum axial load that a column can support when it is on the verge of buckling. This loading represents a case of *neutral equilibrium.*

- An *ideal column* is initially perfectly straight, made of homogeneous material, and the load is applied through the centroid of its cross section.

- A pin-connected column will buckle about the principal axis of the cross section having the *least* moment of inertia.

- The *slenderness ratio* is L/r, where r is the smallest radius of gyration of the cross section. Buckling will occur about the axis where this ratio gives the greatest value.

EXAMPLE 17.1

The A992 steel W8 × 31 member shown in Fig. 17–8 is to be used as a pin-connected column. Determine the largest axial load it can support before it either begins to buckle or the steel yields. Take $\sigma_Y = 50$ ksi.

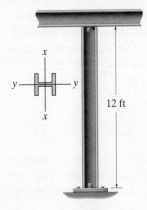

12 ft

Fig. 17–8

SOLUTION

From the table in Appendix B, the column's cross-sectional area and moments of inertia are $A = 9.13$ in^2, $I_x = 110$ in^4, and $I_y = 37.1$ in^4. By inspection, buckling will occur about the y–y axis. Why? Applying Eq. 17–5, we have

$$P_{cr} = \frac{\pi^2 EI}{L^2} = \frac{\pi^2[29(10^3)\ \text{kip/in}^2](37.1\ \text{in}^4)}{[12\ \text{ft}(12\ \text{in.}/\text{ft})]^2} = 512\ \text{kip}$$

When this load is applied, the average compressive stress in the column is

$$\sigma_{cr} = \frac{P_{cr}}{A} = \frac{512\ \text{kip}}{9.13\ \text{in}^2} = 56.1\ \text{ksi}$$

Since this stress exceeds the yield stress (50 ksi), the load P is determined from simple compression.

$$50\ \text{ksi} = \frac{P}{9.13\ \text{in}^2}$$

$$P = 456\ \text{kip} \qquad\qquad Ans.$$

In actual practice, a factor of safety would be placed on this loading.

17

The tubular columns used to support this water tank have been braced at three locations along their length to prevent them from buckling.

17.3 COLUMNS HAVING VARIOUS TYPES OF SUPPORTS

The Euler load in Sec. 17.2 was derived for a column that is pin connected or free to rotate at its ends. Oftentimes, however, columns may be supported in other ways. For example, consider the case of a column fixed at its base and free at the top, Fig. 17–9a. As the column buckles, the load will sidesway and be displaced δ, while at x the displacement is v. From the free-body diagram in Fig. 17–9b, the internal moment at the arbitrary section is $M = P(\delta - v)$, and so the differential equation for the deflection curve is

$$EI\frac{d^2v}{dx^2} = P(\delta - v)$$

$$\frac{d^2v}{dx^2} + \frac{P}{EI}v = \frac{P}{EI}\delta \qquad (17\text{–}7)$$

Unlike Eq. 17–2, this equation is nonhomogeneous because of the nonzero term on the right side. The solution consists of both a complementary and a particular solution, namely,

$$v = C_1\sin\left(\sqrt{\frac{P}{EI}}x\right) + C_2\cos\left(\sqrt{\frac{P}{EI}}x\right) + \delta$$

The constants are determined from the boundary conditions. At $x = 0$, $v = 0$, and so $C_2 = -\delta$. Also,

$$\frac{dv}{dx} = C_1\sqrt{\frac{P}{EI}}\cos\left(\sqrt{\frac{P}{EI}}x\right) - C_2\sqrt{\frac{P}{EI}}\sin\left(\sqrt{\frac{P}{EI}}x\right)$$

and at $x = 0, dv/dx = 0$, then $C_1 = 0$. The deflection curve is therefore

$$v = \delta\left[1 - \cos\left(\sqrt{\frac{P}{EI}}x\right)\right] \qquad (17\text{–}8)$$

Finally, at the top of the column $x = L, v = \delta$, so that

$$\delta\cos\left(\sqrt{\frac{P}{EI}}L\right) = 0$$

The trivial solution $\delta = 0$ indicates that no buckling occurs, regardless of the load P. Instead,

$$\cos\left(\sqrt{\frac{P}{EI}}L\right) = 0 \qquad \text{or} \qquad \sqrt{\frac{P}{EI}}L = \frac{n\pi}{2}, n = 1, 3, 5\ldots$$

The smallest critical load occurs when $n = 1$, so that

$$P_{\text{cr}} = \frac{\pi^2 EI}{4L^2} \qquad (17\text{–}9)$$

By comparison with Eq. 17–5, it is seen that a column fixed supported at its base and free at its top will support only one-fourth the critical load that can be applied to a column pin supported at both ends.

(a)   (b)

Fig. 17–9

17

Other types of supported columns are analyzed in much the same way and will not be covered in detail here.* Instead, we will tabulate the results for the most common types of column support and show how to apply these results by writing Euler's formula in a general form.

Effective Length. To use Euler's formula, Eq. 17–5, for columns having different types of supports, we will modify the column length L to represent the distance between the points of zero moment on the column. This distance is called the column's **effective length**, L_e. Obviously, for a pin-ended column $L_e = L$, Fig. 17–10a. For the fixed-end and free-ended column, the deflection curve is defined by Eq. 17–8. When plotted, its shape is equivalent to a pin-ended column having a length of $2L$, Fig. 17–10b, and so the effective length between the points of zero moment is $L_e = 2L$. Examples for two other columns with different end supports are also shown in Fig. 17–10. The column fixed at its ends, Fig. 17–10c, has inflection points or points of zero moment $L/4$ from each support. The effective length is therefore defined by the middle half of its length, that is, $L_e = 0.5L$. Finally, the pin- and fixed-ended column, Fig. 17–10d, has an inflection point at approximately $0.7L$ from its pinned end, so that $L_e = 0.7L$.

Rather than specifying the column's effective length, many design codes provide column formulas that employ a dimensionless coefficient K called the **effective-length factor**. This factor is defined from

$$L_e = KL \qquad (17\text{–}10)$$

Specific values of K are also given in Fig. 17–10. Based on this generality, we can therefore write Euler's formula as

or

$$P_{cr} = \frac{\pi^2 EI}{(KL)^2} \qquad (17\text{–}11)$$

$$\sigma_{cr} = \frac{\pi^2 E}{(KL/r)^2} \qquad (17\text{–}12)$$

Here (KL/r) is the column's **effective-slenderness ratio**.

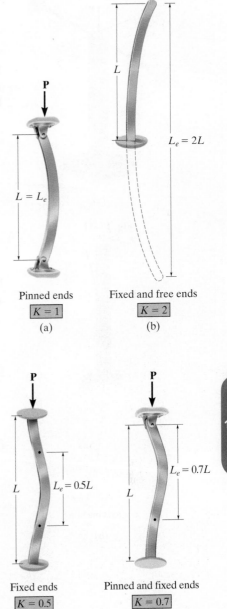

Pinned ends
$K = 1$
(a)

Fixed and free ends
$K = 2$
(b)

$L_e = 2L$

Fixed ends
$K = 0.5$
(c)

$L_e = 0.5L$

Pinned and fixed ends
$K = 0.7$
(d)

$L_e = 0.7L$

Fig. 17–10

*See Problems 17–43, 17–44, and 17–45.

17

EXAMPLE 17.2

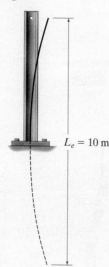

z

P

y

x

5 m

(a)

Fig. 17-11

$L_e = 10$ m

x–x axis buckling

(b)

The aluminum column in Fig. 17–11a is braced at its top by cables so as to prevent movement at the top along the x axis. If it is assumed to be fixed at its base, determine the largest allowable load **P** that can be applied. Use a factor of safety for buckling of F.S. = 3.0. Take $E_{al} = 70$ GPa, $\sigma_Y = 215$ MPa, $A = 7.5(10^{-3})$ m², $I_x = 61.3(10^{-6})$ m⁴, $I_y = 23.2(10^{-6})$ m⁴.

SOLUTION

Buckling about the x and y axes is shown in Figs. 17–11b and 17–11c. Using Fig.17–10a, for x–x axis buckling, $K = 2$, so $(KL)_x = 2(5 \text{ m}) = 10$ m. For y–y axis buckling, $K = 0.7$, so $(KL)_y = 0.7(5 \text{ m}) = 3.5$ m. Applying Eq. 17–11, the critical loads for each case are

$$(P_{cr})_x = \frac{\pi^2 EI_x}{(KL)_x^2} = \frac{\pi^2[70(10^9) \text{ N/m}^2](61.3(10^{-6}) \text{ m}^4)}{(10 \text{ m})^2}$$

$$= 424 \text{ kN}$$

$$(P_{cr})_y = \frac{\pi^2 EI_y}{(KL)_y^2} = \frac{\pi^2[70(10^9) \text{ N/m}^2](23.2(10^{-6}) \text{ m}^4)}{(3.5 \text{ m})^2}$$

$$= 1.31 \text{ MN}$$

By comparison, as P is increased the column will buckle about the x–x axis. The allowable load is therefore

$$P_{allow} = \frac{P_{cr}}{\text{F.S.}} = \frac{424 \text{ kN}}{3.0} = 141 \text{ kN} \qquad Ans.$$

Since

$$\sigma_{cr} = \frac{P_{cr}}{A} = \frac{424 \text{ kN}}{7.5(10^{-3}) \text{ m}^2} = 56.5 \text{ MPa} < 215 \text{ MPa}$$

Euler's equation is valid.

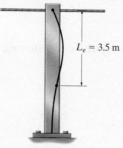

$L_e = 3.5$ m

y–y axis buckling

(c)

EXAMPLE 17.3

A W6 × 15 steel column is 24 ft long and is fixed at its ends as shown in Fig. 17–12a. Its load-carrying capacity is increased by bracing it about the y–y (weak) axis using struts that are assumed to be pin connected at its midheight. Determine the load the column can support so that it does not buckle nor the material exceed the yield stress. Take $E_{st} = 29(10^3)$ ksi and $\sigma_Y = 60$ ksi.

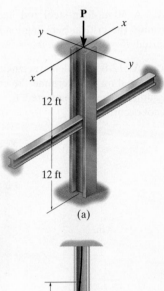

(a)

SOLUTION

The buckling behavior of the column will be *different* about the x–x and y–y axes due to the bracing. The buckled shape for each of these cases is shown in Figs. 17–12b and 17–12c. From Fig. 17–12b, the effective length for buckling about the x–x axis is $(KL)_x = 0.5(24\text{ ft}) = 12\text{ ft} = 144\text{ in.}$, and from Fig. 17–12c, for buckling about the y–y axis $(KL)_y = 0.7(24\text{ ft}/2) = 8.40\text{ ft} = 100.8\text{ in.}$ The moments of inertia for a W6 × 15 are found from the table in Appendix B. We have $I_x = 29.1\text{ in}^4$, $I_y = 9.32\text{ in}^4$.

Applying Eq. 17–11,

$$(P_{cr})_x = \frac{\pi^2 EI_x}{(KL)_x^2} = \frac{\pi^2[29(10^3)\text{ ksi}]29.1\text{ in}^4}{(144\text{ in.})^2} = 401.7\text{ kip} \qquad (1)$$

$$(P_{cr})_y = \frac{\pi^2 EI_y}{(KL)_y^2} = \frac{\pi^2[29(10^3)\text{ ksi}]9.32\text{ in}^4}{(100.8\text{ in.})^2} = 262.5\text{ kip} \qquad (2)$$

x–x axis buckling

(b)

By comparison, buckling will occur about the y–y axis.

The area of the cross section is 4.43 in², so the average compressive stress in the column is

$$\sigma_{cr} = \frac{P_{cr}}{A} = \frac{262.5\text{ kip}}{4.43\text{ in}^2} = 59.3\text{ ksi}$$

Since this stress is less than the yield stress, buckling will occur before the material yields. Thus,

$$P_{cr} = 263\text{ kip} \qquad\qquad Ans.$$

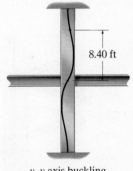

y–y axis buckling

(c)

Fig. 17–12

17

FUNDAMENTAL PROBLEMS

F17–1. A 50-in.-long steel rod has a diameter of 1 in. Determine the critical buckling load if the ends are fixed supported. $E = 29(10^3)$ ksi, $\sigma_Y = 36$ ksi.

F17–2. A 12-ft wooden rectangular column has the dimensions shown. Determine the critical load if the ends are assumed to be pin connected. $E = 1.6(10^3)$ ksi. Yielding does not occur.

F17–4. A steel pipe is fixed supported at its ends. If it is 5 m long and has an outer diameter of 50 mm and a thickness of 10 mm, determine the maximum axial load P that it can carry without buckling. $E_{st} = 200$ GPa, $\sigma_Y = 250$ MPa.

F17–5. Determine the maximum force **P** that can be supported by the assembly without causing member AC to buckle. The member is made of A992 steel and has a diameter of 2 in. Take F.S. = 2 against buckling.

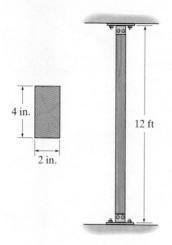

Prob. F17–2

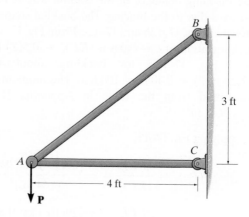

Prob. F17–5

F17–3. The A992 steel column can be considered pinned at its top and bottom and braced against its weak axis at the mid-height. Determine the maximum allowable force **P** that the column can support without buckling. Apply a F.S. = 2 against buckling. Take $A = 7.4(10^{-3})$ m^2, $I_x = 87.3(10^{-6})$ m^4, and $I_y = 18.8(10^{-6})$ m^4.

F17–6. The A992 steel rod BC has a diameter of 50 mm and is used as a strut to support the beam. Determine the maximum intensity w of the uniform distributed load that can be applied to the beam without causing the strut to buckle. Take F.S. = 2 against buckling.

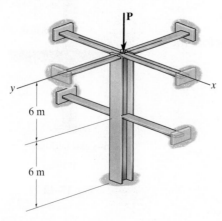

Prob. F17–3

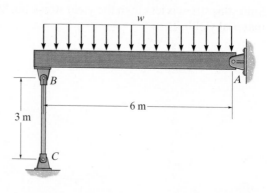

Prob. F17–6

PROBLEMS

17–1. Determine the critical buckling load for the column. The material can be assumed rigid.

17–3. The aircraft link is made from an A992 steel rod. Determine the smallest diameter of the rod, to the nearest $\frac{1}{16}$ in., that will support the load of 4 kip without buckling. The ends are pin connected.

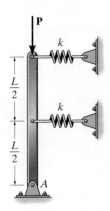

Prob. 17–1

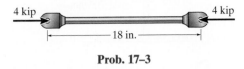

Prob. 17–3

*17–4. Rigid bars AB and BC are pin connected at B. If the spring at D has a stiffness k, determine the critical load P_{cr} that can be applied to the bars.

17–2. The column consists of a rigid member that is pinned at its bottom and attached to a spring at its top. If the spring is unstretched when the column is in the vertical position, determine the critical load that can be placed on the column.

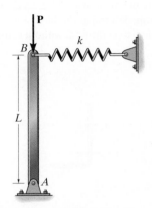

Prob. 17–2

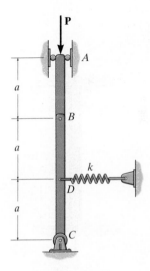

Prob. 17–4

17–5. A 2014-T6 aluminum alloy column has a length of 6 m and is fixed at one end and pinned at the other. If the cross-sectional area has the dimensions shown, determine the critical load. $\sigma_Y = 250$ MPa.

17–6. Solve Prob. 17–5 if the column is pinned at its top and bottom.

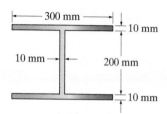

Probs. 17–5/6

17–7. The W12 × 50 is made of A992 steel and is used as a column that has a length of 20 ft. If its ends are assumed pin supported, and it is subjected to an axial load of 150 kip, determine the factor of safety with respect to buckling.

***17–8.** The W12 × 50 is made of A992 steel and is used as a column that has a length of 20 ft. If the ends of the column are fixed supported, can the column support the critical load without yielding?

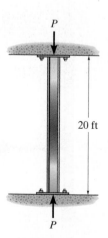

Probs. 17–7/8

17–9. A steel column has a length of 9 m and is fixed at both ends. If the cross-sectional area has the dimensions shown, determine the critical load. $E_{st} = 200$ GPa, $\sigma_Y = 250$ MPa.

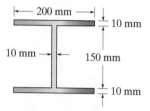

Prob. 17–9

17–10. A steel column has a length of 9 m and is pinned at its top and bottom. If the cross-sectional area has the dimensions shown, determine the critical load. $E_{st} = 200$ GPa, $\sigma_Y = 250$ MPa.

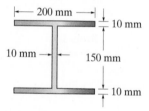

Prob. 17–10

17–11. The A992 steel angle has a cross-sectional area of $A = 2.48$ in^2 and a radius of gyration about the x axis of $r_x = 1.26$ in. and about the y axis of $r_y = 0.879$ in. The smallest radius of gyration occurs about the a–a axis and is $r_a = 0.644$ in. If the angle is to be used as a pin-connected 10-ft-long column, determine the largest axial load that can be applied through its centroid C without causing it to buckle.

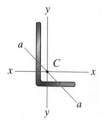

Prob. 17–11

*17–12.   The 50-mm-diameter C86100 bronze rod is fixed supported at A and has a gap of 2 mm from the wall at B. Determine the increase in temperature ΔT that will cause the rod to buckle. Assume that the contact at B acts as a pin.

17–14.   The W8 × 67 wide-flange A-36 steel column can be assumed fixed at its base and pinned at its top. Determine the largest axial force P that can be applied without causing it to buckle.

17–15.   Solve Prob. 17–14 if the column is assumed fixed at its bottom and free at its top.

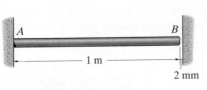

Prob. 17–12

Probs. 17–14/15

17–13.   Determine the maximum load P the frame can support without buckling member AB. Assume that AB is made of steel and is pinned at its ends for y–y axis buckling and fixed at its ends for x–x axis buckling. $E_{st} = 200$ GPa, $\sigma_Y = 360$ MPa.

*17–16.   An A992 steel W200 × 46 column of length 9 m is fixed at one end and free at its other end. Determine the allowable axial load the column can support if F.S. = 2 against buckling.

17–17.   The 10-ft wooden rectangular column has the dimensions shown. Determine the critical load if the ends are assumed to be pin connected. $E_w = 1.6(10^3)$ ksi, $\sigma_Y = 5$ ksi.

17–18.   The 10-ft wooden column has the dimensions shown. Determine the critical load if the bottom is fixed and the top is pinned. $E_w = 1.6(10^3)$ ksi, $\sigma_Y = 5$ ksi.

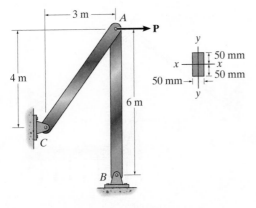

Prob. 17–13

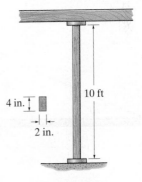

Probs. 17–17/18

17–19. Determine the maximum force P that can be applied to the handle so that the A992 steel control rod AB does not buckle. The rod has a diameter of 1.25 in. It is pin connected at its ends.

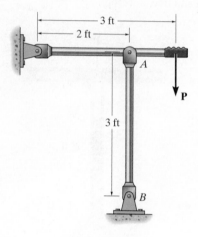

Prob. 17–19

*17–20. The A-36 steel pipe has an outer diameter of 2 in. and a thickness of 0.5 in. If it is held in place by a guywire, determine the largest vertical force P that can be applied without causing the pipe to buckle. Assume that the ends of the pipe are pin connected.

17–21. The A-36 steel pipe has an outer diameter of 2 in. If it is held in place by a guywire, determine its required inner diameter to the nearest $\frac{1}{8}$ in., so that it can support a maximum vertical load of $P = 4$ kip without causing the pipe to buckle. Assume the ends of the pipe are pin connected.

17–22. The deck is supported by the two 40-mm-square columns. Column AB is pinned at A and fixed at B, whereas CD is pinned at C and D. If the deck is prevented from sidesway, determine the greatest weight of the load that can be applied without causing the deck to collapse. The center of gravity of the load is located at $d = 2$ m. Both columns are made from Douglas Fir.

17–23. The deck is supported by the two 40-mm-square columns. Column AB is pinned at A and fixed at B, whereas CD is pinned at C and D. If the deck is prevented from sidesway, determine the position d of the center of gravity of the load and the load's greatest magnitude without causing the deck to collapse. Both columns are made from Douglas Fir.

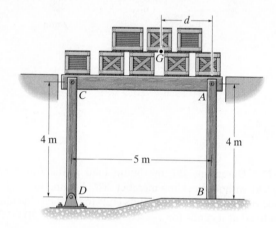

Probs. 17–22/23

*17–24. The beam is supported by the three pin-connected suspender bars, each having a diameter of 0.5 in. and made from A-36 steel. Determine the greatest uniform load w that can be applied to the beam without causing AB or CB to buckle.

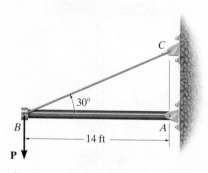

Probs. 17–20/21

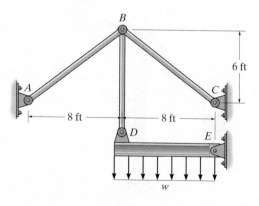

Prob. 17–24

17–25. The W14 × 30 A992 steel column is assumed pinned at both of its ends. Determine the largest axial force P that can be applied without causing it to buckle.

25 ft

Prob. 17–25

17–26. The A992 steel bar AB has a square cross section. If it is pin connected at its ends, determine the maximum allowable load P that can be applied to the frame. Use a factor of safety with respect to buckling of 2.

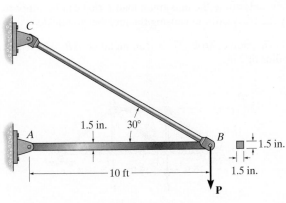

C

1.5 in.   30°

A

10 ft

B

1.5 in.

1.5 in.

P

Prob. 17–26

17–27. The linkage is made using two A992 steel rods, each having a circular cross section. Determine the diameter of each rod to the nearest $\frac{1}{8}$ in. that will support a load of $P = 10$ kip. Assume that the rods are pin connected at their ends. Use a factor of safety with respect to buckling of 1.8.

*17–28. The linkage is made using two A992 steel rods, each having a circular cross section. If each rod has a diameter of 2 in., determine the largest load it can support without causing any rod to buckle if the factor of safety against buckling is 1.8. Assume that the rods are pin connected at their ends.

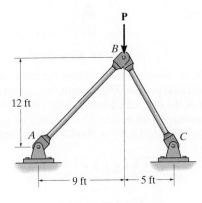

P

B

12 ft

A

C

9 ft   5 ft

Probs. 17–27/28

17–29. The linkage is made using two A-36 steel rods, each having a circular cross section. Determine the diameter of each rod to the nearest $\frac{1}{8}$ in. that will support the 900-lb load. Assume that the rods are pin connected at their ends. Use a factor of safety with respect to buckling of F.S. = 1.8.

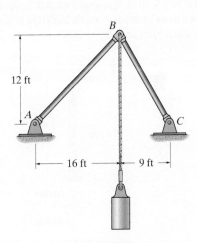

B

12 ft

A

C

16 ft   9 ft

Prob. 17–29

17

17–30. The linkage is made using two A-36 steel rods, each having a circular cross section. If each rod has a diameter of $\frac{3}{4}$ in., determine the largest load it can support without causing any rod to buckle. Assume that the rods are pin connected at their ends.

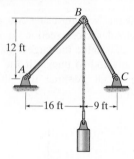

Prob. 17–30

17–31. The steel bar AB has a rectangular cross section. If it is pin connected at its ends, determine the maximum allowable intensity w of the distributed load that can be applied to BC without causing AB to buckle. Use a factor of safety with respect to buckling of 1.5. $E_{st} = 200$ GPa, $\sigma_Y = 360$ MPa.

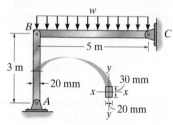

Prob. 17–31

*17–32. Determine if the frame can support a load of $P = 20$ kN if the factor of safety with respect to buckling of member AB is F.S. = 3. Assume that AB is made of steel and is pinned at its ends for x–x axis buckling and fixed at its ends for y–y axis buckling. $E_{st} = 200$ GPa, $\sigma_Y = 360$ MPa.

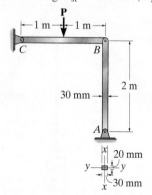

Prob. 17–32

17–33. Determine the maximum allowable load P that can be applied to member BC without causing member AB to buckle. Assume that AB is made of steel and is pinned at its ends for x–x axis buckling and fixed at its ends for y–y axis buckling. Use a factor of safety with respect to buckling of F.S. = 3. $E_{st} = 200$ GPa, $\sigma_Y = 360$ MPa.

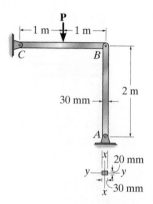

Prob. 17–33

17–34. A 6061-T6 aluminum alloy solid circular rod of length 4 m is pinned at both of its ends. If it is subjected to an axial load of 15 kN and F.S. = 2 against buckling, determine the minimum required diameter of the rod to the nearest mm.

17–35. A 6061-T6 aluminum alloy solid circular rod of length 4 m is pinned at one end while fixed at the other end. If it is subjected to an axial load of 15 kN and F.S. = 2 against buckling, determine the minimum required diameter of the rod to the nearest mm.

*17–36. The members of the truss are assumed to be pin connected. If member BD is an A992 steel rod of radius 2 in., determine the maximum load P that can be supported by the truss without causing the member to buckle.

17–37. Solve Prob. 17–36 for member AB, which has a radius of 2 in.

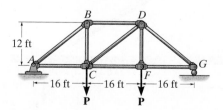

Probs. 17–36/37

17–38. The truss is made from A992 steel bars, each of which has a circular cross section with a diameter of 1.5 in. Determine the maximum force P that can be applied without causing any of the members to buckle. The members are pin connected at their ends.

17–39. The truss is made from A992 steel bars, each of which has a circular cross section. If the applied load $P = 10$ kip, determine the diameter of member AB to the nearest $\frac{1}{8}$ in. that will prevent this member from buckling. The members are pin connected at their ends.

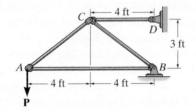

Probs. 17–38/39

*17–40.** The steel bar AB of the frame is assumed to be pin connected at its ends for y–y axis buckling. If $P = 18$ kN, determine the factor of safety with respect to buckling about the y–y axis. $E_{st} = 200$ GPa, $\sigma_Y = 360$ MPa.

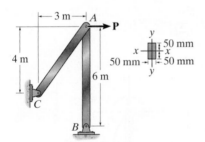

Prob. 17–40

17–41. The ideal column has a weight w (force/length) and is subjected to the axial load P. Determine the maximum moment in the column at midspan. EI is constant. *Hint:* Establish the differential equation for deflection, Eq. 17–1, with the origin at the midspan. The general solution is $v = C_1 \sin kx + C_2 \cos kx + (w/(2P))x^2 - (wL/(2P))x - (wEI/P^2)$ where $k^2 = P/EI$.

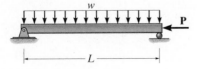

Prob. 17–41

17–42. The ideal column is subjected to the force $\mathbf{F}$ at its midpoint and the axial load $\mathbf{P}$. Determine the maximum moment in the column at midspan. EI is constant. *Hint:* Establish the differential equation for deflection, Eq. 17–1. The general solution is $v = C_1 \sin kx + C_2 \cos kx - c^2 x/k^2$, where $c^2 = F/2EI, k^2 = P/EI$.

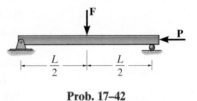

Prob. 17–42

17–43. The column with constant EI has the end constraints shown. Determine the critical load for the column.

Prob. 17–43

*17–44.** Consider an ideal column as in Fig. 17–10c, having both ends fixed. Show that the critical load on the column is $P_{cr} = 4\pi^2 EI/L^2$. *Hint:* Due to the vertical deflection of the top of the column, a constant moment $\mathbf{M}'$ will be developed at the supports. Show that $d^2v/dx^2 + (P/EI)v = M'/EI$. The solution is of the form $v = C_1 \sin(\sqrt{P/EI}x) + C_2 \cos(\sqrt{P/EI}x) + M'/P$.

17–45. Consider an ideal column as in Fig. 17–10d, having one end fixed and the other pinned. Show that the critical load on the column is $P_{cr} = 20.19EI/L^2$. *Hint:* Due to the vertical deflection at the top of the column, a constant moment $\mathbf{M}'$ will be developed at the fixed support and horizontal reactive forces $\mathbf{R}'$ will be developed at both supports. Show that $d^2v/dx^2 + (P/EI)v = (R'/EI)(L - x)$. The solution is of the form $v = C_1 \sin(\sqrt{P/EI}x) + C_2 \cos(\sqrt{P/EI}x) + (R'/P)(L - x)$. After application of the boundary conditions show that $\tan(\sqrt{P/EI}L) = \sqrt{P/EI}L$. Solve numerically for the smallest nonzero root.

17

*17.4 THE SECANT FORMULA

The Euler formula was derived assuming the load P is applied through the centroid of the column's cross-sectional area and that the column is perfectly straight. Actually this is quite unrealistic, since a manufactured column is never perfectly straight, nor is the application of the load known with great accuracy. In reality, then, columns never suddenly buckle; instead they begin to bend, although ever so slightly, immediately upon application of the load. As a result, the actual criterion for load application should be limited, either to a specified sidesway deflection of the column, or by not allowing the maximum stress in the column to exceed an allowable stress.

To study the effect of an eccentric loading, we will apply the load P to the column at a distance e from its centroid, Fig. 17–13a. This loading is statically equivalent to the axial load P and bending moment $M' = Pe$ shown in Fig. 17–13b. In both cases, the ends A and B are supported so that they are free to rotate (pin supported), and as before, we will only consider linear elastic material behavior. Furthermore, the x–v plane is a plane of symmetry for the cross-sectional area.

From the free-body diagram of the arbitrary section, Fig. 17–13c, the internal moment in the column is

$$M = -P(e + v) \tag{17–13}$$

And so the differential equation for the deflection curve becomes

$$EI\frac{d^2v}{dx^2} = -P(e + v)$$

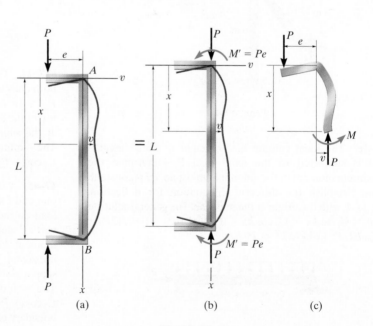

(a)   (b)   (c)

Fig. 17–13

or

$$\frac{d^2v}{dx^2} + \frac{P}{EI}v = -\frac{P}{EI}e$$

This equation is similar to Eq. 17–7, and its solution consists of both complementary and particular solutions, namely,

$$v = C_1 \sin\sqrt{\frac{P}{EI}}x + C_2 \cos\sqrt{\frac{P}{EI}}x - e \qquad (17\text{–}14)$$

To evaluate the constants we must apply the boundary conditions. At $x = 0, v = 0$, so $C_2 = e$. And at $x = L, v = 0$, which gives

$$C_1 = \frac{e[1 - \cos(\sqrt{P/EI}\,L)]}{\sin(\sqrt{P/EI}\,L)}$$

Since $1 - \cos(\sqrt{P/EI}\,L) = 2\sin^2(\sqrt{P/EI}\,L/2)$ and $\sin(\sqrt{P/EI}\,L) = 2\sin(\sqrt{P/EI}\,L/2)\cos(\sqrt{P/EI}\,L/2)$, we have

$$C_1 = e\tan\!\left(\sqrt{\frac{P}{EI}}\frac{L}{2}\right)$$

Hence, the deflection curve, Eq. 17–14, becomes

$$v = e\left[\tan\!\left(\sqrt{\frac{P}{EI}}\frac{L}{2}\right)\sin\!\left(\sqrt{\frac{P}{EI}}x\right) + \cos\!\left(\sqrt{\frac{P}{EI}}x\right) - 1\right] \qquad (17\text{–}15)$$

Maximum Deflection.   Due to symmetry of loading, both the maximum deflection and maximum stress occur at the column's midpoint. Therefore, at $x = L/2$,

$$v_{max} = e\left[\sec\!\left(\sqrt{\frac{P}{EI}}\frac{L}{2}\right) - 1\right] \qquad (17\text{–}16)$$

Notice that if e approaches zero, then v_{max} approaches zero. However, if the terms in the brackets approach infinity as e approaches zero, then v_{max} will have a nonzero value. Mathematically, this represents the behavior of an axially loaded column at failure when subjected to the critical load P_{cr}. Therefore, to find P_{cr} we require

$$\sec\!\left(\sqrt{\frac{P_{cr}}{EI}}\frac{L}{2}\right) = \infty$$

$$\sqrt{\frac{P_{cr}}{EI}}\frac{L}{2} = \frac{\pi}{2}$$

$$P_{cr} = \frac{\pi^2 EI}{L^2} \qquad (17\text{–}17)$$

which is the same result found from the Euler formula, Eq. 17–5.

17

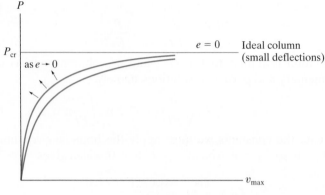

Fig. 17–14

If Eq. 17–16 is plotted for various values of eccentricity e, it results in a family of curves shown in Fig. 17–14. Here the critical load becomes an asymptote to the curves and represents the unrealistic case of an ideal column ($e = 0$). The results developed here apply only for small sidesway deflections, and so they certainly apply if the column is long and slender.

Notice that the curves in Fig. 17–14 show a *nonlinear* relationship between the load P and the deflection v. As a result, the principle of superposition *cannot be used* to determine the total deflection of a column. In other words, the deflection must be determined by applying the *total load* to the column, not a series of component loads. Furthermore, due to this nonlinear relationship, any factor of safety used for design purposes must be applied to the load and not to the stress.

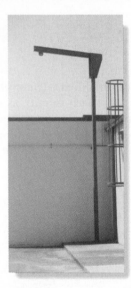

The column supporting this crane is unusually long. It will be subjected not only to uniaxial load, but also a bending moment. To ensure it will not buckle, it should be braced at the roof as a pin connection.

The Secant Formula.   The maximum stress in an eccentrically loaded column is caused by both the axial load and the moment, Fig. 17–15a. Maximum moment occurs at the column's midpoint, and using Eqs. 17–13 and 17–16, it has a magnitude of

$$M = |P(e + v_{max})| \qquad M = Pe \sec\left(\sqrt{\frac{P}{EI}} \frac{L}{2}\right) \qquad (17\text{--}18)$$

As shown in Fig. 17–15b, the maximum stress in the column is therefore

$$\sigma_{max} = \frac{P}{A} + \frac{Mc}{I}; \qquad \sigma_{max} = \frac{P}{A} + \frac{Pec}{I} \sec\left(\sqrt{\frac{P}{EI}} \frac{L}{2}\right)$$

Since the radius of gyration is $r = \sqrt{I/A}$, the above equation can be written in a form called the **secant formula**:

$$\boxed{\sigma_{max} = \frac{P}{A}\left[1 + \frac{ec}{r^2} \sec\left(\frac{L_e}{2r} \sqrt{\frac{P}{EA}}\right)\right]} \qquad (17\text{--}19)$$

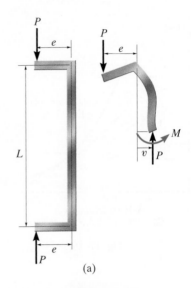

(a)

Here

σ_{max} = maximum *elastic stress* in the column, which occurs at the inner concave side at the column's midpoint. This stress is compressive.

P = vertical load applied to the column. $P < P_{cr}$ unless $e = 0$; then $P = P_{cr}$ (Eq. 17–5).

e = eccentricity of the load P, measured from the centroidal axis of the column's cross-sectional area to the line of action of P

c = distance from the centroidal axis to the outer fiber of the column where the maximum compressive stress σ_{max} occurs

A = cross-sectional area of the column

L_e = unsupported length of the column *in the plane of bending*. Application is restricted to members that are pin connected, $L_e = L$, or have one end free and the other end fixed, $L_e = 2L$.

E = modulus of elasticity for the material

r = radius of gyration, $r = \sqrt{I/A}$, where I is calculated about the centroidal or bending axis

Axial
stress

+

Bending
stress

=

σ_{max}
Resultant
stress

(b)

Fig. 17–15

17

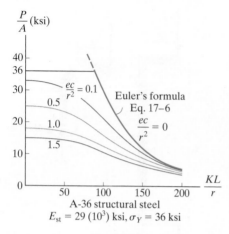

$\frac{P}{A}$ (ksi)

$\dfrac{ec}{r^2} = 0.1$

0.5

Euler's formula
Eq. 17–6

1.0

$\dfrac{ec}{r^2} = 0$

1.5

A-36 structural steel
$E_{st} = 29 \, (10^3)$ ksi, $\sigma_Y = 36$ ksi

Fig. 17–16

Graphs of Eq. 17–19 for various values of the *eccentricity ratio* ec/r^2 are plotted in Fig. 17–16 for a structural-grade A-36 steel. Note that when $e \to 0$, or when $ec/r^2 \to 0$, Eq. 17–16 gives $\sigma_{max} = P/A$, where P is the critical load on the column, defined by Euler's formula. Since the results are valid only for elastic loadings, the stresses shown in the figure cannot exceed $\sigma_Y = 36$ ksi, represented here by the horizontal line.

By inspection, the curves indicate that changes in the eccentricity ratio have a marked effect on the load-carrying capacity of columns with *small* slenderness ratios. However, columns that have large slenderness ratios tend to fail at or near the Euler critical load regardless of the eccentricity ratio, since the curves bunch together. Therefore, when Eq. 17–19 is used for design purposes, it is important to have a somewhat accurate value for the eccentricity ratio for shorter-length columns.

Design.   Once the eccentricity ratio is specified, the column data can be substituted into Eq. 17–19. If a value of $\sigma_{max} = \sigma_Y$ is considered, then the corresponding load P_Y can be determined numerically, since the equation is transcendental and cannot be solved explicitly for P_Y. As a design aid, computer software, or graphs such as those in Fig. 17–16, can also be used to determine P_Y. Realize that due to the eccentric application of P_Y, this load will *always be smaller* than the critical load P_{cr}, which assumes (unrealistically) that the column is axially loaded.

IMPORTANT POINTS

- Due to imperfections in manufacturing or specific application of the load, a column will never suddenly buckle; instead, it begins to bend as it is loaded.

- The load applied to a column is related to its deflection in a nonlinear manner, and so the principle of superposition does not apply.

- As the slenderness ratio increases, eccentrically loaded columns tend to fail at or near the Euler buckling load.

EXAMPLE   17.4

The W8 × 40 A992 steel column shown in Fig. 17–17a is fixed at its base and braced at the top so that it is fixed from displacement, yet free to rotate about the y–y axis. Also, it can sway to the side in the y–z plane. Determine the maximum eccentric load the column can support before it either begins to buckle or the steel yields. Take $\sigma_Y = 50$ ksi.

SOLUTION

From the support conditions it is seen that about the y–y axis the column behaves as if it were pinned at its top and fixed at its bottom, and subjected to an axial load P, Fig. 17–17b. About the x–x axis the column is free at the top and fixed at the bottom, and it is subjected to both an axial load P and moment $M = P(9 \text{ in.})$, Fig. 17–17c.

y–y Axis Buckling. From Fig. 17–10d the effective length factor is $K_y = 0.7$, so $(KL)_y = 0.7(12) \text{ ft} = 8.40 \text{ ft} = 100.8$ in. Using the table in Appendix B to determine I_y for the W8 × 40 section and applying Eq. 17–11, we have

$$(P_{cr})_y = \frac{\pi^2 EI_y}{(KL)_y^2} = \frac{\pi^2[29(10^3) \text{ ksi}](49.1 \text{ in}^4)}{(100.8 \text{ in.})^2} = 1383 \text{ kip}$$

x–x Axis Yielding. From Fig. 17–10b, $K_x = 2$, so $(KL)_x = 2(12) \text{ ft} = 24 \text{ ft} = 288$ in. Again using the table in Appendix B, we have $A = 11.7 \text{ in}^2$, $c = 8.25 \text{ in.}/2 = 4.125$ in., and $r_x = 3.53$ in. Applying the secant formula,

$$\sigma_Y = \frac{P_x}{A}\left[1 + \frac{ec}{r_x^2}\sec\left(\frac{(KL)_x}{2r_x}\sqrt{\frac{P_x}{EA}}\right)\right]$$

$$50 \text{ ksi} = \frac{P_x}{11.7 \text{ in}^2}\left[1 + \frac{9 \text{ in. } (4.125 \text{ in.})}{(3.53 \text{ in.})^2}\sec\left(\frac{(288 \text{ in.})}{2\,(3.53 \text{ in.})}\sqrt{\frac{P_x}{29(10^3) \text{ ksi } (11.7 \text{ in}^2)}}\right)\right]$$

$$585 = P_x[1 + 2.979 \sec(0.0700\sqrt{P_x})]$$

Solving for P_x by trial and error, noting that the argument for the secant is in radians, we get

$$P_x = 115 \text{ kip} \qquad\qquad Ans.$$

Since this value is less than $(P_{cr})_y = 1383$ kip, failure will occur about the x–x axis.

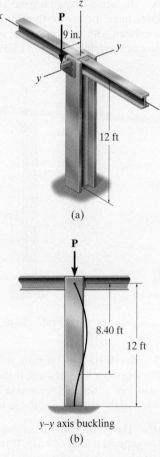

(a)

P

8.40 ft

12 ft

y–y axis buckling
(b)

P

$M = P(9 \text{ in.})$

12 ft

x–x axis yielding
(c)

Fig. 17–17

17

PROBLEMS

17–46. The wood column is fixed at its base and free at its top. Determine the load P that can be applied to the edge of the column without causing the column to fail either by buckling or by yielding. $E_w = 12$ GPa, $\sigma_Y = 55$ MPa.

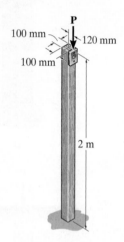

Prob. 17–46

17–47. The W10 × 12 structural A-36 steel column is used to support a load of 4 kip. If the column is fixed at the base and free at the top, determine the sidesway deflection at the top of the column due to the loading.

***17–48.** The W10 × 12 structural A-36 steel column is used to support a load of 4 kip. If the column is fixed at its base and free at its top, determine the maximum stress in the column due to this loading.

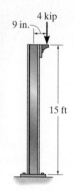

Probs. 17–47/48

17–49. The aluminum column is fixed at the bottom and free at the top. Determine the maximum force P that can be applied at A without causing it to buckle or yield. Use a factor of safety of 3 with respect to buckling and yielding. $E_{al} = 70$ GPa, $\sigma_Y = 95$ MPa.

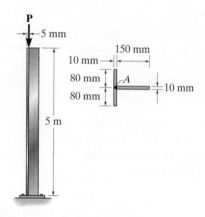

Prob. 17–49

17–50. The aluminum rod is fixed at its base and free at its top. If the eccentric load $P = 200$ kN is applied, determine the greatest allowable length L of the rod so that it does not buckle or yield. $E_{al} = 72$ GPa, $\sigma_Y = 410$ MPa.

17–51. The aluminum rod is fixed at its base and free and at its top. If the length of the rod is $L = 2$ m, determine the greatest allowable load P that can be applied so that the rod does not buckle or yield. $E_{al} = 72$ GPa, $\sigma_Y = 410$ MPa.

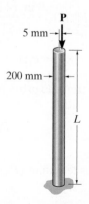

Probs. 17–50/51

*17–52. Assume that the wood column is pin connected at its base and top. Determine the maximum eccentric load P that can be applied without causing the column to buckle or yield. $E_w = 1.8(10^3)$ ksi, $\sigma_Y = 8$ ksi.

17–53. Assume that the wood column is pinned top and bottom for movement about the $x-y$ axis, and fixed at the bottom and free at the top for movement about the $y-y$ axis. Determine the maximum eccentric load P that can be applied without causing the column to buckle or yield. $E_w = 1.8(10^3)$ ksi, $\sigma_Y = 8$ ksi.

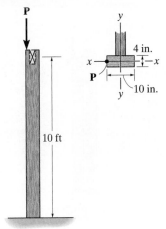

Probs. 17–52/53

17–54. The wood column is pinned at its base and top. If the eccentric force $P = 10$ kN is applied to the column, investigate whether the column is adequate to support this loading without buckling or yielding. Take $E = 10$ GPa and $\sigma_Y = 15$ MPa.

17–55. The wood column is pinned at its base and top. Determine the maximum eccentric force P the column can support without causing it to either buckle or yield. Take $E = 10$ GPa and $\sigma_Y = 15$ MPa.

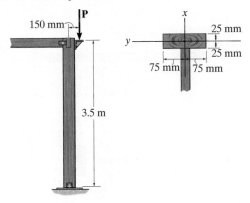

Probs. 17–54/55

*17–56. The A992 steel rectangular hollow section column is pinned at both ends. If it has a length of $L = 14$ ft, determine the maximum allowable eccentric force **P** it can support without causing it to either buckle or yield.

17–57. The A992 steel rectangular hollow section column is pinned at both ends. If it is subjected to the eccentric force $P = 45$ kip, determine its maximum allowable length L without causing it to either buckle or yield.

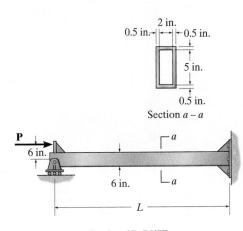

Probs. 17–56/57

17–58. The tube is made of copper and has an outer diameter of 35 mm and a wall thickness of 7 mm. Determine the eccentric load P that it can support without failure. The tube is pin supported at its ends. $E_{cu} = 120$ GPa, $\sigma_Y = 750$ MPa.

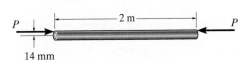

Prob. 17–58

17–59. The wood column is pinned at its base and top. If $L = 5$ ft, determine the maximum eccentric load P that can be applied without causing the column to buckle or yield. $E_w = 1.8(10^3)$ ksi, $\sigma_Y = 8$ ksi.

17–62. The W14 × 53 structural A992 steel column is fixed at its base and free at its top. If $P = 75$ kip, determine the sidesway deflection at its top and the maximum stress in the column.

17–63. The W14 × 53 column is fixed at its base and free at its top. Determine the maximum eccentric load P that it can support without causing it to buckle or yield. $E_{st} = 29(10^3)$ ksi, $\sigma_Y = 50$ ksi.

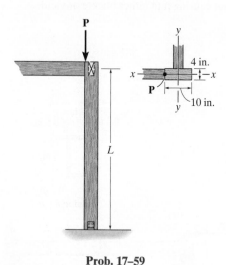

Prob. 17–59

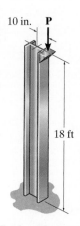

Probs. 17–62/63

*17–60. The brass rod is fixed at one end and free at the other end. If the eccentric load $P = 200$ kN is applied, determine the greatest allowable length L of the rod so that it does not buckle or yield. $E_{br} = 101$ GPa, $\sigma_Y = 69$ MPa.

17–61. The brass rod is fixed at one end and free at the other end. If the length of the rod is $L = 2$ m, determine the greatest allowable load P that can be applied so that the rod does not buckle or yield. Also, determine the largest sidesway deflection of the rod due to the loading. $E_{br} = 101$ GPa, $\sigma_Y = 69$ MPa.

*17–64. Determine the maximum eccentric load P the 2014-T6-aluminum-alloy strut can support without causing it either to buckle or yield. The ends of the strut are pin connected.

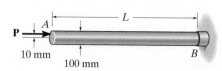

Probs. 17–60/61

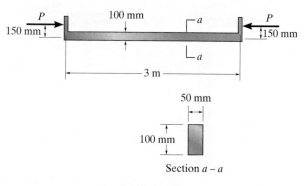

Section $a - a$

Prob. 17–64

CHAPTER REVIEW

Buckling is the sudden instability that occurs in columns or members that support an axial compressive load. The maximum axial load that a member can support just before buckling is called the critical load P_{cr}.		
The critical load for an ideal column is determined from Euler's formula, where $K = 1$ for pin supports, $K = 0.5$ for fixed supports, $K = 0.7$ for a pin and a fixed support, and $K = 2$ for a fixed support and a free end.	$$P_{cr} = \frac{\pi^2 EI}{(KL)^2}$$	
If the axial loading is applied eccentrically to the column, then the secant formula can be used to determine the maximum stress in the column.	$$\sigma_{max} = \frac{P}{A}\left[1 + \frac{ec}{r^2}\sec\left(\frac{L}{2r}\sqrt{\frac{P}{EA}}\right)\right]$$	

17

REVIEW PROBLEMS

R17–1. The wood column is 4 m long and is required to support the axial load of 25 kN. If the cross section is square, determine the dimension a of each of its sides using a factor of safety against buckling of F.S. = 2.5. The column is assumed to be pinned at its top and bottom. Use the Euler equation. $E_w = 11$ GPa, and $\sigma_Y = 10$ MPa.

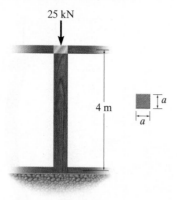

25 kN

4 m

a

a

Prob. R17–1

R17–2. If the torsional springs attached to ends A and C of the rigid members AB and BC have a stiffness k, determine the critical load P_{cr}.

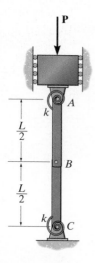

P

$\dfrac{L}{2}$

k

A

B

$\dfrac{L}{2}$

k

C

Prob. R17–2

R17–3. A steel column has a length of 5 m and is free at one end and fixed at the other end. If the cross-sectional area has the dimensions shown, determine the critical load. $E_{st} = 200$ GPa, $\sigma_Y = 360$ MPa.

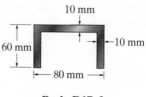

10 mm

60 mm

10 mm

80 mm

Prob. R17–3

*R17–4. The square structural A992 steel tubing has outer dimensions of 8 in. by 8 in. Its cross-sectional area is 14.40 in^2 and its moments of inertia are $I_x = I_y = 131$ in^4. Determine the maximum load P it can support. The column can be assumed fixed at its base and free at its top.

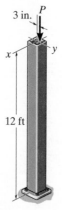

3 in.

P

x

y

12 ft

Prob. R17–4

R17–5. If the A-36 steel solid circular rod *BD* has a diameter of 2 in., determine the allowable maximum force **P** that can be supported by the frame without causing the rod to buckle. Use F.S. = 2 against buckling.

R17–6. If *P* = 15 kip, determine the required minimum diameter of the A992 steel solid circular rod *BD* to the nearest $\frac{1}{16}$ in. Use F.S. = 2 against buckling.

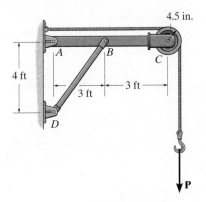

Probs. R17–5/6

R17–7. The steel pipe is fixed supported at its ends. If it is 4 m long and has an outer diameter of 50 mm, determine its required thickness so that it can support an axial load of *P* = 100 kN without buckling. E_{st} = 200 GPa, σ_Y = 250 MPa.

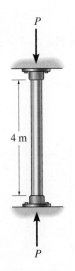

Prob. R17–7

*****R17–8.** The W200 × 46 wide-flange A992-steel column can be considered pinned at its top and fixed at its base. Also, the column is braced at its mid-height against weak axis buckling. Determine the maximum axial load the column can support without causing it to buckle.

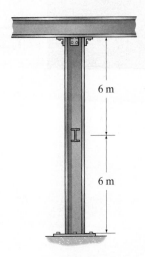

Prob. R17–8

R17–9. The wide-flange A992 steel column has the cross section shown. If it is fixed at the bottom and free at the top, determine the maximum force *P* that can be applied at *A* without causing it to buckle or yield. Use a factor of safety of 3 with respect to buckling and yielding.

R17–10. The wide-flange A992 steel column has the cross section shown. If it is fixed at the bottom and free at the top, determine if the column will buckle or yield when the load *P* = 10 kN is applied at *A*. Use a factor of safety of 3 with respect to buckling and yielding.

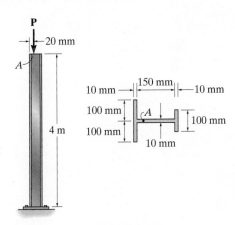

Probs. R17–9/10

MATHEMATICAL REVIEW AND EXPRESSIONS

GEOMETRY AND TRIGONOMETRY REVIEW

The angles θ in Fig. A–1 are equal between the transverse and two parallel lines.

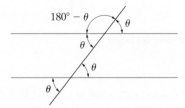

Fig. A–1

For a line and its normal, the angles θ in Fig. A–2 are equal.

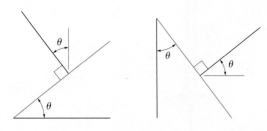

Fig. A–2

For the circle in Fig. A–3 $s = \theta r$, so that when $\theta = 360° = 2\pi$ rad then the circumference is $s = 2\pi r$. Also, since $180° = \pi$ rad, then θ (rad) $= (\pi/180°)\theta°$. The area of the circle is $A = \pi r^2$.

Fig. A–3

The sides of a similar triangle can be obtained by proportion as in Fig. A–4, where $\dfrac{a}{A} = \dfrac{b}{B} = \dfrac{c}{C}$.

For the right triangle in Fig. A–5, the Pythagorean theorem is

$$h = \sqrt{(o)^2 + (a)^2}$$

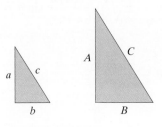

Fig. A–4

The trigonometric functions are

$$\sin \theta = \frac{o}{h}$$

$$\cos \theta = \frac{a}{h}$$

$$\tan \theta = \frac{o}{a}$$

This is easily remembered as "soh, cah, toa," i.e., the sine is the opposite over the hypotenuse, etc. The other trigonometric functions follow from this.

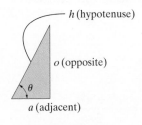

o (opposite)

a (adjacent)

Fig. A–5

$$\csc \theta = \frac{1}{\sin \theta} = \frac{h}{o}$$

$$\sec \theta = \frac{1}{\cos \theta} = \frac{h}{a}$$

$$\cot \theta = \frac{1}{\tan \theta} = \frac{a}{o}$$

Trigonometric Identities

$$\sin^2 \theta + \cos^2 \theta = 1$$

$$\sin(\theta \pm \phi) = \sin \theta \cos \phi \pm \cos \theta \sin \phi$$

$$\sin 2\theta = 2 \sin \theta \cos \theta$$

$$\cos(\theta \pm \phi) = \cos \theta \cos \phi \mp \sin \theta \sin \phi$$

$$\cos 2\theta = \cos^2 \theta - \sin^2 \theta$$

$$\cos \theta = \pm \sqrt{\frac{1 + \cos 2\theta}{2}}, \quad \sin \theta = \pm \sqrt{\frac{1 - \cos 2\theta}{2}}$$

$$\tan \theta = \frac{\sin \theta}{\cos \theta}$$

$$1 + \tan^2 \theta = \sec^2 \theta \qquad 1 + \cot^2 \theta = \csc^2 \theta$$

Quadratic Formula

If $ax^2 + bx + c = 0$, then $x = \dfrac{-b \pm \sqrt{b^2 - 4ac}}{2a}$

Hyperbolic Functions

$$\sinh x = \frac{e^x - e^{-x}}{2},$$

$$\cosh x = \frac{e^x + e^{-x}}{2},$$

$$\tanh x = \frac{\sinh x}{\cosh x}$$

Power-Series Expansions

$$\sin x = x - \frac{x^3}{3!} + \cdots, \quad \cos x = 1 - \frac{x^2}{2!} + \cdots$$

$$\sinh x = x + \frac{x^3}{3!} + \cdots, \quad \cosh x = 1 + \frac{x^2}{2!} + \cdots$$

Derivatives

$$\frac{d}{dx}(u^n) = nu^{n-1}\frac{du}{dx} \qquad \frac{d}{dx}(\sin u) = \cos u \frac{du}{dx}$$

$$\frac{d}{dx}(uv) = u\frac{dv}{dx} + v\frac{du}{dx} \qquad \frac{d}{dx}(\cos u) = -\sin u \frac{du}{dx}$$

$$\frac{d}{dx}\left(\frac{u}{v}\right) = \frac{v\dfrac{du}{dx} - u\dfrac{dv}{dx}}{v^2} \qquad \frac{d}{dx}(\tan u) = \sec^2 u \frac{du}{dx}$$

$$\frac{d}{dx}(\cot u) = -\csc^2 u \frac{du}{dx} \qquad \frac{d}{dx}(\sinh u) = \cosh u \frac{du}{dx}$$

$$\frac{d}{dx}(\sec u) = \tan u \sec u \frac{du}{dx} \quad \frac{d}{dx}(\cosh u) = \sinh u \frac{du}{dx}$$

$$\frac{d}{dx}(\csc u) = -\csc u \cot u \frac{du}{dx}$$

Integrals

$$\int x^n \, dx = \frac{x^{n+1}}{n+1} + C, n \neq -1$$

$$\int \frac{dx}{a + bx} = \frac{1}{b}\ln(a + bx) + C$$

$$\int \frac{dx}{a + bx^2} = \frac{1}{2\sqrt{-ab}}\ln\left[\frac{a + x\sqrt{-ab}}{a - x\sqrt{-ab}}\right] + C, ab < 0$$

$$\int \frac{x \, dx}{a + bx^2} = \frac{1}{2b}\ln(bx^2 + a) + C$$

$$\int \frac{x^2 \, dx}{a + bx^2} = \frac{x}{b} - \frac{a}{b\sqrt{ab}}\tan^{-1}\frac{x\sqrt{ab}}{a} + C, ab > 0$$

$$\int \sqrt{a + bx} \, dx = \frac{2}{3b}\sqrt{(a + bx)^3} + C$$

$$\int x\sqrt{a + bx} \, dx = \frac{-2(2a - 3bx)\sqrt{(a + bx)^3}}{15b^2} + C$$

$$\int x^2\sqrt{a + bx} \, dx =$$
$$\frac{2(8a^2 - 12abx + 15b^2x^2)\sqrt{(a + bx)^3}}{105b^3} + C$$

$$\int \sqrt{a^2 - x^2} \, dx = \frac{1}{2}\left[x\sqrt{a^2 - x^2} + a^2 \sin^{-1}\frac{x}{a}\right] + C,$$
$$a > 0$$

$$\int x\sqrt{a^2 - x^2} \, dx = -\frac{1}{3}\sqrt{(a^2 - x^2)^3} + C$$

$$\int x^2\sqrt{a^2 - x^2} \, dx = -\frac{x}{4}\sqrt{(a^2 - x^2)^3}$$
$$+ \frac{a^2}{8}\left(x\sqrt{a^2 - x^2} + a^2 \sin^{-1}\frac{x}{a}\right) + C, a > 0$$

$$\int \sqrt{x^2 \pm a^2} \, dx =$$
$$\frac{1}{2}\left[x\sqrt{x^2 \pm a^2} \pm a^2 \ln\left(x + \sqrt{x^2 \pm a^2}\right)\right] + C$$

$$\int x\sqrt{x^2 \pm a^2} \, dx = \frac{1}{3}\sqrt{(x^2 \pm a^2)^3} + C$$

$$\int x^2\sqrt{x^2 \pm a^2} \, dx = \frac{x}{4}\sqrt{(x^2 \pm a^2)^3}$$
$$\mp \frac{a^2}{8}x\sqrt{x^2 \pm a^2} - \frac{a^4}{8}\ln\left(x + \sqrt{x^2 \pm a^2}\right) + C$$

$$\int \frac{dx}{\sqrt{a + bx}} = \frac{2\sqrt{a + bx}}{b} + C$$

$$\int \frac{x \, dx}{\sqrt{x^2 \pm a^2}} = \sqrt{x^2 \pm a^2} + C$$

$$\int \frac{dx}{\sqrt{a + bx + cx^2}} = \frac{1}{\sqrt{c}}\ln\left[\sqrt{a + bx + cx^2} +\right.$$
$$\left. x\sqrt{c} + \frac{b}{2\sqrt{c}}\right] + C, c > 0$$

$$= \frac{1}{\sqrt{-c}}\sin^{-1}\left(\frac{-2cx - b}{\sqrt{b^2 - 4ac}}\right) + C, c < 0$$

$$\int \sin x \, dx = -\cos x + C$$

$$\int \cos x \, dx = \sin x + C$$

$$\int x \cos(ax) \, dx = \frac{1}{a^2}\cos(ax) + \frac{x}{a}\sin(ax) + C$$

$$\int x^2 \cos(ax) \, dx = \frac{2x}{a^2}\cos(ax) + \frac{a^2x^2 - 2}{a^3}\sin(ax) + C$$

$$\int e^{ax} \, dx = \frac{1}{a}e^{ax} + C$$

$$\int xe^{ax} \, dx = \frac{e^{ax}}{a^2}(ax - 1) + C$$

$$\int \sinh x \, dx = \cosh x + C$$

$$\int \cosh x \, dx = \sinh x + C$$

Centroid Location	Centroid Location	Area Moment of Inertia
$A = \frac{1}{2}h(a+b)$ — Trapezoidal area, with h, a, b, and $\frac{1}{3}\left(\frac{2a+b}{a+b}\right)h$	$A = \theta r^2$ — Circular sector area, with $\frac{2}{3}\frac{r\sin\theta}{\theta}$	$I_x = \frac{1}{4}r^4\left(\theta - \frac{1}{2}\sin 2\theta\right)$ $I_y = \frac{1}{4}r^4\left(\theta + \frac{1}{2}\sin 2\theta\right)$
$A = \frac{2}{3}ab$ — Semiparabolic area, with $\frac{2}{5}a$, b, $\frac{3}{8}b$, a	$A = \frac{1}{4}\pi r^2$ — Quarter circular area, with $\frac{4r}{3\pi}$	$I_x = \frac{1}{16}\pi r^4$ $I_y = \frac{1}{16}\pi r^4$
$A = \frac{1}{3}ab$ — Exparabolic area, with b, $\frac{3}{4}a$, a, $\frac{3}{10}b$	$A = \frac{\pi r^2}{2}$ — Semicircular area, with r, $\frac{4r}{3\pi}$	$I_x = \frac{1}{8}\pi r^4$ $I_y = \frac{1}{8}\pi r^4$
$A = \frac{4}{3}ab$ — Parabolic area, with a, b, $\frac{2}{5}a$	$A = \pi r^2$ — Circular area, with r	$I_x = \frac{1}{4}\pi r^4$ $I_y = \frac{1}{4}\pi r^4$

Centroid Location	Centroid Location	Area Moment of Inertia

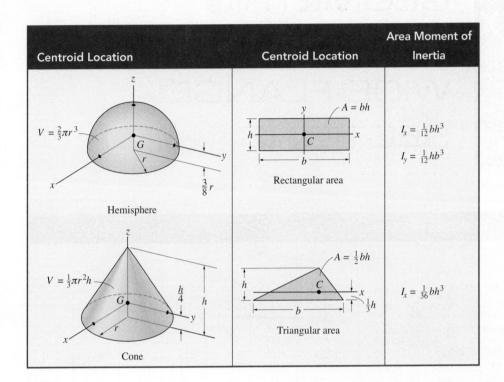

Hemisphere

$V = \frac{2}{3}\pi r^3$

Cone

$V = \frac{1}{3}\pi r^2 h$

Rectangular area

$A = bh$

$I_x = \frac{1}{12}bh^3$

$I_y = \frac{1}{12}hb^3$

Triangular area

$A = \frac{1}{2}bh$

$I_x = \frac{1}{36}bh^3$

B

APPENDIX C

GEOMETRIC PROPERTIES OF WIDE-FLANGE SECTIONS

				Flange		x–x axis			y–y axis		
Designation	Area A	Depth d	Web thickness t_w	width b_f	thickness t_f	I	S	r	I	S	r
in. × lb/ft	in^2	in.	in.	in.	in.	in^4	in^3	in.	in^4	in^3	in.
W24 × 104	30.6	24.06	0.500	12.750	0.750	3100	258	10.1	259	40.7	2.91
W24 × 94	27.7	24.31	0.515	9.065	0.875	2700	222	9.87	109	24.0	1.98
W24 × 84	24.7	24.10	0.470	9.020	0.770	2370	196	9.79	94.4	20.9	1.95
W24 × 76	22.4	23.92	0.440	8.990	0.680	2100	176	9.69	82.5	18.4	1.92
W24 × 68	20.1	23.73	0.415	8.965	0.585	1830	154	9.55	70.4	15.7	1.87
W24 × 62	18.2	23.74	0.430	7.040	0.590	1550	131	9.23	34.5	9.80	1.38
W24 × 55	16.2	23.57	0.395	7.005	0.505	1350	114	9.11	29.1	8.30	1.34
W18 × 65	19.1	18.35	0.450	7.590	0.750	1070	117	7.49	54.8	14.4	1.69
W18 × 60	17.6	18.24	0.415	7.555	0.695	984	108	7.47	50.1	13.3	1.69
W18 × 55	16.2	18.11	0.390	7.530	0.630	890	98.3	7.41	44.9	11.9	1.67
W18 × 50	14.7	17.99	0.355	7.495	0.570	800	88.9	7.38	40.1	10.7	1.65
W18 × 46	13.5	18.06	0.360	6.060	0.605	712	78.8	7.25	22.5	7.43	1.29
W18 × 40	11.8	17.90	0.315	6.015	0.525	612	68.4	7.21	19.1	6.35	1.27
W18 × 35	10.3	17.70	0.300	6.000	0.425	510	57.6	7.04	15.3	5.12	1.22
W16 × 57	16.8	16.43	0.430	7.120	0.715	758	92.2	6.72	43.1	12.1	1.60
W16 × 50	14.7	16.26	0.380	7.070	0.630	659	81.0	6.68	37.2	10.5	1.59
W16 × 45	13.3	16.13	0.345	7.035	0.565	586	72.7	6.65	32.8	9.34	1.57
W16 × 36	10.6	15.86	0.295	6.985	0.430	448	56.5	6.51	24.5	7.00	1.52
W16 × 31	9.12	15.88	0.275	5.525	0.440	375	47.2	6.41	12.4	4.49	1.17
W16 × 26	7.68	15.69	0.250	5.500	0.345	301	38.4	6.26	9.59	3.49	1.12
W14 × 53	15.6	13.92	0.370	8.060	0.660	541	77.8	5.89	57.7	14.3	1.92
W14 × 43	12.6	13.66	0.305	7.995	0.530	428	62.7	5.82	45.2	11.3	1.89
W14 × 38	11.2	14.10	0.310	6.770	0.515	385	54.6	5.87	26.7	7.88	1.55
W14 × 34	10.0	13.98	0.285	6.745	0.455	340	48.6	5.83	23.3	6.91	1.53
W14 × 30	8.85	13.84	0.270	6.730	0.385	291	42.0	5.73	19.6	5.82	1.49
W14 × 26	7.69	13.91	0.255	5.025	0.420	245	35.3	5.65	8.91	3.54	1.08
W14 × 22	6.49	13.74	0.230	5.000	0.335	199	29.0	5.54	7.00	2.80	1.04

Wide-Flange Sections or W Shapes FPS Units

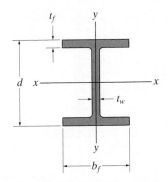

Wide-Flange Sections or W Shapes FPS Units

Designation	Area A	Depth d	Web thickness t_w	Flange width b_f	Flange thickness t_f	x–x axis I	x–x axis S	x–x axis r	y–y axis I	y–y axis S	y–y axis r
in. × lb/ft	in²	in.	in.	in.	in.	in⁴	in³	in.	in⁴	in³	in.
W12 × 87	25.6	12.53	0.515	12.125	0.810	740	118	5.38	241	39.7	3.07
W12 × 50	14.7	12.19	0.370	8.080	0.640	394	64.7	5.18	56.3	13.9	1.96
W12 × 45	13.2	12.06	0.335	8.045	0.575	350	58.1	5.15	50.0	12.4	1.94
W12 × 26	7.65	12.22	0.230	6.490	0.380	204	33.4	5.17	17.3	5.34	1.51
W12 × 22	6.48	12.31	0.260	4.030	0.425	156	25.4	4.91	4.66	2.31	0.847
W12 × 16	4.71	11.99	0.220	3.990	0.265	103	17.1	4.67	2.82	1.41	0.773
W12 × 14	4.16	11.91	0.200	3.970	0.225	88.6	14.9	4.62	2.36	1.19	0.753
W10 × 100	29.4	11.10	0.680	10.340	1.120	623	112	4.60	207	40.0	2.65
W10 × 54	15.8	10.09	0.370	10.030	0.615	303	60.0	4.37	103	20.6	2.56
W10 × 45	13.3	10.10	0.350	8.020	0.620	248	49.1	4.32	53.4	13.3	2.01
W10 × 39	11.5	9.92	0.315	7.985	0.530	209	42.1	4.27	45.0	11.3	1.98
W10 × 30	8.84	10.47	0.300	5.810	0.510	170	32.4	4.38	16.7	5.75	1.37
W10 × 19	5.62	10.24	0.250	4.020	0.395	96.3	18.8	4.14	4.29	2.14	0.874
W10 × 15	4.41	9.99	0.230	4.000	0.270	68.9	13.8	3.95	2.89	1.45	0.810
W10 × 12	3.54	9.87	0.190	3.960	0.210	53.8	10.9	3.90	2.18	1.10	0.785
W8 × 67	19.7	9.00	0.570	8.280	0.935	272	60.4	3.72	88.6	21.4	2.12
W8 × 58	17.1	8.75	0.510	8.220	0.810	228	52.0	3.65	75.1	18.3	2.10
W8 × 48	14.1	8.50	0.400	8.110	0.685	184	43.3	3.61	60.9	15.0	2.08
W8 × 40	11.7	8.25	0.360	8.070	0.560	146	35.5	3.53	49.1	12.2	2.04
W8 × 31	9.13	8.00	0.285	7.995	0.435	110	27.5	3.47	37.1	9.27	2.02
W8 × 24	7.08	7.93	0.245	6.495	0.400	82.8	20.9	3.42	18.3	5.63	1.61
W8 × 15	4.44	8.11	0.245	4.015	0.315	48.0	11.8	3.29	3.41	1.70	0.876
W6 × 25	7.34	6.38	0.320	6.080	0.455	53.4	16.7	2.70	17.1	5.61	1.52
W6 × 20	5.87	6.20	0.260	6.020	0.365	41.4	13.4	2.66	13.3	4.41	1.50
W6 × 16	4.74	6.28	0.260	4.030	0.405	32.1	10.2	2.60	4.43	2.20	0.966
W6 × 15	4.43	5.99	0.230	5.990	0.260	29.1	9.72	2.56	9.32	3.11	1.46
W6 × 12	3.55	6.03	0.230	4.000	0.280	22.1	7.31	2.49	2.99	1.50	0.918
W6 × 9	2.68	5.90	0.170	3.940	0.215	16.4	5.56	2.47	2.19	1.11	0.905

C

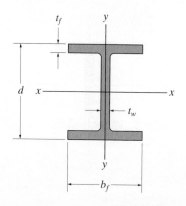

Wide-Flange Sections or W Shapes SI Units

Designation	Area A	Depth d	Web thickness t_w	Flange width b_f	Flange thickness t_f	x–x axis I	x–x axis S	x–x axis r	y–y axis I	y–y axis S	y–y axis r
mm × kg/m	mm²	mm	mm	mm	mm	10^6 mm⁴	10^3 mm³	mm	10^6 mm⁴	10^3 mm³	mm
W610 × 155	19 800	611	12.70	324.0	19.0	1 290	4 220	255	108	667	73.9
W610 × 140	17 900	617	13.10	230.0	22.2	1 120	3 630	250	45.1	392	50.2
W610 × 125	15 900	612	11.90	229.0	19.6	985	3 220	249	39.3	343	49.7
W610 × 113	14 400	608	11.20	228.0	17.3	875	2 880	247	34.3	301	48.8
W610 × 101	12 900	603	10.50	228.0	14.9	764	2 530	243	29.5	259	47.8
W610 × 92	11 800	603	10.90	179.0	15.0	646	2 140	234	14.4	161	34.9
W610 × 82	10 500	599	10.00	178.0	12.8	560	1 870	231	12.1	136	33.9
W460 × 97	12 300	466	11.40	193.0	19.0	445	1 910	190	22.8	236	43.1
W460 × 89	11 400	463	10.50	192.0	17.7	410	1 770	190	20.9	218	42.8
W460 × 82	10 400	460	9.91	191.0	16.0	370	1 610	189	18.6	195	42.3
W460 × 74	9 460	457	9.02	190.0	14.5	333	1 460	188	16.6	175	41.9
W460 × 68	8 730	459	9.14	154.0	15.4	297	1 290	184	9.41	122	32.8
W460 × 60	7 590	455	8.00	153.0	13.3	255	1 120	183	7.96	104	32.4
W460 × 52	6 640	450	7.62	152.0	10.8	212	942	179	6.34	83.4	30.9
W410 × 85	10 800	417	10.90	181.0	18.2	315	1 510	171	18.0	199	40.8
W410 × 74	9 510	413	9.65	180.0	16.0	275	1 330	170	15.6	173	40.5
W410 × 67	8 560	410	8.76	179.0	14.4	245	1 200	169	13.8	154	40.2
W410 × 53	6 820	403	7.49	177.0	10.9	186	923	165	10.1	114	38.5
W410 × 46	5 890	403	6.99	140.0	11.2	156	774	163	5.14	73.4	29.5
W410 × 39	4 960	399	6.35	140.0	8.8	126	632	159	4.02	57.4	28.5
W360 × 79	10 100	354	9.40	205.0	16.8	227	1 280	150	24.2	236	48.9
W360 × 64	8 150	347	7.75	203.0	13.5	179	1 030	148	18.8	185	48.0
W360 × 57	7 200	358	7.87	172.0	13.1	160	894	149	11.1	129	39.3
W360 × 51	6 450	355	7.24	171.0	11.6	141	794	148	9.68	113	38.7
W360 × 45	5 710	352	6.86	171.0	9.8	121	688	146	8.16	95.4	37.8
W360 × 39	4 960	353	6.48	128.0	10.7	102	578	143	3.75	58.6	27.5
W360 × 33	4 190	349	5.84	127.0	8.5	82.9	475	141	2.91	45.8	26.4

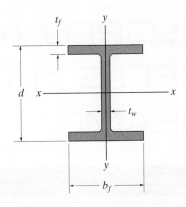

Wide-Flange Sections or W Shapes SI Units

Designation	Area A	Depth d	Web thickness t_w	Flange width b_f	Flange thickness t_f	x–x axis I	x–x axis S	x–x axis r	y–y axis I	y–y axis S	y–y axis r
mm × kg/m	mm²	mm	mm	mm	mm	10^6 mm⁴	10^3 mm³	mm	10^6 mm⁴	10^3 mm³	mm
W310 × 129	16 500	318	13.10	308.0	20.6	308	1940	137	100	649	77.8
W310 × 74	9 480	310	9.40	205.0	16.3	165	1060	132	23.4	228	49.7
W310 × 67	8 530	306	8.51	204.0	14.6	145	948	130	20.7	203	49.3
W310 × 39	4 930	310	5.84	165.0	9.7	84.8	547	131	7.23	87.6	38.3
W310 × 33	4 180	313	6.60	102.0	10.8	65.0	415	125	1.92	37.6	21.4
W310 × 24	3 040	305	5.59	101.0	6.7	42.8	281	119	1.16	23.0	19.5
W310 × 21	2 680	303	5.08	101.0	5.7	37.0	244	117	0.986	19.5	19.2
W250 × 149	19 000	282	17.30	263.0	28.4	259	1840	117	86.2	656	67.4
W250 × 80	10 200	256	9.40	255.0	15.6	126	984	111	43.1	338	65.0
W250 × 67	8 560	257	8.89	204.0	15.7	104	809	110	22.2	218	50.9
W250 × 58	7 400	252	8.00	203.0	13.5	87.3	693	109	18.8	185	50.4
W250 × 45	5 700	266	7.62	148.0	13.0	71.1	535	112	7.03	95	35.1
W250 × 28	3 620	260	6.35	102.0	10.0	39.9	307	105	1.78	34.9	22.2
W250 × 22	2 850	254	5.84	102.0	6.9	28.8	227	101	1.22	23.9	20.7
W250 × 18	2 280	251	4.83	101.0	5.3	22.5	179	99.3	0.919	18.2	20.1
W200 × 100	12 700	229	14.50	210.0	23.7	113	987	94.3	36.6	349	53.7
W200 × 86	11 000	222	13.00	209.0	20.6	94.7	853	92.8	31.4	300	53.4
W200 × 71	9 100	216	10.20	206.0	17.4	76.6	709	91.7	25.4	247	52.8
W200 × 59	7 580	210	9.14	205.0	14.2	61.2	583	89.9	20.4	199	51.9
W200 × 46	5 890	203	7.24	203.0	11.0	45.5	448	87.9	15.3	151	51.0
W200 × 36	4 570	201	6.22	165.0	10.2	34.4	342	86.8	7.64	92.6	40.9
W200 × 22	2 860	206	6.22	102.0	8.0	20.0	194	83.6	1.42	27.8	22.3
W150 × 37	4 730	162	8.13	154.0	11.6	22.2	274	68.5	7.07	91.8	38.7
W150 × 30	3 790	157	6.60	153.0	9.3	17.1	218	67.2	5.54	72.4	38.2
W150 × 22	2 860	152	5.84	152.0	6.6	12.1	159	65.0	3.87	50.9	36.8
W150 × 24	3 060	160	6.60	102.0	10.3	13.4	168	66.2	1.83	35.9	24.5
W150 × 18	2 290	153	5.84	102.0	7.1	9.19	120	63.3	1.26	24.7	23.5
W150 × 14	1 730	150	4.32	100.0	5.5	6.84	91.2	62.9	0.912	18.2	23.0

APPENDIX D

SLOPES AND DEFLECTIONS OF BEAMS

Simply Supported Beam Slopes and Deflections

Beam	Slope	Deflection	Elastic Curve
	$\theta_{max} = \dfrac{-PL^2}{16EI}$	$v_{max} = \dfrac{-PL^3}{48EI}$	$v = \dfrac{-Px}{48EI}(3L^2 - 4x^2)$ $0 \le x \le L/2$
	$\theta_1 = \dfrac{-Pab(L+b)}{6EIL}$ $\theta_2 = \dfrac{Pab(L+a)}{6EIL}$	$v\Big|_{x=a} = \dfrac{-Pba}{6EIL}(L^2 - b^2 - a^2)$	$v = \dfrac{-Pbx}{6EIL}(L^2 - b^2 - x^2)$ $0 \le x \le a$
	$\theta_1 = \dfrac{-M_0 L}{6EI}$ $\theta_2 = \dfrac{M_0 L}{3EI}$	$v_{max} = \dfrac{-M_0 L^2}{9\sqrt{3}EI}$ at $x = 0.5774L$	$v = \dfrac{-M_0 x}{6EIL}(L^2 - x^2)$
	$\theta_{max} = \dfrac{-wL^3}{24EI}$	$v_{max} = \dfrac{-5wL^4}{384EI}$	$v = \dfrac{-wx}{24EI}(x^3 - 2Lx^2 + L^3)$
	$\theta_1 = \dfrac{-3wL^3}{128EI}$ $\theta_2 = \dfrac{7wL^3}{384EI}$	$v\Big|_{x=L/2} = \dfrac{-5wL^4}{768EI}$ $v_{max} = -0.006563\dfrac{wL^4}{EI}$ at $x = 0.4598L$	$v = \dfrac{-wx}{384EI}(16x^3 - 24Lx^2 + 9L^3)$ $0 \le x \le L/2$ $v = \dfrac{-wL}{384EI}(8x^3 - 24Lx^2 + 17L^2x - L^3)$ $L/2 \le x < L$
	$\theta_1 = \dfrac{-7w_0 L^3}{360EI}$ $\theta_2 = \dfrac{w_0 L^3}{45EI}$	$v_{max} = -0.00652\dfrac{w_0 L^4}{EI}$ at $x = 0.5193L$	$v = \dfrac{-w_0 x}{360EIL}(3x^4 - 10L^2x^2 + 7L^4)$

Cantilevered Beam Slopes and Deflections

Beam	Slope	Deflection	Elastic Curve
	$\theta_{max} = \dfrac{-PL^2}{2EI}$	$v_{max} = \dfrac{-PL^3}{3EI}$	$v = \dfrac{-Px^2}{6EI}(3L - x)$
	$\theta_{max} = \dfrac{-PL^2}{8EI}$	$v_{max} = \dfrac{-5PL^3}{48EI}$	$v = \dfrac{-Px^2}{12EI}(3L - 2x) \quad 0 \le x \le L/2$ $v = \dfrac{-PL^2}{48EI}(6x - L) \quad L/2 \le x \le L$
	$\theta_{max} = \dfrac{-wL^3}{6EI}$	$v_{max} = \dfrac{-wL^4}{8EI}$	$v = \dfrac{-wx^2}{24EI}(x^2 - 4Lx + 6L^2)$
	$\theta_{max} = \dfrac{M_0 L}{EI}$	$v_{max} = \dfrac{M_0 L^2}{2EI}$	$v = \dfrac{M_0 x^2}{2EI}$
	$\theta_{max} = \dfrac{-wL^3}{48EI}$	$v_{max} = \dfrac{-7wL^4}{384EI}$	$v = \dfrac{-wx^2}{24EI}\left(x^2 - 2Lx + \tfrac{3}{2}L^2\right)$ $0 \le x \le L/2$ $v = \dfrac{-wL^3}{384EI}(8x - L)$ $L/2 \le x \le L$
	$\theta_{max} = \dfrac{-w_0 L^3}{24EI}$	$v_{max} = \dfrac{-w_0 L^4}{30EI}$	$v = \dfrac{-w_0 x^2}{120EIL}(10L^3 - 10L^2 x + 5Lx^2 - x^3)$

D

Preliminary Problems Solutions

Chapter 2

P2–1.

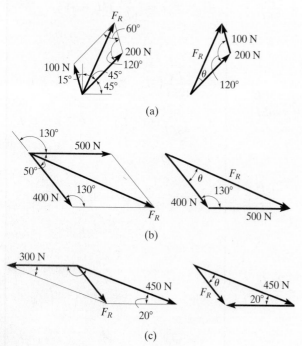

(a)

(b)

(c)

P2–2.

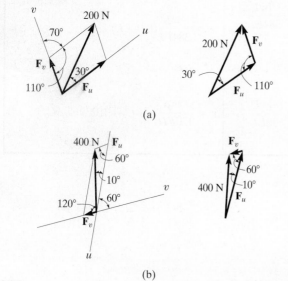

(a)

(b)

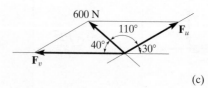

(c)

P2–3.

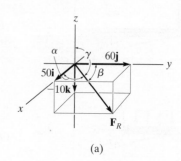

(a)

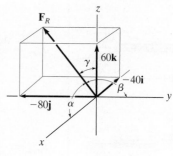

(b)

P2–4. **a.** $\mathbf{F} = \{-4\mathbf{i} - 4\mathbf{j} + 2\mathbf{k}\}$ kN

$F = \sqrt{(4)^2 + (-4)^2 + (2)^2} = 6$ kN

$\cos \beta = \dfrac{-2}{3}$

b. $\mathbf{F} = \{20\mathbf{i} + 20\mathbf{j} - 10\mathbf{k}\}$ N

$F = \sqrt{(20)^2 + (20)^2 + (-10)^2} = 30$ N

$\cos \beta = \dfrac{2}{3}$

P2–5.

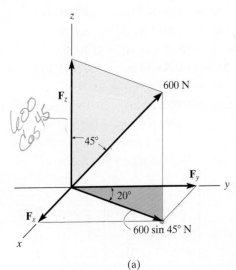

(a)

$$F_x = (600 \sin 45°) \sin 20° \text{ N}$$
$$F_y = (600 \sin 45°) \cos 20° \text{ N}$$
$$F_z = 600 \cos 45° \text{ N}$$

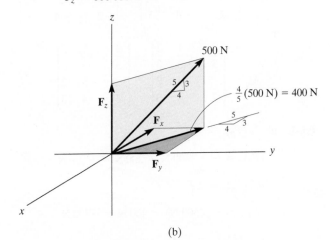

(b)

$$F_x = -\frac{3}{5}(400) \text{ N}$$

$$F_y = \frac{4}{5}(400) \text{ N}$$

$$F_z = \frac{3}{5}(500) \text{ N}$$

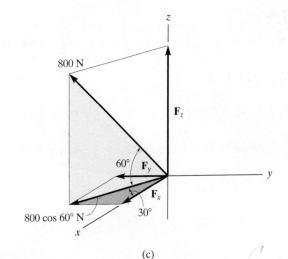

(c)

$$F_x = 800 \cos 60° \cos 30° \text{ N}$$
$$F_y = -800 \cos 60° \sin 30° \text{ N}$$
$$F_z = 800 \sin 60° \text{ N}$$

P2–6. **a.** $\mathbf{r}_{AB} = \{-5\mathbf{i} + 3\mathbf{j} - 2\mathbf{k}\}$ m
   b. $\mathbf{r}_{AB} = \{4\mathbf{i} + 8\mathbf{j} - 3\mathbf{k}\}$ m
   c. $\mathbf{r}_{AB} = \{6\mathbf{i} - 3\mathbf{j} - 4\mathbf{k}\}$ m

P2–7. **a.** $\mathbf{F} = 15 \text{ kN}\left(\dfrac{-3}{5}\mathbf{i} + \dfrac{4}{5}\mathbf{j}\right) = \{-9\mathbf{i} + 12\mathbf{j}\}$ kN

   b. $\mathbf{F} = 600 \text{ N}\left(\dfrac{2}{3}\mathbf{i} + \dfrac{2}{3}\mathbf{j} - \dfrac{1}{3}\mathbf{k}\right)$

   $\qquad = \{400\mathbf{i} + 400\mathbf{j} - 200\mathbf{k}\}$ N

   c. $\mathbf{F} = 300 \text{ N}\left(-\dfrac{2}{3}\mathbf{i} + \dfrac{2}{3}\mathbf{j} - \dfrac{1}{3}\mathbf{k}\right)$

   $\qquad = \{-200\mathbf{i} + 200\mathbf{j} - 100\mathbf{k}\}$ N

P2–8. **a.** $\mathbf{r}_A = \{3\mathbf{k}\}$ m, $r_A = 3$ m
   $\qquad \mathbf{r}_B = \{2\mathbf{i} + 2\mathbf{j} - 1\mathbf{k}\}$ m, $r_B = 3$ m
   $\qquad \mathbf{r}_A \cdot \mathbf{r}_B = 0(2) + 0(2) + (3)(-1) = -3 \text{ m}^2$
   $\qquad \mathbf{r}_A \cdot \mathbf{r}_B = r_A r_B \cos \theta$
   $\qquad\quad -3 = 3(3) \cos \theta$

   b. $\mathbf{r}_A = \{-2\mathbf{i} + 2\mathbf{j} + 1\mathbf{k}\}$ m, $r_A = 3$ m
   $\qquad \mathbf{r}_B = \{1.5\mathbf{i} - 2\mathbf{k}\}$ m, $r_B = 2.5$ m
   $\qquad \mathbf{r}_A \cdot \mathbf{r}_B = (-2)(1.5) + 2(0) + (1)(-2) = -5 \text{ m}^2$
   $\qquad \mathbf{r}_A \cdot \mathbf{r}_B = r_A r_B \cos \theta$
   $\qquad\quad -5 = 3(2.5) \cos \theta$

P2–9. a.

$$\mathbf{F} = 300 \text{ N}\left(\frac{2}{3}\mathbf{i} + \frac{2}{3}\mathbf{j} - \frac{1}{3}\mathbf{k}\right) = \{200\mathbf{i} + 200\mathbf{j} - 100\mathbf{k}\}\text{ N}$$

$$\mathbf{u}_a = -\frac{3}{5}\mathbf{i} + \frac{4}{5}\mathbf{j}$$

$$F_a = \mathbf{F} \cdot \mathbf{u}_a = (200)\left(-\frac{3}{5}\right) + (200)\left(\frac{4}{5}\right) + (-100)\left(0\right)$$

b. $\mathbf{F} = 500 \text{ N}\left(-\frac{4}{5}\mathbf{j} + \frac{3}{5}\mathbf{k}\right) = \{-400\mathbf{j} + 300\mathbf{k}\}\text{ N}$

$$\mathbf{u}_a = -\frac{1}{3}\mathbf{i} + \frac{2}{3}\mathbf{j} + \frac{2}{3}\mathbf{k}$$

$$F_a = \mathbf{F} \cdot \mathbf{u}_a = (0)\left(-\frac{1}{3}\right) + (-400)\left(\frac{2}{3}\right) + (300)\left(\frac{2}{3}\right)$$

Chapter 3

P3–1. a. $M_O = 100 \text{ N}(2 \text{ m}) = 200 \text{ N} \cdot \text{m} \,\circlearrowleft$

b. $M_O = -100 \text{ N}(1 \text{ m}) = 100 \text{ N} \cdot \text{m} \,\circlearrowright$

c. $M_O = -\left(\frac{3}{5}\right)(500 \text{ N})(2 \text{ m}) = 600 \text{ N} \cdot \text{m} \,\circlearrowright$

d. $M_O = \left(\frac{4}{5}\right)(500 \text{ N})(3 \text{ m}) = 1200 \text{ N} \cdot \text{m} \,\circlearrowleft$

e. $M_O = -\left(\frac{3}{5}\right)(100 \text{ N})(5 \text{ m}) = 300 \text{ N} \cdot \text{m} \,\circlearrowright$

f. $M_O = 100 \text{ N}(0) = 0$

g. $M_O = -\left(\frac{3}{5}\right)(500 \text{ N})(2 \text{ m}) + \left(\frac{4}{5}\right)(500 \text{ N})(1 \text{ m})$

$$= 200 \text{ N} \cdot \text{m} \,\circlearrowright$$

h. $M_O = -\left(\frac{3}{5}\right)(500 \text{ N})(3 \text{ m} - 1 \text{ m})$

$$+ \left(\frac{4}{5}\right)(500 \text{ N})(1 \text{ m}) = 200 \text{ N} \cdot \text{m} \,\circlearrowright$$

i. $M_O = \left(\frac{3}{5}\right)(500 \text{ N})(1 \text{ m}) - \left(\frac{4}{5}\right)(500 \text{ N})(3 \text{ m})$

$$= 900 \text{ N} \cdot \text{m} \,\circlearrowright$$

P3–2. $\mathbf{M}_P = \begin{vmatrix} \mathbf{i} & \mathbf{j} & \mathbf{k} \\ 2 & -3 & 0 \\ -3 & 2 & 5 \end{vmatrix}$ $\mathbf{M}_P = \begin{vmatrix} \mathbf{i} & \mathbf{j} & \mathbf{k} \\ 2 & 5 & -1 \\ 2 & -4 & -3 \end{vmatrix}$

$$\mathbf{M}_P = \begin{vmatrix} \mathbf{i} & \mathbf{j} & \mathbf{k} \\ 5 & -4 & -1 \\ -2 & 3 & 4 \end{vmatrix}$$

P3–3. a. $M_x = -(100 \text{ N})(3 \text{ m}) = -300 \text{ N} \cdot \text{m}$

$M_y = -(200 \text{ N})(2 \text{ m}) = -400 \text{ N} \cdot \text{m}$

$M_z = -(300 \text{ N})(2 \text{ m}) = -600 \text{ N} \cdot \text{m}$

b. $M_x = (50 \text{ N})(0.5 \text{ m}) = 25 \text{ N} \cdot \text{m}$

$M_y = (400 \text{ N})(0.5 \text{ m}) - (300 \text{ N})(3 \text{ m}) = -700 \text{ N} \cdot \text{m}$

$M_z = (100 \text{ N})(3 \text{ m}) = 300 \text{ N} \cdot \text{m}$

c. $M_x = (300 \text{ N})(2 \text{ m}) - (100 \text{ N})(2 \text{ m}) = 400 \text{ N} \cdot \text{m}$

$M_y = -(300 \text{ N})(1 \text{ m}) + (50 \text{ N})(1 \text{ m})$

$$+ (400 \text{ N})(0.5 \text{ m}) = 250 \text{ N} \cdot \text{m}$$

$M_z = -(200 \text{ N})(1 \text{ m}) = -200 \text{ N} \cdot \text{m}$

P3–4. a.

$$M_a = \begin{vmatrix} -\frac{4}{5} & -\frac{3}{5} & 0 \\ -5 & 2 & 0 \\ 6 & 2 & 3 \end{vmatrix} = \begin{vmatrix} -\frac{4}{5} & -\frac{3}{5} & 0 \\ -1 & 5 & 0 \\ 6 & 2 & 3 \end{vmatrix}$$

b.

$$M_a = \begin{vmatrix} -\frac{1}{\sqrt{2}} & \frac{1}{\sqrt{2}} & 0 \\ 3 & 4 & -2 \\ 2 & -4 & 3 \end{vmatrix} = \begin{vmatrix} -\frac{1}{\sqrt{2}} & \frac{1}{\sqrt{2}} & 0 \\ 5 & 2 & -2 \\ 2 & -4 & 3 \end{vmatrix}$$

c.

$$M_a = \begin{vmatrix} \frac{2}{3} & -\frac{1}{3} & \frac{2}{3} \\ -5 & -4 & 0 \\ 2 & -4 & 3 \end{vmatrix} = \begin{vmatrix} \frac{2}{3} & -\frac{1}{3} & \frac{2}{3} \\ -3 & -5 & 2 \\ 2 & -4 & 3 \end{vmatrix}$$

P3–5. a. $\xrightarrow{+} (F_R)_x = \Sigma F_x;$

$$(F_R)_x = -\left(\frac{4}{5}\right)500 \text{ N} + 200 \text{ N} = -200 \text{ N}$$

$$+\uparrow (F_R)_y = \Sigma F_y;$$

$$(F_R)_y = -\frac{3}{5}(500 \text{ N}) - 400 \text{ N} = -700 \text{ N}$$

$$\zeta + (M_R)_O = \Sigma M_O;$$

$$(M_R)_O = -\left(\frac{3}{5}\right)(500 \text{ N})(2 \text{ m}) - 400 \text{ N}(4 \text{ m})$$

$$= -2200 \text{ N} \cdot \text{m}$$

b. $\xrightarrow{+} (F_R)_x = \Sigma F_x;$

$$(F_R)_x = \left(\frac{4}{5}\right)(500 \text{ N}) = 400 \text{ N}$$

$$+\uparrow (F_R)_y = \Sigma F_y;$$

$$(F_R)_y = -(300 \text{ N}) - \left(\frac{3}{5}\right)(500 \text{ N}) = -600 \text{ N}$$

$\zeta + (M_R)_O = \Sigma M_O;$

$(M_R)_O = -(300 \text{ N})(2 \text{ m}) - \left(\frac{3}{5}\right)(500 \text{ N})(4 \text{ m})$

$\qquad - 200 \text{ N} \cdot \text{m} = -2000 \text{ N} \cdot \text{m}$

c. $\xrightarrow{+} (F_R)_x = \Sigma F_x;$

$(F_R)_x = \left(\frac{3}{5}\right)(500 \text{ N}) + 100 \text{ N} = 400 \text{ N}$

$+\uparrow (F_R)_y = \Sigma F_y;$

$(F_R)_y = -(500 \text{ N}) - \left(\frac{4}{5}\right)(500 \text{ N}) = -900 \text{ N}$

$\zeta + (M_R)_O = \Sigma M_O;$

$(M_R)_O = -(500 \text{ N})(2 \text{ m}) - \left(\frac{4}{5}\right)(500 \text{ N})(4 \text{ m})$

$\qquad + \left(\frac{3}{5}\right)(500 \text{ N})(2 \text{ m}) = -2000 \text{ N} \cdot \text{m}$

d. $\xrightarrow{+} (F_R)_x = \Sigma F_x;$

$(F_R)_x = -\left(\frac{4}{5}\right)(500 \text{ N}) + \left(\frac{3}{5}\right)(500 \text{ N}) = -100 \text{ N}$

$+\uparrow (F_R)_y = \Sigma F_y;$

$(F_R)_y = -\left(\frac{3}{5}\right)(500 \text{ N}) - \left(\frac{4}{5}\right)(500 \text{ N}) = -700 \text{ N}$

$\zeta + (M_R)_O = \Sigma M_O;$

$(M_R)_O = \left(\frac{4}{5}\right)(500 \text{ N})(4 \text{ m}) + \left(\frac{3}{5}\right)(500 \text{ N})(2 \text{ m})$

$\qquad - \left(\frac{3}{5}\right)(500 \text{ N})(4 \text{ m}) + 200 \text{ N} \cdot \text{m} = 1200 \text{ N} \cdot \text{m}$

P3–6. **a.** $\xrightarrow{+} (F_R)_x = \Sigma F_x;$ $(F_R)_x = 0$

$+\uparrow (F_R)_y = \Sigma F_y;$

$(F_R)_y = -200 \text{ N} - 260 \text{ N} = -460 \text{ N}$

$\zeta + (F_R)_y d = \Sigma M_O;$

$-(460 \text{ N})d = -(200 \text{ N})(2 \text{ m}) - (260 \text{ N})(4 \text{ m})$

$\qquad d = 3.13 \text{ m}$

Note: Although 460 N acts downward, this is *not* why $-(460 \text{ N})d$ is negative. It is because the *moment* of 460 N about O is negative.

b. $\xrightarrow{+} (F_R)_x = \Sigma F_x;$

$(F_R)_x = -\left(\frac{3}{5}\right)(500 \text{ N}) = -300 \text{ N}$

$+\uparrow (F_R)_y = \Sigma F_y;$

$(F_R)_y = -400 \text{ N} - \left(\frac{4}{5}\right)(500 \text{ N}) = -800 \text{ N}$

$\zeta + (F_R)_y d = \Sigma M_O;$

$-(800 \text{ N})d = -(400 \text{ N})(2 \text{ m}) - \left(\frac{4}{5}\right)(500 \text{ N})(4 \text{ m})$

$\qquad d = 3 \text{ m}$

c. $\xrightarrow{+} (F_R)_x = \Sigma F_x;$

$(F_R)_x = \left(\frac{4}{5}\right)(500 \text{ N}) - \left(\frac{4}{5}\right)(500 \text{ N}) = 0$

$+\uparrow (F_R)_y = \Sigma F_y;$

$(F_R)_y = -\left(\frac{3}{5}\right)(500 \text{ N}) - \left(\frac{3}{5}\right)(500 \text{ N}) = -600 \text{ N}$

$\zeta + (F_R)_y d = \Sigma M_O;$

$-(600 \text{ N})d = -\left(\frac{3}{5}\right)(500 \text{ N})(2 \text{ m}) - \left(\frac{3}{5}\right)(500 \text{ N})(4 \text{ m})$

$\qquad - 600 \text{ N} \cdot \text{m}$

$\qquad d = 4 \text{ m}$

P3–7. **a.** $+\downarrow F_R = \Sigma F_z;$

$F_R = 200 \text{ N} + 100 \text{ N} + 200 \text{ N} = 500 \text{ N}$

$(M_R)_x = \Sigma M_x;$

$-(500 \text{ N})y = -(100 \text{ N})(2 \text{ m}) - (200 \text{ N})(2 \text{ m})$

$\qquad y = 1.20 \text{ m}$

$(M_R)_y = \Sigma M_y;$

$(500 \text{ N})x = (100 \text{ N})(2 \text{ m}) + (200 \text{ N})(1 \text{ m})$

$\qquad x = 0.80 \text{ m}$

b. $+\downarrow F_R = \Sigma F_z;$

$F_R = 100 \text{ N} - 100 \text{ N} + 200 \text{ N} = 200 \text{ N}$

$(M_R)_x = \Sigma M_x;$

$-(200 \text{ N})y = (100 \text{ N})(1 \text{ m}) + (100 \text{ N})(2 \text{ m})$

$\qquad - (200 \text{ N})(2 \text{ m})$

$\qquad y = 0.5 \text{ m}$

$(M_R)_y = \Sigma M_y;$

$(200 \text{ N})x = -(100 \text{ N})(2 \text{ m}) + (100 \text{ N})(2 \text{ m})$

$\qquad x = 0$

c. $+\downarrow F_R = \Sigma F_z;$

$F_R = 400 \text{ N} + 300 \text{ N} + 200 \text{ N} + 100 \text{ N} = 1000 \text{ N}$

$(M_R)_x = \Sigma M_x;$

$-(1000 \text{ N})y = -(300 \text{ N})(4 \text{ m}) - (100 \text{ N})(4 \text{ m})$

$\qquad y = 1.6 \text{ m}$

$(M_R)_y = \Sigma M_y;$

$(1000 \text{ N})x = (400 \text{ N})(2 \text{ m}) + (300 \text{ N})(2 \text{ m})$
$\qquad - (200 \text{ N})(2 \text{ m}) - (100 \text{ N})(2 \text{ m})$

$\qquad x = 0.8 \text{ m}$

Chapter 4

P4–1.

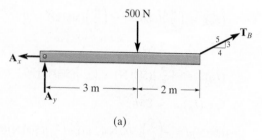

(a)

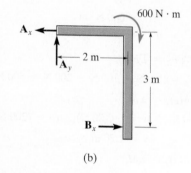

(b)

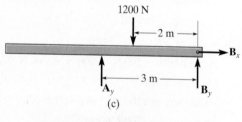

(c)

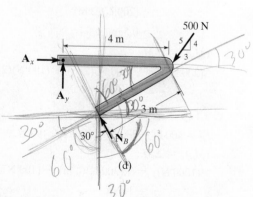

(d)

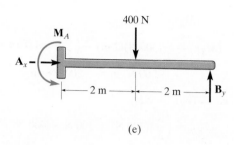

(e)

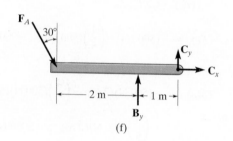

(f)

P4–2.

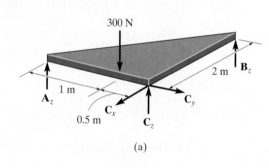

(a)

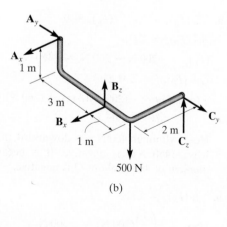

(b)

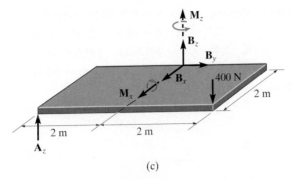

(c)

P4–3.   a. $\Sigma M_x = 0$;

$$-(400 \text{ N})(2 \text{ m}) - (600 \text{ N})(5 \text{ m}) + B_z(5 \text{ m}) = 0$$

$\Sigma M_y = 0$; $-A_z(4 \text{ m}) - B_z(4 \text{ m}) = 0$

$\Sigma M_z = 0$; $B_y(4 \text{ m}) - B_x(5 \text{ m})$

$$+ (300 \text{ N})(5 \text{ m}) = 0$$

b. $\Sigma M_x = 0$; $A_z(4 \text{ m}) + C_z(6 \text{ m}) = 0$

$\Sigma M_y = 0$; $B_z(1 \text{ m}) - C_z(1 \text{ m}) = 0$

$\Sigma M_z = 0$; $-B_y(1 \text{ m}) + (300 \text{ N})(2 \text{ m})$

$$- A_x(4 \text{ m}) + C_y(1 \text{ m}) = 0$$

c. $\Sigma M_x = 0$; $B_z(2 \text{ m}) + C_z(3 \text{ m}) - 800 \text{ N} \cdot \text{m} = 0$

$\Sigma M_y = 0$; $-C_z(1.5 \text{ m}) = 0$

$\Sigma M_z = 0$; $-B_x(2 \text{ m}) + C_y(1.5 \text{ m}) = 0$

P4–4.

a.

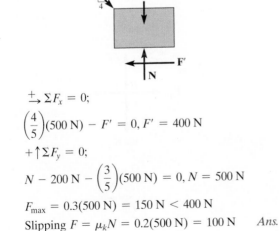

$\xrightarrow{+} \Sigma F_x = 0$;

$$\left(\frac{4}{5}\right)(500 \text{ N}) - F' = 0, F' = 400 \text{ N}$$

$+\uparrow \Sigma F_y = 0$;

$$N - 200 \text{ N} - \left(\frac{3}{5}\right)(500 \text{ N}) = 0, N = 500 \text{ N}$$

$F_{\text{max}} = 0.3(500 \text{ N}) = 150 \text{ N} < 400 \text{ N}$

Slipping $F = \mu_k N = 0.2(500 \text{ N}) = 100 \text{ N}$ *Ans.*

b.

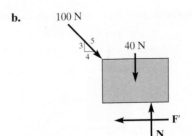

$\xrightarrow{+} \Sigma F_x = 0$;

$$\frac{4}{5}(100 \text{ N}) - F' = 0; F' = 80 \text{ N}$$

$+\uparrow \Sigma F_y = 0$;

$$N - 40 \text{ N} - \left(\frac{3}{5}\right)(100 \text{ N}) = 0; N = 100 \text{ N}$$

$F_{\text{max}} = 0.9(100 \text{ N}) = 90 \text{ N} > 80 \text{ N}$

$F = F' = 80 \text{ N}$ *Ans.*

P4–5.

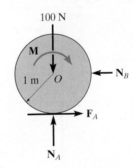

Require $F_A = 0.1 N_A$

$+\uparrow \Sigma F_y = 0$; $N_A - 100 \text{ N} = 0$

$$N_A = 100 \text{ N}$$

$$F_A = 0.1(100 \text{ N}) = 10 \text{ N}$$

$\zeta + \Sigma M_O = 0$; $-M + (10 \text{ N})(1 \text{ m}) = 0$

$$M = 10 \text{ N} \cdot \text{m}$$

P4–6. Slipping must occur between A and B.

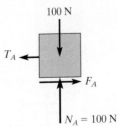

$$F_A = 0.2(100 \text{ N}) = 20 \text{ N}$$

b. Assume B slips on C and C does not slip.

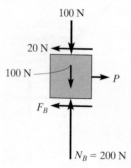

$$F_B = 0.2(200\text{ N}) = 40\text{ N}$$
$$\xrightarrow{+}\ \Sigma F_x = 0; \qquad P - 20\text{ N} - 40\text{ N} = 0$$
$$P = 60\text{ N}$$

c. Assume C slips and B does not slip on C.

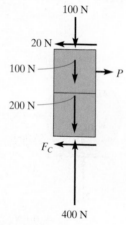

$$F_C = 0.1(400\text{ N}) = 40\text{ N}$$
$$\xrightarrow{+}\ \Sigma F_x = 0; \qquad P - 20\text{ N} - 40\text{ N} = 0$$
$$P = 60\text{ N}$$
Therefore, $\qquad P = 60\text{ N}$ *Ans.*

P4–7.

a.

P

200 N

2 m

0.5 m

O

F

x

$N = 200\text{ N}$

Assume slipping, $\qquad F = 0.3(200\text{ N}) = 60\text{ N}$
$$\xrightarrow{+}\ \Sigma F_x = 0; \quad P - 60\text{ N} = 0; \ P = 60\text{ N}$$
$$\zeta + \Sigma M_O = 0; \quad 200\text{ N}(x) - (60\text{ N})(2\text{ m}) = 0$$
$$x = 0.6\text{ m} > 0.5\text{ m}$$
Block tips, $\qquad x = 0.5\text{ m}$
$$\zeta + \Sigma M_O = 0 \quad (200\text{ N})(0.5\text{ m}) - P(2\text{ m}) = 0$$
$$P = 50\text{ N} \qquad Ans.$$

b.

Assume slipping, $\qquad F = 0.4(100\text{ N}) = 40\text{ N}$
$$\xrightarrow{+}\ \Sigma F_x = 0; \quad P - 40\text{ N} = 0; P = 40\text{ N}$$
$$\zeta + \Sigma M_O = 0; \quad (100\text{ N})(x) - (40\text{ N})(1\text{ m}) = 0$$
$$x = 0.4\text{ m} < 0.5\text{ m}$$
No tipping
$$P = 40\text{ N} \qquad Ans.$$

Chapter 5

P5–1.   **a.** $A_y = 200$ N, $D_x = 0$, $D_y = 200$ N

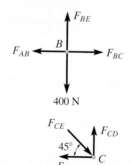

b. $A_y = 300$ N, $C_x = 0$, $C_y = 300$ N

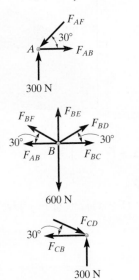

P5–3. **a.**

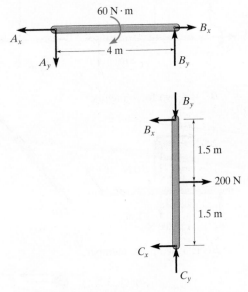

P5–2. **a.**

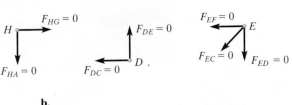

b.

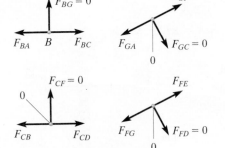

b. *CB* is a two-force member.

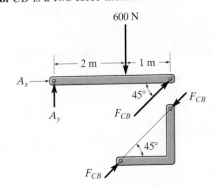

c. *CD* is a two-force member.

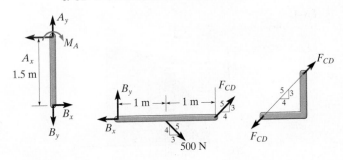

d.

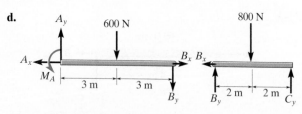

600 N

800 N

A_y

A_x

M_A

3 m

3 m

B_x B_x

B_y

B_y

2 m   2 m

C_y

BC is a two-force member.

e.

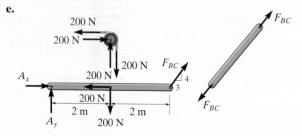

200 N
200 N
200 N F_{BC}
A_x 200 N
200 N
2 m   2 m
A_y
200 N

F_{BC}
4
3
F_{BC}

BC is a two-force member.

f.

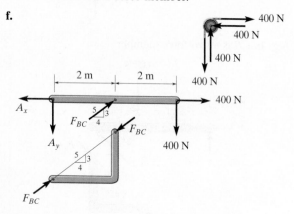

400 N
400 N
400 N
2 m   2 m   400 N
A_x 400 N
F_{BC} 5 3 4
F_{BC}
A_y 5 3 4   400 N
F_{BC}

Chapter 6

P6–1.

a.

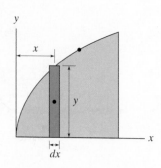

$\tilde{x} = x$

$\tilde{y} = \dfrac{y}{2} = \dfrac{\sqrt{x}}{2}$

$dA = ydx = \sqrt{x}\,dx$

b.

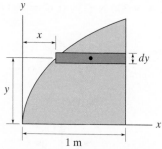

y

x

dy

y

1 m

$\tilde{x} = x + \left(\dfrac{1-x}{2}\right) = \dfrac{1+x}{2} = \dfrac{1+y^2}{2}$

$\tilde{y} = y$

$dA = (1-x)dy = (1-y^2)dy$

c.

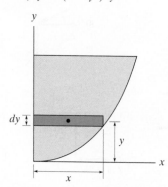

y

dy

y

x

$\tilde{x} = \dfrac{x}{2} = \dfrac{\sqrt{y}}{2}$

$\tilde{y} = y$

$dA = xdy = \sqrt{y}\,dy$

d.

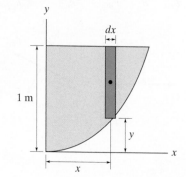

dx

1 m

y

x

$\tilde{x} = x$

$\tilde{y} = y + \left(\dfrac{1-y}{2}\right) = \dfrac{1+y}{2} = \dfrac{1+x^2}{2}$

$dA = (1-y)dx = (1-x^2)dx$

Chapter 7

P7–1a.

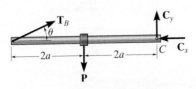

$$\zeta + \Sigma M_C = 0; \text{ get } T_B$$

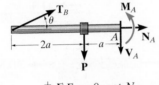

$$\xrightarrow{+} \Sigma F_x = 0; \text{ get } N_A$$
$$+\uparrow \Sigma F_y = 0; \text{ get } V_A$$
$$\zeta + \Sigma M_A = 0; \text{ get } M_A$$

P7–1b.

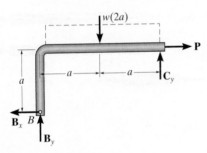

$$\zeta + \Sigma M_B = 0; \text{ get } C_y$$

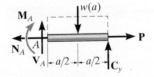

$$\xrightarrow{+} \Sigma F_x = 0; \text{ get } N_A$$
$$+\uparrow \Sigma F_y = 0; \text{ get } V_A$$
$$\zeta + \Sigma M_A = 0; \text{ get } M_A$$

P7–1c.

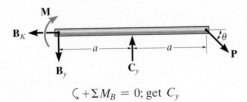

$$\zeta + \Sigma M_B = 0; \text{ get } C_y$$

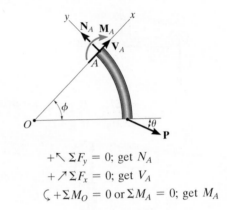

$$\xrightarrow{+} \Sigma F_x = 0; \text{ get } N_A$$
$$+\uparrow \Sigma F_y = 0; \text{ get } V_A$$
$$\zeta + \Sigma M_A = 0; \text{ get } M_A$$

P7–1d.

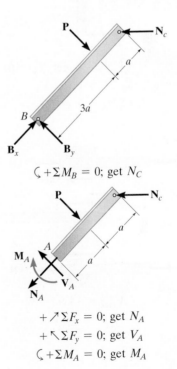

$$+\nwarrow \Sigma F_y = 0; \text{ get } N_A$$
$$+\nearrow \Sigma F_x = 0; \text{ get } V_A$$
$$\zeta + \Sigma M_O = 0 \text{ or } \Sigma M_A = 0; \text{ get } M_A$$

P7–1e.

$$\zeta + \Sigma M_B = 0; \text{ get } N_C$$

$$+\nearrow \Sigma F_x = 0; \text{ get } N_A$$
$$+\nwarrow \Sigma F_y = 0; \text{ get } V_A$$
$$\zeta + \Sigma M_A = 0; \text{ get } M_A$$

P7–1f.

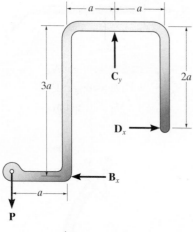

$$+\uparrow \Sigma F_y = 0; \text{ get } C_y \, (= P)$$
$$\circlearrowleft +\Sigma M_B = 0; \text{ get } D_x$$

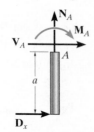

$$+\uparrow \Sigma F_y = 0; \text{ get } N_A \, (= 0)$$
$$\xrightarrow{+} \Sigma F_x = 0; \text{ get } V_A$$
$$\circlearrowleft +\Sigma M_A = 0; \text{ get } M_A$$

P7–2a.

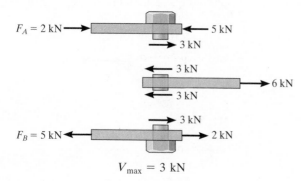

$$V_{\max} = 3 \text{ kN}$$

P7–2b.

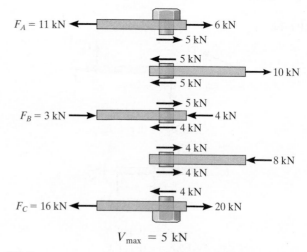

$$V_{\max} = 5 \text{ kN}$$

P7–3.

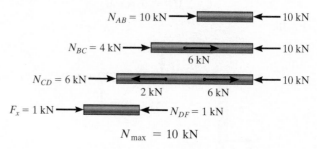

$$N_{\max} = 10 \text{ kN}$$

P7–4.

P7–5.

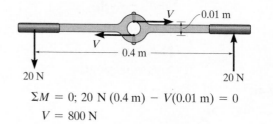

$$\Sigma M = 0; \ 20 \text{ N } (0.4 \text{ m}) - V(0.01 \text{ m}) = 0$$
$$V = 800 \text{ N}$$

P7–6.

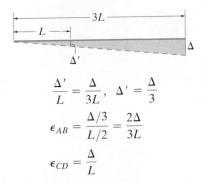

$$N = (5 \text{ kN}) \cos 30° = 4.33 \text{ kN}$$
$$V = (5 \text{ kN}) \sin 30° = 2.5 \text{ kN}$$

P7–7.

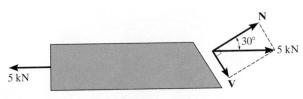

$$\frac{\Delta'}{L} = \frac{\Delta}{3L}, \quad \Delta' = \frac{\Delta}{3}$$

$$\epsilon_{AB} = \frac{\Delta/3}{L/2} = \frac{2\Delta}{3L}$$

$$\epsilon_{CD} = \frac{\Delta}{L}$$

P7–8.

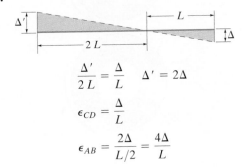

$$\frac{\Delta'}{2L} = \frac{\Delta}{L} \quad \Delta' = 2\Delta$$

$$\epsilon_{CD} = \frac{\Delta}{L}$$

$$\epsilon_{AB} = \frac{2\Delta}{L/2} = \frac{4\Delta}{L}$$

P7–9.

$$\epsilon_{AB} = \frac{L_{A'B} - L_{AB}}{L_{AB}}$$

P7–10.

$$\epsilon_{AB} = \frac{L_{AB'} - L_{AB}}{L_{AB}}, \quad \epsilon_{AC} = \frac{L_{AC'} - L_{AC}}{L_{AC}}$$

$$\epsilon_{BC} = \frac{L_{B'C'} - L_{BC}}{L_{BC}}, \quad (\gamma_A)_{xy} = \left(\frac{\pi}{2} - \theta\right) \text{ rad}$$

P7–11.

$$(\gamma_A)_{xy} = \frac{\pi}{2} - \left(\frac{\pi}{2} + \theta_1\right)$$

$$= (-\theta_1) \text{ rad}$$

$$(\gamma_B)_{xy} = \frac{\pi}{2} - (\pi - \theta_2)$$

$$= \left(-\frac{\pi}{2} + \theta_2\right) \text{ rad}$$

Chapter 9

P9–1a.

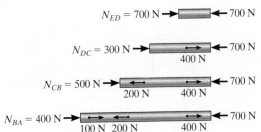

P9–1b.

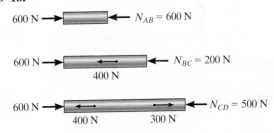

P9–2.

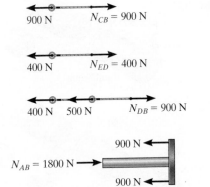

P9–3.

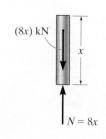

$N = 8x$

P9–4.

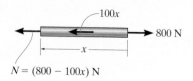

$N = (800 - 100x)$ N

P9–5.

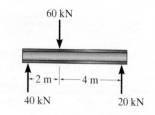

$$\Delta_B = \frac{PL}{AE} = \frac{20(10^3) \text{ N } (3\text{ m})}{2(10^{-3}) \text{ m}^2 (60(10^9) \text{ N/m}^2)}$$
$$= 0.5(10^{-3}) \text{ m} = 0.5 \text{ mm}$$

Chapter 10

P10–1.

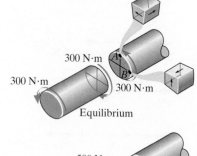

Equilibrium

Equilibrium

P10–2.

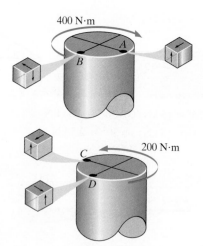

P10–3.

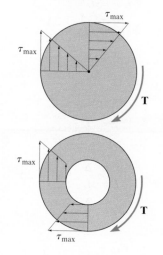

P10–4.

$$P = T\omega$$
$$(10 \text{ hp})\left(\frac{550 \text{ ft} \cdot \text{lb/s}}{1 \text{ hp}}\right) = T\left(1200 \frac{\text{rev}}{\text{min}}\right)\left(\frac{1 \text{ min}}{60 \text{ s}}\right)\frac{2\pi \text{ rad}}{1 \text{ rev}}$$
$$T = 43.8 \text{ lb} \cdot \text{ft}$$

Chapter 11

P11–1a.

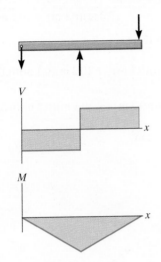

P11–1b.

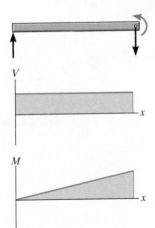

P11–1c.

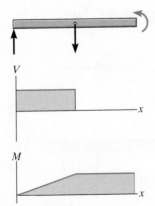

P11–1d.

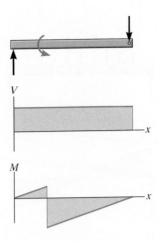

P11–1e.

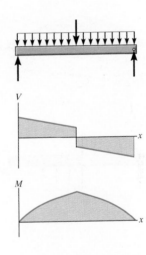

P11–1f.

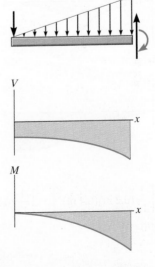

P11–1g.

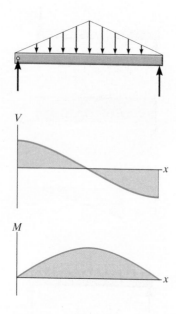

P11–1h.

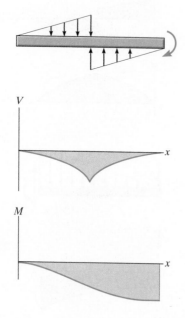

P11–2. $I = \left[\dfrac{1}{12}(0.2 \text{ m})(0.4 \text{ m})^3 \right] - \left[\dfrac{1}{12}(0.1 \text{ m})(0.2 \text{ m})^3 \right]$

$= 1.0\,(10^{-3}) \text{ m}^4$

P11–3.

$$\bar{y} = \frac{\Sigma \tilde{y} A}{\Sigma A} = \frac{(0.05 \text{ m})(0.2 \text{ m})(0.1 \text{ m}) + (0.25 \text{ m})(0.1 \text{ m})(0.3 \text{ m})}{(0.2 \text{ m})(0.1 \text{ m}) + (0.1 \text{ m})(0.3 \text{ m})}$$

$$= 0.17 \text{ m}$$

$$I = \left[\frac{1}{12}(0.2 \text{ m})(0.1 \text{ m})^3 + (0.2 \text{ m})(0.1 \text{ m})(0.17 \text{ m} - 0.05 \text{ m})^2 \right]$$

$$+ \left[\frac{1}{12}(0.1 \text{ m})(0.3 \text{ m})^3 + (0.1 \text{ m})(0.3 \text{ m})(0.25 \text{ m} - 0.17 \text{ m})^2 \right]$$

$$= 0.722\,(10^{-3}) \text{ m}^4$$

P11–4a.

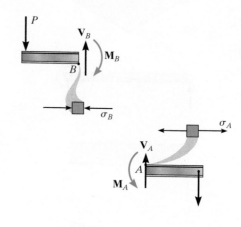

P11–4b.

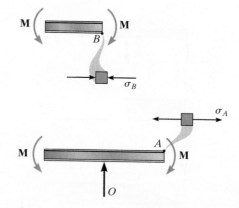

P11–5a.

P11–5b.

Chapter 12

P12–1a.

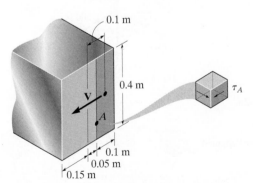

$$Q = \bar{y}'A' = (0.1 \text{ m})(0.1 \text{ m})(0.4 \text{ m}) = 4(10^{-3}) \text{ m}^3$$
$$t = 0.4 \text{ m}$$

P12–1b.

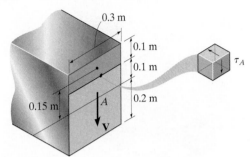

$$Q = \bar{y}'A' = (0.15 \text{ m})(0.3 \text{ m})(0.1 \text{ m}) = 4.5(10^{-3}) \text{ m}^3$$
$$t = 0.3 \text{ m}$$

P12–1c.

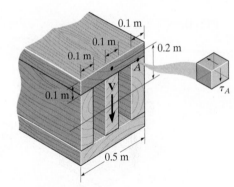

$$Q = \bar{y}'A' = (0.2 \text{ m})(0.1 \text{ m})(0.5 \text{ m}) = 0.01 \text{ m}^3$$
$$t = 3(0.1 \text{ m}) = 0.3 \text{ m}$$

P12–1d.

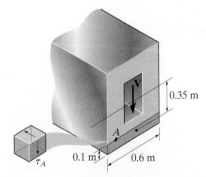

$$Q = \bar{y}'A' = (0.35 \text{ m})(0.6 \text{ m})(0.1 \text{ m}) = 0.021 \text{ m}^3$$
$$t = 0.6 \text{ m}$$

P12–1e.

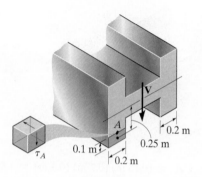

$$Q = \bar{y}'A' = (0.25 \text{ m}) (0.2 \text{ m}) (0.1 \text{ m}) = 5(10^{-3}) \text{ m}^3$$
$$t = 0.2 \text{ m}$$

P12–1f.

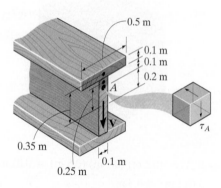

$$Q = \Sigma\bar{y}'A' = (0.25 \text{ m}) (0.1 \text{ m}) (0.1 \text{ m})$$
$$+ (0.35 \text{ m}) (0.1 \text{ m}) (0.5 \text{ m}) = 0.02 \text{ m}^3$$
$$t = 0.1 \text{ m}$$

Chapter 13

P13–1a.

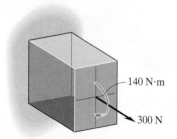

P13–1b.

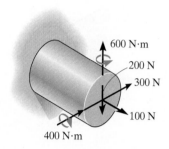

P13–1c.

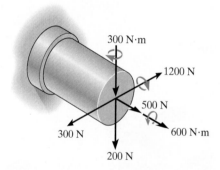

P13–1d.

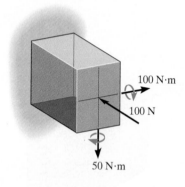

P13–2a.

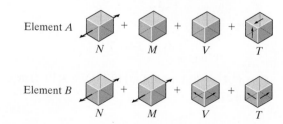

P13–2b.

Element A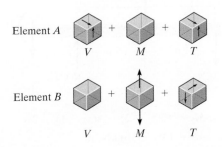

V M T

Element B

V M T

Chapter 14

P14–1.

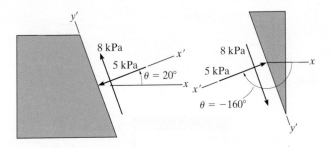

P14–1b.

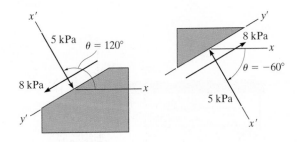

P14–1c.

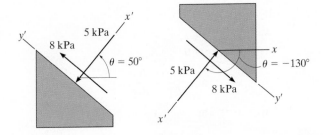

P14–2.

$$\tau_{max} = \sqrt{\left(\frac{\sigma_x - \sigma_y}{2}\right)^2 + \tau_{xy}^2} = \sqrt{\left(\frac{4 - (-4)}{2}\right)^2 + (0)^2}$$

$$= 4 \text{ MPa}$$

$$\sigma_{avg} = \frac{\sigma_x + \sigma_y}{2} = \frac{4 - 4}{2} = 0$$

$$\tan 2\theta_s = -\frac{(\sigma_x - \sigma_y)/2}{\tau_{xy}} = \frac{[4 - (-4)]/2}{0} = -\infty$$

$$\theta_s = -45°$$

$$\tau_{x'y'} = -\frac{\sigma_x - \sigma_y}{2} \sin 2\theta + \tau_{xy} \cos 2\theta$$

$$= -\frac{4 - (-4)}{2} \sin 2(-45°) + 0 = 4 \text{ MPa}$$

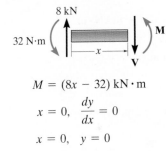

Chapter 16

P16–1a.

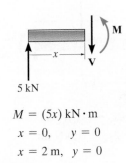

$$M = (8x - 32) \text{ kN} \cdot \text{m}$$

$$x = 0, \quad \frac{dy}{dx} = 0$$

$$x = 0, \quad y = 0$$

P16–1b.

$$M = (5x) \text{ kN} \cdot \text{m}$$

$$x = 0, \quad y = 0$$

$$x = 2 \text{ m}, \quad y = 0$$

P16–1c.

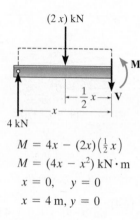

$$M = 4x - (2x)\left(\tfrac{1}{2}x\right)$$
$$M = (4x - x^2)\ \text{kN} \cdot \text{m}$$
$$x = 0, \quad y = 0$$
$$x = 4\ \text{m}, y = 0$$

P16–1d.

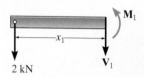

$$M_1 = (-2x_1)\ \text{kN} \cdot \text{m}$$

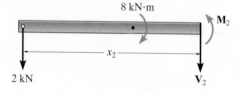

$$M_2 = (-2x + 8)\ \text{kN} \cdot \text{m}$$
$$x_1 = 0, \quad y_1 = 0$$
$$x_2 = 4\ \text{m}, \ y_2 = 0$$
$$x_1 = x_2 = 2\,\text{m}, \quad \frac{dy_1}{dx_1} = \frac{dy_2}{dx_2}$$
$$x_1 = x_2 = 2\,\text{m}, \quad y_1 = y_2$$

P16–1e.

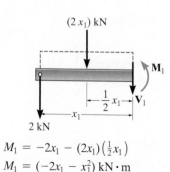

$$M_1 = -2x_1 - (2x_1)\left(\tfrac{1}{2}x_1\right)$$
$$M_1 = (-2x_1 - x_1^2)\ \text{kN} \cdot \text{m}$$

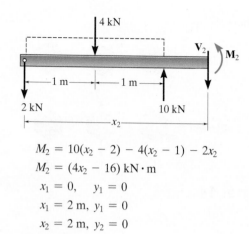

$$M_2 = 10(x_2 - 2) - 4(x_2 - 1) - 2x_2$$
$$M_2 = (4x_2 - 16)\ \text{kN} \cdot \text{m}$$
$$x_1 = 0, \quad y_1 = 0$$
$$x_1 = 2\ \text{m}, \ y_1 = 0$$
$$x_2 = 2\ \text{m}, \ y_2 = 0$$
$$x_1 = x_2 = 2\ \text{m}, \quad \frac{dy_1}{dx_1} = \frac{dy_2}{dx_2}$$

P16–1f.

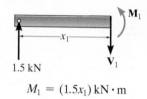

$$M_1 = (1.5x_1)\ \text{kN} \cdot \text{m}$$

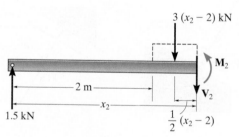

$$M_2 = 1.5x_2 - 3(x_2 - 2)\left(\frac{1}{2}\right)(x_2 - 2)$$
$$M_2 = -1.5x_2^2 + 7.5x_2 - 6$$
$$x_1 = 0, \ y_1 = 0$$
$$x_2 = 4\ \text{m}, \ y_2 = 0$$
$$x_1 = x_2 = 2\,\text{m}, \quad \frac{dy_1}{dx_1} = \frac{dy_2}{dx_2}$$
$$x_1 = x_2 = 2\,\text{m}, \quad y_1 = y_2$$

Fundamental Problems
Solutions and Answers

Chapter 2

F2–1.
$$F_R = \sqrt{(2\text{ kN})^2 + (6\text{ kN})^2 - 2(2\text{ kN})(6\text{ kN})\cos 105°}$$
$$= 6.798\text{ kN} = 6.80\text{ kN} \qquad \textit{Ans.}$$
$$\frac{\sin \phi}{6\text{ kN}} = \frac{\sin 105°}{6.798\text{ kN}}, \quad \phi = 58.49°$$
$$\theta = 45° + \phi = 45° + 58.49° = 103° \qquad \textit{Ans.}$$

F2–2. $\quad F_R = \sqrt{200^2 + 500^2 - 2(200)(500)\cos 140°}$
$$= 666\text{ N} \qquad \textit{Ans.}$$

F2–3. $\quad F_R = \sqrt{600^2 + 800^2 - 2(600)(800)\cos 60°}$
$$= 721.11\text{ N} = 721\text{ N} \qquad \textit{Ans.}$$
$$\frac{\sin \alpha}{800} = \frac{\sin 60°}{721.11}; \quad \alpha = 73.90°$$
$$\phi = \alpha - 30° = 73.90° - 30° = 43.9° \qquad \textit{Ans.}$$

F2–4. $\quad \dfrac{F_u}{\sin 45°} = \dfrac{30}{\sin 105°}; \quad F_u = 22.0\text{ lb} \qquad \textit{Ans.}$

$\qquad\qquad \dfrac{F_v}{\sin 30°} = \dfrac{30}{\sin 105°}; \quad F_v = 15.5\text{ lb} \qquad \textit{Ans.}$

F2–5. $\quad \dfrac{F_{AB}}{\sin 105°} = \dfrac{450}{\sin 30°}$

$\qquad\qquad F_{AB} = 869\text{ lb} \qquad \textit{Ans.}$

$\qquad\qquad \dfrac{F_{AC}}{\sin 45°} = \dfrac{450}{\sin 30°}$

$\qquad\qquad F_{AC} = 636\text{ lb} \qquad \textit{Ans.}$

F2–6. $\quad \dfrac{F}{\sin 30°} = \dfrac{6}{\sin 105°} \quad F = 3.11\text{ kN} \qquad \textit{Ans.}$

$\qquad\qquad \dfrac{F_v}{\sin 45°} = \dfrac{6}{\sin 105°} \quad F_v = 4.39\text{ kN} \qquad \textit{Ans.}$

F2–7. $\quad (F_1)_x = 0 \quad (F_1)_y = 300\text{ N} \qquad \textit{Ans.}$
$\qquad\qquad (F_2)_x = -(450\text{ N})\cos 45° = -318\text{ N} \qquad \textit{Ans.}$
$\qquad\qquad (F_2)_y = (450\text{ N})\sin 45° = 318\text{ N} \qquad \textit{Ans.}$
$\qquad\qquad (F_3)_x = \left(\frac{3}{5}\right)600\text{ N} = 360\text{ N} \qquad \textit{Ans.}$
$\qquad\qquad (F_3)_y = \left(\frac{4}{5}\right)600\text{ N} = 480\text{ N} \qquad \textit{Ans.}$

F2–8. $\quad F_{Rx} = 300 + 400\cos 30° - 250\left(\frac{4}{5}\right) = 446.4\text{ N}$
$\qquad\qquad F_{Ry} = 400\sin 30° + 250\left(\frac{3}{5}\right) = 350\text{ N}$
$\qquad\qquad F_R = \sqrt{(446.4)^2 + 350^2} = 567\text{ N} \qquad \textit{Ans.}$
$\qquad\qquad \theta = \tan^{-1}\frac{350}{446.4} = 38.1° \angle \qquad \textit{Ans.}$

F2–9.
$$\xrightarrow{+}(F_R)_x = \Sigma F_x;$$
$$(F_R)_x = -(700\text{ lb})\cos 30° + 0 + \left(\tfrac{3}{5}\right)(600\text{ lb})$$
$$= -246.22\text{ lb}$$
$$+\uparrow(F_R)_y = \Sigma F_y;$$
$$(F_R)_y = -(700\text{ lb})\sin 30° - 400\text{ lb} - \left(\tfrac{4}{5}\right)(600\text{ lb})$$
$$= -1230\text{ lb}$$
$$F_R = \sqrt{(246.22\text{ lb})^2 + (1230\text{ lb})^2} = 1254\text{ lb} \qquad \textit{Ans.}$$
$$\phi = \tan^{-1}\left(\tfrac{1230\text{ lb}}{246.22\text{ lb}}\right) = 78.68°$$
$$\theta = 180° + \phi = 180° + 78.68° = 259° \qquad \textit{Ans.}$$

F2–10. $\quad \xrightarrow{+}(F_R)_x = \Sigma F_x;$
$$750\text{ N} = F\cos\theta + \left(\tfrac{5}{13}\right)(325\text{ N}) + (600\text{ N})\cos 45°$$
$$+\uparrow(F_R)_y = \Sigma F_y;$$
$$0 = F\sin\theta + \left(\tfrac{12}{13}\right)(325\text{ N}) - (600\text{ N})\sin 45°$$
$$\tan\theta = 0.6190 \quad \theta = 31.76° = 31.8° \angle \qquad \textit{Ans.}$$
$$F = 236\text{ N} \qquad \textit{Ans.}$$

F2–11. $\quad \xrightarrow{+}(F_R)_x = \Sigma F_x;$
$$(80\text{ lb})\cos 45° = F\cos\theta + 50\text{ lb} - \left(\tfrac{3}{5}\right)90\text{ lb}$$
$$+\uparrow(F_R)_y = \Sigma F_y;$$
$$-(80\text{ lb})\sin 45° = F\sin\theta - \left(\tfrac{4}{5}\right)(90\text{ lb})$$
$$\tan\theta = 0.2547 \quad \theta = 14.29° = 14.3° \angle \qquad \textit{Ans.}$$
$$F = 62.5\text{ lb} \qquad \textit{Ans.}$$

F2–12. $\quad (F_R)_x = 15\left(\frac{4}{5}\right) + 0 + 15\left(\frac{4}{5}\right) = 24\text{ kN} \rightarrow$
$\qquad\qquad (F_R)_y = 15\left(\frac{3}{5}\right) + 20 - 15\left(\frac{3}{5}\right) = 20\text{ kN} \uparrow$
$\qquad\qquad F_R = 31.2\text{ kN} \qquad \textit{Ans.}$
$\qquad\qquad \theta = 39.8° \qquad \textit{Ans.}$

F2–13. $\quad F_x = 75\cos 30°\sin 45° = 45.93\text{ lb}$
$\qquad\qquad F_y = 75\cos 30°\cos 45° = 45.93\text{ lb}$
$\qquad\qquad F_z = -75\sin 30° = -37.5\text{ lb}$
$\qquad\qquad \alpha = \cos^{-1}\left(\frac{45.93}{75}\right) = 52.2° \qquad \textit{Ans.}$
$\qquad\qquad \beta = \cos^{-1}\left(\frac{45.93}{75}\right) = 52.2° \qquad \textit{Ans.}$
$\qquad\qquad \gamma = \cos^{-1}\left(\frac{-37.5}{75}\right) = 120° \qquad \textit{Ans.}$

F2–14. $\cos \beta = \sqrt{1 - \cos^2 120° - \cos^2 60°} = \pm 0.7071$

Require $\beta = 135°$.

$\mathbf{F} = F\mathbf{u}_F = (500 \text{ N})(-0.5\mathbf{i} - 0.7071\mathbf{j} + 0.5\mathbf{k})$

$= \{-250\mathbf{i} - 354\mathbf{j} + 250\mathbf{k}\} \text{ N}$ *Ans.*

F2–15. $\cos^2\alpha + \cos^2 135° + \cos^2 120° = 1$

$\alpha = 60°$

$\mathbf{F} = F\mathbf{u}_F = (500 \text{ N})(0.5\mathbf{i} - 0.7071\mathbf{j} - 0.5\mathbf{k})$

$= \{250\mathbf{i} - 354\mathbf{j} - 250\mathbf{k}\} \text{ N}$ *Ans.*

F2–16. $F_z = (50 \text{ lb}) \sin 45° = 35.36 \text{ lb}$

$F' = (50 \text{ lb}) \cos 45° = 35.36 \text{ lb}$

$F_x = \left(\frac{3}{5}\right)(35.36 \text{ lb}) = 21.21 \text{ lb}$

$F_y = \left(\frac{4}{5}\right)(35.36 \text{ lb}) = 28.28 \text{ lb}$

$\mathbf{F} = \{-21.2\mathbf{i} + 28.3\mathbf{j} + 35.4\mathbf{k}\} \text{ lb}$ *Ans.*

F2–17. $F_z = (750 \text{ N}) \sin 45° = 530.33 \text{ N}$

$F' = (750 \text{ N}) \cos 45° = 530.33 \text{ N}$

$F_x = (530.33 \text{ N}) \cos 60° = 265.2 \text{ N}$

$F_y = (530.33 \text{ N}) \sin 60° = 459.3 \text{ N}$

$\mathbf{F}_2 = \{265\mathbf{i} - 459\mathbf{j} + 530\mathbf{k}\} \text{ N}$ *Ans.*

F2–18. $\mathbf{F}_1 = \left(\frac{4}{5}\right)(500 \text{ lb})\mathbf{j} + \left(\frac{3}{5}\right)(500 \text{ lb})\mathbf{k}$

$= \{400\mathbf{j} + 300\mathbf{k}\} \text{ lb}$

$\mathbf{F}_2 = [(800 \text{ lb}) \cos 45°] \cos 30° \mathbf{i}$

$+ [(800 \text{ lb}) \cos 45°] \sin 30°\mathbf{j}$

$+ (800 \text{ lb}) \sin 45° (-\mathbf{k})$

$= \{489.90\mathbf{i} + 282.84\mathbf{j} - 565.69\mathbf{k}\} \text{ lb}$

$\mathbf{F}_R = \mathbf{F}_1 + \mathbf{F}_2 = \{490\mathbf{i} + 683\mathbf{j} - 266\mathbf{k}\} \text{ lb}$ *Ans.*

F2–19. $\mathbf{r}_{AB} = \{-6\mathbf{i} + 6\mathbf{j} + 3\mathbf{k}\} \text{ m}$ *Ans.*

$r_{AB} = \sqrt{(-6 \text{ m})^2 + (6 \text{ m})^2 + (3 \text{ m})^2} = 9 \text{ m}$ *Ans.*

$\alpha = 132°, \quad \beta = 48.2°, \quad \gamma = 70.5°$ *Ans.*

F2–20. $\mathbf{r}_{AB} = \{-4\mathbf{i} + 2\mathbf{j} + 4\mathbf{k}\} \text{ ft}$ *Ans.*

$r_{AB} = \sqrt{(-4 \text{ ft})^2 + (2 \text{ ft})^2 + (4 \text{ ft})^2} = 6 \text{ ft}$ *Ans.*

$\alpha = \cos^{-1}\left(\frac{-4 \text{ ft}}{6 \text{ ft}}\right) = 131.8°$

$\theta = 180° - 131.8° = 48.2°$ *Ans.*

F2–21. $\mathbf{r}_{AB} = \{2\mathbf{i} + 3\mathbf{j} - 6\mathbf{k}\} \text{ m}$

$\mathbf{F}_{AB} = F_{AB}\mathbf{u}_{AB}$

$= (630 \text{ N})\left(\frac{2}{7}\mathbf{i} + \frac{3}{7}\mathbf{j} - \frac{6}{7}\mathbf{k}\right)$

$= \{180\mathbf{i} + 270\mathbf{j} - 540\mathbf{k}\} \text{ N}$ *Ans.*

F2–22. $\mathbf{F} = F\mathbf{u}_{AB} = 900\text{N}\left(-\frac{4}{9}\mathbf{i} + \frac{7}{9}\mathbf{j} - \frac{4}{9}\mathbf{k}\right)$

$= \{-400\mathbf{i} + 700\mathbf{j} - 400\mathbf{k}\} \text{ N}$ *Ans.*

F2–23. $\mathbf{F}_B = F_B\mathbf{u}_B$

$= (840 \text{ N})\left(\frac{3}{7}\mathbf{i} - \frac{2}{7}\mathbf{j} - \frac{6}{7}\mathbf{k}\right)$

$= \{360\mathbf{i} - 240\mathbf{j} - 720\mathbf{k}\} \text{ N}$

$\mathbf{F}_C = F_C\mathbf{u}_C$

$= (420 \text{ N})\left(\frac{2}{7}\mathbf{i} + \frac{3}{7}\mathbf{j} - \frac{6}{7}\mathbf{k}\right)$

$= \{120\mathbf{i} + 180\mathbf{j} - 360\mathbf{k}\} \text{ N}$

$F_R = \sqrt{(480 \text{ N})^2 + (-60 \text{ N})^2 + (-1080 \text{ N})^2}$

$= 1.18 \text{ kN}$ *Ans.*

F2–24. $\mathbf{F}_B = F_B\mathbf{u}_B$

$= (600 \text{ lb})\left(-\frac{1}{3}\mathbf{i} + \frac{2}{3}\mathbf{j} - \frac{2}{3}\mathbf{k}\right)$

$= \{-200\mathbf{i} + 400\mathbf{j} - 400\mathbf{k}\} \text{ lb}$

$\mathbf{F}_C = F_C\mathbf{u}_C$

$= (490 \text{ lb})\left(-\frac{6}{7}\mathbf{i} + \frac{3}{7}\mathbf{j} - \frac{2}{7}\mathbf{k}\right)$

$= \{-420\mathbf{i} + 210\mathbf{j} - 140\mathbf{k}\} \text{ lb}$

$\mathbf{F}_R = \mathbf{F}_B + \mathbf{F}_C = \{-620\mathbf{i} + 610\mathbf{j} - 540\mathbf{k}\} \text{ lb}$ *Ans.*

F2–25. $\mathbf{u}_{AO} = -\frac{1}{3}\mathbf{i} + \frac{2}{3}\mathbf{j} - \frac{2}{3}\mathbf{k}$

$\mathbf{u}_F = -0.5345\mathbf{i} + 0.8018\mathbf{j} + 0.2673\mathbf{k}$

$\theta = \cos^{-1}(\mathbf{u}_{AO} \cdot \mathbf{u}_F) = 57.7°$ *Ans.*

F2–26. $\mathbf{u}_{AB} = -\frac{3}{5}\mathbf{j} + \frac{4}{5}\mathbf{k}$

$\mathbf{u}_F = \frac{4}{5}\mathbf{i} - \frac{3}{5}\mathbf{j}$

$\theta = \cos^{-1}(\mathbf{u}_{AB} \cdot \mathbf{u}_F) = 68.9°$ *Ans.*

F2–27. $\mathbf{u}_{OA} = \frac{12}{13}\mathbf{i} + \frac{5}{13}\mathbf{j}$

$\mathbf{u}_{OA} \cdot \mathbf{j} = u_{OA}(1) \cos \theta$

$\cos \theta = \frac{5}{13}; \quad \theta = 67.4°$ *Ans.*

F2–28. $\mathbf{u}_{OA} = \frac{12}{13}\mathbf{i} + \frac{5}{13}\mathbf{j}$

$\mathbf{F} = F\mathbf{u}_F = [650\mathbf{j}] \text{ N}$

$F_{OA} = \mathbf{F} \cdot \mathbf{u}_{OA} = 250 \text{ N}$

$\mathbf{F}_{OA} = F_{OA} \mathbf{u}_{OA} = \{231\mathbf{i} + 96.2\mathbf{j}\} \text{ N}$ *Ans.*

F2–29. $\quad \mathbf{F} = (400 \text{ N})\dfrac{\{4\,\mathbf{i} + 1\,\mathbf{j} - 6\,\mathbf{k}\}\,\text{m}}{\sqrt{(4\text{ m})^2 + (1\text{ m})^2 + (-6\text{ m})^2}}$

$\qquad\qquad = \{219.78\mathbf{i} + 54.94\mathbf{j} - 329.67\mathbf{k}\}\ \text{N}$

$\qquad \mathbf{u}_{AO} = \dfrac{\{-4\,\mathbf{j} - 6\,\mathbf{k}\}\,\text{m}}{\sqrt{(-4\text{ m})^2 + (-6\text{ m})^2}}$

$\qquad\qquad = -0.5547\mathbf{j} - 0.8321\mathbf{k}$

$\qquad (F_{AO})_{\text{proj}} = \mathbf{F}\cdot\mathbf{u}_{AO} = 244 \text{ N}$ $\qquad\qquad$ *Ans.*

F2–30. $\quad \mathbf{F} = [(-600 \text{ lb}) \cos 60°] \sin 30°\ \mathbf{i}$

$\qquad\qquad + [(600 \text{ lb}) \cos 60°] \cos 30°\ \mathbf{j}$

$\qquad\qquad + [(600 \text{ lb}) \sin 60°]\ \mathbf{k}$

$\qquad\quad = \{-150\mathbf{i} + 259.81\mathbf{j} + 519.62\mathbf{k}\}\ \text{lb}$

$\qquad \mathbf{u}_A = -\tfrac{2}{3}\mathbf{i} + \tfrac{2}{3}\mathbf{j} + \tfrac{1}{3}\mathbf{k}$

$\qquad (F_A)_{\text{par}} = \mathbf{F}\cdot\mathbf{u}_A = 446.41 \text{ lb} = 446 \text{ lb}$ $\qquad$ *Ans.*

$\qquad (F_A)_{\text{per}} = \sqrt{(600 \text{ lb})^2 - (446.41 \text{ lb})^2}$

$\qquad\qquad = 401 \text{ lb}$ $\qquad\qquad\qquad\qquad\qquad\qquad$ *Ans.*

F2–31. $\quad \mathbf{F} = 56\,\text{N}\left(\tfrac{3}{7}\mathbf{i} - \tfrac{6}{7}\mathbf{j} + \tfrac{2}{7}\mathbf{k}\right)$

$\qquad\qquad = \{24\mathbf{i} - 48\mathbf{j} + 16\mathbf{k}\}\ \text{N}$

$\qquad (F_{AO})_{\parallel} = \mathbf{F}\cdot\mathbf{u}_{AO} = (24\mathbf{i} - 48\mathbf{j} + 16\mathbf{k})\cdot\left(\tfrac{3}{7}\mathbf{i} - \tfrac{6}{7}\mathbf{j} - \tfrac{2}{7}\mathbf{k}\right)$

$\qquad\qquad = 46.86 \text{ N} = 46.9 \text{ N}$ $\qquad\qquad\qquad$ *Ans.*

$\qquad (F_{AO})_{\perp} = \sqrt{F^2 - (F_{AO})_{\parallel}} = \sqrt{(56)^2 - (46.86)^2}$

$\qquad\qquad = 30.7 \text{ N}$ $\qquad\qquad\qquad\qquad\qquad\qquad$ *Ans.*

Chapter 3

F3–1. $\quad \zeta + M_O = -\left(\tfrac{4}{5}\right)(100 \text{ N})(2 \text{ m}) - \left(\tfrac{3}{5}\right)(100 \text{ N})(5 \text{ m})$

$\qquad\qquad = -460 \text{ N}\cdot\text{m} = 460 \text{ N}\cdot\text{m}\ \big\downarrow$ $\qquad$ *Ans.*

F3–2. $\quad \zeta + M_O = [(300 \text{ N})\sin 30°][0.4 \text{ m} + (0.3 \text{ m})\cos 45°]$

$\qquad\qquad - [(300 \text{ N})\cos 30°][(0.3 \text{ m})\sin 45°]$

$\qquad\qquad = 36.7 \text{ N}\cdot\text{m}$ $\qquad\qquad\qquad\qquad\qquad$ *Ans.*

F3–3. $\quad \zeta + M_O = (600 \text{ lb})(4 \text{ ft} + (3 \text{ ft})\cos 45° - 1 \text{ ft})$

$\qquad\qquad = 3.07 \text{ kip}\cdot\text{ft}$ $\qquad\qquad\qquad\qquad\qquad$ *Ans.*

F3–4. $\quad \zeta + M_O = -50\sin 60°\,(0.1 + 0.2\cos 45° + 0.1)$

$\qquad\qquad + 50\cos 60°(0.2\sin 45°)$

$\qquad\qquad = -11.2 \text{ N}\cdot\text{m}$ $\qquad\qquad\qquad\qquad\qquad$ *Ans.*

F3–5. $\quad \zeta + M_O = 600\sin 50°\,(5) + 600\cos 50°\,(0.5)$

$\qquad\qquad = 2.49 \text{ kip}\cdot\text{ft}$ $\qquad\qquad\qquad\qquad\qquad$ *Ans.*

F3–6. $\quad \zeta + M_O = 500\sin 45°\,(3 + 3\cos 45°)$

$\qquad\qquad - 500\cos 45°\,(3\sin 45°)$

$\qquad\qquad = 1.06 \text{ kN}\cdot\text{m}$ $\qquad\qquad\qquad\qquad\qquad$ *Ans.*

F3–7. $\quad \zeta + (M_R)_O = \Sigma Fd;$

$\qquad (M_R)_O = -(600 \text{ N})(1 \text{ m})$

$\qquad\qquad + (500 \text{ N})[3 \text{ m} + (2.5 \text{ m})\cos 45°]$

$\qquad\qquad - (300 \text{ N})[(2.5 \text{ m})\sin 45°]$

$\qquad\qquad = 1254 \text{ N}\cdot\text{m} = 1.25 \text{ kN}\cdot\text{m}$ $\qquad$ *Ans.*

F3–8. $\quad \zeta + (M_R)_O = \Sigma Fd;$

$\qquad (M_R)_O = \left[\left(\tfrac{3}{5}\right)500 \text{ N}\right](0.425 \text{ m})$

$\qquad\qquad - \left[\left(\tfrac{4}{5}\right)500 \text{ N}\right](0.25 \text{ m})$

$\qquad\qquad - [(600 \text{ N})\cos 60°](0.25 \text{ m})$

$\qquad\qquad - [(600 \text{ N})\sin 60°](0.425 \text{ m})$

$\qquad\qquad = -268 \text{ N}\cdot\text{m} = 268 \text{ N}\cdot\text{m}\ \big\downarrow$ $\quad$ *Ans.*

F3–9. $\quad \zeta + (M_R)_O = \Sigma Fd;$

$\qquad (M_R)_O = (300\cos 30° \text{ lb})(6 \text{ ft} + 6\sin 30° \text{ ft})$

$\qquad\qquad - (300\sin 30° \text{ lb})(6\cos 30° \text{ ft})$

$\qquad\qquad + (200 \text{ lb})(6\cos 30° \text{ ft})$

$\qquad\qquad = 2.60 \text{ kip}\cdot\text{ft}$ $\qquad\qquad\qquad\qquad\qquad$ *Ans.*

F3–10. $\quad \mathbf{F} = F\mathbf{u}_{AB} = 500\text{N}\left(\tfrac{4}{5}\mathbf{i} - \tfrac{3}{5}\mathbf{j}\right) = \{400\mathbf{i} - 300\mathbf{j}\}\ \text{N}$

$\qquad \mathbf{M}_O = \mathbf{r}_{OA}\times\mathbf{F} = \{3\mathbf{j}\}\ \text{m}\times\{400\mathbf{i} - 300\mathbf{j}\}\ \text{N}$

$\qquad\qquad = \{-1200\mathbf{k}\}\ \text{N}\cdot\text{m}$ $\qquad\qquad\qquad$ *Ans.*

or

$\qquad \mathbf{M}_O = \mathbf{r}_{OB}\times\mathbf{F} = \{4\mathbf{i}\}\ \text{m}\times\{400\mathbf{i} - 300\mathbf{j}\}\ \text{N}$

$\qquad\qquad = \{-1200\mathbf{k}\}\ \text{N}\cdot\text{m}$ $\qquad\qquad\qquad$ *Ans.*

F3–11. $\quad \mathbf{F} = F\mathbf{u}_{BC}$

$\qquad\qquad = 120\,\text{lb}\left[\dfrac{\{4\,\mathbf{i} - 4\,\mathbf{j} - 2\,\mathbf{k}\}\,\text{ft}}{\sqrt{(4\text{ ft})^2 + (-4\text{ ft})^2 + (-2\text{ ft})^2}}\right]$

$\qquad\qquad = \{80\mathbf{i} - 80\mathbf{j} - 40\mathbf{k}\}\ \text{lb}$

$\qquad \mathbf{M}_O = \mathbf{r}_C\times\mathbf{F} = \begin{vmatrix} \mathbf{i} & \mathbf{j} & \mathbf{k} \\ 5 & 0 & 0 \\ 80 & -80 & -40 \end{vmatrix}$

$\qquad\qquad = \{200\mathbf{j} - 400\mathbf{k}\}\ \text{lb}\cdot\text{ft}$ $\qquad\qquad$ *Ans.*

or

$\qquad \mathbf{M}_O = \mathbf{r}_B\times\mathbf{F} = \begin{vmatrix} \mathbf{i} & \mathbf{j} & \mathbf{k} \\ 1 & 4 & 2 \\ 80 & -80 & -40 \end{vmatrix}$

$\qquad\qquad = \{200\mathbf{j} - 400\mathbf{k}\}\ \text{lb}\cdot\text{ft}$ $\qquad\qquad$ *Ans.*

F3–12. $\mathbf{F}_R = \mathbf{F}_1 + \mathbf{F}_2$

$= \{(100 - 200)\mathbf{i} + (-120 + 250)\mathbf{j}$
$+ (75 + 100)\mathbf{k}\}\,\mathrm{lb}$

$= \{-100\mathbf{i} + 130\mathbf{j} + 175\mathbf{k}\}\,\mathrm{lb}$

$(\mathbf{M}_R)_O = \mathbf{r}_A \times \mathbf{F}_R = \begin{vmatrix} \mathbf{i} & \mathbf{j} & \mathbf{k} \\ 4 & 5 & 3 \\ -100 & 130 & 175 \end{vmatrix}$

$= \{485\mathbf{i} - 1000\mathbf{j} + 1020\mathbf{k}\}\,\mathrm{lb}\cdot\mathrm{ft}$ *Ans.*

F3–13. $M_x = \mathbf{i}\cdot(\mathbf{r}_{OB}\times\mathbf{F}) = \begin{vmatrix} 1 & 0 & 0 \\ 0.3 & 0.4 & -0.2 \\ 300 & -200 & 150 \end{vmatrix}$

$= 20\,\mathrm{N}\cdot\mathrm{m}$ *Ans.*

F3–14. $\mathbf{u}_{OA} = \dfrac{\mathbf{r}_A}{r_A} = \dfrac{\{0.3\mathbf{i} + 0.4\mathbf{j}\}\,\mathrm{m}}{\sqrt{(0.3\,\mathrm{m})^2 + (0.4\,\mathrm{m})^2}} = 0.6\,\mathbf{i} + 0.8\,\mathbf{j}$

$M_{OA} = \mathbf{u}_{OA}\cdot(\mathbf{r}_{AB}\times\mathbf{F}) = \begin{vmatrix} 0.6 & 0.8 & 0 \\ 0 & 0 & -0.2 \\ 300 & -200 & 150 \end{vmatrix}$

$= -72\,\mathrm{N}\cdot\mathrm{m}$ *Ans.*

$\left|M_{OA}\right| = 72\,\mathrm{N}\cdot\mathrm{m}$

F3–15. Scalar Analysis

The magnitudes of the force components are

$F_x = |200\cos 120°| = 100\,\mathrm{N}$

$F_y = 200\cos 60° = 100\,\mathrm{N}$

$F_z = 200\cos 45° = 141.42\,\mathrm{N}$

$M_x = -F_y(z) + F_z(y)$

$= -(100\,\mathrm{N})(0.25\,\mathrm{m}) + (141.42\,\mathrm{N})(0.3\,\mathrm{m})$

$= 17.4\,\mathrm{N}\cdot\mathrm{m}$ *Ans.*

Vector Analysis

$M_x = \begin{vmatrix} 1 & 0 & 0 \\ 0 & 0.3 & 0.25 \\ -100 & 100 & 141.42 \end{vmatrix} = 17.4\,\mathrm{N}\cdot\mathrm{m}$ *Ans.*

F3–16. $M_y = \mathbf{j}\cdot(\mathbf{r}_A\times\mathbf{F}) = \begin{vmatrix} 0 & 1 & 0 \\ -3 & -4 & 2 \\ 30 & -20 & 50 \end{vmatrix}$

$= 210\,\mathrm{N}\cdot\mathrm{m}$ *Ans.*

F3–17. $\mathbf{u}_{AB} = \dfrac{\mathbf{r}_{AB}}{r_{AB}} = \dfrac{\{-4\mathbf{i} + 3\mathbf{j}\}\,\mathrm{ft}}{\sqrt{(-4\,\mathrm{ft})^2 + (3\,\mathrm{ft})^2}} = -0.8\mathbf{i} + 0.6\mathbf{j}$

$M_{AB} = \mathbf{u}_{AB}\cdot(\mathbf{r}_{AC}\times\mathbf{F})$

$= \begin{vmatrix} -0.8 & 0.6 & 0 \\ 0 & 0 & 2 \\ 50 & -40 & 20 \end{vmatrix} = -4\,\mathrm{lb}\cdot\mathrm{ft}$

$\mathbf{M}_{AB} = M_{AB}\mathbf{u}_{AB} = \{3.2\mathbf{i} - 2.4\mathbf{j}\}\,\mathrm{lb}\cdot\mathrm{ft}$ *Ans.*

F3–18. Scalar Analysis

The magnitudes of the force components are

$F_x = \left(\frac{3}{5}\right)\left[\frac{4}{5}(500)\right] = 240\,\mathrm{N}$

$F_y = \frac{4}{5}\left[\frac{4}{5}(500)\right] = 320\,\mathrm{N}$

$F_z = \frac{3}{5}(500) = 300\,\mathrm{N}$

$M_x = -320(3) + 300(2) = -360\,\mathrm{N}\cdot\mathrm{m}$ *Ans.*

$M_y = -240(3) - 300(-2) = -120\,\mathrm{N}\cdot\mathrm{m}$ *Ans.*

$M_z = 240(2) - 320(2) = -160\,\mathrm{N}\cdot\mathrm{m}$ *Ans.*

Vector Analysis

$\mathbf{F} = \{-240\mathbf{i} + 320\mathbf{j} + 300\mathbf{k}\}\,\mathrm{N}$

$\mathbf{r}_{OA} = \{-2\mathbf{i} + 2\mathbf{j} + 3\mathbf{k}\}\,\mathrm{m}$

$M_x = \mathbf{i}\cdot(\mathbf{r}_{OA}\times\mathbf{F}) = -360\,\mathrm{N}\cdot\mathrm{m}$

$M_y = \mathbf{j}\cdot(\mathbf{r}_{OA}\times\mathbf{F}) = -120\,\mathrm{N}\cdot\mathrm{m}$

$M_z = \mathbf{k}\cdot(\mathbf{r}_{OA}\times\mathbf{F}) = -160\,\mathrm{N}\cdot\mathrm{m}$

F3–19. $\zeta + M_{C_R} = \Sigma M_A = -400(3) + 400(5) - 300(5)$

$- 200(0.2) = -740\,\mathrm{N}\cdot\mathrm{m}$ *Ans.*

Also,

$\zeta + M_{C_R} = -300(5) + 400(2) - 200(0.2)$

$= -740\,\mathrm{N}\cdot\mathrm{m}$ *Ans.*

F3–20. $\zeta + M_{C_R} = 300(4) + 200(4) + 150(4)$

$= 2600\,\mathrm{lb}\cdot\mathrm{ft}$ *Ans.*

F3–21. $\zeta + (M_B)_R = \Sigma M_B$

$-1.5\,\mathrm{kN}\cdot\mathrm{m} = (2\,\mathrm{kN})(0.3\,\mathrm{m}) - F(0.9\,\mathrm{m})$

$F = 2.33\,\mathrm{kN}$ *Ans.*

F3–22. $\zeta + M_C = 10\left(\frac{3}{5}\right)(2) - 10\left(\frac{4}{5}\right)(4) = -20\,\mathrm{kN}\cdot\mathrm{m}$

$= 20\,\mathrm{kN}\cdot\mathrm{m}\,\circlearrowright$ *Ans.*

F3–23. $\mathbf{u}_1 = \dfrac{\mathbf{r}_1}{r_1} = \dfrac{\{-2\mathbf{i} + 2\mathbf{j} + 3.5\mathbf{k}\}\,\text{ft}}{\sqrt{(-2\,\text{ft})^2 + (2\,\text{ft})^2 + (3.5\,\text{ft})^2}}$

$\qquad\qquad = -\frac{2}{4.5}\mathbf{i} + \frac{2}{4.5}\mathbf{j} + \frac{3.5}{4.5}\mathbf{k}$

$\qquad \mathbf{u}_2 = -\mathbf{k}$

$\qquad \mathbf{u}_3 = \frac{1.5}{2.5}\mathbf{i} - \frac{2}{2.5}\mathbf{j}$

$\qquad (\mathbf{M}_c)_1 = (M_c)_1\mathbf{u}_1$

$\qquad\qquad = (450\,\text{lb}\cdot\text{ft})\left(-\frac{2}{4.5}\mathbf{i} + \frac{2}{4.5}\mathbf{j} + \frac{3.5}{4.5}\mathbf{k}\right)$

$\qquad\qquad = \{-200\mathbf{i} + 200\mathbf{j} + 350\mathbf{k}\}\,\text{lb}\cdot\text{ft}$

$\qquad (\mathbf{M}_c)_2 = (M_c)_2\mathbf{u}_2 = (250\,\text{lb}\cdot\text{ft})(-\mathbf{k})$

$\qquad\qquad = \{-250\mathbf{k}\}\,\text{lb}\cdot\text{ft}$

$\qquad (\mathbf{M}_c)_3 = (M_c)_3\,\mathbf{u}_3 = (300\,\text{lb}\cdot\text{ft})\left(\frac{1.5}{2.5}\mathbf{i} - \frac{2}{2.5}\mathbf{j}\right)$

$\qquad\qquad = \{180\mathbf{i} - 240\mathbf{j}\}\,\text{lb}\cdot\text{ft}$

$\qquad (\mathbf{M}_c)_R = \Sigma M_c;$

$\qquad (\mathbf{M}_c)_R = \{-20\mathbf{i} - 40\mathbf{j} + 100\mathbf{k}\}\,\text{lb}\cdot\text{ft}$ *Ans.*

F3–24. $\mathbf{F}_B = \left(\frac{4}{5}\right)(450\,\text{N})\mathbf{j} - \left(\frac{3}{5}\right)(450\,\text{N})\mathbf{k}$

$\qquad\qquad = \{360\mathbf{j} - 270\mathbf{k}\}\,\text{N}$

$\qquad \mathbf{M}_c = \mathbf{r}_{AB} \times \mathbf{F}_B = \begin{vmatrix} \mathbf{i} & \mathbf{j} & \mathbf{k} \\ 0.4 & 0 & 0 \\ 0 & 360 & -270 \end{vmatrix}$

$\qquad\qquad = \{108\mathbf{j} + 144\mathbf{k}\}\,\text{N}\cdot\text{m}$ *Ans.*

Also,

$\qquad \mathbf{M}_c = (\mathbf{r}_A \times \mathbf{F}_A) + (\mathbf{r}_B \times \mathbf{F}_B)$

$\qquad\quad = \begin{vmatrix} \mathbf{i} & \mathbf{j} & \mathbf{k} \\ 0 & 0 & 0.3 \\ 0 & -360 & 270 \end{vmatrix} + \begin{vmatrix} \mathbf{i} & \mathbf{j} & \mathbf{k} \\ 0.4 & 0 & 0.3 \\ 0 & 360 & -270 \end{vmatrix}$

$\qquad\quad = \{108\mathbf{j} + 144\mathbf{k}\}\,\text{N}\cdot\text{m}$ *Ans.*

F3–25. $\xleftarrow{+} F_{Rx} = \Sigma F_x; \;\; F_{Rx} = 200 - \frac{3}{5}(100) = 140\,\text{lb}$

$\qquad +\!\downarrow F_{Ry} = \Sigma F_y; \;\; F_{Ry} = 150 - \frac{4}{5}(100) = 70\,\text{lb}$

$\qquad\qquad F_R = \sqrt{140^2 + 70^2} = 157\,\text{lb}$ *Ans.*

$\qquad\qquad \theta = \tan^{-1}\left(\frac{70}{140}\right) = 26.6°$ *Ans.*

$\qquad \zeta + M_{A_R} = \Sigma M_A;$

$\qquad\qquad M_{A_R} = -\frac{3}{5}(100)(4) + \frac{4}{5}(100)(6) - 150(3)$

$\qquad\qquad M_{R_A} = -210\,\text{lb}\cdot\text{ft}$ *Ans.*

F3–26. $\xrightarrow{+} F_{Rx} = \Sigma F_x; \;\; F_{Rx} = \frac{4}{5}(50) = 40\,\text{N}$

$\qquad +\!\downarrow F_{Ry} = \Sigma F_y; \;\; F_{Ry} = 40 + 30 + \frac{3}{5}(50)$

$\qquad\qquad = 100\,\text{N}$

$\qquad\qquad F_R = \sqrt{(40)^2 + (100)^2} = 108\,\text{N}$ *Ans.*

$\qquad\qquad \theta = \tan^{-1}\left(\frac{100}{40}\right) = 68.2°$ *Ans.*

$\zeta + M_{A_R} = \Sigma M_A;$

$\qquad M_{A_R} = -30(3) - \frac{3}{5}(50)(6) - 200$

$\qquad\qquad = -470\,\text{N}\cdot\text{m}$ *Ans.*

F3–27. $\xrightarrow{+} (F_R)_x = \Sigma F_x;$

$\qquad (F_R)_x = 900\sin 30° = 450\,\text{N} \;\rightarrow$

$\qquad +\!\uparrow (F_R)_y = \Sigma F_y;$

$\qquad (F_R)_y = -900\cos 30° - 300$

$\qquad\qquad = -1079.42\,\text{N} = 1079.42\,\text{N}\downarrow$

$\qquad F_R = \sqrt{450^2 + 1079.42^2}$

$\qquad\qquad = 1169.47\,\text{N} = 1.17\,\text{kN}$ *Ans.*

$\qquad \theta = \tan^{-1}\left(\frac{1079.42}{450}\right) = 67.4°$ *Ans.*

$\zeta + (M_R)_A = \Sigma M_A;$

$\qquad (M_R)_A = 300 - 900\cos 30° (0.75) - 300(2.25)$

$\qquad\qquad = -959.57\,\text{N}\cdot\text{m}$

$\qquad\qquad = 960\,\text{N}\cdot\text{m}\,\circlearrowright$ *Ans.*

F3–28. $\xrightarrow{+} (F_R)_x = \Sigma F_x;$

$\qquad (F_R)_x = 150\left(\frac{3}{5}\right) + 50 - 100\left(\frac{4}{5}\right) = 60\,\text{lb}\rightarrow$

$\qquad +\!\uparrow (F_R)_y = \Sigma F_y;$

$\qquad (F_R)_y = -150\left(\frac{4}{5}\right) - 100\left(\frac{3}{5}\right)$

$\qquad\qquad = -180\,\text{lb} = 180\,\text{lb}\downarrow$

$\qquad F_R = \sqrt{60^2 + 180^2} = 189.74\,\text{lb} = 190\,\text{lb}$ *Ans.*

$\qquad \theta = \tan^{-1}\left(\frac{180}{60}\right) = 71.6°$ *Ans.*

$\zeta + (M_R)_A = \Sigma M_A;$

$\qquad (M_R)_A = 100\left(\frac{4}{5}\right)(1) - 100\left(\frac{3}{5}\right)(6) - 150\left(\frac{4}{5}\right)(3)$

$\qquad\qquad = -640 = 640\,\text{lb}\cdot\text{ft}\,\circlearrowright$ *Ans.*

F3–29. $\mathbf{F}_R = \Sigma \mathbf{F};$

$\qquad F_R = \mathbf{F}_1 + \mathbf{F}_2$

$\qquad\quad = (-300\mathbf{i} + 150\mathbf{j} + 200\mathbf{k}) + (-450\mathbf{k})$

$\qquad\quad = \{-300\mathbf{i} + 150\mathbf{j} - 250\mathbf{k}\}\,\text{N}$ *Ans.*

$\qquad \mathbf{r}_{OA} = (2 - 0)\mathbf{j} = \{2\mathbf{j}\}\,\text{m}$

$\qquad \mathbf{r}_{OB} = (-1.5-0)\mathbf{i} + (2 - 0)\mathbf{j} + (1 - 0)\mathbf{k}$

$\qquad\qquad = \{-1.5\mathbf{i} + 2\mathbf{j} + 1\mathbf{k}\}\,\text{m}$

$\qquad (\mathbf{M}_R)_O = \Sigma \mathbf{M};$

$\qquad (\mathbf{M}_R)_O = \mathbf{r}_{OB} \times \mathbf{F}_1 + \mathbf{r}_{OA} \times \mathbf{F}_2$

$\qquad\quad = \begin{vmatrix} \mathbf{i} & \mathbf{j} & \mathbf{k} \\ -1.5 & 2 & 1 \\ -300 & 150 & 200 \end{vmatrix} + \begin{vmatrix} \mathbf{i} & \mathbf{j} & \mathbf{k} \\ 0 & 2 & 0 \\ 0 & 0 & -450 \end{vmatrix}$

$\qquad\quad = \{-650\mathbf{i} + 375\mathbf{k}\}\,\text{N}\cdot\text{m}$ *Ans.*

F3–30. $\mathbf{F}_1 = \{-100\mathbf{j}\}$ N

$$\mathbf{F}_2 = (200\text{ N})\left[\frac{\{-0.4\mathbf{i} - 0.3\mathbf{k}\}\text{ m}}{\sqrt{(-0.4\text{ m})^2 + (-0.3\text{ m})^2}}\right]$$

$$= \{-160\mathbf{i} - 120\mathbf{k}\}\text{ N}$$

$\mathbf{M}_c = \{-75\mathbf{i}\}$ N·m

$\mathbf{F}_R = \{-160\mathbf{i} - 100\mathbf{j} - 120\mathbf{k}\}$ N   *Ans.*

$(\mathbf{M}_R)_O = (0.3\mathbf{k}) \times (-100\mathbf{j})$

$$+ \begin{vmatrix} \mathbf{i} & \mathbf{j} & \mathbf{k} \\ 0 & 0.5 & 0.3 \\ -160 & 0 & -120 \end{vmatrix} + (-75\mathbf{i})$$

$$= \{-105\mathbf{i} - 48\mathbf{j} + 80\mathbf{k}\}\text{ N·m}$$ *Ans.*

F3–31. $+\downarrow F_R = \Sigma F_y;$ $F_R = 500 + 250 + 500$

$$= 1250\text{ lb}$$ *Ans.*

$\zeta + F_R x = \Sigma M_O;$

$-1250(x) = -500(3) - 250(6) - 500(9)$

$$x = 6\text{ ft}$$ *Ans.*

F3–32. $\xrightarrow{+} (F_R)_x = \Sigma F_x;$

$(F_R)_x = 100\left(\frac{3}{5}\right) + 50\sin 30° = 85\text{ lb} \rightarrow$

$+\uparrow (F_R)_y = \Sigma F_y;$

$(F_R)_y = 200 + 50\cos 30° - 100\left(\frac{4}{5}\right)$

$$= 163.30\text{ lb}\uparrow$$

$F_R = \sqrt{85^2 + 163.30^2} = 184\text{ lb}$

$\theta = \tan^{-1}\left(\frac{163.30}{85}\right) = 62.5°\ \measuredangle$ *Ans.*

$\zeta + (M_R)_A = \Sigma M_A;$

$163.30(d) = 200(3) - 100\left(\frac{4}{5}\right)(6) + 50\cos 30°(9)$

$$d = 3.12\text{ ft}$$ *Ans.*

F3–33. $\xrightarrow{+} (F_R)_x = \Sigma F_x;$

$(F_R)_x = 15\left(\frac{4}{5}\right) = 12\text{ kN} \rightarrow$

$+\uparrow (F_R)_y = \Sigma F_y;$

$(F_R)_y = -20 + 15\left(\frac{3}{5}\right) = -11\text{ kN} = 11\text{ kN}\downarrow$

$F_R = \sqrt{12^2 + 11^2} = 16.3\text{ kN}$ *Ans.*

$\theta = \tan^{-1}\left(\frac{11}{12}\right) = 42.5°\ \measuredangle$ *Ans.*

$\zeta + (M_R)_A = \Sigma M_A;$

$-11(d) = -20(2) - 15\left(\frac{4}{5}\right)(2) + 15\left(\frac{3}{5}\right)(6)$

$$d = 0.909\text{ m}$$ *Ans.*

F3–34. $\xrightarrow{+} (F_R)_x = \Sigma F_x;$

$(F_R)_x = \left(\frac{3}{5}\right)5\text{ kN} - 8\text{ kN}$

$$= -5\text{ kN} = 5\text{ kN} \leftarrow$$

$+\uparrow (F_R)_y = \Sigma F_y;$

$(F_R)_y = -6\text{ kN} - \left(\frac{4}{5}\right)5\text{ kN}$

$$= -10\text{ kN} = 10\text{ kN}\downarrow$$

$F_R = \sqrt{5^2 + 10^2} = 11.2\text{ kN}$ *Ans.*

$\theta = \tan^{-1}\left(\frac{10\text{ kN}}{5\text{ kN}}\right) = 63.4°\ \measuredangle$ *Ans.*

$\zeta + (M_R)_A = \Sigma M_A;$

$5\text{ kN}(d) = 8\text{ kN}(3\text{ m}) - 6\text{ kN}(0.5\text{ m})$

$$- \left[\left(\frac{4}{5}\right)5\text{ kN}\right](2\text{ m})$$

$$- \left[\left(\frac{3}{5}\right)5\text{ kN}\right](4\text{ m})$$

$$d = 0.2\text{ m}$$ *Ans.*

F3–35. $+\downarrow F_R = \Sigma F_z;$ $F_R = 400 + 500 - 100$

$$= 800\text{ N}$$ *Ans.*

$M_{Rx} = \Sigma M_x;\ -800y = -400(4) - 500(4)$

$$y = 4.50\text{ m}$$ *Ans.*

$M_{Ry} = \Sigma M_y;\ \ 800x = 500(4) - 100(3)$

$$x = 2.125\text{ m}$$ *Ans.*

F3–36. $+\downarrow F_R = \Sigma F_z;$

$F_R = 200 + 200 + 100 + 100$

$$= 600\text{ N}$$ *Ans.*

$M_{Rx} = \Sigma M_x;$

$-600y = 200(1) + 200(1) + 100(3) - 100(3)$

$$y = -0.667\text{ m}$$ *Ans.*

$M_{Ry} = \Sigma M_y;$

$600x = 100(3) + 100(3) + 200(2) - 200(3)$

$$x = 0.667\text{ m}$$ *Ans.*

F3–37. $+\uparrow F_R = \Sigma F_y;$

$-F_R = -6(1.5) - 9(3) - 3(1.5)$

$$F_R = 40.5\text{ kN}\downarrow$$ *Ans.*

$\zeta + (M_R)_A = \Sigma M_A;$

$$-40.5(d) = 6(1.5)(0.75)$$

$$- 9(3)(1.5) - 3(1.5)(3.75)$$

$$d = 1.25\text{ m}$$ *Ans.*

F3–38. $F_R = \frac{1}{2}(6)(150) + 8(150) = 1650\text{ lb}$ *Ans.*

$\zeta + M_{A_R} = \Sigma M_A;$

$-1650d = -\left[\frac{1}{2}(6)(150)\right](4) - [8(150)](10)$

$$d = 8.36\text{ ft}$$ *Ans.*

F3–39. $+\uparrow F_R = \Sigma F_y;$

$-F_R = -\frac{1}{2}(6)(3) - \frac{1}{2}(6)(6)$

$F_R = 27 \text{ kN}\downarrow$ *Ans.*

$\zeta + (M_R)_A = \Sigma M_A;$

$-27(d) = \frac{1}{2}(6)(3)(1) - \frac{1}{2}(6)(6)(2)$

$d = 1 \text{ m}$ *Ans.*

F3–40. $+\downarrow F_R = \Sigma F_y;$

$F_R = \frac{1}{2}(50)(6) + 150(6) + 500$

$= 1550 \text{ lb}$ *Ans.*

$\zeta + M_{A_R} = \Sigma M_A;$

$-1550d = -\left[\frac{1}{2}(50)(6)\right](4) - [150(6)](3) - 500(9)$

$d = 5.03 \text{ ft}$ *Ans.*

F3–41. $+\uparrow F_R = \Sigma F_y;$

$-F_R = -\frac{1}{2}(3)(4.5) - 3(6)$

$F_R = 24.75 \text{ kN}\downarrow$ *Ans.*

$\zeta + (M_R)_A = \Sigma M_A;$

$-24.75(d) = -\frac{1}{2}(3)(4.5)(1.5) - 3(6)(3)$

$d = 2.59 \text{ m}$ *Ans.*

F3–42. $F_R = \int w(x)\, dx = \int_0^4 2.5x^3\, dx = 160 \text{ N}$

$\bar{x} = \dfrac{\int xw(x)\, dx}{\int w(x)\, dx} = \dfrac{\int_0^4 2.5x^4\, dx}{160} = 3.20 \text{ m}$ *Ans.*

Chapter 4

F4–1. $\xrightarrow{+}\Sigma F_x = 0; \quad -A_x + 500\left(\frac{3}{5}\right) = 0$

$A_x = 300 \text{ lb}$ *Ans.*

$\zeta + \Sigma M_A = 0; \quad B_y(10) - 500\left(\frac{4}{5}\right)(5) - 600 = 0$

$B_y = 260 \text{ lb}$ *Ans.*

$+\uparrow \Sigma F_y = 0; \quad A_y + 260 - 500\left(\frac{4}{5}\right) = 0$

$A_y = 140 \text{ lb}$ *Ans.*

F4–2. $\zeta + \Sigma M_A = 0;$

$F_{CD} \sin 45°(1.5 \text{ m}) - 4 \text{ kN}(3 \text{ m}) = 0$

$F_{CD} = 11.31 \text{ kN} = 11.3 \text{ kN}$ *Ans.*

$\xrightarrow{+}\Sigma F_x = 0; \quad A_x + (11.31 \text{ kN}) \cos 45° = 0$

$A_x = -8 \text{ kN} = 8 \text{ kN} \leftarrow$ *Ans.*

$+\uparrow \Sigma F_y = 0;$

$A_y + (11.31 \text{ kN}) \sin 45° - 4 \text{ kN} = 0$

$A_y = -4 \text{ kN} = 4 \text{ kN}\downarrow$ *Ans.*

F4–3. $\zeta + \Sigma M_A = 0;$

$N_B[6 \text{ m} + (6 \text{ m}) \cos 45°]$

$- 10 \text{ kN}[2 \text{ m} + (6 \text{ m}) \cos 45°]$

$- 5 \text{ kN}(4 \text{ m}) = 0$

$N_B = 8.047 \text{ kN} = 8.05 \text{ kN}$ *Ans.*

$\xrightarrow{+}\Sigma F_x = 0;$

$(5 \text{ kN}) \cos 45° - A_x = 0$

$A_x = 3.54 \text{ kN}$ *Ans.*

$+\uparrow \Sigma F_y = 0;$

$A_y + 8.047 \text{ kN} - (5 \text{ kN}) \sin 45° - 10 \text{ kN} = 0$

$A_y = 5.49 \text{ kN}$ *Ans.*

F4–4. $\xrightarrow{+}\Sigma F_x = 0; \quad -A_x + 400 \cos 30° = 0$

$A_x = 346 \text{ N}$ *Ans.*

$+\uparrow \Sigma F_y = 0;$

$A_y - 200 - 200 - 200 - 400 \sin 30° = 0$

$A_y = 800 \text{ N}$ *Ans.*

$\zeta + \Sigma M_A = 0;$

$M_A - 200(2.5) - 200(3.5) - 200(4.5)$

$- 400 \sin 30°(4.5) - 400 \cos 30°(3 \sin 60°) = 0$

$M_A = 3.90 \text{ kN} \cdot \text{m}$ *Ans.*

F4–5. $\zeta + \Sigma M_A = 0;$

$N_C(0.7 \text{ m}) - [25(9.81) \text{ N}](0.5 \text{ m}) \cos 30° = 0$

$N_C = 151.71 \text{ N} = 152 \text{ N}$ *Ans.*

$\xrightarrow{+}\Sigma F_x = 0;$

$T_{AB} \cos 15° - (151.71 \text{ N}) \cos 60° = 0$

$T_{AB} = 78.53 \text{ N} = 78.5 \text{ N}$ *Ans.*

$+\uparrow \Sigma F_y = 0;$

$F_A + (78.53 \text{ N}) \sin 15°$

$+ (151.71 \text{ N}) \sin 60° - 25(9.81) \text{ N} = 0$

$F_A = 93.5 \text{ N}$ *Ans.*

F4–6. $\xrightarrow{+}\Sigma F_x = 0;$

$N_C \sin 30° - (250 \text{ N}) \sin 60° = 0$

$N_C = 433.0 \text{ N} = 433 \text{ N}$ *Ans.*

$\zeta + \Sigma M_B = 0;$

$-N_A \sin 30°(0.15 \text{ m}) - 433.0 \text{ N}(0.2 \text{ m})$

$+ [(250 \text{ N}) \cos 30°](0.6 \text{ m}) = 0$

$N_A = 577.4 \text{ N} = 577 \text{ N}$ *Ans.*

$+\uparrow \Sigma F_y = 0;$

$N_B - 577.4 \text{ N} + (433.0 \text{ N}) \cos 30°$

$- (250 \text{ N}) \cos 60° = 0$

$N_B = 327 \text{ N}$ *Ans.*

F4–7. $\Sigma F_z = 0;$

$$T_A + T_B + T_C - 200 - 500 = 0$$

$\Sigma M_x = 0;$

$$T_A(3) + T_C(3) - 500(1.5) - 200(3) = 0$$

$\Sigma M_y = 0;$

$$-T_B(4) - T_C(4) + 500(2) + 200(2) = 0$$

$T_A = 350$ lb, $T_B = 250$ lb, $T_C = 100$ lb   *Ans.*

F4–8. $\Sigma M_y = 0;$

$$600 \text{ N}(0.2 \text{ m}) + 900 \text{ N}(0.6 \text{ m}) - F_A(1 \text{ m}) = 0$$

$$F_A = 660 \text{ N} \qquad \textit{Ans.}$$

$\Sigma M_x = 0;$

$$D_z(0.8 \text{ m}) - 600 \text{ N}(0.5 \text{ m}) - 900 \text{ N}(0.1 \text{ m}) = 0$$

$$D_z = 487.5 \text{ N} \qquad \textit{Ans.}$$

$\Sigma F_x = 0;$ $D_x = 0$ *Ans.*

$\Sigma F_y = 0;$ $D_y = 0$ *Ans.*

$\Sigma F_z = 0;$

$$T_{BC} + 660 \text{ N} + 487.5 \text{ N} - 900 \text{ N} - 600 \text{ N} = 0$$

$$T_{BC} = 352.5 \text{ N} \qquad \textit{Ans.}$$

F4–9. $\Sigma F_y = 0;$ $400 \text{ N} + C_y = 0;$

$$C_y = -400 \text{ N} \qquad \textit{Ans.}$$

$\Sigma M_y = 0;$ $-C_x (0.4 \text{ m}) - 600 \text{ N} (0.6 \text{ m}) = 0$

$$C_x = -900 \text{ N} \qquad \textit{Ans.}$$

$\Sigma M_x = 0;$ $B_z (0.6 \text{ m}) + 600 \text{ N} (1.2 \text{ m})$

$$+ (-400 \text{ N})(0.4 \text{ m}) = 0$$

$$B_z = -933.3 \text{ N} \qquad \textit{Ans.}$$

$\Sigma M_z = 0;$

$$-B_x (0.6 \text{ m}) - (-900 \text{ N})(1.2 \text{ m})$$

$$+ (-400 \text{ N})(0.6 \text{ m}) = 0$$

$$B_x = 1400 \text{ N} \qquad \textit{Ans.}$$

$\Sigma F_x = 0;$ $1400 \text{ N} + (-900 \text{ N}) + A_x = 0$

$$A_x = -500 \text{ N} \qquad \textit{Ans.}$$

$\Sigma F_z = 0;$ $A_z - 933.3 \text{ N} + 600 \text{ N} = 0$

$$A_z = 333.3 \text{ N} \qquad \textit{Ans.}$$

F4–10. $\Sigma F_x = 0;$ $B_x = 0$ *Ans.*

$\Sigma M_z = 0;$

$C_y(0.4 \text{ m} + 0.6 \text{ m}) = 0$ $C_y = 0$ *Ans.*

$\Sigma F_y = 0;$ $A_y + 0 = 0$ $A_y = 0$ *Ans.*

$\Sigma M_x = 0;$ $C_z(0.6 \text{ m} + 0.6 \text{ m}) + B_z(0.6 \text{ m})$

$$- 450 \text{ N}(0.6 \text{ m} + 0.6 \text{ m}) = 0$$

$$1.2C_z + 0.6B_z - 540 = 0$$

$\Sigma M_y = 0;$ $-C_z(0.6 \text{ m} + 0.4 \text{ m})$

$$- B_z(0.6 \text{ m}) + 450 \text{ N}(0.6 \text{ m}) = 0$$

$$-C_z - 0.6B_z + 270 = 0$$

$C_z = 1350 \text{ N}$ $B_z = -1800 \text{ N}$ *Ans.*

$\Sigma F_z = 0;$

$$A_z + 1350 \text{ N} + (-1800 \text{ N}) - 450 \text{ N} = 0$$

$$A_z = 900 \text{ N} \qquad \textit{Ans.}$$

F4–11. $\Sigma F_y = 0;$ $A_y = 0$ *Ans.*

$\Sigma M_x = 0;$ $-9(3) + F_{CE}(3) = 0$

$$F_{CE} = 9 \text{ kN} \qquad \textit{Ans.}$$

$\Sigma M_z = 0;$ $F_{CF}(3) - 6(3) = 0$

$$F_{CF} = 6 \text{ kN} \qquad \textit{Ans.}$$

$\Sigma M_y = 0;$ $9(4) - A_z (4) - 6(1.5) = 0$

$$A_z = 6.75 \text{ kN} \qquad \textit{Ans.}$$

$\Sigma F_x = 0;$ $A_x + 6 - 6 = 0$ $A_x = 0$ *Ans.*

$\Sigma F_z = 0;$ $F_{DB} + 9 - 9 + 6.75 = 0$

$$F_{DB} = -6.75 \text{ kN} \qquad \textit{Ans.}$$

F4–12. $\Sigma F_x = 0;$ $A_x = 0$ *Ans.*

$\Sigma F_y = 0;$ $A_y = 0$ *Ans.*

$\Sigma F_z = 0;$ $A_z + F_{BC} - 80 = 0$

$\Sigma M_x = 0;$ $(M_A)_x + 6F_{BC} - 80(6) = 0$

$\Sigma M_y = 0;$ $3F_{BC} - 80(1.5) = 0$ $F_{BC} = 40$ lb   *Ans.*

$\Sigma M_z = 0;$ $(M_A)_z = 0$ *Ans.*

$A_z = 40$ lb $(M_A)_x = 240$ lb $\cdot$ ft   *Ans.*

F4–13. a) $+\uparrow \Sigma F_y = 0;$ $N - 50(9.81) - 200\left(\frac{3}{5}\right) = 0$

$$N = 610.5 \text{ N}$$

$\xrightarrow{+} \Sigma F_x = 0;$ $F - 200\left(\frac{4}{5}\right) = 0$

$$F = 160 \text{ N}$$

$F < F_{max} = \mu_s N = 0.3(610.5) = 183.15 \text{ N},$

therefore $F = 160 \text{ N}$ *Ans.*

b) $+\uparrow \Sigma F_y = 0;$ $N - 50(9.81) - 400\left(\frac{3}{5}\right) = 0$

$$N = 730.5 \text{ N}$$

$\xrightarrow{+} \Sigma F_x = 0;$ $F - 400\left(\frac{4}{5}\right) = 0$

$$F = 320 \text{ N}$$

$F > F_{max} = \mu_s N = 0.3(730.5) = 219.15 \text{ N}$

Block slips

$F = \mu_k N = 0.2(730.5) = 146 \text{ N}$ *Ans.*

F4–14. $\zeta + \Sigma M_B = 0;$

$$N_A(3) + 0.2N_A(4) - 30(9.81)(2) = 0$$
$$N_A = 154.89 \text{ N}$$

$\xrightarrow{+} \Sigma F_x = 0; \quad P - 154.89 = 0$

$$P = 154.89 \text{ N} = 155 \text{ N} \qquad Ans.$$

F4–15.   Crate A

$+\uparrow \Sigma F_y = 0; \quad N_A - 50(9.81) = 0$

$$N_A = 490.5 \text{ N}$$

$\xrightarrow{+} \Sigma F_x = 0; \quad T - 0.25(490.5) = 0$

$$T = 122.62 \text{ N}$$

Crate B

$+\uparrow \Sigma F_y = 0; \quad N_B + P \sin 30° - 50(9.81) = 0$

$$N_B = 490.5 - 0.5P$$

$\xrightarrow{+} \Sigma F_x = 0;$

$$P \cos 30° - 0.25(490.5 - 0.5 P) - 122.62 = 0$$
$$P = 247 \text{ N} \qquad Ans.$$

F4–16. $\xrightarrow{+} \Sigma F_x = 0; \quad N_A - 0.3N_B = 0$

$+\uparrow \Sigma F_y = 0;$

$$N_B + 0.3N_A + P - 100(9.81) = 0$$

$\zeta + \Sigma M_O = 0;$

$$P(0.6) - 0.3N_B(0.9) - 0.3 N_A(0.9) = 0$$
$$N_A = 175.70 \text{ N} \qquad N_B = 585.67 \text{ N}$$
$$P = 343 \text{ N} \qquad Ans.$$

F4–17.   If slipping occurs:

$+\uparrow \Sigma F_y = 0; \quad N_c - 250 \text{ lb} = 0; N_c = 250 \text{ lb}$

$\xrightarrow{+} \Sigma F_x = 0; \quad P - 0.4(250) = 0; P = 100 \text{ lb}$

If tipping occurs:

$\zeta + \Sigma M_A = 0; \quad -P(4.5) + 250(1.5) = 0$

$$P = 83.3 \text{ lb} \qquad Ans.$$

F4–18.

$\zeta + \Sigma M_A = 0; \quad 490.5(0.6) - T \cos 60°(0.3 \cos 60° + 0.6)$
$$- T \sin 60° (0.3 \sin 60°) = 0$$
$$T = 490.5 \text{ N}$$

$\xrightarrow{+} \Sigma F_x = 0; \quad 490.5 \sin 60° - N_A = 0; \quad N_A = 424.8 \text{ N}$

$+\uparrow \Sigma F_y = 0; \quad \mu_s(424.8) + 490.5 \cos 60° - 490.5 = 0$

$$\mu_s = 0.577 \qquad Ans.$$

F4–19. A will not move. Assume B is about to slip on C and A, and C is stationary.

$\xrightarrow{+} \Sigma F_x = 0; \quad P - 0.3(50) - 0.4(75); \quad P = 45 \text{ N}$

Assume C is about to slip and B does not slip on C, but is about to slip at A.

$\xrightarrow{+} \Sigma F_x = 0; \quad P - 0.3(50) - 0.35(90) = 0$

$$P = 46.5 \text{ N} > 45 \text{ N}$$
$$P = 45 \text{ N} \qquad Ans.$$

F4–20. A is about to move down the plane and B moves upward.

Block A

$+\nwarrow \Sigma F_y = 0; \quad N = W \cos \theta$

$+\nearrow \Sigma F_x = 0; \quad T + \mu_s(W \cos \theta) - W \sin \theta = 0$

$$T = W \sin \theta - \mu_s W \cos \theta \qquad (1)$$

Block B

$+\nwarrow \Sigma F_y = 0; \quad N' = 2W \cos \theta$

$+\nearrow \Sigma F_x = 0; \quad 2T - \mu_s W \cos \theta - \mu_s(2W \cos \theta)$
$$- W \sin \theta = 0$$

Using Eq.(1);

$$\theta = \tan^{-1} 5\mu_s \qquad Ans.$$

F4–21.   Assume B is about to slip on A, $F_B = 0.3 N_B$.

$\xrightarrow{+} \Sigma F_x = 0; \quad P - 0.3(10)(9.81) = 0$

$$P = 29.4 \text{ N}$$

Assume B is about to tip on A, $x = 0$.

$\zeta + \Sigma M_O = 0; \quad 10(9.81)(0.15) - P(0.4) = 0$

$$P = 36.8 \text{ N}$$

Assume A is about to slip, $F_A = 0.1 N_A$.

$\xrightarrow{+} \Sigma F_x = 0 \quad P - 0.1[7(9.81) + 10(9.81)] = 0$

$$P = 16.7 \text{ N}$$

Choose the smallest result. $P = 16.7 \text{ N} \qquad Ans.$

Chapter 5

F5–1.   *Joint A.*

$+\uparrow \Sigma F_y = 0; \quad 225 \text{ lb} - F_{AD} \sin 45° = 0$

$$F_{AD} = 318.20 \text{ lb} = 318 \text{ lb (C)} \qquad Ans.$$

$\xrightarrow{+} \Sigma F_x = 0; \quad F_{AB} - (318.20 \text{ lb}) \cos 45° = 0$

$$F_{AB} = 225 \text{ lb (T)} \qquad Ans.$$

Joint B.

$\xrightarrow{+} \Sigma F_x = 0;$ $F_{BC} - 225 \text{ lb} = 0$

$F_{BC} = 225 \text{ lb (T)}$ *Ans.*

$+\uparrow \Sigma F_y = 0;$ $F_{BD} = 0$ *Ans.*

Joint D.

$\xrightarrow{+} \Sigma F_x = 0;$

$F_{CD} \cos 45° + (318.20 \text{ lb}) \cos 45° - 450 \text{ lb} = 0$

$F_{CD} = 318.20 \text{ lb} = 318 \text{ lb (T)}$ *Ans.*

F5–2. Joint D.

$+\uparrow \Sigma F_y = 0; \frac{3}{5} F_{CD} - 300 = 0;$

$F_{CD} = 500 \text{ lb (T)}$ *Ans.*

$\xrightarrow{+} \Sigma F_x = 0; -F_{AD} + \frac{4}{5}(500) = 0$

$F_{AD} = 400 \text{ lb (C)}$ *Ans.*

$F_{BC} = 500 \text{ lb (T)}, F_{AC} = F_{AB} = 0$ *Ans.*

F5–3. $D_x = 800 \text{ lb}, D_y = 1400 \text{ lb}, A_x = 800 \text{ lb}$

Joint B.

$\xrightarrow{+} \Sigma F_x = 0; F_{BA} = 0$ *Ans.*

$+\uparrow \Sigma F_y = 0; -600 + F_{BC} = 0; F_{BC} = 600 \text{ lb (T)}$ *Ans.*

Joint C.

$+\uparrow \Sigma F_y = 0; F_{CA}\left(\frac{3}{5}\right) - 600 = 0;$

$F_{CA} = 1000 \text{ lb (C)}$ *Ans.*

$\xrightarrow{+} \Sigma F_x = 0; -F_{CD} + \left(\frac{4}{5}\right)(1000) = 0;$

$F_{CD} = 800 \text{ lb (T)}$ *Ans.*

Joint A.

$+\uparrow \Sigma F_y = 0; -800 - 1000\left(\frac{3}{5}\right) + F_{AD} = 0;$

$F_{AD} = 1400 \text{ lb (T)}$ *Ans.*

F5–4. Joint C.

$+\uparrow \Sigma F_y = 0;$ $2F \cos 30° - P = 0$

$F_{AC} = F_{BC} = F = \frac{P}{2 \cos 30°} = 0.5774P \text{ (C)}$

Joint B.

$\xrightarrow{+} \Sigma F_x = 0; 0.5774P \cos 60° - F_{AB} = 0$

$F_{AB} = 0.2887P \text{ (T)}$

$F_{AB} = 0.2887P = 2 \text{ kN}$

$P = 6.928 \text{ kN}$

$F_{AC} = F_{BC} = 0.5774P = 1.5 \text{ kN}$

$P = 2.598 \text{ kN}$

The *smaller value* of P is chosen,

$P = 2.598 \text{ kN} = 2.60 \text{ kN}$ *Ans.*

F5–5. $F_{CB} = 0$ *Ans.*

$F_{CD} = 0$ *Ans.*

$F_{AE} = 0$ *Ans.*

$F_{DE} = 0$ *Ans.*

F5–6. Joint C.

$+\uparrow \Sigma F_y = 0;$ $259.81 \text{ lb} - F_{CD} \sin 30° = 0$

$F_{CD} = 519.62 \text{ lb} = 520 \text{ lb (C)}$ *Ans.*

$\xrightarrow{+} \Sigma F_x = 0;$ $(519.62 \text{ lb}) \cos 30° - F_{BC} = 0$

$F_{BC} = 450 \text{ lb (T)}$ *Ans.*

Joint D.

$+\nearrow \Sigma F_{y'} = 0;$ $F_{BD} \cos 30° = 0$ $F_{BD} = 0$ *Ans.*

$+\searrow \Sigma F_{x'} = 0;$ $F_{DE} - 519.62 \text{ lb} = 0$

$F_{DE} = 519.62 \text{ lb} = 520 \text{ lb (C)}$ *Ans.*

Joint B.

$\uparrow \Sigma F_y = 0;$ $F_{BE} \sin \phi = 0$ $F_{BE} = 0$ *Ans.*

$\xrightarrow{+} \Sigma F_x = 0;$ $450 \text{ lb} - F_{AB} = 0$

$F_{AB} = 450 \text{ lb (T)}$ *Ans.*

Joint A.

$+\uparrow \Sigma F_y = 0;$ $340.19 \text{ lb} - F_{AE} = 0$

$F_{AE} = 340 \text{ lb (C)}$ *Ans.*

F5–7. $+\uparrow \Sigma F_y = 0; F_{CF} \sin 45° - 600 - 800 = 0$

$F_{CF} = 1980 \text{ lb (T)}$ *Ans.*

$\zeta + \Sigma M_C = 0; F_{FE}(4) - 800(4) = 0$

$F_{FE} = 800 \text{ lb (T)}$ *Ans.*

$\zeta + \Sigma M_F = 0; F_{BC}(4) - 600(4) - 800(8) = 0$

$F_{BC} = 2200 \text{ lb (C)}$ *Ans.*

F5–8. $\zeta + \Sigma M_A = 0;$ $G_y(12 \text{ m}) - 20 \text{ kN}(2 \text{ m})$

$- 30 \text{ kN}(4 \text{ m}) - 40 \text{ kN}(6 \text{ m}) = 0$

$G_y = 33.33 \text{ kN}$

$+\uparrow \Sigma F_y = 0; F_{KC} + 33.33 \text{ kN} - 40 \text{ kN} = 0$

$F_{KC} = 6.67 \text{ kN (C)}$ *Ans.*

$\zeta + \Sigma M_K = 0;$

$33.33 \text{ kN}(8 \text{ m}) - 40 \text{ kN}(2 \text{ m}) - F_{CD}(3 \text{ m}) = 0$

$F_{CD} = 62.22 \text{ kN} = 62.2 \text{ kN (T)}$ *Ans.*

$\xrightarrow{+} \Sigma F_x = 0;$ $F_{LK} - 62.22 \text{ kN} = 0$

$F_{LK} = 62.2 \text{ kN (C)}$ *Ans.*

F5–9. From the geometry of the truss,

$\phi = \tan^{-1}(3 \text{ m}/2 \text{ m}) = 56.31°.$

$\zeta + \Sigma M_K = 0;$

$33.33 \text{ kN}(8 \text{ m}) - 40 \text{ kN}(2 \text{ m}) - F_{CD}(3 \text{ m}) = 0$

$F_{CD} = 62.2 \text{ kN (T)}$ *Ans.*

$\zeta + \Sigma M_D = 0; 33.33 \text{ kN}(6 \text{ m}) - F_{KJ}(3 \text{ m}) = 0$

$F_{KJ} = 66.7 \text{ kN (C)}$ *Ans.*

$+\uparrow\Sigma F_y = 0;$
$33.33 \text{ kN} - 40 \text{ kN} + F_{KD}\sin 56.31° = 0$
$$F_{KD} = 8.01 \text{ kN (T)} \qquad Ans.$$

F5–10. From the geometry of the truss,
$\tan\phi = \frac{(9 \text{ ft}) \tan 30°}{3 \text{ ft}} = 1.732 \quad \phi = 60°$
$\zeta + \Sigma M_C = 0;$
$F_{EF}\sin 30°(6 \text{ ft}) + 300 \text{ lb}(6 \text{ ft}) = 0$
$$F_{EF} = -600 \text{ lb} = 600 \text{ lb (C)} \qquad Ans.$$
$\zeta + \Sigma M_D = 0;$
$300 \text{ lb}(6 \text{ ft}) - F_{CF}\sin 60° (6 \text{ ft}) = 0$
$$F_{CF} = 346.41 \text{ lb} = 346 \text{ lb (T)} \qquad Ans.$$
$\zeta + \Sigma M_F = 0;$
$300 \text{ lb}(9 \text{ ft}) - 300 \text{ lb}(3 \text{ ft}) - F_{BC}(9 \text{ ft})\tan 30° = 0$
$$F_{BC} = 346.41 \text{ lb} = 346 \text{ lb (T)} \qquad Ans.$$

F5–11. From the geometry of the truss,
$\theta = \tan^{-1}(1 \text{ m}/2 \text{ m}) = 26.57°$
$\phi = \tan^{-1}(3 \text{ m}/2 \text{ m}) = 56.31°.$

The location of O can be found using similar triangles.
$$\frac{1 \text{ m}}{2 \text{ m}} = \frac{2 \text{ m}}{2 \text{ m} + x}$$
$$4 \text{ m} = 2 \text{ m} + x$$
$$x = 2 \text{ m}$$
$\zeta + \Sigma M_G = 0;$
$26.25 \text{ kN}(4 \text{ m}) - 15 \text{ kN}(2 \text{ m}) - F_{CD}(3 \text{ m}) = 0$
$$F_{CD} = 25 \text{ kN (T)} \qquad Ans.$$
$\zeta + \Sigma M_D = 0;$
$26.25 \text{ kN}(2 \text{ m}) - F_{GF}\cos 26.57°(2 \text{ m}) = 0$
$$F_{GF} = 29.3 \text{ kN (C)} \qquad Ans.$$
$\zeta + \Sigma M_O = 0; \ 15 \text{ kN}(4 \text{ m}) - 26.25 \text{ kN}(2 \text{ m})$
$\qquad\qquad\qquad - F_{GD}\sin 56.31°(4 \text{ m}) = 0$
$$F_{GD} = 2.253 \text{ kN} = 2.25 \text{ kN (T)} \qquad Ans.$$

F5–12. $\zeta + \Sigma M_H = 0;$
$F_{DC}(12 \text{ ft}) + 1200 \text{ lb}(9 \text{ ft}) - 1600 \text{ lb}(21 \text{ ft}) = 0$
$$F_{DC} = 1900 \text{ lb (C)} \qquad Ans.$$
$\zeta + \Sigma M_D = 0;$
$1200 \text{ lb}(21 \text{ ft}) - 1600 \text{ lb}(9 \text{ ft}) - F_{HI}(12 \text{ ft}) = 0$
$$F_{HI} = 900 \text{ lb (C)} \qquad Ans.$$
$\zeta + \Sigma M_C = 0; \ F_{JI}\cos 45°(12 \text{ ft}) + 1200 \text{ lb}(21 \text{ ft})$
$\qquad\qquad - 900 \text{ lb}(12 \text{ ft}) - 1600 \text{ lb}(9 \text{ ft}) = 0$
$$F_{JI} = 0 \qquad Ans.$$

F5–13. $+\uparrow\Sigma F_y = 0; \quad 3P - 60 = 0$
$$P = 20 \text{ lb} \qquad Ans.$$

F5–14. $\zeta + \Sigma M_C = 0;$
$-\left(\frac{4}{5}\right)(F_{AB})(9) + 400(6) + 500(3) = 0$
$$F_{AB} = 541.67 \text{ lb}$$
$\xrightarrow{+}\Sigma F_x = 0; -C_x + \frac{3}{5}(541.67) = 0$
$$C_x = 325 \text{ lb} \qquad Ans.$$
$+\uparrow\Sigma F_y = 0; C_y + \frac{4}{5}(541.67) - 400 - 500 = 0$
$$C_y = 467 \text{ lb} \qquad Ans.$$

F5–15. $\zeta + \Sigma M_A = 0; 100 \text{ N}(250 \text{ mm}) - N_B(50 \text{ mm}) = 0$
$$N_B = 500 \text{ N} \qquad Ans.$$
$\xrightarrow{+}\Sigma F_x = 0; \quad (500 \text{ N})\sin 45° - A_x = 0$
$$A_x = 353.55 \text{ N}$$
$+\uparrow\Sigma F_y = 0; A_y - 100 \text{ N} - (500 \text{ N})\cos 45° = 0$
$$A_y = 453.55 \text{ N}$$
$F_A = \sqrt{(353.55 \text{ N})^2 + (453.55 \text{ N})^2}$
$$= 575 \text{ N} \qquad Ans.$$

F5–16. $\zeta + \Sigma M_C = 0;$
$400(2) + 800 - F_{BA}\left(\frac{3}{\sqrt{10}}\right)(1)$
$\qquad\qquad\qquad\qquad - F_{BA}\left(\frac{1}{\sqrt{10}}\right)(3) = 0$
$$F_{BA} = 843.27 \text{ N}$$
$\xrightarrow{+}\Sigma F_x = 0; C_x - 843.27\left(\frac{3}{\sqrt{10}}\right) = 0$
$$C_x = 800 \text{ N} \qquad Ans.$$
$+\uparrow\Sigma F_y = 0; C_y + 843.27\left(\frac{1}{\sqrt{10}}\right) - 400 = 0$
$$C_y = 133 \text{ N} \qquad Ans.$$

Chapter 6

F6–1. $\bar{x} = \dfrac{\int_A \tilde{x}\, dA}{\int_A dA} = \dfrac{\frac{1}{2}\int_0^{1\text{ m}} y^{2/3}\, dy}{\int_0^{1\text{ m}} y^{1/3}\, dy} = 0.4 \text{ m} \qquad Ans.$

$\bar{y} = \dfrac{\int_A \tilde{y}\, dA}{\int_A dA} = \dfrac{\int_0^{1\text{ m}} y^{4/3}\, dy}{\int_0^{1\text{ m}} y^{1/3}\, dy} = 0.571 \text{ m} \qquad Ans.$

F6–2. $\bar{x} = \dfrac{\int_A \tilde{x}\, dA}{\int_A dA} = \dfrac{\int_0^{1\text{ m}} x(x^3\, dx)}{\int_0^{1\text{ m}} x^3\, dx}$
$$= 0.8 \text{ m} \qquad Ans.$$

$$\bar{y} = \frac{\int_A \tilde{y}\, dA}{\int_A dA} = \frac{\int_0^{1\,\text{m}} \frac{1}{2} x^3 (x^3\, dx)}{\int_0^{1\,\text{m}} x^3\, dx}$$

$$= 0.286\ \text{m} \qquad \textit{Ans.}$$

F6–3. $\bar{y} = \dfrac{\int_A \tilde{y}\, dA}{\int_A dA} = \dfrac{\int_0^{2\,\text{m}} y\left(2\left(\dfrac{y^{1/2}}{\sqrt{2}}\right)\right) dy}{\int_0^{2\,\text{m}} 2\left(\dfrac{y^{1/2}}{\sqrt{2}}\right) dy}$

$$= 1.2\ \text{m} \qquad \textit{Ans.}$$

F6–4. $\bar{x} = \dfrac{\int_m \tilde{x}\, dW}{\int_m dW} = \dfrac{\int_0^L x\left[W_0\left(1 + \dfrac{x^2}{L^2}\right) dx\right]}{\int_0^L W_0\left(1 + \dfrac{x^2}{L^2}\right) dx}$

$$= \frac{9}{16} L \qquad \textit{Ans.}$$

F6–5. $\bar{y} = \dfrac{\int_V \tilde{y}\, dV}{\int_V dV} = \dfrac{\int_0^{1\,\text{m}} y\left(\dfrac{\pi}{4} y\, dy\right)}{\int_0^{1\,\text{m}} \dfrac{\pi}{4} y\, dy}$

$$= 0.667\ \text{m} \qquad \textit{Ans.}$$

F6–6. $\bar{z} = \dfrac{\int_V \tilde{z}\, dV}{\int_V dV} = \dfrac{\int_0^{2\,\text{ft}} z\left[\dfrac{9\pi}{64}(4-z)^2\, dz\right]}{\int_0^{2\,\text{ft}} \dfrac{9\pi}{64}(4-z)^2\, dz}$

$$= 0.786\ \text{ft} \qquad \textit{Ans.}$$

F6–7. $\bar{x} = \dfrac{\Sigma \bar{x} L}{\Sigma L}$

$$= \frac{150(300) + 300(600) + 300(400)}{300 + 600 + 400}$$

$$= 265\ \text{mm} \qquad \textit{Ans.}$$

$$\bar{y} = \frac{\Sigma \bar{y} L}{\Sigma L}$$

$$= \frac{0(300) + 300(600) + 600(400)}{300 + 600 + 400}$$

$$= 323\ \text{mm} \qquad \textit{Ans.}$$

$$\bar{z} = \frac{\Sigma \bar{z} L}{\Sigma L}$$

$$= \frac{0(300) + 0(600) + (-200)(400)}{300 + 600 + 400}$$

$$= -61.5\ \text{mm} \qquad \textit{Ans.}$$

F6–8. $\bar{y} = \dfrac{\Sigma \bar{y} A}{\Sigma A} = \dfrac{150[300(50)] + 325[50(300)]}{300(50) + 50(300)}$

$$= 237.5\ \text{mm} \qquad \textit{Ans.}$$

F6–9. $\bar{y} = \dfrac{\Sigma \bar{y} A}{\Sigma A} = \dfrac{100[2(200)(50)] + 225[50(400)]}{2(200)(50) + 50(400)}$

$$= 162.5\ \text{mm} \qquad \textit{Ans.}$$

F6–10. $\bar{x} = \dfrac{\Sigma \bar{x} A}{\Sigma A} = \dfrac{0.25[4(0.5)] + 1.75[0.5(2.5)]}{4(0.5) + 0.5(2.5)}$

$$= 0.827\ \text{in.} \qquad \textit{Ans.}$$

$$\bar{y} = \frac{\Sigma \bar{y} A}{\Sigma A} = \frac{2[4(0.5)] + 0.25[(0.5)(2.5)]}{4(0.5) + (0.5)(2.5)}$$

$$= 1.33\ \text{in.} \qquad \textit{Ans.}$$

F6–11. $\bar{x} = \dfrac{\Sigma \bar{x} V}{\Sigma V} = \dfrac{1[2(7)(6)] + 4[4(2)(3)]}{2(7)(6) + 4(2)(3)}$

$$= 1.67\ \text{ft} \qquad \textit{Ans.}$$

$$\bar{y} = \frac{\Sigma \bar{y} V}{\Sigma V} = \frac{3.5[2(7)(6)] + 1[4(2)(3)]}{2(7)(6) + 4(2)(3)}$$

$$= 2.94\ \text{ft} \qquad \textit{Ans.}$$

$$\bar{z} = \frac{\Sigma \bar{z} V}{\Sigma V} = \frac{3[2(7)(6)] + 1.5[4(2)(3)]}{2(7)(6) + 4(2)(3)}$$

$$= 2.67\ \text{ft} \qquad \textit{Ans.}$$

F6–12. $\bar{x} = \dfrac{\Sigma \bar{x} V}{\Sigma V}$

$$= \frac{0.25[0.5(2.5)(1.8)] + 0.25\left[\frac{1}{2}(1.5)(1.8)(0.5)\right] + \left[\frac{1}{2}(1.5)(1.8)(0.5)\right]}{0.5(2.5)(1.8) + \frac{1}{2}(1.5)(1.8)(0.5) + \frac{1}{2}(1.5)(1.8)(0.5)}$$

$$= 0.391\ \text{m} \qquad \textit{Ans.}$$

$$\bar{y} = \frac{\Sigma \bar{y} V}{\Sigma V} = \frac{5.00625}{3.6} = 1.39\ \text{m} \qquad \textit{Ans.}$$

$$\bar{z} = \frac{\Sigma \bar{z} V}{\Sigma V} = \frac{2.835}{3.6} = 0.7875\ \text{m} \qquad \textit{Ans.}$$

F6–13.

$$I_x = \int_A y^2\, dA = \int_0^{1\,\text{m}} y^2\left[\left(1 - y^{3/2}\right) dy\right] = 0.111\ \text{m}^4 \quad \textit{Ans.}$$

F6–14.

$$I_x = \int_A y^2\, dA = \int_0^{1\,\text{m}} y^2\left(y^{3/2} dy\right) = 0.222\ \text{m}^4 \qquad \textit{Ans.}$$

F6–15.

$$I_y = \int_A x^2\, dA = \int_0^{1\,\text{m}} x^2\left(x^{2/3}\right) dx = 0.273\ \text{m}^4 \qquad \textit{Ans.}$$

F6–16.

$$I_y = \int_A x^2 \, dA = \int_0^{1\,m} x^2\left[(1 - x^{2/3})\, dx\right] = 0.0606 \text{ m}^4 \text{ Ans.}$$

F6–17. $I_x = \left[\frac{1}{12}(50)(450^3) + 0\right] + \left[\frac{1}{12}(300)(50^3) + 0\right]$
$$= 383(10^6) \text{ mm}^4 \qquad\qquad Ans.$$
$$I_y = \left[\frac{1}{12}(450)(50^3) + 0\right]$$
$$+ 2\left[\frac{1}{12}(50)(150^3) + (150)(50)(100)^2\right]$$
$$= 183(10^6) \text{ mm}^4 \qquad\qquad Ans.$$

F6–18. $I_x = \frac{1}{12}(360)(200^3) - \frac{1}{12}(300)(140^3)$
$$= 171(10^6) \text{ mm}^4 \qquad\qquad Ans.$$
$$I_y = \frac{1}{12}(200)(360^3) - \frac{1}{12}(140)(300^3)$$
$$= 463(10^6) \text{ mm}^4 \qquad\qquad Ans.$$

F6–19. $I_y = 2\left[\frac{1}{12}(50)(200^3) + 0\right]$
$$+ \left[\frac{1}{12}(300)(50^3) + 0\right]$$
$$= 69.8(10^6) \text{ mm}^4 \qquad\qquad Ans.$$

F6–20.
$$\bar{y} = \frac{\Sigma \bar{y} A}{\Sigma A} = \frac{15(150)(30) + 105(30)(150)}{150(30) + 30(150)} = 60 \text{ mm}$$
$$\bar{I}_{x'} = \Sigma(\bar{I} + Ad^2)$$
$$= \left[\frac{1}{12}(150)(30)^3 + (150)(30)(60 - 15)^2\right]$$
$$+ \left[\frac{1}{12}(30)(150)^3 + 30(150)(105 - 60)^2\right]$$
$$= 27.0(10^6) \text{ mm}^4 \qquad\qquad Ans.$$

Chapter 7

F7–1. Entire beam:

$\zeta + \Sigma M_B = 0;$ $60 - 10(2) - A_y(2) = 0$ $A_y = 20 \text{ kN}$

Left segment:

                                                   Ans.

$\xrightarrow{+} \Sigma F_x = 0;$ $N_C = 0$

$+\uparrow \Sigma F_y = 0;$ $20 - V_C = 0$ $V_C = 20 \text{ kN}$ *Ans.*

$\zeta + \Sigma M_C = 0;$ $M_C + 60 - 20(1) = 0$ $M_C = -40 \text{ kN} \cdot \text{m}$ *Ans.*

F7–2. Entire beam:

$\zeta + \Sigma M_A = 0;$ $B_y(3) - 100(1.5)(0.75) - 200(1.5)(2.25) = 0$

            $B_y = 262.5 \text{ N}$

Right segment:

                                                    Ans.

$\xrightarrow{+} \Sigma F_x = 0;$ $N_C = 0$

$+\uparrow \Sigma F_y = 0;$ $V_C + 262.5 - 200(1.5) = 0$ $V_C = 37.5 \text{ N}$ *Ans.*

$\zeta + \Sigma M_C = 0;$ $262.5(1.5) - 200(1.5)(0.75) - M_C = 0$ $M_C = 169 \text{ N} \cdot \text{m}$ *Ans.*

F7–3. Entire beam:

$\xrightarrow{+} \Sigma F_x = 0;$ $B_x = 0$

$\zeta + \Sigma M_A = 0;$ $20(2)(1) - B_y(4) = 0$ $B_y = 10 \text{ kN}$

Right segment:

                                                    Ans.

$\xrightarrow{+} \Sigma F_x = 0;$ $N_C = 0$

$+\uparrow \Sigma F_y = 0;$ $V_C - 10 = 0$ $V_C = 10 \text{ kN}$ *Ans.*

$\zeta + \Sigma M_C = 0;$ $-M_C - 10(2) = 0$ $M_C = -20 \text{ kN} \cdot \text{m}$ *Ans.*

F7–4. Entire beam:

$\zeta + \Sigma M_B = 0;$ $\frac{1}{2}(10)(3)(2) + 10(3)(4.5) - A_y(6) = 0$ $A_y = 27.5 \text{ kN}$

Left segment:

                                                    Ans.

$\xrightarrow{+} \Sigma F_x = 0;$ $N_C = 0$

$+\uparrow \Sigma F_y = 0;$ $27.5 - 10(3) - V_C = 0$ $V_C = -2.5 \text{ kN}$ *Ans.*

$\zeta + \Sigma M_C = 0;$ $M_C + 10(3)(1.5) - 27.5(3) = 0$ $M_C = 37.5 \text{ kN} \cdot \text{m}$ *Ans.*

F7–5. Entire beam:

$\xrightarrow{+} \Sigma F_x = 0;$ $A_x = 0$

$\zeta + \Sigma M_B = 0;$ $300(6)(3) - \frac{1}{2}(300)(3)(1) - A_y(6) = 0$ $A_y = 825 \text{ lb}$

Left segment:

$\xrightarrow{+} \Sigma F_x = 0;$ $N_C = 0$ *Ans.*

$+\uparrow \Sigma F_y = 0;$ $825 - 300(3) - V_C = 0$ $V_C = -75$ lb   *Ans.*

$\zeta + \Sigma M_C = 0;$ $M_C + 300(3)(1.5) - 825(3) = 0$ $M_C = 1125$ lb · ft   *Ans.*

F7–6.   Entire beam:

$\zeta + \Sigma M_A = 0;$ $F_{BD}\left(\dfrac{3}{5}\right)(4) - 5(6)(3) = 0$ $F_{BD} = 37.5$ kN

$\xrightarrow{+} \Sigma F_x = 0;$ $37.5\left(\dfrac{4}{5}\right) - A_x = 0$ $A_x = 30$ kN

$+\uparrow \Sigma F_y = 0;$ $A_y + 37.5\left(\dfrac{3}{5}\right) - 5(6) = 0$ $A_y = 7.5$ kN

Left segment:

$\xrightarrow{+} \Sigma F_x = 0;$ $N_C - 30 = 0$ $N_C = 30$ kN   *Ans.*

$+\uparrow \Sigma F_y = 0;$ $7.5 - 5(2) - V_C = 0$ $V_C = -2.5$ kN   *Ans.*

$\zeta + \Sigma M_C = 0;$ $M_C + 5(2)(1) - 7.5(2) = 0$ $M_C = 5$ kN · m   *Ans.*

F7–7.   Beam:

$\Sigma M_A = 0; \quad T_{CD} = 2w$

$\Sigma F_y = 0; \quad T_{AB} = w$

Rod AB:

$\sigma = \dfrac{N}{A}; \quad 300(10^3) = \dfrac{w}{10};$

$w = 3$ N/m

Rod CD:

$\sigma = \dfrac{N}{A}; \quad 300(10^3) = \dfrac{2w}{15};$

$w = 2.25$ N/m   *Ans.*

F7–8. $A = \pi(0.1^2 - 0.08^2) = 3.6(10^{-3})\pi$ m²

$\sigma_{avg} = \dfrac{N}{A} = \dfrac{300(10^3)}{3.6(10^{-3})\pi} = 26.5$ MPa   *Ans.*

F7–9. $A = 3[4(1)] = 12$ in²

$\sigma_{avg} = \dfrac{N}{A} = \dfrac{15}{12} = 1.25$ ksi   *Ans.*

F7–10.   Consider the cross section to be a rectangle and two triangles.

$\bar{y} = \dfrac{\Sigma \tilde{y}A}{\Sigma A} = \dfrac{0.15[(0.3)(0.12)] + (0.1)\left[\dfrac{1}{2}(0.16)(0.3)\right]}{0.3(0.12) + \dfrac{1}{2}(0.16)(0.3)}$

$= 0.13$ m $= 130$ mm   *Ans.*

$\sigma_{avg} = \dfrac{N}{A} = \dfrac{600(10^3)}{0.06} = 10$ MPa   *Ans.*

F7–11.

$A_A = A_C = \dfrac{\pi}{4}(0.5^2) = 0.0625\pi$ in², $A_B = \dfrac{\pi}{4}(1^2) = 0.25\pi$ in²

$\sigma_A = \dfrac{N_A}{A_A} = \dfrac{3}{0.0625\pi} = 15.3$ ksi (T)   *Ans.*

$\sigma_B = \dfrac{N_B}{A_B} = \dfrac{-6}{0.25\pi} = -7.64$ ksi = 7.64 ksi (C)   *Ans.*

$\sigma_C = \dfrac{N_C}{A_C} = \dfrac{2}{0.0625\pi} = 10.2$ ksi (T)   *Ans.*

F7–12.   Pin at A:

$F_{AD} = 50(9.81)$ N $= 490.5$ N

$+\uparrow \Sigma F_y = 0;$ $F_{AC}\left(\dfrac{3}{5}\right) - 490.5 = 0$ $F_{AC} = 817.5$ N

$\xrightarrow{+} \Sigma F_x = 0;$ $817.5\left(\dfrac{4}{5}\right) - F_{AB} = 0$ $F_{AB} = 654$ N

$A_{AB} = \dfrac{\pi}{4}(0.008^2) = 16(10^{-6})\pi$ m²

$(\sigma_{AB})_{avg} = \dfrac{F_{AB}}{A_{AB}} = \dfrac{654}{16(10^{-6})\pi} = 13.0$ MPa   *Ans.*

F7–13.   Ring C:

$+\uparrow \Sigma F_y = 0;$ $2F\cos 60° - 200(9.81) = 0$ $F = 1962$ N

$(\sigma_{allow})_{avg} = \dfrac{F}{A}; \quad 150(10^6) = \dfrac{1962}{\dfrac{\pi}{4}d^2}$

$d = 0.00408$ m $= 4.08$ mm

Use $d = 5$ mm.   *Ans.*

F7–14. Entire frame:

$$\Sigma F_y = 0; \quad A_y = 600 \text{ lb}$$

$$\Sigma M_B = 0; \quad A_x = 800 \text{ lb}$$

$$F_A = \sqrt{(600)^2 + (800)^2} = 1000 \text{ lb}$$

$$(\tau_A)_{avg} = \frac{F_A/2}{A} = \frac{1000/2}{\frac{\pi}{4}(0.25)^2} = 10.2 \text{ ksi} \qquad \textit{Ans.}$$

F7–15. Center plate, bolts have double shear:

$$\Sigma F_x = 0; \quad 4V - 10 = 0 \qquad V = 2.5 \text{ kip}$$

$$A = \frac{\pi}{4}\left(\frac{3}{4}\right)^2 = 0.140625\pi \text{ in}^2$$

$$\tau_{avg} = \frac{V}{A} = \frac{2.5}{0.140625\pi} = 5.66 \text{ ksi} \qquad \textit{Ans.}$$

F7–16. Nails have single shear:

$$\Sigma F_x = 0; \quad P - 3V = 0 \qquad V = \frac{P}{3}$$

$$A = \frac{\pi}{4}(0.004^2) = 4(10^{-6})\pi \text{ m}^2$$

$$(\tau_{avg})_{allow} = \frac{V}{A}; \quad 60(10^6) = \frac{\frac{P}{3}}{4(10^{-6})\pi}$$

$$P = 2.262(10^3) \text{ N} = 2.26 \text{ kN} \qquad \textit{Ans.}$$

F7–17. Strut:

$$\xrightarrow{+} \Sigma F_x = 0; \quad V - P \cos 60° = 0 \qquad V = 0.5P$$

$$A = \left(\frac{0.05}{\sin 60°}\right)(0.025) = 1.4434(10^{-3}) \text{ m}^2$$

$$(\tau_{avg})_{allow} = \frac{V}{A}; \quad 600(10^3) = \frac{0.5P}{1.4434(10^{-3})}$$

$$P = 1.732(10^3) \text{ N} = 1.73 \text{ kN} \qquad \textit{Ans.}$$

F7–18. The resultant force on the pin is

$$F = \sqrt{30^2 + 40^2} = 50 \text{ kN}.$$

We have double shear:

$$V = \frac{F}{2} = \frac{50}{2} = 25 \text{ kN}$$

$$A = \frac{\pi}{4}(0.03^2) = 0.225(10^{-3})\pi \text{ m}^2$$

$$\tau_{avg} = \frac{V}{A} = \frac{25(10^3)}{0.225(10^{-3})\pi} = 35.4 \text{ MPa} \qquad \textit{Ans.}$$

F7–19. Eyebolt:

$$\xrightarrow{+} \Sigma F_x = 0; \quad 30 - N = 0 \qquad N = 30 \text{ kN}$$

$$\sigma_{allow} = \frac{\sigma_Y}{\text{F.S.}} = \frac{250}{1.5} = 166.67 \text{ MPa}$$

$$\sigma_{allow} = \frac{N}{A}; \quad 166.67(10^6) = \frac{30(10^3)}{\frac{\pi}{4}d^2}$$

$$d = 15.14 \text{ mm}$$

Use $d = 16$ mm. $\qquad \textit{Ans.}$

F7–20. Right segment through AB:

$$\xrightarrow{+} \Sigma F_x = 0; \quad N_{AB} - 30 = 0 \qquad N_{AB} = 30 \text{ kip}$$

Right segment through CB:

$$\xrightarrow{+} \Sigma F_x = 0; \quad N_{BC} - 15 - 15 - 30 = 0 \qquad N_{BC} = 60 \text{ kip}$$

$$\sigma_{allow} = \frac{\sigma_Y}{\text{F.S.}} = \frac{50}{1.5} = 33.33 \text{ ksi}$$

Segment AB:

$$\sigma_{allow} = \frac{N_{AB}}{A_{AB}}; \quad 33.33 = \frac{30}{h_1(0.5)}$$

$$h_1 = 1.8 \text{ in.}$$

Segment BC:

$$\sigma_{allow} = \frac{N_{BC}}{A_{BC}}; \quad 33.33 = \frac{60}{h_2(0.5)}$$

$$h_2 = 3.6 \text{ in.}$$

Use $h_1 = 1\frac{7}{8}$ in. and $h_2 = 3\frac{5}{8}$ in. $\qquad \textit{Ans.}$

F7–21. $N = P$

$$\sigma_{allow} = \frac{\sigma_Y}{\text{F.S.}} = \frac{250}{2} = 125 \text{ MPa}$$

$$A_r = \frac{\pi}{4}(0.04^2) = 1.2566(10^{-3}) \text{ m}^2$$

$$A_{a-a} = 2(0.06 - 0.03)(0.05) = 3(10^{-3}) \text{ m}^2$$

The rod will fail first.

$$\sigma_{allow} = \frac{N}{A_r}; \quad 125(10^6) = \frac{P}{1.2566(10^{-3})}$$

$$P = 157.08(10^3) \text{ N} = 157 \text{ kN} \qquad \textit{Ans.}$$

F7–22. Pin has double shear:

$$\xrightarrow{+} \Sigma F_x = 0; \quad 80 - 2V = 0 \qquad V = 40 \text{ kN}$$

$$\tau_{allow} = \frac{\tau_{fail}}{\text{F.S.}} = \frac{100}{2.5} = 40 \text{ MPa}$$

$$\tau_{allow} = \frac{V}{A}; \quad 40(10^6) = \frac{40(10^3)}{\frac{\pi}{4}d^2}$$

$$d = 0.03568 \text{ m} = 35.68 \text{ mm}$$

Use $d = 36$ mm. $\qquad \textit{Ans.}$

F7–23.

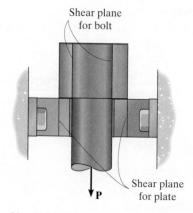

Shear plane for bolt

Shear plane for plate

$V = P$

$$\tau_{\text{allow}} = \frac{\tau_{\text{fail}}}{\text{F.S.}} = \frac{120}{2.5} = 48 \text{ MPa}$$

Area of shear plane for bolt head and plate:

$A_b = \pi dt = \pi(0.04)(0.075) = 0.003\pi \text{ m}^2$

$A_p = \pi dt = \pi(0.08)(0.03) = 0.0024\pi \text{ m}^2$

Since the area of shear plane for the plate is smaller,

$$\tau_{\text{allow}} = \frac{V}{A_p}; \quad 48(10^6) = \frac{P}{0.0024\pi}$$

$P = 361.91(10^3) \text{ N} = 362 \text{ kN}$ *Ans.*

F7–24. Support reaction at A:

$\zeta + \Sigma M_B = 0;$ $\frac{1}{2}(300)(9)(6) - A_y(9) = 0$ $A_y = 900 \text{ lb}$

Each nail has single shear:
$V = 900 \text{ lb}/6 = 150 \text{ lb}$

$$\tau_{\text{allow}} = \frac{\tau_{\text{fail}}}{\text{F.S.}} = \frac{16}{2} = 8 \text{ ksi}$$

$$\tau_{\text{allow}} = \frac{V}{A}; \quad 8(10^3) = \frac{150}{\frac{\pi}{4}d^2}$$

$$d = 0.1545 \text{ in.}$$

Use $d = \frac{3}{16}$ in. *Ans.*

F7–25. $\frac{\delta_C}{600} = \frac{0.2}{400};$ $\delta_C = 0.3 \text{ mm}$

$\epsilon_{CD} = \frac{\delta_C}{L_{CD}} = \frac{0.3}{300} = 0.001 \text{ mm/mm}$ *Ans.*

F7–26.

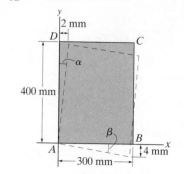

$$\theta = \left(\frac{0.02°}{180°}\right)\pi \text{ rad} = 0.3491(10^{-3}) \text{ rad}$$

$\delta_B = \theta L_{AB} = 0.3491(10^{-3})(600) = 0.2094 \text{ mm}$

$\delta_C = \theta L_{AC} = 0.3491(10^{-3})(1200) = 0.4189 \text{ mm}$

$\epsilon_{BD} = \dfrac{\delta_B}{L_{BD}} = \dfrac{0.2094}{400} = 0.524(10^{-3}) \text{ mm/mm}$ *Ans.*

$\epsilon_{CE} = \dfrac{\delta_C}{L_{CE}} = \dfrac{0.4189}{600} = 0.698(10^{-3}) \text{ mm/mm}$ *Ans.*

F7–27.

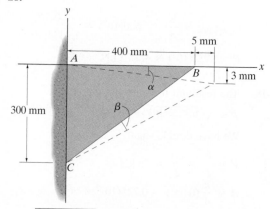

$\alpha = \dfrac{2}{400} = 0.005 \text{ rad}$ $\beta = \dfrac{4}{300} = 0.01333 \text{ rad}$

$(\gamma_A)_{xy} = \dfrac{\pi}{2} - \theta$

$\qquad = \dfrac{\pi}{2} - \left(\dfrac{\pi}{2} - \alpha + \beta\right)$

$\qquad = \alpha - \beta$

$\qquad = 0.005 - 0.01333$

$\qquad = -0.00833 \text{ rad}$ *Ans.*

F7–28.

$L_{BC} = \sqrt{300^2 + 400^2} = 500 \text{ mm}$

$L_{B'C} = \sqrt{(300 - 3)^2 + (400 + 5)^2} = 502.2290 \text{ mm}$

$$\alpha = \frac{3}{405} = 0.007407 \text{ rad}$$

$$(\epsilon_{BC})_{avg} = \frac{L_{B'C} - L_{BC}}{L_{BC}} = \frac{502.2290 - 500}{500}$$

$$= 0.00446 \text{ mm/mm} \qquad \text{Ans.}$$

$$(\gamma_A)_{xy} = \frac{\pi}{2} - \theta = \frac{\pi}{2} - \left(\frac{\pi}{2} - \alpha\right) = -\alpha = -0.00741 \text{ rad} \quad \text{Ans.}$$

F7–29.

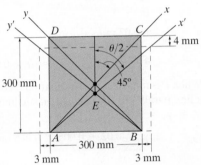

$$L_{AC} = \sqrt{L_{CD}^2 + L_{AD}^2} = \sqrt{300^2 + 300^2} = 424.2641 \text{ mm}$$

$$L_{A'C'} = \sqrt{L_{C'D'}^2 + L_{A'D'}^2} = \sqrt{306^2 + 296^2} = 425.7370 \text{ mm}$$

$$\frac{\theta}{2} = \tan^{-1}\left(\frac{L_{C'D'}}{L_{A'D'}}\right); \theta = 2 \tan^{-1}\left(\frac{306}{296}\right) = 1.6040 \text{ rad}$$

$$(\epsilon_{AC})_{avg} = \frac{L_{A'C'} - L_{AC}}{L_{AC}} = \frac{425.7370 - 424.2641}{424.2641}$$

$$= 0.00347 \text{ mm/mm} \qquad \text{Ans.}$$

$$(\gamma_E)_{xy} = \frac{\pi}{2} - \theta = \frac{\pi}{2} - 1.6040 = -0.0332 \text{ rad} \qquad \text{Ans.}$$

Chapter 8

F8–1. Material has uniform properties throughout. *Ans.*

F8–2. Proportional limit is A.   *Ans.*

       Ultimate stress is D.   *Ans.*

F8–3. The initial slope of the $\sigma - \epsilon$ diagram.   *Ans.*

F8–4. True.   *Ans.*

F8–5. False. Use the *original* cross-sectional area and length.   *Ans.*

F8–6. False. It will normally decrease.   *Ans.*

F8–7.
$$\epsilon = \frac{\sigma}{E} = \frac{N}{AE}$$

$$\delta = \epsilon L = \frac{NL}{AE} = \frac{100(10^3)(0.100)}{\frac{\pi}{4}(0.015)^2 \, 200(10^9)}$$

$$= 0.283 \text{mm} \qquad \text{Ans.}$$

F8–8.
$$\epsilon = \frac{\sigma}{E} = \frac{N}{AE}$$

$$\delta = \epsilon L = \frac{NL}{AE}$$

$$0.003 = \frac{(10\,000)(8)}{12E}$$

$$E = 2.22(10^6) \text{ psi} \qquad \text{Ans.}$$

F8–9.
$$\epsilon = \frac{\sigma}{E} = \frac{N}{AE}$$

$$\delta = \epsilon L = \frac{NL}{AE} = \frac{6(10^3)4}{\frac{\pi}{4}(0.01)^2 100(10^9)}$$

$$= 3.06 \text{ mm} \qquad \text{Ans.}$$

F8–10.
$$\sigma = \frac{N}{A} = \frac{100(10^3)}{\frac{\pi}{4}(0.02)^2} = 318.31 \text{ MPa}$$

Since $\sigma < \sigma_Y = 450$ MPa, Hooke's Law is applicable.

$$E = \frac{\sigma_Y}{\epsilon_Y} = \frac{450(10^6)}{0.00225} = 200 \text{ GPa}$$

$$\epsilon = \frac{\sigma}{E} = \frac{318.31(10^6)}{200(10^9)} = 0.001592 \text{ mm/mm}$$

$$\delta = \epsilon L = 0.001592(50) = 0.0796 \text{ mm} \qquad \text{Ans.}$$

F8–11.
$$\sigma = \frac{N}{A} = \frac{150(10^3)}{\frac{\pi}{4}(0.02^2)} = 477.46 \text{ MPa}$$

Since $\sigma > \sigma_Y = 450$ MPa, Hooke's Law is not applicable. From the geometry of the shaded triangle,

$$\frac{\epsilon - 0.00225}{0.03 - 0.00225} = \frac{477.46 - 450}{500 - 450}$$

$$\epsilon = 0.0174903$$

When the load is removed, the strain recovers along a line AB which is parallel to the original elastic line.

Here $E = \dfrac{\sigma_Y}{\epsilon_Y} = \dfrac{450(10^6)}{0.00225} = 200 \text{ GPa}$.

The elastic recovery is

$$\epsilon_r = \frac{\sigma}{E} = \frac{477.46(10^6)}{200(10^9)} = 0.002387 \text{ mm/mm}$$

$$\epsilon_p = \epsilon - \epsilon_r = 0.017493 - 0.002387$$

$$= 0.01511 \text{ mm/mm}$$

$$\delta_p = \epsilon_p L = 0.0151(50) = 0.755 \text{ mm} \qquad \text{Ans.}$$

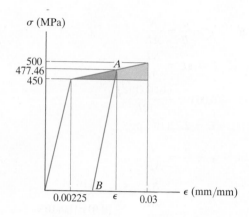

σ (MPa)

F8–12. $\epsilon_{BC} = \dfrac{\delta_{BC}}{L_{BC}} = \dfrac{0.2}{300} = 0.6667(10^{-3})$ mm/mm

$\sigma_{BC} = E\epsilon_{BC} = 200(10^9)[0.6667(10^{-3})]$
$= 133.33$ MPa

Since $\sigma_{BC} < \sigma_Y = 250$ MPa, Hooke's Law is valid.

$\sigma_{BC} = \dfrac{F_{BC}}{A_{BC}};\qquad 133.33(10^6) = \dfrac{F_{BC}}{\dfrac{\pi}{4}(0.003^2)}$

$\qquad\qquad\qquad F_{BC} = 942.48$ N

$\zeta + \Sigma M_A = 0;\qquad 942.48(0.4) - P(0.6) = 0$

$\qquad\qquad\qquad P = 628.31$ N $= 628$ N $\qquad$ *Ans.*

F8–13. $\sigma = \dfrac{P}{A} = \dfrac{10(10^3)}{\frac{\pi}{4}(0.015)^2} = 56.59$ MPa

$\epsilon_{long} = \dfrac{\sigma}{E} = \dfrac{56.59(10^6)}{70(10^9)} = 0.808(10^{-3})$

$\epsilon_{lat} = -\nu\epsilon_{long} = -0.35(0.808(10^{-3}))$
$\qquad = -0.283(10^{-3})$

$\delta d = (-0.283(10^{-3}))(15\text{ mm}) = -4.24(10^{-3})$ mm
$\qquad\qquad\qquad\qquad\qquad\qquad\qquad\qquad$ *Ans.*

F8–14. $\sigma = \dfrac{P}{A} = \dfrac{50(10^3)}{\frac{\pi}{4}(0.02^2)} = 159.15$ MPa

$\epsilon_a = \dfrac{\delta}{L} = \dfrac{1.40}{600} = 0.002333$ mm/mm

$E = \dfrac{\sigma}{\epsilon_a} = \dfrac{159.15(10^6)}{0.002333} = 68.2$ GPa $\qquad$ *Ans.*

$\epsilon_e = \dfrac{d' - d}{d} = \dfrac{19.9837 - 20}{20} = -0.815(10^{-3})$ mm/mm

$\nu = -\dfrac{\epsilon_e}{\epsilon_a} = -\dfrac{-0.815(10^{-3})}{0.002333} = 0.3493 = 0.349$

$G = \dfrac{E}{2(1 + \nu)} = \dfrac{68.21}{2(1 + 0.3493)} = 25.3$ GPa *Ans.*

F8–15. $\alpha = \dfrac{0.5}{150} = 0.003333$ rad

$\gamma = \dfrac{\pi}{2} - \theta = \dfrac{\pi}{2} - \left(\dfrac{\pi}{2} - \alpha\right)$
$\quad = \alpha = 0.003333$ rad

$\tau = G\gamma = [26(10^9)](0.003333) = 86.67$ MPa

$\tau = \dfrac{V}{A};\qquad 86.67(10^6) = \dfrac{P}{0.15(0.02)}$
$\qquad\qquad\qquad P = 260$ kN $\qquad$ *Ans.*

F8–16. $\alpha = \dfrac{3}{150} = 0.02$ rad

$\gamma = \dfrac{\pi}{2} - \theta = \dfrac{\pi}{2} - \left(\dfrac{\pi}{2} - \alpha\right) = \alpha = 0.02$ rad

When P is removed, the shear strain recovers along a line parallel to the original elastic line.

$\gamma_r = \gamma_Y = 0.005$ rad

$\gamma_p = \gamma - \gamma_r = 0.02 - 0.005 = 0.015$ rad $\qquad$ *Ans.*

Chapter 9

F9–1. $A = \dfrac{\pi}{4}(0.02^2) = 0.1(10^{-3})\pi$ m²

$N_{BC} = 40$ kN, $N_{AB} = -60$ kN

$\delta_C = \dfrac{1}{AE}\{40(10^3)(400) + [-60(10^3)(600)]\}$
$\quad = \dfrac{-20(10^6)\text{ N}\cdot\text{mm}}{AE}$
$\quad = -0.318$ mm $\qquad$ *Ans.*

F9–2. $A_{AB} = A_{CD} = \dfrac{\pi}{4}(0.02^2) = 0.1(10^{-3})\pi$ m²

$A_{BC} = \dfrac{\pi}{4}(0.04^2 - 0.03^2) = 0.175(10^{-3})\pi$ m²

$N_{AB} = -10$ kN, $N_{BC} = 10$ kN, $N_{CD} = -20$ kN

$\delta_{D/A} = \dfrac{[-10(10^3)](400)}{[0.1(10^{-3})\pi][68.9(10^9)]}$
$\qquad + \dfrac{[10(10^3)](400)}{[0.175(10^{-3})\pi][68.9(10^9)]}$
$\qquad + \dfrac{[-20(10^3)](400)}{[0.1(10^{-3})\pi][68.9(10^9)]}$
$\qquad = -0.449$ mm $\qquad$ *Ans.*

F9–3. $A = \dfrac{\pi}{4}(0.03^2) = 0.225(10^{-3})\pi$ m²

$N_{BC} = -90$ kN, $N_{AB} = -90 + 2\left(\dfrac{4}{5}\right)(30) = -42$ kN

$$\delta_C = \frac{1}{0.225(10^{-3})\pi[200(10^9)]}\left\{[-42(10^3)(0.4)]\right.$$
$$\left.+ [-90(10^3)(0.6)]\right\}$$
$$= -0.501(10^{-3}) \text{ m} = -0.501 \text{ mm} \quad Ans.$$

F9–4.
$$\delta_{A/B} = \frac{NL}{AE} = \frac{[60(10^3)](0.8)}{[0.1(10^{-3})\pi][200(10^9)]}$$
$$= 0.7639(10^{-3})\text{ m} \downarrow$$
$$\delta_B = \frac{F_{sp}}{k} = \frac{60(10^3)}{50(10^6)} = 1.2(10^{-3})\text{ m} \downarrow$$
$$+\downarrow \quad \delta_A = \delta_B + \delta_{A/B}$$
$$\delta_A = 1.2(10^{-3}) + 0.7639(10^{-3})$$
$$= 1.9639(10^{-3}) \text{ m} = 1.96 \text{ mm} \downarrow \qquad Ans.$$

F9–5.

30 kN/m

N(x)

x

$$A = \frac{\pi}{4}(0.02^2) = 0.1(10^{-3})\pi \text{ m}^2$$

Internal load $N(x) = 30(10^3)x$

$$\delta_A = \int \frac{N(x)dx}{AE}$$
$$= \frac{1}{[0.1(10^{-3})\pi][73.1(10^9)]}\int_0^{0.9 \text{ m}} 30(10^3)x \, dx$$
$$= 0.529(10^{-3}) \text{ m} = 0.529 \text{ mm} \qquad Ans.$$

F9–6.

(50 x) kN/m

N(x)

x

Distributed load $N(x) = \frac{45(10^3)}{0.9}x = 50(10^3)x \text{ N/m}$

Internal load $N(x) = \frac{1}{2}(50(10^3))x(x) = 25(10^3)x^2$

$$\delta_A = \int_0^L \frac{N(x)dx}{AE}$$
$$= \frac{1}{[0.1(10^{-3})\pi][73.1(10^9)]}\int_0^{0.9 \text{ m}} [25(10^3)x^2]dx$$
$$= 0.265 \text{ mm} \qquad Ans.$$

Chapter 10

F10–1. $J = \frac{\pi}{2}(0.04^4) = 1.28(10^{-6})\pi \text{ m}^4$

$$\tau_A = \tau_{max} = \frac{Tc}{J} = \frac{5(10^3)(0.04)}{1.28(10^{-6})\pi} = 49.7 \text{ MPa} \quad Ans.$$

$$\tau_B = \frac{T\rho_B}{J} = \frac{5(10^3)(0.03)}{1.28(10^{-6})\pi} = 37.3 \text{ MPa} \quad Ans.$$

49.7 MPa

A

B

37.3 MPa

F10–2. $J = \frac{\pi}{2}(0.06^4 - 0.04^4) = 5.2(10^{-6})\pi \text{ m}^4$

$$\tau_B = \tau_{max} = \frac{Tc}{J} = \frac{10(10^3)(0.06)}{5.2(10^{-6})\pi} = 36.7 \text{ MPa} \quad Ans.$$

$$\tau_A = \frac{T\rho_A}{J} = \frac{10(10^3)(0.04)}{5.2(10^{-6})\pi} = 24.5 \text{ MPa} \quad Ans.$$

24.5 MPa

A

B

36.7 MPa

F10–3. $J_{AB} = \frac{\pi}{2}(0.04^4 - 0.03^4) = 0.875(10^{-6})\pi \text{ m}^4$

$$J_{BC} = \frac{\pi}{2}(0.04^4) = 1.28(10^{-6})\pi \text{ m}^4$$

$$(\tau_{AB})_{max} = \frac{T_{AB}\,c_{AB}}{J_{AB}} = \frac{[2(10^3)](0.04)}{0.875(10^{-6})\pi} = 29.1 \text{ MPa}$$

$$(\tau_{BC})_{max} = \frac{T_{BC}\,c_{BC}}{J_{BC}} = \frac{[6(10^3)](0.04)}{1.28(10^{-6})\pi}$$
$$= 59.7 \text{ MPa} \qquad Ans.$$

F10–4. $T_{AB} = 0$, $T_{BC} = 600 \text{ N} \cdot \text{m}$, $T_{CD} = 0$

$$J = \frac{\pi}{2}(0.02^4) = 80(10^{-9})\pi \text{ m}^4$$

$$\tau_{max} = \frac{Tc}{J} = \frac{600(0.02)}{80(10^{-9})\pi} = 47.7 \text{ MPa} \qquad Ans.$$

F10–5. $J_{BC} = \dfrac{\pi}{2}(0.04^4 - 0.03^4) = 0.875(10^{-6})\pi \text{ m}^4$

$$(\tau_{BC})_{max} = \dfrac{T_{BC}c_{BC}}{J_{BC}} = \dfrac{2100(0.04)}{0.875(10^{-6})\pi}$$

$$= 30.6 \text{ MPa} \qquad\qquad Ans.$$

F10–6. $t = 5(10^3) \text{ N} \cdot \text{m/m}$

Internal torque is $T = 5(10^3)(0.8) = 4000 \text{ N} \cdot \text{m}$

$$J = \dfrac{\pi}{2}(0.04^4) = 1.28(10^{-6})\pi \text{ m}^4$$

$$\tau_{AB} = \dfrac{T_A c}{J} = \dfrac{4000(0.04)}{1.28(10^{-6})\pi} = 39.8 \text{ MPa} \qquad Ans.$$

F10–7. $T_{AB} = 250 \text{ N} \cdot \text{m}, T_{BC} = 175 \text{ N} \cdot \text{m},$
$T_{CD} = -150 \text{ N} \cdot \text{m}$

Maximum internal torque is in region AB.

$T_{AB} = 250 \text{ N} \cdot \text{m}$

$$\tau_{\substack{abs \\ max}} = \dfrac{T_{AB}c}{J} = \dfrac{250(0.025)}{\dfrac{\pi}{2}(0.025)^4} = 10.2 \text{ MPa} \qquad Ans.$$

F10–8. $P = T\omega;\quad 3(550) \text{ ft} \cdot \text{lb/s} = T\left[150\left(\dfrac{2\pi}{60}\right) \text{ rad/s}\right]$

$T = 105.04 \text{ ft} \cdot \text{lb}$

$$\tau_{allow} = \dfrac{Tc}{J};\quad 12(10^3) = \dfrac{105.04(12)(d/2)}{\dfrac{\pi}{2}(d/2)^4}$$

$$d = 0.812 \text{ in.}$$

$$\text{Use } d = \dfrac{7}{8} \text{ in.} \qquad\qquad Ans.$$

F10–9. $T_{AB} = -2 \text{ kN} \cdot \text{m}, T_{BC} = 1 \text{ kN} \cdot \text{m}$

$$J = \dfrac{\pi}{2}(0.03^4) = 0.405(10^{-6})\pi \text{ m}^4$$

$$\phi_{A/C} = \dfrac{-2(10^3)(0.6) + (10^3)(0.4)}{[0.405(10^{-6})\pi][75(10^9)]}$$

$$= -0.00838 \text{ rad} = -0.480° \qquad Ans.$$

F10–10. $T_{AB} = 600 \text{ N} \cdot \text{m}$

$$J = \dfrac{\pi}{2}(0.02^4) = 80(10^{-9})\pi \text{ m}^4$$

$$\phi_{B/A} = \dfrac{600(0.45)}{[80(10^{-9})\pi][75(10^9)]}$$

$$= 0.01432 \text{ rad} = 0.821° \qquad Ans.$$

F10–11. $J = \dfrac{\pi}{2}(0.04^4 - 0.03^4) = 0.875(10^{-6})\pi \text{ m}^4$

$$\phi_{A/B} = \dfrac{T_{AB}\,L_{AB}}{JG} = \dfrac{3(10^3)(0.9)}{[0.875(10^{-6})\pi][26(10^9)]}$$

$$= 0.03778 \text{ rad}$$

$$\phi_B = \dfrac{T_B}{k_B} = \dfrac{3(10^3)}{90(10^3)} = 0.03333 \text{ rad}$$

$$\phi_A = \phi_B + \phi_{A/B}$$

$$= 0.03333 + 0.03778$$

$$= 0.07111 \text{ rad} = 4.07° \qquad Ans.$$

F10–12. $T_{AB} = 600 \text{ N} \cdot \text{m}, T_{BC} = -300 \text{ N} \cdot \text{m},$
$T_{CD} = 200 \text{ N} \cdot \text{m}, T_{DE} = 500 \text{ N} \cdot \text{m}$

$$J = \dfrac{\pi}{2}(0.02^4) = 80(10^{-9})\pi \text{ m}^4$$

$$\phi_{E/A} = \dfrac{[600 + (-300) + 200 + 500]0.2}{[80(10^{-9})\pi][75(10^9)]}$$

$$= 0.01061 \text{ rad} = 0.608° \qquad Ans.$$

F10–13.

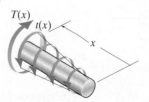

$$J = \dfrac{\pi}{2}(0.04^4) = 1.28(10^{-6})\pi \text{ m}^4$$

$$t = 5(10^3) \text{ N} \cdot \text{m/m}$$

Internal torque is $5(10^3)x \text{ N} \cdot \text{m}$

$$\phi_{A/B} = \int_0^L \dfrac{T(x)dx}{JG}$$

$$= \dfrac{1}{[1.28(10^{-6})\pi][75(10^9)]} \int_0^{0.8\text{ m}} 5(10^3)x\,dx$$

$$= 0.00531 \text{ rad} = 0.304° \qquad Ans.$$

F10–14.

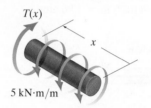

$$J = \dfrac{\pi}{2}(0.04^4) = 1.28(10^{-6})\pi \text{ m}^4$$

Distributed torque is $t = \dfrac{15(10^3)}{0.6}(x)$

$$= 25(10^3)x \text{ N} \cdot \text{m/m}$$

Internal torque in segment AB,

$$T(x) = \frac{1}{2}(25x)(10^3)(x) = 12.5(10^3)x^2 \text{ N} \cdot \text{m}$$

In segment BC,

$$T_{BC} = \frac{1}{2}[25(10^3)(0.6)](0.6) = 4500 \text{ N} \cdot \text{m}$$

$$\phi_{A/C} = \int_0^L \frac{T(x)dx}{JG} + \frac{T_{BC}L_{BC}}{JG}$$

$$= \frac{1}{[1.28(10^{-6})\pi][75(10^9)]}\left[\int_0^{0.6 \text{ m}} 12.5(10^3)x^2 \, dx + 4500(0.4)\right]$$

$$= 0.008952 \text{ rad} = 0.513° \qquad\qquad Ans.$$

Chapter 11

F11–1.

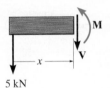

5 kN

$$\zeta + \Sigma M_B = 0; \qquad A_y(6) - 30 = 0 \qquad A_y = 5 \text{ kN}$$

$$+\uparrow \Sigma F_y = 0; \qquad -V - 5 = 0 \qquad V = -5 \text{ kN}$$
$$\qquad\qquad\qquad\qquad\qquad\qquad\qquad Ans.$$

$$\zeta + \Sigma M_0 = 0; \quad M + 5x = 0 \quad M = \{-5x\} \text{ kN} \cdot \text{m}$$
$$\qquad\qquad\qquad\qquad\qquad\qquad\qquad Ans.$$

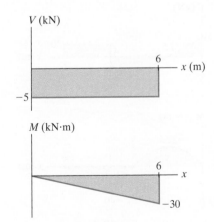

F11–2. 9 kN

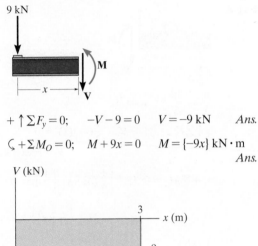

$$+\uparrow \Sigma F_y = 0; \qquad -V - 9 = 0 \qquad V = -9 \text{ kN} \qquad Ans.$$

$$\zeta + \Sigma M_O = 0; \quad M + 9x = 0 \qquad M = \{-9x\} \text{ kN} \cdot \text{m}$$
$$\qquad\qquad\qquad\qquad\qquad\qquad\qquad Ans.$$

F11–3.

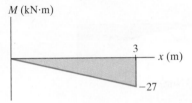

$$+\uparrow \Sigma F_y = 0; \quad -V - 2x = 0 \quad V = \{-2x\} \text{ kip} \qquad Ans.$$

$$\zeta + \Sigma M_O = 0; \quad M + 2x\left(\frac{x}{2}\right) - 18 = 0$$

$$M = \{18 - x^2\} \text{ kip} \cdot \text{ft} \quad Ans.$$

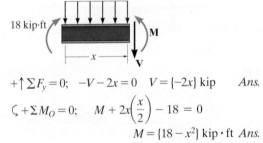

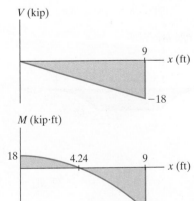

F11–4.

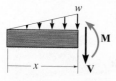

$$\frac{w}{x} = \frac{12}{3} \qquad w = 4x$$

$$+\uparrow\Sigma F_y = 0; \quad -V - \frac{1}{2}(4x)(x) = 0$$

$$V = \{-2x^2\} \text{ kN} \quad Ans.$$

$$\circlearrowleft +\Sigma M_O = 0; \; M + \left[\frac{1}{2}(4x)(x)\right]\left(\frac{x}{3}\right) = 0$$

$$M = \left\{-\frac{2}{3}x^3\right\} \text{ kN} \cdot \text{m} \quad Ans.$$

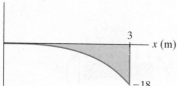

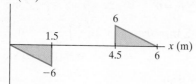

F11–5.

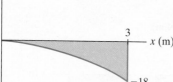

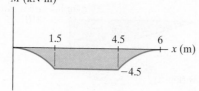

F11–6.

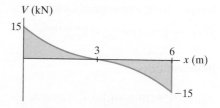

F11–7.

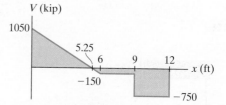

F11–8.

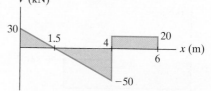

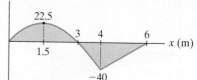

F11–9. Consider two vertical rectangles and a horizontal rectangle.

$$I = 2\left[\frac{1}{12}(0.02)(0.2^3)\right] + \frac{1}{12}(0.26)(0.02^3)$$

$$= 26.84(10^{-6}) \text{ m}^4$$

$$\sigma_{max} = \frac{Mc}{I} = \frac{20(10^3)(0.1)}{26.84(10^{-6})} = 74.5 \text{ MPa} \qquad Ans.$$

F11–10. See inside front cover.

$$\bar{y} = \frac{0.3}{3} = 0.1 \text{ m}$$

$$I = \frac{1}{36}(0.3)(0.3^3) = 0.225(10^{-3}) \text{ m}^4$$

$$(\sigma_{max})_c = \frac{Mc}{I} = \frac{50(10^3)(0.3 - 0.1)}{0.225(10^{-3})}$$

$$= 44.4 \text{ MPa (C)} \qquad Ans.$$

$$(\sigma_{max})_t = \frac{My}{I} = \frac{50(10^3)(0.1)}{0.225(10^{-3})} = 22.2 \text{ MPa (T)} \quad Ans.$$

F11–11. Consider large rectangle minus the two side rectangles.

$$I = \frac{1}{12}(0.2)(0.3^3) - (2)\frac{1}{12}(0.09)(0.26^3)$$

$$= 0.18636(10^{-3}) \text{ m}^4$$

$$\sigma_{max} = \frac{Mc}{I} = \frac{50(10^3)(0.15)}{0.18636(10^{-3})} = 40.2 \text{ MPa} \qquad Ans.$$

F11–12. Consider two vertical rectangles and two horizontal rectangles.

$$I = 2\left[\frac{1}{12}(0.03)(0.4^3)\right] + 2\left[\frac{1}{12}(0.14)(0.03^3) + 0.14(0.03)(0.15^2)\right]$$

$$= 0.50963(10^{-3}) \text{ m}^4$$

$$\sigma_{max} = \frac{Mc}{I} = \frac{10(10^3)(0.2)}{0.50963(10^{-3})} = 3.92 \text{ MPa} \qquad Ans.$$

$$\sigma_A = 3.92 \text{ MPa (C)}$$

$$\sigma_B = 3.92 \text{ MPa (T)}$$

F11–13. Consider center rectangle and two side rectangles.

$$I = \frac{1}{12}(0.05)(0.4)^3 + 2\left[\frac{1}{12}(0.025)(0.3)^3\right]$$

$$= 0.37917(10^{-3}) \text{ m}^4$$

$$\sigma_A = \frac{My_A}{I} = \frac{5(10^3)(-0.15)}{0.37917(10^{-3})} = 1.98 \text{ MPa (T)} \qquad Ans.$$

F11–14. $M_y = 50\left(\frac{4}{5}\right) = 40 \text{ kN} \cdot \text{m}$

$$M_z = 50\left(\frac{3}{5}\right) = 30 \text{ kN} \cdot \text{m}$$

$$I_y = \frac{1}{12}(0.3)(0.2^3) = 0.2(10^{-3}) \text{ m}^4$$

$$I_z = \frac{1}{12}(0.2)(0.3^3) = 0.45(10^{-3}) \text{ m}^4$$

$$\sigma = -\frac{M_z y}{I_z} + \frac{M_y z}{I_y}$$

$$\sigma_A = -\frac{[30(10^3)](-0.15)}{0.45(10^{-3})} + \frac{[40(10^3)](0.1)}{0.2(10^{-3})}$$

$$= 30 \text{ MPa (T)} \qquad Ans.$$

$$\sigma_B = -\frac{[30(10^3)](0.15)}{0.45(10^{-3})} + \frac{[40(10^3)](0.1)}{0.2(10^{-3})}$$

$$= 10 \text{ MPa (T)} \qquad Ans.$$

$$\tan \alpha = \frac{I_z}{I_y} \tan \theta$$

$$\tan \alpha = \left[\frac{0.45(10^{-3})}{0.2(10^{-3})}\right]\left(\frac{4}{3}\right)$$

$$\alpha = 71.6° \qquad Ans.$$

F11–15. Maximum stress occurs at D or A.

$$(\sigma_{max})_D = \frac{(50\cos 30°)12(3)}{\frac{1}{12}(4)(6)^3} + \frac{(50\sin 30°)12(2)}{\frac{1}{12}(6)(4)^3}$$

$$= 40.4 \text{ psi} \qquad Ans.$$

Chapter 12

F12–1. Consider two vertical rectangles and a horizontal rectangle.

$$I = 2\left[\frac{1}{12}(0.02)(0.2^3)\right] + \frac{1}{12}(0.26)(0.02^3)$$

$$= 26.84(10^{-6}) \text{ m}^4$$

Take two rectangles above A.

$$Q_A = 2[0.055(0.09)(0.02)] = 198(10^{-6}) \text{ m}^3$$

$$\tau_A = \frac{VQ_A}{It} = \frac{100(10^3)[198(10^{-6})]}{[26.84(10^{-6})]2(0.02)}$$

$$= 18.4 \text{ MPa} \qquad Ans.$$

F12–2. Consider a vertical rectangle and two squares.

$$I = \frac{1}{12}(0.1)(0.3^3) + (2)\frac{1}{12}(0.1)(0.1^3)$$

$$= 0.24167(10^{-3}) \text{ m}^4$$

Take top half of area (above A).

$$Q_A = y'_1 A'_1 + y'_2 A'_2$$

$$= \left[\frac{1}{2}(0.05)\right](0.05)(0.3) + 0.1(0.1)(0.1)$$

$$= 1.375(10^{-3}) \text{ m}^3$$

$$\tau_A = \frac{VQ}{It} = \frac{600(10^3)[1.375(10^{-3})]}{[0.24167(10^{-3})](0.3)} = 11.4 \text{ MPa } \textit{Ans.}$$

Take top square (above B).

$$Q_B = y'_2 A'_2 = 0.1(0.1)(0.1) = 1(10^{-3}) \text{ m}^3$$

$$\tau_B = \frac{VQ}{It} = \frac{600(10^3)[1(10^{-3})]}{[0.24167(10^{-3})](0.1)} = 24.8 \text{ MPa } \textit{Ans.}$$

F12–3. $V_{max} = 4.5$ kip

$$I = \frac{1}{12}(3)(6^3) = 54 \text{ in}^4$$

Take top half of area.

$$Q_{max} = y'A' = 1.5(3)(3) = 13.5 \text{ in}^3$$

$$(\tau_{max})_{abs} = \frac{V_{max} Q_{max}}{It} = \frac{4.5(10^3)(13.5)}{54(3)} = 375 \text{ psi}$$

$$\textit{Ans.}$$

F12–4. Consider two vertical rectangles and two horizontal rectangles.

$$I = 2\left[\frac{1}{12}(0.03)(0.4^3)\right] + 2\left[\frac{1}{12}(0.14)(0.03^3)\right.$$

$$\left. + 0.14(0.03)(0.15^2)\right] = 0.50963(10^{-3}) \text{ m}^4$$

Take the top half of area.

$$Q_{max} = 2y'_1 A'_1 + y'_2 A'_2 = 2(0.1)(0.2)(0.03)$$

$$+ (0.15)(0.14)(0.03) = 1.83(10^{-3}) \text{ m}^3$$

$$\tau_{max} = \frac{VQ_{max}}{It} = \frac{20(10^3)[1.83(10^{-3})]}{0.50963(10^{-3})[2(0.03)]} = 1.20 \text{ MPa}$$

$$\textit{Ans.}$$

F12–5. Consider one large vertical rectangle and two side rectangles.

$$I = \frac{1}{12}(0.05)(0.4)^3 + 2\left[\frac{1}{12}(0.025)(0.3)^3\right]$$

$$= 0.37917(10^{-3}) \text{ m}^4$$

Take the top half of area.

$$Q_{max} = 2y'_1 A'_1 + y'_2 A'_2 = 2(0.075)(0.025)(0.15)$$

$$+ (0.1)(0.05)(0.2) = 1.5625(10^{-3}) \text{ m}^3$$

$$\tau_{max} = \frac{VQ_{max}}{It} = \frac{20(10^3)[1.5625(10^{-3})]}{[0.37917(10^{-3})][2(0.025)]}$$

$$= 1.65 \text{ MPa} \qquad \textit{Ans.}$$

F12–6. $I = \frac{1}{12}(0.3)(0.2^3) = 0.2(10^{-3}) \text{ m}^4$

Top (or bottom) board

$$Q = y'A' = 0.05(0.1)(0.3) = 1.5(10^{-3}) \text{ m}^3$$

Two rows of nails

$$q_{allow} = 2\left(\frac{F}{s}\right) = \frac{2[15(10^3)]}{s} = \frac{30(10^3)}{s}$$

$$q_{allow} = \frac{VQ}{I}; \quad \frac{30(10^3)}{s} = \frac{50(10^3)[1.5(10^{-3})]}{0.2(10^{-3})}$$

$$s = 0.08 \text{ m} = 80 \text{ mm} \qquad \textit{Ans.}$$

F12–7. Consider large rectangle minus two side rectangles.

$$I = \frac{1}{12}(0.2)(0.34^3) - (2)\frac{1}{12}(0.095)(0.28^3)$$

$$= 0.3075(10^{-3}) \text{ m}^4$$

Top plate

$$Q = y'A' = 0.16(0.02)(0.2) = 0.64(10^{-3}) \text{ m}^3$$

Two rows of bolts

$$q_{allow} = 2\left(\frac{F}{s}\right) = \frac{2[30(10^3)]}{s} = \frac{60(10^3)}{s}$$

$$q_{allow} = \frac{VQ}{I}; \quad \frac{60(10^3)}{s} = \frac{300(10^3)[0.64(10^{-3})]}{0.3075(10^{-3})}$$

$$s = 0.09609 \text{ m} = 96.1 \text{ mm}$$

Use $s = 96$ mm $\qquad \textit{Ans.}$

F12–8. Consider two large rectangles and two side rectangles.

$$I = 2\left[\frac{1}{12}(0.025)(0.3^3)\right] + 2\left[\frac{1}{12}(0.05)(0.2^3) + 0.05(0.2)(0.15^2)\right]$$

$$= 0.62917(10^{-3}) \text{ m}^4$$

Top center board is held onto beam by the top row of bolts.

$$Q = y'A' = 0.15(0.2)(0.05) = 1.5(10^{-3}) \text{ m}^3$$

Each bolt has two shearing surfaces.

$$q_{allow} = 2\left(\frac{F}{s}\right) = \frac{2[8(10^3)]}{s} = \frac{16(10^3)}{s}$$

$$q_{allow} = \frac{VQ}{I}; \quad \frac{16(10^3)}{s} = \frac{20(10^3)[1.5(10^{-3})]}{0.62917(10^{-3})}$$

$$s = 0.3356 \text{ m} = 335.56 \text{ mm}$$

Use $s = 335$ mm $\qquad \textit{Ans.}$

F12–9. Consider center board and four side boards.

$$I = \frac{1}{12}(1)(6^3) + 4\left[\frac{1}{12}(0.5)(4^3) + 0.5(4)(3^2)\right]$$

$$= 100.67 \text{ in}^4$$

Top-right board is held onto beam by a row of bolts.

$$Q = y'A' = 3(4)(0.5) = 6 \text{ in}^3$$

Bolts have one shear surface.

$$q_{allow} = \frac{F}{s} = \frac{6}{s}$$

$$q_{allow} = \frac{VQ}{I}; \qquad \frac{6}{s} = \frac{15(6)}{100.67}$$

$$s = 6.711 \text{ in.}$$

Use $s = 6\frac{5}{8}$ in. *Ans.*

Also, can consider the top *two* boards held onto beam by a row of bolts with two shearing surfaces.

Chapter 13

F13–1. $+\uparrow \Sigma F_z = (F_R)_z;$ $-500 - 300 = P$

$P = -800 \text{ kN}$

$\Sigma M_x = 0;$ $300(0.05) - 500(0.1) = M_x$

$M_x = -35 \text{ kN} \cdot \text{m}$

$\Sigma M_y = 0;$ $300(0.1) - 500(0.1) = M_y$

$M_y = -20 \text{ kN} \cdot \text{m}$

$A = 0.3(0.3) = 0.09 \text{ m}^2$

$I_x = I_y = \frac{1}{12}(0.3)(0.3^3) = 0.675(10^{-3}) \text{ m}^4$

$$\sigma_A = \frac{-800(10^3)}{0.09} + \frac{[20(10^3)](0.15)}{0.675(10^{-3})} + \frac{[35(10^3)](0.15)}{0.675(10^{-3})}$$

$$= 3.3333 \text{ MPa} = 3.33 \text{ MPa (T)} \qquad Ans.$$

$$\sigma_B = \frac{-800(10^3)}{0.09} + \frac{[20(10^3)](0.15)}{0.675(10^{-3})} - \frac{[35(10^3)](0.15)}{0.675(10^{-3})}$$

$$= -12.22 \text{ MPa} = 12.2 \text{ MPa (C)} \qquad Ans.$$

F13–2. $+\uparrow \Sigma F_y = 0;$ $V - 400 = 0$ $V = 400 \text{ kN}$

$\zeta + \Sigma M_A = 0; -M - 400(0.5) = 0$ $M = -200 \text{ kN} \cdot \text{m}$

$I = \frac{1}{12}(0.1)(0.3^3) = 0.225(10^{-3}) \text{ m}^4$

Bottom segment:

$$\sigma_A = \frac{My}{I} = \frac{[200(10^3)](-0.05)}{0.225(10^{-3})}$$

$$= -44.44 \text{ MPa} = 44.4 \text{ MPa (C)} \qquad Ans.$$

$$Q_A = y'A' = 0.1(0.1)(0.1) = 1(10^{-3}) \text{ m}^3$$

$$\tau_A = \frac{VQ}{It} = \frac{400(10^3)[1(10^{-3})]}{0.225(10^{-3})(0.1)} = 17.8 \text{ MPa} \qquad Ans.$$

17.8 MPa — 44.4 MPa

F13–3. Left reaction is 20 kN.

Left segment:

$+\uparrow \Sigma F_y = 0;$ $20 - V = 0$ $V = 20 \text{ kN}$

$\zeta + \Sigma M_s = 0;$ $M - 20(0.5) = 0$ $M = 10 \text{ kN} \cdot \text{m}$

Consider large rectangle minus two side rectangles.

$$I = \frac{1}{12}(0.1)(0.2^3) - (2)\frac{1}{12}(0.045)(0.18^3)$$

$$= 22.9267(10^{-6}) \text{ m}^4$$

Top segment above A:

$Q_A = y'_1A'_1 + y'_2A'_2 = 0.07(0.04)(0.01)$

$+ 0.095(0.1)(0.01) = 0.123(10^{-3}) \text{ m}^3$

$$\sigma_A = -\frac{My_A}{I} = -\frac{[10(10^3)](0.05)}{22.9267(10^{-6})}$$

$$= -21.81 \text{ MPa} = 21.8 \text{ MPa (C)} \qquad Ans.$$

$$\tau_A = \frac{VQ_A}{It} = \frac{20(10^3)[0.123(10^{-3})]}{[22.9267(10^{-6})](0.01)}$$

$$= 10.7 \text{ MPa} \qquad Ans.$$

10.7 MPa — 21.8 MPa

F13–4. At the section through centroidal axis:

$N = P$

$V = 0$

$M = (2 + 1)P = 3P$

$\sigma = \frac{N}{A} + \frac{Mc}{I}$

$$30 = \frac{P}{2(0.5)} + \frac{(3P)(1)}{\frac{1}{12}(0.5)(2)^3}$$

$$P = 3 \text{ kip} \qquad Ans.$$

F13–5. At section through B:

$N = 500$ lb, $V = 400$ lb

$M = 400(10) = 4000$ lb $\cdot$ in.

Axial load:

$$\sigma_x = \frac{N}{A} = \frac{500}{4(3)} = 41.667 \text{ psi (T)}$$

Shear load:

$$\tau_{xy} = \frac{VQ}{It} = \frac{400[(1.5)(3)(1)]}{[\frac{1}{12}(3)(4)^3]3} = 37.5 \text{ psi}$$

Bending moment:

$$\sigma_x = \frac{My}{I} = \frac{4000(1)}{\frac{1}{12}(3)(4)^3} = 250 \text{ psi (C)}$$

Thus

$\sigma_x = 41.667 - 250 = 208 \text{ psi (C)}$ *Ans.*

$\sigma_y = 0$ *Ans.*

$\tau_{xy} = 37.5 \text{ psi}$ *Ans.*

37.5 psi

208 psi

F13–6. Top segment:

$\Sigma F_y = 0; \quad V_y + 1000 = 0 \qquad V_y = -1000 \text{ N}$

$\Sigma F_x = 0; \quad V_x - 1500 = 0 \qquad V_x = 1500 \text{ N}$

$\Sigma M_z = 0; \quad T_z - 1500(0.4) = 0 \quad T_z = 600 \text{ N} \cdot \text{m}$

$\Sigma M_y = 0; \quad M_y - 1500(0.2) = 0 \quad M_y = 300 \text{ N} \cdot \text{m}$

$\Sigma M_x = 0; \quad M_x - 1000(0.2) = 0 \quad M_x = 200 \text{ N} \cdot \text{m}$

$$I_y = I_x = \frac{\pi}{4}(0.02^4) = 40(10^{-9})\pi \text{ m}^4$$

$$J = \frac{\pi}{2}(0.02^4) = 80(10^{-9})\pi \text{ m}^4$$

$$(Q_y)_A = \frac{4(0.02)}{3\pi}\left[\frac{\pi}{2}(0.02^2)\right] = 5.3333(10^{-6}) \text{ m}^3$$

$$\sigma_A = \frac{M_x y}{I_x} - \frac{M_y x}{I_y} = \frac{-200(0)}{40(10^{-9})\pi} - \frac{-300(0.02)}{40(10^{-9})\pi}$$

$= 47.7 \text{ MPa (T)}$ *Ans.*

$$[(\tau_{zy})_T]_A = \frac{T_z c}{J} = \frac{600(0.02)}{80(10^{-9})\pi} = 47.746 \text{ MPa}$$

$$[(\tau_{zy})_V]_A = \frac{V_y(Q_y)_A}{I_x t} = \frac{1000[5.3333(10^{-6})]}{[40(10^{-9})\pi](0.04)}$$

$= 1.061 \text{ MPa}$

Combining these two shear stress components,

$(\tau_{zy})_A = 47.746 + 1.061 = 48.8 \text{ MPa}$ *Ans.*

47.7 MPa

48.8 MPa

F13–7. Right Segment:

$\Sigma F_z = 0; \qquad V_z - 6 = 0 \qquad V_z = 6 \text{ kN}$

$\Sigma M_y = 0; \qquad T_y - 6(0.3) = 0 \qquad T_y = 1.8 \text{ kN} \cdot \text{m}$

$\Sigma M_x = 0; \qquad M_x - 6(0.3) = 0 \qquad M_x = 1.8 \text{ kN} \cdot \text{m}$

$$I_x = \frac{\pi}{4}(0.05^4 - 0.04^4) = 0.9225(10^{-6})\pi \text{ m}^4$$

$$J = \frac{\pi}{2}(0.05^4 - 0.04^4) = 1.845(10^{-6})\pi \text{ m}^4$$

$(Q_z)_A = y_2' A_2' - y_1' A_1'$

$$= \frac{4(0.05)}{3\pi}\left[\frac{\pi}{2}(0.05^2)\right] - \frac{4(0.04)}{3\pi}\left[\frac{\pi}{2}(0.04^2)\right]$$

$= 40.6667(10^{-6}) \text{ m}^3$

$$\sigma_A = \frac{M_x z}{I_x} = \frac{1.8(10^3)(0)}{0.9225(10^{-6})\pi} = 0$$ *Ans.*

$$[(\tau_{yz})_T]_A = \frac{T_y c}{J} = \frac{[1.8(10^3)](0.05)}{1.845(10^{-6})\pi} = 15.53 \text{ MPa}$$

$$[(\tau_{yz})_V]_A = \frac{V_z(Q_z)_A}{I_x t} = \frac{6(10^3)[40.6667(10^{-6})]}{[0.9225(10^{-6})\pi](0.02)}$$

$= 4.210 \text{ MPa}$

Combining these two shear stress components,

$(\tau_{yz})_A = 15.53 - 4.210 = 11.3 \text{ MPa}$ *Ans.*

11.3 MPa

F13–8. Left Segment:

$\Sigma F_z = 0; \quad V_z - 900 - 300 = 0 \qquad V_z = 1200 \text{ N}$

$\Sigma M_y = 0; \quad T_y + 300(0.1) - 900(0.1) = 0 \quad T_y = 60 \text{ N} \cdot \text{m}$

$\Sigma M_x = 0; \quad M_x + (900 + 300)0.3 = 0 \qquad M_x = -360 \text{ N} \cdot \text{m}$

$$I_x = \frac{\pi}{4}(0.025^4 - 0.02^4) = 57.65625(10^{-9})\pi \text{ m}^4$$

$$J = \frac{\pi}{2}(0.025^4 - 0.02^4) = 0.1153125(10^{-6})\pi \text{ m}^4$$

$(Q_y)_A = 0$

$$\sigma_A = \frac{M_x y}{I_x} = \frac{(360)(0.025)}{57.65625(10^{-9})\pi} = 49.7 \text{ MPa} \quad Ans.$$

$$[(\tau_{xy})_T]_A = \frac{T_y \rho_A}{J} = \frac{60(0.025)}{0.1153125(10^{-6})\pi} = 4.14 \text{ MPa } Ans.$$

$$[(\tau_{yz})_V]_A = \frac{V_z(Q_z)_A}{I_x t} = 0 \qquad\qquad Ans.$$

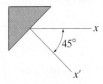

4.14 MPa

49.7 MPa

Chapter 14

F14–1. $\theta = 120°$ $\sigma_x = 500 \text{ kPa}$ $\sigma_y = 0$ $\tau_{xy} = 0$

Apply Eqs. 14–1, 14–2.

$\sigma_{x'} = 125 \text{ kPa}$ *Ans.*

$\tau_{x'y'} = 217 \text{ kPa}$ *Ans.*

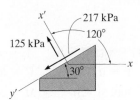

217 kPa

120°

125 kPa

30°

x

x'

y'

F14–2.

x

45°

x'

$\theta = -45°$ $\sigma_x = 0$ $\sigma_y = -400 \text{ kPa}$

$\tau_{xy} = -300 \text{ kPa}$

Apply Eqs.14–1, 14–3, 14–2.

$\sigma_{x'} = 100 \text{ kPa}$ *Ans.*

$\sigma_{y'} = -500 \text{ kPa}$ *Ans.*

$\tau_{x'y'} = 200 \text{ kPa}$ *Ans.*

F14–3. $\theta_x = 80 \text{ kPa}$ $\sigma_y = 0$ $\tau_{xy} = 30 \text{ kPa}$

Apply Eqs. 14–5, 14–4.

$\sigma_1 = 90 \text{ kPa}$ $\sigma_2 = -10 \text{ kPa}$ *Ans.*

$\theta_p = 18.43°$ and $108.43°$

From Eq. 14–1,

$$\sigma_{x'} = \frac{80 + 0}{2} + \frac{80 - 0}{2}\cos 2 \,(18.43°)$$

$$+ 30 \sin 2(18.43°)$$

$$= 90 \text{ kPa} = \sigma_1$$

Thus,

$(\theta_p)_1 = 18.4°$ for σ_1 *Ans.*

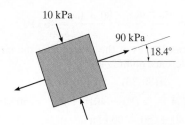

10 kPa

90 kPa

18.4°

F14–4. $\sigma_x = 100 \text{ kPa}$ $\sigma_y = 700 \text{ kPa}$

$\tau_{xy} = -400 \text{ kPa}$

Apply Eqs. 14–7, 14–8.

$\underset{\text{in-plane}}{\tau_{\max}} = 500 \text{ kPa}$ *Ans.*

$\sigma_{\text{avg}} = 400 \text{ kPa}$ *Ans.*

F14–5. At the cross section through B:

$N = 4 \text{ kN}$ $V = 2 \text{ kN}$

$M = 2(2) = 4 \text{ kN} \cdot \text{m}$

$$\sigma_B = \frac{P}{A} + \frac{Mc}{I} = \frac{4(10^3)}{0.03(0.06)} + \frac{4(10^3)(0.03)}{\frac{1}{12}(0.03)(0.06)^3}$$

$$= 224 \text{ MPa (T)}$$

Note $\tau_B = 0$ since $Q = 0$.

Thus

$\sigma_1 = 224 \text{ MPa}$ *Ans.*

$\sigma_2 = 0$

F14–6. $A_y = B_y = 12 \text{ kN}$

Segment AC:

$V_C = 0$ $M_C = 24 \text{ kN} \cdot \text{m}$

$\tau_C = 0$ (since $V_C = 0$)

$\sigma_C = 0$ (since C is on neutral axis)

$\sigma_1 = \sigma_2 = 0$ *Ans.*

F14–7. $\sigma_{\text{avg}} = \dfrac{\sigma_x + \sigma_y}{2} = \dfrac{500 + 0}{2} = 250 \text{ kPa}$

The coordinates of the center C of the circle and the reference point A are

$$A(500, 0) \qquad C(250, 0)$$

$R = CA = 500 - 250 = 250 \text{ kPa}$

$\theta = 120°$ (counterclockwise). Rotate the radial line CA counterclockwise $2\theta = 240°$ to the coordinates of point $P(\sigma_{x'}, \tau_{x'y'})$.

$\alpha = 240° - 180° = 60°$

$\sigma_{x'} = 250 - 250 \cos 60° = 125 \text{ kPa}$ *Ans.*

$\tau_{x'y'} = 250 \sin 60° = 217 \text{ kPa}$ *Ans.*

F14–8. $\sigma_{\text{avg}} = \dfrac{\sigma_x + \sigma_y}{2} = \dfrac{80 + 0}{2} = 40 \text{ kPa}$

The coordinates of the center C of the circle and the reference point A are

$$A(80, 30) \qquad C(40, 0)$$

$R = CA = \sqrt{(80 - 40)^2 + 30^2} = 50 \text{ kPa}$

$\sigma_1 = 40 + 50 = 90 \text{ kPa}$ *Ans.*

$\sigma_2 = 40 - 50 = -10 \text{ kPa}$ *Ans.*

$\tan 2(\theta_p)_1 = \dfrac{30}{80 - 40} = 0.75$

$(\theta_p)_1 = 18.4° \text{ (counterclockwise)}$ *Ans.*

F14–9. The coordinates of the reference point A and the center C of the circle are

$$A(30, 40) \qquad C(0, 0)$$

$R = CA = 50 \text{ MPa}$

$\sigma_1 = 50 \text{ MPa}$

$\sigma_2 = -50 \text{ MPa}$

F14–10. $J = \dfrac{\pi}{2}(0.04^4 - 0.03^4) = 0.875(10^{-6})\pi \text{ m}^4$

$\tau = \dfrac{Tc}{J} = \dfrac{4(10^3)(0.04)}{0.875(10^{-6})\pi} = 58.21 \text{ MPa}$

$\sigma_x = \sigma_y = 0$ and $\tau_{xy} = -58.21 \text{ MPa}$

$\sigma_{\text{avg}} = \dfrac{\sigma_x + \sigma_y}{2} = 0$

The coordinates of the reference point A and the center C of the circle are

$$A(0, -58.21) \qquad C(0, 0)$$

$R = CA = 58.21 \text{ MPa}$

$\sigma_1 = 0 + 58.21 = 58.2 \text{ MPa}$ *Ans.*

$\sigma_2 = 0 - 58.21 = -58.2 \text{ MPa}$ *Ans.*

F14–11.

$+\uparrow\Sigma F_y = 0; \qquad V - 30 = 0 \qquad V = 30 \text{ kN}$

$\zeta+\Sigma M_O = 0; \qquad -M - 30(0.3) = 0 \qquad M = -9 \text{ kN} \cdot \text{m}$

$I = \dfrac{1}{12}(0.05)(0.15^3) = 14.0625(10^{-6}) \text{ m}^4$

Segment above A,

$Q_A = y'A' = 0.05(0.05)(0.05) = 0.125(10^{-3}) \text{ m}^3$

$\sigma_A = -\dfrac{My_A}{I} = \dfrac{[-9(10^3)](0.025)}{14.0625(10^{-6})} = 16 \text{ MPa (T)}$

$\tau_A = \dfrac{VQ_A}{It} = \dfrac{30(10^3)[0.125(10^{-3})]}{14.0625(10^{-6})(0.05)} = 5.333 \text{ MPa}$

$\sigma_x = 16 \text{ MPa}, \sigma_y = 0,$ and $\tau_{xy} = -5.333 \text{ MPa}$

$\sigma_{\text{avg}} = \dfrac{\sigma_x + \sigma_y}{2} = \dfrac{16 + 0}{2} = 8 \text{ MPa}$

The coordinates of the reference point A and the center C of the circle are

$$A(16, -5.333) \qquad C(8, 0)$$

$R = CA = \sqrt{(16 - 8)^2 + (-5.333)^2} = 9.615 \text{ MPa}$

$\sigma_1 = 8 + 9.615 = 17.6 \text{ MPa}$ *Ans.*

$\sigma_2 = 8 - 9.615 = -1.61 \text{ MPa}$ *Ans.*

F14–12.

$\zeta+\Sigma M_B = 0; \qquad 60(1) - A_y(1.5) = 0 \qquad A_y = 40 \text{ kN}$

$+\uparrow\Sigma F_y = 0; \qquad 40 - V = 0 \qquad V = 40 \text{ kN}$

$\zeta+\Sigma M_O = 0; \qquad M - 40(0.5) = 0 \qquad M = 20 \text{ kN} \cdot \text{m}$

Consider large rectangle minus two side rectangles.

$I = \dfrac{1}{12}(0.1)(0.2^3) - (2)\dfrac{1}{12}(0.045)(0.18^3) = 22.9267(10^{-6}) \text{ m}^4$

Top rectangle,

$Q_A = y'A' = 0.095(0.01)(0.1) = 95(10^{-6}) \text{ m}^3$

$\sigma_A = -\dfrac{My_A}{I} = -\dfrac{[20(10^3)](0.09)}{22.9267(10^{-6})} = -78.51 \text{ MPa}$

$\quad = 78.51 \text{ MPa (C)}$

$\tau_A = \dfrac{VQ_A}{It} = \dfrac{40(10^3)[95(10^{-6})]}{[22.9267(10^{-6})](0.01)} = 16.57 \text{ MPa}$

$\sigma_x = -78.51 \text{ MPa}, \sigma_y = 0,$ and $\tau_{xy} = -16.57 \text{ MPa}$

$\sigma_{\text{avg}} = \dfrac{\sigma_x + \sigma_y}{2} = \dfrac{-78.51 + 0}{2} = -39.26 \text{ MPa}$

The coordinates of the reference point A and the center C of the circle are

$$A(-78.51, -16.57) \quad C(-39.26, 0)$$

$R = CA = \sqrt{[-78.51 - (-39.26)]^2 + (-16.57)^2}$

$\quad = 42.61 \text{ MPa}$

$\tau_{\substack{\text{max} \\ \text{in-plane}}} = |R| = 42.6 \text{ MPa}$ *Ans.*

Chapter 15

F15–1.

At support,

$V_{max} = 12 \text{ kN}$ $M_{max} = 18 \text{ kN} \cdot \text{m}$

$I = \dfrac{1}{12}(a)(2a)^3 = \dfrac{2}{3}a^4$

$\sigma_{allow} = \dfrac{M_{max}c}{I};$ $10(10^6) = \dfrac{18(10^3)(a)}{\dfrac{2}{3}a^4}$

$a = 0.1392 \text{ m} = 139.2 \text{ mm}$

Use $a = 140 \text{ mm}$ *Ans.*

$I = \dfrac{2}{3}(0.14^4) = 0.2561(10^{-3}) \text{ m}^4$

$Q_{max} = \dfrac{0.14}{2}(0.14)(0.14) = 1.372(10^{-3}) \text{ m}^3$

$\tau_{max} = \dfrac{V_{max}Q_{max}}{It} = \dfrac{12(10^3)[1.372(10^{-3})]}{[0.2561(10^{-3})](0.14)}$

$= 0.459 \text{ MPa} < \tau_{allow} = 1 \text{ MPa (OK)}$

F15–2.

At support,

$V_{max} = 3 \text{ kip}$ $M_{max} = 12 \text{ kip} \cdot \text{ft}$

$I = \dfrac{\pi}{4}\left(\dfrac{d}{2}\right)^4 = \dfrac{\pi d^4}{64}$

$\sigma_{allow} = \dfrac{M_{max}c}{I};$ $20 = \dfrac{12(12)\left(\dfrac{d}{2}\right)}{\dfrac{\pi d^4}{64}}$

$d = 4.19 \text{ in.}$

Use $d = 4\dfrac{1}{4} \text{ in.}$ *Ans.*

$I = \dfrac{\pi}{64}(4.25^4) = 16.015 \text{ in}^4$

Semicircle,

$Q_{max} = \dfrac{4(4.25/2)}{3\pi}\left[\dfrac{1}{2}\left(\dfrac{\pi}{4}\right)(4.25^2)\right] = 6.397 \text{ in}^3$

$\tau_{max} = \dfrac{V_{max}Q_{max}}{It} = \dfrac{3(6.397)}{16.015(4.25)}$

$= 0.282 \text{ ksi} < \tau_{allow} = 10 \text{ ksi (OK)}$

F15–3.

At the supports,

$V_{max} = 10 \text{ kN}$

Under 15-kN load,

$M_{max} = 5 \text{ kN} \cdot \text{m}$

$I = \dfrac{1}{12}(a)(2a)^3 = \dfrac{2}{3}a^4$

$\sigma_{allow} = \dfrac{M_{max}c}{I};$ $12(10^6) = \dfrac{5(10^3)(a)}{\dfrac{2}{3}a^4}$

$a = 0.0855 \text{ m} = 85.5 \text{ mm}$

Use $a = 86 \text{ mm}$ *Ans.*

$I = \dfrac{2}{3}(0.086^4) = 36.4672(10^{-6}) \text{ m}^4$

Top half of rectangle,

$Q_{max} = \dfrac{0.086}{2}(0.086)(0.086)$

$= 0.318028(10^{-3}) \text{ m}^3$

$\tau_{max} = \dfrac{V_{max}Q_{max}}{It} = \dfrac{10(10^3)[0.318028(10^{-3})]}{[36.4672(10^{-6})](0.086)}$

$= 1.01 \text{ MPa} < \tau_{allow} = 1.5 \text{ MPa (OK)}$

F15–4.

At the supports,

$V_{max} = 4.5 \text{ kip}$

At the center,

$M_{max} = 6.75 \text{ kip} \cdot \text{ft}$

$I = \dfrac{1}{12}(4)(h^3) = \dfrac{h^3}{3}$

$\sigma_{allow} = \dfrac{M_{max}c}{I};$ $2 = \dfrac{6.75(12)\left(\dfrac{h}{2}\right)}{\dfrac{h^3}{3}}$

$h = 7.794 \text{ in.}$

Top half of rectangle,

$Q_{max} = y'A' = \dfrac{h}{4}\left(\dfrac{h}{2}\right)(4) = \dfrac{h^2}{2}$

$\tau_{max} = \dfrac{V_{max}Q_{max}}{It}; 0.2 = \dfrac{4.5\left(\dfrac{h^2}{2}\right)}{\dfrac{h^3}{3}(4)}$

$h = 8.4375 \text{ in. (controls)}$

Use $h = 8\dfrac{1}{2} \text{ in.}$ *Ans.*

F15–5.

At the supports,

$V_{max} = 25$ kN

At the center,

$M_{max} = 20$ kN $\cdot$ m

$I = \dfrac{1}{12}(b)(3b)^3 = 2.25b^4$

$\sigma_{allow} = \dfrac{M_{max}c}{I}; \quad 12(10^6) = \dfrac{20(10^3)(1.5b)}{2.25b^4}$

$b = 0.1036$ m $= 103.6$ mm

Use $b = 104$ mm   *Ans.*

$I = 2.25(0.104^4) = 0.2632(10^{-3})$ m^4

Top half of rectangle,

$Q_{max} = 0.75(0.104)[1.5(0.104)(0.104)] = 1.2655(10^{-3})$ m^3

$\tau_{max} = \dfrac{V_{max} Q_{max}}{It} = \dfrac{25(10^3)[1.2655(10^{-3})]}{[0.2632(10^{-3})](0.104)}$

$\quad = 1.156$ MPa $< \tau_{allow} = 1.5$ MPa (OK).

F15–6.

Within the overhang,

$V_{max} = 150$ kN

At B,

$M_{max} = 150$ kN $\cdot$ m

$S_{reqd} = \dfrac{M_{max}}{\sigma_{allow}} = \dfrac{150(10^3)}{150(10^6)} = 0.001$ m$^3 = 1000(10^3)$ mm^3

Select W410 × 67 [$S_x = 1200(10^3)$ mm^3, $d = 410$ mm, and $t_w = 8.76$ mm].   *Ans.*

$\tau_{max} = \dfrac{V}{t_w d} = \dfrac{150(10^3)}{0.00876(0.41)}$

$\quad = 41.76$ MPa $< \tau_{allow} = 75$ MPa (OK)

Chapter 16

F16–1.

Use left segment,

$M(x) = 30$ kN $\cdot$ m

$EI\dfrac{d^2v}{dx^2} = 30$

$EI\dfrac{dv}{dx} = 30x + C_1$

$EIv = 15x^2 + C_1 x + C_2$

At $x = 3$ m, $\dfrac{dv}{dx} = 0$.

$C_1 = -90$ kN $\cdot$ m^2

At $x = 3$ m, $v = 0$.

$C_2 = 135$ kN $\cdot$ m^3

$\dfrac{dv}{dx} = \dfrac{1}{EI}(30x - 90)$

$v = \dfrac{1}{EI}(15x^2 - 90x + 135)$

For end A, $x = 0$

$\theta_A = \dfrac{dv}{dx}\Big|_{x=0} = -\dfrac{90(10^3)}{200(10^9)[65.0(10^{-6})]} = -0.00692$ rad

Ans.

$v_A = v|_{x=0} = \dfrac{135(10^3)}{200(10^9)[65.0(10^{-6})]} = 0.01038$ m $= 10.4$ mm

Ans.

F16–2.

Use left segment,

$M(x) = (-10x - 10)$ kN $\cdot$ m

$EI\dfrac{d^2v}{dx^2} = -10x - 10$

$EI\dfrac{dv}{dx} = -5x^2 - 10x + C_1$

$EIv = -\dfrac{5}{3}x^3 - 5x^2 + C_1 x + C_2$

At $x = 3$ m, $\dfrac{dv}{dx} = 0$.

$EI(0) = -5(3^2) - 10(3) + C_1 \qquad C_1 = 75$ kN $\cdot$ m^2

At $x = 3$ m, $v = 0$.

$EI(0) = -\dfrac{5}{3}(3^3) - 5(3^2) + 75(3) + C_2 \quad C_2 = -135$ kN $\cdot$ m^3

$\dfrac{dv}{dx} = \dfrac{1}{EI}(-5x^2 - 10x + 75)$

$v = \dfrac{1}{EI}\left(-\dfrac{5}{3}x^3 - 5x^2 + 75x - 135\right)$

For end A, $x = 0$

$\theta_A = \dfrac{dv}{dx}\Big|_{x=0} = \dfrac{1}{EI}[-5(0) - 10(0) + 75]$

$\quad = \dfrac{75(10^3)}{200(10^9)[65.0(10^{-6})]} = 0.00577$ rad   *Ans.*

$v_A = v|_{x=0} = \dfrac{1}{EI}\left[-\dfrac{5}{3}(0^3) - 5(0^2) + 75(0) - 135\right]$

$\quad = -\dfrac{135(10^3)}{200(10^9)[65.0(10^{-6})]} = -0.01038$ m $= -10.4$ mm *Ans.*

F16–3.

Use left segment,

$M(x) = \left(-\dfrac{3}{2}x^2 - 10x\right)$ kN $\cdot$ m

$$EI\frac{d^2v}{dx^2} = -\frac{3}{2}x^2 - 10x$$

$$EI\frac{dv}{dx} = -\frac{1}{2}x^3 - 5x^2 + C_1$$

At $x = 3$ m, $\dfrac{dv}{dx} = 0$.

$$EI(0) = -\frac{1}{2}(3^3) - 5(3^2) + C_1 \qquad C_1 = 58.5 \text{ kN} \cdot \text{m}^2$$

$$\frac{dv}{dx} = \frac{1}{EI}\left(-\frac{1}{2}x^3 - 5x^2 + 58.5\right)$$

For end $A, x = 0$

$$\theta_A = \frac{dv}{dx}\Big|_{x=0} = \frac{58.5(10^3)}{200(10^9)[65.0(10^{-6})]} = 0.0045 \text{ rad} \qquad Ans.$$

F16–4.

$A_y = 600$ lb

Use left segment,

$$M(x) = (600x - 50x^2) \text{ lb} \cdot \text{ft}$$

$$EI\frac{d^2v}{dx^2} = 600x - 50x^2$$

$$EI\frac{dv}{dx} = 300x^2 - 16.667x^3 + C_1$$

$$EIv = 100x^3 - 4.1667x^4 + C_1x + C_2$$

At $x = 0, v = 0$.

$$EI(0) = 100(0^3) - 4.1667(0^4) + C_1(0) + C_2 \qquad C_2 = 0$$

At $x = 12$ ft, $v = 0$.

$$EI(0) = 100(12^3) - 4.1667(12^4) + C_1(12)$$

$$C_1 = -7200 \text{ lb} \cdot \text{ft}^2$$

$$\frac{dv}{dx} = \frac{1}{EI}(300x^2 - 16.667x^3 - 7200)$$

$$v = \frac{1}{EI}(100x^3 - 4.1667x^4 - 7200x)$$

v_{max} occurs where $\dfrac{dv}{dx} = 0$.

$$300x^2 - 16.667x^3 - 7200 = 0$$

$$x = 6 \text{ ft} \qquad\qquad\qquad Ans.$$

$$v = \frac{1}{EI}[100(6^3) - 4.1667(6^4) - 7200(6)]$$

$$= \frac{-27\,000(12 \text{ in.}/\text{ft})^3}{1.5(10^6)\left[\frac{1}{12}(3)(6^3)\right]}$$

$$= -0.576 \text{ in.} \qquad\qquad Ans.$$

F16–5.

$A_y = -5$ kN

Use left segment,

$$M(x) = (40 - 5x) \text{ kN} \cdot \text{m}$$

$$EI\frac{d^2v}{dx^2} = 40 - 5x$$

$$EI\frac{dv}{dx} = 40x - 2.5x^2 + C_1$$

$$EIv = 20x^2 - 0.8333x^3 + C_1x + C_2$$

At $x = 0, v = 0$.

$$EI(0) = 20(0^2) - 0.8333(0^3) + C_1(0) + C_2 \qquad C_2 = 0$$

At $x = 6$ m, $v = 0$.

$$EI(0) = 20(6^2) - 0.8333(6^3) + C_1(6) + 0$$

$$C_1 = -90 \text{ kN} \cdot \text{m}^2$$

$$\frac{dv}{dx} = \frac{1}{EI}(40x - 2.5x^2 - 90)$$

$$v = \frac{1}{EI}(20x^2 - 0.8333x^3 - 90x)$$

v_{max} occurs where $\dfrac{dv}{dx} = 0$.

$$40x - 2.5x^2 - 90 = 0$$

$$x = 2.7085 \text{ m}$$

$$v = \frac{1}{EI}[20(2.7085^2) - 0.83333(2.7085^3) - 90(2.7085)]$$

$$= -\frac{113.60(10^3)}{200(10^9)[39.9(10^{-6})]} = -0.01424 \text{ m} = -14.2 \text{ mm } Ans.$$

F16–6.

$A_y = 10$ kN

Use left segment,

$$M(x) = (10x + 10) \text{ kN} \cdot \text{m}$$

$$EI\frac{d^2v}{dx^2} = 10x + 10$$

$$EI\frac{dv}{dx} = 5x^2 + 10x + C_1$$

Due to symmetry, $\dfrac{dv}{dx} = 0$ at $x = 3$ m.

$$EI(0) = 5(3^2) + 10(3) + C_1 \quad C_1 = -75 \text{ kN} \cdot \text{m}^2$$

$$\frac{dv}{dx} = \frac{1}{EI}[5x^2 + 10x - 75]$$

At $x = 0$,

$$\frac{dv}{dx} = \frac{-75(10^3)}{200(10^9)(39.9(10^{-6}))} = -9.40(10^{-3}) \text{ rad} \qquad Ans.$$

F16–7.

Remove B_y,

$$(v_B)_1 = \frac{Px^2}{6EI}(3L - x) = \frac{40(4^2)}{6EI}[3(6) - 4] = \frac{1493.33}{EI} \downarrow$$

Apply B_y,

$$(v_B)_2 = \frac{PL^3}{3EI} = \frac{B_y(4^3)}{3EI} = \frac{21.33B_y}{EI} \uparrow$$

$$(+\uparrow) \ v_B = 0 = (v_B)_1 + (v_B)_2$$

$$0 = -\frac{1493.33}{EI} + \frac{21.33B_y}{EI}$$

$B_y = 70 \text{ kN}$ *Ans.*

For the beam,

$\xrightarrow{+} \Sigma F_x = 0; \qquad A_x = 0$ *Ans.*

$+\uparrow \Sigma F_y = 0; \qquad 70 - 40 - A_y = 0 \qquad A_y = 30 \text{ kN}$ *Ans.*

$\zeta + \Sigma M_A = 0; \qquad 70(4) - 40(6) - M_A = 0$

$M_A = 40 \text{ kN} \cdot \text{m}$ *Ans.*

F16–8.

Remove B_y,

To use the deflection tables, consider loading as a superposition of uniform distributed load minus a triangular load.

$$(v_B)_1 = \frac{w_0 L^4}{8EI} \downarrow \qquad (v_B)_2 = \frac{w_0 L^4}{30EI} \uparrow$$

Apply B_y,

$$(+\uparrow) \ (v_B)_3 = \frac{B_y L^3}{3EI} \uparrow \quad v_B = 0 = (v_B)_1 + (v_B)_2 + (v_B)_3$$

$$0 = -\frac{w_0 L^4}{8EI} + \frac{w_0 L^4}{30EI} + \frac{B_y L^3}{3EI}$$

$$B_y = \frac{11w_0 L}{40}$$ *Ans.*

For the beam,

$\xrightarrow{+} \Sigma F_x = 0; \qquad A_x = 0$ *Ans.*

$$+\uparrow \Sigma F_y = 0; \qquad A_y + \frac{11w_0 L}{40} - \frac{1}{2}w_0 L = 0$$

$$A_y = \frac{9w_0 L}{40}$$ *Ans.*

$$\zeta + \Sigma M_A = 0; \qquad M_A + \frac{11w_0 L}{40}(L) - \frac{1}{2}w_0 L\left(\frac{2}{3}L\right) = 0$$

$$M_A = \frac{7w_0 L^2}{120}$$ *Ans.*

F16–9.

Remove B_y,

$$(v_B)_1 = \frac{wL^4}{8EI} = \frac{[10(10^3)](6^4)}{8[200(10^9)][65.0(10^{-6})]} = 0.12461 \text{ m} \downarrow$$

Apply B_y,

$$(v_B)_2 = \frac{B_y L^3}{3EI} = \frac{B_y(6^3)}{3[200(10^9)][65.0(10^{-6})]} = 5.5385(10^{-6})B_y \uparrow$$

$$(+\downarrow) \qquad v_B = (v_B)_1 + (v_B)_2$$

$$0.002 = 0.12461 - 5.5385(10^{-6})B_y$$

$$B_y = 22.314(10^3) \text{ N} = 22.1 \text{ kN}$$ *Ans.*

For the beam,

$\xrightarrow{+} \Sigma F_x = 0; \qquad A_x = 0$ *Ans.*

$+\uparrow \Sigma F_y = 0; \qquad A_y + 22.14 - 10(6) = 0 \qquad A_y = 37.9 \text{ kN}$ *Ans.*

$\zeta + \Sigma M_A = 0; \qquad M_A + 22.14(6) - 10(6)(3) = 0$

$M_A = 47.2 \text{ kN} \cdot \text{m}$ *Ans.*

F16–10.

Remove B_y,

$$(v_B)_1 = \frac{M_O L}{6EI(2L)}[(2L)^2 - L^2] = \frac{M_O L^2}{4EI} \downarrow$$

Apply B_y,

$$(v_B)_2 = \frac{B_y(2L)^3}{48EI} = \frac{B_y L^3}{6EI} \uparrow$$

$$(+\uparrow) \qquad v_B = 0 = (v_B)_1 + (v_B)_2$$

$$0 = -\frac{M_O L^2}{4EI} + \frac{B_y L^3}{6EI}$$

$$B_y = \frac{3M_O}{2L}$$ *Ans.*

F16–11.

Remove B_y,

$$(v_B)_1 = \frac{Pbx}{6EIL}(L^2 - b^2 - x^2) = \frac{50(4)(6)}{6EI(12)}(12^2 - 4^2 - 6^2)$$

$$= \frac{1533.3 \text{ kN} \cdot \text{m}^3}{EI} \downarrow$$

Apply B_y,

$$(v_B)_2 = \frac{B_y L^3}{48EI} = \frac{B_y(12^3)}{48EI} = \frac{36B_y}{EI} \uparrow$$

$$(+\uparrow) \qquad v_B = 0 = (v_B)_1 + (v_B)_2$$

$$0 = -\frac{1533.3 \text{ kN} \cdot \text{m}^3}{EI} + \frac{36B_y}{EI}$$

$$B_y = 42.6 \text{ kN}$$ *Ans.*

F16–12.

Remove B_y,

$$(v_B)_1 = \frac{5wL^4}{384EI} = \frac{5[10(10^3)](12^4)}{384[200(10^9)][65.0(10^{-6})]} = 0.20769 \downarrow$$

Apply B_y,

$$(v_B)_2 = \frac{B_y L^3}{48EI} = \frac{B_y (12^3)}{48[200(10^9)][65.0(10^{-6})]}$$
$$= 2.7692(10^{-6})B_y \uparrow$$

$$(+\uparrow) \qquad v_B = (v_B)_1 + (v_B)_2$$
$$-0.005 = -0.20769 + 2.7692(10^{-6})B_y$$
$$B_y = 73.19(10^3) \text{ N} = 73.2 \text{ kN} \qquad\qquad Ans.$$

Chapter 17

F17–1.

$$P = \frac{\pi^2 EI}{(KL)^2} = \frac{\pi^2 \left[29(10^3)\right]\left[\dfrac{\pi}{4}(0.5)^4\right]}{[0.5(50)]^2} = 22.5 \text{ kip} \qquad Ans.$$

$$\sigma = \frac{P}{A} = \frac{22.5}{\pi(0.5)^2} = 28.6 \text{ ksi} < \sigma_Y \quad \text{OK}$$

F17–2.

$$P = \frac{\pi^2 EI}{(KL)^2} = \frac{\pi^2 \left[1.6(10^3)\right]\left[\dfrac{1}{12}(4)(2)^3\right]}{[1(12)(12)]^2}$$
$$= 2.03 \text{ kip} \qquad\qquad Ans.$$

F17–3.

For buckling about the x axis, $K_x = 1$ and $L_x = 12$ m.

$$P_{cr} = \frac{\pi^2 EI_x}{(K_x L_x)^2} = \frac{\pi^2 [200(10^9)][87.3(10^{-6})]}{[1(12)]^2} = 1.197(10^6) \text{ N}$$

For buckling about the y axis, $K_y = 1$ and $L_y = 6$ m.

$$P_{cr} = \frac{\pi^2 EI_y}{(K_y L_y)^2} = \frac{\pi^2 [200(10^9)][18.8(10^{-6})]}{[1(6)]^2}$$
$$= 1.031(10^6) \text{ N (controls)}$$

$$P_{allow} = \frac{P_{cr}}{\text{F.S.}} = \frac{1.031(10^6)}{2} = 515 \text{ kN} \qquad Ans.$$

$$\sigma_{cr} = \frac{P_{cr}}{A} = \frac{1.031(10^6)}{7.4(10^{-3})} = 139.30 \text{ MPa} < \sigma_Y = 345 \text{ MPa (OK)}$$

F17–4.

$$A = \pi[(0.025)^2 - (0.015)^2] = 1.257(10^{-3}) \text{ m}^2$$

$$I = \frac{1}{4}\pi[(0.025)^4 - (0.015)^4] = 267.04(10^{-9}) \text{ m}^4$$

$$P = \frac{\pi^2 EI}{(KL)^2} = \frac{\pi^2 [200(10^9)][267.04(10^{-9})]}{[0.5(5)]^2} = 84.3 \text{ kN} \quad Ans.$$

$$\sigma = \frac{P}{A} = \frac{84.3(10^3)}{1.257(10^{-3})} = 67.1 \text{ MPa} < 250 \text{ MPa} \quad \text{(OK)}$$

F17–5.

Joint A,

$$+\uparrow \Sigma F_y = 0; \qquad F_{AB}\left(\frac{3}{5}\right) - P = 0 \qquad F_{AB} = 1.6667P \text{ (T)}$$

$$\xrightarrow{+} \Sigma F_x = 0; \qquad 1.6667P\left(\frac{4}{5}\right) - F_{AC} = 0$$
$$F_{AC} = 1.3333P \text{ (C)}$$

$$A = \frac{\pi}{4}(2^2) = \pi \text{ in}^2 \qquad\qquad I = \frac{\pi}{4}(1^4) = \frac{\pi}{4} \text{ in}^4$$

$$P_{cr} = F(\text{F.S.}) = 1.3333P(2) = 2.6667P$$

$$P_{cr} = \frac{\pi^2 EI}{(KL)^2}$$

$$2.6667P = \frac{\pi^2 [29(10^3)]\left(\dfrac{\pi}{4}\right)}{[1(4)(12)]^2}$$

$$P = 36.59 \text{ kip} = 36.6 \text{ kip} \qquad\qquad Ans.$$

$$\sigma_{cr} = \frac{P_{cr}}{A} = \frac{2.6667(36.59)}{\pi} = 31.06 \text{ ksi} < \sigma_Y = 50 \text{ ksi}$$
$$\text{(OK)}$$

F17–6.

Beam AB,

$$\zeta + \Sigma M_A = 0; \qquad w(6)(3) - F_{BC}(6) = 0 \qquad F_{BC} = 3w$$

Strut BC,

$$A_{BC} = \frac{\pi}{4}(0.05^2) = 0.625(10^{-3})\pi \text{ m}^2 \qquad I = \frac{\pi}{4}(0.025^4)$$
$$= 97.65625(10^{-9})\pi \text{ m}^4$$

$$P_{cr} = F_{BC}(\text{F.S.}) = 3w(2) = 6w$$

$$P_{cr} = \frac{\pi^2 EI}{(KL)^2}$$

$$6w = \frac{\pi^2 [200(10^9)][97.65625(10^{-9})\pi]}{[1(3)]^2}$$

$$w = 11.215(10^3) \text{ N/m} = 11.2 \text{ kN/m} \qquad\qquad Ans.$$

$$\sigma_{cr} = \frac{P_{cr}}{A} = \frac{6[11.215(10^3)]}{0.625(10^{-3})\pi} = 34.27 \text{ MPa} < \sigma_Y = 345 \text{ MPa}$$
$$\text{(OK)}$$

Selected Answers

Chapter 1

1–1.
    a. 78.5 N
    b. 0.392 mN
    c. 7.46 MN

1–2.
    a. GN/s
    b. Gg/N
    c. $GN/(kg \cdot s)$

1–3.
    a. Gg/s
    b. kN/m
    c. $kN/(kg \cdot s)$

1–5.
    a. 45.3 MN
    b. 56.8 km
    c. 5.63 μg

1–6.
    a. 58.3 km
    b. 68.5 s
    c. 2.55 kN
    d. 7.56 mg

1–7.
    a. 0.431 g
    b. 35.3 kN
    c. 5.32 m

1–9.
    a. km/s
    b. mm
    c. Gs/kg
    d. $mm \cdot N$

1–10.
    a. $kN \cdot m$
    b. Gg/m
    c. $\mu N/s^2$
    d. GN/s

1–11.
    a. 8.653 s
    b. 8.368 kN
    c. 893 g

1–13.
    a. 27.1 $N \cdot m$
    b. 70.7 kN/m^3
    c. 1.27 mm/s

1–14.
    a. $44.9(10)^{-3} N^2$
    b. $2.79(10^3) s^2$
    c. 23.4 s

1–15. 7.41 μN

1–17.
    a. 98.1 N
    b. 4.90 mN
    c. 44.1 kN

1–18.
    a. 0.447 $kg \cdot m/N$
    b. 0.911 $kg \cdot s$
    c. 18.8 GN/m

1–19. 1.04 kip

1–21.
    a. 70.2 kg
    b. 689 N
    c. 25.5 lb
    d. 70.2 kg

Chapter 2

2–1. $F_R = 497$ N, $\phi = 155°$

2–2. $F = 960$ N, $\theta = 45.2°$

2–3. $F_R = 393$ lb, $\phi = 353°$

2–5. $F_{AB} = 314$ lb, $F_{AC} = 256$ lb

2–6. $F_R = 8.03$ kN, $\phi = 1.22°$

2–7. $(F_1)_v = 2.93$ kN, $(F_1)_u = 2.07$ kN

2–9. $F = 616$ lb, $\theta = 46.9°$

2–10. $F_R = 980$ lb, $\phi = 19.4°$

2–11. $F_R = 10.8$ kN, $\phi = 3.16°$

2–13. $F_a = 30.6$ lb, $F_b = 26.9$ lb

2–14. $F = 19.6$ lb, $F_b = 26.4$ lb

2–15. $F = 917$ lb, $\theta = 31.8°$

2–17. $\theta = 36.3°$, $\phi = 26.4°$

2–18. $\theta = 54.3°$, $F_A = 686$ N

2–19. $F_R = 1.23$ kN, $\theta = 6.08°$

2–21. $F_B = 1.61$ kN, $\theta = 38.3°$

2–22. $F_R = 4.01$ kN, $\phi = 16.2°$

2–23. $\theta = 90°$, $F_B = 1$ kN, $F_R = 1.73$ kN

2–25. $F_R = 983$ N, $\theta = 21.8°$

2–26. $\mathbf{F}_1 = \{200\mathbf{i} + 346\mathbf{j}\}$ N, $\mathbf{F}_2 = \{177\mathbf{i} - 177\mathbf{j}\}$ N

2–27. $F_R = 413$ N, $\theta = 24.2°$

2–29. $F_R = 1.96$ kN, $\theta = 4.12°$

2–30. $\mathbf{F}_1 = \{30\mathbf{i} + 40\mathbf{j}\}$ N, $\mathbf{F}_2 = \{-20.7\mathbf{i} - 77.3\mathbf{j}\}$ N,
$\mathbf{F}_3 = \{30\mathbf{i}\}$, $F_R = 54.2$ N, $\theta = 43.5°$

2–31. $F_{1x} = 141$ N, $F_{1y} = 141$ N, $F_{2x} = -130$ N,
$F_{2y} = 75$ N

2–33. $F_R = 12.5$ kN, $\theta = 64.1°$

2–34. $\mathbf{F}_1 = \{680\mathbf{i} - 510\mathbf{j}\}$ N, $\mathbf{F}_2 = \{-312\mathbf{i} - 541\mathbf{j}\}$ N,
$\mathbf{F}_3 = \{-530\mathbf{i} + 530\mathbf{j}\}$ N

2–35. $F_R = 546$ N, $\theta = 253°$

2–37. $F_R = \sqrt{F_1^2 + F_2^2 + 2F_1F_2 \cos \phi}$,
$\theta = \tan^{-1}\left(\dfrac{F_1 \sin \phi}{F_2 + F_1 \cos \phi}\right)$

2–38. $F_x = 40.0$ lb, $F_y = 56.6$ lb, $F_z = 40.0$ lb

2–39. $F_x = 40$ N, $F_y = 40$ N, $F_z = 56.6$ N

2–41. $F_R = 114$ lb, $\alpha = 62.1°$, $\beta = 113°$, $\gamma = 142°$

2–42. $\mathbf{F}_1 = \{53.1\mathbf{i} - 44.5\mathbf{j} + 40\mathbf{k}\}$ lb, $\alpha_1 = 48.4°$,
$\beta_1 = 124°$, $\gamma_1 = 60°$, $\mathbf{F}_2 = \{-130\mathbf{k}\}$ lb,
$\alpha_2 = 90°$, $\beta_2 = 90°$, $\gamma_2 = 180°$

2–43. $\mathbf{F}_1 = \{-106\mathbf{i} + 106\mathbf{j} + 260\mathbf{k}\}$ N,
$\mathbf{F}_2 = \{250\mathbf{i} + 354\mathbf{j} - 250\mathbf{k}\}$ N,
$\mathbf{F}_R = \{144\mathbf{i} + 460\mathbf{j} + 9.81\mathbf{k}\}$ N, $F_R = 482$ N,
$\alpha = 72.6°$, $\beta = 17.4°$, $\gamma = 88.8°$

2–45. $F_3 = 428$ lb, $\alpha = 88.3°$, $\beta = 20.6°$, $\gamma = 69.5°$

2–46. $F_3 = 250$ lb, $\alpha = 87.0°$, $\beta = 143°$, $\gamma = 53.1°$

2–47. $F_R = 430$ N, $\alpha = 28.9°$, $\beta = 67.3°$, $\gamma = 107°$

2–49. $F_1 = 429$ lb, $\alpha_1 = 62.2°$, $\beta_1 = 110°$, $\gamma_1 = 145°$

2–50. $F_R = 116$ lb, $\cos \alpha_2 = 130°$, $\cos \beta_2 = 81.9°$,
$\cos \gamma_2 = 41.4°$

2–51. $\mathbf{F}_1 = \{72.0\mathbf{i} + 54.0\mathbf{k}\}$ N,
$\mathbf{F}_2 = \{53.0\mathbf{i} + 53.0\mathbf{j} + 130\mathbf{k}\}$ N, $\mathbf{F}_3 = \{200\mathbf{k}\}$

2–53. $\mathbf{F}_1 = \{14.0\mathbf{j} - 48.0\mathbf{k}\}$ lb,
$\mathbf{F}_2 = \{90\mathbf{i} - 127\mathbf{j} + 90\mathbf{k}\}$ lb

2–54. $F_{Rx} = 90$, $F_{Ry} = -113$, $F_{Rz} = 42$,
$\mathbf{F}_R = \{90\mathbf{i} - 113\mathbf{j} + 42\mathbf{k}\}$ lb

2–55. $F_R = 610$ N, $\alpha = 19.4°$, $\beta = 77.5°$, $\gamma = 105°$

2–57. $\mathbf{F} = \{59.4\mathbf{i} - 88.2\mathbf{j} - 83.2\mathbf{k}\}$ lb, $\alpha = 63.9°$,
$\beta = 131°$, $\gamma = 128°$

2–58. $\mathbf{F}_1 = \{-26.2\mathbf{i} - 41.9\mathbf{j} + 62.9\mathbf{k}\}$ lb,
$\mathbf{F}_2 = \{13.4\mathbf{i} - 26.7\mathbf{j} - 40.1\mathbf{k}\}$ lb,
$F_R = 73.5$ lb, $\alpha = 100°$, $\beta = 159°$, $\gamma = 71.9°$

2–59. $x = -5.06$ m, $y = 3.61$ m, $z = 6.51$ m

2–61. $x = y = 4.42$ m

2–62. $\mathbf{F}_{AB} = \{97.3\mathbf{i} - 129\mathbf{j} - 191\mathbf{k}\}$ N,
$\mathbf{F}_{AC} = \{221\mathbf{i} - 27.7\mathbf{j} - 332\mathbf{k}\}$ N,
$F_R = 620$ N, $\cos \alpha = 59.1°$, $\cos \beta = 80.6°$,
$\cos \gamma = 147°$

2–63. $F_R = 1.17$ kN, $\alpha = 66.9°$, $\beta = 92.0°$, $\gamma = 157°$

2–65. $\mathbf{F}_{BA} = \{-109\mathbf{i} + 131\mathbf{j} + 306\mathbf{k}\}$ lb,
$\mathbf{F}_{CA} = \{103\mathbf{i} + 103\mathbf{j} + 479\mathbf{k}\}$ lb,
$\mathbf{F}_{DA} = \{-52.1\mathbf{i} - 156\mathbf{j} + 365\mathbf{k}\}$ lb

2–66. $\mathbf{F}_C = \{-324\mathbf{i} - 130\mathbf{j} + 195\mathbf{k}\}$ N,
$\mathbf{F}_B = \{-324\mathbf{i} + 130\mathbf{j} + 195\mathbf{k}\}$ N,
$\mathbf{F}_E = \{-194\mathbf{i} + 291\mathbf{j}\}$ N

2–67. $F_R = 757$ N, $\alpha = 149°$, $\beta = 90.0°$, $\gamma = 59.0°$

2–69. $\mathbf{F} = \{13.4\mathbf{i} + 23.2\mathbf{j} + 53.7\mathbf{k}\}$ lb

2–70. $F_R = 194$ N, $\cos \alpha = 90.9°$, $\cos \beta = 76.4°$,
$\cos \gamma = 166°$

2–73. $\theta = 142°$

2–74. $F_\| = 0.182$ kN

2–75. $\theta = 36.4°$

2–77. $\theta = 70.5°$

2–78. $F_u = 246$ N

2–79. $F_\| = 10.5$ lb

2–81. $\theta = 82.9°$

2–82. Proj $\mathbf{F}_{AB} = \{0.229\mathbf{i} - 0.916\mathbf{j} + 1.15\mathbf{k}\}$ lb

2–83. $\theta = 74.4°$, $\phi = 55.4°$

2–85. $F_{AC} = 25.87$ lb,
$\mathbf{F}_{AC} = \{-18.0\mathbf{i} - 15.4\mathbf{j} + 10.3\mathbf{k}\}$ lb

2–86. $\theta = 132°$

2–87. $\theta = 23.4°$

2–89. $F_x = 75$ N, $F_y = 260$ N

2–90. $F_{OA} = 242$ N

2–91. The magnitude is $(F_1)_{F_2} = 5.44$ lb

R2–1. $F_R = \sqrt{(300)^2 + (500)^2 - 2(300)(500) \cos 95°}$
$= 605.1 = 605$ N
$\dfrac{605.1}{\sin 95°} = \dfrac{500}{\sin \theta}$
$\theta = 55.40°$
$\phi = 55.40° + 30° = 85.4°$

R2–2. $\dfrac{F_{1v}}{\sin 30°} = \dfrac{250}{\sin 105°}$ $F_{1v} = 129$ N
$\dfrac{F_{1u}}{\sin 45°} = \dfrac{250}{\sin 105°}$ $F_{1u} = 183$ N

R2–3. $F_{Rx} = F_{1x} + F_{2x} + F_{3x} + F_{4x}$
$F_{Rx} = -200 + 320 + 180 - 300 = 0$
$F_{Ry} = F_{1y} + F_{2y} + F_{3y} + F_{4y}$
$F_{Ry} = 0 - 240 + 240 + 0 = 0$
Thus, $F_R = 0$

R2–5. $\mathbf{r} = \{50 \sin 20°\mathbf{i} + 50 \cos 20°\mathbf{j} - 35\mathbf{k}\}$ ft
$r = \sqrt{(17.10)^2 + (46.98)^2 + (-35)^2} = 61.03$ ft
$\mathbf{u} = \dfrac{\mathbf{r}}{r} = (0.280\mathbf{i} + 0.770\mathbf{j} - 0.573\mathbf{k})$
$\mathbf{F} = F\mathbf{u} = \{98.1\mathbf{i} + 269\mathbf{j} - 201\mathbf{k}\}$ lb

R2–6. $\mathbf{F}_1 = 600\left(\dfrac{4}{5}\right)\cos 30°(+\mathbf{i}) + 600\left(\dfrac{4}{5}\right)\sin 30°(-\mathbf{j})$
$+ 600\left(\dfrac{3}{5}\right)(+\mathbf{k})$
$= \{415.69\mathbf{i} - 240\mathbf{j} + 360\mathbf{k}\}$ N
$\mathbf{F}_2 = 0\mathbf{i} + 450 \cos 45°(+\mathbf{j}) + 450 \sin 45°(+\mathbf{k})$
$= \{318.20\mathbf{j} + 318.20\mathbf{k}\}$ N

R2–7. $\mathbf{r}_1 = \{400\mathbf{i} + 250\mathbf{k}\}$ mm; $r_1 = 471.70$ mm

$\mathbf{r}_2 = \{50\mathbf{i} + 300\mathbf{j}\}$ mm; $r_2 = 304.14$ mm

$\mathbf{r}_1 \cdot \mathbf{r}_2 = (400)(50) + 0(300) + 250(0) = 20\,000$

$\theta = \cos^{-1}\left(\dfrac{\mathbf{r}_1 \cdot \mathbf{r}_2}{r_1 r_2}\right) = \cos^{-1}\left(\dfrac{20\,000}{(471.70)(304.14)}\right)$

$= 82.0°$

Chapter 3

3–5. $(M_{F_1})_B = 4.125$ kip·ft $\circlearrowright$,

$(M_{F_2})_B = 2.00$ kip·ft $\circlearrowright$,

$(M_{F_3})_B = 40.0$ lb·ft $\circlearrowright$

3–6. $M_P = 341$ in.·lb $\circlearrowright$

$M_F = 403$ in.·lb $\circlearrowleft$

Not sufficient

3–7. $(M_{F_1})_A = 433$ N·m $\circlearrowleft$

$(M_{F_2})_A = 1.30$ kN·m $\circlearrowleft$

$(M_{F_3})_A = 800$ N·m $\circlearrowleft$

3–9. $M_B = 90.6$ lb·ft $\circlearrowleft$, $M_C = 141$ lb·ft $\circlearrowright$

3–10. $M_A = 195$ lb·ft $\circlearrowright$

3–11. $(M_O)_{max} = 48.0$ kN·m $\circlearrowright$, $x = 9.81$ m

3–13. $\mathbf{M}_B = \{-3.36\mathbf{k}\}$ N·m, $\alpha = 90°$, $\beta = 90°$,

$\gamma = 180°$

3–14. $\mathbf{M}_O = \{0.5\mathbf{i} + 0.866\mathbf{j} - 3.36\mathbf{k}\}$ N·m,

$\alpha = 81.8°$, $\beta = 75.7°$, $\gamma = 163°$

3–15. $(M_A)_C = 768$ lb·ft $\circlearrowleft$

$(M_A)_B = 636$ lb·ft $\circlearrowright$

Clockwise

3–17. $m = \left(\dfrac{l}{d + l}\right) M$

3–18. $M_P = (537.5 \cos\theta + 75 \sin\theta)$ lb·ft

3–19. $F = 239$ lb

3–21. $F = 27.6$ lb

3–22. $r = 13.3$ mm

3–23. $(M_R)_A = (M_R)_B = 76.0$ kN·m $\circlearrowright$

3–25. $(M_{AB})_A = 3.88$ kip·ft $\circlearrowleft$,

$(M_{BCD})_A = 2.05$ kip·ft $\circlearrowleft$,

$(M_{man})_A = 2.10$ kip·ft $\circlearrowleft$

3–26. $(M_R)_A = 8.04$ kip·ft $\circlearrowleft$

3–27. $\mathbf{M}_O = \{-40\mathbf{i} - 44\mathbf{j} - 8\mathbf{k}\}$ kN·m

3–29. $\mathbf{M}_O = \{-25\mathbf{i} + 6200\mathbf{j} - 900\mathbf{k}\}$ lb·ft

3–30. $\mathbf{M}_A = \{-175\mathbf{i} + 5600\mathbf{j} - 900\mathbf{k}\}$ lb·ft

3–31. $\mathbf{M}_P = \{-24\mathbf{i} + 24\mathbf{j} + 8\mathbf{k}\}$ kN·m

3–33. $F = 18.6$ lb

3–34. $M_O = 4.27$ N·m, $\alpha = 95.2°$, $\beta = 110°$, $\gamma = 20.6°$

3–35. $\mathbf{M}_A = \{-5.39\mathbf{i} + 13.1\mathbf{j} + 11.4\mathbf{k}\}$ N·m

3–37. $y = 2$ m, $z = 1$ m

3–38. $y = 1$ m, $z = 3$ m, $d = 1.15$ m

3–39.   No, Yes

3–41. $M_{y'} = 464$ lb·ft

3–42. $M_x = 440$ lb·ft

3–43. $M_x = 15.0$ lb·ft, $M_y = 4.00$ lb·ft,

$M_z = 36.0$ lb·ft

3–45. $M_x = 21.7$ N·m

3–46. $F = 139$ N

3–47. $M_y = 282$ lb·ft

3–49. $M_{BC} = 165$ N·m

3–50. $M_{CA} = 226$ N·m

3–51. $M_z = \{35.4\,\mathbf{k}\}$ N·m

3–53. $M_a = 4.37$ N·m, $\alpha = 33.7°$, $\beta = 90°$, $\gamma = 56.3°$,

$M = 5.41$ N·m

3–54. $R = 28.9$ N

3–55. $F = 133$ N, $P = 800$ N

3–57. $(M_R)_C = 435$ lb·ft $\circlearrowright$

3–58. $F = 139$ lb, anywhere

3–59. $M_R = 64.0$ lb·ft, $\alpha = 94.7°$, $\beta = 13.2°$, $\gamma = 102°$

3–61. $M_R = 576$ lb·in., $\alpha = 37.0°$, $\beta = 111°$, $\gamma = 61.2°$

3–62. $\mathbf{M}_C = \{-65.0\mathbf{i} - 37.5\mathbf{k}\}$ N·m

3–63. $F = 15.4$ N

3–65. $F = 832$ N

3–66. $M_C = 40.8$ N·m, $\alpha = 11.3°$, $\beta = 101°$, $\gamma = 90°$

3–67. $F = 98.1$ N

3–69. $(M_C)_R = 71.9$ N·m, $\alpha = 44.2°$, $\beta = 131°$, $\gamma = 103°$

3–70. $F_2 = 112$ N, $F_1 = 87.2$ N, $F_3 = 100$ N

3–71. $F_R = 365$ N, $\theta = 70.8°$ $\nearrow$, $(M_R)_O = 2364$ N·m $\circlearrowright$

3–73. $F_R = 5.93$ kN, $\theta = 77.8°$ $\nearrow$, $M_{R_A} = 34.8$ kN·m $\circlearrowleft$

3–74. $F_R = 5.93$ kN, $\theta = 77.8°$ $\nearrow$,

$M_B = 11.6$ kN·m (**Counterclockwise**)

3–75. $F_R = 294$ N, $\theta = 40.1°$ $\nearrow$,

$M_{RO} = 39.6$ N·m $\circlearrowleft$

3–77. $F_R = 1.30$ kN, $\theta = 86.7°$ $\searrow$,

$(M_R)_B = 10.1$ kN·m $\circlearrowright$

3–78. $F_R = 416$ lb, $\theta = 35.2°$ $\nearrow$,

$(M_R)_A = 1.48$ kip·ft (**Clockwise**)

3–79. $F_R = 938$ N, $\theta = 35.9°$ $\searrow$, $(M_R)_A = 680$ N·m $\circlearrowright$

3–81. $\mathbf{F}_R = \{270\mathbf{k}\}$ N, $\mathbf{M}_{RO} = \{-2.22\mathbf{i}\}$ N·m

3–82. $\mathbf{F}_R = \{-200\mathbf{i} + 700\mathbf{j} - 600\mathbf{k}\}$ N,

$(\mathbf{M}_R)_O = \{-1200\mathbf{i} + 450\mathbf{j} + 1450\mathbf{k}\}$ N·m

3–83. $\mathbf{F}_R = \{6\mathbf{i} + 5\mathbf{j} - 5\mathbf{k}\}$ kN,

$(\mathbf{M}_R)_O = \{2.5\mathbf{i} - 7\mathbf{j}\}$ kN·m

3–85. $F_R = \{-40\mathbf{j} - 40\mathbf{k}\}$ N,
$M_{RA} = \{-12\mathbf{j} + 12\mathbf{k}\}$ N·m

3–86. $F_R = \{-28.3\mathbf{j} - 68.3\mathbf{k}\}$ N,
$M_{RA} = \{-20.5\mathbf{j} + 8.49\mathbf{k}\}$ N·m

3–87. $F_R = 10.75$ kip $\downarrow$, $d = 13.7$ ft

3–89. $F = 798$ lb, $\theta = 67.9°$ ↗, $x = 7.43$ ft

3–90. $F = 798$ lb, $\theta = 67.9°$ ↗, $x = 6.57$ ft

3–91. $F = 1302$ N, $\theta = 84.5°$ ↗, $x = 7.36$ m

3–93. $F_R = 1000$ N, $\theta = 53.1°$ ↘, $d = 2.17$ m

3–94. $F_R = 356$ N, $\theta = 51.8°$, $d = b = 3.32$ m

3–95. $F_R = 356$ N, $\theta = 51.8°$, $d = 1.75$ m

3–97. $F_R = 542$ N, $\theta = 10.6°$ ↖, $d = 2.17$ m

3–98. $F_R = -10$ kN, $x = 1$ m, $z = 1.4$ m

3–99. $F_R = 197$ lb, $\theta = 42.6°$∡, $d = 5.24$ ft

3–101. $F_R = 26$ kN, $y = 82.7$ mm, $x = 3.85$ mm

3–102. $F_A = 18.0$ kN, $F_B = 16.7$ kN, $F_R = 48.7$ kN

3–103. $F_C = 600$ N, $F_D = 500$ N

3–105. $F_1 = 27.6$ kN, $F_2 = 24.0$ kN

3–106. $F_R = 215$ kN, $y = 3.68$ m, $x = 3.54$ m

3–107. $F_A = 30$ kN, $F_B = 20$ kN, $F_R = 190$ kN

3–109. $F_R = 6.75$ kN, $\bar{x} = 2.5$ m

3–110. $F_R = 21.0$ kN, $d = 3.43$ m

3–111. $F_R = 7$ lb, $\bar{x} = 0.268$ ft

3–113. $F_R = 12.0$ kN, $\theta = 48.4°$ ↗, $d = 3.28$ m

3–114. $F_R = 12.0$ kN, $\theta = 48.4°$ ↗, $d = 3.69$ m

3–115. $a = 1.26$ m, $b = 2.53$ m

3–117. $F_R = 43.6$ lb, $x = 3.27$ ft

R3–1. $20(10^3) = 800(16\cos 30°) + W(30\cos 30° + 2)$
$W = 319$ lb

R3–2. $F_R = 50$ lb $\left[\dfrac{(10\mathbf{i} + 15\mathbf{j} - 30\mathbf{k})}{\sqrt{(10)^2 + (15)^2 + (-30)^2}}\right]$

$F_R = \{14.3\mathbf{i} + 21.4\mathbf{j} - 42.9\mathbf{k}\}$ lb

$(M_R)_C = \mathbf{r}_{CB} \times \mathbf{F} = \begin{vmatrix} \mathbf{i} & \mathbf{j} & \mathbf{k} \\ 10 & 45 & 0 \\ 14.29 & 21.43 & -42.86 \end{vmatrix}$

$= \{-1929\mathbf{i} + 428.6\mathbf{j} - 428.6\mathbf{k}\}$ lb·ft

R3–3. $\mathbf{r} = \{4\mathbf{i}\}$ ft

$\mathbf{F} = 24$ lb $\left(\dfrac{-2\mathbf{i} + 2\mathbf{j} + 4\mathbf{k}}{\sqrt{(-2)^2 + (2)^2 + (4)^2}}\right)$

$= \{-9.80\mathbf{i} + 9.80\mathbf{j} + 19.60\mathbf{k}\}$ lb

$M_y = \begin{vmatrix} 0 & 1 & 0 \\ 4 & 0 & 0 \\ -9.80 & 9.80 & 19.60 \end{vmatrix} = -78.4$ lb·ft

$\mathbf{M}_y = \{-78.4\mathbf{j}\}$ lb·ft

R3–5. $\xrightarrow{+}\Sigma F_{Rx} = \Sigma F_x;$ $F_{Rx} = 6\left(\dfrac{5}{13}\right) - 4\cos 60°$
$= 0.30769$ kN

$+\uparrow\Sigma F_{Ry} = \Sigma F_y;$ $F_{Ry} = 6\left(\dfrac{12}{13}\right) - 4\sin 60°$
$= 2.0744$ kN

$F_R = \sqrt{(0.30769)^2 + (2.0744)^2} = 2.10$ kN

$\theta = \tan^{-1}\left[\dfrac{2.0744}{0.30769}\right] = 81.6°$ ∡

$\zeta + M_P = \Sigma M_P;\ M_P = 8 - 6\left(\dfrac{12}{13}\right)(7) + 6\left(\dfrac{5}{13}\right)(5)$
$- 4\cos 60°(4) + 4\sin 60°(3)$
$M_P = -16.8$ kN·m
$= 16.8$ kN·m ↻

R3–6. $\xrightarrow{+}\Sigma(F_R)_x = \Sigma F_x;\ (F_R)_x = 200\cos 45° - 250\left(\dfrac{4}{5}\right)$
$- 300 = -358.58$ lb $= 358.58$ lb $\leftarrow$

$+\uparrow(F_R)_y = \Sigma F_y;\ (F_R)_y = -200\sin 45° - 250\left(\dfrac{3}{5}\right)$
$= -291.42$ lb $= 291.42$ lb $\downarrow$

$F_R = \sqrt{(F_R)_x^2 + (F_R)_y^2} = \sqrt{358.58^2 + 291.42^2}$
$= 462.07$ lb $= 462$ lb

$\theta = \tan^{-1}\left[\dfrac{(F_R)_y}{(F_R)_x}\right] = \tan^{-1}\left[\dfrac{291.42}{358.58}\right] = 39.1°$ ↗

$\zeta + (M_R)_A = \Sigma M_A;\ 358.58(d) = 250\left(\dfrac{3}{5}\right)(2.5) + 250\left(\dfrac{4}{5}\right)(4)$
$+ 300(4) - 200\cos 45°(6) - 200\sin 45°(3)$
$d = 3.07$ ft

R3–7. $+\uparrow F_R = \Sigma F_z;$ $F_R = -20 - 50 - 30 - 40$
$= -140$ kN $= 140$ kN $\downarrow$

$(M_R)_x = \Sigma M_x;$ $-140y = -50(3) - 30(11) - 40(13)$
$y = 7.14$ m

$(M_R)_y = \Sigma M_y;$ $140x = 50(4) + 20(10) + 40(10)$
$x = 5.71$ m

Chapter 4

4–1. $A_x = 3.46$ kN, $A_y = 8$ kN, $M_A = 20.2$ kN·m

4–2. $N_A = 750$ N, $B_y = 600$ N, $B_x = 450$ N

4–3. $N_B = 3.46$ kN, $A_x = 1.73$ kN, $A_y = 1.00$ kN

4–5. $N_A = 3.33$ kN, $B_x = 2.40$ kN, $B_y = 133$ N

4–6. $A_y = 5.00$ kN, $N_B = 9.00$ kN, $A_x = 5.00$ kN

4–7. $F_{BC} = 1.82$ kip, $F_A = 2.06$ kip

4–9. $F = 14.0$ kN

4–10. $N_A = 173$ N, $N_C = 416$ N, $N_B = 69.2$ N

4–11. $\theta = 3.82°$

4–13. $P = 660$ N, $N_A = 442$ N, $\theta = 48.0°$ ⬊

4–14. $d = \dfrac{3a}{4}$

4–15. $F_{BC} = 80$ kN, $A_x = 54$ kN, $A_y = 16$ kN

4–17. $F_C = 10$ mN

4–18. $k = 250$ N/m

4–19. $w_B = 2.19$ kip/ft, $w_A = 10.7$ kip/ft

4–21. $\alpha = 10.4°$

4–22. $w_1 = \dfrac{2P}{L}, w_2 = \dfrac{4P}{L}$

4–23. $\theta = 23.2°, 85.2°$

4–25. $N_A = 346$ N, $N_B = 693$ N, $a = 0.650$ m

4–26. $T = 1.84$ kN, $F = 6.18$ kN

4–27. $R_D = 22.6$ kip, $R_E = 22.6$ kip, $R_F = 13.7$ kip

4–29. $N_A = 28.6$ lb, $N_B = 10.7$ lb, $N_C = 10.7$ lb

4–30. $T_{BC} = 43.9$ N, $N_B = 58.9$ N, $A_x = 58.9$ N, $A_y = 39.2$ N, $A_z = 177$ N

4–31. $T_C = 14.8$ kN, $T_B = 16.5$ kN, $T_A = 7.27$ kN

4–33. $F_{AB} = 467$ N, $F_{AC} = 674$ N, $D_x = 1.04$ kN, $D_y = 0, D_z = 0$

4–34. $A_x = 633$ lb, $A_y = -141$ lb, $B_x = -721$ lb, $B_z = 895$ lb, $C_y = 200$ lb, $C_z = -506$ lb

4–35. $F_2 = 674$ lb

4–37. $C_z = 10.6$ lb, $D_y = -0.230$ lb, $C_y = 0.230$ lb, $D_x = 5.17$ lb, $C_x = 5.44$ lb, $M = 0.459$ lb·ft

4–38. $F_{BD} = 294$ N, $F_{BC} = 589$ N, $A_x = 0$, $A_y = 589$ N, $A_z = 490.5$ N

4–39. $T = 58.0$ N, $C_z = 87.0$ N, $C_y = 28.8$ N, $D_x = 0$, $D_y = 79.2$ N, $D_z = 58.0$ N

4–41. $F_{BC} = 0, A_y = 0, A_z = 800$ lb, $(M_A)_x = 4.80$ kip·ft, $(M_A)_y = 0, (M_A)_z = 0$

4–42. $P = 12.8$ kN

4–43. $N_B = 2.43$ kip, $N_C = 1.62$ kip, $F = 200$ lb

4–45. $T = 3.67$ kip

4–46. $F = 2.76$ kN

4–47. $F = 5.79$ kN

4–49. (a) No, (b) Yes

4–51. $\phi = \theta$, $P = W \sin(\alpha + \theta)$

4–53. $\theta = 52.0°$

4–54. $\mu_s = 0.231$

4–55. $P = 1350$ lb

4–57. $P = 14.4$ N

4–58. $P = 50$ lb

4–59. No

4–61. $P = 8.18$ lb

4–62. $P = 286$ N

4–63. $O_y = 400$ N, $O_x = 46.4$ N

R4–1. $\zeta + \Sigma M_A = 0$: $F(6) + F(4) + F(2) - 3 \cos 45°(2) = 0$

$$F = 0.3536 \text{ kN} = 354 \text{ N}$$

R4–2. $\zeta + \Sigma M_A = 0$; $N_B(7) - 1400(3.5) - 300(6) = 0$

$$N_B = 957.14 \text{ N} = 957 \text{ N}$$

$+\uparrow \Sigma F_y = 0$; $A_y - 1400 - 300 + 957 = 0$　$A_y = 743$ N

$\xrightarrow{+} \Sigma F_x = 0$; 　$A_x = 0$

R4–3. $\zeta + \Sigma M_A = 0$; 　$10(0.6 + 1.2 \cos 60°) + 6(0.4)$
$$- N_A(1.2 + 1.2 \cos 60°) = 0$$
$$N_A = 8.00 \text{ kN}$$

$\xrightarrow{+} \Sigma F_x = 0$; 　$B_x - 6 \cos 30° = 0$; 　$B_x = 5.20$ kN

$+\uparrow \Sigma F_y = 0$; 　$B_y + 8.00 - 6 \sin 30° - 10 = 0$
$$B_y = 5.00 \text{ kN}$$

R4–5. $\Sigma F_x = 0$; 　　　　$A_x = 0$

$\Sigma F_y = 0$; 　　　　$A_y + 200 = 0$

　　　　　　　　　$A_y = -200$ N

$\Sigma F_z = 0$; 　　　　$A_z - 150 = 0$

　　　　　　　　　$A_z = 150$ N

$\Sigma M_x = 0$; 　　$-150(2) + 200(2) - (M_A)_x = 0$

　　　　　　　　$(M_A)_x = 100$ N·m

$\Sigma M_y = 0$; 　　$(M_A)_y = 0$

$\Sigma M_z = 0$; 　　$200(2.5) - (M_A)_z = 0$

　　　　　　　　$(M_A)_z = 500$ N·m

R4–6.

$\Sigma M_y = 0$; 　$P(8) - 80(10) = 0$		$P = 100$ lb
$\Sigma M_x = 0$; 　$B_z(28) - 80(14) = 0$		$B_z = 40$ lb
$\Sigma M_z = 0$; 　$-B_x(28) - 100(10) = 0$		$B_x = -35.7$ lb
$\Sigma F_x = 0$; 　$A_x + (-35.7) - 100 = 0$		$A_x = 136$ lb
$\Sigma F_y = 0$; 　$B_y = 0$		
$\Sigma F_z = 0$; 　$A_z + 40 - 80 = 0$		$A_z = 40$ lb

R4–7.   Assume that the ladder slips at A:

$$F_A = 0.4\,N_A$$
$$+\uparrow \Sigma F_y = 0; \quad N_A - 20 = 0$$
$$N_A = 20\text{ lb}$$
$$F_A = 0.4(20) = 8\text{ lb}$$
$$\zeta + \Sigma M_B = 0; \quad P(4) - 20(3) + 20(6) - 8(8) = 0$$
$$P = 1\text{ lb}$$
$$\xrightarrow{+} \Sigma F_x = 0; \quad N_B + 1 - 8 = 0$$
$$N_B = 7\text{ lb} > 0$$

The ladder will remain in contact with the wall.

Chapter 5

5–1. $F_{CB} = 0$, $F_{CD} = 20.0$ kN (C),
$F_{DB} = 33.3$ kN (T), $F_{DA} = 36.7$ kN (C)

5–2. $F_{CB} = 0$, $F_{CD} = 45.0$ kN (C),
$F_{DB} = 75.0$ kN (T), $F_{DA} = 90.0$ kN (C)

5–3. $F_{AC} = 150$ lb (C), $F_{AB} = 140$ lb (T),
$F_{BD} = 140$ lb (T), $F_{BC} = 0$, $F_{CD} = 150$ lb (T),
$F_{CE} = 180$ lb (C), $F_{DE} = 120$ lb (C),
$F_{DF} = 230$ lb (T), $F_{EF} = 300$ lb (C)

5–5. $F_{CD} = 5.21$ kN (C), $F_{CB} = 4.17$ kN (T),
$F_{AD} = 1.46$ kN (C), $F_{AB} = 4.17$ kN (T),
$F_{BD} = 4$ kN (T)

5–6. $F_{CD} = 5.21$ kN (C), $F_{CB} = 2.36$ kN (T),
$F_{AD} = 1.46$ kN (C), $F_{AB} = 2.36$ kN (T),
$F_{BD} = 4$ kN (T)

5–7. $F_{DE} = 16.3$ kN (C), $F_{DC} = 8.40$ kN (T),
$F_{EA} = 8.85$ kN (C), $F_{EC} = 6.20$ kN (C),
$F_{CF} = 8.77$ kN (T), $F_{CB} = 2.20$ kN (T),
$F_{BA} = 3.11$ kN (T), $F_{BF} = 6.20$ kN (C),
$F_{FA} = 6.20$ kN (T)

5–9. $P_{max} = 849$ lb

5–10. $P_{max} = 849$ lb

5–11. $F_{DC} = 9.24$ kN (T), $F_{DE} = 4.62$ kN (C),
$F_{CE} = 9.24$ kN (C), $F_{CB} = 9.24$ kN (T),
$F_{BE} = 9.24$ kN (C), $F_{BA} = 9.24$ kN (T),
$F_{EA} = 4.62$ kN (C)

5–13. $F_{DE} = 8.94$ kN (T), $F_{DC} = 4.00$ kN (C),
$F_{CB} = 4.00$ kN (C), $F_{CE} = 0$,
$F_{EB} = 11.3$ kN (C), $F_{EF} = 12.0$ kN (T),
$F_{BA} = 12.0$ kN (C), $F_{BF} = 18.0$ kN (T),
$F_{FA} = 20.1$ kN (C), $F_{FG} = 21.0$ kN (T)

5–14. $F_{DE} = 13.4$ kN (T), $F_{DC} = 6.00$ kN (C),
$F_{CB} = 6.00$ kN (C), $F_{CE} = 0$, $F_{EB} = 17.0$ kN (C),

$F_{EF} = 18.0$ kN (T), $F_{BA} = 18.0$ kN (C),
$F_{BF} = 20.0$ kN (T), $F_{FA} = 22.4$ kN (C),
$F_{FG} = 28.0$ kN (T)

5–15. $F_{CD} = F_{DE} = F_{AF} = 0$,
$F_{CE} = 16.9$ kN (C), $F_{CB} = 10.1$ kN (T),
$F_{BA} = 10.1$ kN (T), $F_{BE} = 15.0$ kN (T),
$F_{AE} = 1.875$ kN (C), $F_{FE} = 9.00$ kN (C)

5–17. $F_{HI} = 42.5$ kN (T), $F_{HC} = 100$ kN (T),
$F_{DC} = 125$ kN (C)

5–18. $F_{GH} = 76.7$ kN (T), $F_{ED} = 100$ kN (C),
$F_{EH} = 29.2$ kN (T)

5–19. $F_{HG} = 1125$ lb (T), $F_{DE} = 3375$ lb (C),
$F_{EH} = 3750$ lb (T)

5–21. $F_{KJ} = 11.25$ kip (T), $F_{CD} = 9.375$ kip (C),
$F_{CJ} = 3.125$ kip (C), $F_{DJ} = 0$

5–22. $F_{JI} = 7.50$ kip (T), $F_{EI} = 2.50$ kip (C)

5–23. $F_{GF} = 12.5$ kN (C), $F_{CD} = 6.67$ kN (T), $F_{GC} = 0$

5–25. $F_{BC} = 5.33$ kN (C), $F_{EF} = 5.33$ kN (T),
$F_{CF} = 4.00$ kN (T)

5–26. $F_{AF} = 21.3$ kN (T), $F_{BC} = 5.33$ kN (C),
$F_{BF} = 20.0$ kN (C)

5–27. $F_{EF} = 14.0$ kN (C), $F_{BC} = 13.0$ kN (T),
$F_{BE} = 1.41$ kN (T), $F_{BF} = 8.00$ kN (T)

5–29. $F_{BC} = 10.4$ kN (C), $F_{HG} = 9.16$ kN (T),
$F_{HC} = 2.24$ kN (T)

5–30. $F_{CD} = 11.2$ kN (C), $F_{CF} = 3.21$ kN (T),
$F_{CG} = 6.80$ kN (C)

5–31. $F_{BC} = 18.0$ kN (T), $F_{FE} = 15.0$ kN (C),
$F_{EB} = 5.00$ kN (C)

5–33.   **a.** $P = 25.0$ lb, **b.** $P = 33.3$ lb, **c.** $P = 11.1$ lb

5–34. $P = 18.9$ N

5–35. $P = 368$ N

5–37. $F_C = 572$ N, $F_A = 572$ N, $F_B = 478$ N

5–38. $N_E = 3.60$ kN, $N_B = 900$ N, $A_x = 0$,
$A_y = 2.70$ kN, $M_A = 8.10$ kN $\cdot$ m

5–39. $T = 350$ lb, $A_y = 700$ lb, $A_x = 1.88$ kip,
$D_x = 1.70$ kip, $D_y = 1.70$ kip

5–41. $A_x = 0$, $A_y = 2.025$ kN, $B_x = 1.80$ kN,
$B_y = 2.025$ kN

5–42. $F_{FD} = 20.1$ kN, $F_{BD} = 25.5$ kN,
Member EDC: $C_x' = 18.0$ kN, $C_y' = 12.0$ kN
Member ABC: $C_y'' = 12.0$ kN, $C_x'' = 18.0$ kN,

5–43. $A_x = 294$ N, $A_y = 196$ N, $N_C = 147$ N,
$N_E = 343$ N

5–45. $M = 314$ lb $\cdot$ ft

5–46. $F_C = 19.6$ kN

5–47. $F = 6.93$ kN

5–49. $N_B = N_C = 49.5$ N

5–50. $F_{EF} = 8.18$ kN (T), $F_{AD} = 158$ kN (C)

5–51. $F_{CA} = 12.9$ kip, $F_{AB} = 11.9$ kip, $F_{AD} = 2.39$ kip

5–53. $F = 66.1$ lb

5–54. $d = 0.638$ ft

5–55. $N_A = 11.1$ kN (Both wheels), $F'_{CD} = 6.47$ kN,
$F'_E = 5.88$ kN

5–57. $\theta = 23.7°$

5–58. $m = 26.0$ kg

5–59. $M = 14.2$ lb $\cdot$ ft

5–61. $m_L = 106$ kg

R5–1. Joint B:

$\xrightarrow{+} \Sigma F_x = 0;$ $F_{BC} = 3$ kN (C)

$+\uparrow\Sigma F_y = 0;$ $F_{BA} = 8$ kN (C)

Joint A:

$+\uparrow\Sigma F_y = 0;$ $8.875 - 8 - \dfrac{3}{5}F_{AC} = 0$

$F_{AC} = 1.458 = 1.46$ kN (C)

$\xrightarrow{+} \Sigma F_x = 0;$ $F_{AF} - 3 - \dfrac{4}{5}(1.458) = 0$

$F_{AF} = 4.17$ kN (T)

Joint C:

$\xrightarrow{+} \Sigma F_x = 0;$ $3 + \dfrac{4}{5}(1.458) - F_{CD} = 0$

$F_{CD} = 4.167 = 4.17$ kN (C)

$+\uparrow\Sigma F_y = 0;$ $F_{CF} - 4 + \dfrac{3}{5}(1.458) = 0$

$F_{CF} = 3.125 = 3.12$ kN (C)

Joint E:

$\xrightarrow{+} \Sigma F_x = 0;$ $F_{EF} = 0$

$+\uparrow\Sigma F_y = 0;$ $F_{ED} = 13.125 = 13.1$ kN (C)

Joint D:

$+\uparrow\Sigma F_y = 0;$ $13.125 - 10 - \dfrac{3}{5}F_{DF} = 0$

$F_{DF} = 5.21$ kN (T)

R5–2. Joint A:

$\xrightarrow{+} \Sigma F_x = 0;$ $F_{AB} - F_{AG}\cos 45° = 0$

$+\uparrow\Sigma F_y = 0;$ $333.3 - F_{AG}\sin 45° = 0$

$F_{AG} = 471$ lb (C)

$F_{AB} = 333.3 = 333$ lb (T)

Joint B:

$\xrightarrow{+} \Sigma F_x = 0;$ $F_{BC} = 333.3 = 333$ lb (T)

$+\uparrow\Sigma F_y = 0;$ $F_{GB} = 0$

Joint D:

$\xrightarrow{+} \Sigma F_x = 0;$ $-F_{DC} + F_{DE}\cos 45° = 0$

$+\uparrow\Sigma F_y = 0;$ $666.7 - F_{DE}\sin 45° = 0$

$F_{DE} = 942.9$ lb $= 943$ lb (C)

$F_{DC} = 666.7$ lb $= 667$ lb (T)

Joint E:

$\xrightarrow{+} \Sigma F_x = 0;$ $-942.9\sin 45° + F_{EG} = 0$

$+\uparrow\Sigma F_y = 0;$ $-F_{EC} + 942.9\cos 45° = 0$

$F_{EC} = 666.7$ lb $= 667$ lb (T)

$F_{EG} = 666.7$ lb $= 667$ lb (C)

Joint C:

$+\uparrow\Sigma F_y = 0;$ $F_{GC}\cos 45° + 666.7 - 1000 = 0$

$F_{GC} = 471$ lb (T)

R5–3. $\zeta + \Sigma M_C = 0;$ $-1000(10) + 1500(20)$
$- F_{GJ}\cos 30°(20\tan 30°) = 0$

$F_{GJ} = 2.00$ kip (C)

$+\uparrow\Sigma F_y = 0;$ $-1000 + 2(2000\cos 60°) - F_{GC} = 0$

$F_{GC} = 1.00$ kip (T)

R5–5. CB is a two force member.

Member AC:

$\zeta + \Sigma M_A = 0;$ $-600(0.75) + 1.5(F_{CB}\sin 75°) = 0$

$F_{CB} = 310.6$

$B_x = B_y = 310.6\left(\dfrac{1}{\sqrt{2}}\right) = 220$ N

$\xrightarrow{+} \Sigma F_x = 0;$ $-A_x + 600\sin 60° - 310.6\cos 45° = 0$

$A_x = 300$ N

$+\uparrow\Sigma F_y = 0;$ $A_y - 600\cos 60° + 310.6\sin 45° = 0$

$A_y = 80.4$ N

R5–6. Member AB:

$\zeta + \Sigma M_A = 0;$ $-750(2) + B_y(3) = 0$

$B_y = 500$ N

Member BC:

$\zeta + \Sigma M_C = 0;$ $-1200(1.5) - 900(1) + B_x(3) - 500(3) = 0$

$B_x = 1400$ N

$+\uparrow\Sigma F_y = 0;$ $A_y - 750 + 500 = 0$

$A_y = 250$ N

Member AB:

$\xrightarrow{+} \Sigma F_x = 0;$ $-A_x + 1400 = 0$

$A_x = 1400$ N $= 1.40$ kN

Member BC:

$\xrightarrow{+} \Sigma F_x = 0;$ $C_x + 900 - 1400 = 0$

$C_x = 500$ N

$+\uparrow\Sigma F_y = 0;$ $-500 - 1200 + C_y = 0$

$C_y = 1700$ N $= 1.70$ kN

R5–7. $F_D = 20.8$ lb, $F_F = 14.7$ lb, $F_A = 24.5$ lb

Chapter 6

6–1. $\bar{x} = \dfrac{3}{8}a$

6–2. $\bar{y} = \dfrac{\pi}{8}a, \bar{x} = 0$

6–3. $\bar{x} = \dfrac{3}{2}$ m

6–5. $\bar{x} = \dfrac{3}{4}b$

6–6. $\bar{y} = \dfrac{3}{10}h$

6–7. $\bar{x} = 6$ m

6–9. $\bar{x} = 0.398$ m

6–10. $\bar{y} = 1.00$ m

6–11. $\bar{y} = 1.43$ in.

6–13. $\bar{y} = \dfrac{hn}{2n + 1}$

6–14. $\bar{y} = \dfrac{hn + 1}{2(2n + 1)}$

6–15. $\bar{x} = 1\dfrac{3}{5}$ ft

6–17. $\bar{x} = \dfrac{3}{8}a$

6–18. $\bar{y} = \dfrac{2}{5}h$

6–19. $\bar{x} = 3.20$ ft, $\bar{y} = 3.20$ ft, $T_A = 384$ lb, $T_C = 384$ lb, $T_B = 1.15$ kip

6–21. $\bar{y} = \dfrac{6}{5}$ ft

6–22. $\bar{x} = 50.0$ mm

6–23. $\bar{y} = 40.0$ mm

6–25. $\bar{y} = \dfrac{h}{3}$

6–26. $\bar{x} = \dfrac{\pi}{2}a$

6–27. $\bar{y} = \dfrac{\pi a}{8}$

6–29. $\bar{x} = \left[\dfrac{2(n + 1)}{3(n + 2)}\right]a$

6–30. $\bar{y} = \left[\dfrac{(4n + 1)}{3(2n + 1)}\right]h$

6–31. $\bar{y} = 2.61$ ft

6–33. $\bar{z} = 12.8$ in.

6–34. $\bar{z} = \dfrac{4}{3}$ m

6–35. $\bar{y} = \dfrac{3}{8}b, \bar{x} = \bar{z} = 0$

6–37. $\bar{y} = 249$ mm

6–38. $\bar{y} = 79.7$ mm

6–39. $\bar{x} = -1.00$ in., $\bar{y} = 4.625$ in.

6–41. $\bar{x} = -1.18$ in., $\bar{y} = 1.39$ in.

6–42. $\bar{x} = 1.57$ in., $\bar{y} = 1.57$ in.

6–43. $\bar{y} = 2$ in.

6–45. $\bar{y} = 0.75$ in.

6–46. $\bar{z} = 1.625$ in.

6–47. $\bar{z} = 122$ mm

6–49. $\bar{x} = 5.07$ ft, $\bar{y} = 3.80$ ft

6–50. $h = 323$ mm

6–51. $\bar{z} = 128$ mm

6–53. $\bar{x} = 19.0$ ft, $\bar{y} = 11.0$ ft

6–54. $\bar{x} = 8.22$ in.

6–55. $\Sigma m = 16.4$ kg, $\bar{x} = 153$ mm, $\bar{y} = -15$ mm, $\bar{z} = 111$ mm

6–57. $I_y = \dfrac{a^3 b}{n + 3}$

6–58. $I_x = 457(10^6)$ mm^4

6–59. $I_y = 53.3(10^6)$ mm^4

6–61. $I_y = 0.286$ m^4

6–62. $I_x = 0.267$ m^4

6–63. $I_y = 1.22$ m^4

6–65. $I_x = \dfrac{2}{15}bh^3$

6–66. $I_x = 614$ m^4

6–67. $I_y = 85.3$ m^4

6–69. $I_y = \dfrac{\pi}{2}$ m^4

6–70. $I_x = 205$ in^4

6–71. $I_y = 780$ in^4

6–73. $I_y = \dfrac{b^3 h}{6}$

6–74. $I_x = 0.267$ m^4

6–75. $I_y = 0.305$ m^4

6–77. $I_y = 0.571$ m^4

6–78. $I_x = \dfrac{3ab^3}{35}$

6–79. $I_y = \dfrac{3a^3 b}{35}$

6–81. $I_y = 533$ in^4

6–82. $A = 14.0(10^3)$ mm^2

6–83. $\bar{y} = 91.7$ mm, $I_x = 216(10^6)$ mm^4

6–85. $I_x = 182$ in^4

6–86. $I_y = 966$ in^4

6–87. $I_x = 1.72(10^9)$ mm^4

6–89. $I_y = 115(10^6) \text{ mm}^4$

6–90. $\bar{y} = 207 \text{ mm}, \bar{I}_{x'} = 222(10^6) \text{ mm}^4$

6–91. $I_x = 511(10^6) \text{ mm}^4$

R6–1. Using an element of thickness dx,

$$\bar{x} = \frac{\int_A \tilde{x}\,dA}{\int_A dA} = \frac{\int_a^b x\left(\dfrac{c^2}{x}\,dx\right)}{c^2 \ln \dfrac{b}{a}} = \frac{\int_a^b c^2\,dx}{c^2 \ln \dfrac{b}{a}} = \frac{c^2 x \Big|_a^b}{c^2 \ln \dfrac{b}{a}} = \frac{b-a}{\ln \dfrac{b}{a}}$$

R6–2. Using an element of thickness dx,

$$\bar{y} = \frac{\int_A y\,dA}{\int_A dA} = \frac{\int_a^b \left(\dfrac{c^2}{2x}\right)\left(\dfrac{c^2}{x}\,dx\right)}{c^2 \ln \dfrac{b}{a}} = \frac{\int_a^b \dfrac{c^4}{2x^2}\,dx}{c^2 \ln \dfrac{b}{a}}$$

$$= \frac{-\dfrac{c^4}{2x}\Big|_a^b}{c^2 \ln \dfrac{b}{a}} = \frac{c^2(b-a)}{2ab \ln \dfrac{b}{a}}$$

R6–3. $\Sigma \tilde{x}L = 0(4) + 2(\pi)(2) = 12.5664 \text{ ft}^2$

$$\Sigma \tilde{y}L = 0(4) + \frac{2(2)}{\pi}(\pi)(2) = 8 \text{ ft}^2$$

$$\Sigma \tilde{z}L = 2(4) + 0(\pi)(2) = 8 \text{ ft}^2$$

$$\Sigma L = 4 + \pi(2) = 10.2832 \text{ ft}$$

$$\bar{x} = \frac{\Sigma \tilde{x}L}{\Sigma L} = \frac{12.5664}{10.2832} = 1.22 \text{ ft}$$

$$\bar{y} = \frac{\Sigma \tilde{y}L}{\Sigma L} = \frac{8}{10.2832} = 0.778 \text{ ft}$$

$$\bar{z} = \frac{\Sigma \tilde{z}L}{\Sigma L} = \frac{8}{10.2832} = 0.778 \text{ ft}$$

R6–5.

$$I_x = \int_A y^2\,dA = \int_0^2 y^2(4-x)\,dy = \int_0^2 y^2\left(4 - (32)^{\frac{1}{3}}y^{\frac{1}{3}}\right)dy$$

$$= 1.07 \text{ in}^4$$

R6–6.

$$I_y = \int_A x^2\,dA = 2\int_0^2 x^2(y\,dx) = 2\int_0^2 x^2(1 - 0.25\,x^2)\,dx$$

$$= 2.13 \text{ ft}^4$$

R6–7. $I_x = \left[\dfrac{1}{12}(d)(d^3) + 0\right] + 4\left[\dfrac{1}{36}(0.2887d)\left(\dfrac{d}{2}\right)^3\right.$

$$\left. + \dfrac{1}{2}(0.2887d)\left(\dfrac{d}{2}\right)\left(\dfrac{d}{6}\right)^2\right]$$

$$= 0.0954d^4$$

Chapter 7

7–1. $N_E = 0, V_E = -200 \text{ lb}, M_E = -2.40 \text{ kip} \cdot \text{ft}$

7–2. (a) $N_a = 500 \text{ lb}, V_a = 0$,
(b) $N_b = 433 \text{ lb}, V_b = 250 \text{ lb}$

7–3. $V_{b-b} = 2.475 \text{ kip}, N_{b-b} = 0.390 \text{ kip}$,
$M_{b-b} = 3.60 \text{ kip} \cdot \text{ft}$

7–5. $N_B = 0, V_B = 288 \text{ lb}$,
$M_B = -1.15 \text{ kip} \cdot \text{ft}$

7–6. $N_D = 0.703 \text{ kN}, V_D = 0.3125 \text{ kN}$,
$M_D = 0.3125 \text{ kN} \cdot \text{m}$

7–7. $N_F = 1.17 \text{ kN}, V_F = 0, M_F = 0, N_E = 0.703 \text{ kN}$,
$V_E = -0.3125 \text{ kN}, M_E = 0.3125 \text{ kN} \cdot \text{m}$

7–9. $N_D = 0, V_D = -3.25 \text{ kN}, M_D = 5.625 \text{ kN} \cdot \text{m}$

7–10. $N_A = 0, V_A = 450 \text{ lb}, M_A = -1.125 \text{ kip} \cdot \text{ft}$,
$N_B = 0, V_B = 850 \text{ lb}, M_B = -6.325 \text{ kip} \cdot \text{ft}$,
$V_C = 0, N_C = -1.20 \text{ kip}, M_C = -8.125 \text{ kip} \cdot \text{ft}$

7–11. $N_D = -527 \text{ lb}, V_D = -373 \text{ lb}, M_D = -373 \text{ lb} \cdot \text{ft}$,
$N_E = 75.0 \text{ lb}, V_E = 355 \text{ lb}, M_E = -727 \text{ lb} \cdot \text{ft}$

7–13. $N_{a-a} = -100 \text{ N}, V_{a-a} = 0, M_{a-a} = -15 \text{ N} \cdot \text{m}$

7–14. $N_{b-b} = -86.6 \text{ N}, V_{b-b} = 50 \text{ N}, M_{b-b} = -15 \text{ N} \cdot \text{m}$

7–15. $N_C = 0, V_C = -1.40 \text{ kip}, M_C = 8.80 \text{ kip} \cdot \text{ft}$

7–17. $(N_D)_x = 0, (V_D)_y = 154 \text{ N}, (V_D)_z = -171 \text{ N}$,
$(T_D)_x = 0, (M_D)_y = -94.3 \text{ N} \cdot \text{m}$,
$(M_D)_z = -149 \text{ N} \cdot \text{m}$

7–18. $(N_C)_x = 0, (V_C)_y = -246 \text{ N}, (V_C)_z = -171 \text{ N}$,
$(T_C)_x = 0, (M_C)_y = -154 \text{ N} \cdot \text{m}$,
$(M_C)_z = -123 \text{ N} \cdot \text{m}$

7–19. $(V_A)_x = 0, (N_A)_y = -25 \text{ lb}, (V_A)_z = 43.3 \text{ lb}$,
$(M_A)_x = 303 \text{ lb} \cdot \text{in.}, (T_A)_y = -130 \text{ lb} \cdot \text{in.}$,
$(M_A)_z = -75 \text{ lb} \cdot \text{in.}$

7–21. $N_E = -2.94 \text{ kN}, V_E = -2.94 \text{ kN}$,
$M_E = -2.94 \text{ kN} \cdot \text{m}$

7–22. $F_{BC} = 1.39 \text{ kN}, F_A = 1.49 \text{ kN}, N_D = 120 \text{ N}$,
$V_D = 0, M_D = 36.0 \text{ N} \cdot \text{m}$

7–23. $N_E = 0, V_E = 120 \text{ N}, M_E = 48.0 \text{ N} \cdot \text{m}$,
Short link: $V = 0, N = 1.39 \text{ kN}, M = 0$

7–25. $V_B = 496 \text{ lb}, N_B = 59.8 \text{ lb}, M_B = 480 \text{ lb} \cdot \text{ft}$,
$N_C = 495 \text{ lb}, V_C = 70.7 \text{ lb}, M_C = 1.59 \text{ kip} \cdot \text{ft}$

7–26. $\tau_{\text{avg}} = 119 \text{ MPa}$

7–27. $w = 16.0 \text{ kN/m}$

7–29. $F = 22.5 \text{ kip}, d = 0.833 \text{ in.}$

7–30. $P_{\text{allow}} = 9.12 \text{ kip}$

7–31. $\sigma = 78.9 \text{ psi}$

7–33. $\sigma = 2.92 \text{ psi}, \tau = 8.03 \text{ psi}$

7–34. $\sigma_{BC} = \left\{ \dfrac{1.528 \cos \theta}{\sin (45° + \theta/2)} \right\}$ ksi

7–35. $(\sigma_{avg})_{BC} = 159$ MPa, $(\sigma_{avg})_{AC} = 95.5$ MPa
$(\sigma_{avg})_{AB} = 127$ MPa

7–37. $(\tau_{avg})_A = 50.9$ MPa

7–38. $\tau_B = \tau_C = 81.9$ MPa, $\tau_A = 88.1$ MPa

7–39. $P = 4.54$ kN

7–41. $\sigma_{AB} = 333$ MPa, $\sigma_{CD} = 250$ MPa

7–42. $d = 1.20$ m

7–43. $\tau_B = 324$ MPa, $\tau_A = 324$ MPa

7–45. $\sigma_{a-a} = 90.0$ kPa, $\tau_{a-a} = 52.0$ kPa

7–46. $\sigma = 4.69$ MPa, $\tau = 8.12$ MPa

7–47. $\sigma = \{46.9 - 7.50x^2\}$ MPa

7–49. $\sigma = 66.7$ psi, $\tau = 115$ psi

7–50. $\sigma_{AB} = 127$ MPa, $\sigma_{AC} = 129$ MPa

7–51. $d_{AB} = 11.9$ mm

7–53. $d = 5.71$ mm

7–54. $P = 0.491$ kip

7–55. $P = 3.26$ kip

7–57. $F_H = 20.0$ kN, $F_{BF} = F_{AG} = 15.0$ kN,
$d_{EF} = d_{CG} = 11.3$ mm

7–58. For A': Use a 3 in. × 3 in. plate,
For B': Use a $4\frac{1}{2}$ in. × $4\frac{1}{2}$ in. plate

7–59. $P_{allow} = 1.16$ kip

7–61. $d_{AB} = 4.81$ mm, $d_{AC} = 5.22$ mm

7–62. $P = 5.83$ kN

7–63. $d = 13.8$ mm, $t = 7.00$ mm

7–65. $(F.S.)_{st} = 2.14$, $(F.S.)_{con} = 3.53$

7–66. $W = 680$ lb

7–67. Use $= \dfrac{7}{8}$ in.

7–69. $d_{AB} = 15.5$ mm, $d_{AC} = 13.0$ mm

7–70. $P = 7.54$ kN

7–71. $\epsilon = 0.167$ in./in.

7–73. $\epsilon_{CE} = 0.00250$ mm/mm, $\epsilon_{BD} = 0.00107$ mm/mm

7–74. $(\epsilon_{avg})_{AH} = 0.0349$ mm/mm,
$(\epsilon_{avg})_{CG} = 0.0349$ mm/mm,
$(\epsilon_{avg})_{DF} = 0.0582$ mm/mm

7–75. $\gamma_{xy} = -0.0200$ rad

7–77. $(\epsilon_{avg})_{AC} = 6.04(10^{-3})$ mm/mm

7–78. $\epsilon_{AB} = 0.0343$

7–79. $\epsilon_{AB} = \dfrac{0.5\Delta L}{L}$

7–81. $(\gamma_{xy})_C = 25.5(10^{-3})$ rad, $(\gamma_{xy})_D = 18.1(10^{-3})$ rad

7–82. $(\epsilon_x)_A = 0$, $(\epsilon_y)_A = 1.80(10^{-3})$ mm/mm,
$(\gamma_{xy})_A = 0.0599$ rad, $\epsilon_{BE} = -0.0198$ mm/mm

7–83. $\epsilon_{AD} = 0.0566$ mm/mm, $\epsilon_{CF} = -0.0255$ mm/mm

7–85. $(\gamma_C)_{xy} = 11.6(10^{-3})$ rad, $(\gamma_D)_{xy} = 11.6(10^{-3})$ rad

7–86. $\epsilon_{AC} = 1.60(10^{-3})$ mm/mm,
$\epsilon_{DB} = 12.8(10^{-3})$ mm/mm

7–87. $\epsilon_x = -0.03$ in./in., $\epsilon_y = 0.02$ in./in.

7–89. $\epsilon_x = 0.00443$ mm/mm

7–90. $\epsilon_{x'} = 0.00884$ mm/mm

7–91. $\gamma_A = 0$, $\gamma_B = 0.199$ rad

R7–1. $N_D = -2.16$ kip, $V_D = 0$, $M_D = 2.16$ kip · ft
$V_E = 0.540$ kip, $N_E = 4.32$ kip, $M_E = 2.16$ kip · ft

R7–2. $\sigma_s = 208$ MPa, $(\tau_{avg})_a = 4.72$ MPa,
$(\tau_{avg})_b = 45.5$ MPa

R7–3. $t = \dfrac{1}{4}$ in., $d_A = 1\dfrac{1}{8}$ in., $d_B = \dfrac{13}{16}$ in.

R7–5. $\tau_{avg} = 25.5$ MPa, $\sigma_b = 4.72$ MPa

R7–6. $\sigma_{a-a} = 200$ kPa, $\tau_{a-a} = 115$ kPa

R7–7. $(\epsilon_{avg})_{CA} = -5.59(10^{-3})$ mm/mm

R7–9. At $(b/2, a/2)$: $\gamma_{xy} = \tan^{-1}\left[\dfrac{3}{4}\left(\dfrac{v_0}{b}\right)\right]$,
At (b, a): $\gamma_{xy} = \tan^{-1}\left[3\left(\dfrac{v_0}{b}\right)\right]$

Chapter 8

8–1. $(\sigma_u)_{approx} = 110$ ksi, $(\sigma_f)_{approx} = 93.1$ ksi,
$(\sigma_Y)_{approx} = 55$ ksi, $E_{approx} = 32.0(10^3)$ ksi

8–2. $E = 55.3(10^3)$ ksi, $u_r = 9.96 \dfrac{\text{in} \cdot \text{lb}}{\text{in}^3}$

8–3. $(u_t)_{approx} = 85.0 \dfrac{\text{in} \cdot \text{lb}}{\text{in}^3}$

8–5. Elastic recovery = 0.00350 in.,
Permanent elongation = 0.1565 in.

8–6. $(u_r)_{approx} = 20.0 \dfrac{\text{in} \cdot \text{lb}}{\text{in}^3}$, $(u_t)_{approx} = 18.0 \dfrac{\text{in} \cdot \text{kip}}{\text{in}^3}$

8–7. $\delta_{AB} = 0.152$ in.

8–9. $\delta = 0.979$ in.

8–10. $E_{approx} = 10.0(10^3)$ ksi, $P_Y = 9.82$ kip,
$P_u = 13.4$ kip

8–11. Elastic recovery = 0.012 in.,
Permanent elongation = 0.0680 in.

8–13. $E = 28.6(10^3)$ ksi

8–14. $\delta_{BD} = 0.0632$ in.

8–15. $P = 570$ lb

8–17. $\delta_{AB} = 0.0913$ in.

8–18. $w = 228$ lb/ft

8–19. $\sigma_{YS} = 2.03$ MPa

8–21. $p = 741$ kPa, $\delta = 7.41$ mm

8–22. $\nu = 0.350$

8–23. $L = 51.14$ mm, $d = 12.67$ mm

8–25. $\gamma = 3.06\,(10^{-3})$ rad

8–26. $\gamma_P = 0.0189$ rad

8–27. $\epsilon_x = 0.0075$ in./in., $\epsilon_y = -0.00375$ in./in.,

        $\gamma_{xy} = 0.0122$ rad

8–29. $\delta = \dfrac{Pa}{2bhG}$

R8–1. $G_{al} = 4.31(10^3)$ ksi

R8–2. $d' = 0.4989$ in.

R8–3. $x = 1.53$ m, $d_A' = 30.008$ mm

R8–5. $P = 6.48$ kip

R8–6. $\epsilon = 0.000999$ in./in., $\epsilon_{unscr} = 0$

R8–7. $L = 10.17$ in.

R8–9. $\epsilon_b = 0.00227$ mm/mm, $\epsilon_s = 0.000884$ mm/mm

R8–10. $G = 5$ MPa

Chapter 9

9–1. $\delta_B = 2.93$ mm $\downarrow$, $\delta_A = 3.55$ mm $\downarrow$

9–2. $\delta_{A/D} = 0.111$ in. away from end D

9–3. $\sigma_{AB} = 22.2$ ksi (T), $\sigma_{BC} = 41.7$ ksi (C),

      $\sigma_{CD} = 25.0$ ksi (C),

      $\delta_{A/D} = 0.00157$ in. towards end D

9–5. $\delta_{A/E} = 0.697$ mm

9–6. $\sigma_A = 13.6$ ksi, $\sigma_B = 10.3$ ksi, $\sigma_C = 3.2$ ksi,

      $\delta_D = 2.99$ ft

9–7. $\delta_C = 0.0975$ mm $\rightarrow$

9–9. $\delta_F = 0.453$ mm

9–10. $P = 4.97$ kN

9–11. $\delta_l = 0.0260$ in.

9–13. $\delta_D = 17.3$ mm

9–14. $F = 8.00$ kN, $\delta_{A/B} = -0.311$ mm

9–15. $F = 4.00$ kN, $\delta_{A/B} = -0.259$ mm

9–17. $P = \dfrac{F_{max}\,L}{2}$, $\delta = \dfrac{F_{max}\,L^2}{3AE}$

9–18. $\delta_B = 0.262$ in.

9–19. $P = 57.3$ kip

9–21. $d_{AB} = 0.841$ in., $d_{CD} = 0.486$ in.

9–22. $x = 4.24$ ft, $w = 1.02$ kip/ft

9–23. $\delta_{A/D} = 0.129$ mm,

      $h' = 49.9988$ mm, $w' = 59.9986$ mm

9–25. $\delta_C = 0.00843$ in., $\delta_E = 0.00169$ in., $\delta_B = 0.0333$ in.

9–26. $P = 6.80$ kip

9–27. $P = 11.8$ kip

9–29. $\delta = 2.37$ mm

9–30. $\delta = \dfrac{2.63\,P}{\pi r E}$

9–31. $\sigma_{con} = 2.29$ ksi, $\sigma_{st} = 15.8$ ksi

9–33. $\sigma_{al} = 27.5$ MPa, $\sigma_{st} = 79.9$ MPa

9–34. $\sigma_{con} = 1.64$ ksi, $\sigma_{st} = 11.3$ ksi

9–35. $P = 114$ kip

9–37. $F_C = \left[\dfrac{9(8ka + \pi d^2 E)}{136ka + 18\pi d^2 E}\right]P,$

      $F_A = \left(\dfrac{64ka + 9\pi d^2 E}{136ka + 18\pi d^2 E}\right)P$

9–38. $T_{AC} = 0.806$ kip, $T_{AB} = 1.19$ kip

9–39. $A_{AB} = 0.0144$ in^2

9–41. $P = 126$ kN

9–42. $\sigma_{st} = 102$ MPa, $\sigma_{br} = 50.9$ MPa

9–43. $\sigma_{AB} = \sigma_{CD} = 26.5$ MPa, $\sigma_{EF} = 33.8$ MPa

9–45. $P_b = 14.4$ kN

9–46. $F_D = 20.4$ kN, $F_A = 180$ kN

9–47. $P = 198$ kN

9–49. $\delta_B = 0.0733$ in.

9–50. $\theta = 0.0875°$

9–51. $\delta_B = 0.00257$ in.

9–53. $F_D = 71.4$ kN, $F_C = 329$ kN

9–54. $F_D = 219$ kN, $F_C = 181$ kN

9–55. $\sigma_{BE} = 96.3$ MPa, $\sigma_{AD} = 79.6$ MPa,

      $\sigma_{CF} = 113$ MPa

9–57. $\sigma_{al} = 2.46$ ksi, $\sigma_{br} = 5.52$ ksi, $\sigma_{st} = 22.1$ ksi

9–58. $F = 0.510$ kip

9–59. $T_2 = 112°F$, $\sigma_{al} = \sigma_{cu} = 25.6$ ksi

9–61. $\sigma = 19.1$ ksi

9–62. $F = 7.60$ kip

9–63. $\delta = 0.348$ in., $F = 19.5$ kip

9–65. $F = \dfrac{\alpha A E}{2}(T_B - T_A)$

9–66. $\sigma = 180$ MPa

9–67. $\sigma = 105$ MPa

9–69. $F = 904$ N

9–70. $T_2 = 244°C$

9–71. $F_{AC} = F_{AB} = 10.0$ lb, $F_{AD} = 136$ lb

9–73. $F_{AB} = F_{EF} = 1.85$ kN

9–74. $d = \left(\dfrac{2E_2 + E_1}{3(E_2 + E_1)} \right) w$

R9–1. $\sigma_b = 33.5$ MPa, $\sigma_r = 16.8$ MPa

R9–2. $T = 507°C$

R9–3. $F_{AB} = F_{AC} = F_{AD} = 58.9$ kN (C)

R9–5. $F_B = 2.13$ kip, $F_A = 2.14$ kip

R9–6. $P = 4.85$ kip

R9–7. $F_B = 86.6$ lb, $F_C = 195$ lb

Chapter 10

10–1. $r' = 0.841r$

10–2. $r' = 0.707\ r$

10–3. $T = 19.6$ kN·m, $T' = 13.4$ kN·m

10–5. $\tau_A = 3.45$ ksi, $\tau_B = 2.76$ ksi

10–6. $(T_1)_{max} = 2.37$ kN·m, $(\tau_{max})_{CD} = 35.6$ MPa, $(\tau_{max})_{DE} = 23.3$ MPa

10–7. $\tau^{abs}_{max} = 44.8$ MPa

10–9. $\tau_B = 6.79$ MPa, $\tau_A = 7.42$ MPa

10–10. $\tau_{max} = 14.5$ MPa

10–11. $\tau_{AB} = 7.82$ ksi, $\tau_{BC} = 2.36$ ksi

10–13. $\tau_i = 34.5$ MPa, $\tau_o = 43.1$ MPa

10–14. Use $d = 1\dfrac{3}{4}$ in.

10–15. $(\tau_{AB})_{max} = 23.9$ MPa, $(\tau_{BC})_{max} = 15.9$ MPa

10–17. $\tau_{max} = 4.89$ ksi

10–18. $\tau_{max} = 7.33$ ksi

10–19. $\tau_A = 1.31$ ksi, $\tau_B = 2.62$ ksi

10–21. $\tau_A = 9.43$ MPa, $\tau_B = 14.1$ MPa

10–22. $\tau^{abs}_{max} = 0$ occurs at $x = 0.700$ m, $\tau^{abs}_{max} = 33.0$ MPa occurs at $x = 0$.

10–23. $d = 34.4$ mm

10–25. $d = 46.7$ mm

10–26. $\tau_{max} = 3.44$ MPa

10–27. $\tau_{max} = 856$ psi

10–29. $\tau_{max} = 1.07$ ksi

10–30. Use $d = \dfrac{7}{8}$ in.

10–31. Use $d = \dfrac{11}{16}$ in.

10–33. $(\tau_{max})_{CF} = 12.5$ MPa $(\tau_{max})_{BC} = 7.26$ MPa

10–34. $\tau^{abs}_{max} = 12.5$ MPa

10–35. $\tau_{max} = 44.3$ MPa, $\phi = 11.9°$

10–37. $\phi_{B/A} = 0.730°$ ↶

10–38. $\tau^{abs}_{max} = 10.2$ MPa

10–39. $T = 5.09$ kN·m, $\phi_{A/C} = 3.53°$

10–41. $\phi_A = 1.57°$ ↻

10–42. Use $d = 22$ mm, $\phi_{A/D} = 2.54°$ ↻

10–43. Use $d = 25$ mm

10–45. $\tau_{max} = 9.12$ MPa, $\phi_{E/B} = 0.585°$

10–46. $\tau_{max} = 14.6$ MPa, $\phi_{B/E} = 1.11°$

10–47. $\phi_{B/D} = 1.15°$

10–49. $\tau^{abs}_{max} = 20.4$ MPa,
For $0 \le x < 0.5$ m,
$\phi(x) = \{ 0.005432 \ (x^2 + x) \}$ rad
For 0.5 m $< x \le 1$ m,
$\phi(x) = \{ -0.01086x^2 + 0.02173\ x - 0.004074 \}$ rad

10–50. $\tau^{abs}_{max} = 24.3$ MPa, $\phi_{D/A} = 0.929°$

10–51. $\phi_{A/C} = 5.45°$

10–53. $\phi_{D/C} = 0.0823$ rad, $\tau^{abs}_{max} = 34.0$ ksi

10–54. $t = 7.53$ mm

10–55. $\omega = 131$ rad/s

10–57. $k = 1.20(10^6)$ N/m², $\phi = 3.56°$

10–58. $k = 12.3(10^3)$ N/m$^{2/3}$, $\phi = 2.97°$

10–59. $d_t = 201$ mm, $\phi = 3.30°$

10–61. $\phi_{F/E} = 0.999(10)^{-3}$ rad, $\phi_{F/D} = 0.999(10)^{-3}$ rad, $\tau_{max} = 3.12$ MPa

10–62. $(\tau_{AC})_{max} = 14.3$ MPa, $(\tau_{CB})_{max} = 9.55$ MPa

10–63. $\tau^{abs}_{max} = 9.77$ MPa

10–65. $F = 23.4$ lb

10–66. $\tau_{max} = 389$ psi

10–67. $(\tau_{max})_{AC} = 68.2$ MPa, $(\tau_{max})_{BC} = 90.9$ MPa

10–69. $T = 4.34$ kN·m, $\phi_A = 2.58°$

10–70. $(\tau_{st})_{max} = 86.5$ MPa, $(\tau_{mg})_{max} = 41.5$ MPa, $(\tau_{mg})|_{\rho=0.02\text{m}} = 20.8$ MPa

10–71. $T_B = 22.2$ N·m, $T_A = 55.6$ N·m

10–73. $d = 42.7$ mm

10–74. $\tau^{abs}_{max} = 64.1$ MPa

10–75. $(\tau_{BD})_{max} = 4.35$ ksi, $(\tau_{AC})_{max} = 2.17$ ksi

10–77. $\tau^{abs}_{max} = 93.1$ MPa

10–78. $T_B = \dfrac{37}{189}T$, $T_A = \dfrac{152}{189}T$

10–79. $T_B = \dfrac{7t_0 L}{12}$, $T_A = \dfrac{3t_0 L}{4}$

R10–1. Use $d = 26$ mm, $\phi_{A/C} = 2.11°$

R10–2. Use $d = 28$ mm.

R10–3. $\tau = 88.3$ MPa, $\phi = 4.50°$

R10–5. Use $d = 1\dfrac{3}{8}$ in.

R10–6. $(\tau_{max})_{AB} = 3.60$ ksi, $(\tau_{max})_{BC} = 10.7$ ksi

R10–7. $P = 2.80$ kip

R10–9. $P = 1.10$ kW, $\tau_{max} = 825$ kPa

Chapter 11

11–1. For $0 \leq x < 3$ ft:
$V = 170$ lb,
$M = \{170x\}$ lb·ft,
For 3 ft $< x < 5$ ft:
$V = -630$ lb,
$M = \{-630x + 2400\}$ lb·ft,
For 5 ft $< x \leq 6$ ft:
$V = 500$ lb,
$M = \{500x - 3250\}$ lb·ft

11–2. For $0 < x \leq 4$ ft:
$V = -250$ lb,
$M = \{-250\,x\}$ lb·ft,
For 4 ft $\leq x \leq 10$ ft: $V = \{1050 - 150\,x\}$ lb,
$M = \{-75x^2 + 1050x - 4000\}$ lb·ft,
For 10 ft $< x \leq 14$ ft: $V = 250$ lb,
$M = \{250x - 3500\}$ lb·ft

11–3. For $0 \leq x < 6$ ft:
$V = \{30.0 - 2x\}$ kip,
$M = \{-x^2 + 30.0x - 216\}$ kip·ft,
For 6 ft $< x \leq 10$ ft:
$V = 8.00$ kip,
$M = \{8.00x - 120\}$ kip·ft

11–5. $V = \{-300 - 16.67x^2\}$ lb,
$M = \{-300x - 5.556x^3\}$ lb·ft

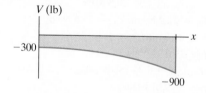

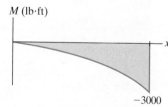

11–6. $V = 15.6$ N, $M = \{15.6x + 100\}$ N·m

11–7. For $0 \leq x < \dfrac{L}{2}$: $V = \dfrac{w_0 L}{24}$, $M = \dfrac{w_0 L}{24}x$,

For $\dfrac{L}{2} < x \leq L$: $V = \dfrac{w_0}{24L}\left[L^2 - 6(2x - L)^2\right]$,

$M = \dfrac{w_0}{24L}\left[L^2 x - (2x - L)^3\right]$

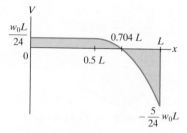

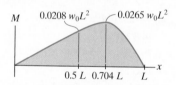

11–9. $T_1 = 250$ lb, $T_2 = 200$ lb

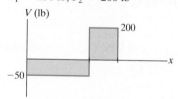

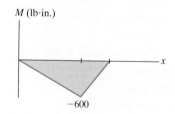

11–10.

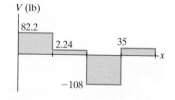

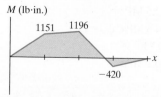

11–11.

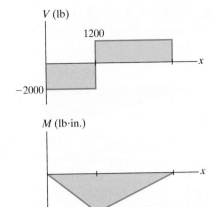

11–13. $V = -\dfrac{M_0}{L},$

For $0 \le x < \dfrac{L}{2}, M = M_0 - \left(\dfrac{M_0}{L}\right)x,$

For $\dfrac{L}{2} < x \le L, M = -\left(\dfrac{M_0}{L}\right)x$

11–14. For $0 \le x < 5$ ft:

$V = \{-2x\}$ kip,

$M = \{-x^2\}$ kip $\cdot$ ft,

For 5 ft $< x < 10$ ft:

$V = -0.5$ kip,

$M = \{-22.5 - 0.5x\}$ kip $\cdot$ ft,

For 10 ft $< x \le 15$ ft:

$V = -0.5$ kip,

$M = \{7.5 - 0.5x\}$ kip $\cdot$ ft

11–15.

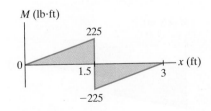

11–17. $V = 1050 - 150x$

$M = -75x^2 + 1050x - 3200$

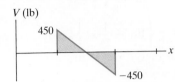

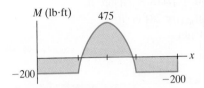

11–18.

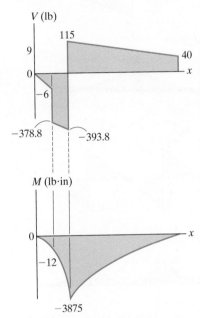

11–19. $a = 0.866\,L, M_{max} = 0.134\,PL$

11–21. For $0 \le x < 5$ ft:

$V = \{-2x\}$ kip,

$M = \{-20.0 - x^2\}$ kip $\cdot$ ft,

For 5 ft $< x < 10$ ft:

$V = 3.00$ kip,

$M = \{-20.0 + 3x\}$ kip $\cdot$ ft,

For 10 ft $< x \le 15$ ft:

$V = \{23 - 2x\}$ kip,

$M = \{-120 + 23x - x^2\}$ kip $\cdot$ ft

11–22. For $0 \le x < 12$ ft:

$$V = \left\{10 - \frac{1}{8}x^2\right\} \text{kip},$$

$$M = \left\{10x - \frac{1}{24}x^3\right\} \text{kip} \cdot \text{ft},$$

For 12 ft $< x \le 18$ ft:

$V = -8$ kip,

$M = [8(18 - x)]$ kip $\cdot$ ft

11–23. $M_{max} = 281$ lb $\cdot$ ft

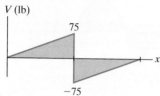

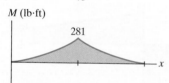

11–25.

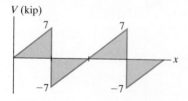

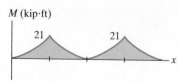

11–26. $V_{AB} = -1.625$ kip, $M_B = -18$ kip $\cdot$ ft

11–27.

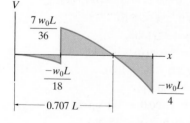

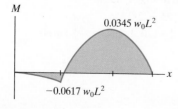

11–29. $V_B = -45$ kN, $M_B = -63$ kN $\cdot$ m

11–30. $V|_{x=15 \text{ ft}} = 1.12$ kip, $M|_{x=15} = -1.95$ kip $\cdot$ ft

11–31.

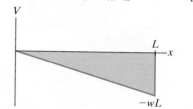

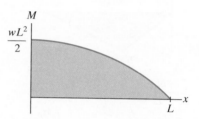

11–33.

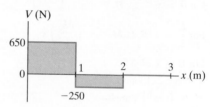

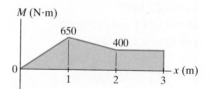

11–34.

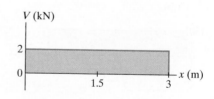

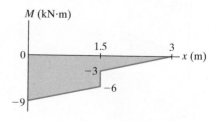

11–35. For $0 \leq x < 3$ m: $V = 200$ N, $M = \{200x\}$ N·m,

For 3 m $< x \leq 6$ m: $V = \left\{ -\dfrac{100}{3}x^2 + 500 \right\}$ N,

$M = \left\{ -\dfrac{100}{9}x^3 + 500x - 600 \right\}$ N·m

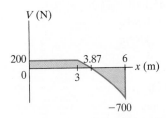

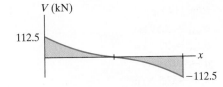

11–37.

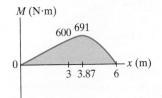

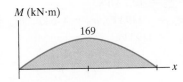

11–38.

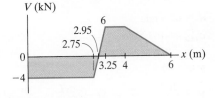

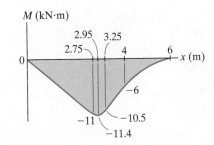

11–39.

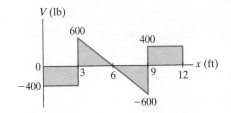

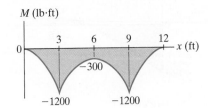

11–41.

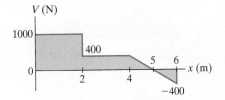

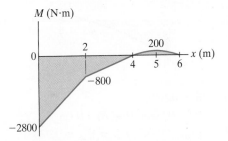

11–42.

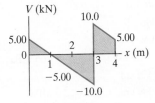

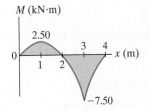

11–43.

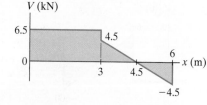

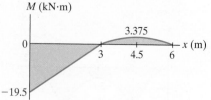

11–45.

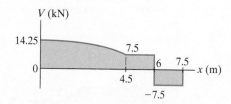

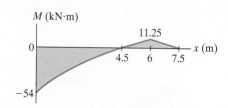

11–46. $a = 0.207L$

11–47. $r = 909$ mm, $M = 61.9$ N·m

11–49. $(\sigma_t)_{max} = 3.72$ ksi, $(\sigma_c)_{max} = 1.78$ ksi

11–50. $\sigma_{max} = 1.46$ ksi

11–51. $F = 10.5$ kip

11–53. $F = 4.56$ kN

11–54. $(\sigma_{max})_c = 78.1$ MPa, $(\sigma_{max})_t = 165$ MPa

11–55. $M = 50.3$ kN·m

11–57. $M = 15.6$ kN·m, $\sigma_{max} = 12.0$ MPa

11–58. $\sigma_{max} = 93.0$ psi

11–59. $F = 753$ lb

11–61. % of moment carried by web = 22.6%

11–62. $\sigma_A = 199$ MPa, $\sigma_B = 66.2$ MPa

11–63. $\sigma_{max} = 20.4$ ksi

11–65. (a) $\sigma_{max} = 497$ kPa, (b) $\sigma_{max} = 497$ kPa

11–66. (a) $\sigma_{max} = 249$ kPa, (b) $\sigma_{max} = 249$ kPa

11–67. $\sigma_{max} = 158$ MPa

11–69. $\sigma_{max} = 12.2$ ksi

11–70. $\sigma_{max} = 2.70$ ksi

11–71. $\sigma_{max} = 21.1$ ksi

11–73. $d = 1.28$ in.

11–74. $\sigma_{max} = 45.1$ ksi

11–75. $\sigma_{max} = 52.8$ MPa

11–77. $w = 18.75$ kN/m

11–78. Use $b = 3\frac{5}{8}$ in.

11–79. $w_0 = 415$ lb/ft

11–81. $a = 66.9$ mm

11–82. $\sigma_{max} = \dfrac{23w_0 L^2}{36\, bh^2}$

11–83. $\sigma_{max} = 119$ MPa

11–85. $\sigma_{abs}^{max} = 24.0$ ksi

11–86. $\sigma_{abs}^{max} = 6.88$ ksi

11–87. Use $d = 1\frac{3}{8}$ in.

11–89. $P = 114$ kip

11–90. $\sigma_{max} = 7.59$ ksi

11–91. $\sigma_{max} = 22.1$ ksi

11–93. $d = 410$ mm

11–94. $w = 937.5$ N/m

11–95. $\sigma_{abs}^{max} = 10.7$ MPa

11–97. $\sigma_A = -119$ kPa, $\sigma_B = 446$ kPa, $\sigma_D = -446$ kPa, $\sigma_E = 119$ kPa

11–98. $a = 0, b = -\left(\dfrac{M_z I_y + M_y I_{yz}}{I_y I_z - I_{yz}^2}\right), c = \dfrac{M_y I_z + M_z I_{yz}}{I_y I_z - I_{yz}^2}$

11–99. $\sigma_A = 21.0$ ksi (C)

11–101. $d = 62.9$ mm

11–102. $\sigma_{max} = 163$ MPa

11–103. $\sigma_A = 20.6$ MPa (C)

R11–1. $k = 1.22$

R11–2. $V = \dfrac{2wL}{27} - \dfrac{w}{2L}x^2, M = \dfrac{2wL}{27}x - \dfrac{w}{6L}x^3$

R11–3. $\sigma_{max} = 0.410$ MPa

R11–5. $V = 20 - 2x, M = -x^2 + 20x - 166$

R11–6. Case (a), $\Delta\sigma_{max} = 2.49\left(\dfrac{M}{a^3}\right)$

R11–7. $V|_{x=600\ mm^-} = -233$ N, $M|_{x=600\ mm} = -50$ N·m

Chapter 12

12–1. $\tau_A = 2.56$ MPa

12–2. $\tau_{max} = 3.46$ MPa

12–3. $V_w = 19.0$ kN

12–5. $\tau_{max} = 3.91$ MPa

12–6. $V_{max} = 100$ kN

12–7. $\tau_{max} = 17.9$ MPa

12–9. $V_{max} = 32.1$ kip

12–10. $\tau_{max} = 4.48$ ksi

12–11. $\tau_{max} = 45.0$ MPa

12–13. $\tau_{max} = 4.22$ MPa

12–14. $V_{max} = 190$ kN

12–15. $V = 9.96$ kip

12–17. $\tau_A = 1.99$ MPa, $\tau_B = 1.65$ MPa

12–18. $\tau_{max} = 4.62$ MPa

12–19. $V_w = 27.1$ kN

12–21. $P_{max} = 1.28$ kip

12–22. $w_{max} = 2.15$ kip/ft

12–23. $\tau_{max} = 298$ psi

12–25. $\tau_{max} = 4.85$ MPa

12–26. $w = 11.3$ kip/ft, $\tau_{max} = 531$ psi

12–27. $\tau_{max} = 22.0$ MPa, $(\tau_{max})_s = 66.0$ MPa

12–29. $\tau_B = 4.41$ MPa

12–30. $\tau_{max} = 3.67$ MPa

12–31. $\alpha = 1.27$ in.

12–33. $F = 300$ lb

12–34. $s_t = 1.42$ in., $s_b = 1.69$ in.

12–35. $V = 4.97$ kip, $s_t = 1.14$ in.,
$s_b = 1.36$ in.

12–37. $V = 499$ kN

12–38. $P = 4.97$ kip.
For regions AB and CD,
$s_{top} = 1.14$ in., $s_{bottom} = 1.36$ in.
For region CD, theoretically no nails are required.

12–39. $F = 12.5$ kN

12–41. $s = 71.3$ mm

12–42. $V_{max} = 8.82$ kip, use $s = 1\frac{1}{8}$ in.

12–43. $P_{max} = 317$ lb

12–45. $\tau_{max} = 1.83$ ksi

12–46. $w_0 = 983$ lb/ft

12–47. $s = 8.66$ in., $s' = 1.21$ in.

R12–1. $F_C = 197$ lb, $F_D = 1.38$ kip

R12–2. $V = 131$ kN

R12–3. $q_A = 0$, $q_B = 1.21$ kN/m, $q_C = 3.78$ kN/m

R12–5. $\tau_B = 795$ psi, $\tau_C = 596$ psi

R12–6. $\tau_{max} = 928$ psi

R12–7. $V_{AB} = 1.47$ kip

R12–9. $\tau_{max} = 7.38$ ksi

Chapter 13

13–1. $t = 18.8$ mm

13–2. $r_o = 75.5$ in.

13–3. (a) $\sigma_1 = 1.04$ ksi, $\sigma_2 = 0$,
(b) $\sigma_1 = 1.04$ ksi, $\sigma_2 = 520$ psi

13–5. $\sigma_1 = 7.07$ MPa, $\sigma_2 = 0$

13–6. $P = 848$ N

13–7. (a) $\sigma_1 = 127$ MPa,
(b) $\sigma_1' = 79.1$ MPa,
(c) $(\tau_{avg})_b = 322$ MPa

13–9. $\sigma_{hoop} = 7.20$ ksi, $\sigma_{long} = 3.60$ ksi

13–10. $\sigma_1 = 1.60$ ksi, $p = 25$ psi, $\delta = 0.00140$ in.

13–11. $\sigma_2 = 11.5$ ksi, $\sigma_1 = 24$ ksi

13–13. $T_1 = 128°F$, $\sigma_1 = 12.1$ ksi, $p = 252$ psi

13–14. $\delta r_i = \dfrac{pr_i^2}{E(r_o - r_i)}$

13–15. $p = \dfrac{E(r_2 - r_3)}{\dfrac{r_2^2}{r_2 - r_1} + \dfrac{r_3^2}{r_4 - r_3}}$

13–17. $\sigma_{fil} = \dfrac{pr}{t + t'w/L} + \dfrac{T}{wt'}$, $\sigma_w = \dfrac{pr}{t + t'w/L} - \dfrac{T}{Lt}$

13–18. $d = 66.7$ mm

13–19. $d = 133$ mm

13–21. $\sigma_{max} = \sigma_B = 13.9$ ksi (T), $\sigma_A = 13.6$ ksi (C)

13–22. $P_{max} = 2.01$ kip

13–23. $\sigma_{max} = 22.4$ ksi (T)

13–25. $P_{max} = 128$ kN

13–26. $\sigma_{max} = 44.0$ ksi (T)

13–27. $\sigma_{max} = 44.0$ ksi (T), $\sigma_{min} = 41.3$ ksi (C)

13–29. $\sigma_A = 0.800$ ksi (T), $\sigma_B = 5.20$ ksi (C),
$\tau_A = 1.65$ ksi, $\tau_B = 0$

13–30. $\sigma_A = 25$ MPa (C), $\sigma_B = 0$, $\tau_A = 0$, $\tau_B = 5$ MPa

13–31. $\sigma_A = 28.8$ ksi, $\tau_A = 0$

13–33. $\sigma_A = 70.0$ MPa (C), $\sigma_B = 10.0$ MPa (C)

13–34. $\sigma_A = 70.0$ MPa (C), $\sigma_B = 10.0$ MPa (C),
$\sigma_C = 50.0$ MPa (T), $\sigma_D = 10.0$ MPa (C)

13–35. $\sigma_A = 27.3$ ksi (T), $\sigma_B = 0.289$ ksi (T),
$\tau_A = 0$, $\tau_B = 0.750$ ksi

13–37. $\sigma_B = 1.53$ MPa (C), $(\tau_{xz})_B = 0$,
$(\tau_{xy})_B = 100$ MPa

13–38. $\sigma_D = -88.0$ MPa, $\tau_D = 0$

13–39. $\sigma_E = 57.8$ MPa, $\tau_E = 864$ kPa

13–41. $\sigma_B = 27.5$ MPa (C), $(\tau_{xz})_B = -8.81$ MPa,
$(\tau_{xy})_B = 0$

13–42. $(\sigma_A)_y = 16.2$ ksi (T), $(\tau_A)_{yx} = -2.84$ ksi,
$(\tau_A)_{yz} = 0$

13–43. $(\sigma_B)_y = 7.80$ ksi (T), $(\tau_B)_{yz} = 3.40$ ksi,
$(\tau_B)_{yx} = 0$

13–45. $\sigma_A = 1.00$ ksi (C), $\sigma_B = 3.00$ ksi (C),
$\sigma_C = 1.00$ ksi (C), $\sigma_D = 1.00$ ksi (T)

13–46. $P = \dfrac{\delta_{max}\pi\left(r_0^4 - r_i^4\right)}{r_0^2 + r_i^2 + 4er_0}$

13–47. $\sigma_A = 224$ MPa (T), $(\tau_{xz})_A = -30.7$ MPa,
$(\tau_{xy})_A = 0$

13–49. $\sigma_C = 295$ MPa (C), $(\tau_{xy})_C = 25.9$ MPa,
$(\tau_{xz})_C = 0$

13–50. $(\sigma_{max})_t = 106$ MPa, $(\sigma_{max})_c = -159$ MPa

13–51. $P_{max} = 9.08$ kN

13–53. $\sigma_F = 17.7$ ksi (C), $\sigma_E = 125$ ksi (C),
$(\tau_{xz})_E = -62.4$ ksi, $(\tau_{xy})_E = 0$, $(\tau_{xy})_F = 67.2$ ksi,
$(\tau_{xz})_F = 0$

R13–1. $(\sigma_t)_{max} = 15.8$ ksi, $(\sigma_c)_{max} = -10.5$ ksi

R13–2. $\sigma_E = 802$ kPa, $\tau_E = 69.8$ kPa

R13–3. $\sigma_F = 695$ kPa (C), $\tau_A = 31.0$ kPa

R13–5. $\sigma_{max} = 236$ psi (C)

R13–6. $\theta = 0.286°$

R13–7. $\sigma_C = 11.6$ ksi, $\tau_C = 0$, $\sigma_D = -23.2$ ksi, $\tau_D = 0$

Chapter 14

14–2. $\sigma_{x'} = 31.4$ MPa, $\tau_{x'y'} = 38.1$ MPa

14–3. $\sigma_{x'} = -388$ psi, $\tau_{x'y'} = 455$ psi

14–5. $\sigma_{x'} = 1.45$ ksi, $\tau_{x'y'} = 3.50$ ksi

14–6. $\sigma_{x'} = -4.05$ ksi, $\tau_{x'y'} = -0.404$ ksi

14–7. $\sigma_{x'} = -61.5$ MPa, $\tau_{x'y'} = 62.0$ MPa

14–9. $\sigma_{x'} = 36.0$ MPa, $\tau_{x'y'} = -37.0$ MPa

14–10. $\sigma_{x'} = 36.0$ MPa, $\tau_{x'y'} = -37.0$ MPa

14–11. $\sigma_{x'} = 47.5$ MPa, $\sigma_{y'} = 202$ MPa,
$\tau_{x'y'} = -15.8$ MPa

14–13. $\sigma_{x'} = -62.5$ MPa, $\tau_{x'y'} = -65.0$ MPa

14–14. $\sigma_1 = 319$ MPa, $\sigma_2 = -219$ MPa, $\theta_{p1} = 10.9°$,
$\theta_{p2} = -79.1°$, $\tau_{\substack{max \\ \text{in-plane}}} = 269$ MPa,
$\theta_s = -34.1°$ and $55.9°$, $\sigma_{avg} = 50.0$ MPa

14–15. $\sigma_1 = 53.0$ MPa, $\sigma_2 = -68.0$ MPa, $\theta_{p1} = 14.9°$,
$\theta_{p2} = -75.1°$, $\tau_{\substack{max \\ \text{in-plane}}} = 60.5$ MPa,
$\sigma_{avg} = -7.50$ MPa, $\theta_s = -30.1°$ and $59.9°$

14–17. $\sigma_1 = 137$ MPa, $\sigma_2 = -86.8$ MPa,
$\theta_{p1} = -13.3°$, $\theta_{p2} = 76.7°$, $\tau_{\substack{max \\ \text{in-plane}}} = 112$ MPa,
$\theta_s = 31.7°$ and $122°$, $\sigma_{avg} = 25$ MPa

14–18. $\sigma_x = 33.0$ MPa, $\sigma_y = 137$ MPa, $\tau_{xy} = -30$ MPa

14–19. $\sigma_1 = 5.90$ MPa, $\sigma_2 = -106$ MPa,
$\theta_{p1} = 76.7°$ and $\theta_{p2} = -13.3°$,
$\tau_{\substack{max \\ \text{in-plane}}} = 55.9$ MPa, $\sigma_{avg} = -50$ MPa,
$\theta_s = 31.7°$ and $122°$

14–21. $\tau_a = -1.96$ ksi, $\sigma_1 = 80.1$ ksi, $\sigma_2 = 19.9$ ksi

14–22. $\sigma_{x'} = -63.3$ MPa, $\tau_{x'y'} = 35.7$ MPa

14–23. $\sigma_1 = 2.29$ MPa, $\sigma_2 = -7.20$ kPa, $\theta_p = -3.21°$

14–25. $\sigma_1 = 16.6$ MPa, $\sigma_2 = 0$, $\tau_{\substack{max \\ \text{in-plane}}} = 8.30$ MPa

14–26. $\sigma_1 = 14.2$ MPa, $\sigma_2 = -8.02$ MPa,
$\tau_{\substack{max \\ \text{in-plane}}} = 11.1$ MPa

14–27. $M = 8.73$ kip $\cdot$ in.

14–29. Point A: $\sigma_1 = \sigma_y = 0$, $\sigma_2 = \sigma_x = -30.5$ MPa,
Point B: $\sigma_1 = 0.541$ MPa, $\sigma_2 = -1.04$ MPa,
$\theta_{p1} = -54.2°$, $\theta_{p2} = 35.8°$

14–30. $\tau_{x'y'} = 160$ kPa

14–31. $\sigma_{x'} = -191$ kPa

14–33. Point A: $\sigma_1 = 37.8$ kPa, $\sigma_2 = -10.8$ MPa
Point B: $\sigma_1 = 42.0$ MPa, $\sigma_2 = -10.6$ kPa

14–34. $\sigma_1 = 233$ psi, $\sigma_2 = -774$ psi, $\tau_{\substack{max \\ \text{in-plane}}} = 503$ psi

14–35. $\sigma_1 = 382$ psi, $\sigma_2 = -471$ psi, $\tau_{\substack{max \\ \text{in-plane}}} = 427$ psi

14–37. $\sigma_1 = 838$ psi, $\sigma_2 = -37.8$ psi

14–38. $\sigma_1 = 628$ psi, $\sigma_2 = -166$ psi

14–39. $\sigma_1 = 198$ MPa, $\sigma_2 = -1.37$ MPa

14–41. $\sigma_1 = 111$ MPa, $\sigma_2 = 0$

14–42. $\sigma_1 = 2.40$ MPa, $\sigma_2 = -6.68$ MPa

14–43. $\sigma_{x'} = 31.4$ MPa, $\tau_{x'y'} = 38.1$ MPa

14–45. $\sigma_{x'} = -4.05$ ksi, $\tau_{x'y'} = -0.404$ ksi

14–46. $\sigma_{x'} = 47.5$ MPa, $\tau_{x'y'} = -15.8$ MPa,
$\sigma_{y'} = 202$ MPa

14–47. $\sigma_1 = 53.0$ MPa, $\sigma_2 = -68.0$ MPa,
$\theta_{p1} = 14.9°$ counterclockwise, $\tau_{\substack{max \\ in\text{-}plane}} = 60.5$ MPa,
$\sigma_{avg} = -7.50$ MPa, $\theta_{s1} = 30.1°$ clockwise

14–50. $\sigma_{avg} = -40.0$ MPa, $\sigma_1 = 32.1$ MPa,
$\sigma_2 = -112$ MPa, $\theta_{p1} = 28.2°$, $\tau_{\substack{max \\ in\text{-}plane}} = 72.1$ MPa,
$\theta_s = -16.8°$

14–51. $\sigma_{avg} = -4.00$ ksi, $\sigma_1 = 14.9$ ksi,
$\sigma_2 = -22.9$ ksi, $\theta_{p1} = -74.0°$, $\tau_{\substack{max \\ in\text{-}plane}} = -18.9$ ksi,
$\theta_s = -29.0°$

14–55. $\sigma_{avg} = 60.0$ MPa, $\sigma_1 = 117$ MPa,
$\sigma_2 = 3.43$ MPa, $\tau_{\substack{max \\ in\text{-}plane}} = -56.6$ MPa, $\theta_s = 22.5°$,
$\theta_{p1} = 22.5°$ **(Clockwise)**

14–57. $\sigma_1 = 64.1$ MPa, $\sigma_2 = -14.1$ MPa, $\theta_P = 25.1°$,
$\sigma_{avg} = 25.0$ MPa, $\tau_{\substack{max \\ in\text{-}plane}} = 39.1$ MPa, $\theta_s = -19.9°$

14–58. $\theta_P = -14.9°$, $\sigma_1 = 227$ MPa, $\sigma_2 = -177$ MPa,
$\tau_{\substack{max \\ in\text{-}plane}} = 202$ MPa, $\sigma_{avg} = 25$ MPa, $\theta_s = 30.1°$

14–59. $\sigma_1 = 12.3$ ksi, $\sigma_2 = -17.3$ ksi, $\theta_P = -16.3°$,
$\tau_{\substack{max \\ in\text{-}plane}} = 14.8$ ksi, $\sigma_{avg} = -2.5$ ksi, $\theta_s = 28.7°$

14–61. $\sigma_{x'} = 19.5$ kPa, $\tau_{x'y'} = -53.6$ kPa

14–62. $\tau_{\substack{max \\ in\text{-}plane}} = 41.0$ psi, $\sigma_1 = 0.976$ psi, $\sigma_2 = -81.0$ psi

14–63. $\sigma_{avg} = 5$ MPa, $\sigma_1 = 88.8$ MPa, $\sigma_2 = -78.8$ MPa,
$\theta_P = 36.3°$ *(Counterclockwise)*, $\tau_{\substack{max \\ in\text{-}plane}} = 83.8$ MPa,
$\theta_s = 8.68°$ *(Clockwise)*

14–65. $\sigma_{x'} = 75.3$ kPa, $\tau_{x'y'} = -78.5$ kPa

14–66. $\sigma_{x'} = -45.0$ kPa, $\tau_{x'y'} = 45.0$ kPa

14–67. $\sigma_1 = 3.85$ ksi, $\sigma_2 = -2.08$ ksi, $\tau_{\substack{max \\ in\text{-}plane}} = 2.96$ ksi

14–69. Mohr's circle is a point located at (4.80, 0).

14–70. $\sigma_{x'} = 500$ MPa, $\tau_{x'y'} = -167$ MPa

14–71. $\sigma_{x'} = 470$ kPa, $\tau_{x'y'} = 592$ kPa

14–73. $\sigma_1 = 2.97$ ksi, $\sigma_2 = -2.97$ ksi, $\theta_{p1} = 45.0°$,
$\theta_{p2} = -45.0°$, $\tau_{\substack{max \\ in\text{-}plane}} = 2.97$ ksi, $\theta_s = 0$

14–74. $\sigma_1 = 2.59$ ksi, $\sigma_2 = -3.61$ ksi, $\theta_{p1} = -40.3°$,
$\theta_{p2} = -49.7°$, $\tau_{\substack{max \\ in\text{-}plane}} = 3.10$ ksi, $\theta_s = 4.73°$

14–75. $\tau_{\substack{max \\ in\text{-}plane}} = 322$ psi, $\sigma_1 = 639$ psi, $\sigma_2 = -5.50$ psi

14–79. $\sigma_{max} = 158$ psi, $\sigma_{min} = -8.22$ psi,
$\sigma_{int} = 0$ psi, $\tau_{\substack{abs \\ max}} = 83.2$ psi

14–81. $\sigma_1 = 222$ MPa, $\sigma_2 = -102$ MPa, $\tau_{\substack{abs \\ max}} = 162$ MPa

14–82. $\sigma_1 = 6.73$ ksi, $\sigma_2 = -4.23$ ksi, $\tau_{\substack{abs \\ max}} = 5.48$ ksi

14–85. $\epsilon_{x'} = 248(10^{-6})$, $\gamma_{x'y'} = -233(10^{-6})$,
$\epsilon_{y'} = -348(10^{-6})$

14–86. $\epsilon_{x'} = 55.1(10^{-6})$, $\gamma_{x'y'} = 133(10^{-6})$,
$\epsilon_{y'} = 325(10^{-6})$

14–87. $\epsilon_{x'} = 325(10^{-6})$, $\gamma_{x'y'} = -133(10^{-6})$,
$\epsilon_{y'} = 55.1(10^{-6})$

14–89. $\epsilon_{x'} = -116(10^{-6})$, $\epsilon_{y'} = 466(10^{-6})$,
$\gamma_{x'y'} = 393(10^{-6})$

14–90. $\epsilon_{x'} = 466(10^{-6})$, $\epsilon_{y'} = -116(10^{-6})$,
$\gamma_{x'y'} = -393(10^{-6})$

14–91. $\epsilon_1 = 385(10^{-6})$, $\epsilon_2 = 195(10^{-6})$,
$\theta_{p1} = 54.2°$, $\theta_{p1} = -35.8°$,
$\gamma_{\substack{max \\ in\text{-}plane}} = 190(10^{-6})$,
$\theta_s = 9.22°$ and $-80.8°$, $\epsilon_{avg} = 290(10^{-6})$

14–93. (a) $\epsilon_1 = 713(10^{-6})$, $\epsilon_2 = 36.6(10^{-6})$, $\theta_{p1} = 133°$,
(b) $\gamma_{\substack{max \\ in\text{-}plane}} = 677(10^{-6})$, $\epsilon_{avg} = 375(10^{-6})$,
$\theta_s = -2.12°$

14–94. $\epsilon_{x'} = 649(10^{-6})$, $\gamma_{x'y'} = -85.1(10^{-6})$,
$\epsilon_{y'} = 201(10^{-6})$

14–95. $\epsilon_{x'} = -365(10^{-6})$, $\gamma_{x'y'} = -271(10^{-6})$,
$\epsilon_{y'} = -35.0(10^{-6})$

14–97. $\epsilon_1 = 368(10^{-6})$, $\epsilon_2 = 182(10^{-6})$,
$\theta_{p1} = -52.8°$ and $\theta_{p2} = 37.2°$,
$\gamma_{\substack{max \\ in\text{-}plane}} = 187(10^{-6})$, $\theta_s = -7.76°$ and $82.2°$,
$\epsilon_{avg} = 275(10^{-6})$

14–99. $\epsilon_1 = 114(10^{-6})$, $\epsilon_2 = -314(10^{-6})$,
$(\theta_p)_1 = 79.7°$, $(\theta_p)_2 = -10.3°$,
$\gamma_{\substack{max \\ in\text{-}plane}} = 427(10^{-6})$,
$\theta_s = 34.7°$ and $125°$, $\epsilon_{avg} = -100(10^{-6})$

14–101. $\epsilon_{x'} = 325(10^{-6})$, $\gamma_{x'y'} = -133(10^{-6})$,
$\epsilon_{y'} = 55.1(10^{-6})$

14–102. $\epsilon_{x'} = 77.4(10^{-6})$, $\gamma_{x'y'} = 1279(10^{-6})$,
$\epsilon_{y'} = 383(10^{-6})$

14–103. $\epsilon_{avg} = 290(10^{-6})$, $\epsilon_1 = 385(10^{-6})$,
$\epsilon_2 = 195(10^{-6})$, $\theta_{P1} = 54.2°$ **(Counterclockwise)**,
$\gamma_{\substack{max \\ in\text{-}plane}} = 190(10^{-6})$,
$\theta_s = 9.22°$ **(Counterclockwise)**

14–105. (a) $\epsilon_1 = 773(10^{-6})$, $\epsilon_2 = 76.8(10^{-6})$,
(b) $\gamma_{\substack{max \\ in\text{-}plane}} = 696(10^{-6})$, (c) $\gamma_{\substack{abs \\ max}} = 773(10^{-6})$

14–106. $\epsilon_1 = 870(10^{-6})$, $\epsilon_2 = 405(10^{-6})$,
$\gamma_{\substack{max \\ \text{in-plane}}} = 465(10^{-6})$, $\gamma_{\substack{abs \\ max}} = 870(10^{-6})$

14–107. $\epsilon_1 = 946(10^{-6})$, $\epsilon_2 = 254(10^{-6})$,
$\gamma_{\substack{max \\ \text{in-plane}}} = 693(10^{-6})$, $\gamma_{\substack{abs \\ max}} = 946(10^{-6})$

14–111. $E = 30.7(10^3)$ ksi, $\nu = 0.291$

14–113. $\Delta\theta = -0.0103°$

14–114. $\nu_{pvc} = 0.164$

14–115. $p = 3.43$ MPa, $\tau_{\substack{max \\ \text{in-plane}}} = 0$, $\tau_{\substack{abs \\ max}} = 85.7$ MPa

14–117. $p = 0.967$ ksi, $\gamma_{\substack{max \\ \text{in-plane}}} = 1.30(10^{-3})$

14–118. $\sigma_1 = 10.2$ ksi, $\sigma_2 = 7.38$ ksi

14–119. $P = 26.1$ kip, $\gamma_{xy} = -27.5(10^{-6})$ rad

14–121. $\epsilon_x = 2.35(10^{-3})$, $\epsilon_y = -0.972(10^{-3})$,
$\epsilon_z = -2.44(10^{-3})$

14–122. $\sigma_1 = 8.37$ ksi, $\sigma_2 = 6.26$ ksi

14–123. $a' = 2.00302$ in., $b' = 2.00553$ in.,
$t' = 0.24964$ in.

R14–1. $\sigma_1 = 119$ psi, $\sigma_2 = -119$ psi

R14–2. $\sigma_1 = 329$ psi, $\sigma_2 = -72.1$ psi

R14–3. $\sigma_{x'} = -0.611$ ksi, $\tau_{x'y'} = 7.88$ ksi, $\sigma_{y'} = -3.39$ ksi

R14–6. $\delta_a = 0.367$ mm, $\delta_b = -0.255$ mm,
$\delta_t = -0.00167$ mm

R14–7. $\epsilon_1 = 0.0243$, $\epsilon_2 = -0.0311$

R14–9. $\epsilon_{avg} = 83.3(10^{-6})$, $\epsilon_1 = 880(10^{-6})$,
$\epsilon_2 = -713(10^{-6})$, $\theta_p = 54.8°$ (clockwise),
$\gamma_{\substack{max \\ \text{in-plane}}} = -1593(10^{-6})$,
$\theta_s = 9.78°$ (clockwise)

Chapter 15

15–1. $b = 211$ mm, $h = 264$ mm

15–2. Use $b = 4$ in.

15–3. Use $b = 5$ in.

15–5. Use W12 × 16.

15–6. Yes

15–7. Use W12 × 22.

15–9. Use W360 × 45.

15–10. Yes, it can.

15–11. Use s = $s'' = 2$ in.,
Use $s' = 1$ in.

15–13. $h = 8.0$ in., $P = 3.20$ kip

15–14. $s = 1.93$ in., $s' = 2.89$ in.,
$s'' = 5.78$ in.

15–15. Use $s = 1\frac{5}{8}$ in., $s' = 1\frac{1}{8}$ in.,
$s'' = 3\frac{1}{8}$ in.

15–17. $P = 103$ kN

15–18. Use $a = 3\frac{1}{8}$ in.

15–19. $P = 750$ lb

15–21. $P = 85.9$ N

15–22. $b = 3.40$ in.

15–23. Use W16 × 31.

15–25. Use $d = 3$ in.

Chapter 16

16–1. $\sigma = 3.02$ ksi

16–2. $\sigma = 75.5$ ksi

16–3. $\sigma = 582$ MPa

16–5. $v_C = -6.11$ mm

16–6. $\theta_{max} = -\dfrac{M_0 L}{EI}$,
$v = -\dfrac{M_0 x^2}{2EI}$,
$v_{max} = -\dfrac{M_0 L^2}{2EI}$

16–7. $\rho = 336$ ft,
$\theta_{max} = \dfrac{M_0 L}{EI}\curvearrowright$,
$v_{max} = -\dfrac{M_0 L^2}{2EI}$

16–9. $v_1 = \dfrac{P}{12EI}(2x_1^3 - 3Lx_1^2)$,
$v_2 = \dfrac{PL^2}{48EI}(-6x_2 + L)$

16–10. $v_1 = \dfrac{wax_1}{12EI}(2x^2 - 9ax_1)$,
$v_2 = \dfrac{w}{24EI}(-x_2^4 + 28a^3 x_2 - 41a^4)$,
$\theta_C = -\dfrac{wa^3}{EI}$, $v_B = -\dfrac{41wa^4}{24EI}$

16–11. $v_1 = \dfrac{wax_1}{12EI}(2x^2 - 9ax_1)$,
$v_3 = \dfrac{w}{24EI}(-x_3^4 + 8ax_3^3 - 24a^2 x_3^2 + 4a^3 x_3 - a^4)$,
$\theta_B = -\dfrac{7wa^3}{6EI}$, $v_C = -\dfrac{7wa^4}{12EI}$

16–13. $\theta_A = -\dfrac{M_0 a}{2EI}$, $v_{max} = -\dfrac{5M_0 a^2}{8EI}$

16–14. $v_{max} = -\dfrac{3PL^3}{256EI}$

16–15. $P = 40.0$ lb, $s = 0.267$ in.

16–17. $v_1 = \dfrac{Px_1}{12EI}\left(-x_1^2 + L^2\right)$,

$v_2 = \dfrac{P}{24EI}\left(-4x_2^3 + 7L^2 x_2 - 3L^3\right)$,

$v_{max} = \dfrac{PL^3}{8EI}$

16–18. $\theta_A = -\dfrac{3PL^2}{8EI}$, $v_C = -\dfrac{PL^3}{6EI}$

16–19. $v_B = -\dfrac{11PL^3}{48EI}$

16–21. $v_{max} = -11.5$ mm

16–22. $v = \dfrac{1}{EI}(2.25x^3 - 0.002778x^5 - 40.5x^2)$ kip·ft³,

$\theta_{max} = -0.00466$ rad, $v_{max} = -0.369$ in.

16–23. $\theta_C = \dfrac{4M_0 L}{3EI}\nearrow$, $v_1 = \dfrac{M_0}{6EIL}\left(-x_1^3 + L^2 x_1\right)$,

$v_2 = \dfrac{M_0}{6EIL}\left(-3Lx_2^2 + 8L^2 x_2 - 5L^3\right)$,

$v_C = -\dfrac{5M_0 L^2}{6EI}$

16–25. $v_{max} = -\dfrac{18.8 \text{ kip·ft}^3}{EI}$

16–26. $v_{max} = -0.396$ in.

16–27. $\theta_A = \dfrac{2\gamma L^3}{3t^2 E}$,

$v_A = -\dfrac{\gamma L^4}{2t^2 E}$

16–29. $\theta_B = -\dfrac{wa^3}{6EI}$,

$v_1 = \dfrac{w}{24EI}\left(-x_1^4 + 4ax_1^3 - 6a^2 x_1^2\right)$,

$v_2 = \dfrac{wa^3}{24EI}(-4x_2 + a)$,

$v_B = \dfrac{wa^3}{24EI}(-4L + a)$

16–30. $\theta_B = -\dfrac{wa^3}{6EI}$, $v_1 = \dfrac{wx_1^2}{24EI}\left(-x_1^2 + 4ax_1 - 6a^2\right)$,

$v_2 = \dfrac{wa^3}{24EI}(4x_3 + a - 4L)$, $v_B = \dfrac{wa^3}{24EI}(a - 4L)$

16–31. $v = \dfrac{1}{EI}\left[-\dfrac{Pb}{6a}x^3 + \dfrac{P(a+b)}{6a}\langle x - a\rangle^3 + \dfrac{Pab}{6}x\right]$

16–33. $E = \dfrac{Pa}{24\Delta I}(3L^2 - 4a^2)$

16–34. $v = \dfrac{P}{12EI}\left[-2\langle x - a\rangle^3 + 4\langle x - 2a\rangle^3 + a^2 x\right]$,

$(v_{max})_{AB} = \dfrac{0.106Pa^3}{EI}$, $v_C = -\dfrac{3Pa^3}{4EI}$

16–35. $v = \dfrac{1}{EI}\left[-2.5x^2 + 2\langle x - 4\rangle^3 - \dfrac{1}{8}\langle x - 4\rangle^4\right.$

$\left. + 2\langle x - 12\rangle^3 + \dfrac{1}{8}\langle x - 12\rangle^4\right.$

$\left. - 24x + 136\right]$ kip·ft³

16–37. $v = \dfrac{M_0}{6EI}\left[3\left\langle x - \dfrac{L}{3}\right\rangle^2 - 3\left\langle x - \dfrac{2}{3}L\right\rangle^2 - Lx\right]$,

$v_{max} = -\dfrac{5M_0 L^2}{72EI}$

16–38. $v = \dfrac{1}{EI}[-8.33x^3 + 17.1\langle x - 12\rangle^3$

$- 13.3\langle x - 36\rangle^3 + 1680x - 5760]$ lb·in³

16–39. $v_{max} = -12.9$ mm

16–41. $(v_{max})_{AB} = 0.0867$ in.

16–42. $\theta_A = -\dfrac{1920}{EI}$, $\theta_B = \dfrac{6720 \text{ lb·in}^2}{EI}$

16–43. $v = \dfrac{1}{EI}\left[-0.0833x^3 + 3\langle x - 8\rangle^2\right.$

$\left. + 3\langle x - 16\rangle^2 + 8.00x\right]$ kip·ft³

16–45. $\theta_B = -0.00778$ rad, $v_B = 0.981$ in.↓

16–46. $\theta_B = 2.08°$, $v_B = 3.61$ in.↓

16–47. $v_C = 1.20$ in.↓

16–49. $v_C = 0.429$ in.↓

16–50. $\theta_A = -0.283°$, $\theta_B = 0.427°$

16–51. $\theta_A = 0.00458$ rad, $v_C = 0.187$ in.↓

16–53. $(v_A)_v = 0.0737$ in., $(v_A)_k = 0.230$ in.

16–54. $v_A = 0.593$ in.↓

16–55. $v = PL^2\left(\dfrac{1}{k} + \dfrac{L}{3EI}\right)$

16–57. $v_A = PL^3\left(\dfrac{1}{12EI} + \dfrac{1}{8GJ}\right)\downarrow$

16–58. $\Delta_D = 3.23$ in.↓

16–59. $F = 0.349$ N, $a = 0.800$ mm

16–61. $B_y = 550$ N, $A_y = 125$ N, $C_y = 125$ N

16–62. $A_x = 0, B_y = \dfrac{5wL}{4}, C_y = \dfrac{3wL}{8}$

16–63. $B_y = \dfrac{5}{8}wL \uparrow, C_y = \dfrac{wL}{16} \downarrow, A_y = \dfrac{7}{16}wL \uparrow$

16–65. $A_x = 0, B_y = 35.0 \text{ kip}, A_y = 15.0 \text{ kip},$
$M_A = 40.0 \text{ kip} \cdot \text{ft}$

16–66. $A_x = 0, B_y = \dfrac{7P}{4}, A_y = \dfrac{3P}{4}, M_A = \dfrac{PL}{4}$

16–67. $A_x = 0, B_y = \dfrac{7wL}{128}, A_y = \dfrac{57wL}{128}, M_A = \dfrac{9wL^2}{128}$

16–69. $M_A = M_B = \dfrac{1}{24}PL, A_y = B_y = \dfrac{1}{6}P,$
$C_y = D_y = \dfrac{1}{3}P, D_x = 0$

16–70. $T_{AC} = \dfrac{3wA_2E_2L_1^4}{8(3E_1I_1L_2 + A_2E_2L_1^3)}$

16–71. $M = \dfrac{PL}{8} - \dfrac{2EI}{L}\alpha, \Delta_{\max} = \dfrac{PL^3}{192EI} + \dfrac{\alpha L}{4}$

16–73. $a = L - \left(\dfrac{72\Delta EI}{w_0}\right)^{1/4}$

R16–1. $v = \dfrac{1}{EI}\left[-30x^3 + 46.25\langle x - 12\rangle^3\right.$
$\left. - 11.7\langle x - 24\rangle^3 + 38{,}700x - 412{,}560\right] \text{lb} \cdot \text{in}^3$

R16–2. $v_1 = \dfrac{1}{EI}(4.44x_1^3 - 640x_1) \text{ lb} \cdot \text{in}^3,$
$v_2 = \dfrac{1}{EI}(-4.44x_2^3 + 640x_2) \text{ lb} \cdot \text{in}^3$

R16–3. $M_B = \dfrac{w_0L^2}{30}, M_A = \dfrac{w_0L^2}{20}$

R16–5. $(v_2)_{\max} = \dfrac{wL^4}{18\sqrt{3}EI}$

R16–6. $v_A = \dfrac{w_0L^4}{Eth_0^3} \downarrow$

R16–7. $\Delta_G = 5.82 \text{ in.} \downarrow$

R16–9. $\Delta_C = 0.644 \text{ in.} \downarrow$

Chapter 17

17–1. $P_{cr} = \dfrac{5kL}{4}$

17–2. $P_{cr} = kL$

17–3. Use $d = \dfrac{9}{16}$ in.

17–5. $P_{cr} = 1.84 \text{ MN}$

17–6. $P_{cr} = 902 \text{ kN}$

17–7. F.S. $= 1.87$

17–9. $P_{cr} = 1.30 \text{ MN}$

17–10. $P_{cr} = 325 \text{ kN}$

17–11. $P_{cr} = 20.4 \text{ kip}$

17–13. $P = 42.8 \text{ kN}$

17–14. $P_{cr} = 575 \text{ kip}$

17–15. $P_{cr} = 70.4 \text{ kip}$

17–17. $P_{cr} = 2.92 \text{ kip}$

17–18. $P_{cr} = 5.97 \text{ kip}$

17–19. $P = 17.6 \text{ kip}$

17–21. Use $d_i = 1\dfrac{1}{8}$ in.

17–22. $W = 4.31 \text{ kN}$

17–23. $W = 5.24 \text{ kN}, d = 1.64 \text{ m}$

17–25. $P = 62.3 \text{ kip}$

17–26. $P = 2.42 \text{ kip}$

17–27. Use $d_{AB} = 2\dfrac{1}{8}$ in., $d_{BC} = 2\dfrac{1}{4}$ in.

17–29. Use $d_{AB} = 1\dfrac{1}{2}$ in., $d_{BC} = 1\dfrac{3}{8}$ in.

17–30. $P = 129 \text{ lb}$

17–31. $w = 1.17 \text{ kN/m}$

17–33. $P = 14.8 \text{ kN}$

17–34. Use $d = 62 \text{ mm.}$

17–35. Use $d = 52 \text{ mm.}$

17–37. $P = 37.5 \text{ kip}$

17–38. $P = 5.79 \text{ kip}$

17–39. Use $d = 1\dfrac{3}{4}$ in.

17–41. $M_{\max} = -\dfrac{wEI}{P}\left[\sec\left(\dfrac{L}{2}\sqrt{\dfrac{P}{EI}}\right) - 1\right]$

17–42. $M_{\max} = -\dfrac{F}{2}\sqrt{\dfrac{EI}{P}}\tan\left(\dfrac{L}{2}\sqrt{\dfrac{P}{EI}}\right)$

17–43. $P_{cr} = \dfrac{\pi^2EI}{4L^2}$

17–46. $P = 31.4 \text{ kN}$

17–47. $v_{\max} = 0.387 \text{ in.}$

17–49. $P_{\text{allow}} = 7.89 \text{ kN}$

17–50. $L = 8.34 \text{ m}$

17–51. $P = 3.20 \text{ MN}, v_{\max} = 70.5 \text{ mm}$

17–53. $P = 45.7 \text{ kip}$

17–54. Yes

17–55. $P_{cr} = 12.6 \text{ kN}$

17–57. $L = 21.2$ ft
17–58. $P_{max} = 16.9$ kN
17–59. $P = 76.6$ kip
17–61. $P = 174$ kN, $v_{max} = 16.5$ mm
17–62. $v_{max} = 1.23$ in., $\sigma_{max} = 15.6$ ksi
17–63. $P = 88.5$ kip
R17–1. $a = 103$ mm
R17–2. $P_{cr} = \dfrac{2k}{L}$

R17–3. $P_{cr} = 12.1$ kN
R17–5. $P = 12.5$ kip
R17–6.   Use $d = 2\dfrac{1}{8}$ in.
R17–7. $t = 5.92$ mm
R17–9. $P_{allow} = 77.2$ kN
R17–10. It does not buckle or yield.

INDEX